Algebra: Abstract and Concrete (Stressing Symmetry) 2e
Fred Goodman
0-13-067342-0

Linear Algebra 4e
S. Friedberg, A. Insel, and L. Spence
0-13-008451-4

Applied Linear Algebra
Peter Olver and Cheri Shakiban
0-13-147382-4

Linear Algebra 2e
Ken Hoffman and Ray Kunze
0-13-536797-2

Applied Linear Algebra
Lorenzo Sadun
0-13-085645-2

Applied Linear Algebra 3e
Ben Noble and Jim Daniel
0-13-041260-0

Introduction to Linear Programming
Leonid Vaserstein
0-13-035917-3

Introduction to Mathematical Programming
Russ Walker
0-13-263765-0

Applied Algebra
Darel Hardy and Carol Walker
0-13-067464-8

The Mathematics of Coding Theory
Paul Garrett
0-13-101967-8

Introduction to Cryptography with Coding Theory
Wade Trappe and Larry Washington
0-13-061814-4

Making, Breaking Codes
Paul Garrett
0-13-030369-0

Invitation to Cryptology
Tom Barr
0-13-088976-8

An Introduction to Dynamical Systems
R. Clark Robinson
0-13-143140-4

Linear Algebra for Engineers and Scientists Using MATLAB®

KENNETH HARDY
Carleton University, Ottawa

PEARSON
Prentice Hall

Upper Saddle River, New Jersey 07458

Library of Congress Cataloging-in-Publication Data
Hardy, Kenneth
Linear algebra for engineers and scientists using MATLAB/ Kenneth Hardy.
p. cm.
Includes bibliographical references and index.
ISBN 0-13-906728-0
1. Algebras, Linear. 2. Science—Mathematics. 3. Engineering mathematics.
4. MATLAB. I. Title.
QA184.2.H37 2005
512'.5—dc22
2004044517

Editor in Chief: *Sally Yagan*
Executive Acquisitions Editor: *George Lobell*
Production Editor: *Lynn Savino Wendel*
Vice President/Director of Production and Manufacturing: *David W. Riccardi*
Senior Managing Editor: *Linda Mihatov Behrens*
Executive Managing Editor: *Kathleen Schiaparelli*
Assistant Manufacturing Manager/Buyer: *Michael Bell*
Manufacturing Manager: *Trudy Pisciotti*
Marketing Manager: *Halee Dinsey*
Marketing Assistant: *Rachel Beckman*
Creative Director: *Jayne Conte*
Editorial Assistant: *Jennifer Brady*
Director, Image Resource Center: *Melinda Reo*
Manager, Rights and Permissions: *Zina Arabia*
Manager, Visual Research: *Beth Brenzel*
Manager, Cover Visual Research & Permissions: *Karen Sanatar*
Image Permission Coordinator: *Joanne Dippel*
Photo Researcher: *Melinda Alexander*
Cover Designer: *Kiwi Design*
Cover Photo Credits: *Fra Luca Pacioli, Museo Archeologico Nazionale di Napoli*
Art Studio: *Laserwords*

The painting on the cover shows Fra Luca Pacioli (1445–1514), the famous Renaissance teacher of mathematics. If you visit Italy you can see the painting in the Museo di Capodimonte in Naples. Historians are confident about many aspects of this work, including the fact that it was painted in Venice in 1495 by Jacopo de'Barbari. In the painting on the right is a regular dodecahedron standing on top of the *Summa*, a 600-page text published by Pacioli in 1494 in which he compiled the then known facts of arithmetic, algebra, and geometry. Pacioli is working from a first printed edition (Venice 1482) of Euclid's *Elements*. He is pointing to a diagram illustrating Euclid's Proposition 12 in Book XIII of *Elements*. In the top left is suspended a glass model of a rhombicuboctahedron (half-full of water), one of the 13 semi regular solids discovered by Archimedes. Some art historians believe that this item was added to the painting at a later time by Leonard da Vinci.

Pearson Prentice Hall
Pearson Education, Inc.,
Upper Saddle River, New Jersey 07458

Printed in the United States of America

ISBN 0-13-906728-0

Pearson Education Ltd., London
Pearson Education Australia Pty. Limited, Sydney
Pearson Education Singapore Pte. Ltd.
Pearson Education North Asia, Ltd, Hong Kong
Pearson Education Canada, Ltd., Toronto
Pearson Educación de Mexico, S.A.,de C.V.
Pearson Education, Japan, Tokyo
Pearson Education Malaysia, Pte. Ltd.

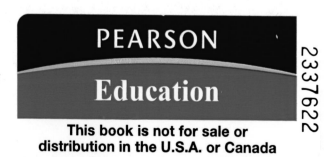

To Professor Guido Lorenz

Here on the level sand,
Between the sea and land,
What shall I build or write
Against the fall of night?

Tell me of runes to grave
That hold the bursting wave,
Or bastions to design
For longer date than mine.

From *A Mathematician's Apology*
by Godfrey Harold Hardy (1877–1947)

ABOUT THE AUTHOR

Kenneth Hardy was born in Derbyshire, England and educated at Leicester University (B.Sc.) and McGill University in Montréal, Quebec, Canada (M.Sc. and Ph.D.). He taught mathematics at Carleton University, Ottawa, Canada for 31 years, maintaining a special interest in courses for applied audiences, particularly engineering and science. During his career he has contributed to a number of special projects in mathematics education, including the following:

- Development and teaching of a course in the mathematics of Bloodstain Pattern Analysis (BPA) offered to members of the Royal Canadian Mounted Police (RCMP) and to officers and forensic scientists from the United States and Canada. BPA is a branch of Forensic Science dealing with the analysis of bloodstain patterns found at the scene of violent crimes. BPA is now used in courtroom crime-scene reconstruction.

- A founding member of The Canadian Center for Creative Technology's *Shad Valley Program* (1981—). This summer co-op program, co-sponsored by industry and hosted by universities across Canada, is offered to bright high school students with the aim of fostering creativity, invention, and entrepreneurship in engineering and science. He taught in the program for 14 years.

- Co-author (with K.S. Williams) of *The Green Book* and *The Red Book* each containing 100 practice problems with hints and solutions for students training for the prestigious Putnam Competition in the United States.

- Special projects in mathematics education undertaken for the Commonwealth of Learning (COL), an organization of the British Commonwealth dedicated to the improvement of education throughout the Commonwealth using distance learning methods.

In 1995 he received a Teaching Achievement Award from Carleton University *for outstanding work in teaching innovation and development.* His interests include painting, poetry, cooking, golf, running, and hill-walking. He lives in Ottawa.

CONTENTS

PREFACE

Mathematics is the invisible culture of our age. Although frequently hidden from public view, mathematics ... influences the way we live and the way we work.

—

[Everybody Counts: A Report to the Nation on the Future of Mathematics Education, National Academic Press, Washington, D.C. (1989)]

To describe some of the challenges of modern mathematics education in just a few short lines is, of course, an impossible task. High on the list must be the challenge that faces every instructor—that of demonstrating the elegance of pure mathematical ideas side-by-side with their application to the real world. In recent years, a new layer of complexity has been added: Incorporating the use of machine computation in a way that enhances, and in no way detracts from, the traditional pedagogy of understanding theory and working problems by hand. Meeting the challenges of mathematics education requires creativity and innovation in the classroom together with progressive textbooks that provide a platform for instruction. After many years of teaching mathematics, both in university and to other applied audiences, I was invited to prepare a text in the area of linear algebra—this book is a contribution toward meeting the challenges that we face.

The Text

This text provides an introduction to linear algebra and its applications suitable for students in their freshman or sophomore year. All the standard theory for an introductory level course is included here, together with complete explanations of some current applications particularly appropriate for engineers and scientists. The theory is presented in a rigorous way and the language, notation, illustrations, and examples have been carefully chosen so as to make the exposition as clear as possible. A knowledge of calculus is not a prerequisite although some understanding of functions, derivatives and differential equations would be beneficial in a few sections.

This text encourages the use of technology but is not bound by it. Reference to MATLAB calculation and code within the text body has been kept to a minimum and the use of MATLAB exercises and projects that are included at the end of exercise sets is left to the discretion of the instructor. These exercises vary considerably in difficulty. Some exercises are routine, intended for a combination of hand and machine computation, while others require more time. Exercises labeled *Project* are aimed at getting started with MATLAB M-file programming and those labeled *Experiment* introduce the idea of discovery through experimentation—for example, observing the consequences of changing the parameters in an age-structured population model or using Monte Carlo simulation to estimate the probability that a certain phenomenon will occur.

Distinctive Features

The core course material is contained in Chapters 1 through 7. I have tried to order the chapters and topics in such a way that the transition from topic to topic appears to the student both timely and logical. Particular special features of the text are mentioned next.

- The chapter introductions are key elements in the text—they give motivation and meaning; show directions and make connections.
- Linear systems appear first because they are a foundation for the rest of the theory and are of interest in their own right through a multitude of applications. The rank of a matrix appears first in Section 1.2 and plays a central role in the chapters that follow.
- Some major applications involving matrix multiplication are discussed in Chapter 2. Discrete dynamical systems and age-structured population models are topics that are revisited as eigenvalue problems in Chapter 6.
- Whether vector spaces are introduced early or late in a course is an important question. My approach has been to develop the key vector space concepts early in Chapter 3 in the context of column vectors and extend these concepts to real vectors spaces in Chapter 7. This approach, although attracting a certain amount of repetition, is designed to help the student make the transition from concrete to abstract. Proofs of many important theorems, stated in Chapter 3, are postponed to Chapter 7 allowing time in Chapter 3 to concentrate more on the implications of the theorems in the context of $\mathbb{R}^n$.
- Chapter 8 provides a substantial introduction to complex numbers which I hope will be a useful reference for students in engineering and science. A section in this chapter outlines linear algebra over the complex field.
- Linear Programming is used to solve applied problems in a wide variety of situations, from economic to scientific, and is a favorite topic with some instructors and their students. Chapter 9 is an introduction to this subject.
- Appendix A: MATLAB gives a brief introduction to interactive computation, M-file programming and 2-D graphics.
- Appendix B: Toolbox contains a few theoretical items, including an important discussion of methods of proof and a description of the principle of mathematical induction.

Course Design

Many instructors have definite ideas about the order in which course material is presented and I will therefore give only my suggestion of time-allocation in chapters 1 through 7 for a typical 35 lecture course.

Chapter 1	4 lectures	Chapter 5	2 lectures
Chapter 2	8 lectures	Chapter 6	4 lectures
Chapter 3	7 lectures	Chapter 7	3 lectures
Chapter 4	7 lectures		
		35 lectures	

Use of Computers

Incorporating computer activity into courses requires more organization and effort, especially when time is a priority. However, in my experience, the level of activity can vary considerably from light to heavy. For example, an instructor may choose to devote all lecture time to theoretical work, leaving computer exercises and projects to be done in-lab or outside class time. Short computer presentations from time to time by the instructor, either in-class or in-lab, are excellent for experimenting with models that are numerically challenging (such as discrete dynamical systems) and for stimulating interest in what the software can do.

Commercial software packages for mathematical computing provide both students and instructors with exciting possibilities that enhance and complement the traditional methods of mathematics education, including

- Fast and accurate computations
- Visualization using engineering and scientific graphics
- Modeling, experimentation, exploration, and discovery
- The ability to handle more complicated problems
- Checking hand calculations

I chose to use MATLAB[1] for two reasons—first, because it is an industry standard in mathematical computing for engineers and scientists, and second, because, in my view, it offers students the most trouble-free introduction to machine computation and the greatest potential for future use in their professional lives.

History

Only recently has it become popular to mention history in mathematics texts. In the past, students (myself included) were no doubt left wondering if the great body of knowledge we call mathematics happened over night. In years gone by, mathematicians were often part scientist, engineer, artist, writer, linguist,

[1] MATLAB® is a registered trademark of The MathWorks, Inc.

economist, and so on, all rolled into one. Leonardo da Vinci and Bertrand Russell are prominent examples. Through history we gain an appreciation for the length of time (sometimes hundreds of years) that it took for concepts to grow from mere ideas into finished form.

Each chapter in this text opens with a profile of a prominent mathematician connected with its subject matter—the idea being to draw the reader in and bring the subject alive by linking mathematical ideas more closely to the life and times of the mathematicians who created them. For portraits of mathematicians and other historical information, search the web for *The MacTutor History of Mathematics*, School of Mathematics and Statistics, University of St. Andrews, Scotland.

Graphics

The production and use of accurately drawn graphical images is fundamental to engineering and science. Whenever possible I have illustrated concepts with either schematic images or scaled figures. Schematic images were conceived and executed in Adobe Illustrator 8.0. Scaled figures were often created using MATLAB graphics and modified in Adobe Illustrator. Throughout the text students are encouraged to learn the graphic and important color capabilities that MATLAB has to offer.

To Students

Becoming proficient in any language requires four basic skills—reading, writing, listening and speaking. The same basic skills also apply to learning the language of mathematics, which has become indispensable to professional people in all applied fields for understanding and communicating technical ideas and for critical and deductive thinking. Learning the language usually begins with first courses in linear algebra and calculus and these act as a foundation on which later mathematical training is built. Being successful in mathematics involves dedication and patience—learning definitions, knowing the accurate statement of theorems and understanding their proofs—but it also involves the thrill of finding solutions! You will find the theory of linear algebra very elegant in its structure and very profound in its applications. I wish you well on your journey of discovery.

Supplements

The following publications complement this textbook and are available through the publisher:

- *Student's Solutions Manual* containing complete solutions to odd–numbered exercises (ISBN 0-13-061962-0)
- *Instructor's Solutions Manual*, containing complete solutions to all exercises (ISBN 0-13-061950-7)

- *ATLAST Computer Exercises for Linear Algebra*, S. Leon, G. Hermann, R. Faulkenberry (editors), Prentice Hall, 1997 (ISBN 0-13-270273-8)
- *Linear Algebra with MATLAB*, 2th ed., D. Hill, D. Zitarelli, Prentice Hall, 1996 (ISBN 0-13-505439-7)

Reviewers

I wish to express my deep appreciation to the following people for their careful reading and helpful advice during the development of this text.

Anjan Biswas, *Tennessee State University*
Saroj Biswas, *Temple University*
Carlos Borges, *Naval Postgraduate School*
David Boyles, *University of Wisconsin-Platteville*
Juergen Guerlach, *Radford University*
Dawn A. Lott, *New Jersey Institute of Technology*
Mohsen Maesumi, *Lamar University*
Curtis Olson, *University of South Dakota*
John Palmer, *University of Arizona*
Alexander Shibakov, *Tennessee Technological University*
Marie Vitulli, *University of Oregon*
Rebecca G. Wahl, *Butler University*
Bruno Welfert, *Arizona State University*
John Woods, *Southwest Oklahoma State University*
Xian Wu, *University of South Carolina*

Acknowledgments

The following individuals and groups provided support and encouragement during the writing of this text, and I am indebted to them:

- Carleton University, Ottawa, Canada. Special thanks to Dr. M. J. Moore, School of Mathematics and Statistics for invaluable help with exercises and solutions and Lal Sebastian for computer-related technical assistance.
- In Ottawa, Mary Frances Huband, for editorial advice.
- In Montréal, Shauna Eliot Hardy at Straight to the Point Productions for editorial advice and Angelica Kate Hardy at Endurance Studio for graphics design.
- Special thanks to George Lobell for his unflagging patience and encouragement, and to the whole production team at Pearson Prentice Hall, U.S.A.
- The MathWorks, Inc., Natick, Mass., U.S.A. for supply of software and for technical support.
- LINDO SYSTEMS, Inc., Chicago, IL 60622 for supply of software.
- In France, Dr. Alan W. Hewat, Diffraction Group Leader at the Institute Laue-Langevin, Grenoble, for permission to use the graphic "topaz" in Chapter 5.

- In Italia, Professoressa Clara Silvia Roero, Università di Torino, e Dottoressa Anna Merlo, per la ricerca su Giuseppe Peano e fotographie storiche. Professor Dr. Tommaso Ruggeri e colleghi, Centro Interdipartimentale di Ricerca per le Applicazioni della Matematica (CIRAM), Università di Bologna, Italia, per l'assistenza generale.

Kenneth Hardy, Ottawa, June, 2004.
kenneth_hardy@carleton.ca

PHOTO CREDITS

Chapter 1 Openers: Left: Photo Researchers, Inc.; Right: Dr. Edward G. Nawy, D. Eng., P.E.

Chapter 2 Openers: Left: Corbis/Bettmann; Right: Portland Cement Association

Chapter 3 Openers: Left: CORBIS-NY; Right: Courtesy of Figg Engineering Group

Chapter 4 Openers: Left: The Image Works; Right: Bethlehem Steel Corporation

Chapter 5 Openers: Left: Corbis/Bettmann; Right: Courtesy Dr. Alan Hewat, Grenoble, France Source: http://www.ill.fr/dif/3D–crystals/minerals.html

Chapter 6 Opener: Left: Sovfoto/Eastfoto

Chapter 7 Openers: Left: Clara Silvia Roero; Right: SolidWorks Corporation

Chapter 8 Openers: Left: Corbis/Bettmann; Right: Getty Images Inc.—Hulton Archive Photos

Chapter 9 Openers: George B. Dantzig; Right: Portland Cement Association

Cover Photo: Museo Archeologico Nazionale di Napoli

Karl Friedrich Gauss (1777–1855)

G auss was a child prodigy with a phenomenal memory. At an early age he amazed his school teacher by rapidly summing the integers $1, 2, \ldots, 99, 100$. He noted that each of the 50 pairs $(1, 100), (2, 99), \ldots, (50, 51)$ has sum 101 giving a total of $50 \cdot 101 = 5050$.

Gauss dominated 19th century mathematics and science. He started his famous scientific diary on March 30, 1796 with an entry that recorded his solution to an ancient problem of constructing a regular 17-side polygon using only ruler and compass. The Italian astronomer Giuseppe Piazzi (1746–1826) discovered the asteroid Ceres on January 1, 1801, but Piazzi was only able to obtain a few observations of its orbit before it was hidden by the sun. Gauss became famous for his calculations that correctly predicted its position when it appeared on December 7, 1801. The methods are still in use today. The basis of his orbit calculations, which he revealed only at a later date, was his *method of least squares* approximation. Gauss became director of the Göttingen Observatory in 1807, a position he kept throughout his life. He is now universally acclaimed by the scientific community as the *Prince of Mathematicians*, an equal of Archimedes and Newton.

Gauss made important discoveries in number theory, analysis, statistics, geometry, astronomy, geodesy, electricity and magnetism, and optics. His scientific inventions include the bifilar magnetometer, the heliotrope, and an electrotelegraph. In this chapter we remember Gauss particularly for the *elimination algorithms* that are used today for solving linear systems.

Chapter 1

LINEAR SYSTEMS

Introduction

The simplest possible linear equation

$$ax = b, \tag{1.1}$$

where a and b are given real or complex numbers and x is an unknown, is solved in the following way. There are three cases to consider:

(a) If $a = 0$ and $b \neq 0$ in (1.1), then $0x = 0 = b$, which is false for any value x, and so there are no solutions;
(b) If $a \neq 0$, then (1.1) has the unique solution $x = b/a$ for any value of b;
(c) If $a = b = 0$ in (1.1), then $0x = 0$ for any value of x and so there are infinitely many solutions.

We say that $ax = b$ has *zero, one*, or *infinitely many* solutions (only one of these three possibilities can occur).

Equation (1.3) in Section 1.1 shows the most general form of a linear equation. A *linear system* is an ordered list of linear equations that we treat as a single entity. For example, the following linear system (S) is an ordered list of four linear equations E_1, E_2, E_3, E_4 in five unknowns x_1, x_2, x_3, x_4, x_5.

$$(S)\quad \left\{\begin{array}{rl} x_2 - 2x_3 + 2x_4 - x_5 = 2 & E_1 \\ 2x_1 + 3x_2 + 6x_3 + 3x_4 + 6x_5 = 11 & E_2 \\ -4x_1 - 13x_2 + 2x_3 + x_4 + 2x_5 = -15 & E_3 \\ - x_2 + 2x_3 + x_4 - 3x_5 = 1 & E_4 \end{array}\right. \tag{1.2}$$

Linear systems arise in a great many theoretical and practical situations and in many applied fields, including engineering, science, business, and economics. Some modern applications require the solution of linear systems having millions of equations and unknowns. These are called *large-scale systems*. By comparison, (1.2) is a *small-scale* linear system. It may be surprising that any linear system no matter how large or small-scale still has the same three possibilities for solution as the single equation (1.1), namely zero, one or infinitely many solutions.

Our motivation in choosing linear systems as the first topic of study in this text comes from two distinct yet equally important considerations:

(a) Linear systems and their associated vector and matrix forms play a key role in linear algebra. The ideas developed in this chapter are used extensively throughout the text;

(b) Given any linear system, a primary goal is to find solutions when they exist or to determine that there are none.

These days, linear systems are almost always solved with the aid of a computer. Machine computation introduces further complications which arise from the *floating point arithmetic* on which machine computation is based. An understanding of the possible errors that may arise through machine computation is therefore an important consideration.

Keep in mind that most of the examples and exercises in this text involve small-scale linear systems with *integer coefficients*, such as (1.2), which are designed to illustrate theoretical ideas. MATLAB will be used to help solve more numerically challenging problems and to explore other ideas through experimentation.

1.1 Solving Linear Systems

1.1.1 Linear Equations

An equation in *unknowns*, or *variables*, $x_1, x_2, \ldots, x_n$ is called *linear* if it can be written in the *standard form*

$$a_1x_1 + a_2x_2 + \cdots + a_nx_n = b. \tag{1.3}$$

The numbers $a_1, a_2, \ldots, a_n$ are called *coefficient* for the unknowns and the number b is called the *constant term*. For example, the equation

$$3.2x_1 - 5.4x_2 = 7.9 \tag{1.4}$$

is linear. Comparing (1.4) with (1.3) tells us that in (1.4) there are $n = 2$ unknowns x_1, x_2 and that $a_1 = 3.2$ is the coefficient of x_1, $a_2 = -5.4$ is the coefficient of x_2 and $b = 7.9$ is the constant term.

When all the terms in a linear equation are real numbers, as in (1.4), we call it a *real* linear equation. On the other hand, the linear equation

$$(1 + 2i)x_1 + 4x_2 = 3 - 3i, \qquad \text{where } i = \sqrt{-1},$$

which involves complex numbers of the form $a + ib$, where a and b are real numbers, is called a *complex* linear equation. The theory of complex numbers and an outline of linear algebra in the *complex case* are given in Chapter 8.

The variables $x_1, x_2, x_3, \ldots$ form an ordered list: $x_1 =$ "first variable," $x_2 =$ "second variable," and so on. In some situations it may be more convenient or simpler to use other symbols to denote variables, such as

$p, q, r, \ldots, x, y, z, \ldots$ or Greek letters α (alpha), β (beta), γ (gamma), ...

EXAMPLE 1

Linear and Nonlinear Equations

The equations

$$x_1^2 + x_2^2 = 1 \qquad \sqrt{x_1} - x_2 + \sqrt[3]{x_3} = 0$$

$$\log x - \log y = \log z \qquad e^{x+y} - \tan(x - y) = 0$$

cannot be rearranged to fit the standard form (1.3) and are therefore not linear. We call these *nonlinear* equations. In some cases rearrangement or simplification of an equation will reveal that it is in fact linear. For example, rearranging and canceling, we have

$$3x_1 + 4x_2 - 6 = 2x_3 + 4x_2 \quad \Rightarrow \quad 3x_1 + 0x_2 - 2x_3 = 6, \quad \text{which is linear.}$$

Of course, the term $0x_2$ ($= 0$ for any value of x_2) could be omitted, or included in order to show x_2 with its coefficient 0. The following equation simplifies, with the nonlinear terms canceling:

$$(x_1 + 2)^2 - x_1^2 + 3x_2 = 9 \quad \Rightarrow \quad 4x_1 + 3x_2 = 5, \text{ which is linear.}$$

In some calculations it may be convenient to express a nonlinear equation as a linear equation by renaming expressions in the equation. For example, substituting $p = e^x$ and $q = e^y$ into the nonlinear equation $2e^x - 3e^y = 4$ gives $2p - 3q = 4$, which is linear in the new unknowns p and q. ■

1.1.2 Linear Systems

An ordered list of m linear equations in the same n unknowns $x_1, x_2, \ldots, x_n$ is called an $m \times n$ (read "m by n") linear system. For example, the linear system (1.2) on page 1 is 4×5.

A *solution* to a linear system (S) in unknowns $x_1, x_2, \ldots, x_n$ is a set of n numbers $c_1, c_2, \ldots, c_n$ such that the assignment

$$x_1 = c_1, \ x_2 = c_2, \ \ldots, \ x_n = c_n \tag{1.5}$$

satisfies *every* equation in (S) simultaneously. The *solution set* or *general solution* to (S) is the set of all solutions to (S). A *particular solution* is any solution in the solution set.

Definition 1.1

Inconsistent, consistent, solving

If a linear system (S) has no solutions, we call (S) *inconsistent*. If (S) has one or more solutions, we call (S) *consistent*. To *solve* (S) means to find the solution set to (S) or to determine that (S) is inconsistent.

The simple, yet powerful, method of *elimination* for solving linear systems will now be explained. Although some forms of these techniques were known as early as the 13th century, they are traditionally linked today with the names of Karl Friedrich Gauss and the German engineer Wilhelm Jordan (1842–1899), who developed and used them during the 19th century.

1.1.3 Forward Elimination

The first phase of solving a linear system is called *forward elimination.* This is accomplished using *elementary equation operations* (EEOs). There are three types of operation, called *replacement*, *interchange*, and *scaling*, and we now introduce these in a series of illustrations.

■ ILLUSTRATION 1.1

Forward Elimination, Replacement

Consider the problem of finding a solution (assuming one exists) to the linear system (S) in (1.6). The most direct approach is by *substitution.*

$$(S)\begin{cases} x_1 + x_2 + x_3 = 1 & E_1 \\ 2x_1 - 2x_2 - x_3 = 0 & E_2 \\ -x_1 + 3x_2 + 7x_3 = 0 & E_3 \end{cases} \tag{1.6}$$

Suppose equation E_1 is *solved* for the *leading unknown* x_1, as follows:

$$x_1 + x_2 + x_3 = 1 \quad \Rightarrow \quad x_1 = 1 - x_2 - x_3$$

and then the expression for x_1 is substituted into all equations below E_1. We obtain a new system

$$(S)'\begin{cases} x_1 + x_2 + x_3 = 1 & E_1 \\ -4x_2 - 3x_3 = -2 & E_2\ (\text{new})\ . \\ 4x_2 + 8x_3 = 1 & E_3\ (\text{new}) \end{cases} \tag{1.7}$$

The effect of the two substitutions is to *eliminate* the unknown x_1 from all equations below E_1 in (S).

The process of solving and substituting just described can be combined into a single *elementary equation operation*, called *replacement*, which achieves the same goal. Replacement uses the operations of subtraction and multiplication. For example, in order to eliminate x_1 from the equations below E_1 in (S), we use the following two replacements:

Equation E_2 minus 2 times equation E_1	replaces	Equation E_2	
$E_2 - 2E_1$	$\rightarrow$	new E_2	
			(1.8)
Equation E_3 minus (-1) times equation E_1	replaces	Equation E_3	
$E_3 - (-1)E_1$	$\rightarrow$	new E_3	

■ NOTATION Replacement

Suppose an $m \times n$ linear system (S) contains equations E_i and E_j, where i and j are *distinct* positive integers. The elementary equation operation which replaces equation E_i with equation "E_i minus m times equation E_j" is denoted by

$$E_i - mE_j \rightarrow E_i \tag{1.9}$$

and the number m is called the *multiplier* for the operation. Only equation E_i is new in the resulting system $(S)'$. ■

Rewriting (1.8) using the preceding notation, we have

$$E_2 - 2E_1 \rightarrow E_2 \qquad i = 2, \quad j = 1, \quad m = 2$$

$$E_3 - (-1)E_1 \rightarrow E_3 \qquad i = 3, \quad j = 1, \quad m = -1. \tag{1.10}$$

Of course, the replacement (1.10) may be written $E_3 + E_1 \rightarrow E_3$, but the notation in (1.9), which uses a minus sign before the multiplier, is chosen purposely to anticipate later theory.[1]

Now return to (1.7) and use equation E_2 in replacements to eliminate x_2 from all equations below E_2. Only one replacement is required in this case, namely

$E_3 - (-1)E_2 \rightarrow E_3 \quad (m = -1)$ and we obtain a new system

$$(S)'' \begin{cases} x_1 + x_2 + x_3 = 1 & E_1 \\ -4x_2 - 3x_3 = -2 & E_2 \\ 5x_3 = -1 & E_3 \text{ (new)} \end{cases} \tag{1.11}$$

There are no more unknowns below equation E_3 to eliminate and we stop. Forward elimination changes (S) into a new system $(S)''$, which is said to be in *upper triangular form.*[2] ■

Back-Substitution

The system $(S)''$ in (1.11) can now be solved by *back-substitution.* Move upward through $(S)''$, solving and substituting, in the following way:

Solve E_3 for x_3: $\quad x_3 = -0.2$

Solve E_2 for x_2 using x_3: $\quad 4x_2 = 2 - 3x_3 = 2 + 0.6 = 2.6 \quad \Rightarrow x_2 = 0.65$

Solve E_1 for x_1 using x_2, x_3: $\quad x_1 = 1 - x_2 - x_3 = 1 - 0.45 = 0.55.$

Hence, the unique (single) solution to $(S)''$ in (1.11) is

$$x_1 = 0.55, \qquad x_2 = 0.65, \qquad x_3 = -0.2. \tag{1.12}$$

But what is the solution set to the original system (S) in (1.6)? Here is the key!

Definition 1.2 *Equivalent linear systems*

Two linear systems (S) and $(S)'$ are called *equivalent* if $(S)'$ is the result of performing one or more EEOs on (S) and we write $(S) \sim (S)'$ to designate equivalence.

Thus, the systems (S) in (1.6) and $(S)'$ in (1.7) are equivalent because $(S)'$ is the result of performing two replacements on (S). Furthermore, $(S) \sim (S)''$

[1] See Section 2.3.

[2] A precise definition is given at the end of Section 1.2.

(explain). We shall see on page 9 that equivalent linear systems have exactly the same solution set. Thus, (1.12) is the unique solution to (1.6).

Caution! It is always good practice to cross-check your answers by substituting the solution set back into *all* equations in the original system. As an exercise, verify that the solution (1.12) satisfies (1.6).

■ NOTATION Recording Solutions, n-tuples

An *n-tuple* is an *ordered list* of n numbers enclosed in round brackets. For example, $(4, -5, 3)$ is a 3-tuple of real numbers and $4, -5, 3$ are called *components*. The solution (1.5) can be written compactly as an *equality* of n-tuples in the form

$$(x_1, x_2, \ldots, x_n) = (c_1, c_2, \ldots, c_n), \tag{1.13}$$

with the understanding that corresponding components are equal, as in (1.5). ■

Using 3-tuples, the solution (1.12) can be written $(x_1, x_2, x_3) = (0.55, 0.65, -0.2)$.

There are cases when elimination using replacements alone breaks down and dealing with this eventuality leads to a second type of EEO which we explain next.

■ ILLUSTRATION 1.2

Interchange

Apply forward elimination to the 4×3 linear system (S) in (1.14).

$$(S)\begin{cases} x_1 + x_2 + x_3 = 1 & E_1 \\ x_1 + x_2 = 0 & E_2 \\ -x_1 + 3x_2 + 7x_3 = 0 & E_3 \\ 2x_1 - 6x_2 - 14x_3 = 0 & E_4 \end{cases} \tag{1.14}$$

Use equation E_1 to eliminate x_1 from all equations below E_1. The required replacements, shown next, have multipliers $m = 1, -1, 2$, respectively.

$$E_2 - E_1 \to E_2, \quad E_3 + E_1 \to E_3, \quad E_4 - 2E_1 \to E_4$$

$$(S) \sim (S)'\begin{cases} x_1 + x_2 + x_3 = 1 & E_1 \\ -x_3 = -1 & E_2 \text{ (new)} \\ 4x_2 + 8x_3 = 1 & E_3 \text{ (new)} \\ -8x_2 - 16x_3 = -2 & E_4 \text{ (new)} \end{cases} \tag{1.15}$$

The next step would normally be to use the nonzero coefficient of x_2 from equation E_2 in (1.15) to eliminate x_2 from all equations below E_2, but this is obviously impossible! It is possible, however, in this case to *interchange* E_2 with an equation further down the list in order to obtain a nonzero coefficient for x_2 in the desired position. By the way, if the coefficient of x_2 is zero in all equations below E_1, we would go on to eliminate the next available variable in the list. Example 2 on page 13 illustrates this situation. ■

■ NOTATION Interchange

Suppose an $m \times n$ linear system (S) contains equations E_i and E_j, where i and j are distinct positive integers. The elementary equation operation which swaps equations E_i and E_j in (S) is called an *interchange*. The notation is

$$E_i \leftrightarrow E_j \tag{1.16}$$

and the two equations in the resulting system $(S)'$ are relabeled. ■

Performing an interchange on a linear system changes the order of equations in the system but clearly does not change the solution set. Returning to (1.15), we eliminate x_2 by performing the following EEOs:

$$E_2 \leftrightarrow E_3, \quad E_4 + 2E_2 \to E_4 \quad (m = -2)$$

$$(S)' \sim (S)'' \begin{cases} x_1 + x_2 + x_3 = 1 & E_1 \\ 4x_2 + 8x_3 = 1 & E_2 \text{ (new)} \\ -x_3 = -1 & E_3 \text{ (new)} \\ 0x_1 + 0x_2 + 0x_3 = 0 & E_4 \text{ (new)} \end{cases} . \tag{1.17}$$

Equation E_4 in $(S)''$ is called a *zero equation*—it is true for any values of the unknowns x_1, x_2, x_3. Zero equations appear during forward elimination when some equation in the original system (S) is a *linear combination*[3] of other equations in (S). In (1.14), note that $E_4 = -2E_3$, that is, $2E_3 + E_4$ is the zero equation $0x_1 + 0x_2 + 0x_3 = 0$.

The system $(S)''$ is in upper triangular form and back-substitution can be applied to yield the unique solution $x_1 = 1.75$, $x_2 = -1.75$, $x_3 = 1$, or

$$(x_1, x_2, x_3) = (1.75, -1.75, 1). \tag{1.18}$$

The systems (S) and $(S)''$ are equivalent, and so (1.18) is the unique solution to (1.14).

Forward elimination can be carried out using only replacements and interchanges. However, there is one more EEO, called *scaling*, that is often used during forward elimination.

■ NOTATION Scaling

Suppose an $m \times n$ linear system (S) contains an equation E_i for some positive integer i. The elementary equation operation that multiplies E_i throughout by a *nonzero* scalar c is called *scaling*. The notation is

$$cE_i \to E_i \tag{1.19}$$

and equation E_i is new in the resulting system $(S)'$. ■

[3] *Linear combinations* of vectors and *linear independence* of vectors are important concepts that will be explained in Chapter 3.

Scaling a linear equation in a linear system does not change the solution set.

The first and second illustrations describe the basic ideas of forward elimination. In both cases it was found that the solution consisted of a *fixed numeric value* for each of the unknowns. The unknowns are said to be *uniquely determined*[4] in such cases. In other systems certain unknowns are not uniquely determined, meaning that they are expressed in terms of one or more *auxiliary variables*, which we call *parameters*.

■ ILLUSTRATION 1.3

Parameters, Pivot Variables, Free Variables

Consider solving the linear system

$$(S)\quad \begin{cases} 2x_1 + 4x_2 - 6x_3 - 14x_4 = 11 & E_1 \\ -4x_1 - 8x_2 + 11x_3 + 26x_4 = -22 & E_2 \end{cases}. \tag{1.20}$$

Forward elimination on (S) requires the single EEO $E_2+2E_1 \to E_2$ $(m=-2)$ and we obtain the equivalent upper triangular form $(S)'$, where

$$(S)'\quad \begin{cases} 2\textcircled{x_1} + 4x_2 - 6x_3 - 14x_4 = 11 & E_1 \\ \qquad\qquad -\textcircled{x_3} - 2x_4 = 0 & E_2 \end{cases}. \tag{1.21}$$

To write down the solution to $(S)'$, consider each equation in (1.21) from left to right. The variable x_1 (circled) is the first variable we encounter in E_1 with nonzero coefficient, and similarly for x_3 in E_2. The variables x_1 and x_3 are called *pivot* (or *leading*) variables and are determined by the process of forward elimination. The remaining variables x_2 and x_4 in $(S)'$ are called *free* (or *nonpivot*) variables. The standard method of solving $(S)'$ is to express the leading variables in terms of the free variables using back-substitution. Move upward through the system, giving each free variable an independent parametric value, and solve for the pivot variable.

Solve E_2 for x_3: Let $x_4 = t$, where t is a real parameter. Scale using the EEO $(-1)E_2 \to E_2$ and then $x_3 = 0 - 2x_4 = -2t$.

Solve E_1 for x_1: Let $x_2 = s$, where s is a real parameter. Scale using the EEO $0.5\,E_1 \to E_1$ and then

$$\begin{aligned} x_1 &= 5.5 - 2x_2 + 3x_3 + 7x_4 \\ &= 5.5 - 2s + 3(-2t) + 7t \qquad \Rightarrow \qquad x_1 = 5.5 - 2s + t. \end{aligned}$$

Now $(S) \sim (S)'$ and so (S) has *infinitely many* solutions. Expressed as an equality of 4-tuples, the solution set is

$$(x_1, x_2, x_3, x_4) = (5.5 - 2s + t, s, -2t, t), \tag{1.22}$$

---

[4] See [1] for some recent work on variables that are uniquely determined. References appear at the end of each section.

for all real numbers s and t. Assigning particular values to s and t in (1.22) gives a particular solution for (1.20). Choosing $s = 2$ and $t = 1$, for example, we have

$$(x_1, x_2, x_3, x_4) = (2.5, 2, -2, 1).$$

From (1.22) we see that no unknown in (S) is uniquely determined. In other systems it may happen that some (perhaps all) pivot variables are uniquely determined and any remaining pivot variables are determined in terms of free variables. For example, all three pivot variables in (1.11) and (1.17) are uniquely determined and there are no free variables in those cases. ■

Recall from page 5 that two linear systems (S) and $(S)'$ are called equivalent, written $(S) \sim (S)'$, if $(S)'$ is the result of performing one or more EEOs on (S).

Theorem 1.1 **Fundamental Property of Equivalent Linear Systems**

If (S) and $(S)'$ are two equivalent linear systems, then (S) and $(S)'$ have exactly the same solution set.

Proof It is easy to show (Exercises 1.1) that the action of any EEO on a linear system does not change the solution set to the linear system.[5] Thus, if $(S) \sim (S)'$, then every solution to (S) is also a solution to $(S)'$. However, for every EEO there is a corresponding *inverse* EEO (of the same type) and so (S) can be recovered from $(S)'$ as follows:

	EEO			Inverse EEO		
(S)	$E_i - mE_j \to E_i$	$\to$	$(S)'$	$E_i + mE_j \to E_i$	$\to$	(S)
(S)	$E_i \leftrightarrow E_j$	$\to$	$(S)'$	$E_i \leftrightarrow E_j$	$\to$	(S)
(S)	$cE_i \to E_i$	$\to$	$(S)'$	$\frac{1}{c} E_i \to E_i$	$\to$	(S)

where i, j are distinct positive integers and m, c are nonzero scalars. Thus, $(S)' \sim (S)$ and so every solution to $(S)'$ is also a solution to (S). The solution sets to (S) and $(S)'$ are therefore identical.

Inconsistent Linear Systems

Whenever b is nonzero, the equation

$$0x_1 + 0x_2 + \cdots + 0x_n = b \tag{1.23}$$

has no solution for *any* values $x_1, x_2, \ldots, x_n$ because (1.23) reduces to $0 = b$, which is false.[6]

Suppose the systems (S) and $(S)'$ are equivalent. Then, by Theorem 1.1, (S) and $(S)'$ are either both consistent or both inconsistent. If $(S)'$ contains an equation of type (1.23) with b nonzero, then $(S)'$ and (S) are both inconsistent.

[5] Caution! Due to round-off error, this statement may fail to be true when performing EEOs using a computer.

[6] On the other hand, note that the equation $0x_1 + 0x_2 + \cdots + 0x_n = 0$ has infinitely many solutions, as does the equation $0x_1 + 0x_2 + \cdots + 0x_{n-1} + x_n = 0$, which is solved by choosing any values for the unknowns $x_1, x_2, \ldots, x_{n-1}$ and setting $x_n = 0$.

■ ILLUSTRATION 1.4

Inconsistency

Consider solving the linear system

$$(S)\quad \begin{cases} x_1 - 2x_2 + 5x_3 = 1 & E_1 \\ 2x_1 - 4x_2 + 8x_3 = 9 & E_2 \\ -3x_1 + 6x_2 - 7x_3 = 1 & E_3 \end{cases}.$$

We will apply forward elimination to (S) and display all zero coefficients in this case.

Eliminate x_1 below E_1: $E_2 - 2E_1 \to E_2$ $(m = 2)$, $E_3 + 3E_1 \to E_3$ $(m = -3)$.

$$(S) \quad \sim \quad (S)' \begin{cases} x_1 - 2x_2 + 5x_3 = 1 & E_1 \\ 0x_1 + 0x_2 - 2x_3 = 7 & E_2\ (\text{new}) \\ 0x_1 + 0x_2 + 8x_3 = 4 & E_3\ (\text{new}) \end{cases}$$

Eliminate x_3 below E_2: $E_3 + 4E_2 \to E_3$ $(m = -4)$.

$$(S)' \quad \sim \quad (S)'' \begin{cases} x_1 - 2x_2 + 5x_3 = 1 & E_1 \\ 0x_1 + 0x_2 - 2x_3 = 7 & E_2 \\ 0x_1 + 0x_2 + 0x_3 = 32 & E_3\ (\text{new}) \end{cases}$$

Equation E_3 in $(S)''$ is false for all x_1, x_2, x_3. Therefore, $(S)''$ is inconsistent and so is (S) because $(S) \sim (S)''$. ■

1.1.4 Backward Elimination

Consider the upper triangular forms (1.11), (1.17), and (1.21). As an alternative to back-substitution, note that in each case the elimination process can be continued further by moving upward and backward through the system.

■ ILLUSTRATION 1.5

Backward Elimination

Suppose a linear system has been changed by forward elimination into an upper triangular form. Begin with the last nonzero equation. Scale it if necessary to give the pivot (leading) variable a coefficient 1 and then use the new equation in replacements to eliminate the pivot variable from all equations above it. Move to the left and upward through the system, scaling and eliminating above each pivot variable. The details for the system $(S)''$ in (1.11) on page 5 are shown next.

Eliminate x_3 above E_3: $0.2E_3 \to E_3$, $E_2 + 3E_3 \to E_2$, $E_1 - E_3 \to E_1$.

$$(S)'' \begin{cases} x_1 + x_2 + x_3 = 1 & E_1 \\ -4x_2 - 3x_3 = -2 & E_2 \\ 5x_3 = -1 & E_3 \end{cases} \sim \begin{cases} x_1 + x_2 = 1.2 & E_1 \\ -4x_2 = -2.6 & E_2 \\ x_3 = -0.2 & E_3 \end{cases}$$

Eliminate x_2 above E_2: $-0.25E_2 \to E_2$, $E_1 - E_2 \to E_1$.

$$\begin{cases} x_1 + x_2 = 1.2 & E_1 \\ -4x_2 = -2.6 & E_2 \\ x_3 = -0.2 & E_3 \end{cases} \sim (S)^* \begin{cases} x_1 = 0.55 & E_1 \\ x_2 = 0.65 & E_2 \\ x_3 = -0.2 & E_3 \end{cases}$$

The system (S) in (1.6) is equivalent to the upper triangular form $(S)^*$, which we will call the *reduced upper triangular form*[7] for (S). The solution to $(S)^*$ can be read off immediately and agrees with (1.12). ■

Gauss Elimination, Gauss–Jordan Elimination

Applying forward elimination to a linear system (S) results in an equivalent system that is in upper triangular form and that is in general not unique because the choice of EEOs used is not unique. Forward elimination will reveal inconsistency in the system. Solving a linear system by forward elimination, followed by back-substitution when the system is consistent, is called *Gauss* (or *Gaussian*) *elimination*. Applying forward elimination followed by backward elimination results in a unique equivalent system that is in reduced upper triangular form. Solving by the latter method is called *Gauss–Jordan elimination*.

Computational Note

Consider the cost of solving an $n \times n$ linear system in n unknowns. The number of arithmetic operations (additions, subtractions, multiplications, divisions) required for forward elimination is about $2n^3/3$. Back-substitution requires a further n^2 operations which, for large n, is negligible compared to the count for forward elimination. Gauss–Jordan elimination requires about n^3 operations and hence Gaussian elimination is considerably more efficient than Gauss–Jordan elimination.

1.1.5 Matrix Representation of a Linear System

At this point we consider a more efficient way of representing a linear system using matrix notation. The *matrix form* of a linear system provides both a welcomed simplification of notation and a means of entering a linear system into a computer for solution by machine.

Definition 1.3

Matrix, row vector, column vector

Let m and n be positive integers. An $m \times n$ (read "m by n") *matrix* is a rectangular array of mn numbers arranged in m rows and n columns and enclosed in square brackets. The numbers in a matrix are called *entries*. An $m \times 1$ matrix (m rows, one column) is called a *column vector*, or *m-vector*.

Matrices are denoted by boldfaced capital letters: $\mathbf{A}, \mathbf{B}, \mathbf{C}, \ldots, \mathbf{M}, \mathbf{N}, \ldots$ and vectors by lower case boldfaced letters: $\mathbf{a}, \mathbf{b}, \mathbf{c}, \ldots, \mathbf{u}, \mathbf{v}, \ldots$.

[7] A precise definition is given at the end of Section 1.2.

Let us develop the matrix form of the 4×4 linear system (S) shown next.

$$(S)\quad \begin{cases} \phantom{2x_1 +{}} x_2 - 2x_3 + 2x_4 = 2 & E_1 \\ 2x_1 + 3x_2 + 6x_3 + 3x_4 = 11 & E_2 \\ -4x_1 - 13x_2 + 2x_3 + x_4 = -15 & E_3 \\ - x_2 + 2x_3 + x_4 = 1 & E_4 \end{cases} \tag{1.24}$$

The coefficients from the left side of (S) are stored in a 4×4 *coefficient matrix* **A** and the constant terms from the right side of (S) are stored in a 4×1 matrix (column vector) **b**. We call **b** the *column of constant terms* in (S).

$$\mathbf{A} = \begin{bmatrix} 0 & 1 & -2 & 2 \\ 2 & 3 & 6 & 3 \\ -4 & -13 & 2 & 1 \\ 0 & -1 & 2 & 1 \end{bmatrix} \qquad \mathbf{b} = \begin{bmatrix} 2 \\ 11 \\ -15 \\ 1 \end{bmatrix} \tag{1.25}$$

The zero entries in **A** correspond to the unknowns in (S) that have coefficient zero. The numerical data that define (S) can now be stored in a 4×5 matrix **M**, shown in (1.26), which has $m = 4$ rows, $n = 5$ columns, and $mn = 20$ entries in all.

$$\mathbf{M} = [\mathbf{A}\,|\,\mathbf{b}] = \left[\begin{array}{rrrr|r} 0 & 1 & -2 & 2 & 2 \\ 2 & 3 & 6 & 3 & 11 \\ -4 & -13 & 2 & 1 & -15 \\ 0 & -1 & 2 & 1 & 1 \end{array}\right] \tag{1.26}$$

The entry that lies at the intersection of row i and column j in a matrix **M** is called the (i, j)*-entry* in **M**. In (1.26), the (1, 1)-entry is 0 and the (2, 5)-entry is 11. We also say that -4 is in the (3, 1)*-position* in **M**.

The Augmented Matrix of a Linear System

The matrix $\mathbf{M} = [\mathbf{A}\,|\,\mathbf{b}]$ in (1.26) is called the *augmented matrix*[8] for (S) because it is formed by adjoining the column vector **b** to **A** as an extra column on the right. The vertical line that separates coefficients from constant terms in (1.26) is called a *partition* of **M**. More ideas concerning partitioning of matrices and block multiplication appear in Section 2.1.

Row Vectors of a Matrix

A $1 \times n$ matrix (one row, n columns) is called a *row vector*. The entries in an $m \times n$ matrix define m row vectors, each with n entries. For example, the 4×4 matrix **A** in (1.25) defines four row vectors, denoted by $\mathbf{r}_1$, $\mathbf{r}_2$, $\mathbf{r}_3$, $\mathbf{r}_4$, as follows:

$$\begin{aligned} \mathbf{r}_1 &= [\;0 \quad 1 \quad -2 \quad 2] \\ \mathbf{r}_2 &= [\;2 \quad 3 \quad 6 \quad 3] \\ \mathbf{r}_3 &= [-4 \quad -13 \quad 2 \quad 1] \\ \mathbf{r}_4 &= [\;0 \quad -1 \quad 2 \quad 1] \end{aligned} \tag{1.27}$$

[8] Augment means "to enlarge or add to."

Sometimes we use commas to separate the entries in row vectors (in order to avoid ambiguity), as in $\mathbf{r} = [0, 1, -2, 2]$, which is also acceptable input into MATLAB.

Elementary Row Operations

A linear system (S) is solved by performing a sequence of EEOs on the equations in (S). When (S) is represented by its augmented matrix $\mathbf{M}$, each EEO corresponds to an *elementary row operation* (ERO) that is performed on row vectors in $\mathbf{M}$.

■ NOTATION Elementary Row Operations

Suppose the row vectors in a matrix $\mathbf{M}$, ordered top to bottom, are denoted by $\mathbf{r}_1, \mathbf{r}_2, \ldots$. The EROs and their inverses that correspond to the EEOs and their inverses are as follows:

	ERO	**Inverse ERO**
(row) replacement	$\mathbf{r}_i - m\mathbf{r}_j \to \mathbf{r}_i$	$\mathbf{r}_i + m\mathbf{r}_j \to \mathbf{r}_i$
(row) interchange	$\mathbf{r}_i \leftrightarrow \mathbf{r}_j$	$\mathbf{r}_i \leftrightarrow \mathbf{r}_j$
(row) scaling	$c\mathbf{r}_i \to \mathbf{r}_i$	$\frac{1}{c}\mathbf{r}_i \to \mathbf{r}_i$

where i, j are distinct positive integers and m, c are nonzero scalars. ■

In practice, we perform forward elimination and backward elimination on a linear system using the associated augmented matrices. We illustrate this process after introducing the following terminology.

Definition 1.4

Equivalent matrices

Two $m \times n$ matrices $\mathbf{M}$ and $\mathbf{M}'$ are called *(row) equivalent* if there is a finite sequence of elementary row operations (EROs) that changes one matrix into the other matrix. We write $\mathbf{M} \sim \mathbf{M}'$ when this is the case.

EXAMPLE 2

Solving a Linear System Using Matrices

We will apply Gauss–Jordan elimination to solve the 3×4 linear system (S) shown. Beginning with the augmented matrix $\mathbf{M}$, the matrix entries (corresponding to coefficients) used in the elimination processes are circled. We call these entries *pivots*.

$$(S) \quad \begin{cases} x_1 + 2x_2 + x_3 + 3x_4 = 4 \\ 3x_1 + 6x_2 \quad 5x_3 + 10x_4 = 0 \\ 5x_1 + 10x_2 + 7x_3 + 17x_4 = 23 \end{cases}$$

$$\mathbf{M} = [\mathbf{A} \,|\, \mathbf{b}] = \left[\begin{array}{cccc|c} 1 & 2 & 1 & 3 & 4 \\ 3 & 6 & 5 & 10 & 0 \\ 5 & 10 & 7 & 17 & 23 \end{array}\right]$$

Forward elimination:

$$\mathbf{M} = \left[\begin{array}{cccc|c} \textcircled{1} & 2 & 1 & 3 & 4 \\ 3 & 6 & 5 & 10 & 0 \\ 5 & 10 & 7 & 17 & 23 \end{array}\right] \underset{\substack{\mathbf{r}_2 - 3\mathbf{r}_1 \to \mathbf{r}_2 \\ \mathbf{r}_3 - 5\mathbf{r}_1 \to \mathbf{r}_3}}{\sim} \left[\begin{array}{cccc|c} \textcircled{1} & 2 & 1 & 3 & 4 \\ 0 & 0 & \textcircled{2} & 1 & -12 \\ 0 & 0 & 2 & 2 & 3 \end{array}\right]$$

$$\underset{\mathbf{r}_3 - \mathbf{r}_2 \to \mathbf{r}_3}{\sim} \left[\begin{array}{cccc|c} \textcircled{1} & 2 & 1 & 3 & 4 \\ 0 & 0 & \textcircled{2} & 1 & -12 \\ 0 & 0 & 0 & \textcircled{1} & 15 \end{array}\right] = \mathbf{M}'$$

It is clear from $\mathbf{M}'$ that (S) is consistent. Inconsistency would be revealed during forward elimination by the appearance of an equation $0x_1 + \cdots + 0x_4 = b$, where b is nonzero, and this would correspond to a row vector in the corresponding augmented matrix of the form $[0\ 0\ 0\ 0 \mid b]$.

Backward elimination (all pivots are scaled to 1 during this phase):

$$\mathbf{M}' = \left[\begin{array}{cccc|c} \textcircled{1} & 2 & 1 & 3 & 4 \\ 0 & 0 & \textcircled{2} & 1 & -12 \\ 0 & 0 & 0 & \textcircled{1} & 15 \end{array}\right] \underset{\substack{\mathbf{r}_2 - \mathbf{r}_3 \to \mathbf{r}_2 \\ \mathbf{r}_1 - 3\mathbf{r}_3 \to \mathbf{r}_1}}{\sim} \left[\begin{array}{cccc|c} \textcircled{1} & 2 & 1 & 0 & -41 \\ 0 & 0 & \textcircled{2} & 0 & -27 \\ 0 & 0 & 0 & \textcircled{1} & 15 \end{array}\right]$$

$$\underset{\substack{0.5\mathbf{r}_2 \to \mathbf{r}_2 \\ \mathbf{r}_1 - \mathbf{r}_2 \to \mathbf{r}_1}}{\sim} \left[\begin{array}{cccc|c} \textcircled{1} & 2 & 0 & 0 & -27.5 \\ 0 & 0 & \textcircled{1} & 0 & -13.5 \\ 0 & 0 & 0 & \textcircled{1} & 15 \end{array}\right] = \mathbf{M}^*$$

$\mathbf{M}^*$ is the augmented matrix of the system $(S)^*$ shown next. Note that $(S) \sim (S)^*$ and $\mathbf{M} \sim \mathbf{M}^*$.

$$(S)^* \begin{cases} \textcircled{x_1} + 2x_2 & = -27.5 \\ \qquad\qquad \textcircled{x_3} & = -13.5 \\ \qquad\qquad\qquad \textcircled{x_4} & = 15 \end{cases}$$

The variables x_1, x_3, x_4 are pivot variables—x_3, x_4 are uniquely determined and x_1 is determined in terms of the free variable x_2. There are infinitely many solutions to (S) given by

$$(x_1, x_2, x_3, x_4) = (-27.5 - 2t, t, -13.5, 15),$$

where $x_2 = t$ is a real parameter. ▪

Visualizing Linear Systems

Each linear equation in two variables is represented by a *line* in $\mathbb{R}^2$ (the plane), and conversely. Consider the 4×2 linear system

$$(S)\left\{\begin{array}{rlll} -x+3y &= 1 & E_1 & (L_1) \\ x+\ y &= 1 & E_2 & (L_2) \\ x+\ y &= 2 & E_3 & (L_3) \\ 4x+4y &= 8 & E_4 & (L_4) \end{array}\right. . \tag{1.28}$$

A MATLAB plot[9] of the four lines L_1, L_2, L_3, L_4 that represent the equations in (S) is shown in Figure 1.1. Note that the lines L_3 and L_4 are coincident.

The system in (1.28) is consistent if and only if there is at least one point that lies simultaneously on all four lines. Clearly, (S) is inconsistent. Three *subsystems* of (S) are shown next. The systems have, respectively, zero, one, and infinitely many solutions (check this geometrically).

$$(S_1)\left\{\begin{array}{ll} x+y=1 & E_2 \\ x+y=2 & E_3 \end{array}\right. \qquad (S_2)\left\{\begin{array}{ll} -x+3y=1 & E_1 \\ x+\ y=1 & E_2 \end{array}\right.$$

$$(S_3)\left\{\begin{array}{ll} x+\ y=2 & E_3 \\ 4x+4y=8 & E_4 \end{array}\right.$$

Each linear equation in three variables is represented by a *plane* in $\mathbb{R}^3$ (3-space), and conversely. Consider the 2×3 linear system

$$(S)\left\{\begin{array}{rlll} x+\ y+z &= 2 & E_1 & (P_1) \\ 3x-2y+z &= 1 & E_2 & (P_2) \end{array}\right. . \tag{1.29}$$

A MATLAB plot[10] of the planes P_1, P_2 that represent the equations in (1.29) is shown in Figure 1.2. Two planes in $\mathbb{R}^3$ are either parallel or, if not, intersect in a line L. The system (1.29) will be inconsistent if P_1 and P_2 are parallel.

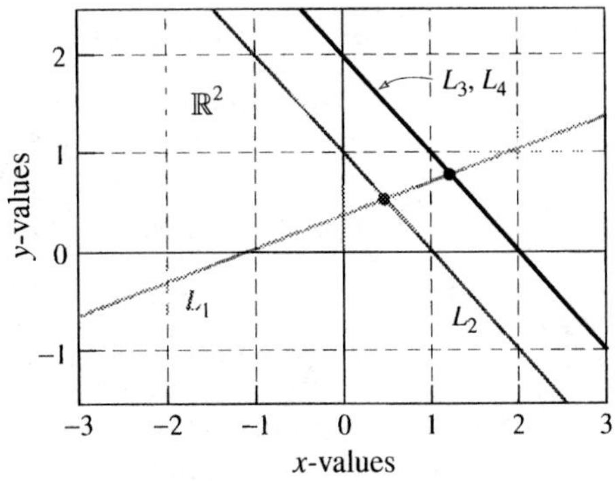

Figure 1.1 Lines in $\mathbb{R}^2$.

[9] MATLAB graphics have been modified using Adobe Illustrator®.

[10] Some practice may be required in reading MATLAB 3-D graphics due to the orientation and placement of the axes.

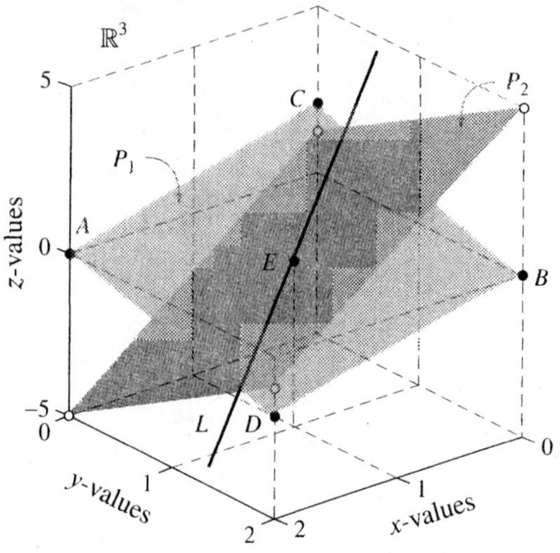

Figure 1.2 Planes in $\mathbb{R}^3$.

However, elimination shows that (S) is consistent with infinitely many solutions given by

$$(x, y, z) = (1 - 0.6t, 1 - 0.4t, t), \quad \text{where } t \text{ is a real parameter.} \tag{1.30}$$

Hence, P_1 and P_2 intersect in a line L whose equation in *parametric form* is obtained by equating components in (1.30), that is,

$$x = 1 - 0.6t, \qquad y = 1 - 0.4t, \qquad z = t. \tag{1.31}$$

Any point (x, y, z) on L is obtained by assigning t a value in (1.31). The value $t = 0$, for example, corresponds to $E = (1, 1, 0)$, which is the point where L passes through the xy-plane ($z = 0$).

Higher Dimensions, Hyperplanes

We rely on analogy to interpret geometrically linear equations in four or more variables. Observe that a linear equation in two variables is visualized as a line (a one-dimensional space) in $\mathbb{R}^2$, and a linear equation in three variables is visualized as a plane (a two-dimensional space) in $\mathbb{R}^3$. Consider the linear equation

$$2x_1 - 3x_2 + 5x_3 - x_4 = 10. \tag{1.32}$$

The set of all points $P = (x_1, x_2, x_3, x_4)$ in $\mathbb{R}^4$ that satisfy equation (1.32) is called *hyperplane*, and is a three-dimensional space in $\mathbb{R}^4$. The hyperplane passes through the point $P = (5, 0, 0, 0)$ because the coordinates of P satisfy (1.32). The points $Q = (0, 0, 2, 0)$ and $R = (2.1, 4.3, 3.6, -0.7)$ also lie in the hyperplane. This interpretation is perfectly general. For example, the equation

$$x_1 - 2x_2 + 5x_3 - 2x_4 - x_5 = 1$$

has four *degrees of freedom* and represents a four-dimensional hyperplane in $\mathbb{R}^5$. The points $P = (0, 0, 0, 0, -1)$ and $Q = (1, 1, 1, 1, 1)$ lie in this hyperplane.

REFERENCES

1. Hardy, K., Williams, K. S., and Spearman, B. K., "Uniquely Determined Unknowns in Systems of Linear Equations," *Mathematics Magazine*, 75 (1), 53–57 (2002).

Historical Notes

Qin Jiu-Shao (1202–1261) Chinese mathematician. Versions of the elimination algorithms to solve linear systems have been known for centuries, including one which appeared in the work of Qin Jiu-Shao.

Wilhelm Jordan (1842–1899) German engineer, specializing in geodesy, whose name is associated with the Gauss–Jordan algorithm. He should not be confused with the famous French mathematician **Camille Jordan** (1838–1922).

EXERCISES 1.1

Exercises 1–4. When possible, simplify each equation and write in standard linear form.

1. $(x+2)^2 + (y-1)^2 = 3 + x^2 + y^2$

2. $(x-y)^2 = x^2 + y^2$

3. $(x-y)(x+y) = x^2 + y^2$

4. $x + (x-1)(y-1) = xy + y - 1$

Exercises 5–8. The linear systems are in upper triangular form. In each case, identify the pivot variables and free (nonpivot) variables. Solve by back-substitution and state which variables (if any) are uniquely determined.

5. $\begin{cases} x_1 - 2x_2 \qquad\quad - 10x_4 = 0 \\ \quad - x_2 + x_3 - 2x_4 = 1 \\ \qquad\qquad 2x_3 + 4x_4 = 6 \end{cases}$

6. $\begin{cases} x_1 + 2x_2 + x_3 - x_4 = 2 \\ \qquad x_2 - x_3 + 7x_4 = 1 \\ \qquad\qquad\qquad 5x_4 = 12 \end{cases}$

7. $\begin{cases} 4.5x_1 + 2.25x_2 \qquad\qquad\qquad = 18.0 \\ \qquad\qquad - 2.2x_3 + 1.1x_4 = 3.3 \end{cases}$

8. $\begin{cases} x_1 - x_2 + 2x_3 + x_4 \qquad = 7.0 \\ \qquad\quad 3x_3 \qquad\qquad\quad = 6.3 \\ \qquad\qquad\qquad x_4 + x_5 = 2.0 \end{cases}$

Exercises 9–10. The linear systems are in lower triangular form. *Solve by* forward-substitution. *Plot the system in Exercise 9.*

9. $\begin{cases} 3x_1 \qquad\quad = 6.0 \\ x_1 - x_2 = 5.5 \end{cases}$

10. $\begin{cases} 3x_1 \qquad\qquad\quad = 2 \\ x_1 + x_2 \qquad\quad = 0 \\ 3x_1 - 3x_2 + x_3 = 1 \end{cases}$

Exercises 11–16. Represent each linear system in matrix form. Solve by Gauss elimination when the system is consistent and cross-check by substituting your solution set back into all equations. Interpret the solution geometrically in terms of planes in $\mathbb{R}^3$.

11. $\begin{cases} 2x_1 + 3x_2 - x_3 = 1 \\ 4x_1 + 7x_2 + x_3 = 3 \\ 7x_1 + 10x_2 - 4x_3 = 4 \end{cases}$

12. $\begin{cases} 3x_1 + 3x_2 + x_3 = -4.5 \\ x_1 + x_2 + x_3 = 0.5 \\ -2x_1 - 2x_2 \qquad = 5.0 \end{cases}$

13. $\begin{cases} x_1 + 2x_2 - 3x_3 = 1 \\ 3x_1 + 6x_2 + x_3 = 13 \\ 4x_1 + 8x_2 - 2x_3 = 9 \end{cases}$

14. $\begin{cases} \qquad x_2 - 5x_3 = -3 \\ x_1 + 3x_2 \qquad = 1 \end{cases}$

15. $$\begin{cases} 5x_1 - 2x_2 + x_3 = 2 \\ 3x_1 + 2x_2 + 7x_3 = 3 \\ x_1 + x_2 + 3x_3 = 2 \\ 9x_1 + x_2 + 11x_3 = 7 \end{cases}$$

16. $$\begin{cases} x_1 - x_2 + x_3 = 1 \\ 2x_1 + 3x_2 - x_3 = 4 \\ -x_1 - 2x_2 + 5x_3 = 2 \end{cases}$$

Exercises 17–20. Before solving the given system, try to predict the number of solutions: zero, one, or infinitely many, then solve using matrix methods.

17. $$\begin{cases} x_1 + 4x_2 - 4x_3 + 4x_4 = 5 \\ 2x_1 - x_2 + x_3 - x_4 = 1 \\ x_1 + x_2 - x_3 + x_4 = 2 \end{cases}$$

18. $$\begin{cases} x_1 - 2x_2 + x_3 = 2 \\ 2x_1 - 5x_2 + 3x_3 = 6 \\ x_1 + 2x_2 + 2x_3 = 4 \\ 2x_1 \qquad + 3x_3 = 6 \end{cases}$$

19. $$\begin{cases} -x_1 + 3x_2 - 4x_3 + x_4 = 2 \\ x_1 - x_2 + x_3 - x_4 = 3 \end{cases}$$

20. $$\begin{cases} x_1 + 2x_2 \qquad + 2x_4 + 2x_5 = 10 \\ 2x_1 + 4x_2 + 2x_3 + 2x_4 - 2x_5 = 14 \\ x_3 - x_4 - x_5 = -1 \\ x_1 + 2x_2 - x_3 + 3x_4 + x_5 = 3 \end{cases}$$

21. Find the general solution to the linear system and find a particular solution for which $x_3 = 0$.

$$\begin{cases} 3x_1 + x_2 - 2x_3 \qquad = 3 \\ x_1 + 2x_2 + x_3 - 2x_4 = 4 \\ -5x_2 \qquad + x_4 = 3 \end{cases}$$

22. Regard x_1 as a pivot variable and solve the single equation $x_1 + 2x_2 + 3x_3 = 4$. Is $(3, 2, -1)$ a particular solution? Solve the equation again by letting $x_1 = p$ and $x_3 = q$. Show that the two solution sets you have obtained are identical.

Exercises 23–24. Write down the augmented matrix $\mathbf{M} = [\mathbf{A}|\mathbf{b}]$ *of each system. State the sizes of* $\mathbf{A}$, $\mathbf{b}$, *and* $\mathbf{M}$. *Solve by Gauss–Jordan elimination. Find all particular solutions in nonnegative integers.*

23. $$\begin{cases} 2x_1 + x_2 + 13x_3 = 28 \\ -x_1 + x_2 - 2x_3 = -8 \\ x_1 + x_2 + 8x_3 = 16 \end{cases}$$

24. $$\begin{cases} 7x_1 + 14x_2 - 2x_3 + 5x_4 = 15 \\ 3x_1 + 6x_2 - x_3 + 2x_4 = 6 \end{cases}$$

Exercises 25–26. Find the value(s) of α *such that each system has (a) no solutions, (b) one solution, (c) infinitely many solutions.*

25. $$\begin{cases} \alpha x_1 + x_2 + x_3 = 1 \\ x_1 + \alpha x_2 + x_3 = 1 \\ x_1 + x_2 + \alpha x_3 = 1 \end{cases}$$

26. $$\begin{cases} x_1 + 2x_2 + 2x_3 = 1 \\ x_2 + \alpha x_3 = 1 \\ -x_1 + x_2 + \alpha x_3 = \alpha \end{cases}$$

27. Find the general solution to the linear system (S) and all particular solutions in positive integers.

$$(S) \begin{cases} x_1 + x_2 - x_3 + x_4 = -1 \\ x_1 + x_2 \qquad - x_4 = 0 \\ 2x_1 + 2x_2 - 2x_3 + 5x_4 = 1 \end{cases}$$

28. Explain why $(-1, 1, 0)$ is not a particular solution for (S), where

$$(S) \begin{cases} -3x_1 + 5x_2 - 7x_3 = -2 & E_1 \\ 4x_1 - 6x_2 + 8x_3 = 2 & E_2 \end{cases}.$$

Solve (S) and give a geometric interpretation. Find particular solutions in positive integers such that $x_1 < 3$.

Exercises 29–32. Each matrix $\mathbf{M} = [\mathbf{A}|\mathbf{b}]$ *is the augmented matrix of a linear system* (S). *State the sizes of* $\mathbf{M}$ *and* $\mathbf{A}$. *Write down* (S). *Use* $\mathbf{M}$ *to solve* (S) *by Gauss elimination.*

29. $$\left[\begin{array}{ccc|c} 1 & 2 & 4 & 3 \\ 2 & 4 & 8 & 6 \\ -1 & 6 & 8 & -15 \\ 4 & 2 & 7 & 21 \end{array}\right]$$

30. $$\left[\begin{array}{ccccc|c} 1 & -1 & 0 & 1 & 1 & 2 \\ -1 & 2 & -1 & 1 & 1 & 2 \\ 5 & -8 & 3 & -1 & -1 & 0 \end{array}\right]$$

31. $$\left[\begin{array}{cccc|c} 0 & 2 & 2 & 0 & 0 \\ 0 & 0 & 3 & 3 & 3 \\ 1 & 0 & 0 & -1 & 2 \\ -5 & -5 & 0 & 0 & 15 \end{array}\right]$$

32. $\left[\begin{array}{rrrrr|r} 1 & -1 & 0 & 1 & 1 & 2 \\ -1 & 2 & -1 & 1 & 1 & 2 \end{array}\right]$

Exercises 33–36. Solve each linear system by Gauss–Jordan elimination. Draw an accurate plot of each system in $\mathbb{R}^2$.

33. $\begin{cases} x + 5y = 2 \\ -3x - 9y = 6 \\ 2x + 16y = 16 \end{cases}$

34. $\begin{cases} 3x + 4y + 6 = x \\ 7x + 4y - 6 = -y \end{cases}$

35. $\begin{cases} 2y = 4 \\ 2x - 4y = -5 \end{cases}$

36. $\begin{cases} 6x - 9y = 0 \\ -x + 1.5y = 2.5 \end{cases}$

Exercises 37–38. Find the value(s) of c for which each linear system is consistent. Solve the system for these values of c and plot in $\mathbb{R}^2$.

37. $\begin{cases} 2x - 4y = -1 & E_1 \\ x + 3y = 2 & E_2 \\ 4x + 5y = c & E_3 \end{cases}$

38. $\begin{cases} x + y = 1 & E_1 \\ -x + 2y = 2 & E_2 \\ 2x + 3y = c^2 & E_3 \end{cases}$

39. Find the equation of the line $y = mx + b$ in $\mathbb{R}^2$ that passes through the points $(1, 2)$ and $(-1, 0.5)$.

40. Find the equation of the plane $z = ax + by + c$ in $\mathbb{R}^3$ that passes through the points $(-1, 1, 0)$, $(0, 2, 1)$, $(4, 5, 2)$.

41. Show that $(x_1, x_2) = (0.7, 0)$ is not a solution to the system (S). Determine if $(x_1, x_2) = (1.2, -1)$ is a particular solution. Plot the corresponding lines in $\mathbb{R}^2$.

$$(S)\quad \begin{cases} x_1 + 1.5x_2 = -0.3 & E_1 \\ 2x_1 + x_2 = 1.4 & E_2 \\ 3x_1 + 2.5x_2 = 1.1 & E_3 \end{cases}$$

42. Perform forward elimination on the given system. (a) Give an algebraic condition in a, b, c that will guarantee that the system (S) is consistent. (b) How many solutions will (S) have? (c) Interpret the condition geometrically.

$$(S)\quad \begin{cases} x_1 + 2x_2 + 3x_3 = a \\ 2x_1 + 4x_2 + 7x_3 = b \\ 5x_1 + 10x_2 + 16x_3 = c \end{cases}$$

Exercises 43–44. Give a geometric argument that shows that three distinct, noncollinear points in the plane determine a unique circle passing through the points. Use Gauss–Jordan elimination to find the equation of the circle $x^2 + y^2 + ax + by + c = 0$ *that passes through the given points. Express the equation in the form* $(x - p)^2 + (y - q)^2 = r^2$ *and so find the center* (p, q) *and radius* r *of the circle.*

43. $(0, -1)$, $(1, 0)$, $(2, 2)$

44. $(1, 1)$, $(-1, 1)$, $(-1, 2)$

45. Solve the system of nonlinear equations

$$\begin{cases} xy + y^2 + 2yz = 9 \\ 2xy + 4y^2 - 3yz = 1 \\ 3xy + 6y^2 - 5yz = 0 \end{cases}.$$

46. Find the solution set of angles (α, β, γ), where $0 \le \alpha, \gamma \le \pi$ and $0 \le \beta \le 2\pi$ (π radians $= 180$ degrees). Greek letters: alpha (α), beta (β), gamma (γ).

$$\begin{cases} \sin\alpha + 4\cos\beta + \tan\gamma = 2 \\ \sin\alpha + 6\cos\beta - \tan\gamma = 5 \\ 4\sin\alpha + 18\cos\beta + \tan\gamma = 12 \end{cases}$$

47. Find all angles α that satisfy the nonlinear system

$$\begin{cases} 2x\cos\alpha + 3x = 8 \\ 3x\cos\alpha + 4x = 11 \end{cases}.$$

48. To illustrate that the action of EEOs does not change solution sets, suppose that $(x_1, x_2, x_3) = (a_1, a_2, a_3)$ is a particular solution to the linear system

$$(S_1)\quad \begin{cases} p_1x_1 + p_2x_2 + p_3x_3 = b_1 & E_1 \\ q_1x_1 + q_2x_2 + q_3x_3 = b_2 & E_2 \end{cases}.$$

Apply each EEO shown to (S_1) to obtain an equivalent system (S_2). In each case show that the particular solution to (S_1) is also a particular solution to (S_2).

(a) Replacement $E_1 - mE_2 \to E_1$
(b) Interchange $E_1 \leftrightarrow E_2$
(c) Scaling $cE_1 \to E_1$

49. Refer to Figure 1.2. Find the points in $\mathbb{R}^3$ where the line L passes through the plane (a) $x = 0$ and (b) $y = 0$. Find coordinates of points A, B, C, D.

50. The Chinese and Babylonians were acquainted with systems of linear equations and historians of mathematics believe that the *Nine Chapters of the Mathematical Art*, written during the Han Dynasty between 200 and 100 B.C., contains the first recorded use of matrix methods. The author of *Nine Chapters* describes a practical problem that in modern language requires the solution of the 3×3 linear system (S). Solve the system by forward elimination.

$$(S)\quad \begin{cases} 3x_1 + 2x_2 + x_3 = 39 \\ 2x_1 + 3x_2 + x_3 = 34 \\ x_1 + 2x_2 + 3x_3 = 26 \end{cases}$$

Exercises 51–52. Find the parametric form of the equation of the line of intersection of the given planes.

51. $\begin{cases} 2x_1 + 3x_2 + x_3 = 1 \\ 4x_1 + 7x_2 + 5x_3 = 7 \end{cases}$ **52.** $\begin{cases} 2x_1 - 3x_2 \quad = 2 \\ 4x_1 - 5x_2 + x_3 = 6 \end{cases}$

Exercises 53–65. Label each statement as either true (T) or false (F). If a false statement can be made true by modifying the wording, then do so.

53. If a consistent linear system has no free variables, then each unknown is uniquely determined.

54. A single linear equation has one pivot variable.

55. A 2×2 linear system has two pivot variables.

56. A 3×3 linear system cannot have two free variables.

57. A linear system is inconsistent whenever there are more equations than unknowns.

58. Forward elimination can be carried out using only replacements and scaling.

59. Forward elimination can be carried out using only replacements and interchanges.

60. If a linear system (S) is inconsistent, then every subsystem formed using equations from (S) is also inconsistent.

61. Gauss–Jordan elimination is the most efficient method of solving any linear system.

62. An $m \times n$ matrix has mn entries and n columns.

63. A linear equation in eight variables is represented by a hyperplane in $\mathbb{R}^8$.

64. An $m \times n$ linear system is more likely to be inconsistent whenever $m > n$.

65. A linear system with two equations in four unknowns has infinitely many solutions when the constant terms are both zero.

USING MATLAB

Read Appendix A: MATLAB. *Download the M-file* **gauss.m** *from the Web site*

www.prenhall.com/hardy

If the matrix **M** *is in the* MATLAB *workspace, the command* **gauss(M)** *performs EROs of your choice on the matrix* **M**.

66. Use **gauss.m** to solve the linear system

$$(S) \quad \begin{cases} x_1 + 2x_2 = 3 \\ 4x_1 - 5x_2 = 6 \end{cases}$$

67. Use **gauss.m** to solve the linear system (1.2) on page 1. Use your solution to solve the system (1.24) on page 12.

68. Use **gauss.m** to solve the linear system (S) whose augmented matrix **M** is given by

$$\mathbf{M} = \left[\begin{array}{ccccc|c} 1 & 1 & 2 & 12 & 48 & 100 \\ 2 & 2 & 6 & 24 & 96 & 50 \\ 0 & 0 & 3 & 4 & 6 & 24 \\ -1 & -1 & 0 & -12 & -48 & -250 \\ 2 & 2 & 11 & 28 & 102 & -76 \end{array}\right].$$

69. *Plotting Lines.* Consult online help for the topics **hold on** and **plot**. Build an M-file containing the commands

```
hold on;
x = -2:0.1:5;
y1 = 1-0.5*x;
y2 = 3-x;
plot (x,y1,x,y2)
```

Run the M-file in order to plot the lines $x + 2y = 2$, $x + y = 3$ in the same Figure Window.

70. *Project.* Reproduce the plot in Figure 1.1.

71. *Project.* Reproduce the plot in Figure 1.2.

72. *Project.* Let $n = abc$ be a three-digit integer. The sum of the digits in n is 14 and the middle digit is the sum of the other two. Let $m = acb$ be the result of interchanging the last two digits in n.

(a) If $n - m = 27$, find n.

(b) Use hand calculation to find all other integers n with $n - m = k$, where k is a positive integer. Check your hand calculations by writing an M-file to step through positive integer values k such that $n - m = k$, solving the problem for each value of k. What is the maximum value of k which should be tested?

73. Open the M-file **gauss.m** using the MATLAB editor and look at the structure of the program. Make improvements.

74. Use **gauss.m** to cross-check your hand calculations in this exercise set.

1.2 Echelon Forms, Rank

1.2.1 Forward Reduction

Forward and backward elimination are algorithms (step-by-step procedures) that are used to solve linear systems. In this section we focus on matrices in general and formulate forward and backward elimination as algorithms that may be applied to any matrix. In this context we use the terms *forward reduction* and *backward reduction* of the matrix.

Algorithm 1.1 Forward reduction of a matrix

Let **M** be an $m \times n$ matrix.

Step 1 Locate the first nonzero column[11] from the left in **M** and call it column j. This is a *pivot column.*

Step 2 Choose a nonzero entry p_j in column j. The number p_j is called the *pivot* for column j and the row in which the pivot appears is called a *pivot row*. If necessary, use an interchange to move the pivot row into first position and then use replacements to reduce any nonzero entries below p_j to zero.

Step 3 Cover up the pivot row to form a *submatrix* of the current matrix with one less row. If the submatrix has no rows or all zero rows, then stop; else apply Steps 1 and 2 to the submatrix. Continue in this way, always covering up the pivot row in the current submatrix to form a new submatrix with one less row.

■ ILLUSTRATION 1.6

Forward Reduction of a Matrix

Recall from page 13 that two matrices **M** and **M′** are (row) equivalent, written $\mathbf{M} \sim \mathbf{M}'$, if $\mathbf{M}'$ is the result of performing a sequence of EROs on **M**. We apply Algorithm 1.1 to forward reduce the matrix **M** shown next. The chosen pivots are circled.

Step 1 Column 1, $p_1 = 2$.

Step 2 $\mathbf{r}_1 \leftrightarrow \mathbf{r}_2$, reduce below the pivot, $\mathbf{r}_3 + 2\mathbf{r}_1 \to \mathbf{r}_3$.

$$\mathbf{M} = \begin{bmatrix} 0 & 1 & -2 & 2 & 2 \\ \textcircled{2} & 3 & 6 & 3 & 11 \\ -4 & -13 & 2 & 1 & -15 \\ 0 & -1 & 2 & 1 & 1 \end{bmatrix} \sim \begin{bmatrix} \textcircled{2} & 3 & 6 & 3 & 11 \\ \hline 0 & 1 & -2 & 2 & 2 \\ 0 & -7 & 14 & 7 & 7 \\ 0 & -1 & 2 & 1 & 1 \end{bmatrix} = \mathbf{M}'$$

Step 3 Cover up row 1 in **M′** to form a 3×5 submatrix of **M′**.

Step 1 Column 2, $p_2 = 1$.

Step 2 Reduce below the pivot, $\mathbf{r}_3 + 7\mathbf{r}_2 \to \mathbf{r}_3$, $\mathbf{r}_4 + \mathbf{r}_2 \to \mathbf{r}_4$.

$$\begin{bmatrix} \textcircled{2} & 3 & 6 & 3 & 11 \\ \hline 0 & \textcircled{1} & -2 & 2 & 2 \\ 0 & -7 & 14 & 7 & 7 \\ 0 & -1 & 2 & 1 & 1 \end{bmatrix} \sim \begin{bmatrix} \textcircled{2} & 3 & 6 & 3 & 11 \\ 0 & \textcircled{1} & -2 & 2 & 2 \\ \hline 0 & 0 & 0 & \textcircled{21} & 21 \\ 0 & 0 & 0 & 3 & 3 \end{bmatrix} = \mathbf{M}''$$

Step 3 Cover up rows 1 and 2 in **M″** to form a 2×5 submatrix of **M″**.

---

[11] A column (row) vector **v** is called a *zero* column (row) if all entries in **v** are zero and so **v** is nonzero if **v** has *at least one* nonzero entry.

Step 1 Column 4, $p_4 = 21$.
Step 2 Reduce below the pivot, $\mathbf{r}_4 - \frac{3}{21}\mathbf{r}_3 \to \mathbf{r}_4$.

$$\left[\begin{array}{ccccc} \textcircled{2} & 3 & 6 & 3 & 11 \\ 0 & \textcircled{1} & -2 & 2 & 2 \\ \hline 0 & 0 & 0 & \textcircled{21} & 21 \\ 0 & 0 & 0 & 3 & 3 \end{array}\right] \sim \left[\begin{array}{ccccc} \textcircled{2} & 3 & 6 & 3 & 11 \\ 0 & \textcircled{1} & -2 & 2 & 2 \\ 0 & 0 & 0 & \textcircled{21} & 21 \\ \hline 0 & 0 & 0 & 0 & 0 \end{array}\right] = \mathbf{U}$$

Step 3 Cover up rows 1 through 3 in **U**. The 1×5 submatrix of **U** has all zero rows. Stop. ■

Remarks

(a) The nonzero rows in **U** lie at the top of the matrix and the first nonzero entry in each nonzero row of **U** marks a *pivot position* in **U** and in **M**. The (1, 1), (2, 2) and (3, 4)-positions in **U** are pivot positions and the pivot values are 2, 1, 21, respectively. The columns in **U** that contain a pivot are *pivot columns* in **U** and in **M**.

(b) The matrix **U** is not in general unique because there may be a choice of pivots in each pivot column. Machine algorithms will select the largest nonzero pivot (in absolute value) in each pivot column in order to minimize possible *error* caused by finite-precision arithmetic. Thus, in Illustration 1.6, a machine would select $p_1 = -4$ as the pivot in column 1 and proceed according to the algorithm.

No matter which pivots are chosen in each pivot column, forward reduction of **M** in Illustration 1.6 will end with a matrix **U** which has the *form* shown on the right side of (1.33).

$$\mathbf{M} = \begin{bmatrix} 0 & 1 & -2 & 2 & 2 \\ 2 & 3 & 6 & 3 & 11 \\ -4 & -13 & 2 & 1 & -15 \\ 0 & -1 & 2 & 1 & 1 \end{bmatrix} \sim \begin{bmatrix} \textcircled{p_1} & * & * & * & * \\ 0 & \textcircled{p_2} & * & * & * \\ 0 & 0 & 0 & \textcircled{p_4} & * \\ 0 & 0 & 0 & 0 & 0 \end{bmatrix} = \mathbf{U} \qquad (1.33)$$

The main features of **U** are identified as follows:

(a) The pivot values, p_1, p_2, p_4, are nonzero numbers which depend on the choice of pivots;
(b) Each star represents a number (zero accepted) that also depends on the choice of pivots;
(c) The pivot positions are independent of the choice of pivots and descend from left to right in a staircase or *echelon* pattern;
(d) The zero entries shown in **U** are independent of the choice of pivots;
(e) The number of zero rows[12] in **U** (one in this case) and the number of nonzero rows in **U** (three in this case) are independent of the choice of pivots. The last fact is not obvious and requires proof.

[12] In parallel with the remark made on page 7, the reduction process will create a zero row in **U** whenever a row vector in **M** is a linear combination of other row vectors in **M**.

1.2.2 Echelon Forms

Forward reduction on a matrix **M** will result in a matrix **U** whose properties are now defined.

Definition 1.5

Row echelon form

A matrix **U** is in *row echelon form* (briefly, *echelon form*) if two conditions are met.

(a) The nonzero rows in **U** lie above any zero rows.
(b) The first nonzero entry (pivot) in a nonzero row lies to the right of the first nonzero entry (pivot) in the row immediately above it.

The definition implies that all entries below each pivot in **U** are zero. It is not a requirement at this stage that pivots have value 1, although scaling each pivot row will accomplish this. A matrix that is already forward reduced is said to be *in echelon form.* You may verify that each matrix shown in (1.34) is in echelon form (pivots are circled).

$$\begin{bmatrix} 0 & 0 & 0 \\ 0 & 0 & 0 \end{bmatrix}, \quad \begin{bmatrix} \textcircled{2} & 2 \\ 0 & \textcircled{5} \end{bmatrix}, \quad \begin{bmatrix} \textcircled{3} & -2 & 9 \\ 0 & 0 & 0 \end{bmatrix}, \quad \begin{bmatrix} \textcircled{3} & 0 & 9 & -4 \\ 0 & \textcircled{2} & 4 & 6 \\ 0 & 0 & 0 & \textcircled{2} \\ 0 & 0 & 0 & 0 \end{bmatrix} \tag{1.34}$$

Each matrix shown next is not in echelon form. Explain why.

$$\begin{bmatrix} 5 & 0 & 3 \\ 0 & 0 & 0 \\ 0 & -2 & 5 \end{bmatrix} \quad \begin{bmatrix} 5 & 0 & 3 \\ 0 & 0 & 1 \\ 0 & 2 & 0 \end{bmatrix} \quad \begin{bmatrix} 5 & 0 & 3 \\ 0 & -2 & 5 \\ 3 & 0 & 0 \end{bmatrix}$$

EXAMPLE 1

Row Echelon Forms are not Unique

This example underlines the fact that a matrix can have more than one echelon form. We therefore speak of the echelon forms of a matrix. Apply forward reduction to the matrix **M** in two different ways. The chosen pivot for column 1 is circled.

$$\mathbf{M} = \begin{bmatrix} \textcircled{2} & 4 & 6 \\ 1 & 1 & 1 \end{bmatrix} \quad \underset{\mathbf{r}_2 - 0.5\mathbf{r}_1 \to \mathbf{r}_2}{\sim} \quad \begin{bmatrix} \textcircled{2} & 4 & 6 \\ 0 & -1 & -2 \end{bmatrix} = \mathbf{U}_1$$

$$\mathbf{M} = \begin{bmatrix} 2 & 4 & 6 \\ \textcircled{1} & 1 & 1 \end{bmatrix} \quad \underset{\substack{\mathbf{r}_1 \leftrightarrow \mathbf{r}_2 \\ \mathbf{r}_2 - 2\mathbf{r}_1 \to \mathbf{r}_2}}{\sim} \quad \begin{bmatrix} \textcircled{1} & 1 & 1 \\ 0 & 2 & 4 \end{bmatrix} = \mathbf{U}_2$$

$\mathbf{U}_1$ and $\mathbf{U}_2$ are both echelon forms for **M** that are equivalent matrices. Note that $\mathbf{U}_1 \sim \mathbf{U}_2$ using the sequence of EROs:

$$0.5\mathbf{r}_1 \to \mathbf{r}_1, \qquad \mathbf{r}_1 + \mathbf{r}_2 \to \mathbf{r}_1, \qquad -2\mathbf{r}_2 \to \mathbf{r}_2,$$

and also $\mathbf{U}_2 \sim \mathbf{U}_1$ by applying the inverse of each of the above EROs in reverse order, namely

$$-0.5\mathbf{r}_2 \to \mathbf{r}_2, \qquad \mathbf{r}_1 - \mathbf{r}_2 \to \mathbf{r}_1, \qquad 2\mathbf{r}_1 \to \mathbf{r}_1. \qquad ▪$$

1.2.3 Backward Reduction

The process of backward elimination, introduced in the previous section, reduces a linear system in upper triangular form to a reduced upper triangular form. Backward elimination is now formulated as an algorithm that can be applied to any echelon form **U** of a given matrix. We call this process *backward reduction* of **U**.

Algorithm 1.2 **Backward reduction**

Let **U** be an echelon form for an $m \times n$ matrix. Select the last pivot column to the right in **U**.

Step 1 If necessary, scale the pivot row to give pivot value $p = 1$ and use the pivot row in replacements to reduce all entries above the pivot to zero.
Step 2 Repeat Step 1 for each pivot column in **U** to the left of the one just processed. Stop after the last pivot column has been processed.

▪ ILLUSTRATION 1.7

Backward Reduction

Perform backward reduction on the echelon form **U** in Illustration 1.6.

Step 1 Column 4, scale, $\frac{1}{21}\mathbf{r}_3 \to \mathbf{r}_3$, clear column 4, $\mathbf{r}_2 - 2\mathbf{r}_3 \to \mathbf{r}_2$, $\mathbf{r}_1 - 3\mathbf{r}_3 \to \mathbf{r}_1$.
Step 2 Repeat Step 1 for column 2, clear column 2, $\mathbf{r}_1 - 3\mathbf{r}_2 \to \mathbf{r}_1$.
Step 2 Repeat Step 1 for column 1, scale, $0.5\mathbf{r}_1 \to \mathbf{r}_1$. Stop.

$$\mathbf{U} = \begin{bmatrix} \textcircled{2} & 3 & 6 & 3 & 11 \\ 0 & \textcircled{1} & -2 & 2 & 2 \\ 0 & 0 & 0 & \textcircled{21} & 21 \\ 0 & 0 & 0 & 0 & 0 \end{bmatrix} \sim \begin{bmatrix} \textcircled{1} & 0 & 6 & 0 & 4 \\ 0 & \textcircled{1} & -2 & 0 & 0 \\ 0 & 0 & 0 & \textcircled{1} & 1 \\ 0 & 0 & 0 & 0 & 0 \end{bmatrix} = \mathbf{M}^*$$

▪

Remarks

(a) Backward reduction does not change the pivot positions and pivot columns in **U**. In fact, $\mathbf{M}^*$ is also an echelon form for **M** in Illustration 1.6.
(b) All entries in each pivot column of $\mathbf{M}^*$ are zero except for the pivot (= 1).
(c) The matrix $\mathbf{M}^*$ is the final form in the reduction process on **M**.

1.2.4 Reduced Row Echelon Form

Backward reduction on an echelon form $\mathbf{U}$ for $\mathbf{M}$ results in an echelon form $\mathbf{M}^*$ for $\mathbf{M}$ whose properties are now defined.

Definition 1.6

Reduced row echelon form

A matrix is in *reduced row echelon form* (briefly, *reduced form*) if two conditions are met.

(a) The matrix is already in echelon form.
(b) In every pivot column, the pivot has value 1 and all other entries in the column are zero.

The following important fact is not obvious and we refer the interested reader to a proof in [1] that uses mathematical induction.

Theorem 1.2

Uniqueness of Reduced Row Echelon Forms

Every matrix $\mathbf{M}$ *has a reduced row echelon form* $\mathbf{M}^*$ *that is unique.*

Thus, forward and backward reduction on a matrix $\mathbf{M}$ results in a unique echelon form $\mathbf{M}^*$ for $\mathbf{M}$. We call $\mathbf{M}^*$ the *reduced form* for $\mathbf{M}$.

You may verify that $\mathbf{M}^*$ in Illustration 1.7 is in reduced form. The reduced forms for the matrices in (1.34) are shown next. Note that pivot positions remain unchanged.

$$\begin{bmatrix} 0 & 0 & 0 \\ 0 & 0 & 0 \end{bmatrix}, \quad \begin{bmatrix} \textcircled{1} & 0 \\ 0 & \textcircled{1} \end{bmatrix}, \quad \begin{bmatrix} \textcircled{1} & -\frac{2}{3} & 3 \\ 0 & 0 & 0 \end{bmatrix}, \quad \begin{bmatrix} \textcircled{1} & 0 & 3 & 0 \\ 0 & \textcircled{1} & 2 & 0 \\ 0 & 0 & 0 & \textcircled{1} \\ 0 & 0 & 0 & 0 \end{bmatrix}$$

EXAMPLE 2

Uniqueness of Reduced Row Echelon Forms

Refer to Example 1 on page 23. The matrices $\mathbf{U}_1$ and $\mathbf{U}_2$ are two different echelon forms for $\mathbf{M}$. Backward reduction on $\mathbf{U}_1$ and on $\mathbf{U}_2$ results in $\mathbf{M}^*$, which is the unique reduced form for all the matrices $\mathbf{U}_1$, $\mathbf{U}_2$, and $\mathbf{M}$.

$$\mathbf{M} \sim \mathbf{U}_1 = \begin{bmatrix} 2 & 4 & 6 \\ 0 & -1 & -2 \end{bmatrix} \underset{\substack{-\mathbf{r}_2 \to \mathbf{r}_2 \\ \mathbf{r}_1 - 4\mathbf{r}_2 \to \mathbf{r}_1 \\ -\frac{1}{2}\mathbf{r}_1 \to \mathbf{r}_1}}{\sim} \begin{bmatrix} 1 & 0 & -1 \\ 0 & 1 & 2 \end{bmatrix} = \mathbf{M}^*$$

$$\mathbf{M} \sim \mathbf{U}_2 = \begin{bmatrix} 1 & 1 & 1 \\ 0 & 2 & 4 \end{bmatrix} \underset{\substack{\frac{1}{2}\mathbf{r}_2 \to \mathbf{r}_2 \\ \mathbf{r}_1 - \mathbf{r}_2 \to \mathbf{r}_1}}{\sim} \begin{bmatrix} 1 & 0 & -1 \\ 0 & 1 & 2 \end{bmatrix} = \mathbf{M}^*$$ ▪

Clearing Columns

Note that forward and backward reduction can be performed simultaneously. Having located a pivot p_j in row i of column j, scale the pivot (if necessary) to give pivot value 1 and use row i in replacements to reduce all entries in column j below the pivot to zero (forward reduction). After this is done, use row i in replacements to reduce all the entries above the pivot to zero (backward reduction). When this is done, we say that column j has been *cleared*. The zeros to the left of the pivot in row i ensure that no change is made to the entries above them during backward reduction.

EXAMPLE 3

Clearing a Column

Choose a pivot $p_2 = 4$ in column $j = 2$ of $\mathbf{M}$. Row $i = 2$ is used in replacements to reduce all entries below the pivot to zero, and then above it, thus clearing column 2.

$$\mathbf{M} = \begin{bmatrix} 3 & 1 & 5 & 4 \\ 0 & \textcircled{4} & 4 & 8 \\ 0 & 2 & 7 & 8 \end{bmatrix} \quad \underset{\substack{\frac{1}{4}\mathbf{r}_2 \to \mathbf{r}_2 \\ \mathbf{r}_3 - 2\mathbf{r}_2 \to \mathbf{r}_3 \\ \mathbf{r}_1 - \mathbf{r}_2 \to \mathbf{r}_1}}{\sim} \quad \begin{bmatrix} 3 & 0 & 4 & 2 \\ 0 & \textcircled{1} & 1 & 2 \\ 0 & 0 & 5 & 4 \end{bmatrix}$$

Column 3 is a pivot column that can be cleared (above) by backward reduction. Column 4 is not a pivot column. ▪

The terminology used for linear systems on page 5 and page 11 is now clarified by the following definition.

Definition 1.7

Linear systems in upper triangular form

A linear system (S) is said to be in *upper triangular form* if its augmented matrix is in echelon form and (S) is in *reduced upper triangular form* if its augmented matrix is in reduced row echelon form.

Every linear system is therefore equivalent to a unique reduced upper triangular form.

1.2.5 Rank

Forward reduction of a matrix to an echelon form will reveal an "inner structure" of the matrix, namely the number and location of its pivots, which may not be immediately obvious without reduction. For example, the matrix $\mathbf{M}$ in (1.33) has three pivot positions which are located using the pivot positions in $\mathbf{U}$. It can be proved that any echelon form for a given matrix contains exactly the same number of nonzero rows and this integer coincides with the number of pivots.

Definition 1.8

Rank of a matrix

Let $\mathbf{A}$ be an $m \times n$ matrix. The *rank* of $\mathbf{A}$, denoted by rank $\mathbf{A}$, is the number of nonzero rows in an echelon form for $\mathbf{A}$. We write rank $\mathbf{A} = r$.

EXAMPLE 4

Finding the Rank

For any *zero matrix* $\mathbf{O}$, which has all entries equal to zero, we have rank $\mathbf{O} = 0$. The rank of a nonzero matrix (having at least one nonzero entry) is a positive integer. Computing the reduced form in each case, we have

$$\mathbf{A} = \begin{bmatrix} 1 & 2 \\ 2 & 4 \\ 0 & 0 \end{bmatrix} \sim \begin{bmatrix} \textcircled{1} & 2 \\ 0 & 0 \\ 0 & 0 \end{bmatrix} \quad \text{and} \quad \mathbf{B} = \begin{bmatrix} 1 & 2 \\ 3 & 4 \end{bmatrix} \sim \begin{bmatrix} \textcircled{1} & 0 \\ 0 & \textcircled{1} \end{bmatrix},$$

which shows that rank $\mathbf{A} = 1$ and rank $\mathbf{B} = 2$. Note that the rank of a matrix does not necessarily coincide with the number of nonzero rows in the matrix. ■

Remarks

Suppose $\mathbf{A}$ is an $m \times n$ matrix with rank $\mathbf{A} = r$ and let $\mathbf{U}$ be an echelon form for $\mathbf{A}$. The nonzero rows in $\mathbf{U}$ lie at the top of the matrix and the first nonzero entry in each of these rows determines a pivot position and a pivot column in $\mathbf{U}$. Thus

$$\begin{aligned} r &= \text{the number of nonzero rows in } \mathbf{U} \\ &= \text{the number of pivot columns in } \mathbf{U} \text{ and in } \mathbf{A}. \end{aligned} \tag{1.35}$$

Observe that

$$\operatorname{rank} \mathbf{A} \le m \quad \text{and} \quad \operatorname{rank} \mathbf{A} \le n \tag{1.36}$$

because $\mathbf{U}$ has m rows and the number of pivot columns cannot exceed the number of columns in $\mathbf{U}$. From (1.36) we conclude that

$$\operatorname{rank} \mathbf{A} \le \min\{m, n\} = \text{the smallest of the integers } m, n. \tag{1.37}$$

The inequality (1.37) implies, for example, that the rank of a 3×5 matrix, or a 20×3 matrix, cannot exceed 3.

The number of solutions to a linear system can be classified using the ranks of the coefficient and augmented matrices of the system.

Theorem 1.3 **Classifying Solutions**

Let (S) be an $m \times n$ linear system represented by its $m \times (n + 1)$ augmented matrix $\mathbf{M} = [\mathbf{A} | \mathbf{b}]$. Then one and only one of the following cases can occur.

Case 1 If rank $\mathbf{A} <$ rank $\mathbf{M}$, then (S) is inconsistent.

Case 2 If rank $\mathbf{A} =$ rank $\mathbf{M} = n$, then (S) has a unique solution.

Case 3 If rank $\mathbf{A} =$ rank $\mathbf{M} < n$, then (S) has infinitely many solutions.

Notice that the conditions in Theorem 1.3 do not involve m explicitly.

Proof and Examples

First, the reduction of $\mathbf{M}$ to an echelon form reduces $\mathbf{A}$ at the same time. Hence

$$0 \le \operatorname{rank} \mathbf{A} \le \operatorname{rank} \mathbf{M} \le m. \tag{1.38}$$

The inequality rank $\mathbf{A} \le$ rank $\mathbf{M}$ in (1.38) gives us two possibilities:

$$\text{either} \qquad \operatorname{rank} \mathbf{A} < \operatorname{rank} \mathbf{M} \qquad \text{or} \qquad \operatorname{rank} \mathbf{A} = \operatorname{rank} \mathbf{M}. \tag{1.39}$$

Case 1 rank **A** < rank **M**.

Let r = rank **A**. Suppose that **M** = [**A**| **b**] is reduced to $\left[\mathbf{U} \mid \mathbf{b}'\right]$, where **U** is an echelon form for **A**. The general pattern is shown immediately below: circles denote (nonzero) pivots and the stars denote numerical (possibly zero) values.

$$[\mathbf{A}|\,\mathbf{b}] \sim [\,\mathbf{U}\,|\,\mathbf{b}'\,] = \left[\begin{array}{cccccccc|c} \vdots & \vdots & \ddots & \vdots & \vdots & \vdots & \vdots & \vdots & \vdots \\ 0 & 0 & \cdots & 0 & \bigcirc & * & \cdots & * & * \\ 0 & 0 & \cdots & 0 & 0 & 0 & \cdots & 0 & \bigcirc \\ 0 & 0 & \cdots & 0 & 0 & 0 & \cdots & 0 & 0 \\ \vdots & \vdots & \ddots & \vdots & \vdots & \vdots & \vdots & \vdots & \vdots \end{array}\right] \begin{array}{l} \\ \text{row } r \\ \text{row } (r+1) \\ \\ \\ \end{array}$$

Row r is the last nonzero row in **U**. This row contains a pivot located somewhere between columns r and n inclusive. If r < rank **M**, then **M** can only have one more nonzero row than **A** with a pivot in the $(r+1, n+1)$-position. The last column of **M** is therefore a pivot column. We have rank **M** = rank **A** + 1, and row $(r+1)$ in $\left[\mathbf{U} \mid \mathbf{b}'\right]$ implies that (S) is inconsistent.

EXAMPLE 5

An Inconsistent Linear System

Consider a 4×2 linear system (S) with augmented matrix

$$\mathbf{M} = [\mathbf{A}|\,\mathbf{b}] = \left[\begin{array}{rr|r} 1 & 1 & 1 \\ -1 & 1 & 1 \\ 2 & 4 & 5 \\ 3 & 3 & 6 \end{array}\right].$$

Reducing **M** to an echelon form, we have

$$\left[\begin{array}{rr|r} 1 & 1 & 1 \\ -1 & 1 & 1 \\ 2 & 4 & 5 \\ 3 & 3 & 6 \end{array}\right] \sim \left[\begin{array}{rr|r} 1 & 1 & 1 \\ 0 & 2 & 2 \\ 0 & 2 & 3 \\ 0 & 0 & 3 \end{array}\right] \sim \left[\begin{array}{rr|r} 1 & 1 & 1 \\ 0 & 2 & 2 \\ 0 & 0 & 1 \\ 0 & 0 & 3 \end{array}\right]$$

$$\sim \left[\begin{array}{cc|c} 1 & 1 & 1 \\ 0 & \textcircled{2} & 2 \\ 0 & 0 & \textcircled{1} \\ 0 & 0 & 0 \end{array}\right] = \left[\mathbf{U} \mid \mathbf{b}'\right].$$

The crucial consecutive rows are row 2 and row 3. Comparing the ranks of **A** and **M**, we have rank **A** = 2 < 3 = rank **M**, and so (S) is inconsistent. ▩

The second possibility in (1.38) is rank **A** = rank **M**. Under this condition, (S) is consistent and there are two subcases to consider:

either rank **A** = n (Case 2) or rank **A** < n (Case 3).

Case 2 rank $\mathbf{A}$ = rank $\mathbf{M} = n$.

The system is consistent and row reduction must accommodate n pivots in n columns of $\mathbf{A}$. Hence $\mathbf{M}$ has the reduced form $\mathbf{M}^*$ with the following general pattern. There will be no zero rows in $\mathbf{M}^*$ when $m = n$.

$$\mathbf{M} \sim \left[\begin{array}{ccccc|c} \textcircled{1} & 0 & 0 & \cdots & 0 & k_1 \\ 0 & \textcircled{1} & 0 & \cdots & 0 & k_2 \\ \vdots & \vdots & \vdots & \ddots & \vdots & \vdots \\ 0 & 0 & 0 & \cdots & \textcircled{1} & k_n \\ 0 & 0 & 0 & \cdots & 0 & 0 \\ \vdots & \vdots & \vdots & \ddots & \vdots & \vdots \\ 0 & 0 & 0 & \cdots & 0 & 0 \end{array}\right] = \mathbf{M}^* \tag{1.40}$$

The unique solution is $(x_1, x_2, \ldots, x_n) = (k_1, k_2, \ldots, k_n)$.

EXAMPLE 6

A Consistent Linear System, Unique Solution

Consider a 4×2 linear system (S) with augmented matrix $\mathbf{M} = [\mathbf{A}\,|\,\mathbf{b}]$ shown in (1.41). Note that (S) was obtained by modifying the column $\mathbf{b}$ in the system in Example 5. Reducing $\mathbf{M}$ to its reduced form, we have

$$\mathbf{M} = \left[\begin{array}{cc|c} 1 & 1 & 1 \\ -1 & 1 & 1 \\ 2 & 4 & 4 \\ 3 & 3 & 3 \end{array}\right] \sim \left[\begin{array}{cc|c} \textcircled{1} & 0 & 0 \\ 0 & \textcircled{1} & 1 \\ 0 & 0 & 0 \\ 0 & 0 & 0 \end{array}\right] = \mathbf{M}^* \tag{1.41}$$

and we see that rank $\mathbf{A}$ = rank $\mathbf{M} = 2$ and (S) has the unique solution $(x_1, x_2) = (0, 1)$. Geometrically, the equations in (S) represent four lines in $\mathbb{R}^2$ that intersect in a single point. ▩

Case 3 rank $\mathbf{A}$ = rank $\mathbf{M} < n$.

The system is consistent and rank $\mathbf{A} = r < n$. The r pivot columns in $\mathbf{A}$ each determine a pivot (leading) variable, and the $n - r$ nonpivot columns each determined a free variable that is given an independent parametric value. There are thus infinitely many solutions.

Caution! It can still happen that some variables are uniquely determined.

EXAMPLE 7

A Consistent Linear System, Infinitely Many Solutions

Consider a 3×3 linear system (S) with augmented matrix $\mathbf{M} =[\mathbf{A}\,|\,\mathbf{b}]$ shown in (1.42). Reducing $\mathbf{M}$ to its reduced form, we have

$$\mathbf{M} = \left[\begin{array}{ccc|c} 1 & -2 & 4 & 1 \\ -2 & 5 & 5 & -1 \\ 5 & -12 & -6 & 3 \end{array}\right] \sim \left[\begin{array}{ccc|c} \textcircled{1} & 0 & 30 & 3 \\ 0 & \textcircled{1} & 13 & 1 \\ 0 & 0 & 0 & 0 \end{array}\right] = \mathbf{M}^*. \tag{1.42}$$

Then $r = \operatorname{rank}\mathbf{A} = \operatorname{rank}\mathbf{M} = 2$, showing that (S) is consistent. Also, $r = 2 < 3 = n$ shows that there is $n - r = 3 - 2 = 1$ parameter required. The solution set for (S) is

$$(x_1, x_2, x_3) = (3 - 30t, 1 - 13t, t),$$

where t is a real parameter. ■

Some Observations and Deductions

The consistency or inconsistency of a linear system (S) depends on both its coefficient matrix **A** and the column vector **b** of constant terms. If **A** is fixed and **b** is allowed to vary, then (S) may be inconsistent for some **b** and consistent for some other **b** (see Example 5 and Example 6). The cases when (S) is consistent for every column vector **b** are of particular interest and Theorems 1.4 and 1.5 give conditions for this to happen.

A linear system with more unknowns than equations is sometimes called *underdetermined*. Likewise, an *overdetermined* linear system has more equations than unknowns and a linear system is called *square* if the number of equations equals the number of unknowns.

Theorem 1.4 **Underdetermined Linear Systems**

An $m \times n$ linear system (S) for which $m < n$ is either inconsistent or has infinitely many solutions. If $\operatorname{rank}\mathbf{A} = m$, *then (S) has infinitely many solutions for every column of constants* **b**.

Proof Let $r = \operatorname{rank}\mathbf{A}$ and note that $r \le m < n$. Then (S) is either inconsistent, or consistent with $n - r$ (≥ 1) free variables. The condition $r = m$ forces $\operatorname{rank}\mathbf{A} = \operatorname{rank}[\mathbf{A}\,|\,\mathbf{b}]$ for every **b**.

Theorem 1.5 **Square Linear Systems**

An $n \times n$ linear system (S) for which $\operatorname{rank}\mathbf{A} = n$ *has a unique solution for every column of constants* **b**.

Proof Refer to the remarks that precede (1.40) in Case 2 above. For every column **b**, the augmented matrix $\mathbf{M} = [\mathbf{A}\,|\,\mathbf{b}]$ for (S) will have reduced form $\mathbf{M}^*$ as in (1.40) but with no zero rows in the case $m = n$.

EXAMPLE 8 **Consistency for Every Column of Constants**

Consider the 2×2 linear system

$$(S)\begin{cases} x_1 + 2x_2 = s \\ 3x_1 + 4x_2 = t \end{cases} \qquad \mathbf{A} = \begin{bmatrix} 1 & 2 \\ 3 & 4 \end{bmatrix} \qquad \mathbf{b} = \begin{bmatrix} s \\ t \end{bmatrix},$$

of the $m \times n$ coefficient matrix **A** *and the number of free variables.*

26. $\begin{cases} x_1 + 2x_2 = 0 \\ 3x_1 + 8x_2 = 0 \\ 5x_1 - 6x_2 = 0 \end{cases}$

27. $\begin{cases} x_1 + 2x_2 + 3x_3 = 0 \\ 4x_1 + 5x_2 + 6x_3 = 0 \end{cases}$

28. $\begin{cases} 4.8x_1 - 3.2x_2 = 0 \\ -2.4x_1 + 1.6x_2 = 0 \\ 12x_1 - 8x_2 = 0 \end{cases}$

29. $\begin{cases} 2x_1 + 3x_2 - x_3 + 5x_4 = 0 \\ 3x_1 - x_2 + 2x_3 - 7x_4 = 0 \\ 4x_1 + x_2 - 3x_3 + 6x_4 = 0 \\ x_1 - 2x_2 + 4x_3 - 7x_4 = 0 \end{cases}$

30. $\begin{cases} 0.6x_1 + 0.2x_2 = x_1 \\ 0.4x_1 + 0.8x_2 = x_2 \end{cases}$

Exercises 31–34. Find a matrix **E** *in reduced form for which the number of solutions to the linear system with augmented matrix* [**E** | **b**] *is as stated, for various* **b**.

31. Zero or one, depending on **b**.

32. Zero or infinitely many, depending on **b**.

33. One, for any **b**.

34. Infinitely many, for any **b**.

Exercises 35–40. Find the rank r of each matrix.

35. $\begin{bmatrix} 1 & -1 \\ -2 & 2 \end{bmatrix}$

36. $\begin{bmatrix} 1 & -1 \\ -2 & 3 \end{bmatrix}$

37. $\begin{bmatrix} 0 & 0 \\ 0 & 1 \end{bmatrix}$

38. $\begin{bmatrix} 1 & 0 & 0 \\ 5 & 5 & 5 \end{bmatrix}$

39. $\begin{bmatrix} 1 & -2 & 3 \\ -5 & 10 & -15 \\ 4 & -8 & 12 \end{bmatrix}$

40. $\begin{bmatrix} 1 & 2 & 3 \\ 4 & 5 & 6 \\ 7 & 8 & 9 \end{bmatrix}$

Exercises 41–42. For each matrix **A**, *find the value(s) of α for which (a)* rank **A** = 1, *(b)* rank **A** = 2.

41. $\begin{bmatrix} 1-\alpha & 2 \\ 2 & 1-\alpha \end{bmatrix}$

42. $\begin{bmatrix} 2 & 6 & 4 \\ 1 & 3 & \alpha \end{bmatrix}$

43. Find the value(s) of α for which (a) rank **A** = 2, (b) rank **A** = 3.

$$\mathbf{A} = \begin{bmatrix} 1 & 1 & 1 & -1 \\ 1 & 2 & \alpha & 2\alpha \\ 1 & \alpha & 2 & -2 \end{bmatrix}$$

44. Find the value(s) of α and β for which (a) rank **A** = 2, (b) rank **A** = 3.

$$\mathbf{A} = \begin{bmatrix} 1 & 1 & 3 & 4 \\ 1 & 2 & 4 & 5 \\ 1 & -1 & \alpha & \beta \end{bmatrix}$$

Exercises 45–49. Each matrix represents the augmented matrix $\mathbf{M} = [\mathbf{A}|\,\mathbf{b}]$ *of an $m \times n$ linear system (S). Find* rank **A** *and* rank **M**. *Use Theorem 1.3 to classify the solution to (S). State the number of free variables in terms of r =* rank **A** *and n. Give the solution set when (S) is consistent.*

45. $\left[\begin{array}{ccc|c} 3 & 2 & 1 & -1 \\ 2 & 3 & 1 & 1 \\ 2 & 2 & 2 & -1 \end{array}\right]$

46. $\left[\begin{array}{ccc|c} 5 & 6 & 6 & 13 \\ 2 & 2 & 4 & 6 \\ 6 & 8 & 4 & 0 \end{array}\right]$

47. $\left[\begin{array}{cccc|c} 1 & -2 & 1 & -1 & 1 \\ 2 & 1 & -1 & 2 & -3 \\ 3 & -1 & 0 & 1 & 2 \end{array}\right]$

48. $\left[\begin{array}{cccc|c} 1 & -2 & 1 & 1 & -1 \\ 2 & 1 & -1 & -1 & 1 \\ 1 & 7 & -5 & -5 & 5 \end{array}\right]$

49. $\left[\begin{array}{ccc|c} 2 & -3 & 1 & 7 \\ 5 & 2 & 4 & 8 \\ 1 & 1 & -1 & 1 \\ 3 & -2 & 2 & 8 \\ -2 & 7 & 0 & -11 \end{array}\right]$

50. Let $\mathbf{M} = [\mathbf{A}|\,\mathbf{b}]$ the augmented matrix of a 4×6 linear system (S). Give an example of the reduced form $\mathbf{M}^*$ of **M** for which rank **A** = 3 and exactly one variable in (S) is uniquely determined.

51. Consider the linear system

$$(S)\ \begin{cases} x_1 + x_2 + 3x_3 = \alpha \\ 2x_1 + 2x_2 + 7x_3 = \beta \end{cases}, \quad \text{where}$$

$$\mathbf{A} = \begin{bmatrix} 1 & 1 & 3 \\ 2 & 2 & 7 \end{bmatrix}, \quad \mathbf{b} = \begin{bmatrix} \alpha \\ \beta \end{bmatrix}.$$

Solve (S) and explain how Theorem 1.4 applies in this case.

Exercises 52–53. Each matrix is the augmented matrix $\mathbf{M} = [\mathbf{A} \,|\, \mathbf{b}]$ *of a square linear system* (S)*. Find* rank **A** *and* rank **M** *and use Theorem 1.3 to determine when* (S) *will have zero, one, or infinitely many solutions.*

52. $\begin{bmatrix} 1 & 2 & 3 & a \\ 0 & 2 & 1 & b \\ 1 & 4 & 5 & c \end{bmatrix}$ **53.** $\begin{bmatrix} 1 & 1 & 1 & a \\ 1 & 2 & 2 & b \\ 2 & 3 & 3 & c \end{bmatrix}$

Exercises 54–55. Each matrix **A** *is the coefficient matrix of an overdetermined homogeneous linear system* (S)*. Find* rank **A** *and state the number of solutions in* (S)*. Solve* (S)*.*

54. $\begin{bmatrix} 1 & 2 & 3 \\ 4 & 9 & 13 \\ 7 & 14 & 22 \\ 9 & 18 & 27 \end{bmatrix}$ **55.** $\begin{bmatrix} 3 & 21 & 144 \\ 5 & 34 & 233 \\ 8 & 55 & 377 \\ 13 & 89 & 610 \end{bmatrix}$

Exercises 56–59. For each given value λ*, find all nonzero solutions to the linear system. Solve the system again when* $\lambda = 0$.

56. $(S_\lambda) \begin{cases} (-\lambda)x_1 + 3x_2 = 0 \\ x_1 + (2-\lambda)x_2 = 0 \end{cases}$

$$(\lambda = -1, 3)$$

57. $(S_\lambda) \begin{cases} (3-\lambda)x_1 + 5x_2 = 0 \\ -x_1 + (-3-\lambda)x_2 = 0 \end{cases}$

$$(\lambda = -2, 2)$$

58. $(S_\lambda) \begin{cases} (1-\lambda)x_1 + 2x_2 + 3x_3 = 0 \\ (2-\lambda)x_2 + x_3 = 0 \\ (3-\lambda)x_3 = 0 \end{cases}$

$$(\lambda = 1, 2, 3)$$

59. $(S_\lambda) \begin{cases} (1-\lambda)x_1 + x_3 = 0 \\ (1-\lambda)x_2 + x_3 = 0 \\ (1-\lambda)x_3 = 0 \end{cases}$

$$(\lambda = 1, 1, 1)$$

60. A consistent 3×5 linear system (S) in unknowns $x_1, \ldots, x_5$ has an augmented matrix **M**. The pivot positions in **M** are $(1, 1)$, $(2, 2)$, $(3, 4)$. Use rank to determine the number of free variables and the number of solutions to (S).

61. Consider the matrix **M** in Illustration 1.6. Why did the zero row appear in the echelon form **U**? Find a row vector **r** in **M** such that $\mathbf{r} = s_1\mathbf{r}_1 + s_2\mathbf{r}_2 + s_3\mathbf{r}_3$, for some scalars s_1, s_2, s_3 and rows vectors $\mathbf{r}_1, \mathbf{r}_2, \mathbf{r}_3$ distinct from **r**. *Hint*: Solve a linear system.

62. List the possible reduced forms of a general 2×2 matrix **M**.

63. Show that the row interchange $\mathbf{r}_1 \leftrightarrow \mathbf{r}_2$ can be performed using the other two elementary row operations. The same property applies to elementary equation operations.

Exercises 64–68. Label each statement as either true (T) or false (F). If a false statement can be made true by modifying the wording, then do so.

64. If two matrices of the same size have equal rank, then their reduced forms must be identical.

65. There exists a 3×2 matrix with rank 3.

66. Two row equivalent matrices have the same rank.

67. If a matrix **A** has a row of zeros, then so does its reduced form $\mathbf{A}^*$.

68. If the reduced form $\mathbf{A}^*$ has a row of zeros, then so does **A**.

USING MATLAB

Consult online help for the MATLAB *commands:* **rref**, **rank**, **rrefmovie**, **rand**, *and* **magic**.

Exercises 69–70. Use MATLAB *commands to find the reduced form and the rank of each matrix* **M**.

69. $\begin{bmatrix} 1 & 2 & 3 \\ 4 & 5 & 6 \\ 7 & 8 & 9 \end{bmatrix}$

70. $\begin{bmatrix} 1 & -1 & 1 & -1 & 1 \\ -1 & 1 & -1 & 1 & -1 \\ 1 & -1 & 1 & -1 & 1 \\ -1 & 1 & -1 & 1 & -1 \end{bmatrix}$

71. Refer to Exercises 69–70. Use the command **rrefmovie** to step through the reduction process on **M**. Note how the program chooses the largest pivot (in absolute value) in each pivot column, moves the pivot row into position, and then clears the column.

72. *Experiment.* Consider an $n \times (n+1)$ matrix **M** with real entries chosen at random. What is the probability that the associated $n \times n$ linear system (S) will be consistent? Write an M-file to explore this question using the command **rand(n,n+1)**. Report on your results.

73. *Experiment.* Your goal is to use Monte Carlo *simulation* to estimate the probability p that a $n \times n$ linear system (S) will be consistent when the coefficients and constant terms are digits 0, 1, ... , 9. Write an M-file with the following features:

(a) Input the chosen value n from the keyboard;

(b) Use the code **floor(rand(n,n+1)*10)** to define a random augmented matrix with the desired entries;

(c) Loop through a positive integer k of cases in order to find the number c of systems that are consistent. Set $k = 1, 2, 3$ in order to test your program and then increase k.

The number $p = c/k$ is the desired estimate. The value $1 - c/k$ estimates the probability of obtaining an inconsistent linear system.

74. *Experiment.* Extend the previous exercise to $m \times n$ linear systems for various values of m and n. Use matrices that have integer entries between e and $-e$ inclusive, for certain values, $e = 1, 2, 3, \ldots$. Report on your results.

75. *Experiment.* The magic (square) matrix **A** of order $n = 4$ displayed on the front cover of this textbook is obtained using the command: **magic(4)**. Investigate the ranks and reduced forms of magic matrices **A** of order $n = 3, 4, \ldots$. Conjecture whether a square linear system (S) with magic coefficient matrix **A** will be consistent for all columns of constant terms **b**.

76. Compare the matrix **magic(3)** with the matrix in Exercise 69. Compare the character of the solution set to linear systems whose coefficient matrix is either of these two matrices.

77. Use MATLAB to cross-check hand calculations in this exercise set.

1.3 Applications

We are restricted somewhat in the presentation of the many interesting applications of linear systems by the need for lengthy explanation of the problem's background and the accompanying mathematical theory (some applications require a knowledge of calculus, for example). The following applications are some of the most accessible.

1.3.1 Linear Models

Using mathematics to solve an applied problem involves *translation* of the salient features of the problem into *mathematical language* (terminology, symbols, equations, and so on). We refer to this translation process as *building a mathematical model* of the problem. The chemical equations presented in Section 1.3.4, for example, are *modeled* by homogeneous linear systems. Mathematical models that involve some form of linearity are called *linear models.*

Illustration 1.9 is typical of a broad class of problems called *scheduling problems*, which deal with such things as movement of people or equipment, organization of work forces, project analysis, queuing analysis, and so on. When these problems have many solutions, we are often interested in finding the *optimal* or *best possible* solution.

■ ILLUSTRATION 1.9

Transportation Schedules

A construction company has trucks of four sizes, labeled A, B, C, D, which can carry heavy machinery according to the following table:

Truck Size	A	B	C	D	On Site
Forklift	1	2	1	0	38
Drills	2	1	3	2	53
Compressors	2	4	1	4	78

The meaning is that a truck of type A can carry 1 forklift, 2 drills, and 2 compressors. The company must clear a large construction site and haul away 38 forklifts, 53 drills, and 78 compressors as efficiently as possible. Build a mathematical model that analyzes this problem.

Some typical steps in the model building process are itemized as follows:

Step 1 Defining Variables. Be precise here! Let x_1, x_2, x_3, x_4 represent, respectively, the number of trucks of types A, B, C, D to be used in clearing the construction site.

Step 2 Identifying Constraints. In this problem there are natural physical restrictions. We cannot have a negative number or a fractional number of trucks, and it could happen that no trucks of a certain type are required. The variables are therefore constrained to be nonnegative integers $(0, 1, 2, 3, \dots)$.

Step 3 Building the Model. Translating the problem into mathematics we obtain the following 3×4 linear system:

$$(S)\quad \begin{cases} x_1 + 2x_2 + x_3 \phantom{{}+2x_4} = 38 & \text{(counts forklifts)} \\ 2x_1 + x_2 + 3x_3 + 2x_4 = 53 & \text{(counts drills)} \\ 2x_1 + 4x_2 + x_3 + 4x_4 = 78 & \text{(counts compressors)} \end{cases}.$$

Computing the reduced form $\mathbf{M}^*$ of $\mathbf{M} = [\mathbf{A}|\,\mathbf{b}]$, we have

$$\mathbf{M} = \left[\begin{array}{cccc|c} 1 & 2 & 1 & 0 & 38 \\ 2 & 1 & 3 & 2 & 53 \\ 2 & 4 & 1 & 4 & 78 \end{array}\right] \sim \left[\begin{array}{cccc|c} \textcircled{1} & 0 & 0 & 8 & 26 \\ 0 & \textcircled{1} & 0 & -2 & 7 \\ 0 & 0 & \textcircled{1} & -4 & -2 \end{array}\right] = \mathbf{M}^*.$$

The pivot variables are x_1, x_2, x_3 corresponding to the pivots shown in $\mathbf{M}^*$ and x_4 is free. Setting $x_4 = t$, where t is a real parameter, there are infinitely many solutions for (S) given by

$$(x_1, x_2, x_3, x_4) = (26 - 8t, 7 + 2t, -2 + 4t, t) \tag{1.46}$$

Step 4 Applying Constraints. The solutions in (1.46) must be nonnegative integers, and so the following inequalities must *all* be satisfied, where t $(= x_4)$ is a nonnegative integer.

$$0 \le 26 - 8t, \qquad 0 \le 7 + 2t, \qquad 0 \le 4t - 2 \tag{1.47}$$

The inequality $0 \le 26 - 8t$ in (1.47) gives $0 \le t \le 3.25$ so that $t \le 3$. The inequality $0 \le 4t - 2$ gives $0.5 \le t$ and so $1 \le t$. Thus t is restricted in the range $t = 1, 2, 3$. From (1.46) there are three possible solutions:

$$(x_1, x_2, x_3, x_4) = (18, 9, 2, 1), \quad (10, 11, 6, 2), \quad (2, 13, 10, 3). \tag{1.48}$$

Step 5 Drawing Conclusions. The condition that the variables be nonnegative effectively reduces the number of solutions from infinitely many to just three in this case. The solutions in (1.48) are *optimal* in the sense that the construction site can be completely cleared without trucks making extra trips while being partially filled. The last solution in (1.48) uses the smallest number of trucks. ■

Illustration 1.9 can be extended to include some optimization of costs. Suppose the construction company must meet overhead costs for each truck amounting to, respectively, 1, 3, 1, and 2 units of a fixed hourly running cost (suppose 1 unit is \$100, for example) in order to keep the trucks on the road. What is the most economical way to clear the building site? A simple calculation yields

$$18(1) + \ 9(3) + 2(1) + 1(2) = 49 \text{ units}$$
$$10(1) + 11(3) + 6(1) + 2(2) = 53 \text{ units}$$
$$2(1) + 13(3) + 10(1) + 3(2) = 57 \text{ units}$$

and so the solution (18, 9, 2, 1) is the most economical.

1.3.2 General Networks

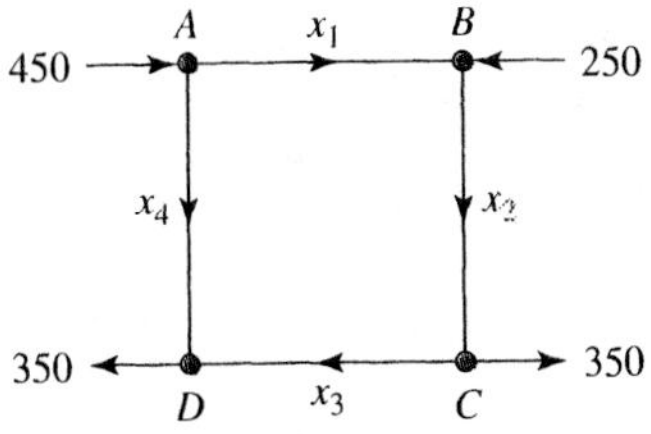

Figure 1.3 An abstract network.

An abstract *network* consists of a system of *directed line segments* that meet at *junctions* or *nodes*. The network shown in Figure 1.3 has nodes A, B, C, D, and the arrows indicate the direction of *flow* along the line segments. Networks are used to model a great variety of physical problems such as traffic flow, electrical networks, and fluid flow in pipeline networks.

EXAMPLE 1

Traffic flow

Suppose that Figure 1.3 represents a network of one-way roads, with arrows indicating direction of traffic flow. The numbers of vehicles entering or leaving the network (per hour) at nodes A, B, C, D are as shown.

Let x_1, x_2, x_3, x_4 denote the number of vehicles per hour passing through branches AB, BC, CD, AD, respectively. Assume the network is in a *steady state* (*equilibrium*), meaning that no new vehicles appear within the network, none are lost, and *input* must equal *output* at each node A, B, C, D. The network is modeled by a linear system (S), which is built as follows.

Node	Input		Output
A	450	=	$x_1 + x_4$
B	$x_1 + 250$	=	x_2
C	x_2	=	$x_3 + 350$
D	$x_3 + x_4$	=	350

$$\Rightarrow \quad (S) \quad \left\{ \begin{aligned} x_1 \qquad\qquad\quad + x_4 &= 450 \\ -x_1 + x_2 \qquad\qquad &= 250 \\ x_2 - x_3 \qquad &= 350 \\ x_3 + x_4 &= 350 \end{aligned} \right.$$

Computing the reduced form $\mathbf{M}^*$ of $\mathbf{M} = [\mathbf{A} \,|\, \mathbf{b}]$, we have

$$\mathbf{M} = \left[\begin{array}{rrrr|r} 1 & 0 & 0 & 1 & 450 \\ -1 & 1 & 0 & 0 & 250 \\ 0 & 1 & -1 & 0 & 350 \\ 0 & 0 & 1 & 1 & 350 \end{array}\right] \sim \left[\begin{array}{cccc|r} \textcircled{1} & 0 & 0 & 1 & 450 \\ 0 & \textcircled{1} & 0 & 1 & 700 \\ 0 & 0 & \textcircled{1} & 1 & 350 \\ 0 & 0 & 0 & 0 & 0 \end{array}\right] = \mathbf{M}^*.$$

We have rank $\mathbf{A} = 3 = \text{rank}\,\mathbf{M}$, showing that (S) is consistent, and $r = \text{rank}\,\mathbf{A} = 3 < n = 4$ indicates that there is $n - r = 1$ free variable and infinitely many solutions. Let $x_4 = t$, where t is a real parameter. Then

$$(x_1, x_2, x_3, x_4) = (450 - t, 700 - t, 350 - t, t) \tag{1.49}$$

is the general solution to (S).

Network Constraints: Integer Solutions

In this application we will suppose that the roads are one way as indicated and that each vehicle is treated as an indivisible unit. The variables are therefore nonnegative integers. From (1.49) we have $0 \leq x_1$ implies $t \leq 450$, $0 \leq x_2$ implies $t \leq 700$, and $0 \leq x_3$ implies $t \leq 350$. All inequalities must be satisfied and so $0 \leq t \leq 350$, where $t = x_4$ is an integer.

The basic analysis is complete. Other constraints could be imposed. For example, if we suppose that section AD is closed for repair, then $t = 0$, giving the unique solution $(x_1, x_2, x_3, x_4) = (450, 700, 350, 0)$. ▩

1.3.3 Electrical Networks

The electrical network shown in Figure 1.5 contains *batteries* (voltage sources) and *resistors* (light bulbs or heaters, for example) joined together by *conductors* (wires). A voltage source provides an *electromotive force* E, measured in *volts*, which moves electrons through the network. The rate at which electrons flow along a conductor is called *current* I and is measured in *amperes* (amps). The resistors act to retard the flow of electrons, using up the current (in the form of heat) and lowering the voltage.

Refer to Figure 1.4. The potential difference between A and B across a battery is E volts, where the sign of E is determined by the direction shown.

Ohm's Law

According to Ohm, the potential difference E volts across a resistor is given by $E = IR$, where I is the current measured in amps and R is the resistance measured in ohms (Ω). Referring to Figure 1.4, the potential difference between C and D (voltage drop) is $+5I$ (in the direction of the current flow) and the potential difference between D and C is $-5I$ (against the current flow).

Network analysis is based on Kirchhoff's two laws that apply to the *closed voltage loops* and *nodes* in the network. A node is a junction of three or more conductors. The network in Figure 1.5 has two nodes (B and D) and three closed voltage loops defined as follows:

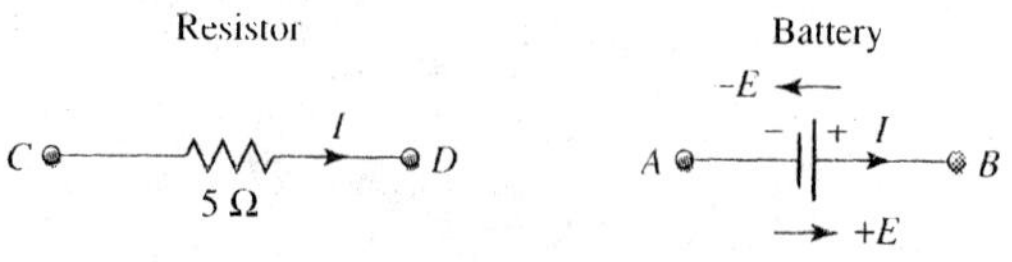

Figure 1.4 Potential difference.

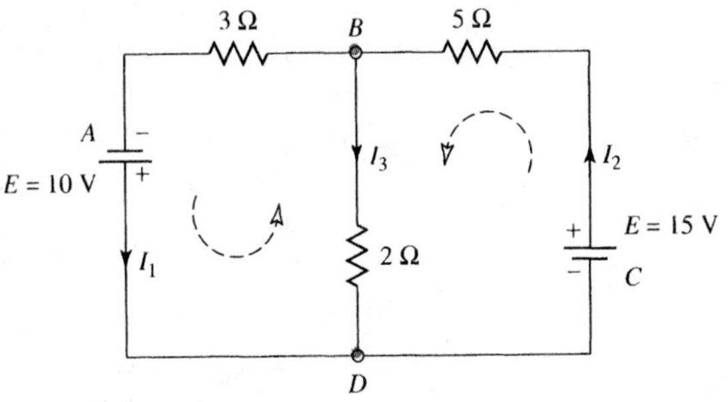

Figure 1.5 An electrical network.

$$\begin{array}{rl} A \to D \to B \to A & \text{(Loop } ADBA\text{)} \\ C \to B \to D \to C & \text{(Loop } CBDC\text{)} \\ A \to D \to C \to B \to A & \text{(Loop } ADCBA\text{)} \end{array}$$

Kirchhoff's Laws

Conservation of Energy: Around any closed voltage loop in the network, the algebraic sum of potential differences is zero (voltage drops caused by resistors and electromotive forces provided by voltage sources must balance).
Conservation of Charge: At each node, the total inflow of current equals the total outflow of current.

EXAMPLE 2

Network Analysis

Refer to Figure 1.5. Our goal is to find values for the currents in each branch of the network. Assign currents I_1, I_2, I_3 as shown. Applying Kirchhoff's laws, we obtain the 4×3 linear system

$$(S) \quad \left\{ \begin{array}{rcll} 3I_1 \qquad\quad - 2I_3 &=& 10 & E_1 \quad \text{(Loop } ADBA\text{)} \\ 5I_2 + 2I_3 &=& 15 & E_2 \quad \text{(Loop } CBDC\text{)} \\ 3I_1 + 5I_2 \qquad\quad &=& 25 & E_3 \quad \text{(Loop } ABCDA\text{)} \\ I_1 - \ I_2 + \ I_3 &=& 0 & E_4 \quad \text{(Nodes B, D)} \end{array} \right. .$$

Note that equation $E_3 = E_1 + E_2$, and so we expect at least one zero row in any echelon form for the augmented matrix for (S). We leave the details of solution as an exercise and record the unique solution to (S), which is

$$(I_1, I_2, I_3) = \left(\frac{100}{31}, \frac{95}{31}, -\frac{5}{31} \right) \simeq (3.23, 3.06, -0.16) \text{ amps.}$$

Current directions may be assigned arbitrarily. The negative value for I_3 indicates that the flow in branch BD is in the opposite direction. ■

Historical Notes

Georg Simon Ohm (1789–1854) German physicist and mathematician. Ohm taught in Cologne and later became a professor at Nüremberg and Munich. In 1826, Ohm gave a mathematical description of conduction in networks that used Fourier's theory of heat conduction. Ohm's law appeared in his famous book (1827), which gave a complete theory of electricity from a mathematical point of view. In 1841, Ohm was awarded the Copley Medal by the Royal Society of Great Britain.

Gustav Robert Kirchhoff (1824–1887) German physicist. Born in Königsberg, East Prussia, Kirchhoff was a student of Gauss. He was appointed Professor of Physics at Heidelberg in 1854 (his two fundamental laws for electrical networks appeared in the same year) and later at Berlin. He collaborated with Robert Bunsen, developed the spectroscope, and used it to discover the elements cesium and rubidium (1860). His fundamental work on spectrum analysis of light from the sun and on black body radiation led to the development of quantum theory.

André Marie Ampère (1775–1836) French physicist. Ampère laid the foundation of thermodynamics, formulated independently Avogadro's law (1814), and his work between 1820 and 1827 explored the connection between current and magnetism (an electrical current produces a magnetic field whose direction is determined by the direction of the current).

1.3.4 Balancing Chemical Equations

A chemical equation is an expression of chemical change. It tells us the kinds and numbers of molecules entering the reaction and the kinds and number of product molecules formed as a result of the reaction. Homogeneous linear systems can be applied in finding all possible ways to *balance* a chemical equation.

Our illustration, taken from biochemistry, uses the elements[14] carbon (C), hydrogen (H), and oxygen (O).

EXAMPLE 3

Balancing Chemical Equations

Catabolism (the breakdown of carbohydrates) provides energy for animals. In the catabolic process carbohydrates combine with oxygen (O_2) to produce carbon dioxide (CO_2), water (H_2O), and energy. Glucose (chemical formula $C_6H_{12}O_6$) is just one of many carbohydrates used in this process. A *mole*[15] is a standard unit of measure for substances. We will suppose that x_1 moles of glucose and x_2 moles of oxygen combine to produce x_3 moles of carbon dioxide and x_4 moles of water. The chemical equation for this reaction is given by

$$x_1 C_6H_{12}O_6 + x_2 O_2 \quad \Longrightarrow \quad x_3 CO_2 + x_4 H_2O \quad + \quad \boxed{\text{ENERGY}} \qquad (1.50)$$

The subscripts on the elements give the number of atoms of each element present in the molecule. The symbol H_{12} in glucose, for example, indicates that there are 12 atoms of hydrogen present. The reaction (1.50) must balance

---

[14] An element is a substance that cannot be broken down chemically into simpler substances. All elements are listed in the *periodic table*.

[15] *Mole* (mol), unit of substance defined as the amount of substance containing the same number of entities (atoms, ions, molecules, etc.) as there are in 0.012 kilograms of carbon-12. This number is 6.022×10^{23} and is called *Avogadro's number*. The mass of 1 mol of a substance is called its *molecular* or *molar* mass.

numbers of atoms *before* and *after* combustion, resulting in a homogeneous linear system (S) as follows.

$$\begin{array}{ll} \text{C: } 6x_1 & = x_3 \\ \text{H: } 12x_1 & = 2x_4 \\ \text{O: } 6x_1 + 2x_2 & = 2x_3 + x_4 \end{array} \quad \Rightarrow \quad (S) \left\{ \begin{array}{rrrrl} 6x_1 & & - x_3 & & = 0 \\ 12x_1 & & & - 2x_4 & = 0 \\ 6x_1 & + 2x_2 & - 2x_3 & - x_4 & = 0 \end{array} \right. \tag{1.51}$$

Gauss–Jordan elimination on the coefficient matrix $\mathbf{A}$ of (S) gives

$$\mathbf{A} \sim \begin{bmatrix} \textcircled{1} & 0 & 0 & -\frac{1}{6} \\ 0 & \textcircled{1} & 0 & -1 \\ 0 & 0 & \textcircled{1} & -1 \end{bmatrix} = \mathbf{A}^*.$$

Column 4 in $\mathbf{A}^*$ indicates that x_4 is free. Let $x_4 = t$, where t is a real parameter, and then (S) has infinitely many solutions given by

$$(x_1, x_2, x_3, x_4) = (t/6, t, t, t)$$

A solution in positive integers is obtained by letting $t = 6k$, where k is a positive integer. When $t = 6$, we have

$$C_6H_{12}O_6 + 6\,O_2 \quad \Rightarrow \quad 6\,CO_2 + 6\,H_2 \quad + \quad \boxed{\text{ENERGY}}$$

and other solutions correspond to scalings of the (stochiometric) coefficients of this reaction. We will refer to the solution (1, 6, 6, 6) as the *smallest solution* that balances the chemical reaction (1.50). ▩

In practice, simple chemical equations are balanced by inspection, one element at a time, or by using other techniques. Homogeneous linear systems are useful in balancing more complicated equations.

Historical Notes

Amedeo Avogadro (1776–1856) Italian physicist and chemist. Avogadro's law states that equal volumes of gases at the same temperature and pressure contain the same number of molecules.

1.3.5 Fitting Curves

The expression

$$p(x) = a_n x^n + a_{n-1} x^{n-1} + \cdots + a_1 x + a_0$$

is called a *polynomial*[16] of degree n. The positive integer n is the highest power of the variable x appearing in $p(x)$ and $a_n, a_{n-1}, \ldots, a_1, a_0$ are the coefficients in $p(x)$ of the monomials $x^n, x^{n-1}, \ldots, x, 1$, respectively. For example,

$$n = 2, \quad a_2 = 3, \quad a_1 = -2, \quad a_0 = 1 \quad \Rightarrow \quad p(x) = 3x^2 - 2x + 1$$

defines a polynomial $p(x)$ of second degree (quadratic). The equation [17]

$$a_n x^n + a_{n-1} x^{n-1} + \cdots + a_1 x + a_0 = y \tag{1.52}$$

[16] This format for a polynomial is used in MATLAB.

[17] Writing $p(x) = y$ here, rather than $y = p(x)$, to fit the form in (1.54).

defines a *polynomial curve* in the xy-plane. The curve passes through m given points

$$P_1 = (x_1, y_1), \quad P_2 = (x_2, y_2), \quad \ldots \quad , P_m = (x_m, y_m) \tag{1.53}$$

in the plane if and only if the $m \times (n+1)$ linear system (S) in unknowns $a_n, a_{n-1}, \ldots, a_0$ given by

$$(S) \begin{cases} a_n x_1^n + a_{n-1} x_1^{n-1} + \cdots + a_1 x_1 + a_0 = y_1, & \text{using } (x_1, y_1) \\ a_n x_2^n + a_{n-1} x_2^{n-1} + \cdots + a_1 x_2 + a_0 = y_2, & \text{using } (x_2, y_2) \\ \vdots \qquad\qquad\qquad\qquad \vdots & \vdots \\ a_n x_m^n + a_{n-1} x_m^{n-1} + \cdots + a_1 x_m + a_0 = y_m, & \text{using } (x_m, y_m) \end{cases} \tag{1.54}$$

is consistent. The $m \times (n+2)$ augmented matrix $\mathbf{M}$ for (S) is

$$\mathbf{M} = [\mathbf{A} | \, \mathbf{y}] = \left[\begin{array}{cccccc|c} x_1^n & x_1^{n-1} & \cdots & x_1 & 1 & & y_1 \\ x_2^n & x_2^{n-1} & \cdots & x_2 & 1 & & y_2 \\ \vdots & \vdots & \ddots & \vdots & \vdots & & \vdots \\ x_m^n & x_m^{n-1} & \cdots & x_m & 1 & & y_m \end{array}\right]. \tag{1.55}$$

The coefficient matrix $\mathbf{A}$ is called a *Vandermonde* matrix due to its special form. The solution to (S) depends on m, n, rank $\mathbf{A}$, rank $\mathbf{M}$ and is described next.

Theorem 1.7 **Fitting Curves**

Consider m points $P_i = (x_i, y_i)$, $1 \leq i \leq m$, such that the x-values are distinct; that is, $x_i \neq x_j$ whenever $i \neq j$.

(a) If $m < n+1$, then rank $\mathbf{A} = m$ and (S) has infinitely many solutions. There is a family of curves passing through all the points.
(b) If $m = n+1$, then rank $\mathbf{A} = m$ and (S) has a unique solution. There is a unique curve passing through all the points.
(c) If $m > n+1$, then either rank $\mathbf{A} <$ rank $\mathbf{M}$, in which case (S) is inconsistent or rank $\mathbf{A} = n+1$, in which case there is a unique solution for (S). There is either no curve or a unique curve passing through all the points.

EXAMPLE 4 **Parabolic Curves**

Consider the points P_1, P_2 and the quadratic $p(x)$ given by

$$P_1 = (0, 1), \quad P_2 = (1, 2), \qquad p(x) = a_2 x^2 + a_1 x + a_0 = y.$$

The x-values are distinct and $m = n = 2 < 3 = n+1$. Theorem 1.7(a) applies and there is a family of quadratic curves (parabolas) passing through

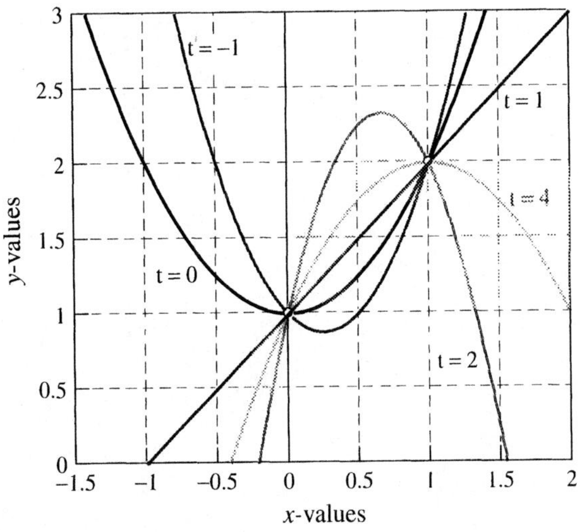

Figure 1.6 A family of parabolas.

both points. The reduced form $\mathbf{M}^*$ for the augmented matrix $\mathbf{M}$ corresponding to (1.55) is as follows (pivots are circled):

$$\mathbf{M} = \left[\begin{array}{ccc|c} 0 & 0 & 1 & 1 \\ 1 & 1 & 1 & 2 \end{array}\right] \sim \left[\begin{array}{ccc|c} \textcircled{1} & 1 & 0 & 1 \\ 0 & 0 & \textcircled{1} & 1 \end{array}\right].$$

Let $a_1 = t$, where t is a real parameter. Then $a_2 = 1 - t$ and $a_0 = 1$. The family of parabolas is given by

$$p(x) = (1-t)x^2 + tx + 1 = y, \quad \text{for any real } t \neq 1. \tag{1.56}$$

Equation (1.56) is linear when $t = 1$, giving the line (degenerate case) that passes through P_1 and P_2. Four curves from the family in (1.56) are plotted in Figure 1.6. ▩

EXAMPLE 5

Fitting Cubic Curves

Consider four points in the plane given by

$$P_1 = (-3, 6.7), \quad P_2 = (-1, -2.3), \quad P_3 = (1, 4.2), \quad P_4 = (2, 1.2).$$

The x-values are all distinct and $m = 4 = 3 + 1$. With $n = 3$, Theorem 1.7(b) applies and there is a unique cubic curve (polynomial of degree 3)

$$y = p(x) = a_3x^3 + a_2x^2 + a_1x + a_0 \tag{1.57}$$

passing through all the points. Using MATLAB, the augmented matrix $\mathbf{M} = [\mathbf{A}|\ \mathbf{y}]$ has reduced form $\mathbf{M}^*$, where

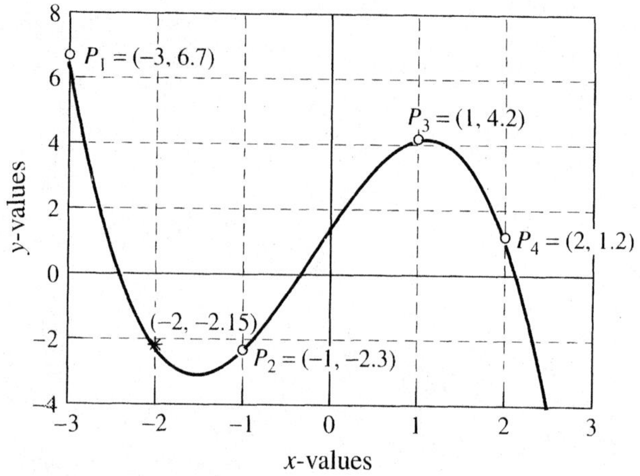

Figure 1.7 Fitting a cubic curve.

$$\mathbf{M} = \left[\begin{array}{rrrr|r} -27 & 9 & -3 & 1 & 6.7 \\ -1 & 1 & -1 & 1 & -2.3 \\ 1 & 1 & 1 & 1 & 4.2 \\ 8 & 4 & 2 & 1 & 1.2 \end{array}\right]$$

$$\sim \left[\begin{array}{rrrr|r} \textcircled{1} & 0 & 0 & 0 & -0.8042 \\ 0 & \textcircled{1} & 0 & 0 & -0.4750 \\ 0 & 0 & \textcircled{1} & 0 & 4.0542 \\ 0 & 0 & 0 & \textcircled{1} & 1.4250 \end{array}\right] = \mathbf{M}^*.$$

Distinct x-values ensure that $\operatorname{rank} \mathbf{A} = 4 = \operatorname{rank} \mathbf{M}$ and the unique coefficients for $p(x)$ in (1.57) are

$$a_3 \simeq -0.8042, \quad a_2 \simeq -0.4750, \quad a_1 \simeq 4.0542, \quad a_0 \simeq 1.4250.$$

A MATLAB plot of this cubic curve appears in Figure 1.7.

Given a value of x between the data points, say $x = -2$, the corresponding value $y = -2.15$ can now be *interpolated*. ▩

Historical Notes

Alexandre Théophile Vandermonde (1735–1796) French mathematician. Although his first love was music, Vandermonde turned to mathematics in his mid-thirties, working mainly in the theory of equations and determinants. Four memoirs published between 1771 and 1772 contain important results and methods. The Vandermonde determinant, named after him by Henri Léon Lebesgue (1875–1941), did not appear in his published work.

EXERCISES 1.3

1.3.1 LINEAR MODELS

1. An engineering company has three divisions (Design, Production, Testing) with a combined annual budget of US $1.2 million. Production has an annual budget equal to the combined annual budgets of Design and Testing. Testing requires a budget of at least $75,500. What is the Production budget and the maximum possible budget for the Design division?
2. A pharmaceutical company spends $6 million on radio, magazine, and television advertising. If the company spends as much on television advertising as on magazines and radio together, and the amount spent on magazines and television combined equals five times that spent on radio, what is the amount spent on each type of advertising?
3. Two products (labeled P_1 and P_2) are produced using two machines (labeled M_1 and M_2). P_1 requires 6 minutes on M_1 and 5 minutes on M_2, and P_2 requires 5 minutes on M_1 and 4 minutes on M_2. Before the machines need to be serviced, M_1 can run for 48 hours and M_2 can run for 39 hours. How many products can be produced before the M_1 is serviced? Can all the machine time be utilized?
4. The volume and weight of items A_1, A_2, A_3, A_4 are shown in the following table.

Item	Volume (in^3)	Weight (lb)
A_1	80	300
A_2	20	400
A_3	100	200
A_4	60	400

How can a shipment of 16 items be made if the shipment has total volume 1200 in^3 and total weight 4800 lb? Is there any way to ship exactly 5 items of type A_3?

5. *A Typical Diet Problem.* The following table shows the vitamin content of one serving of various foods (1 mg $= 10^{-3}$ g [gram]).

Food	Vitamin B_1 (mg)	Iron (mg)	Protein (g)
Skim Milk	0.07	0.1	8.5
Whole Wheat Bread	0.18	1.1	3.0
Hamburger	0.14	3.4	22.0

Make up a daily diet that includes 1.14 mg of Vitamin B_1, 11.4 mg of iron, and 73 g of protein.

6. A trucking company has three depots labeled A, B, C. On a particular day a number of trucks left each depot for the other two depots. The following (partial) information is known.

Depot	Total Trucks Leaving	Total Trucks Arriving
A	5	8
B	10	11
C	7	Unknown

We know that more trucks traveled $A \to C$ than $C \to A$. Build a linear model to find the exact number of trucks traveling between the depots.

7. An industrial city has four heavy industries (denoted by A, B, C, D) each of which burns coal to manufacture its products. By law, no industry can burn more than 45 units of coal per day. Each industry produces the pollutants Pb (lead), SO_2 (sulfur dioxide), and NO_2 (nitrogen dioxide) at (different) daily rates per unit of coal burned and these are released into the atmosphere. The rates are shown in the following table.

Industry	A	B	C	D
Pb	1	0	1	7
SO_2	2	1	2	9
NO_2	0	2	2	0

The CAAG (Clean Air Action Group) has just leaked a government report that claims that during one weekday one year ago, 250 units of Pb, 550 units of SO_2, and 400 units of NO_2 were measured in the atmosphere. An inspector reported that C did not break the law on that day. Which industry (industries) broke the law on that day?

8. A nut company sells three brands of mixed nuts labeled Mister Peanut (MP), Cocktail Plus (CP), and Auld Salt (AS). The consumer price (per pound) of each brand is MP $3.60, CP $8.00, AS $8.80. The brands are composed of three types of nuts in the following proportions:

Brand	Spanish Nuts	Cocktail Nuts	Cashews
MP	40%	60%	
CP	40%	40%	20%
AS	20%	50%	30%

The company buys its nuts at one-half of the consumer price. What is the dollar cost (per pound) of each type of nut?

9. A steel company sells flat steel in 8-foot lengths which come in three types of bundles (labeled A, B, C). There are four gauges[18] of the steel: (G_1) 2 mm, (G_2) 4 mm, (G_3) 6 mm, (G_4) 8 mm. The bundles are composed of a number of lengths of different gauges as shown in the following table.

Bundle	G_1	G_2	G_3	G_4
A	2	1	3	3
B	3	1	1	2
C	5	2	4	5

A bundle of type A sells for \$18, one of type B for \$16, and one of type C for \$21. The company receives an order for 37 lengths of 2-mm gauge, 14 lengths of 4-mm gauge, 24 lengths of 6-mm gauge, and 33 lengths of 8-mm gauge. In how many ways can the order be filled? In how many ways can the order be filled with exactly two types of bundles? Find the cheapest order.

10. A chemical company is testing four brands of fertilizer labeled B_1, B_2, B_3, B_4. Each bag of fertilizer contains Potash (Po) (Potassium Hydroxide), Nitrate (N), and Phosphate (Pho) in the numbers of units shown in the table.

Chemical	B_1	B_2	B_3	B_4
Po	2	1	0	1
N	1	1	2	0
Pho	0	1	1	2

[18] gauge: a standard measure, especially width (railway track) and thickness (steel sheet or wire).

Company research indicates that the best crop yield from one mile^2 ($\simeq 2.59$ km^2) of land requires 12 units of Po, 8 units of N, and 4 units of Pho. Your goal is to find the number of bags of each type of fertilizer that should be applied to one mile^2 in order to test the hypothesis. Let x_i be the number of bags of B_i required, where $1 \le i \le 4$. Set up a 3×4 linear system and solve it.

11. A bakery displays the number of ounces of yogurt, wheat germ, and butter used in the production of one batch of its major products. It uses 10 oz of yogurt, 10 oz of wheat germ and 10 oz of butter in a batch of rolls; 15 oz of wheat germ and 15 oz of butter in a batch of cookies; 10 oz of butter in a batch of buns; and 20 oz of yogurt and 15 oz of butter in a batch of bread. The bakery is supplied with 400 lb of yogurt, 337.5 lb of wheat germ, and 562.5 lb of butter, which must be used up completely to avoid spoiling. What is the maximum number of batches of rolls that can be made to completely use up all the supplies?

1.3.2 GENERAL NETWORKS

12. Refer to Figure 1.3 in this section. Impose an extra constraint that the total flow per hour in branches AB and AD equals the total flow in branches BC and CD. Find the flow in each branch.

13.

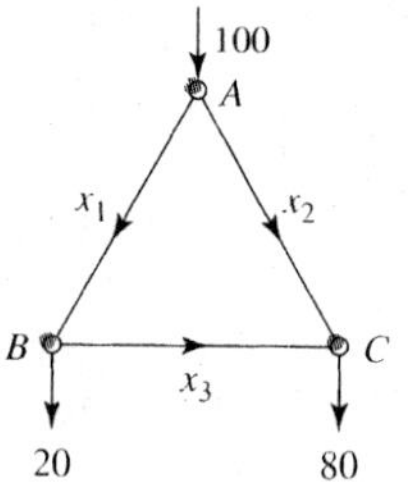

The network shown represents fluid flow in a system of pipes. The system is in equilibrium, with one-way flows in units per minute shown by the arrows.

(a) Find the unknown flows in each branch of the network.

(b) Find the minimum flow in branch AB.

(c) If the flow in branch AB is k (> 0) times the flow in branch AC, determine the unknown flows in terms of k.

14.

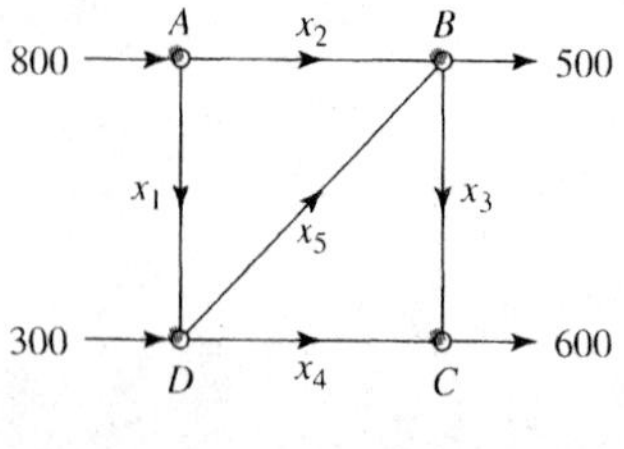

The network shown is in a steady state. Flows are in units per minute (units are indivisible).

(a) Find the flows in all branches.

(b) If there is no flow in branch BC, find the maximum flow in branch BD.

(c) If the flow in branch BD is twice the flow in branch DC, find the maximum flow in branch DC.

15.

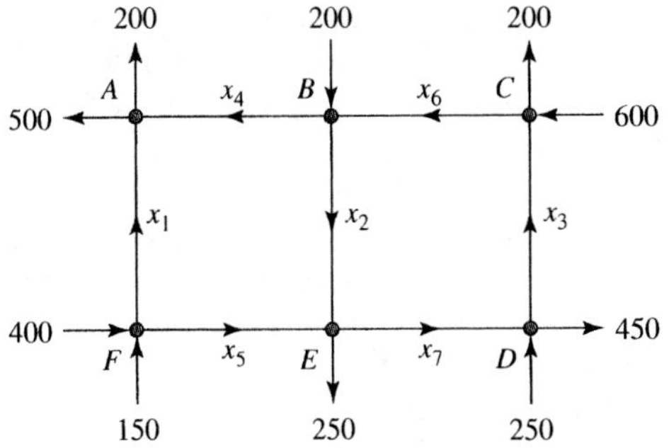

The network shown is in a steady state. Flows are in units per minute (units are indivisible).

(a) Write down the input–output equations for the network and solve the resulting linear system.

(b) Find the minimum flow in branch ED.

(c) If the flow in branch ED is 200 units, find the flow in branch DC.

(d) If there is zero flow in section FE, find the flow in branch BA.

16.

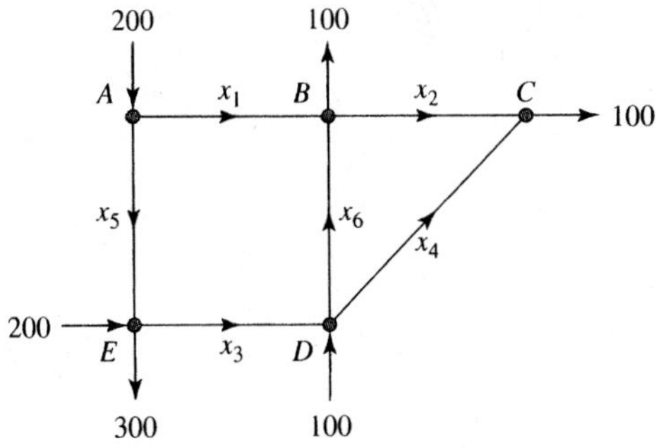

Find the flow in all branches of the network. If the flow in AB is zero, find the maximum flow in BD. Assuming the flows in AB and BC are zero, find the rest of the flows.

1.3.3 ELECTRICAL NETWORKS

Exercises 17–20. Find the unknown currents in each circuit.

17.

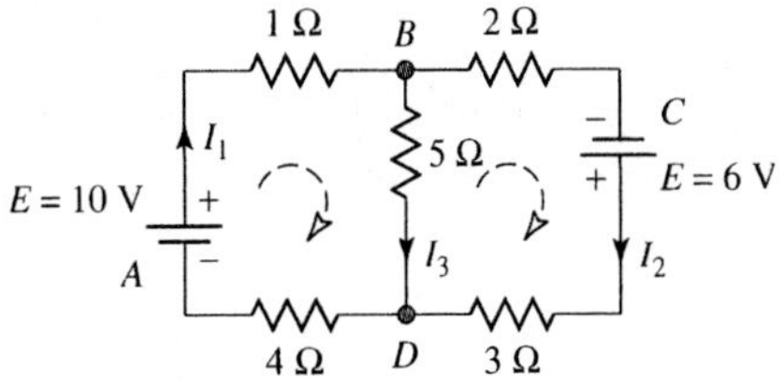

18.

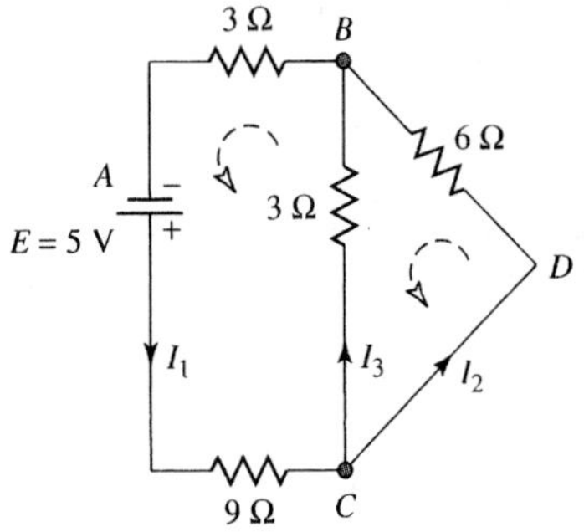

19.

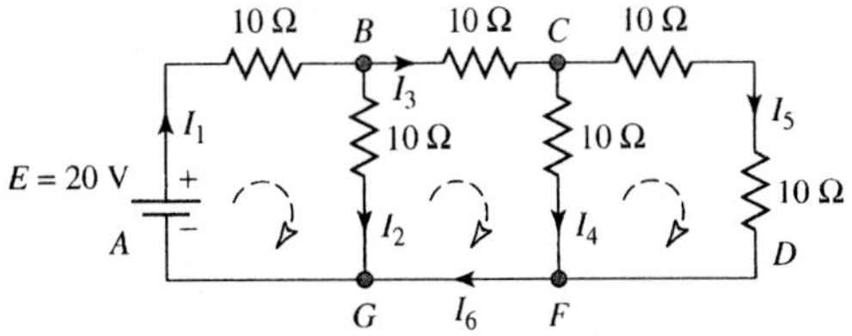

20.

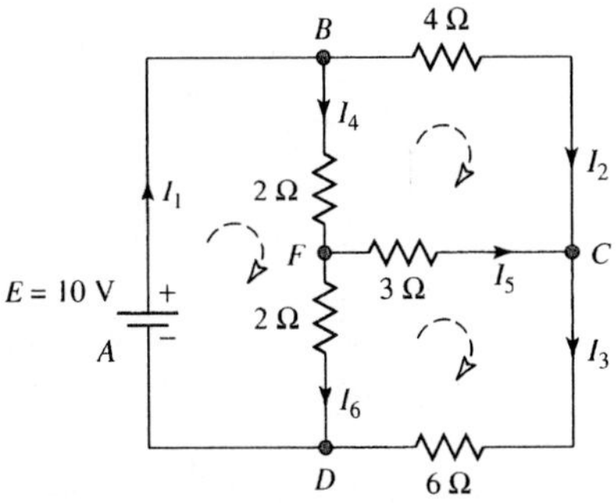

1.3.4 BALANCING CHEMICAL EQUATIONS

21. The action of sunlight on green plants powers a process called *photosynthesis*, in which plants capture energy from light and store energy in chemical form. During photosynthesis, x_1 moles of carbon dioxide (CO_2) and x_2 moles of water (H_2O) are converted into x_3 moles of oxygen (O_2) and x_4 moles of complex carbohydrates, such as glucose ($C_6H_{12}O_6$). Find the smallest positive integer solution that balances the chemical equation

$$x_1CO_2 + x_2H_2O \overset{\boxed{\text{LIGHT}}}{\Longrightarrow} x_3O_2 + x_4C_6H_{12}O_6.$$

22. Gasoline is a mixture of hydrocarbons containing from 6 to 8 carbon atoms per molecule. Octane (C_8H_{18}), a constituent of petroleum, burns in oxygen (O_2) to produce carbon dioxide (CO_2) and water (H_2O). Find the smallest solution (x_1, x_2, x_3, x_4) in moles for the chemical equation

$$x_1 C_8H_{18} + x_2 O_2 \overset{\boxed{\text{COMBUSTION}}}{\Longrightarrow} x_3 CO_2 + x_4 H_2O.$$

23. $C_6H_{10}O_5$ (a monomer of cellulose) is the main constituent of the cell walls of plants and of numerous fibrous products, including paper and cloth. It is by far the most abundant organic substance found in nature. It burns in oxygen (O_2) to form carbon dioxide (CO_2) and water (H_2O) plus k moles of carbon (C). Find the smallest solutions that will balance the equation for each value $k = 0, 1, 2$.

24. Trinitrotoluene ($C_7H_5O_6N_3$), also known as TNT, is produced from toluene (C_7H_8) and nitric acid (HNO_3), with water (H_2O) as the byproduct. Find all solutions in positive integers that will balance this chemical reaction.

1.3.5 FITTING CURVES

25. Use Gauss elimination to find the equation of the line $y = mx + c$ that passes through the points $(-1, 1.1)$, $(2, 3.2)$.

26. Use Gauss–Jordan elimination to find the equation of the parabola $y = a_2x^2 + a_1x + a_0$ passing through the points $(1, 1)$, $(-1, 1)$, $(2, 2)$.

27. Prove that there is a unique parabola passing through the four points $(-1, -1)$, $(0, 1)$, $(1, 1)$, $(2, -1)$. Explain how Theorem 1.7 applies in this case.

28. Show that no parabola passes through the points $(-1, 0)$, $(0, 1)$, $(1, 0)$, $(2, -5)$. Explain how Theorem 1.7 applies in this case.

29. Find the family of polynomial curves of second degree passing through the points $(2, 2)$, $(4, 2)$. Find the equation of the specific curve that passes through $(3, 6)$.

USING MATLAB

30. *E-Bond.com* offers four bond portfolios (labeled A, B, C, D) over the Internet. Each portfolio consists of blocks of federal bonds (F), regional bonds (R), provincial bonds (P), and municipal bonds (M). No fraction of portfolios are sold. A Web page contains the following table showing the numbers of blocks of each type of bond in each portfolio.

Type	F	R	P	M
A	1	2	3	5
B	1	1	2	4
C	0	3	3	3
D	3	1	4	10

A customer requires 11 blocks of federal, 17 blocks of regional, 28 blocks of provincial, and 50 blocks of municipal bonds. Find all possible ways in which the order can be filled using the portfolios. Determine the number of ways of filling the order using exactly two types of portfolios.

31. *E-Stock.com* offers four types of portfolios (labeled A, B, C, D) over the Internet. Each portfolio consists of blocks (100 shares per block) of oil, mining, gold, and commodity stocks. The composition of the portfolios is shown in the table.

Type	Oil	Mining	Gold	Commodity
A	11	6	4	4
B	3	3	1	2
C	5	2	2	1
D	3	1	1	1

The company receives an order for 69 blocks of oil, 32 blocks of mining, 26 blocks of gold, and 20 blocks of commodity stocks. The portfolios A, B, C, D each cost, respectively, \$1600, \$600, \$1200, and \$300. *E-Stock.com* guarantees that each portfolio will be worth (respectively) at least \$2400, \$1600, \$1800, \$900 after one year. No fraction of portfolios are to be bought or sold. Determine all the ways of filling the order with the portfolios. Determine the number of ways of filling the order using exactly three types of portfolios. Determine the order that will make the biggest gain and the one that will make the smallest gain.

32.

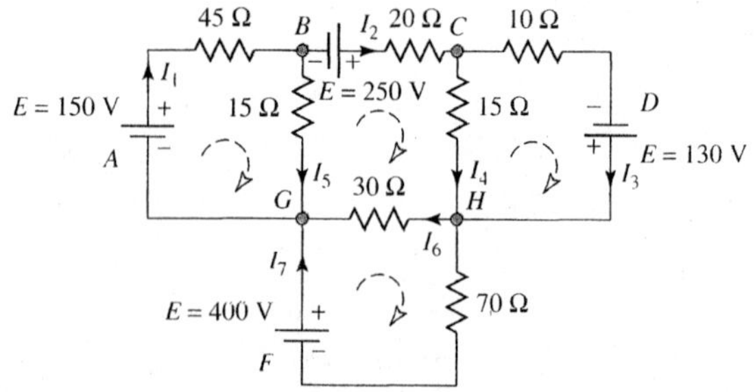

Find the unknown currents in the circuit shown.

33. Revisit Exercise 17. Set up the linear system when the battery at C has variable voltage V. Find the currents in the cases (a) $E = 9$ V, (b) $E = 23$ V. Vary the voltage E in order to explore how the value of current changes. What is the voltage at C when no current flows through branch BD?

34. In this chemical reaction each manganese (Mn) atom is reduced and gains 5 electrons while each atom of iron (Fe) loses 1 electron. Thus potassium permanganate ($KMnO_4$) is reduced and iron sulphate ($FeSO_4$) is oxidized. Find the smallest solution that will balance the chemical reaction

$$x_1 FeSO_4 + x_2 KMnO_4 + x_3 H_2SO_4 \Longrightarrow x_4 Fe_2(SO_4)_3 + x_5 MnSO_4 + x_6 K_2SO_4 + x_7 H_2O.$$

35. Ethanol (C_2H_5OH) is oxidized by sodium dichromate ($Na_2Cr_2O_7$) in the presence of hydrochloric acid (HCl) to produce acetic acid ($C_2H_4O_2$), chromium–III chloride ($CrCl_3$), water, and sodium chloride (NaCl). Find the smallest solution that will balance the chemical reaction

$$x_1 C_2H_5OH + x_2 Na_2Cr_2O_7 + x_3 HCl \Longrightarrow x_4 C_2H_4O_2 + x_5 CrCl_3 + x_6 H_2O + x_7 NaCl.$$

This reaction takes place during a police breathalyzer test!

Exercises 36–37. *Vandermonde matrices.* Refer to Example 5. Define a vector

$$\mathbf{x} = [\,-3 \;\; -1 \;\; 1 \;\; 2\,]$$

whose components are the x-values of four points in the plane. The command **A = vander(x)** returns the required 4×4 Vandermonde coefficient matrix and the commands

$$\mathbf{y} = [\,6.7;\; -2.3;\; 4.2;\; 1.2\,], \quad \mathbf{M} = [\,\mathbf{A}, \mathbf{y}\,]$$

define the column of constant terms $\mathbf{y}$ (y-values) and augmented matrix $\mathbf{M}$. Use the methods just described in the following exercises.

36. Find the equation of the cubic curve $a_3x^3 + a_2x^2 + a_1x + a_0 = y$ that passes through the points $(2, 2)$, $(-1, 9)$, $(-2, 2)$, $(1, 2)$. Use MATLAB to plot the curve and extrapolate the value of y when $x = -3.37$.

37. *Project.* Write an M-file that calls the coordinates of four points in the plane as input, solves the required linear system for values a_3, a_2, a_1, a_0, and plots the points and the cubic curve $a_3x^3 + a_2x^2 + a_1x + a_0 = y$ passing through all four points. Extend the program to interpolate or extrapolate the value of y for any given value of x.

38. Use MATLAB to check hand calculation in this exercise set.

CHAPTER 1 REVIEW

State whether each statement is true or false as stated. Provide a clear reason for your answer. If a false statement can be modified to make it true, then do so.

1. The operation $4E_2 - 5E_1 \to E_2$ is an elementary equation operation.

2. The inverse of the ERO $\mathbf{r}_3 - \mathbf{r}_2 \to \mathbf{r}_3$ is $\mathbf{r}_2 + \mathbf{r}_3 \to \mathbf{r}_2$.

3. The inverse of the ERO $\mathbf{r}_1 \leftrightarrow \mathbf{r}_2$ is $\mathbf{r}_1 \leftrightarrow \mathbf{r}_2$.

4. A linear system must contain at least two linear equations.

5. Forward elimination on a linear system will reveal which variables are uniquely determined.

6. If a variable in a linear system is not free, then it is uniquely determined.

7. A consistent $n \times n$ linear system has a unique solution.

8. Two linear systems are equivalent if and only if their associated augmented matrices are row equivalent.

9. If $\mathbf{M} = [\mathbf{A}\,|\,\mathbf{b}]$ is the augmented matrix of an $n \times n$ linear system (S) and $\operatorname{rank} \mathbf{A} < n$, then (S) is inconsistent for some $\mathbf{b}$.

10. A consistent linear system with less equations than unknowns is consistent for every column of constants $\mathbf{b}$.

11. An $n \times n$ homogeneous linear system has a unique solution.

12. If a 3×3 linear system (S) is inconsistent, then deleting some equation from (S) will result in a consistent subsystem.

13. Any two echelon forms for a given matrix $\mathbf{A}$ have the same number of zero rows.

14. If a 3×3 matrix $\mathbf{A}$ has a zero row, then $\operatorname{rank} \mathbf{A} = 2$.

15. The rank of a matrix $\mathbf{A}$ is the number of nonzero rows in $\mathbf{A}$.

16. There exists a 6×4 matrix of rank 5.

17. There exists a 3×2 matrix with rank 2.

18. If $\mathbf{A} = \begin{bmatrix} -1 & 2 \\ 2 & 1 - c \end{bmatrix}$, then $\operatorname{rank} \mathbf{A} = 2$ if and only if $c \neq 5$.

19. A consistent linear system (S) with less equations than unknowns (underdetermined) never has a unique solution.

20. If the coefficient matrix $\mathbf{A}$ of a linear system (S) is 4×5 and has four pivots, then (S) is consistent for any column of constants $\mathbf{b}$.

Arthur Cayley was one of the leading mathematicians of the 19th century. He was educated at Trinity College, Cambridge, England, and graduated in 1842 with the first Smith Medal. He taught for the next four years at Cambridge and published 28 research papers.

Cayley then took up law and over the next 14 years became a successful lawyer in London, England. While training as a lawyer, Cayley went to Dublin to hear a lecture by Sir William Rowan Hamilton (1805–1865) on quaternions—a concept that Cayley would later use in his research on groups. During his law career, he met and discussed mathematics with James Joseph Sylvester (1814–1897), who was also working in the law courts. Cayley and Sylvester had a very productive collaboration, making great discoveries in modern algebra that would soon find application in the development of relativity and quantum mechanics. In 1863 Cayley gave up law and accepted the inaugural chair of Sadlerian Professor at Cambridge, a position he kept for the next 30 years.

Cayley was truly a pure and applied mathematician, equally at home with abstract theories and scientific applications. He defined the notion of an abstract group and developed the theory of invariance. His work on n-dimensional geometry was applied to space-time continuum in physics. His work on matrices became a foundation for the theory of quantum mechanics. The list of his contributions goes on.

Cayley ranks with Euler and Cauchy as one of the three most prolific writers of mathematics of all time. He published 966 papers which fill 13 large volumes. The work of Arthur Cayley has been characterized by its clarity and extreme elegance of form.

Chapter 2

MATRICES

Introduction

A *matrix* is a general data structure that is used in diverse ways to tabulate and process information. Recall from Section 1.1 how matrices (plural for matrix) were used to represent and solve linear systems. A linear system is solved by performing elementary row operations on the augmented matrix of the system. The concept of rank (of a matrix) was introduced in Section 1.2 and used to determine the character of the solution set, namely, zero, one or infinitely many solutions. In this chapter we focus on *matrix algebra* and its applications.

The term matrix was introduced into mathematical literature by Sylvester in 1850. However, the major development in matrix algebra occurred in 1858 when Arthur Cayley published his *Memoir on the Theory of Matrices*. In this work he introduced matrix notation and defined matrix addition, scalar multiplication, matrix multiplication, and the inverse of a square matrix. Matrix addition and scalar multiplication are operations that extend the arithmetic laws of real and complex numbers to matrices. However, the operation of matrix multiplication is of a different character. This operation is *noncommutative* in general, meaning that $\mathbf{AB} \neq \mathbf{BA}$, for some pairs of matrices $\mathbf{A}$ and $\mathbf{B}$, and in this respect matrix algebra differs from the algebra of real and complex numbers in which $ab = ba$ for any numbers a and b.

Inverses of square matrices play an important role in the theory of linear algebra and its applications. Not every square matrix has an inverse and it is therefore necessary to know whether a given square matrix is *invertible* (the inverse exists) or *singular* (no inverse exists). There are many conditions that indicate *invertibility*. Several of these conditions are given in this chapter and others appear later in the text. There are several methods for calculating the inverse of an invertible matrix. Two methods appear in this chapter and another is given in Chapter 5.

There are situations in theory and in practice when it is useful to be able to write a given matrix as a product of some special matrices. This general problem, called *matrix factorization*, will enter into the discussion from time to time. Section 2.3 deals with LU-*factorization*, which uses triangular matrices. Other factorizations of a matrix appear in later chapters.

The applications in Section 2.4 illustrate the use of matrix multiplication. A deeper analysis of some of these models continues in Chapter 6 which introduces *eigenvalue problems*.

2.1 Matrix Algebra

Let m and n be positive integers. Recall from Section 1.1 that an $m \times n$ (read "m by n") matrix $\mathbf{A}$, as shown in (2.1), is a rectangular array of mn numbers arranged in m rows and n columns.

$$\mathbf{A} = \begin{bmatrix} a_{11} & a_{12} & \cdots & a_{1j} & \cdots & a_{1n} \\ a_{21} & a_{22} & \cdots & a_{2j} & \cdots & a_{2n} \\ \vdots & \vdots & & \vdots & & \vdots \\ a_{i1} & a_{i2} & \cdots & \boxed{a_{ij}} & \cdots & a_{in} \\ \vdots & \vdots & & \vdots & & \vdots \\ a_{m1} & a_{m2} & \cdots & a_{mj} & \cdots & a_{mn} \end{bmatrix} \begin{matrix} \\ \\ \\ \leftarrow \text{row } i \\ \\ \\ \end{matrix} \qquad (2.1)$$

$$\uparrow \text{ column } j$$

We say that $\mathbf{A}$ has *size* $m \times n$. The number lying in row i and column j (circled above) is called the (i, j)-*entry* in $\mathbf{A}$ and is denoted by a_{ij} (lowercase a corresponding to uppercase $\mathbf{A}$). The shorthand notation for (2.1) is $\mathbf{A} = [a_{ij}]$, or sometimes $\mathbf{A} = [a_{ij}]_{m \times n}$ when the size of $\mathbf{A}$ needs to be mentioned.

Submatrices

A *submatrix* of a matrix $\mathbf{A}$ is formed by deleting some of its rows and/or some of its columns to form a new matrix. For example, suppose

$$\mathbf{A} = \begin{bmatrix} 1 & 2 & 3 & 4 \\ 5 & 6 & 7 & 8 \\ 9 & 1 & 2 & 3 \end{bmatrix}.$$

Then $\mathbf{A}_1$, $\mathbf{A}_2$, $\mathbf{A}_3$ defined by

Delete from **A**: rows 1, 3; cols 1, 2	Delete from **A**: row 1; cols 2, 4	Delete from **A**: cols 1, 2, 3
$\mathbf{A}_1 = \begin{bmatrix} 7 & 8 \end{bmatrix}$	$\mathbf{A}_2 = \begin{bmatrix} 5 & 7 \\ 9 & 2 \end{bmatrix}$	$\mathbf{A}_3 = \begin{bmatrix} 4 \\ 8 \\ 3 \end{bmatrix}$

are three submatrices of $\mathbf{A}$.

and so

$$\mathbf{X} = 0.25\begin{bmatrix} -8 & 4 \\ 12 & -4 \end{bmatrix} = \begin{bmatrix} -2 & 1 \\ 3 & -1 \end{bmatrix}.$$ ■

2.1.2 Matrix Multiplication

The definition of matrix multiplication, given by Arthur Cayley in 1858, may appear at first sight rather complicated and unnatural. However, the rationale for defining matrix multiplication in this particular way will soon become evident in Illustration 2.1 and in applications presented in Section 2.4.

Definition 2.4

Matrix multiplication

Let $\mathbf{A} = [a_{ij}]$ be an $m \times n$ matrix and let $\mathbf{B} = [b_{ij}]$ be an $n \times p$ matrix. The product $\mathbf{AB}$ is an $m \times p$ matrix $\mathbf{C} = [c_{ij}]$, where

$$\underset{\mathbf{A}}{\overset{m \times n}{\begin{bmatrix} * & * & \cdots & * \\ \vdots & \vdots & & \vdots \\ a_{i1} & a_{i2} & \cdots & a_{in} \\ \vdots & \vdots & & \vdots \\ * & * & \cdots & * \end{bmatrix}}} \underset{\mathbf{B}}{\overset{n \times p}{\begin{bmatrix} * & \cdots & b_{1j} & \cdots & * \\ * & \cdots & b_{2j} & \cdots & * \\ \vdots & & \vdots & & \vdots \\ * & \cdots & b_{nj} & \cdots & * \end{bmatrix}}} = \underset{\mathbf{C}}{\overset{m \times p}{\begin{bmatrix} * & \cdots & * & \cdots & * \\ \vdots & & \vdots & & \vdots \\ * & \cdots & \boxed{c_{ij}} & \cdots & * \\ \vdots & & \vdots & & \vdots \\ * & \cdots & * & \cdots & * \end{bmatrix}}}$$

The (i, j)-entry c_{ij} in $\mathbf{C}$ is calculated using the entries from row i in $\mathbf{A}$ and the entries from column j in $\mathbf{B}$, and defined as follows:

$$c_{ij} = a_{i1}b_{1j} + a_{i2}b_{2j} + \cdots + a_{in}b_{nj}. \tag{2.3}$$

Computing the product $\mathbf{AB}$ using (2.3) is called *row-by-column multiplication*.[1]

Remarks

First, note that the product $\mathbf{AB}$ can only be formed if the number of columns (n) in $\mathbf{A}$ is equal to the number of rows (n) in $\mathbf{B}$, and when this is the case we say that $\mathbf{A}$ and $\mathbf{B}$ (in that order) are *conformable* for multiplication. When multiplying, always check sizes:

$$m \times n \quad \leftrightarrow \quad n \times p \quad \text{(inside dimensions must agree!)}.$$

When the product $\mathbf{AB}$ exists, we say that $\mathbf{A}$ *pre-multiplies* $\mathbf{B}$ and that $\mathbf{B}$ *post-multiplies* $\mathbf{A}$. Observe that there are m rows in $\mathbf{A}$ and p columns in $\mathbf{B}$ and so

[1] It has been said (by a pianist perhaps?) that matrix multiplication is a two-finger exercise: To obtain the entry c_{ij} in $\mathbf{C}$, locate row i in $\mathbf{A}$ with the left-index finger and column j in $\mathbf{B}$ with the right-index finger. Move simultaneously along row i and down column j, multiplying corresponding entries and adding.

2.1.1 Matrix Operations

■ **NOTATION** Sets of Matrices of the Same Size

We will use the symbol $\mathbb{R}^{m \times n}$ to denote the set of all $m \times n$ matrices. ■

Definition 2.1

Equality

Two $m \times n$ matrices $\mathbf{A} = [a_{ij}]$ and $\mathbf{B} = [b_{ij}]$ are *equal* whenever each pair of corresponding entries in $\mathbf{A}$ and $\mathbf{B}$ are equal; that is, $a_{ij} = b_{ij}$, for each pair of subscripts (i, j). We write $\mathbf{A} = \mathbf{B}$ when this is the case.

EXAMPLE 1

Equality of Matrices

The matrices

$$\mathbf{A} = \begin{bmatrix} x & 2 & 3 \\ 4 & 5 & y \end{bmatrix} \quad \text{and} \quad \mathbf{B} = \begin{bmatrix} 1 & 2 & z \\ 4 & 5 & 6 \end{bmatrix} \tag{2.2}$$

are both 2×3 and so $\mathbf{A} = \mathbf{B}$ if and only if $x = 1$, $y = 6$, $z = 3$.

Note that a pair of $m \times n$ matrices are not equal if *at least one pair* of corresponding entries are not equal. If, for example, $x \neq 1$ in (2.2) then $\mathbf{A} \neq \mathbf{B}$. ■

Definition 2.2

Matrix addition

The *sum* of two $m \times n$ matrices $\mathbf{A} = [a_{ij}]$ and $\mathbf{B} = [b_{ij}]$ is the $m \times n$ matrix $\mathbf{A}+\mathbf{B} = [a_{ij}+b_{ij}]$ obtained by adding each pair of corresponding entries in $\mathbf{A}$ and $\mathbf{B}$. The sum $\mathbf{A}+\mathbf{B}$ is not defined if the sizes of $\mathbf{A}$ and $\mathbf{B}$ are different.

EXAMPLE 2

Matrix Addition

The matrices

$$\mathbf{A} = \begin{bmatrix} 2 & 1 \\ 0 & 1 \end{bmatrix} \quad \text{and} \quad \mathbf{B} = \begin{bmatrix} 1 & -1 \\ 2 & 0 \end{bmatrix}$$

are both 2×2 and so $\mathbf{A}+\mathbf{B}$ is defined and has size 2×2. Adding corresponding entries, we have

$$\mathbf{A}+\mathbf{B} = \begin{bmatrix} 2 & 1 \\ 0 & 1 \end{bmatrix} + \begin{bmatrix} 1 & -1 \\ 2 & 0 \end{bmatrix} = \begin{bmatrix} 2+1 & 1+(-1) \\ 0+2 & 1+0 \end{bmatrix} = \begin{bmatrix} 3 & 0 \\ 2 & 1 \end{bmatrix}.$$ ■

Definition 2.3

Scalar multiplication

Let $\mathbf{A} = [a_{ij}]$ be an $m \times n$ matrix. The matrix $s\mathbf{A} = [sa_{ij}]$ obtained by multiplying all entries in $\mathbf{A}$ by s is called a *scalar multiple* of $\mathbf{A}$. We say that $s\mathbf{A}$ is the result of *scaling* $\mathbf{A}$ by s.

EXAMPLE 3

Scaling a Matrix

Using the matrix **A** shown, forming $s\mathbf{A}$ when $s = -2$ and $s = 0$, respectively, we have

$$\mathbf{A} = \begin{bmatrix} -2 & 3 \\ 0 & 1 \end{bmatrix} \quad \Rightarrow \quad -2\mathbf{A} = \begin{bmatrix} 4 & -6 \\ 0 & -2 \end{bmatrix}, \qquad 0\,\mathbf{A} = \begin{bmatrix} 0 & 0 \\ 0 & 0 \end{bmatrix}$$ ▪

Zero Matrices

Each set $\mathbb{R}^{m\times n}$ contains a unique *zero matrix* $\mathbf{O}_{m\times n}$ that has *all* of its entries equal to zero. For example,

$$\mathbf{O}_{2\times 3} = \begin{bmatrix} 0 & 0 & 0 \\ 0 & 0 & 0 \end{bmatrix} \quad \text{and} \quad \mathbf{O}_{2\times 2} = \begin{bmatrix} 0 & 0 \\ 0 & 0 \end{bmatrix}$$

are, respectively, the zero matrices in the sets $\mathbb{R}^{2\times 3}$ and $\mathbb{R}^{2\times 2}$.

The notation **O** (without size) is also used when the size of **O** is clear from the context. The zero matrix **O** in $\mathbb{R}^{m\times n}$ behaves like the number 0 in the real or complex number system; that is, for any **A** in $\mathbb{R}^{m\times n}$ we have

$$\mathbf{A} + \mathbf{O} = \mathbf{A} = \mathbf{O} + \mathbf{A}.$$

Note also that $0\mathbf{A} = \mathbf{O}$ (scaling **A** by $s = 0$).

Subtraction

For each real or complex number a there is a number $-a$ such that $a + (-a) = 0$. This property is extended to the set $\mathbb{R}^{m\times n}$ in the following way. For each matrix $\mathbf{A} = [a_{ij}]$ in $\mathbb{R}^{m\times n}$ there is a matrix $-\mathbf{A} = [-a_{ij}]$ in $\mathbb{R}^{m\times n}$ called the *additive inverse* **A** such that $\mathbf{A} + (-\mathbf{A}) = \mathbf{O}_{m\times n}$. To find $-\mathbf{A}$, negate all the entries in **A**. For example,

$$\mathbf{A} = \begin{bmatrix} 1 & -2 \\ -3 & 4 \end{bmatrix}, \qquad -\mathbf{A} = \begin{bmatrix} -1 & 2 \\ 3 & -4 \end{bmatrix} \quad \Rightarrow \quad \mathbf{A} + (-\mathbf{A}) = \mathbf{O}_{2\times 2}.$$

For any $m \times n$ matrix **A**, we have $-\mathbf{A} = (-1)\mathbf{A}$. Subtraction $\mathbf{A} - \mathbf{B}$ is defined ...eration $\mathbf{A} + (-\mathbf{B})$.

...eview

... algebraic rules stated in Theorem 2.1 ...d here in the context of the ...old in particular for ...onents). In

Theorem 2.1

Algebraic Rules for Matrix Addition and Scalar Multiplication

Let **A**, **B**, **C** *be matrices in the set* $\mathbb{R}^{m\times n}$. *Let* **O** *be the* $m \times n$ *zero matrix* ... $\mathbb{R}^{m\times n}$ *and let s and t be any real scalars. Then the following rules apply.*

Matrix Addition		Scalar Multiplication	
(A1)	$\mathbf{A} + \mathbf{B} = \mathbf{B} + \mathbf{A}$	**(A5)**	$s(\mathbf{A} + \mathbf{B}) = s\mathbf{A}$...
(A2)	$(\mathbf{A} + \mathbf{B}) + \mathbf{C} = \mathbf{A} + (\mathbf{B} + \mathbf{C})$	**(A6)**	$(s + t)\mathbf{A} = s$...
(A3)	$\mathbf{A} + \mathbf{O} = \mathbf{A}$	**(A7)**	$s(t\mathbf{A}) = t(s$...
(A4)	$\mathbf{A} + (-\mathbf{A}) = \mathbf{O} = -\mathbf{A} + \mathbf{A}$	**(A8)**	$1\mathbf{A} = \mathbf{A}$

Remarks

Matrix operations are performed entry by entry and s... generalizations of corresponding algebraic rules for ... For example, the *commutative law* **(A1)** generali... $a + b = b + a$ for real or complex numbers a ... computing $\mathbf{A} + \mathbf{B}$, we may interchange the ord... **(A2)** is the *associative law* for addition. Addi... on two matrices at a time. Computing th... find $\mathbf{A} + \mathbf{B}$ and then compute $(\mathbf{A} + \mathbf{B}) +$ ... $\mathbf{A} + (\mathbf{B} + \mathbf{C})$. **(A2)** tells us that in eithe...

Matrix Equations

The linear equation $2x + a = b$, w... has solution $x = \frac{1}{2}(b - a)$. Th... $2\mathbf{X} + \mathbf{A} = \mathbf{B}$, in which **A**, **B** ... equation "formally" by caref...

$$(2\mathbf{X} + \mathbf{A}) + (-\mathbf{A} \ldots$$
$$2\mathbf{X} + (\mathbf{A} - \mathbf{A}) \ldots$$
$$2\mathbf{X} + \mathbf{O} \ldots$$
$$2\mathbf{X} \ldots$$

and scaling both ... equations are s...

EXAMPLE 4

Solving M...

Conside...

Note that the u... in **X** to the left sid...

$$5\mathbf{X} - \mathbf{X} = 4\mathbf{X} = 2\Big[\ldots$$

computing all entries in **C** requires mp row-by-column multiplications. According to (2.3), each row-by-column multiplication requires n multiplications and $n-1$ additions. Therefore, the total number of flops (floating point operations) to compute **C** using row-by-column multiplication is $mp(2n-1)$.

EXAMPLE 5

Computing a Product

Let **A** be a 2×3 matrix and **B** be a 3×2 matrix (such as those shown). The product **AB** exists and the result is a 2×2 matrix **C**. There are 4 calculations using 20 flops required to find all entries in **C**.

$$\underset{\mathbf{A}}{\begin{bmatrix} 1 & 2 & 5 \\ 3 & 0 & 4 \end{bmatrix}} \underset{\mathbf{B}}{\begin{bmatrix} 2 & 1 \\ 3 & 6 \\ 1 & 7 \end{bmatrix}} = \underset{\mathbf{C}}{\begin{bmatrix} c_{11} & c_{12} \\ c_{21} & c_{22} \end{bmatrix}}$$

$$\begin{array}{ll} i=1,\ j=1 & c_{11} = 1(2) + 2(3) + 5(1) = 13 \\ i=1,\ j=2 & c_{12} = 1(1) + 2(6) + 5(7) = 48 \\ i=2,\ j=1 & c_{21} = 3(2) + 0(3) + 4(1) = 10 \\ i=2,\ j=2 & c_{22} = 3(1) + 0(6) + 4(7) = 31 \end{array} \quad \Rightarrow \quad \mathbf{C} = \begin{bmatrix} 13 & 48 \\ 10 & 31 \end{bmatrix}$$

■

■ ILLUSTRATION 2.1

Matrix Multiplication, Cost of Materials

This simple illustration explores the use of matrix multiplication in a more practical setting. Three mining companies, C_1, C_2, C_3, sell three metals at different (dollar) costs per pound. The costs are displayed in the following table.

Metal	C_1	C_2	C_3
Lithium (L)	1.2	1.4	1.6
Nickel (N)	2.3	2.5	2.7
Zinc (Z)	3.4	3.6	3.8

Costs associated with individual companies are stored in three 3×1 matrices (column vectors), denoted by $\mathbf{c}_1$, $\mathbf{c}_2$, $\mathbf{c}_3$, and all costs are stored in the 3×3 cost matrix **C**, as shown.

$$\begin{matrix} & C_1 & & C_2 & & C_3 \\ \mathbf{c}_1 = & \begin{bmatrix} 1.2 \\ 2.3 \\ 3.4 \end{bmatrix} & \mathbf{c}_2 = & \begin{bmatrix} 1.4 \\ 2.5 \\ 3.6 \end{bmatrix} & \mathbf{c}_3 = & \begin{bmatrix} 1.6 \\ 2.7 \\ 3.8 \end{bmatrix} \end{matrix}$$

$$\mathbf{C} = [\, \mathbf{c}_1 \;\; \mathbf{c}_2 \;\; \mathbf{c}_3 \,] = \begin{bmatrix} 1.2 & 1.4 & 1.6 \\ 2.3 & 2.5 & 2.7 \\ 3.4 & 3.6 & 3.8 \end{bmatrix}$$

Consider computing the total cost of buying quantities of each of the three metals. An order for 4 lb of L, 5 lb of N, and 6 lb of Z is stored as a 1×3 matrix (row vector) denoted by $\mathbf{q}_1 = [4 \;\; 5 \;\; 6]$. The total cost of filling this order (Order 1) from company C_1 is given by the following row-by-column computation:

$$4(1.2) + 5(2.3) + 6(3.4) = 36.7.$$

The total cost of filling a second order (Order 2), defined by the row vector $\mathbf{q}_2 = [7 \;\; 8 \;\; 9]$, from company C_1 is computed similarly:

$$7(1.2) + 8(2.3) + 9(3.4) = 57.4.$$

Computing costs of various orders (in pounds) from various companies, we have

$$\underset{\text{Order 1}}{[4 \;\; 5 \;\; 6]} \underset{C_1}{\begin{bmatrix} 1.2 \\ 2.3 \\ 3.4 \end{bmatrix}} = \begin{bmatrix} 36.7 \end{bmatrix} \qquad \underset{\text{Order 1}}{[4 \;\; 5 \;\; 6]} \begin{bmatrix} C_1 & C_2 \\ 1.2 & 1.4 \\ 2.3 & 2.5 \\ 3.4 & 3.6 \end{bmatrix} = [\, 36.7 \;\; 39.7 \,]$$

$$\underset{\text{Order 1, 2}}{\begin{bmatrix} 4 & 5 & 6 \\ 7 & 8 & 9 \end{bmatrix}} \underset{C_1}{\begin{bmatrix} 1.2 \\ 2.3 \\ 3.4 \end{bmatrix}} = \begin{bmatrix} 36.7 \\ 57.4 \end{bmatrix} \qquad \underset{\text{Order 1, 2}}{\begin{bmatrix} 4 & 5 & 6 \\ 7 & 8 & 9 \end{bmatrix}} \underset{C_1 \;\; C_2}{\begin{bmatrix} 1.2 & 1.4 \\ 2.3 & 2.5 \\ 3.4 & 3.6 \end{bmatrix}} = \begin{bmatrix} 36.7 & 39.7 \\ 57.4 & 62.2 \end{bmatrix}$$

Multiplying the 2×3 matrix $\mathbf{Q} = \begin{bmatrix} \mathbf{q}_1 \\ \mathbf{q}_2 \end{bmatrix}$, whose row vectors $\mathbf{q}_1$, $\mathbf{q}_2$ define each order, with the 3×3 cost matrix $\mathbf{C}$ results in a 2×3 matrix $\mathbf{T}$ that displays the cost of filling all orders. We have

$$\mathbf{QC} = \begin{bmatrix} 4 & 5 & 6 \\ 7 & 8 & 9 \end{bmatrix} \begin{bmatrix} 1.2 & 1.4 & 1.6 \\ 2.3 & 2.5 & 2.7 \\ 3.4 & 3.6 & 3.8 \end{bmatrix} = \begin{bmatrix} 36.7 & 39.7 & 41.7 \\ 57.4 & 62.2 & 67.0 \end{bmatrix} = \mathbf{T}.$$

This illustration extends to accommodate any number of metals and mining companies. The only restriction in forming the product **QC** (in that order) is that the number of columns in **Q** is equal to the number of rows in **C**, that is, **Q** must be conformable to **C** for multiplication. ■

Existence of Products, Order of Multiplication

(a) If **A** is $m \times n$ and **B** is $n \times p$, then **AB** exists, but **BA** does not exist unless $m = p$. For example, if **A** is 3×3 and **B** is 3×4, then **AB** exists and is 3×4 but **BA** is not defined because the inside dimensions ($3 \times 4 \leftrightarrow 3 \times 3$) do not agree.

(b) If **A** is $m \times n$ and **B** is $n \times m$, then both products **AB** and **BA** exist and have size $m \times m$ and $n \times n$, respectively.

(c) If **A** and **B** are both $n \times n$ matrices, the products **AB** and **BA** exist and are $n \times n$, yet it may happen that $\mathbf{AB} \neq \mathbf{BA}$.

EXAMPLE 6

Changing Order of Multiplication

We illustrate item (c) above when $n = 2$. Consider

$$\mathbf{A} = \begin{bmatrix} 1 & 2 \\ 0 & 1 \end{bmatrix}, \quad \mathbf{B} = \begin{bmatrix} 1 & 3 \\ -1 & 0 \end{bmatrix}, \quad \mathbf{AB} = \begin{bmatrix} -1 & 3 \\ -1 & 0 \end{bmatrix}, \quad \mathbf{BA} = \begin{bmatrix} 1 & 5 \\ -1 & -2 \end{bmatrix}.$$

Then $\mathbf{AB} \neq \mathbf{BA}$ because (at least) one pair of corresponding entries in **AB** and **BA** are not equal. ▪

Caution! Never change order of multiplication unless you know it is permissible to do so.

If it happens that $\mathbf{AB} = \mathbf{BA}$ for $n \times n$ matrices **A** and **B**, we say that **A** and **B** *commute*.

EXAMPLE 7

Some Pairs of Matrices Commute

The 2×2 matrices

$$\mathbf{A} = \begin{bmatrix} 1 & 2 \\ 3 & 4 \end{bmatrix} \quad \text{and} \quad \mathbf{B} = \begin{bmatrix} -2 & 1 \\ 1.5 & -0.5 \end{bmatrix}$$

commute because

$$\mathbf{AB} = \begin{bmatrix} 1 & 2 \\ 3 & 4 \end{bmatrix}\begin{bmatrix} -2 & 1 \\ 1.5 & -0.5 \end{bmatrix} = \begin{bmatrix} 1 & 0 \\ 0 & 1 \end{bmatrix} = \begin{bmatrix} -2 & 1 \\ 1.5 & -0.5 \end{bmatrix}\begin{bmatrix} 1 & 2 \\ 3 & 4 \end{bmatrix} = \mathbf{BA}.$$

In Section 2.2 it will be seen that every *invertible* (square) matrix commutes with its own inverse. Actually, **B** shown above is the inverse of **A** (and vice versa). ▪

Suppose a, b, c are real or complex numbers. Then the following properties are true.

(a) If $ab = 0$, then either $a = 0$ or $b = 0$.
(b) If $ab = ac$ and $a \neq 0$, then $b = c$ (canceling a).

The corresponding properties when stated for matrices are false in general!

EXAMPLE 8

Failure of Properties (a) and (b) for Matrices

Consider the calculation:

$$\underset{\mathbf{A}}{\begin{bmatrix} 1 & -1 & 2 \\ 0 & 0 & 1 \end{bmatrix}} \underset{\mathbf{B}}{\begin{bmatrix} -1 & 2 \\ -1 & 2 \\ 0 & 0 \end{bmatrix}} = \underset{\mathbf{O}}{\begin{bmatrix} 0 & 0 \\ 0 & 0 \end{bmatrix}} = \underset{\mathbf{A}}{\begin{bmatrix} 1 & -1 & 2 \\ 0 & 0 & 1 \end{bmatrix}} \underset{\mathbf{C}}{\begin{bmatrix} 3 & 2 \\ 3 & 2 \\ 0 & 0 \end{bmatrix}}$$

Then $\mathbf{AB} = \mathbf{O}$ but $\mathbf{A} \neq \mathbf{O}$ and $\mathbf{B} \neq \mathbf{O}$. Also, $\mathbf{AB} = \mathbf{AC}$ and $\mathbf{A} \neq \mathbf{O}$ but $\mathbf{B} \neq \mathbf{C}$. ▩

The lesson from Example 8 is stated in the next two rules.

(M1) For some pairs of matrices $\mathbf{A}$ and $\mathbf{B}$, the equation $\mathbf{AB} = \mathbf{O}$ does not imply $\mathbf{A} = \mathbf{O}$ or $\mathbf{B} = \mathbf{O}$.

(M2) For some matrices $\mathbf{A}$, $\mathbf{B}$, $\mathbf{C}$, the equation $\mathbf{AB} = \mathbf{AC}$ does not imply $\mathbf{B} = \mathbf{C}$.

Caution! However, for some pairs of matrices $\mathbf{A}$ and $\mathbf{B}$, $\mathbf{AB} = \mathbf{O}$ does imply $\mathbf{A} = \mathbf{O}$ or $\mathbf{B} = \mathbf{O}$ (take $\mathbf{A}$ to be a zero matrix, for example) and for certain $\mathbf{A}$, $\mathbf{B}$, $\mathbf{C}$, $\mathbf{AB} = \mathbf{AC}$ does imply $\mathbf{B} = \mathbf{C}$ (see Exercises 2.2).

Associative Law for Matrix Multiplication

Suppose $\mathbf{A}$ is $m \times p$, $\mathbf{B}$ is $p \times q$ and $\mathbf{C}$ is $q \times n$. Then

(M3) $(\mathbf{AB})\mathbf{C} = \mathbf{A}(\mathbf{BC})$ associative law

Note that matrix multiplication is a *binary* operation, acting on two matrices at a time. The sizes of $\mathbf{A}$, $\mathbf{B}$, and $\mathbf{C}$ imply that the products $\mathbf{AB}$, $\mathbf{BC}$, $(\mathbf{AB})\mathbf{C}$, $\mathbf{A}(\mathbf{BC})$ all exist and **(M3)** tells us that the computations $(\mathbf{AB})\mathbf{C}$ and $\mathbf{A}(\mathbf{BC})$ result in the same matrix, which is denoted by $\mathbf{ABC}$. It should be pointed out that the truth of **(M3)** can be established by (laboriously!) checking that the (i, j)-entries in the $m \times n$ matrices $(\mathbf{AB})\mathbf{C}$ and $\mathbf{A}(\mathbf{BC})$ are in fact equal.

EXAMPLE 9

Applying the Associative Law

To compute the product $\mathbf{ABCD} = \mathbf{E}$, where

$$\mathbf{A} = \begin{bmatrix} 2 \\ -3 \\ 1 \end{bmatrix}, \quad \mathbf{B} = [4 \;\; 5 \;\; -1], \quad \mathbf{C} = \begin{bmatrix} 1 \\ 0 \\ 2 \end{bmatrix}, \quad \mathbf{D} = [2 \;\; 1 \;\; -3]$$

we can apply the associative law in several ways. Computing (**AB**)(**CD**), which is a product of two 3×3 matrices, is less efficient than computing **A**(**BC**)**D**, which contains the product **BC** = [2]. The calculation goes as follows:

$$\mathbf{E} = \begin{bmatrix} 2 \\ -3 \\ 1 \end{bmatrix} [2] \begin{bmatrix} 2 & 1 & -3 \end{bmatrix} = \begin{bmatrix} 2 \\ -3 \\ 1 \end{bmatrix} \begin{bmatrix} 4 & 2 & -6 \end{bmatrix}$$

$$= \begin{bmatrix} 8 & 4 & -12 \\ -12 & -6 & 18 \\ 4 & 2 & -6 \end{bmatrix}.$$

▪

Distributive Laws for Multiplication

Suppose **B** and **C** are $m \times n$, **A** is $p \times m$ and **D** is $n \times q$.

(**M4**) $\mathbf{A}(\mathbf{B}+\mathbf{C}) = \mathbf{AB}+\mathbf{AC}$ (premultiplying by **A**)
(**M5**) $(\mathbf{B}+\mathbf{C})\mathbf{D} = \mathbf{BD}+\mathbf{CD}$ (postmultiplying by **D**)

EXAMPLE 10

Applying the Distributive Laws, Carefully

Let **A** and **B** be $n \times n$ matrices. Then

$$\begin{aligned}(\mathbf{A}+\mathbf{B})^2 &= (\mathbf{A}+\mathbf{B})(\mathbf{A}+\mathbf{B}) \\ &= (\mathbf{A}+\mathbf{B})\mathbf{A} + (\mathbf{A}+\mathbf{B})\mathbf{B} \quad \text{(premultiplication by } (\mathbf{A}+\mathbf{B})) \\ &= \mathbf{AA}+\mathbf{BA}+\mathbf{AB}+\mathbf{BB} \quad \text{(postmultiplication by } \mathbf{A}, \mathbf{B}) \\ &= \mathbf{A}^2+\mathbf{BA}+\mathbf{BA}+\mathbf{B}^2.\end{aligned}$$

However, $\mathbf{BA}+\mathbf{AB} \neq 2\mathbf{AB}$, unless **A** and **B** commute, and so $(\mathbf{A}+\mathbf{B})^2$ and $\mathbf{A}^2+2\mathbf{AB}+\mathbf{B}^2$ are not equal in general. ▪

Scalar Multiplication

Suppose **A** is conformable to **B** for multiplication and s is any scalar.

(**M6**) $\mathbf{A}(s\mathbf{B}) = (s\mathbf{A})\mathbf{B} = s(\mathbf{AB})$ scalar multiplication

2.1.3 Square Matrices

An $n \times n$ matrix $\mathbf{A} = [a_{ij}]$ is called a *square matrix* of *order* n. The entries a_{ii}, $1 \leq i \leq n$, lie on the *main diagonal* of **A** (passing though **A** from upper left corner to lower right corner), as follows:

$$\mathbf{A} = \begin{bmatrix} a_{11} & * & \cdots & * \\ * & a_{22} & \cdots & * \\ \vdots & \vdots & \ddots & \vdots \\ * & * & \cdots & a_{nn} \end{bmatrix} \tag{2.4}$$

An $n \times n$ matrix $\mathbf{A}$ is called a *diagonal matrix* if the off-diagonal entries, denoted by stars in (2.4), are all zero.[2] Note that two diagonal $n \times n$ matrices commute. A diagonal matrix is called a *scalar matrix* if each entry on the main diagonal is the same scalar s. For example, if

$$\mathbf{D} = \begin{bmatrix} 2.5 & 0 \\ 0 & -4.6 \end{bmatrix}, \qquad \mathbf{S} = \begin{bmatrix} 2 & 0 & 0 \\ 0 & 2 & 0 \\ 0 & 0 & 2 \end{bmatrix} \quad (s = 2),$$

then $\mathbf{D}$ is a diagonal matrix, and $\mathbf{S}$ is a scalar matrix.

An $n \times n$ scalar matrix with all diagonal entries equal to 1 is called the *identity matrix* of order n, denoted as follows:

$$\mathbf{I}_n = \begin{bmatrix} 1 & 0 & \cdots & 0 \\ 0 & 1 & \cdots & 0 \\ \vdots & \vdots & \ddots & \vdots \\ 0 & 0 & \cdots & 1 \end{bmatrix}.$$

If $\mathbf{A}$ is $m \times n$, then $\mathbf{I}_m\mathbf{A} = \mathbf{A} = \mathbf{A}\mathbf{I}_n$. We may write $\mathbf{I}$ in place of $\mathbf{I}_n$ when the order of $\mathbf{I}$ is clear from the context.

The matrix $\mathbf{I}_n$ in the set $\mathbb{R}^{n\times n}$ plays the same role as the multiplicative identity 1 in the real or complex number system ($1r = r1 = r$, for any number r). For any $n \times n$ matrix $\mathbf{A}$ we have $\mathbf{AI} = \mathbf{IA} = \mathbf{A}$ ($\mathbf{A}$ commutes with the identity). The next result shows that there can be no other matrix $\mathbf{J}$ with the property that $\mathbf{JA} = \mathbf{AJ} = \mathbf{A}$ for all $\mathbf{A}$ in $\mathbb{R}^{n\times n}$.

Theorem 2.2 **Uniqueness of Identity Matrices**

There is only one identity $\mathbf{I}_n$ in each set $\mathbb{R}^{n\times n}$ of square matrices of order n. We call $\mathbf{I}_n$ the multiplicative identity *in $\mathbb{R}^{n\times n}$.*

Proof The identity $\mathbf{I}$ in $\mathbb{R}^{n\times n}$ satisfies $\mathbf{AI} = \mathbf{A} = \mathbf{IA}$ for every $\mathbf{A}$ in $\mathbb{R}^{n\times n}$. Suppose a matrix $\mathbf{J}$ in $\mathbb{R}^{n\times n}$ satisfies $\mathbf{AJ} = \mathbf{A} = \mathbf{JA}$ for every $\mathbf{A}$ in $\mathbb{R}^{n\times n}$. Then, in particular, $\mathbf{I} = \mathbf{JI} = \mathbf{J}$ and so $\mathbf{I} = \mathbf{J}$.

2.1.4 Powers and Roots of Square Matrices

Let $\mathbf{A}$ be an $n \times n$ matrix. The products $\mathbf{A}^2 = \mathbf{AA}$, $\mathbf{A}^3 = \mathbf{AAA}$, ... all exist and each one is a square matrix of order n. The associative law (**M3**) ensures that $(\mathbf{AA})\mathbf{A} = \mathbf{A}(\mathbf{AA})$ and so $\mathbf{A}^3 = \mathbf{A}^2\mathbf{A} = \mathbf{AA}^2$. The power $\mathbf{A}^k$, where k is a nonnegative integer, is defined *inductively* by the rule

$$\mathbf{A}^0 = \mathbf{I}_n, \qquad \mathbf{A}^k = \mathbf{A}^{k-1}\mathbf{A}, \qquad k = 1, 2, \ldots \tag{2.5}$$

and for positive integers p and q, we have

$$\mathbf{A}^p\mathbf{A}^q = \mathbf{A}^{(p+q)}. \tag{2.6}$$

If $\mathbf{A} = \mathbf{B}^n$, then $\mathbf{B} = \sqrt[n]{\mathbf{A}}$ is called an nth *root* of $\mathbf{A}$. The matrix $\mathbf{B}$ is in general not unique (see Exercises 2.1).

[2] Caution! Do not conclude that the entries on the main diagonal are necessarily all nonzero.

2.1.5 Block Matrices, Block Multiplication

There are some practical situations in which it is necessary to *partition* a matrix **A** into smaller submatrices, each of which is handled as a separate entity. A partition is created by drawing vertical or horizontal lines (or both) in **A**, as in (2.7). The resulting submatrices defined by the partition are called *blocks*. Why is it necessary to partition matrices? Here are two possible answers. First, there are large-scale applications[3] in which partitioning becomes necessary due to storage limitations within the computer. The partitioned matrix is stored as separate blocks. The second answer is more theoretical. Row-by-column multiplication on **AB** is not the only way to multiply matrices **A** and **B**. We will soon see that there are several other possibilities that are defined by partitioning **A** or **B** (or both) in certain ways. Seeing the product in its partitioned form gives greater insight into the meaning of the product **AB**.

Blocks are treated as the *entries* in a new matrix and labeled with double subscripts. One possible partition of a 6×5 matrix **A** is shown.

$$\mathbf{A} = \left[\begin{array}{ccc|cc} 1 & 2 & 3 & 4 & 5 \\ 3 & -2 & 1 & 1 & 0 \\ \hline 0 & 0 & 1 & 1 & 1 \\ \hline 0 & 0 & 0 & 0 & 5 \\ 0 & 0 & 0 & 1 & -1 \\ 0 & 0 & 0 & 0 & 0 \end{array}\right] = \begin{bmatrix} \mathbf{A}_{11} & \mathbf{A}_{12} \\ \mathbf{A}_{21} & \mathbf{A}_{22} \\ \mathbf{A}_{31} & \mathbf{A}_{32} \end{bmatrix} \tag{2.7}$$

The matrix on the right side of (2.7) is called a *block matrix* and its size is 3×2.

The blocks arising from the chosen partition of **A** are

$$\mathbf{A}_{11} = \begin{bmatrix} 1 & 2 & 3 \\ 3 & -2 & 1 \end{bmatrix} \quad \mathbf{A}_{12} = \begin{bmatrix} 4 & 5 \\ 1 & 0 \end{bmatrix} \quad \mathbf{A}_{21} = \begin{bmatrix} 0 & 0 & 1 \end{bmatrix}$$

$$\mathbf{A}_{22} = \begin{bmatrix} 1 & 1 \end{bmatrix} \quad \mathbf{A}_{31} = \begin{bmatrix} 0 & 0 & 0 \\ 0 & 0 & 0 \\ 0 & 0 & 0 \end{bmatrix} \quad \mathbf{A}_{32} = \begin{bmatrix} 0 & 5 \\ 1 & -1 \\ 0 & 0 \end{bmatrix}. \tag{2.8}$$

Yet another reason for partitioning is to reduce computer storage space. Treating special submatrices as blocks, such as identity or zero submatrices, can result in such a saving. The square block matrix $\mathbf{A}_{31}$ in (2.8), for example, only requires storage of the positive integers 3, 3, 0, representing its size and common entry.

Row-by-column multiplication can be performed on two block matrices provided their respective blocks are conformable for *block multiplication*.

Definition 2.5

Block product

Let **A** be an $m \times n$ matrix and let **B** be an $n \times p$ matrix. Suppose the n columns of **A** and the n rows of **B** be partitioned in *exactly the same*

[3] See [1] for an application involving an $n \times n$ matrix, where n is currently about 2.7 billion and increases on a daily basis.

way, and the rows of **A** and columns of **B** are partitioned in *any way* we please, then the *block product* **AB** exists and is found using row-by-column multiplication on blocks.

EXAMPLE 11

Block Multiplication

The matrix **A** in (2.7) is 6×5 and the matrix **B** shown is 5×4. Hence, **A** is conformable to **B** for ordinary row-by-column multiplication and **AB** is 6×4.

$$\mathbf{B} = \left[\begin{array}{ccc|c} 1 & 0 & 1 & 1 \\ 0 & 2 & -1 & 0 \\ 0 & 3 & 4 & -1 \\ \hline 0 & 0 & 0 & 1 \\ 0 & 0 & 0 & 2 \end{array}\right] = \begin{bmatrix} \mathbf{B}_{11} & \mathbf{B}_{12} \\ \mathbf{B}_{21} & \mathbf{B}_{22} \end{bmatrix}.$$

The columns of **A** are partitioned into two sets (3 columns, then 2 columns) and this agrees with the partitioning on the rows of **B** (3 rows, then 2 rows). This ensures that matrix products such as $\mathbf{A}_{11}\mathbf{B}_{11}$, $\mathbf{A}_{12}\mathbf{B}_{21}$ exist. If the rows of **A** are partitioned into p sets and the columns of **B** are partitioned into q sets, then **AB** will have block size $p \times q$. Computing **AB** using row-by-column multiplication on blocks, we have

$$\begin{bmatrix} \mathbf{A}_{11} & \mathbf{A}_{12} \\ \mathbf{A}_{21} & \mathbf{A}_{22} \\ \mathbf{A}_{31} & \mathbf{A}_{32} \end{bmatrix} \begin{bmatrix} \mathbf{B}_{11} & \mathbf{B}_{12} \\ \mathbf{B}_{21} & \mathbf{B}_{22} \end{bmatrix} = \begin{bmatrix} \mathbf{A}_{11}\mathbf{B}_{11} + \mathbf{A}_{12}\mathbf{B}_{21} & \mathbf{A}_{11}\mathbf{B}_{12} + \mathbf{A}_{12}\mathbf{B}_{22} \\ \mathbf{A}_{21}\mathbf{B}_{11} + \mathbf{A}_{22}\mathbf{B}_{21} & \mathbf{A}_{21}\mathbf{B}_{12} + \mathbf{A}_{22}\mathbf{B}_{22} \\ \mathbf{A}_{31}\mathbf{B}_{11} + \mathbf{A}_{32}\mathbf{B}_{21} & \mathbf{A}_{31}\mathbf{B}_{12} + \mathbf{A}_{32}\mathbf{B}_{22} \end{bmatrix}$$

$$= \begin{bmatrix} \mathbf{C}_{11} & \mathbf{C}_{12} \\ \mathbf{C}_{21} & \mathbf{C}_{22} \\ \mathbf{C}_{31} & \mathbf{C}_{32} \end{bmatrix} = \mathbf{C},$$

where **C** is a 3×2 block matrix, and

$$\mathbf{C} = \begin{bmatrix} \mathbf{C}_{11} & \mathbf{C}_{12} \\ \mathbf{C}_{21} & \mathbf{C}_{22} \\ \mathbf{C}_{31} & \mathbf{C}_{32} \end{bmatrix} = \left[\begin{array}{ccc|c} 1 & 13 & 11 & 12 \\ 3 & -1 & 9 & 3 \\ \hline 0 & 3 & 4 & 2 \\ \hline 0 & 0 & 0 & 10 \\ 0 & 0 & 0 & -1 \\ 0 & 0 & 0 & 0 \end{array}\right].$$

The products involving $\mathbf{A}_{31}$ or $\mathbf{B}_{21}$ (such as $\mathbf{A}_{31}\mathbf{B}_{12}$ or $\mathbf{A}_{32}\mathbf{B}_{21}$) are zero matrices. Hence, we have the immediate simplification $\mathbf{C}_{11} = \mathbf{A}_{11}\mathbf{B}_{11}$, $\mathbf{C}_{21} = \mathbf{A}_{21}\mathbf{B}_{11}$, $\mathbf{C}_{31} = \mathbf{O}_{3\times 3}$, $\mathbf{C}_{32} = \mathbf{A}_{32}\mathbf{B}_{22}$. ■

Note that row-by-column multiplication on a product **AB** is really block multiplication, where the blocks are defined by the rows of **A** and the columns of **B**.

2.1.6 Three Special Cases of Matrix Multiplication

I Outer Products

Suppose **u** is a column vector with m entries and **v** is a row vector with n entries. Then **u** is conformable to **v** for matrix multiplication; that is, $\mathbf{uv} = \mathbf{C}$ exists and **C** is $m \times n$. For example, with $m = 2$ and $n = 3$, we have

$$\mathbf{uv} = \begin{bmatrix} 2 \\ 3 \end{bmatrix} [-4 \;\; 5 \;\; 2] = \begin{bmatrix} -8 & 10 & 4 \\ -12 & 15 & 6 \end{bmatrix} = \mathbf{C}.$$

Observe that the rows of **C** (ordered top to bottom) are copies of **v** scaled by each component in **u** (ordered top to bottom). The columns of **C** (ordered left to right) are copies of **u** scaled by the components in **v**. **C** is called the *outer product* of **u** and **v**.

The next two cases, which involve partitioning by rows and by columns, are illustrated using the product

$$\underset{\mathbf{A}}{\begin{bmatrix} 1 & 2 & 5 \\ 3 & 0 & 4 \end{bmatrix}} \underset{\mathbf{B}}{\begin{bmatrix} 2 & 1 \\ 3 & 6 \\ 1 & 7 \end{bmatrix}} = \underset{\mathbf{C}}{\begin{bmatrix} 13 & 48 \\ 10 & 31 \end{bmatrix}}. \tag{2.9}$$

II Row Vector-Matrix Products

Suppose **u** is a $1 \times n$ row vector and **B** is an $n \times p$ matrix. Then **u** is conformable to **B** for multiplication; that is, $\mathbf{uB} = \mathbf{w}$ exists and **w** is a $1 \times p$ row vector.

Using the row vectors $\mathbf{a}_1$ and $\mathbf{a}_2$ from **A** in (2.9), we have

$$\mathbf{a}_1 = [1 \;\; 2 \;\; 5] \Rightarrow \mathbf{a}_1\mathbf{B} = [1 \;\; 2 \;\; 5] \begin{bmatrix} 2 & 1 \\ 3 & 6 \\ 1 & 7 \end{bmatrix} = [13 \;\; 48], \tag{2.10}$$

$$\mathbf{a}_2 = [3 \;\; 0 \;\; 4] \Rightarrow \mathbf{a}_2\mathbf{B} = [3 \;\; 0 \;\; 4] \begin{bmatrix} 2 & 1 \\ 3 & 6 \\ 1 & 7 \end{bmatrix} = [10 \;\; 31]. \tag{2.11}$$

Hence, the product **AB** can be obtained by partitioning **A** into rows and forming two row vector-matrix products, as follows:

$$\mathbf{AB} = \begin{bmatrix} \mathbf{a}_1 \\ \hline \mathbf{a}_2 \end{bmatrix} \mathbf{B} = \begin{bmatrix} \mathbf{a}_1\mathbf{B} \\ \hline \mathbf{a}_2\mathbf{B} \end{bmatrix} = \mathbf{C}.$$

In general, if **A** is $m \times n$ and **B** is $n \times p$, then the product **AB** is an $m \times p$ matrix **C** whose rows are the row vector-matrix products $\mathbf{a}_1\mathbf{B}$, $\mathbf{a}_2\mathbf{B}$, $\ldots$, $\mathbf{a}_m\mathbf{B}$

and we have

$$\mathbf{AB} = \begin{bmatrix} \mathbf{a}_1 \\ \hline \vdots \\ \hline \mathbf{a}_m \end{bmatrix} \mathbf{B} = \begin{bmatrix} \mathbf{a}_1\mathbf{B} \\ \hline \vdots \\ \hline \mathbf{a}_m\mathbf{B} \end{bmatrix} = \mathbf{C}. \tag{2.12}$$

III Matrix-Column Vector Products

Pay special attention to this case. These products are used in representing linear systems in matrix form and in defining matrix (linear) transformations on $\mathbb{R}^n$ (Section 3.4).

Suppose $\mathbf{A}$ is an $m \times n$ matrix and $\mathbf{v}$ is an $n \times 1$ column matrix. Then $\mathbf{A}$ is conformable to $\mathbf{v}$ for matrix multiplication. In other words, $\mathbf{Av} = \mathbf{w}$ exists and $\mathbf{w}$ is an $m \times 1$ column vector.

Partitioning $\mathbf{B}$ from (2.9) into columns $\mathbf{b}_1$, $\mathbf{b}_2$, we have

$$\mathbf{b}_1 = \begin{bmatrix} 2 \\ 3 \\ 1 \end{bmatrix} \Rightarrow \mathbf{Ab}_1 = \begin{bmatrix} 1 & 2 & 5 \\ 3 & 0 & 4 \end{bmatrix} \begin{bmatrix} 2 \\ 3 \\ 1 \end{bmatrix} = \begin{bmatrix} 13 \\ 10 \end{bmatrix},$$

$$\mathbf{b}_2 = \begin{bmatrix} 1 \\ 6 \\ 7 \end{bmatrix} \Rightarrow \mathbf{Ab}_2 = \begin{bmatrix} 1 & 2 & 5 \\ 3 & 0 & 4 \end{bmatrix} \begin{bmatrix} 1 \\ 6 \\ 7 \end{bmatrix} = \begin{bmatrix} 48 \\ 31 \end{bmatrix}.$$

Hence, the product $\mathbf{AB}$ can be obtained by partitioning $\mathbf{B}$ into columns and forming two matrix-column vector products, as follows:

$$\mathbf{AB} = \mathbf{A}\,[\,\mathbf{b}_1 \;\; \mathbf{b}_2\,] = [\,\mathbf{Ab}_1 \;\; \mathbf{Ab}_2\,] = \mathbf{C}.$$

In general, if $\mathbf{A}$ is $m \times n$ and $\mathbf{B}$ is $n \times p$, then the product $\mathbf{AB}$ is an $m \times p$ matrix $\mathbf{C}$ whose columns are the matrix-column vector products $\mathbf{Ab}_1$, $\mathbf{Ab}_2$, $\ldots$, $\mathbf{Ab}_p$ and we have

$$\mathbf{AB} = \mathbf{A}\left[\mathbf{b}_1 \;\; \mathbf{b}_2 \;\; \cdots \;\; \mathbf{b}_p\right] = \left[\mathbf{Ab}_1 \;\; \mathbf{Ab}_2 \;\; \cdots \;\; \mathbf{Ab}_p\right] = \mathbf{C}. \tag{2.13}$$

Note that the product of two matrices $\mathbf{AB}$ (when it exists) can now be computed in three different ways: row-by-column, row vector-matrix, matrix-column vector.

2.1.7 The Matrix Form of a Linear System

The matrix-column vector product described immediately above enables us to write any linear system (S) in the form $\mathbf{Ax} = \mathbf{b}$, which is called the *matrix form* for (S).

EXAMPLE 12

The Matrix Form of a Linear System

Consider the linear system (S) shown.

$$(S)\quad \left\{ \begin{array}{rcr} 3x_1 + x_2 + 8x_3 \phantom{{}- 2x_4} & = & 1.9 \\ x_1 + 2x_2 + x_3 - 2x_4 & = & -2.2 \end{array} \right. \tag{2.14}$$

The coefficient matrix **A** and the 4×1 column vector **x** of unknowns are conformable for matrix-column vector multiplication, and we have

$$\mathbf{Ax} = \begin{bmatrix} 3 & 1 & 8 & 0 \\ 1 & 2 & 1 & -2 \end{bmatrix} \begin{bmatrix} x_1 \\ x_2 \\ x_3 \\ x_4 \end{bmatrix} = \begin{bmatrix} 3x_1 + x_2 + 8x_3 + 0x_4 \\ x_1 + 2x_2 + x_3 - 2x_4 \end{bmatrix}.$$

The constant terms on the right side of (2.14) are used to form the 2×1 column vector **b**. Using the equations from (S) and equating corresponding entries, we have

$$\begin{bmatrix} 3x_1 + x_2 + 8x_3 + 0x_4 \\ x_1 + 2x_2 + x_3 - 2x_4 \end{bmatrix} = \begin{bmatrix} 1.9 \\ -2.2 \end{bmatrix} \quad \Rightarrow \quad \mathbf{Ax} = \mathbf{b}, \text{ where } \mathbf{b} = \begin{bmatrix} 1.9 \\ -2.2 \end{bmatrix}.$$

■

Example 12 extends in a natural way to any $m \times n$ linear system (S). The matrix form for (S) is shown on the left in (2.15) and the short-hand notation on the right:

$$\begin{bmatrix} a_{11} & a_{12} & \cdots & a_{1n} \\ a_{21} & a_{22} & \cdots & a_{2n} \\ \vdots & \vdots & \ddots & \vdots \\ a_{m1} & a_{m2} & \cdots & a_{mn} \end{bmatrix} \begin{bmatrix} x_1 \\ x_2 \\ \vdots \\ x_n \end{bmatrix} = \begin{bmatrix} b_1 \\ b_2 \\ \vdots \\ b_m \end{bmatrix} \quad \Longleftrightarrow \quad \mathbf{Ax} = \mathbf{b}. \qquad (2.15)$$

Conversely, the equation $\mathbf{Ax} = \mathbf{b}$, where **A** is an $m \times n$ matrix, **x** is an $n \times 1$ column vector of unknowns, and **b** is an $m \times 1$ column vector of constants, represents an $m \times n$ linear system (S).

2.1.8 The Transpose

The *transpose* of the $m \times n$ matrix **A** is the $n \times m$ matrix $\mathbf{A}^\mathsf{T}$ formed by writing the columns of **A** (in order, left to right) as the rows of $\mathbf{A}^\mathsf{T}$ (in order, top to bottom). Equivalently, $\mathbf{A}^\mathsf{T}$ can be formed by writing the rows of **A** as the columns of $\mathbf{A}^\mathsf{T}$ (keeping the same order). Transposition is a *unary operation* because it acts on one matrix at a time. For example

$$\mathbf{A} = \begin{bmatrix} \mathbf{1} & 2 \\ \mathbf{3} & 4 \\ \mathbf{5} & 6 \end{bmatrix} \Rightarrow \mathbf{A}^\mathsf{T} = \begin{bmatrix} \mathbf{1} & \mathbf{3} & \mathbf{5} \\ 2 & 4 & 6 \end{bmatrix}, \quad \mathbf{B} = \begin{bmatrix} \mathbf{1} & 3 \\ \mathbf{5} & 4 \end{bmatrix} \Rightarrow \mathbf{B}^\mathsf{T} = \begin{bmatrix} \mathbf{1} & \mathbf{5} \\ 3 & 4 \end{bmatrix}$$

A is 3×2 $\qquad$ $\mathbf{A}^\mathsf{T}$ is 2×3 $\qquad$ **B** is 2×2 $\qquad$ $\mathbf{B}^\mathsf{T}$ is 2×2.

The (i, j)-entry in $\mathbf{A}^\mathsf{T}$ is the (j, i)-entry in **A**. For example, the $(2, 1)$-entry in **A** is $a_{21} = 3$, which is the $(1, 2)$-entry in $\mathbf{A}^\mathsf{T}$. We write $a_{ij}^\mathsf{T} = a_{ji}$ and $\mathbf{A}^\mathsf{T} = [a_{ij}]^\mathsf{T} = [a_{ji}]$. Row vectors become column vectors under transposition, and vice versa.

$$\mathbf{a} = [1\ \ 2] \quad \Rightarrow \quad \mathbf{a}^{\mathsf{T}} = \begin{bmatrix} 1 \\ 2 \end{bmatrix}, \qquad \mathbf{b} = \begin{bmatrix} 1 \\ 0 \\ 0 \end{bmatrix} \quad \Rightarrow \quad \mathbf{b}^{\mathsf{T}} = [1\ \ 0\ \ 0]$$

Laws of Transposition

If **A** and **B** are matrices in $\mathbb{R}^{m \times n}$ and s is a scalar, then $\mathbf{A}^{\mathsf{T}}$ and $\mathbf{B}^{\mathsf{T}}$ are matrices in $\mathbb{R}^{n \times m}$ and the following properties are easily established.

(T1) $(\mathbf{A}^{\mathsf{T}})^{\mathsf{T}} = \mathbf{A}$
(T2) $(\mathbf{A} + \mathbf{B})^{\mathsf{T}} = \mathbf{A}^{\mathsf{T}} + \mathbf{B}^{\mathsf{T}}$
(T3) $(s\mathbf{A})^{\mathsf{T}} = s(\mathbf{A}^{\mathsf{T}})$

If **A** is $m \times r$ and **B** is $r \times n$, then we have

(T4) $(\mathbf{AB})^{\mathsf{T}} = \mathbf{B}^{\mathsf{T}}\mathbf{A}^{\mathsf{T}}$ (note the order of $\mathbf{A}^{\mathsf{T}}$ and $\mathbf{B}^{\mathsf{T}}$)

Proof Checking sizes, $\mathbf{B}^{\mathsf{T}}$ is $n \times r$ and $\mathbf{A}^{\mathsf{T}}$ is $r \times m$ so that $\mathbf{B}^{\mathsf{T}}\mathbf{A}^{\mathsf{T}}$ exists and is $n \times m$, which is also the size of $(\mathbf{AB})^{\mathsf{T}}$. The (i, j)-entry in $(\mathbf{AB})^{\mathsf{T}}$ is the (j, i)-entry in **AB**, and the latter is obtained from row j in **A** (column j in $\mathbf{A}^{\mathsf{T}}$) and column i in **B** (row i in $\mathbf{B}^{\mathsf{T}}$). Thus, the (i, j)-entry in $(\mathbf{AB})^{\mathsf{T}}$ is obtained from row i in $\mathbf{B}^{\mathsf{T}}$ and column j in $\mathbf{A}^{\mathsf{T}}$, which is the (i, j)-entry in $\mathbf{B}^{\mathsf{T}}\mathbf{A}^{\mathsf{T}}$.

Definition 2.6 *Symmetric, skew-symmetric*

An $n \times n$ matrix **A** is called *symmetric* if $\mathbf{A}^{\mathsf{T}} = \mathbf{A}$ and is called *skew-symmetric* if $\mathbf{A}^{\mathsf{T}} = -\mathbf{A}$.

For example, with $n = 3$,

$$\begin{bmatrix} 1 & -2 & 3 \\ -2 & 0 & 4 \\ 3 & 4 & -7 \end{bmatrix} \text{ is symmetric,} \qquad \begin{bmatrix} 0 & -2 & 3 \\ 2 & 0 & -4 \\ -3 & 4 & 0 \end{bmatrix} \text{ is skew-symmetric.}$$

REFERENCES

1. Moler, C., "The World's Largest Matrix Computation," MATLAB *News & Notes*, The MathWorks, October 2002.

EXERCISES 2.1

Exercises 1–6. For the given ordered pair of positive integers (m, n), write down the $m \times n$ matrix $\mathbf{A} = [a_{ij}]$ whose entries are defined by the given formula for a_{ij}.

1. $(2, 2)$, $a_{ij} = i + j$

2. $(3, 2)$, $a_{ij} = ij$

3. $(3, 3)$, $a_{ij} = i - j$

4. $(1, 3)$, $a_{ij} = 3i + 4j$

5. $(2, 4)$, $a_{ij} = \begin{cases} i^2 + j^2 & \text{if } i + j \text{ even} \\ 0 & \text{otherwise} \end{cases}$

6. $(3, 4)$, $a_{ij} = \begin{cases} 0 & \text{if } i + j \text{ is odd} \\ 1 & \text{otherwise} \end{cases}$

Exercises 7–12. Find the values of the variables that make each equation valid.

7. $\begin{bmatrix} x \\ y \end{bmatrix} = \begin{bmatrix} x^2 - y \\ x \end{bmatrix}$

8. $3\begin{bmatrix} x & 4 \\ 2 & z \end{bmatrix} = \begin{bmatrix} y & z \\ 2x & 0 \end{bmatrix} + \begin{bmatrix} -2 & y \\ z & -6 \end{bmatrix}$

9. $\begin{bmatrix} x & y \\ z & 0 \end{bmatrix} = 2\begin{bmatrix} 1 & x - z \\ -3 & 0 \end{bmatrix}$

10. $\begin{bmatrix} 1 & x & 2y \end{bmatrix} = k\begin{bmatrix} 2 & 0 & -4 \end{bmatrix}$

11. $\begin{bmatrix} x^2 - 2x \\ y \\ z \end{bmatrix} = \begin{bmatrix} -y \\ 1 + z \\ 1 - y \end{bmatrix}$

12. $[1 \;\; x \;\; 1]\begin{bmatrix} 1 \\ x \\ 2x \end{bmatrix} = \begin{bmatrix} 5 \end{bmatrix}$

Exercises 13–14. State the size of the unknown matrix $\mathbf{X} = [x_{ij}]$ *and find* x_{12}.

13. $5\mathbf{X} - 3\begin{bmatrix} -2 & 0 \\ 4 & -2 \end{bmatrix} = 2\begin{bmatrix} -1 & 2 \\ 0 & 1 \end{bmatrix} + \mathbf{X}$

14. $-2\begin{bmatrix} 2 & -1 & 3 \\ 0 & -1 & 2 \end{bmatrix} + 3\mathbf{X} = \begin{bmatrix} -3 & 2 & 1 \\ 4 & -2 & -1 \end{bmatrix} + 5\mathbf{X}$

Exercises 15–16. State the size of each product and then compute the product.

15. $\mathbf{A} = \begin{bmatrix} 3 & 2 & 1 \\ 4 & -2 & 2 \end{bmatrix}\begin{bmatrix} 2 \\ 1 \\ -1 \end{bmatrix}\begin{bmatrix} 1 & -3 & 2 \end{bmatrix}$

16. $\mathbf{B} = \begin{bmatrix} 1 \\ 2 \\ 3 \end{bmatrix}\begin{bmatrix} 3 & 2 & 1 \end{bmatrix}\begin{bmatrix} 1 \\ -1 \\ 4 \end{bmatrix}$

Exercises 17–18. Verify that each numerical entry shown in the product $\mathbf{C} = \mathbf{AB}$ *is correct. Compute the entries* c_{12} *and* c_{33}.

17. $\mathbf{AB} = \begin{bmatrix} 1 & 2 & -2 \\ -1 & 3 & 0 \\ 0 & -2 & 4 \end{bmatrix}\begin{bmatrix} 6 & 0 & 6 \\ 2 & 2 & 2 \\ 1 & 3 & -1 \end{bmatrix} = \begin{bmatrix} 8 & * & 12 \\ 0 & 6 & * \\ * & 8 & * \end{bmatrix} = \mathbf{C}$

18. $\mathbf{AB} = \begin{bmatrix} 1.1 & -1 \\ 2 & 3.5 \\ -1 & 5 \end{bmatrix}\begin{bmatrix} 3 & 4 & -8 \\ 6 & 7 & 10 \end{bmatrix} = \begin{bmatrix} -2.7 & * & -18.8 \\ * & * & * \\ 27 & 31 & * \end{bmatrix} = \mathbf{C}$

Exercises 19–20. Use your knowledge of row-by-column matrix multiplication to compute the given entries w_{ij} *in* $\mathbf{W} = [w_{ij}]$ *without computing the whole product.*

19. $\mathbf{W} = \begin{bmatrix} 1 & 3 & -2 & 1 \\ 2 & 1 & -1 & 0 \end{bmatrix}\begin{bmatrix} 1 & 2 \\ -3 & 5 \\ 4 & 6 \\ -1 & 0 \end{bmatrix}\begin{bmatrix} 2 & -1 & 0 \\ 1 & 0 & 2 \end{bmatrix}.$

Find w_{13}, w_{21}.

20. $\mathbf{W} = \begin{bmatrix} 1 & -2 & 3 & 1 \\ 2 & 0 & 1 & -1 \end{bmatrix}\begin{bmatrix} 1 & 2 \\ 3 & 4 \\ 5 & 6 \\ 7 & 8 \end{bmatrix}\begin{bmatrix} 3 & 2 & -1 & 0 \\ 2 & 0 & 4 & 1 \end{bmatrix}.$

Find w_{14}, w_{22}.

Exercises 21–22. Compute the $(2, 2)$*-entry in Exercise 21, and the* $(1, 3)$*-entry in Exercise 22. Do not multiply out the whole product.*

21. $\begin{bmatrix} 1 & -2 & 1 \\ 3 & 5 & -4 \\ -1 & -1 & 1 \end{bmatrix}\begin{bmatrix} -2 & -1 & 2 \\ -8 & -3 & 6 \\ -13 & -7 & 11 \end{bmatrix}\begin{bmatrix} 1 & 1 & 3 \\ 1 & 2 & 7 \\ 2 & 3 & 11 \end{bmatrix}$

22. $\begin{bmatrix} 1 & 2 & -1 \\ -2 & -1 & 1 \\ 1 & 0 & 2 \end{bmatrix}\begin{bmatrix} 3 & -3 & -1 \\ -3 & 5 & 2 \\ -1 & 2 & 2 \end{bmatrix}\begin{bmatrix} 1 & -2 & 1 \\ 2 & -1 & 0 \\ -1 & 1 & 2 \end{bmatrix}$

Exercises 23–30. For the matrices **A**, **B**, **C** *shown, compute the matrix expressions, when defined, and give a reason for failure when not defined.*

$$\mathbf{A} = \begin{bmatrix} 2 & 1 \\ 3 & -1 \end{bmatrix}, \quad \mathbf{B} = \begin{bmatrix} 2 & -1 & 0 \\ 3 & 4 & 1 \end{bmatrix},$$

$$\mathbf{C} = \begin{bmatrix} 3 \\ 1 \\ 2 \end{bmatrix}$$

23. **AB**, **BA** **24.** **A(BC)**, **(AB)C** **25.** **BC**, **CB**

26. $\mathbf{AA}^{\mathsf{T}}$, $\mathbf{A}^{\mathsf{T}}\mathbf{A}$ **27.** $\mathbf{BB}^{\mathsf{T}}$, $\mathbf{B}^{\mathsf{T}}\mathbf{B}$ **28.** $\mathbf{CC}^{\mathsf{T}}$, $\mathbf{C}^{\mathsf{T}}\mathbf{C}$

29. $(\mathbf{BC})\mathbf{A}$, $(\mathbf{BC})^{\mathsf{T}}(\mathbf{BC})$ **30.** $(\mathbf{BC})^{\mathsf{T}}\mathbf{A}$, $\mathbf{A}(\mathbf{BC})^{\mathsf{T}}$

31. *The Action of Diagonal Matrices.* Consider the matrices

$$\mathbf{F} = \begin{bmatrix} p & 0 \\ 0 & q \end{bmatrix}, \quad \mathbf{G} = \begin{bmatrix} r & 0 & 0 \\ 0 & s & 0 \\ 0 & 0 & t \end{bmatrix},$$

$$\mathbf{A} = \begin{bmatrix} 1 & 2 & 3 \\ 4 & 5 & 6 \end{bmatrix}.$$

Compute **FA** and **AG**, and observe the effect **F** and **G** have on **A**. Describe the effect of premultiplying and postmultiplying any $m \times n$ matrix **A** by a diagonal matrix.

32. Show that $\mathbf{A} = \begin{bmatrix} a & b \\ -b & a \end{bmatrix}$ and $\mathbf{B} = \begin{bmatrix} c & d \\ -d & c \end{bmatrix}$ commute for any real or complex numbers a, b, c, d.

33. The following matrices **A** and **B** commute. Confirm this fact for only the (1, 3)- and (3, 2)-entries in the products **AB** and **BA**. Do not multiply out.

$$\mathbf{A} = \begin{bmatrix} 1 & 3 & 4 \\ -1 & 1 & 3 \\ -2 & -3 & 1 \end{bmatrix}, \quad \mathbf{B} = \begin{bmatrix} -1 & 3 & 4 \\ -1 & -1 & 3 \\ -2 & -3 & -1 \end{bmatrix}$$

34. Let **A** be any $m \times n$ matrix. Show that $\mathbf{AA}^{\mathsf{T}}$ and $\mathbf{A}^{\mathsf{T}}\mathbf{A}$ are symmetric. Give the size of each product.

Exercises 35–36. Concerning linear systems in matrix form. Express each matrix equation $\mathbf{Ax} = \mathbf{b}$ *as a linear system* (S)*. Solve* (S)*.*

35. $\mathbf{A} = \begin{bmatrix} 1 & 3 & -2 \\ -2 & 0 & 1 \end{bmatrix}, \quad \mathbf{x} = \begin{bmatrix} x_1 \\ x_2 \\ x_3 \end{bmatrix},$

$$\mathbf{b} = \begin{bmatrix} 2 \\ 3 \end{bmatrix}$$

36. $\mathbf{A} = \begin{bmatrix} 5 & 4 \\ -2 & 0 \\ 6 & -1 \end{bmatrix}, \quad \mathbf{x} = \begin{bmatrix} x_1 \\ x_2 \end{bmatrix}, \quad \mathbf{b} = \begin{bmatrix} 2 \\ 3 \\ 8 \end{bmatrix}$

37. Use the matrices **A** and **B** to show that a product of two symmetric matrices is not generally symmetric, where

$$\mathbf{A} = \begin{bmatrix} 1 & 2 \\ 2 & 1 \end{bmatrix}, \quad \mathbf{B} = \begin{bmatrix} 1 & 3 \\ 3 & 4 \end{bmatrix}.$$

Give a condition under which a product of two symmetric matrices of order n will be symmetric.

38. Find all nonzero 2×2 matrices **A** for which $\mathbf{BA} = \mathbf{O}$, where $\mathbf{B} = \begin{bmatrix} 1 & 2 \\ 2 & 4 \end{bmatrix}$.

39. If $\mathbf{A} = \begin{bmatrix} 1 & 2 \\ 3 & 4 \end{bmatrix}$, find $\mathbf{A}^2 - 5\mathbf{A}$.

40. If $\mathbf{A} = \begin{bmatrix} 2 & -5 \\ 3 & 1 \end{bmatrix}$, find $\mathbf{A}^2 - 3\mathbf{A}$.

41. If $\mathbf{A} = \begin{bmatrix} 1 & -1 & 1 \\ 2 & -1 & 0 \\ 1 & 0 & 0 \end{bmatrix}$, find $\mathbf{A}^3$.

42. If $\mathbf{R} = \begin{bmatrix} \cos\theta & \sin\theta \\ -\sin\theta & \cos\theta \end{bmatrix}$, find $\mathbf{R}^2$ in terms of $\cos 2\theta$, $\sin 2\theta$.

43. Simplify the expression $(\mathbf{A} + \mathbf{B})^2 + (\mathbf{A} - \mathbf{B})^2$. Do not change order of multiplication.

44. Show that an $m \times n$ matrix **A** has $(2^m - 1)(2^n - 1)$ submatrices.

Hint: Designate the rows of **A** as either "on" (= 1, not deleted) or "off" (= 0, deleted). There are two possibilities for one row, four possibilities for two rows, and so on.

45. Count the number of entries below the main diagonal in an 8×4 matrix. Find a formula for the number of entries on or above the main diagonal of an $m \times n$ matrix when $m < n$.

46. Suppose $\mathbf{A}$ is a square matrix of order n. How many entries in $\mathbf{A}$ lie off the main diagonal? How many entries lie (a) above the main diagonal, (b) on or below the main diagonal?

Exercises 47–50. Suppose an $m \times n$ matrix $\mathbf{A}$ has e entries, for some positive integer e. For each given value e, give all possible sizes for $\mathbf{A}$.

47. $e = 12$ **48.** $e = 9$ **49.** $e = 13$

50. $e = 90$ (use prime factors)

Exercises 51–52. An expression of the type $\mathbf{x}^\mathsf{T}\mathbf{A}\mathbf{x}$, where $\mathbf{A}$ is an $n \times n$ matrix and $\mathbf{x}$ is a column vector of unknowns, is called a binary quadratic form. *Expand the product $\mathbf{x}^\mathsf{T}\mathbf{A}\mathbf{x}$ in each case.*

51. $\begin{bmatrix} x_1 & x_2 \end{bmatrix} \begin{bmatrix} 1 & 2 \\ 3 & 4 \end{bmatrix} \begin{bmatrix} x_1 \\ x_2 \end{bmatrix}$

52. $\begin{bmatrix} x_1 & x_2 & x_3 \end{bmatrix} \begin{bmatrix} 2 & 3 & 1 \\ 1 & -2 & 0 \\ 1 & 0 & 1 \end{bmatrix} \begin{bmatrix} x_1 \\ x_2 \\ x_3 \end{bmatrix}$

Exercises 53–56. An $n \times n$ matrix $\mathbf{A}$ is called nilpotent *if $\mathbf{A}^k = \mathbf{O}$, for some positive integer k. The smallest such integer k is called the* index of nilpotency *for $\mathbf{A}$. Show that each given matrix is nilpotent and state the index of nilpotency k in each case (x and y are real or complex numbers).*

53. $\begin{bmatrix} -2 & 1 \\ -4 & 2 \end{bmatrix}$

54. $\begin{bmatrix} -1 & 1 & 0 \\ 0 & -1 & 1 \\ 1 & -3 & 2 \end{bmatrix}$

55. $\begin{bmatrix} 1 & 5 & -2 \\ 1 & 2 & -1 \\ 3 & 6 & -3 \end{bmatrix}$

56. $\begin{bmatrix} xy & x^2 \\ -y^2 & -xy \end{bmatrix}$

57. Find three examples of nonzero 2×2 nilpotent matrices $\mathbf{A}$ with index of nilpotency $k = 2$.

58. Show that the matrix $\mathbf{B}$ is nilpotent, where

$$\mathbf{B} = \begin{bmatrix} 0 & 1 & 1 \\ 0 & 0 & 1 \\ 0 & 0 & 0 \end{bmatrix}$$

Let $\mathbf{A} = \mathbf{I} + \mathbf{B}$. Use the binomial theorem to expand $\mathbf{A}^5$ and show that $\mathbf{A}^5 = \mathbf{I} + 5\mathbf{B} + 10\mathbf{B}^2$. Compute $\mathbf{A}^5$.

59. An $n \times n$ matrix $\mathbf{A}$ is called *idempotent* if $\mathbf{A}^2 = \mathbf{A}$. Show that the transpose of an idempotent matrix is idempotent. Find four examples of 2×2 idempotent matrices.

60. If $\mathbf{A}^2 = \mathbf{I}$, for some $n \times n$ matrix $\mathbf{A}$, show that $\mathbf{A}^\mathsf{T}$ satisfies the same equation. Find two different 2×2 matrices with nonzero integer entries such that $\mathbf{A}^2 = \mathbf{I}$.

Exercises 61–62. Show that each matrix $\mathbf{A}$, where x is a real or complex number, is a square root of the identity $\mathbf{I}$.

61. $\begin{bmatrix} x & 1+x \\ 1-x & -x \end{bmatrix}$

62. $\begin{bmatrix} 1-2x & -2x & -2x \\ x & 1+x & x \\ -2+x & -2+x & -1+x \end{bmatrix}$

63. For the following matrices $\mathbf{A}$ and $\mathbf{J}$, prove that $\mathbf{A}^2 = \mathbf{J}$ for no matrix $\mathbf{A}$. Hence $\mathbf{J}$ has no square root.

$$\mathbf{A} = \begin{bmatrix} a & b \\ c & d \end{bmatrix}, \qquad \mathbf{J} = \begin{bmatrix} 0 & 1 \\ 0 & 0 \end{bmatrix}$$

64. Show that $\mathbf{A}\mathbf{A}^\mathsf{T} = \mathbf{O}$ implies $\mathbf{A} = \mathbf{O}$ for any 2×2 matrix $\mathbf{A}$. Prove the result for any square matrix $\mathbf{A}$ of order n.

65. Suppose $\mathbf{C}$ is a square matrix of order n such that $\mathbf{C}^2 = \mathbf{I}$. Define $\mathbf{A} = \mathbf{C}\mathbf{D}_1\mathbf{C}$ and $\mathbf{B} = \mathbf{C}\mathbf{D}_2\mathbf{C}$, where $\mathbf{D}_1$ and $\mathbf{D}_2$ are diagonal matrices. Show that $\mathbf{AB} = \mathbf{BA}$.

66. Suppose $\mathbf{A} + \mathbf{I} = \begin{bmatrix} 1 & 3 & 4 \\ -1 & 1 & 3 \\ -2 & -3 & 1 \end{bmatrix}$. Evaluate $(\mathbf{A} + \mathbf{I})(\mathbf{A} - \mathbf{I})$.

67. Consider the matrix $\mathbf{A} = \begin{bmatrix} 2 & -5 \\ 3 & 1 \end{bmatrix}$ and let $\mathbf{B} = \mathbf{A}^2 - 3\mathbf{A} + 17\mathbf{I}$. Show that $\mathbf{B} = \mathbf{O}$.

68. If $\mathbf{A}$ and $\mathbf{B}$ are square matrices of order n, show that

$$(\mathbf{AB} - 2\mathbf{I})\mathbf{A}(\mathbf{BA} + \mathbf{I}) = (\mathbf{AB} + \mathbf{I})\mathbf{A}(\mathbf{BA} - 2\mathbf{I}).$$

69. Prove properties (**T1**), (**T2**), (**T3**) for transposes.

Exercises 70–73. Concerning symmetric and skew-symmetric matrices. Let $\mathbf{A}$ *be an* $n \times n$ *matrix.*

70. If $\mathbf{A}$ is skew-symmetric, show that all the entries on the main diagonal of $\mathbf{A}$ are necessarily zero.

71. Let

$$\mathbf{S} = \frac{1}{2}(\mathbf{A} + \mathbf{A}^{\mathsf{T}}) \quad \text{and} \quad \mathbf{K} = \frac{1}{2}(\mathbf{A} - \mathbf{A}^{\mathsf{T}})$$

Show that $\mathbf{S}$ is symmetric, $\mathbf{K}$ is skew-symmetric, and $\mathbf{A} = \mathbf{S} + \mathbf{K}$. Hence, any square matrix can be *decomposed* into a sum of a symmetric and a skew-symmetric matrix.

Find the decomposition $\mathbf{A} = \mathbf{S} + \mathbf{K}$ when $\mathbf{A} = \begin{bmatrix} 1 & 2 \\ 3 & 4 \end{bmatrix}$.

72. Let $\mathbf{A}$ be a symmetric $m \times m$ matrix and let $\mathbf{B}$ be an $m \times n$ matrix. Show that $\mathbf{B}^{\mathsf{T}}\mathbf{AB}$ is symmetric. Verify this fact for the matrices

$$\mathbf{A} = \begin{bmatrix} 1 & -1 \\ -1 & 2 \end{bmatrix}, \qquad \mathbf{B} = \begin{bmatrix} 1 & 0 & -1 \\ 2 & 3 & 0 \end{bmatrix}.$$

73. Let $\mathbf{B}$ be an $n \times n$ skew-symmetric matrix and let $\mathbf{A} = \mathbf{I} + \mathbf{B}$. Show that $\mathbf{AA}^{\mathsf{T}} = (\mathbf{I} + \mathbf{B})(\mathbf{I} - \mathbf{B}) = \mathbf{A}^{\mathsf{T}}\mathbf{A}$.

Exercises 74–77. Use mathematical induction to prove each identity for all positive integers n.

74. $$\begin{bmatrix} 1 & 1 \\ 0 & 1 \end{bmatrix}^n = \begin{bmatrix} 1 & n \\ 0 & 1 \end{bmatrix}$$

75. $$\begin{bmatrix} 7 & 4 \\ -9 & -5 \end{bmatrix}^n = \begin{bmatrix} 1 + 6n & 4n \\ -9n & 1 - 6n \end{bmatrix}$$

76. If $\mathbf{A}$ is an idempotent matrix ($\mathbf{A}^2 = \mathbf{A}$) of order n, show that

$$(\mathbf{I} + \mathbf{A})^n = \mathbf{I} + (2^n - 1)\mathbf{A}.$$

77. Given matrices $\mathbf{A}$ and $\mathbf{B}$ such that $\mathbf{AB} - \mathbf{BA} = \mathbf{I}$, show that

$$\mathbf{A}^n\mathbf{B} - \mathbf{BA}^n = n\mathbf{A}^{n-1}.$$

Exercises 78–79. Let $\mathbf{A}$ *be an* $n \times n$ *matrix and define an* $n \times n$ *matrix* $\mathbf{B}$ *as follows:*

$$\mathbf{B} = x_0\mathbf{A}^0 + x_1\mathbf{A}^1 + x_2\mathbf{A}^2 + \cdots + x_k\mathbf{A}^k, \tag{2.16}$$

where k is a positive integer and $x_0, x_1, x_2, \ldots, x_k$ *are scalars. The expression on the right side of (2.16) is called a* polynomial *in* $\mathbf{A}$. *For each matrix* $\mathbf{A}$, *compute the entries in the* 2×2 *matrix* $\mathbf{B}$.

78. $\mathbf{A} = \begin{bmatrix} 1 & 1 \\ 0 & 1 \end{bmatrix}$ *Hint: See Exercise 74.*

79. $\mathbf{A} = \begin{bmatrix} 1 & 0 \\ 0 & -1 \end{bmatrix}$

80. Partition the matrices $\mathbf{A}$ and $\mathbf{B}$ into blocks so that $\mathbf{A}$ is conformable to $\mathbf{B}$ for block multiplication and then compute $\mathbf{AB}$ using block products. Check your answer by computing $\mathbf{AB}$ using row-by-column multiplication.

$$\mathbf{A} = \begin{bmatrix} 1 & 0 & 0 & 1 & 0 \\ 0 & 1 & 0 & 0 & 1 \\ 0 & 0 & 1 & 0 & 0 \\ 0 & 0 & 0 & 1 & 2 \\ 0 & 0 & 0 & 3 & -1 \end{bmatrix}, \quad \mathbf{B} = \begin{bmatrix} 2 & 2 \\ -1 & 3 \\ 1 & 4 \\ 5 & 0 \\ 0 & 5 \end{bmatrix}$$

81. In each part find $\mathbf{AB}$ by block multiplication using the indicated partitions. Find $\mathbf{AB}$ directly using row-by-column multiplication.

(a) $$\mathbf{A} = \left[\begin{array}{cc|cc} 1 & -2 & 1 & 2 \\ 0 & 3 & 3 & 4 \\ \hline 0 & 0 & -1 & 2 \end{array}\right], \quad \mathbf{B} = \left[\begin{array}{cc|c} 1 & 0 & 1 \\ 0 & 1 & 1 \\ \hline 1 & -2 & 2 \\ 0 & 3 & -1 \end{array}\right]$$

(b) $$\mathbf{A} = \left[\begin{array}{ccc|c} 1 & 2 & 0 & -1 \\ 0 & 1 & 4 & 5 \end{array}\right], \quad \mathbf{B} = \left[\begin{array}{ccc} 1 & -1 & 1 \\ 2 & 0 & 4 \\ 3 & 1 & 0 \\ \hline 2 & 2 & 1 \end{array}\right]$$

82. Show how block multiplication can be used to compute $\mathbf{AB}$, where $\mathbf{A}$ is as shown and $\mathbf{B} = [b_{ij}]$ is a 5×3 matrix with entries defined by $b_{ij} = (-1)^{i+j}$.

$$\mathbf{A} = \begin{bmatrix} 5 & 0 & 0 & 0 & 0 \\ 0 & 5 & 0 & 0 & 0 \\ 1 & 0 & 1 & 1 & 1 \\ 0 & 1 & 1 & 1 & 1 \end{bmatrix}$$

83. Use block multiplication to find $\mathbf{A}^k$ for all positive integers k, where

$$\mathbf{A} = \begin{bmatrix} 1 & 0 & 0 & 0 \\ 0 & 1 & 0 & 0 \\ a_1 & a_2 & -1 & 0 \\ b_1 & b_2 & 0 & -1 \end{bmatrix}.$$

84. The definition of matrix multiplication given by Arthur Cayley in 1858 is based on the following ideas.

The linear system (2.17) defines unknowns z_1, z_2 in terms of unknowns y_1, y_2 and given constants a_{ij}, $1 \le i, j \le 2$.

$$\begin{cases} z_1 = a_{11}y_1 + a_{12}y_2 \\ z_2 = a_{21}y_1 + a_{22}y_2 \end{cases} \tag{2.17}$$

The linear system (2.18) defines y_1, y_2 in terms of unknowns x_1, x_2 and given constants b_{ij}, $1 \le i \le 2$, $1 \le j \le 3$.

$$\begin{cases} y_1 = b_{11}x_1 + b_{12}x_2 + b_{13}x_3 \\ y_2 = b_{21}x_1 + b_{22}x_2 + b_{23}x_3 \end{cases} \tag{2.18}$$

Write (2.17) as a matrix equation $\mathbf{z} = \mathbf{Ay}$ and (2.18) as a matrix equation $\mathbf{y} = \mathbf{Bx}$. Substitute the unknowns y_1, y_2 from (2.18) into (2.17) to obtain a linear system that defines z_1, z_2 in terms of x_1, x_2. Show that this system can be written as a matrix equation $\mathbf{z} = \mathbf{ABx}$.

85. Rework the previous exercise using the relationships

$$\begin{cases} z_1 = 4y_1 - 5y_2 \\ z_2 = 3y_1 + 7y_2 \end{cases} \qquad \begin{cases} y_1 = x_1 - 3x_2 + x_3 \\ y_2 = 6x_1 - 2x_2 - 9x_3 \end{cases}$$

USING MATLAB

86. Write an M-file that will generate the matrices in Exercises 1–6.

87. Use the Symbolic Math Toolbox to verify that $\mathbf{AB} = \mathbf{BA} = \mathbf{O}$, where

$$\mathbf{A} = \begin{bmatrix} 0 & c & -b \\ -c & 0 & a \\ b & -a & 0 \end{bmatrix}, \qquad \mathbf{B} = \begin{bmatrix} a^2 & ab & ac \\ ba & b^2 & bc \\ ca & cb & c^2 \end{bmatrix}.$$

Hence, **A** and **B** commute. What rule in Section 2.1 does this problem illustrate?

88. Write an M-file that will partition the matrices

$$\mathbf{A} = \begin{bmatrix} 1 & 0 & 2 & 4 & 1 \\ 0 & 1 & 6 & 2 & -2 \\ 0 & 0 & -1 & 2 & 9 \\ 0 & 0 & 2 & 1 & 3 \end{bmatrix}, \qquad \mathbf{B} = \begin{bmatrix} -1 & 1 & 6 \\ 0 & 4 & -3 \\ 1 & 0 & 0 \\ 0 & 1 & 0 \\ 0 & 0 & 1 \end{bmatrix}$$

into convenient blocks so that **A** is conformable to **B** for block multiplication. Your program should compute **AB** by block multiplication and the result should then be compared with **AB** computed by row-by-column multiplication.

89. Use MATLAB to cross-check hand calculations in this exercise set.

2.2 Inverses

2.2.1 Definition and Computation

A number x is called a *multiplicative inverse* for a nonzero real or complex number a if $xa = ax = 1$. The number x is unique. For example, $x = \frac{1}{3}$ is the unique multiplicative inverse for the real number $a = 3$ because $(\frac{1}{3})\,3 = 3(\frac{1}{3}) = 1$. In this section we extend the concept of a multiplicative inverse to square matrices.

Caution! Not every square matrix has a multiplicative inverse!

Definition 2.7 *Inverse of a square matrix*

An $n \times n$ matrix **A** is called *invertible* (*nonsingular*) if there is an $n \times n$ matrix **X** which satisfies the following two equations:

$$\mathbf{AX} = \mathbf{I}_n \quad \text{and} \quad \mathbf{XA} = \mathbf{I}_n. \tag{2.19}$$

We call **X** *an inverse* for **A**. If no matrix **X** exists satisfying (2.19), then **A** is said to be *singular*[4] (*noninvertible*).

Note that the equations (2.19) are symmetric in **A** and **X**. Hence, if **X** is an inverse for **A**, then **A** is an inverse for **X** and conversely.

EXAMPLE 1

An Invertible Matrix of Order 2

Consider the matrices

$$\mathbf{A} = \begin{bmatrix} 2 & 1 \\ 5 & 3 \end{bmatrix} \quad \text{and} \quad \mathbf{X} = \begin{bmatrix} 3 & -1 \\ -5 & 2 \end{bmatrix}.$$

As an exercise, verify that $\mathbf{AX} = \mathbf{I}_2$ and $\mathbf{XA} = \mathbf{I}_2$. Equations (2.19) are satisfied and so **X** is an inverse for **A** and **A** is an inverse for **X**; in other words, **A** and **X** are both invertible matrices. ■

Theorem 2.3

Inverses are Unique

Let **X** *be an inverse for the* $n \times n$ *matrix* **A**. *Then* **X** *is the only matrix in the set* $\mathbb{R}^{n \times n}$ *that satisfies equations (2.19).*

Proof Suppose **A** is invertible and that **X** and **Y** are two inverses for **A**. Then, in particular, $\mathbf{AX} = \mathbf{I}$ and $\mathbf{YA} = \mathbf{I}$. Using associativity of matrix multiplication, we have

$$\mathbf{Y} = \mathbf{YI} = \mathbf{Y(AX)} = \mathbf{(YA)X} = \mathbf{IX} = \mathbf{X} \quad \Rightarrow \quad \mathbf{Y} = \mathbf{X},$$

showing that **X** and **Y** are identical. Hence, an invertible matrix has a unique inverse.

■ **NOTATION**

An invertible matrix **A** has a unique inverse that we denote by $\mathbf{A}^{-1}$. When **A** is singular we say that $\mathbf{A}^{-1}$ *does not exist.* ■

Returning to Example 1, we were careful to call **X** *an* inverse for **A** at that stage. But we now know that **X** is unique and we can write $\mathbf{X} = \mathbf{A}^{-1}$ and $\mathbf{A} = \mathbf{X}^{-1}$.

Two Questions

When presented with an $n \times n$ matrix **A**, two basic questions need to be addressed:

Question How can we determine whether or not **A** is invertible? (2.20)

Question If **A** is invertible, how do we compute $\mathbf{A}^{-1}$ as efficiently as possible? (2.21)

Question (2.20) can be answered by testing **A** against one of several conditions, each of which characterizes invertibility. Many (but not all) of these conditions

4 The word *singular* in this context means "unusual" or "strange."

are listed in Section 2.2.4. One condition in particular is that rank $\mathbf{A} = n$, the order of the matrix. Illustration 2.2 explains the standard computational approach to finding the inverse of a square matrix. It answers questions (2.20) and (2.21) simultaneously and shows how rank $\mathbf{A}$ enters the picture. But before we begin, we look at the equations (2.19) again and introduce the following terminology.

Definition 2.8

Right-inverse and left-inverse

A matrix $\mathbf{X}$ is called a *right-inverse* for the $n \times n$ matrix $\mathbf{A}$ if $\mathbf{AX} = \mathbf{I}_n$ and $\mathbf{X}$ is called a *left-inverse* for $\mathbf{A}$ if $\mathbf{XA} = \mathbf{I}_n$.

Caution! Matrix multiplication is noncommutative in general and so you should not deduce that a right-inverse is automatically a left-inverse (or vice versa) without some proof. Definition 2.7 can now be rephrased: $\mathbf{A}$ is invertible exactly when there is a matrix $\mathbf{X}$ which is both a right-inverse and a left-inverse for $\mathbf{A}$. We call $\mathbf{X}$ a *two-sided inverse* in this case.

■ ILLUSTRATION 2.2

Computational Approach to Finding Inverses

Consider the square matrix $\mathbf{A}$ in the set $\mathbb{R}^{3\times 3}$ shown on the left.

$$\mathbf{A} = \begin{bmatrix} 1 & -2 & 1 \\ 2 & 1 & 2 \\ 3 & 0 & -1 \end{bmatrix} \qquad \mathbf{I}_3 = \begin{bmatrix} 1 & 0 & 0 \\ 0 & 1 & 0 \\ 0 & 0 & 1 \end{bmatrix}$$

As a first step, we will look for a right-inverse for $\mathbf{A}$; that is, we wish to find a 3×3 matrix $\mathbf{X}$ such that $\mathbf{AX} = \mathbf{I}_3$. Write the unknown matrix $\mathbf{X}$ and the identity $\mathbf{I}_3$ in terms of column vectors, $\mathbf{X} = [\,\mathbf{x}_1\ \mathbf{x}_2\ \mathbf{x}_3\,]$ and $\mathbf{I} = [\,\mathbf{e}_1\ \mathbf{e}_2\ \mathbf{e}_3\,]$, and consider solving $\mathbf{AX} = \mathbf{I}$ for $\mathbf{X}$. Applying matrix-column vector multiplication (Section 2.1) to the product $\mathbf{AX}$, we have

$$\mathbf{AX} = \mathbf{A}[\,\mathbf{x}_1\ \mathbf{x}_2\ \mathbf{x}_3\,] = [\,\mathbf{Ax}_1\ \mathbf{Ax}_2\ \mathbf{Ax}_3\,] = [\,\mathbf{e}_1\ \mathbf{e}_2\ \mathbf{e}_3\,] = \mathbf{I}_3$$

and equating columns in the third equality, we have

$$\mathbf{Ax}_1 = \mathbf{e}_1, \qquad \mathbf{Ax}_2 = \mathbf{e}_2, \qquad \mathbf{Ax}_3 = \mathbf{e}_3. \tag{2.22}$$

Solving $\mathbf{AX} = \mathbf{I}_3$ amounts to solving three linear systems, each with the same coefficient matrix $\mathbf{A}$, for solutions $\mathbf{x}_1$, $\mathbf{x}_2$, $\mathbf{x}_3$ (we have called this a multi-system in Exercises 1.2). The matrix equations (2.22) are solved simultaneously by forming the partitioned matrix $[\,\mathbf{A} \mid \mathbf{I}_3\,]$ and reducing $\mathbf{A}$ to its reduced form

$\mathbf{A}^*$. Clearing columns, we have

$$\left[\begin{array}{rrr|rrr} 1 & -2 & 1 & 1 & 0 & 0 \\ 2 & 1 & 2 & 0 & 1 & 0 \\ 3 & 0 & -1 & 0 & 0 & 1 \end{array}\right] \underset{\substack{\mathbf{r}_2 - 2\mathbf{r}_1 \to \mathbf{r}_2 \\ \mathbf{r}_3 - 3\mathbf{r}_1 \to \mathbf{r}_3}}{\sim} \left[\begin{array}{rrr|rrr} 1 & -2 & 1 & 1 & 0 & 0 \\ 0 & 5 & 0 & -2 & 1 & 0 \\ 0 & 6 & -4 & -3 & 0 & 1 \end{array}\right]$$

$$\underset{\substack{\frac{1}{5}\mathbf{r}_2 \to \mathbf{r}_2 \\ \mathbf{r}_3 - 6\mathbf{r}_2 \to \mathbf{r}_3 \\ \mathbf{r}_1 + 2\mathbf{r}_2 \to \mathbf{r}_1}}{\sim} \left[\begin{array}{rrr|rrr} 1 & 0 & 1 & \frac{1}{5} & \frac{2}{5} & 0 \\ 0 & 1 & 0 & -\frac{2}{5} & \frac{1}{5} & 0 \\ 0 & 0 & -4 & -\frac{3}{5} & -\frac{6}{5} & 1 \end{array}\right]$$

$$\underset{\substack{-\frac{1}{4}\mathbf{r}_3 \to \mathbf{r}_3 \\ \mathbf{r}_1 - \mathbf{r}_3 \to \mathbf{r}_1}}{\sim} \left[\begin{array}{rrr|rrr} 1 & 0 & 0 & \frac{1}{20} & \frac{2}{20} & \frac{5}{20} \\ 0 & 1 & 0 & -\frac{8}{20} & -\frac{4}{20} & 0 \\ 0 & 0 & 1 & \frac{3}{20} & \frac{6}{20} & -\frac{5}{20} \end{array}\right].$$

In this case $[\mathbf{A} \mid \mathbf{I}_3] \sim [\mathbf{I}_3 \mid \mathbf{X}]$, where

$$\mathbf{X} = [\,\mathbf{x}_1 \;\; \mathbf{x}_2 \;\; \mathbf{x}_3\,] = \begin{bmatrix} \frac{1}{20} & \frac{2}{20} & \frac{5}{20} \\ -\frac{8}{20} & -\frac{4}{20} & 0 \\ \frac{3}{20} & \frac{6}{20} & -\frac{5}{20} \end{bmatrix}.$$

The computation shows that $\mathbf{A}^* = \mathbf{I}_3$ and so rank $\mathbf{A} = 3$, which implies that each system in (2.22) has a unique solution (Section 1.2). Hence, $\mathbf{X}$ is a *unique* right-inverse for $\mathbf{A}$. ■

The computation in Illustration 2.2 can be carried out for a square matrix of any order, and we therefore have the following general method.

> **Computation** Given an $n \times n$ matrix $\mathbf{A}$ such that rank $\mathbf{A} = n$, the reduction
>
> $$[\mathbf{A} \mid \mathbf{I}_n] \sim [\mathbf{I}_n \mid \mathbf{X}] \tag{2.23}$$
>
> results in a unique right-inverse $\mathbf{X}$ for $\mathbf{A}$.

Only one more piece of information is required to complete the current discussion. It is this: A unique right-inverse for $\mathbf{A}$ turns out to be also a left-inverse for $\mathbf{A}$ and so the computation (2.23) actually gives us $\mathbf{X} = \mathbf{A}^{-1}$. Here is the proof.

Theorem 2.4 A Unique Right-Inverse is also a Left-Inverse

If $\mathbf{A}$ is $n \times n$ and there is a unique matrix $\mathbf{X}$ such that $\mathbf{AX} = \mathbf{I}_n$, then $\mathbf{XA} = \mathbf{I}_n$ and so $\mathbf{X} = \mathbf{A}^{-1}$.

Proof Using the matrix $\mathbf{X}$, define a matrix $\mathbf{Y} = \mathbf{XA} + \mathbf{X} - \mathbf{I}_n$ and consider the product $\mathbf{AY}$. We have

$$\begin{aligned} \mathbf{AY} &= \mathbf{A}(\mathbf{XA} + \mathbf{X} - \mathbf{I}_n) = (\mathbf{AX})\mathbf{A} + \mathbf{AX} - \mathbf{AI}_n \\ &= \mathbf{I}_n\mathbf{A} + \mathbf{I}_n - \mathbf{A} = \mathbf{I}_n \quad \Rightarrow \quad \mathbf{AY} = \mathbf{I}_n. \end{aligned}$$

But, by assumption, $\mathbf{X}$ is a unique right-inverse for $\mathbf{A}$ and so $\mathbf{Y} = \mathbf{X}$. Hence, $\mathbf{X} = \mathbf{XA} + \mathbf{X} - \mathbf{I}_n$, which simplifies to give $\mathbf{XA} = \mathbf{I}_n$.

The following sufficient condition for invertibility is now clear.

Theorem 2.5 **Rank and Invertibility**

Let $\mathbf{A}$ *be an* $n \times n$ *matrix. If* $\operatorname{rank} \mathbf{A} = n$, *then* $\mathbf{A}$ *is invertible.*

Rank Deficient Matrices

The next short discussion explores the reason why some square matrices fail to be invertible. We begin with some terminology.

An $n \times n$ matrix $\mathbf{A}$ with $\operatorname{rank} \mathbf{A} = n$ is said to have *full rank*—every column in $\mathbf{A}$ is a pivot column. Theorem 2.5 says that matrices with full rank are invertible.

When $\operatorname{rank} \mathbf{A} < n$, then $\mathbf{A}$ is said to be *rank deficient*. What happens in the reduction process in (2.23) when $\mathbf{A}$ is rank deficient? Reducing $\mathbf{A}$ to its reduced form $\mathbf{A}^*$ looks like this:

$$[\mathbf{A} \mid \mathbf{I}_n] \sim [\mathbf{A}^* \mid \mathbf{B}] = [\mathbf{A}^* \mid \mathbf{b}_1 \ \mathbf{b}_2 \ \cdots \ \mathbf{b}_n], \tag{2.24}$$

for some matrix $\mathbf{B} = [\mathbf{b}_1 \ \mathbf{b}_2 \ \cdots \ \mathbf{b}_n]$. Row n in $\mathbf{A}^*$ is certainly a zero row because $\operatorname{rank} \mathbf{A} < n$. But $\operatorname{rank} \mathbf{I}_n = n$, and so the EROs that reduce $\mathbf{A}$ to $\mathbf{A}^*$ also reduce $\mathbf{I}_n$ to $\mathbf{B}$, which also has rank n and all its rows are therefore nonzero! Let p be the first nonzero entry in row n of $\mathbf{B}$ occurring in column j, say.

$$[\mathbf{A} \mid \mathbf{I}_n] \sim \left[\begin{array}{ccc|cccccc} & & & & & & * & * & \\ \vdots & & \vdots & & & & \vdots & \vdots & \\ 0 & \cdots & 0 & 0 & \cdots & 0 & p & * & \cdots \end{array}\right] = [\mathbf{A}^* \mid \mathbf{B}]$$

$$\uparrow$$
$$\text{column } j$$

It follows that the jth linear system $\mathbf{Ax}_j = \mathbf{e}_j$ in the multi-system defined by $\mathbf{AX} = \mathbf{I}_n$ is inconsistent because $\operatorname{rank} \mathbf{A} < \operatorname{rank}[\mathbf{A} \mid \mathbf{e}_j]$. Hence, column j of the unknown matrix $\mathbf{X} = [\mathbf{x}_1 \ \mathbf{x}_2 \ \cdots \ \mathbf{x}_n]$ cannot be computed and so no right-inverse for $\mathbf{A}$ exists.

EXAMPLE 2 **A rank deficient matrix**

The matrix shown has $\operatorname{rank} \mathbf{A} = 1$.

$$\mathbf{A} = \begin{bmatrix} 1 & 2 & 3 \\ 1 & 2 & 3 \\ 2 & 4 & 6 \end{bmatrix}$$

Reducing $\mathbf{A}$ to its reduced form $\mathbf{A}^*$, we have

$$[\mathbf{A} \mid \mathbf{e}_1 \ \mathbf{e}_2 \ \mathbf{e}_3] \sim \left[\begin{array}{ccc|ccc} 1 & 2 & 3 & 1 & 0 & 0 \\ 0 & 0 & 0 & -1 & 1 & 0 \\ 0 & 0 & 0 & -2 & 0 & 1 \end{array}\right] = [\mathbf{A}^* \mid \mathbf{b}_1 \ \mathbf{b}_2 \ \mathbf{b}_3]$$

Referring to the explanation proceeding the example, we have $p = -2$, $j = 1$ and $\text{rank}\,\mathbf{A} = 1 < 2 = \text{rank}[\mathbf{A} \mid \mathbf{e}_1]$, indicating that the system $\mathbf{Ax} = \mathbf{e}_1$ has no solution. In this example it turns out that $\mathbf{Ax} = \mathbf{e}_j$ is inconsistent for $j = 2, 3$ because $\text{rank}\,\mathbf{A} = 1 < 2 = \text{rank}[\mathbf{A} \mid \mathbf{e}_2]$ and $\text{rank}\,\mathbf{A} = 1 < 2 = \text{rank}[\mathbf{A} \mid \mathbf{e}_3]$, and so no columns of the unknown matrix $\mathbf{X} = [\,\mathbf{x1}\ \ \mathbf{x2}\ \ \mathbf{x3}\,]$ can be computed. ■

We are now able to recognize invertible matrices using the following important result.

Theorem 2.6 **Invertibility and Rank**

An $n \times n$ matrix $\mathbf{A}$ is invertible if and only if $\text{rank}\,\mathbf{A} = n$.

Proof We have shown immediately above that $\text{rank}\,\mathbf{A} < n$ implies $\mathbf{A}$ is singular. The *contrapositive* (Appendix B: Toolbox) of the latter statement gives the necessary condition: If $\mathbf{A}$ is invertible, then $\text{rank}\,\mathbf{A} = n$. Theorem 2.5 already establishes that $\text{rank}\,\mathbf{A} = n$ implies $\mathbf{A}$ is invertible.

Appealing to Theorem 2.6, each of the matrices

$$\mathbf{A} = \begin{bmatrix} 3 & -2 \\ -6 & 4 \end{bmatrix} \quad (\text{rank}\,\mathbf{A} = 1), \tag{2.25}$$

$$\mathbf{B} = \begin{bmatrix} 1 & 2 & 3 \\ 4 & 5 & 6 \\ 7 & 8 & 9 \end{bmatrix} \quad (\text{rank}\,\mathbf{B} = 2), \quad \mathbf{M} = \begin{bmatrix} 16 & 2 & 3 & 13 \\ 5 & 11 & 10 & 8 \\ 9 & 7 & 6 & 12 \\ 4 & 14 & 15 & 1 \end{bmatrix} \quad (\text{rank}\,\mathbf{M} = 3)$$

is singular because its rank is less than the order of the matrix. The matrix $\mathbf{M}$ is **magic(4)**, the regular magic (square) matrix of order 4 that appears on the front cover of this text book. It has been proved (see [1]) that all regular magic matrices of even order are singular and those of odd order are invertible.

Inverses of 2×2 *Matrices, Determinants*

We will now give the formula for the inverse of a invertible 2×2 matrix. Suppose that

$$\mathbf{A} = \begin{bmatrix} a & b \\ c & d \end{bmatrix}. \tag{2.26}$$

The number $\det(\mathbf{A}) = ad - bc$ is called the *determinant* of $\mathbf{A}$. The method for computing $\det(\mathbf{A})$ for a general $n \times n$ matrix $\mathbf{A}$ is given in Chapter 5, and it is shown there that

$$\mathbf{A} \text{ is invertible if and only } \det(\mathbf{A}) \text{ is nonzero.}$$

It follows that the matrix $\mathbf{A}$ in (2.25) is singular because $\det(\mathbf{A}) = 3(4) - (-2)(-6) = 0$. If the matrix $\mathbf{A}$ in (2.26) has a nonzero determinant, then its inverse is given by

$$\mathbf{A}^{-1} = \frac{1}{\det(\mathbf{A})}\begin{bmatrix} d & -b \\ -c & a \end{bmatrix}. \tag{2.27}$$

The rule is—interchange the entries on the main diagonal of **A**, negate the off-diagonal entries in **A**, and divide throughout by det(**A**).

EXAMPLE 3

Finding the Inverse of a Square Matrix of Order 2

The matrix **A** shown below is invertible because $\det(\mathbf{A}) = (2)(5) - (3)(-4) = 22 \neq 0$. Hence,

$$\mathbf{A} = \begin{bmatrix} 2 & 3 \\ -4 & 5 \end{bmatrix} \quad \Rightarrow \quad \mathbf{A}^{-1} = \frac{1}{22}\begin{bmatrix} 5 & -3 \\ 4 & 2 \end{bmatrix} = \begin{bmatrix} 5/22 & -3/22 \\ 2/11 & 1/11 \end{bmatrix}.$$ ■

One important comment before we change topics. Although pairs of square matrices of the same order do not commute in general, the situation is different with invertible matrices. Theorem 2.9 in Section 2.2.4 will show that a right-inverse for a square matrix **A** is automatically a left-inverse for **A** and vice versa. Thus the existence of a matrix **X** such that *either* $\mathbf{AX} = \mathbf{I}$ *or* $\mathbf{XA} = \mathbf{I}$ implies that **A** is invertible and its inverse is $\mathbf{X} = \mathbf{A}^{-1}$.

2.2.2 Elementary Matrices

This subsection outlines the theory of *elementary matrices*. The highlights of the theory are these. Performing an elementary row operation on a matrix **A** can be accomplished by premultiplying **A** by an elementary matrix. Each invertible matrix **A** is a product of elementary matrices, and so is its inverse $\mathbf{A}^{-1}$. Elementary matrices are used in Section 2.3 (LU-factorization) and they appear in other situations where row operations need to be interpreted as matrix multiplications. We now begin the discussion.

Definition 2.9

Elementary matrix

An $m \times m$ matrix **E** is called *elementary* if it is the result of performing a single elementary row operation (ERO) on the identity matrix $\mathbf{I}_m$.

For example, the matrices $\mathbf{E}_1$, $\mathbf{E}_2$, $\mathbf{E}_3$ shown next are elementary. Each of them is obtained by performing a single ERO on the appropriate identity matrix.

$$\mathbf{E}_1 = \begin{bmatrix} 0 & 1 \\ 1 & 0 \end{bmatrix}, \quad \mathbf{E}_2 = \begin{bmatrix} 1 & 0 & 0 \\ 0 & -5 & 0 \\ 0 & 0 & 1 \end{bmatrix}, \quad \mathbf{E}_3 = \begin{bmatrix} 1 & 0 & 0 & 0 \\ 0 & 1 & 0 & 0 \\ 0 & 0 & 1 & 0 \\ 0 & 0 & 2 & 1 \end{bmatrix}$$

$$\mathbf{r}_1 \leftrightarrow \mathbf{r}_2 \qquad -5\mathbf{r}_2 \to \mathbf{r}_2 \qquad \mathbf{r}_4 + 2\mathbf{r}_3 \to \mathbf{r}_4$$

None of the following matrices are elementary (explain why).

$$\begin{bmatrix} 0 & 1 \\ 0 & 0 \end{bmatrix}, \quad \begin{bmatrix} 1 & 0 & 0 \\ -1 & 1 & 0 \\ 0 & 0 & 2 \end{bmatrix}, \quad \begin{bmatrix} 0 & 0 & 0 & 1 \\ 0 & 0 & 1 & 0 \\ 0 & 1 & 0 & 0 \\ 1 & 0 & 0 & 0 \end{bmatrix} \tag{2.28}$$

Theorem 2.7 **Elementary Row Operations and Matrix Multiplication**

Suppose a single ERO is performed on an $m \times n$ *matrix* $\mathbf{A}$*, resulting in a matrix* $\mathbf{B}$*. If* $\mathbf{E}$ *is the elementary matrix obtained by performing the same ERO on* $\mathbf{I}_m$*, then* $\mathbf{EA} = \mathbf{B}$.

Proof Exercises 2.2.

■ **NOTATION**

An elementary matrix is one of three types, depending on which type of ERO was used on $\mathbf{I}_m$. To keep notation as simple as possible, let $\mathbf{E}$ denote an elementary matrix (without specifying its type) and attach subscripts, $\mathbf{E}_1, \mathbf{E}_2, \ldots$ when there is a sequence of them. ■

Reduction of an $m \times n$ matrix $\mathbf{A}$ to a row echelon form can be done in one step by premultiplying $\mathbf{A}$ by an $m \times m$ matrix $\mathbf{F}$ that is a product of elementary matrices. Here is an example.

EXAMPLE 4 **Row Reduction using Elementary Matrices**

We will reduce the matrix

$$\mathbf{A} = \begin{bmatrix} 4 & 5 & 6 \\ 1 & 2 & 3 \end{bmatrix}$$

to its reduced form $\mathbf{A}^*$. The sequence of EROs (appropriate for hand calculation) used to obtain $\mathbf{A}^*$ is shown in the left column, the corresponding elementary matrices are shown in the second column, and each matrix multiplication is in the third column.

$$\mathbf{r}_1 \leftrightarrow \mathbf{r}_2 \qquad \mathbf{E}_1 = \begin{bmatrix} 0 & 1 \\ 1 & 0 \end{bmatrix} \qquad \mathbf{E}_1\mathbf{A} = \begin{bmatrix} 1 & 2 & 3 \\ 4 & 5 & 6 \end{bmatrix} = \mathbf{B}_1$$

$$\mathbf{r}_2 - 4\mathbf{r}_1 \rightarrow \mathbf{r}_2 \qquad \mathbf{E}_2 = \begin{bmatrix} 1 & 0 \\ -4 & 1 \end{bmatrix} \qquad \mathbf{E}_2\mathbf{B}_1 = \begin{bmatrix} 1 & 2 & 3 \\ 0 & -3 & -6 \end{bmatrix} = \mathbf{B}_2$$

$$-\tfrac{1}{3}\mathbf{r}_2 \rightarrow \mathbf{r}_2 \qquad \mathbf{E}_3 = \begin{bmatrix} 1 & 0 \\ 0 & -\frac{1}{3} \end{bmatrix} \qquad \mathbf{E}_3\mathbf{B}_2 = \begin{bmatrix} 1 & 2 & 3 \\ 0 & 1 & 2 \end{bmatrix} = \mathbf{B}_3$$

$$\mathbf{r}_1 - 2\mathbf{r}_2 \rightarrow \mathbf{r}_1 \qquad \mathbf{E}_4 = \begin{bmatrix} 1 & -2 \\ 0 & 1 \end{bmatrix} \qquad \mathbf{E}_4\mathbf{B}_3 = \begin{bmatrix} 1 & 0 & -1 \\ 0 & 1 & 2 \end{bmatrix} = \mathbf{A}^*$$

Define

$$\mathbf{F} = \mathbf{E}_4\mathbf{E}_3\mathbf{E}_2\mathbf{E}_1 = \frac{1}{3}\begin{bmatrix} 2 & -5 \\ -1 & 4 \end{bmatrix}$$

Check that $\mathbf{FA} = \mathbf{A}^*$. Hence, forward and backward reduction can be realized by premultiplying $\mathbf{A}$ by $\mathbf{F}$, which is a product of elementary matrices. ■

Theorem 2.8 **Elementary Matrices are Invertible**

For every elementary matrix $\mathbf{E}$ *there is a corresponding elementary matrix* $\mathbf{F}$ *(of the same type) such that* $\mathbf{FE} = \mathbf{EF} = \mathbf{I}$. *Hence* $\mathbf{F} = \mathbf{E}^{-1}$.

Proof Exercises 2.2.

EXAMPLE 5 **Inverses of Elementary Matrices**

We will illustrate Theorem 2.8 when $n = 2$. The ERO used to obtain each $\mathbf{E}$ is shown on the extreme left and the inverse of this ERO, which is actually used to create $\mathbf{F}$, is shown on the extreme right. Check that $\mathbf{FE} = \mathbf{EF} = \mathbf{I}_2$ in each case.

$$\mathbf{r}_1 \leftrightarrow \mathbf{r}_2 \qquad \mathbf{E} = \begin{bmatrix} 0 & 1 \\ 1 & 0 \end{bmatrix} \qquad \mathbf{F} = \begin{bmatrix} 0 & 1 \\ 1 & 0 \end{bmatrix} \qquad \mathbf{r}_1 \leftrightarrow \mathbf{r}_2$$

$$2\mathbf{r}_2 \to \mathbf{r}_2 \qquad \mathbf{E} = \begin{bmatrix} 1 & 0 \\ 0 & 2 \end{bmatrix} \qquad \mathbf{F} = \begin{bmatrix} 1 & 0 \\ 0 & \frac{1}{2} \end{bmatrix} \qquad \tfrac{1}{2}\mathbf{r}_2 \to \mathbf{r}_2$$

$$\mathbf{r}_1 + 3\mathbf{r}_2 \to \mathbf{r}_1 \qquad \mathbf{E} = \begin{bmatrix} 1 & 3 \\ 0 & 1 \end{bmatrix} \qquad \mathbf{F} = \begin{bmatrix} 1 & -3 \\ 0 & 1 \end{bmatrix} \qquad \mathbf{r}_1 - 3\mathbf{r}_2 \to \mathbf{r}_1$$

■

The next illustration explores the connection between elementary matrices and invertible matrices. We will see that each invertible matrix is a product of elementary matrices and, in fact, the latter condition characterizes invertible matrices.

■ ILLUSTRATION 2.3

Inverses and Elementary Matrices

The 2×2 matrix

$$\mathbf{A} = \begin{bmatrix} 1 & 3 \\ 5 & 7 \end{bmatrix}$$

is invertible because its determinant $\det(\mathbf{A}) = 1(7) - 3(5) = -8 \neq 0$. To find $\mathbf{A}^{-1}$, reduce $[\,\mathbf{A} \mid \mathbf{I}_2\,] \sim [\,\mathbf{I}_2 \mid \mathbf{X}\,]$ using a sequence of EROs (the sequence is in general not unique). We have

$$
[\mathbf{A} \mid \mathbf{I}_2] = \left[\begin{array}{cc|cc} 1 & 3 & 1 & 0 \\ 5 & 7 & 0 & 1 \end{array}\right] \underset{\mathbf{r}_2 - 5\mathbf{r}_1 \to \mathbf{r}_2}{\sim} \left[\begin{array}{cc|cc} 1 & 3 & 1 & 0 \\ 0 & -8 & -5 & 1 \end{array}\right]
$$

$$
\underset{-\frac{1}{8}\mathbf{r}_2 \to \mathbf{r}_2}{\sim} \left[\begin{array}{cc|cc} 1 & 3 & 1 & 0 \\ 0 & 1 & \frac{5}{8} & -\frac{1}{8} \end{array}\right]
$$

$$
\underset{\mathbf{r}_1 - 3\mathbf{r}_2 \to \mathbf{r}_1}{\sim} \left[\begin{array}{cc|cc} 1 & 0 & -\frac{7}{8} & \frac{3}{8} \\ 0 & 1 & \frac{5}{8} & -\frac{1}{8} \end{array}\right] = [\mathbf{I}_2 \mid \mathbf{X}]
$$

and so $\mathbf{X} = \mathbf{A}^{-1} = \frac{1}{8}\begin{bmatrix} -7 & 3 \\ 5 & -1 \end{bmatrix}$.

The sequence of three EROs used to reduce $\mathbf{A}$ to $\mathbf{I}_2$ correspond to three elementary matrices

$$
\mathbf{E}_1 = \begin{bmatrix} 1 & 0 \\ -5 & 1 \end{bmatrix}, \quad \mathbf{E}_2 = \begin{bmatrix} 1 & 0 \\ 0 & -\frac{1}{8} \end{bmatrix}, \quad \mathbf{E}_3 = \begin{bmatrix} 1 & -3 \\ 0 & 1 \end{bmatrix}
$$

(replacement) (scaling) (replacement)

and we have

$$
\mathbf{E}_3\mathbf{E}_2\mathbf{E}_1\mathbf{A} = \mathbf{I}_2. \tag{2.29}
$$

Therefore, $\mathbf{X} = \mathbf{E}_3\mathbf{E}_2\mathbf{E}_1$ is a left-inverse for $\mathbf{A}$. We now apply Theorem 2.8. Starting with (2.29), premultiply by the inverse of each elementary matrix in turn and use associativity of multiplication, as follows:

$$
\begin{aligned}
\mathbf{E}_3^{-1}(\mathbf{E}_3\mathbf{E}_2\mathbf{E}_1\mathbf{A}) = \mathbf{E}_3^{-1}\mathbf{I}_2 \quad &\Rightarrow (\mathbf{E}_3^{-1}\mathbf{E}_3)\mathbf{E}_2\mathbf{E}_1\mathbf{A} = \mathbf{E}_3^{-1} \\
&\Rightarrow \mathbf{E}_2\mathbf{E}_1\mathbf{A} = \mathbf{E}_3^{-1} \\
\mathbf{E}_2^{-1}(\mathbf{E}_2\mathbf{E}_1\mathbf{A}) = \mathbf{E}_2^{-1}\mathbf{E}_3^{-1} \quad &\Rightarrow (\mathbf{E}_2^{-1}\mathbf{E}_2)\mathbf{E}_1\mathbf{A} = \mathbf{E}_2^{-1}\mathbf{E}_3^{-1} \\
&\Rightarrow \mathbf{E}_1\mathbf{A} = \mathbf{E}_2^{-1}\mathbf{E}_3^{-1} \\
\mathbf{E}_1^{-1}(\mathbf{E}_1\mathbf{A}) = \mathbf{E}_1^{-1}\mathbf{E}_2^{-1}\mathbf{E}_3^{-1} \quad &\Rightarrow (\mathbf{E}_1^{-1}\mathbf{E}_1)\mathbf{A} = \mathbf{E}_1^{-1}\mathbf{E}_2^{-1}\mathbf{E}_3^{-1} \\
&\Rightarrow \mathbf{A} = \mathbf{E}_1^{-1}\mathbf{E}_2^{-1}\mathbf{E}_3^{-1}
\end{aligned}
$$

Hence $\mathbf{A}$ is a product of elementary matrices. Note the order in which the elementary matrices appear in $\mathbf{X}$ compared with the reverse order in which their inverses appear in $\mathbf{A}$. Now postmultiply $\mathbf{A}$ by $\mathbf{X}$, as follows:

$$
\begin{aligned}
\mathbf{AX} &= (\mathbf{E}_1^{-1}\mathbf{E}_2^{-1}\mathbf{E}_3^{-1})(\mathbf{E}_3\mathbf{E}_2\mathbf{E}_1) = \mathbf{E}_1^{-1}\mathbf{E}_2^{-1}(\mathbf{E}_3^{-1}\mathbf{E}_3)\mathbf{E}_2\mathbf{E}_1 \\
&= \mathbf{E}_1^{-1}(\mathbf{E}_2^{-1}\mathbf{E}_2)\mathbf{E}_1) = \mathbf{E}_1^{-1}\mathbf{E}_1 = \mathbf{I}_2
\end{aligned}
$$

Hence, $\mathbf{X}$ is also a right-inverse for $\mathbf{A}$ and so $\mathbf{A}^{-1} = \mathbf{E}_3\mathbf{E}_2\mathbf{E}_1$. ■

The computations in Illustration 2.3 apply to any $n \times n$ matrix $\mathbf{A}$ that has $\operatorname{rank} \mathbf{A} = n$, and the general statement is this:

Let $\mathbf{A}$ be an invertible $n \times n$ matrix. Then $\operatorname{rank} \mathbf{A} = n$ and there is a sequence of elementary matrices $\mathbf{E}_1, \mathbf{E}_2, \ldots, \mathbf{E}_k$ that correspond to the sequence of EROs that change $\mathbf{A}$ into $\mathbf{I}_n$. We have

$$\mathbf{E}_k \mathbf{E}_{k-1} \cdots \mathbf{E}_2 \mathbf{E}_1 \mathbf{A} = \mathbf{I}_n. \tag{2.30}$$

The inverse of $\mathbf{A}$ is given as the product of elementary matrices

$$\mathbf{A}^{-1} = \mathbf{E}_k \mathbf{E}_{k-1} \cdots \mathbf{E}_2 \mathbf{E}_1.$$

Each elementary matrix is invertible, and so

$$\mathbf{A} = \mathbf{E}_1^{-1} \mathbf{E}_2^{-1} \cdots \mathbf{E}_k^{-1} \tag{2.31}$$

and because the inverse of an elementary matrix is again an elementary matrix it follows from (2.31) that $\mathbf{A}$ is a product of elementary matrices. Conversely, if $\mathbf{A} = \mathbf{F}_1 \mathbf{F}_2 \cdots \mathbf{F}_k$, for elementary matrices $\mathbf{F}_1, \mathbf{F}_2, \ldots, \mathbf{F}_k$, then $\mathbf{A}$ is invertible, with inverse $\mathbf{A}^{-1} = \mathbf{F}_k^{-1} \mathbf{F}_{k-1}^{-1} \cdots \mathbf{F}_1^{-1}$.

The preceding discussion shows that a square matrix $\mathbf{A}$ is invertible if and only if $\mathbf{A}$ is a product of elementary matrices.

2.2.3 Operations on Inverses

Let $\mathbf{A}$ and $\mathbf{B}$ be $n \times n$ matrices.

(P1) If $\mathbf{A}$ is invertible, then $\mathbf{A}^{-1}$ is invertible and $(\mathbf{A}^{-1})^{-1} = \mathbf{A}$.
(P2) If $\mathbf{A}$ and $\mathbf{B}$ are invertible, then $\mathbf{AB}$ is invertible and $(\mathbf{AB})^{-1} = \mathbf{B}^{-1}\mathbf{A}^{-1}$.
(P3) If $\mathbf{A}$ is invertible, then $s\mathbf{A}$ is invertible for any scalar $s \neq 0$ and

$$(s\mathbf{A})^{-1} = s^{-1}\mathbf{A}^{-1}.$$

(P4) $\mathbf{A}$ is invertible if and only if $\mathbf{A}^{\mathsf{T}}$ is invertible and $(\mathbf{A}^{\mathsf{T}})^{-1} = (\mathbf{A}^{-1})^{\mathsf{T}}$.

Proof

(P1) If $\mathbf{A}$ is invertible, then $\mathbf{A}^{-1}\mathbf{A} = \mathbf{A}\mathbf{A}^{-1} = \mathbf{I}$ and so $\mathbf{A}$ is a right- and left-inverse for $\mathbf{A}^{-1}$, proving the formula.
(P2) Using associativity, we have

$$(\mathbf{AB})(\mathbf{B}^{-1}\mathbf{A}^{-1}) = \mathbf{A}(\mathbf{B}\mathbf{B}^{-1})\mathbf{A}^{-1} = \mathbf{A}\mathbf{I}\mathbf{A}^{-1} = \mathbf{A}\mathbf{A}^{-1} = \mathbf{I}$$

and similarly, $(\mathbf{B}^{-1}\mathbf{A}^{-1})(\mathbf{AB}) = \mathbf{I}$, showing that $\mathbf{B}^{-1}\mathbf{A}^{-1}$ is a right- and left-inverse for $\mathbf{AB}$, proving the formula.
(P3) We have $(s\mathbf{A})(s^{-1}\mathbf{A}^{-1}) = (ss^{-1})\mathbf{A}\mathbf{A}^{-1} = \mathbf{I} = (s^{-1}\mathbf{A}^{-1})(s\mathbf{A})$, showing that $s^{-1}\mathbf{A}^{-1}$ is a right- and left-inverse for $s\mathbf{A}$, proving the formula.
(P4) Exercises 2.2.

Powers of Invertible Matrices

We can now extend formula (2.5) in Section 2.1.4. Let $\mathbf{A}$ be an $n \times n$ invertible matrix and recall that we defined $\mathbf{A}^0 = \mathbf{I}_n$. If k is a positive integer, define

$$\mathbf{A}^{-k} = \mathbf{A}^{-1} \cdots \mathbf{A}^{-1} \quad (k \text{ factors})$$

Suppose p and q are integers (positive, negative, or zero). Then

$$\mathbf{A}^p\mathbf{A}^q = \mathbf{A}^{p+q}, \quad (\mathbf{A}^p)^q = (\mathbf{A}^q)^p = \mathbf{A}^{pq}, \quad (\mathbf{A}^\mathsf{T})^p = (\mathbf{A}^p)^\mathsf{T}. \tag{2.32}$$

To illustrate the first equation in (2.32), take $p = -2$ and $q = 3$ and, using associativity of matrix multiplication, we obtain

$$\mathbf{A}^{-2}\mathbf{A}^3 = (\mathbf{A}^{-1}\mathbf{A}^{-1})\,\mathbf{AAA} = \mathbf{A}^{-1}(\mathbf{A}^{-1}\mathbf{A})\,\mathbf{AA} = (\mathbf{A}^{-1}\mathbf{A})\,\mathbf{A} = \mathbf{A} = \mathbf{A}^{-2+3}$$

2.2.4 Conditions for invertibility

We now list many conditions (others will be added later) that guarantee that a given square matrix $\mathbf{A}$ is invertible. The conditions are all "equivalent," meaning that any condition implies any other condition. In particular, if any condition (b) through (h) fails to be true, then $\mathbf{A}$ is singular.

In the following theorem, $\mathbf{Ax} = \mathbf{b}$ can be regarded as a matrix equation or the matrix form of an $n \times n$ linear system (S) in which $\mathbf{A}$ is the $n \times n$ coefficient matrix, $\mathbf{x}$ is the column vector of unknowns and $\mathbf{b}$ is the column vector of constant terms.

Theorem 2.9 **Matrix Invertibility**

Let $\mathbf{A}$ *be an* $n \times n$ *matrix and let* $\mathbf{I}$ *be the identity of order* n. *The following statements are equivalent:*

(*a*) $\mathbf{A}$ is invertible.
(*b*) There is a matrix $\mathbf{X}$ such that $\mathbf{XA} = \mathbf{I}$ ($\mathbf{A}$ has a left-inverse).
(*c*) The equation $\mathbf{Ax} = \mathbf{b}$ has a unique solution $\mathbf{x}$ for every $\mathbf{b}$.
(*d*) The equation $\mathbf{Ax} = \mathbf{0}$ has only the zero solution $\mathbf{x} = \mathbf{0}$.
(*e*) $\operatorname{rank} \mathbf{A} = n$.
(*f*) The reduced echelon form for $\mathbf{A}$ is $\mathbf{I}$.
(*g*) $\mathbf{A}$ is a product of elementary matrices.
(*h*) There is a matrix $\mathbf{X}$ such that $\mathbf{AX} = \mathbf{I}$ ($\mathbf{A}$ has a right-inverse).

Proof We prove two chains of implications

First chain: $(a) \Rightarrow (b) \Rightarrow (c) \Rightarrow (d) \Rightarrow (e) \Rightarrow (f) \Rightarrow (g) \Rightarrow (a)$

Second chain: $(a) \Rightarrow (h) \Rightarrow (a)$

First chain

(a) $\Rightarrow$ (b)

$\mathbf{A}$ has a two-sided inverse $\mathbf{X}$, and so in particular, $\mathbf{XA} = \mathbf{I}$.

(b) $\Rightarrow$ (c)

Suppose there exists $\mathbf{X}$ with $\mathbf{XA} = \mathbf{I}$. Using associativity, we have

$$\mathbf{Ax} = \mathbf{b} \;\Rightarrow\; \mathbf{X}(\mathbf{Ax}) = \mathbf{Xb} \;\Rightarrow\; (\mathbf{XA})\mathbf{x} = \mathbf{Xb} \quad \mathbf{Ix} = \mathbf{Xb} \;\Rightarrow\; \mathbf{x} = \mathbf{Xb},$$

showing that $\mathbf{x} = \mathbf{Xb}$ is a solution to $\mathbf{Ax} = \mathbf{b}$. To show $\mathbf{x}$ is unique, suppose that $\mathbf{y}$ satisfies $\mathbf{Ay} = \mathbf{b}$. Then $\mathbf{X}(\mathbf{Ax}) = \mathbf{Xb} = \mathbf{X}(\mathbf{Ay})$ implies $(\mathbf{XA})\mathbf{x} = (\mathbf{XA})\mathbf{y}$ implies $\mathbf{Ix} = \mathbf{Iy}$. Hence, $\mathbf{x} = \mathbf{y}$.

(c) $\Rightarrow$ (d)

Take $\mathbf{b} = \mathbf{0}$ in (c).

$$\text{(d)} \Rightarrow \text{(e)}$$

If rank $\mathbf{A} < n$, then the reduced form $\mathbf{A}^*$ has at least one zero row and so $\mathbf{Ax} = \mathbf{0}$ has infinitely many solutions. Now take the contrapositive (Appendix B: Toolbox) of the latter implication.

$$\text{(e)} \Rightarrow \text{(f)}$$

The matrix $\mathbf{A}$ has n pivots and n pivot columns. Hence, the reduced form $\mathbf{A}^* = \mathbf{I}$.

$$\text{(f)} \Rightarrow \text{(g)}$$

If $\mathbf{A}^* = \mathbf{I}$, then there is a sequence of k EROs that reduce $\mathbf{A}$ to $\mathbf{I}$. The corresponding elementary matrices have the property that

$$\mathbf{E}_k \cdots \mathbf{E}_1 \mathbf{A} = \mathbf{I} \quad \Rightarrow \quad \mathbf{A} = \mathbf{E}_1^{-1} \cdots \mathbf{E}_k^{-1}$$

and hence (g) because each elementary matrix is invertible and its inverse is elementary.

$$\text{(g)} \Rightarrow \text{(a)}$$

If $\mathbf{A} = \mathbf{E}_1 \cdots \mathbf{E}_k$ for elementary matrices $\mathbf{E}_1, \ldots, \mathbf{E}_k$, then $\mathbf{X} = \mathbf{E}_k^{-1} \cdots \mathbf{E}_1^{-1}$ has the property $\mathbf{XA} = \mathbf{AX} = \mathbf{I}$.

Second chain

$$\text{(a)} \Rightarrow \text{(h)}$$

$\mathbf{A}$ has a two-sided inverse and so $\mathbf{AX} = \mathbf{I}$ for some matrix $\mathbf{X}$.

$$\text{(h)} \Rightarrow \text{(a)}$$

Suppose there is a matrix $\mathbf{X}$ such that $\mathbf{AX} = \mathbf{I}$. Taking transposes, $(\mathbf{AX})^\mathsf{T} = \mathbf{I}^\mathsf{T}$ implies $\mathbf{X}^\mathsf{T}\mathbf{A}^\mathsf{T} = \mathbf{I}$ and so $\mathbf{X}^\mathsf{T}$ is a left-inverse for $\mathbf{A}^\mathsf{T}$. But (b) $\Rightarrow$ (a) and so $\mathbf{A}^\mathsf{T}$ is invertible. Property (**P4**) on page 85 shows that $\mathbf{A}$ is invertible.

▬

Note that any two conditions in Theorem 2.9, say (c) and (g), are equivalent; that is, (c) $\Rightarrow$ (g) and (g) $\Rightarrow$ (c), due to the two chains of implications:

$$\text{(c)} \Rightarrow \text{(d)} \Rightarrow \text{(e)} \Rightarrow \text{(f)} \Rightarrow \text{(g)} \quad \text{and} \quad \text{(g)} \Rightarrow \text{(h)} \Rightarrow \text{(a)} \Rightarrow \text{(b)} \Rightarrow \text{(c)}$$

Note also that the contrapositive of the implication (d) $\Rightarrow$ (a) has the logical form

$$\text{not (a)} \Rightarrow \text{not (d)},$$

which translates into the following statement: If $\mathbf{A}$ is singular, then the homogeneous linear system $\mathbf{Ax} = \mathbf{0}$ has infinitely many solutions.

Solving Square Systems

If $\mathbf{A}$ is square and invertible, the linear system defined by $\mathbf{Ax} = \mathbf{b}$ is solved by premultiplying both sides of the equation by $\mathbf{A}^{-1}$, as follows:

$$\mathbf{Ax} = \mathbf{b} \quad \Rightarrow \quad (\mathbf{A}^{-1}\mathbf{A})\mathbf{x} = \mathbf{A}^{-1}\mathbf{b} \quad \Rightarrow \quad \mathbf{x} = \mathbf{A}^{-1}\mathbf{b}.$$

EXAMPLE 6

Solving Square Linear Systems

Consider solving the 2×2 linear system (S).

$$(S)\ \begin{cases} 5x_1 - 3x_2 = \ \ 6 \\ -x_1 + \ x_2 = -2 \end{cases} \qquad \mathbf{A} = \begin{bmatrix} 5 & -3 \\ -1 & 1 \end{bmatrix}, \quad \mathbf{x} = \begin{bmatrix} x_1 \\ x_2 \end{bmatrix},$$

$$\mathbf{b} = \begin{bmatrix} 3 \\ -1 \end{bmatrix}$$

The determinant $\det(\mathbf{A}) = (5)(1) - (-3)(-1) = 2 \neq 0$ and so $\mathbf{A}$ is invertible. Using the formula (2.27) for $\mathbf{A}^{-1}$, we have

$$\mathbf{A}^{-1} = \begin{bmatrix} \frac{1}{2} & \frac{3}{2} \\ \frac{1}{2} & \frac{5}{2} \end{bmatrix} \quad \text{and} \quad \mathbf{x} = \begin{bmatrix} x_1 \\ x_2 \end{bmatrix} = \mathbf{A}^{-1}\mathbf{b} = \begin{bmatrix} \frac{1}{2} & \frac{3}{2} \\ \frac{1}{2} & \frac{5}{2} \end{bmatrix} \begin{bmatrix} 6 \\ -2 \end{bmatrix} = \begin{bmatrix} 0 \\ -2 \end{bmatrix},$$

giving the unique solution $(x_1, x_2) = (0, -2)$. Geometrically, the pair of lines that represent the equations in (S) intersect in the single point $(0, -2)$ in $\mathbb{R}^2$. ▩

Computational Note

Consider the cost of solving a large-scale square linear system that has a unique solution. Computation of $\mathbf{A}^{-1}$ requires about n^3 flops (forward and backward reduction on $\mathbf{A}$) and multiplication of $\mathbf{A}^{-1}$ by $\mathbf{b}$ requires a further n^2 flops. This count does not compare favorably with the $2n^3/3$ flops required for forward reduction on $[\mathbf{A}|\mathbf{b}]$ followed by back-substitution. Thus Gauss elimination is cost-effective.

2.2.5 General Inverses

The idea of a multiplicative inverse can be extended to nonsquare matrices. Here is an insight into the theory. Consider a general $m \times n$ matrix $\mathbf{A}$ with $m \neq n$ and an $n \times m$ matrix $\mathbf{X}$. Then $\mathbf{X}$ is a right-inverse for $\mathbf{A}$ if $\mathbf{AX} = \mathbf{I}_m$ and $\mathbf{X}$ is a left-inverse for $\mathbf{A}$ if $\mathbf{XA} = \mathbf{I}_n$.

Caution! A matrix may have a right-inverse but no left-inverse, and vice versa. Right- and left-inverses may not be unique.

A simple way to determine whether or not a right- or a left-inverse exists is by solving a system of linear equations.

EXAMPLE 7

Left-Inverse

Consider the following unknown 2×3 matrix $\mathbf{X}$ and the given 3×2 matrix $\mathbf{A}$.

$$\mathbf{X} = \begin{bmatrix} x_1 & x_2 & x_3 \\ x_4 & x_5 & x_6 \end{bmatrix}, \qquad \mathbf{A} = \begin{bmatrix} 1 & 0 \\ -1 & 2 \\ 3 & 1 \end{bmatrix}$$

Then $\mathbf{X}$ and $\mathbf{A}$ are conformable for multiplication and we have

$$\mathbf{XA} = \begin{bmatrix} x_1 - x_2 + 3x_3 & 2x_2 + x_3 \\ x_4 - x_5 + 3x_6 & 2x_5 + x_6 \end{bmatrix} = \begin{bmatrix} 1 & 0 \\ 0 & 1 \end{bmatrix} = \mathbf{I}_2$$

Then $\mathbf{X}$ is a left-inverse for $\mathbf{A}$ if and only if the 4×6 linear system

$$(S) \quad \begin{cases} x_1 - x_2 + 3x_3 = 1 \\ 2x_2 + x_3 = 0 \\ x_4 - x_5 + 3x_6 = 0 \\ 2x_5 + x_6 = 1 \end{cases} \tag{2.33}$$

is consistent. The coefficient matrix $\mathbf{B}$ for (S) is a block matrix in which the block pattern is easily recognized:

$$\mathbf{B} = \begin{bmatrix} \mathbf{A}^{\mathsf{T}} & \mathbf{O}_{2\times 3} \\ \mathbf{O}_{2\times 3} & \mathbf{A}^{\mathsf{T}} \end{bmatrix}.$$

The reduced form of the augmented matrix for (S) is

$$\left[\begin{array}{cccccc|c} 1 & 0 & 3.5 & 0 & 0 & 0 & 1 \\ 0 & 1 & 0.5 & 0 & 0 & 0 & 0 \\ 0 & 0 & 0 & 1 & 0 & 3.5 & 0.5 \\ 0 & 0 & 0 & 0 & 1 & 0.5 & 0.5 \end{array}\right]$$

and there are infinitely many solutions given by

$$(x_1, x_2, x_3, x_4, x_5, x_6) = (1 - 3.5s, -0.5s, s, 0.5 - 3.5t, 0.5 - 0.5t, t),$$

where $x_3 = s$ and $x_6 = t$ are independent real parameters. For a particular left-inverse $\mathbf{X}$ with integer entries, set $s = -2$ and $t = 3$, to obtain

$$\begin{array}{lll} x_1 = 8 & x_2 = 1 & x_3 = -2 \\ x_4 = -10 & x_5 = -1 & x_6 = 3 \end{array} \quad \text{giving} \quad \mathbf{X} = \begin{bmatrix} 8 & 1 & -2 \\ -10 & -1 & 3 \end{bmatrix}.$$

This example is continued in Exercises 2.2. ■

The existence of inverses in general is summed up in the next result, which is stated without proof (see [2]).

Theorem 2.10

Existence Criteria for Right- and Left-Inverses

Let $\mathbf{A}$ *be an* $m \times n$ *matrix such that* rank $\mathbf{A} = r$.

(a) $\mathbf{A}$ has a right-inverse if and only if $r = m$ and $m \le n$.
(b) $\mathbf{A}$ has a left-inverse if and only if $r = n$ and $n \le m$.

REFERENCES

1. Mattingly, R. Bruce, "Even Order Magic Squares are Singular," *Mathematics Magazine*, November 2000.
2. Noble, B. and Daniel, J. W., *Applied Linear Algebra*, 3rd ed., Prentice Hall, Englewood Cliffs, NJ 1988.

EXERCISES 2.2

Exercises 1–8. Reduce the matrix $[\mathbf{A}|\mathbf{I}]$ *in order to determine if* $\mathbf{A}$ *is invertible. Compute* $\mathbf{A}^{-1}$ *when* $\mathbf{A}$ *is invertible and check that* $\mathbf{A}\mathbf{A}^{-1} = \mathbf{I}$. *Verify that the* $(1, 1)$*-entry in the product* $\mathbf{A}^{-1}\mathbf{A}$ *is 1.*

1. $\begin{bmatrix} 2 & 1 \\ 5 & 3 \end{bmatrix}$ **2.** $\begin{bmatrix} 2.5 & -3 \\ -7.5 & 9 \end{bmatrix}$ **3.** $\begin{bmatrix} a & b \\ b & a \end{bmatrix}$

4. $\begin{bmatrix} 1 & 2 & 3 \\ 1 & 3 & 5 \\ 1 & 3 & 4 \end{bmatrix}$ **5.** $\begin{bmatrix} 1 & -2 & 1 \\ -2 & 1 & 3 \\ -1 & -4 & 9 \end{bmatrix}$ **6.** $\begin{bmatrix} 1 & 1 & 0 \\ 0 & 1 & 1 \\ 0 & 0 & 1 \end{bmatrix}$

7. $\begin{bmatrix} 2 & 0 & 0 \\ 0 & 4 & 0 \\ 0 & 0 & -2 \end{bmatrix}$ **8.** $\begin{bmatrix} 0 & 1 & 0 \\ 1 & 0 & 0 \\ 0 & 0 & 1 \end{bmatrix}$

Exercises 9–10. Find the value(s) of k for which the given matrix is singular.

9. $\begin{bmatrix} k^2 & 2k \\ 8 & k \end{bmatrix}$ **10.** $\begin{bmatrix} 1 & k & 1 \\ 0 & 1 & 2 \\ 1 & 1 & 3 \end{bmatrix}$

Exercises 11–12. Let $\mathbf{A}$ *and* $\mathbf{B}$ *be invertible* $n \times n$ *matrices such that* $\mathbf{AXB} = \mathbf{C}$, *for some matrix* $\mathbf{X}$. *Show that* $\mathbf{X} = \mathbf{A}^{-1}\mathbf{C}\mathbf{B}^{-1}$. *Compute the first row of* $\mathbf{X}$ *for the given matrices.*

11.

$$\mathbf{A} = \begin{bmatrix} 1 & 1 & -1 \\ 2 & 0 & 3 \\ 4 & 3 & -2 \end{bmatrix}, \quad \mathbf{B} = \begin{bmatrix} -9 & -1 & 3 \\ 16 & 2 & -5 \\ 6 & 1 & -2 \end{bmatrix},$$

$\mathbf{C} = \mathbf{I}_3$

12.

$$\mathbf{A} = \begin{bmatrix} 1 & 0 & 0 \\ 3 & 1 & -2 \\ 1 & 0 & 1 \end{bmatrix}, \quad \mathbf{B} = \begin{bmatrix} 1 & 1 & 3 \\ 0 & 2 & -2 \\ 0 & 0 & 1 \end{bmatrix},$$

$$\mathbf{C} = \begin{bmatrix} 1 & 1 & 1 \\ 0 & 1 & 1 \\ 0 & 0 & 1 \end{bmatrix}$$

13. Find examples of invertible 2×2 matrices $\mathbf{A}$ and $\mathbf{B}$ such that $\mathbf{A} + \mathbf{B}$ is (a) invertible, (b) singular.

14. Your goal is to show that the 3×3 matrix

$$\mathbf{A} = \begin{bmatrix} 1 & 2 & 3 \\ 1 & 1 & 3 \\ 1 & 1 & 3 \end{bmatrix}$$

is singular. Let $\mathbf{X} = [\mathbf{x}_1\ \mathbf{x}_2\ \mathbf{x}_3]$ be a matrix such that $\mathbf{AX} = \mathbf{I}_3$. Reduce $[\mathbf{A}|\mathbf{I}]$ and deduce that $\operatorname{rank}\mathbf{A} < 3$. Show that the system $\mathbf{A}\mathbf{x}_1 = \mathbf{e}_1$ is consistent and has infinitely many solutions but that the systems $\mathbf{A}\mathbf{x}_2 = \mathbf{e}_2$ and $\mathbf{A}\mathbf{x}_3 = \mathbf{e}_3$ are inconsistent. Which columns of $\mathbf{X}$ cannot be computed?

Exercises 15–16. Let $\mathbf{A}$ *and* $\mathbf{B}$ *be an* $n \times n$ *matrices.*

15. Solve the equation $\mathbf{A}^{\mathsf{T}}\mathbf{X}\mathbf{A} = \mathbf{I}_n$ for $\mathbf{X}$. Compute the entries on the main diagonal of $\mathbf{X} = [x_{ij}]$ when

$$\mathbf{A} = \begin{bmatrix} 1 & 0 & 1 \\ 0 & 1 & 3 \\ 2 & -1 & 0 \end{bmatrix}.$$

16. Solve the equation $\mathbf{AX} = \mathbf{B}$ for $\mathbf{X}$. Compute $\mathbf{X}$ when

$$\mathbf{A} = \begin{bmatrix} 1 & 2 & 3 \\ 1 & -1 & -1 \\ 1 & 0 & 1 \end{bmatrix} \text{ and } \mathbf{B} = \begin{bmatrix} 1 & 2 & 3 \\ 4 & 6 & 7 \\ 5 & 8 & 9 \end{bmatrix}.$$ If $\mathbf{BY} = \mathbf{A}$, compute the first column of $\mathbf{Y}$.

17. Prove property (**P4**) in Section 2.2.3.

18. Let $\mathbf{A} = \dfrac{1}{3}\begin{bmatrix} -1 & 2 & -2 \\ -2 & 1 & 2 \\ 2 & 2 & 1 \end{bmatrix}$. Compute $\mathbf{A}^{-1}$ using property (**P3**) in Section 2.2.3 and verify that $\mathbf{A}^{\mathsf{T}} = \mathbf{A}^{-1}$.

19. Let $\mathbf{D} = [d_{ij}]$ be an $n \times n$ diagonal matrix. What condition is sufficient for $\mathbf{D}$ to be invertible? Describe $\mathbf{D}^{-1}$ in general.

20. If $\mathbf{A}$ is invertible and $\mathbf{A}\mathbf{A}^{\mathsf{T}} = \mathbf{A}^{\mathsf{T}}\mathbf{A}$, show that $(\mathbf{A}^{\mathsf{T}}\mathbf{A}^{-1})^{\mathsf{T}} = (\mathbf{A}^{\mathsf{T}}\mathbf{A}^{-1})^{-1}$.

21. The $n \times n$ matrix $\mathbf{B}$ is *similar* (Section 7.3) to the $n \times n$ matrix $\mathbf{C}$ if $\mathbf{B} = \mathbf{P}\mathbf{C}\mathbf{P}^{-1}$, for some invertible $n \times n$ matrix $\mathbf{P}$. Use mathematical induction on k to prove that $\mathbf{B}^k = \mathbf{P}\mathbf{C}^k\mathbf{P}^{-1}$, for all positive integers k. What can you deduce from this exercise?

22. Prove that $(\mathbf{I}-\mathbf{B}\mathbf{A}^{-1})(\mathbf{I}+\mathbf{B}(\mathbf{A}-\mathbf{B})^{-1})=\mathbf{I}$, where all the indicated inverses exist.

23. Let $\mathbf{P}$ and $\mathbf{Q}$ be invertible matrices such that $\mathbf{A}=\mathbf{PLP}^{-1}$ and $\mathbf{B}=\mathbf{QMQ}^{-1}$. If a matrix $\mathbf{X}$ exists such that $\mathbf{AX}=\mathbf{XB}$, then there is a matrix $\mathbf{Y}$ such that $\mathbf{LY}=\mathbf{YM}$. *Hint*: Show that $\mathbf{Y}=\mathbf{PXQ}^{-1}$.

24. Find the inverse of the rotation matrix

$$\begin{bmatrix}\cos\theta & -\sin\theta\\ \sin\theta & \cos\theta\end{bmatrix}.$$

25. If $\mathbf{A}$ is an invertible matrix and $\mathbf{B}$ is (row) equivalent to $\mathbf{A}$, is $\mathbf{B}$ necessarily invertible?

Exercises 26–27. Refer to Exercises 2.1 for definitions.

26. Let $\mathbf{A}$ be a nonzero square nilpotent matrix with index of nilpotency k. Show that $\mathbf{A}$ is singular. *Hint*: Consider the equation $\mathbf{A}^k=\mathbf{O}$ and assume that $\mathbf{A}$ is invertible.

27. Describe the invertible idempotent matrices.

28. Consider the quadratic matrix equation $\mathbf{A}^2+b\mathbf{A}+c\mathbf{I}=\mathbf{O}$, where $\mathbf{A}$ is an $n\times n$ invertible matrix and b, c are real numbers. Find a formula for $\mathbf{A}^{-1}$ in terms of $\mathbf{A}$. To test your result, show that the matrix $\mathbf{A}$ below on the left satisfies the quadratic equation on the right. Find $\mathbf{A}^{-1}$ using your formula and cross-check using the formula for inverse of 2×2 matrices.

$$\mathbf{A}=\begin{bmatrix}1 & 2\\ 2 & 5\end{bmatrix} \qquad \mathbf{A}^2-6\mathbf{A}+\mathbf{I}=\mathbf{O}$$

29. If $\mathbf{A}$ is an $m\times n$ matrix of rank n, show that every solution of $(\mathbf{AB})\mathbf{x}=\mathbf{0}$ is a solution of $\mathbf{Bx}=\mathbf{0}$, and conversely. Hence, show that $\operatorname{rank}(\mathbf{AB})=\operatorname{rank}\mathbf{B}$ in this case.

30. Let $\mathbf{A}$ be an $n\times n$ matrix. If $\mathbf{Ax}=\mathbf{0}$ has only the zero solution $\mathbf{x}=\mathbf{0}$, prove that $\mathbf{A}^k\mathbf{x}=\mathbf{0}$ has only the zero solution for all positive integers k.

Exercises 31–36. Solve each square linear system by finding the inverse of the coefficient matrix.

31. $\begin{cases}6x_1+2x_2=4\\ -x_1+3x_2=-4\end{cases}$ **32.** $\begin{cases}5x_1+4x_2=9\\ 6x_1+4x_2=10\end{cases}$

33. $\begin{cases}x_1+3x_2+3x_3=12\\ x_1+4x_2+3x_3=-10\\ x_1+3x_2+4x_3=16\end{cases}$

34. $\begin{cases}x_1+x_2-x_3=5\\ 2x_1+3x_3=-7\\ 4x_1+3x_2-2x_3=4\end{cases}$

35. $\begin{cases}x_1+2x_2-x_3-x_4=1\\ x_2-x_3+x_4=2\\ x_1-x_2+x_3+x_4=-1\\ x_1+x_3-x_4=0\end{cases}$

36. $\begin{cases}x_1+2x_2-x_3-x_4=1\\ x_2+x_4=0\\ x_2+x_3+x_4=0\\ x_1+x_2-x_4=0\end{cases}$

37. Refer to properties (**M1**) and (**M2**) in Section 2.1. Let $\mathbf{A}$, $\mathbf{B}$, $\mathbf{C}$, $\mathbf{D}$ be $n\times n$ matrices. Prove each statement.

(a) If $\mathbf{A}$ is invertible and $\mathbf{AB}=\mathbf{O}$, then $\mathbf{B}=\mathbf{O}$.

(b) If $\mathbf{A}$ and $\mathbf{D}$ are invertible, then either of the equations $\mathbf{AB}=\mathbf{AC}$ and $\mathbf{BD}=\mathbf{CD}$ implies $\mathbf{B}=\mathbf{C}$.

Exercises 38–39 are related. Let $\mathbf{A}$ and $\mathbf{X}$ be $n\times n$ matrices such that $\mathbf{AX}=\mathbf{O}$.

38. If $\mathbf{A}$ is singular, show that $\mathbf{X}$ can be chosen so that each column has at least one nonzero entry. *Hint*: Consider $\mathbf{AX}=\mathbf{O}$ as a multi-system.

39. Find such a matrix $\mathbf{X}$ when $\mathbf{A}=\begin{bmatrix}1 & 2 & 3\\ 4 & 5 & 6\\ 7 & 8 & 9\end{bmatrix}$.

Exercises 40–41 are related.

40. Let $\mathbf{A}$ be an $n\times n$ matrix such that the homogeneous system defined by $\mathbf{Ax}=\mathbf{0}$ is satisfied by every $n\times 1$ column $\mathbf{x}$. Show that $\mathbf{A}=\mathbf{0}$.

41. If $\mathbf{Ax}=\mathbf{Bx}$ for every $n\times 1$ column matrix $\mathbf{x}$, then $\mathbf{A}=\mathbf{B}$.

Exercises 42–44 are related. Define an $n\times n$ matrix $\mathbf{A}=[a_{ij}]$ by the formula

$$a_{ij}=\begin{cases}(-1)^{i+j}, & \text{if } i\le j\\ 0, & \text{if } i>j\end{cases}. \tag{2.34}$$

42. Write down $\mathbf{A}$ when $n=2,3,4$ and find $\mathbf{A}^{-1}$ in each case.

43. Use Exercise 42 to conjecture (guess) the general form of $\mathbf{A}^{-1}$ for any n.

44. Use $\mathbf{A}^{-1}$ to solve the linear system $\mathbf{Ax}=\mathbf{b}$, where $\mathbf{A}$ is defined by (2.34), $\mathbf{x}=[x_i]$, $\mathbf{b}=[b_i]$.

45. If the entries in $\mathbf{A} = \begin{bmatrix} a & b \\ c & d \end{bmatrix}$ are all nonzero, show that $\mathbf{A}$ is singular if and only if $\dfrac{a}{c} = \dfrac{b}{d}$.

Exercises 46–53. State whether each matrix **A** *is elementary or not. For each elementary matrix, give the ERO that formed it and show that* $\mathbf{A}^{\mathsf{T}}$ *is also elementary.*

46. $\begin{bmatrix} 1 & 0 \\ -5 & 1 \end{bmatrix}$ **47.** $\begin{bmatrix} 0 & 0 \\ 0 & 0 \end{bmatrix}$ **48.** $\begin{bmatrix} 2 & 0 \\ 0 & 2 \end{bmatrix}$

49. $\begin{bmatrix} 0 & 1 \\ 1 & 0 \end{bmatrix}$ **50.** $\begin{bmatrix} 1 & 0 & 0 \\ 0 & 0 & 1 \end{bmatrix}$ **51.** $\begin{bmatrix} 1 & 0 & 0 \\ 0 & 0 & 1 \\ 0 & 1 & 0 \end{bmatrix}$

52. $\begin{bmatrix} 1 & 0 & 0 \\ 0 & 1 & 0 \\ 0 & 3.4 & 1 \end{bmatrix}$ **53.** $\begin{bmatrix} 1 & 0 & 1 \\ 0 & 0 & 0 \\ 0 & 1 & 0 \end{bmatrix}$

54. Consider the matrix

$$\mathbf{A} = \begin{bmatrix} 1 & 2 & 0 \\ 3 & 4 & 1 \\ 5 & 6 & 2 \end{bmatrix}.$$

Perform each ERO on **A** to obtain a matrix **B**. Find an elementary matrix **E** corresponding to the ERO and check that **EA** = **B**.

(a) $\mathbf{r}_1 \leftrightarrow \mathbf{r}_3$, (b) $-2\mathbf{r}_2 \to \mathbf{r}_2$, (c) $\mathbf{r}_2 - 3\mathbf{r}_1 \to \mathbf{r}_2$

55. A sequence of three EROs is used to change **A** into **B**, as follows

$$\mathbf{A} = \begin{bmatrix} 0 & 2 & -2 \\ 1 & 3 & -1 \end{bmatrix} \sim \begin{bmatrix} 1 & 3 & -1 \\ 0 & 2 & -2 \end{bmatrix} \sim \begin{bmatrix} 1 & -1 & 3 \\ 0 & 2 & -2 \end{bmatrix} \sim \begin{bmatrix} 1 & -1 & 3 \\ 0 & 1 & -1 \end{bmatrix} = \mathbf{B}$$

(a) Find elementary matrices $\mathbf{E}_1$, $\mathbf{E}_2$, $\mathbf{E}_3$ that correspond exactly to this sequence of EROs and show that $\mathbf{E}_3\mathbf{E}_2\mathbf{E}_1\mathbf{A} = \mathbf{B}$.

(b) Find the inverse of each elementary matrix.

(c) Find a matrix **F** such that **FA** = **B** and check your answer by multiplication.

Exercises 56–64. For the given matrix **A**, *find a matrix* **F** *such that* $\mathbf{FA} = \mathbf{A}^*$, *where* $\mathbf{A}^*$ *is the reduced form for* **A**. *Compute* **FA** *to verify your answer.*

56. $\begin{bmatrix} 4 & 8 \\ -3 & -6 \end{bmatrix}$ **57.** $\begin{bmatrix} 2 & 4 \\ 0 & 2 \end{bmatrix}$

58. $\begin{bmatrix} 4 & -8 \\ 0 & 1 \\ 3 & 0 \end{bmatrix}$ **59.** $\begin{bmatrix} 1 & 1 \\ -1 & 2 \\ 5 & 1 \end{bmatrix}$

60. $\begin{bmatrix} 1 & 2 & 1 & -1 \\ -1 & -1 & 0 & 2 \\ 0 & 1 & 1 & 3 \end{bmatrix}$ **61.** $\begin{bmatrix} 0 & 1 & -3 & 1 \\ 1 & 0 & -1 & 2 \\ 1 & 2 & 0 & -1 \end{bmatrix}$

62. $\begin{bmatrix} 1 & 1 & 0 \\ 1 & 2 & 1 \\ 2 & 3 & 1 \end{bmatrix}$ **63.** $\begin{bmatrix} 1 & 0 & 3 \\ 0 & -2 & 1 \\ 5 & 1 & 15 \end{bmatrix}$

64. $\begin{bmatrix} 1 & 1 & 1 & 0 \\ -1 & 0 & 2 & 0 \\ 0 & 1 & 3 & 2 \\ 1 & -1 & 4 & 2 \end{bmatrix}$

Exercises 65–68. Each matrix **E** *shown is elementary. In each case find an elementary matrix* **F** *such that* $\mathbf{FE} = \mathbf{I}_n$. *Describe how* **F** *is formed in each case.*

65. $\begin{bmatrix} 1 & 2 \\ 0 & 1 \end{bmatrix}$ **66.** $\begin{bmatrix} -5 & 0 \\ 0 & 1 \end{bmatrix}$

67. $\begin{bmatrix} 1 & 0 & 0 \\ 0 & 1 & 0 \\ 0 & 0 & 3 \end{bmatrix}$ **68.** $\begin{bmatrix} 1 & 0 & -1 \\ 0 & 1 & 0 \\ 0 & 0 & 1 \end{bmatrix}$

69. Prove that every elementary matrix is invertible and its inverse is an elementary matrix of the same type. There are three types of EROs to consider.

70. Prove that the transpose of an elementary matrix is an elementary matrix of the same type. There are three types of EROs to consider. Show that two types of elementary matrix are symmetric.

71. Let **A** and **B** be square matrices of order n. Prove that if any two matrices from the list **A**, **B**, **AB** are invertible, then the third matrix is necessarily invertible.

72. Show that a square matrix having a row (column) of zeros is necessarily singular.

73. If $\mathbf{A}$, $\mathbf{B}$, $\mathbf{A}+\mathbf{B}$ are all invertible matrices of order n, prove that $\mathbf{A}^{-1}+\mathbf{B}^{-1}$ is invertible and that

$$(\mathbf{A}^{-1}+\mathbf{B}^{-1})^{-1} = \mathbf{A}(\mathbf{A}+\mathbf{B})^{-1}\mathbf{B} = \mathbf{B}(\mathbf{A}+\mathbf{B})^{-1}\mathbf{A}.$$

74. Let $\mathbf{A}$ and $\mathbf{B}$ be invertible matrices of order n such that $\mathbf{A}$ commutes with $\mathbf{B}$. Show that $\mathbf{A}$ commutes with $\mathbf{B}^{-1}$ and $\mathbf{B}$ commutes with $\mathbf{A}^{-1}$.

Exercises 75–76. Prove each statement.

75. If a symmetric $n \times n$ matrix is invertible, then its inverse is also symmetric.

76. The inverse of an invertible upper (lower) triangular matrix is upper (lower) triangular.

Exercises 77–78. Refer to Example 7 in Section 2.2.

77. Check that $\mathbf{XA} = \mathbf{I}_2$. Explain why the equation $\mathbf{XA} = \mathbf{AX}$ cannot be formed. Show that $\mathbf{X}$ is not a right-inverse for $\mathbf{A}$. Explain why $\mathbf{A}$ has no right-inverse.

78. Explain how Theorem 2.10 applies to Example 7.

79. Find all right-inverses for $\mathbf{A} = \begin{bmatrix} 1 & 0 & 1 \\ 2 & 1 & 1 \end{bmatrix}$ and explain how Theorem 2.10 applies in this case.

80. Solve a linear system to show that the matrix $\mathbf{A} = \begin{bmatrix} 1 & 2 \\ 2 & 4 \\ 3 & 6 \end{bmatrix}$ has no left-inverse. How does Theorem 2.10 apply to this case? Does $\mathbf{A}$ have a right-inverse?

81. Assuming all inverses exits, show that

$$\begin{bmatrix} \mathbf{A} & \mathbf{B} \\ \mathbf{O} & \mathbf{C} \end{bmatrix}^{-1} = \begin{bmatrix} \mathbf{A}^{-1} & -\mathbf{A}^{-1}\mathbf{B}\mathbf{C}^{-1} \\ \mathbf{O} & \mathbf{C}^{-1} \end{bmatrix}$$

and hence find the inverse of $\begin{bmatrix} 1 & 1 & 1 & 2 \\ 1 & 2 & 1 & 3 \\ 0 & 0 & 0 & -1 \\ 0 & 0 & -1 & 0 \end{bmatrix}$.

USING MATLAB

Consult online help for the command **inv**, and related commands.

82. *Project.* Write an M-file that uses simulation to estimate the probability p that a random 2×2 matrix with digit entries $(0, 1, 2, \ldots, 9)$ will be invertible. Extend the scope of the project to higher-order square matrices.

83. *Project.* The goal is to explore how certain matrices exacerbate the potential for round-off error in machine calculation. Write a (function) M-file **hilbert** that calls a positive integer n and returns the $n \times n$ Hilbert matrix $\mathbf{H} = [h_{ij}]$, where $h_{ij} = 1/(i+j-1)$. Compare your output with hand calculation of $\mathbf{H}$ for the cases $n = 2, 3, 4, 5$ and then compare your output with output from **hilb(n)**, which is a built-in MATLAB command. Comment on how the products **hilb(n)*inv(hilb(n))** differ from **eye(n)** as n increases in value. Determine if accuracy is improved by using the command **hilb(n)*invhilb(n)**, where **invhilb(n)** is the special MATLAB command for finding the inverse of $\mathbf{H}$. Now consider the random matrix **rand(n)** and compare the product **rand(n)*inv(rand(n))** with **eye(n)**. Report on your findings.

84. Use MATLAB to cross-check hand calculations in this exercise set.

2.3 LU-Factorization

2.3.1 Triangular and Permutation Matrices

We will start out by defining some special matrices that are important in their own right and that are required later in this section.

Definition 2.10 *Upper triangular, lower triangular*

An $m \times n$ matrix $\mathbf{A} = [a_{ij}]$ is called *upper triangular* if $a_{ij} = 0$ whenever $j < i$; that is, all entries below the main diagonal are zero. $\mathbf{A}$ is called *lower triangular* if $a_{ij} = 0$ whenever every $i < j$; that is, all entries above the main diagonal are zero.

EXAMPLE 1 **Triangular Matrices**

Consider the matrices

$$\mathbf{A} = \begin{bmatrix} * & * & * \\ 0 & * & * \\ 0 & 0 & * \end{bmatrix}, \quad \mathbf{B} = \begin{bmatrix} * & * & * & * \\ 0 & * & * & * \\ 0 & 0 & * & * \end{bmatrix}, \quad \mathbf{C} = \begin{bmatrix} * & 0 & 0 \\ * & * & 0 \\ * & * & * \\ * & * & * \end{bmatrix},$$

where each $*$ represents a number (which may be zero). Then **A** and **B** are upper triangular matrices and **C** is lower triangular. Note that an echelon form **U** is upper triangular and a diagonal matrix **D** is both upper and lower triangular. ▪

If **A** is $m \times n$ upper triangular, then the transpose $\mathbf{A}^{\mathsf{T}}$ is $n \times m$ lower triangular, and if **A** is $m \times n$ lower triangular, then $\mathbf{A}^{\mathsf{T}}$ is $n \times m$ upper triangular.

Theorem 2.11 **Products of Triangular Matrices**

Let **A** *and* **B** *be upper (lower) triangular square matrices of order n. Then* **AB** *is upper (lower) triangular.*

Proof Suppose $\mathbf{A} = [a_{ij}]$ and $\mathbf{B} = [b_{ij}]$ are upper triangular square matrices of order n. Let c_{ij} be the (i, j)-entry in **AB**. Then c_{ij} is computed using row i from **A** and column j from **B**. Suppose $j < i$. Then $a_{i1} = \cdots = a_{ij} = 0$ and $b_{(j+1)j} = \cdots = b_{nj} = 0$. Hence

$$\begin{aligned} c_{ij} &= a_{i1}b_{1j} + \cdots + a_{ij}b_{jj} + a_{i(j+1)}b_{(j+1)j} + \cdots a_{in}b_{nj} \\ &= 0\ b_{1j} + \cdots + 0\ b_{jj} + a_{i(j+1)}\ 0 + \cdots + a_{in}\ 0 = 0, \end{aligned}$$

proving $c_{ij} = 0$ whenever $j < i$, and so **AB** is upper triangular. If **A** and **B** are lower triangular, then $(\mathbf{AB})^{\mathsf{T}} = \mathbf{B}^{\mathsf{T}}\mathbf{A}^{\mathsf{T}}$ is upper triangular, because $\mathbf{A}^{\mathsf{T}}$ and $\mathbf{B}^{\mathsf{T}}$ are upper triangular. Hence **AB** is lower triangular. ▬

Definition 2.11 *Permutation matrix*

A square matrix **P** of order n is called a *permutation matrix* if **P** is obtained by rearranging the rows (columns) in the identity $\mathbf{I}_n$. We say that the rows (columns) of $\mathbf{I}_n$ are *permuted* to obtain **P**.

▪ ILLUSTRATION 2.4

Defining Permutation Matrices

Consider the identity matrix $\mathbf{I}_3$ in (2.35). The natural order of the rows (top to bottom) in $\mathbf{I}_3$ is denoted by the 3-tuple $(1, 2, 3)$. A *permutation* of the rows of $\mathbf{I}_3$ can be defined by reordering the integers $1, 2, 3$ to form a new 3-tuple, say $(2, 3, 1)$, which means that row 2 in $\mathbf{I}_3$ is row 1 of **P**, row 3 in $\mathbf{I}_3$ is row

2 of **P**, and row 1 in $\mathbf{I}_3$ is row 3 of **P**. Reordering the rows of $\mathbf{I}_3$ according to (2, 3, 1) results in a permutation matrix **P**.

$$\mathbf{I}_3 = \begin{bmatrix} 1 & 0 & 0 \\ 0 & 1 & 0 \\ 0 & 0 & 1 \end{bmatrix} \quad \begin{matrix} \text{permute rows} \\ (1,2,3) \;\rightarrow\; (2,3,1) \end{matrix} \quad \mathbf{P} = \begin{bmatrix} 0 & 1 & 0 \\ 0 & 0 & 1 \\ 1 & 0 & 0 \end{bmatrix}. \tag{2.35}$$

Note that if (1, 2, 3) denotes the natural order of the columns (left to right) in $\mathbf{I}_3$, then the permutation (3, 1, 2) of the columns results in the same permutation matrix **P** in (2.35). Observe that a permutation matrix **P** has the property that there is one (and only one) entry 1 in any row and in any column of **P**, and all other entries are 0. The ideas here extend to permutation matrices of any order. ■

Theorem 2.12 **Permutation Matrices**

Let **P** *be an* $n \times n$ *permutation matrix. Then* $\mathbf{P}^{\mathsf{T}}$ *is a permutation matrix. Furthermore,* **P** *is invertible and* $\mathbf{P}^{-1} = \mathbf{P}^{\mathsf{T}}$.

Proof See Exercises 2.3 for proofs and illustrations.

The Action of Permutation Matrices

Suppose **A** is an $m \times n$ matrix and let **P** and **Q** be permutation matrices. Premultiplication of **A** by **P** of order m permutes the rows of **A**. Postmultiplication of **A** by **Q** of order n permutes the columns of **A**.

EXAMPLE 2 **Permuting Rows or Columns in a Matrix**

Consider permuting the rows and columns of the 2×3 matrix **A**.

$$\underset{\mathbf{Q}}{\begin{bmatrix} 0 & 1 \\ 1 & 0 \end{bmatrix}} \underset{\mathbf{A}}{\begin{bmatrix} 1 & 2 & 3 \\ 4 & 5 & 6 \end{bmatrix}} = \underset{\mathbf{B}}{\begin{bmatrix} 4 & 5 & 6 \\ 1 & 2 & 3 \end{bmatrix}}, \quad \underset{\mathbf{A}}{\begin{bmatrix} 1 & 2 & 3 \\ 4 & 5 & 6 \end{bmatrix}} \underset{\mathbf{P}}{\begin{bmatrix} 0 & 1 & 0 \\ 0 & 0 & 1 \\ 1 & 0 & 0 \end{bmatrix}} = \underset{\mathbf{C}}{\begin{bmatrix} 3 & 1 & 2 \\ 6 & 4 & 5 \end{bmatrix}}$$

Note that the rows (columns) of **A** are permuted in exactly the same way that the rows (columns) of the corresponding identity matrix were permuted. ■

We are now ready to begin the theory of this section.

2.3.2 Factorization

Consider the following generic problem, which arises in a variety of applications. Let (S) be a large-scale square linear system represented by $\mathbf{Ax} = \mathbf{b}$, where **A** is invertible. Suppose that (S) must be solved repeatedly after a given interval of time, say every hour or every day. On each occasion, the coefficient matrix **A** is the same but the column vector **b** of constant terms is new, updated as a result of new data or information. Each time, Gauss elimination (forward elimination followed by back-substitution) is used to solve (S). The goal is to lower the cost (in time) of solving (S) by removing the obvious computational repetition—the EROs used on each occasion are the same. We will see how the process of forward elimination is equivalent to writing $\mathbf{A}$ as a product of special matrices. Using this *factorization* of $\mathbf{A}$, the system (S) can be rewritten as two triangular systems, each of which is easily solved by substitution. This method provides a significant saving in computation.

In what follows, the matrix $\mathbf{L}$ denotes a square lower triangular matrix and $\mathbf{U}$ denotes a square upper triangular matrix. If $\mathbf{L}$ or $\mathbf{U}$ has all unit entries $(= 1)$ on the main diagonal, we call it a *unit triangular matrix*.

EXAMPLE 3

Factorizations

We give two examples of factorization $\mathbf{A} = \mathbf{LU}$, both with $\mathbf{L}$ unit lower triangular.

$$\underset{\mathbf{A}}{\begin{bmatrix} 4 & 5 \\ 1 & 2 \end{bmatrix}} = \underset{\mathbf{L}}{\begin{bmatrix} 1 & 0 \\ 0.25 & 1 \end{bmatrix}} \underset{\mathbf{U}}{\begin{bmatrix} 4 & 5 \\ 0 & 0.75 \end{bmatrix}} \tag{2.36}$$

$$\underset{\mathbf{A}}{\begin{bmatrix} 2 & 6 & 5 \\ -1 & 4 & -2 \\ 1 & 2 & 3 \end{bmatrix}} = \underset{\mathbf{L}}{\begin{bmatrix} 1 & 0 & 0 \\ -\frac{1}{2} & 1 & 0 \\ \frac{1}{2} & -\frac{1}{7} & 1 \end{bmatrix}} \underset{\mathbf{U}}{\begin{bmatrix} 2 & 6 & 5 \\ 0 & 7 & \frac{1}{2} \\ 0 & 0 & \frac{4}{7} \end{bmatrix}} \tag{2.37}$$

▪

Forward Reduction, Replacements Only

Suppose $\mathbf{A}$ is an invertible $n \times n$ matrix. There are n pivot columns in $\mathbf{A}$ and so forward reduction, using replacements and interchanges, will reduce $\mathbf{A}$ to an echelon form $\mathbf{U}$ that has nonzero entries on its main diagonal. Suppose first that $\mathbf{A}$ can be reduced to $\mathbf{U}$ using *only* replacements. This is not possible for every $\mathbf{A}$. The case when interchanges are also required is handled later in the section.

Our notation for row replacement that appears in Section 1.1 is modified here to read

$$\mathbf{r}_i - m_{ij}\mathbf{r}_j \to \mathbf{r}_i \tag{2.38}$$

writing m_{ij} instead of just m in order to record the rows i and j. The multiplier m_{ij} is a nonzero number. We insist on the format in (2.38), with a minus sign before the multiplier.

EXAMPLE 4

Forward Reduction, Replacement and Multipliers

Consider reducing the $(2, 1)$-entry in the following matrix $\mathbf{A}$ to zero. Here $i = 2$ and $j = 1$ in (2.38) and we subtract $-\frac{3}{4}$ row 1 from row 2.

$$\mathbf{A} = [a_{ij}] = \begin{bmatrix} \textcircled{4} & 2 \\ -3 & 5 \end{bmatrix} \underset{\mathbf{r}_2 - (-\frac{3}{4})\mathbf{r}_1 \to \mathbf{r}_2}{\sim} \begin{bmatrix} 4 & 2 \\ 0 & 6.5 \end{bmatrix} \tag{2.39}$$

The multiplier for this replacement is $m_{21} = \dfrac{a_{21}}{a_{11}} = \dfrac{-3}{4} = -\dfrac{3}{4}$. ▪

▪ ILLUSTRATION 2.5

LU-Factorization, Replacements Only

Consider the matrix

$$\mathbf{A} = \begin{bmatrix} 2 & 4 & 2 \\ -3 & -5 & 5 \\ 4 & 7 & 2 \end{bmatrix}. \tag{2.40}$$

Three replacements (shown on the left side below) are required to reduce $\mathbf{A}$ to an echelon form $\mathbf{U}$. Each ERO is performed using premultiplication by a corresponding elementary matrix (Section 2.2), denoted by $\mathbf{E}$. Note that the *negative* of each multiplier m_{ij} appears in the (i, j)-position in each elementary matrix. Pivots are circled.

$$\mathbf{r}_2 - \left(-\tfrac{3}{2}\right)\mathbf{r}_1 \to \mathbf{r}_2 \qquad m_{21} = \tfrac{-3}{2} = -\tfrac{3}{2} \qquad \mathbf{E}_1 = \begin{bmatrix} 1 & 0 & 0 \\ \frac{3}{2} & 1 & 0 \\ 0 & 0 & 1 \end{bmatrix}, \quad \mathbf{E}_1\mathbf{A} = \begin{bmatrix} \textcircled{2} & 4 & 2 \\ 0 & 1 & 8 \\ 4 & 7 & 2 \end{bmatrix}$$

$$\mathbf{r}_3 - \left(\tfrac{4}{2}\right)\mathbf{r}_1 \to \mathbf{r}_3 \qquad m_{31} = \tfrac{4}{2} = 2 \qquad \mathbf{E}_2 = \begin{bmatrix} 1 & 0 & 0 \\ 0 & 1 & 0 \\ -2 & 0 & 1 \end{bmatrix}, \quad \mathbf{E}_2\mathbf{E}_1\mathbf{A} = \begin{bmatrix} \textcircled{2} & 4 & 2 \\ 0 & \textcircled{1} & 8 \\ 0 & -1 & -2 \end{bmatrix}$$

$$\mathbf{r}_3 - (-1)\mathbf{r}_2 \to \mathbf{r}_3 \qquad m_{32} = \tfrac{-1}{1} = -1 \qquad \mathbf{E}_3 = \begin{bmatrix} 1 & 0 & 0 \\ 0 & 1 & 0 \\ 0 & 1 & 1 \end{bmatrix}, \quad \mathbf{E}_3\mathbf{E}_2\mathbf{E}_1\mathbf{A} = \begin{bmatrix} \textcircled{2} & 4 & 2 \\ 0 & \textcircled{1} & 8 \\ 0 & 0 & \textcircled{6} \end{bmatrix} = \mathbf{U}$$

Premultiplying $\mathbf{E}_3\mathbf{E}_2\mathbf{E}_1\mathbf{A}$ by the inverse of each elementary matrix (note the order of factors) gives

$$\mathbf{E}_3\mathbf{E}_2\mathbf{E}_1\mathbf{A} = \mathbf{U} \qquad \Rightarrow \qquad \mathbf{A} = \mathbf{E}_1^{-1}\mathbf{E}_2^{-1}\mathbf{E}_3^{-1}\mathbf{U},$$

where

$$\mathbf{E}_1^{-1} = \begin{bmatrix} 1 & 0 & 0 \\ -\frac{3}{2} & 1 & 0 \\ 0 & 0 & 1 \end{bmatrix}, \quad \mathbf{E}_2^{-1} = \begin{bmatrix} 1 & 0 & 0 \\ 0 & 1 & 0 \\ 2 & 0 & 1 \end{bmatrix}, \quad \mathbf{E}_3^{-1} = \begin{bmatrix} 1 & 0 & 0 \\ 0 & 1 & 0 \\ 0 & -1 & 1 \end{bmatrix}.$$

Each elementary matrix is unit lower triangular and its inverse is also unit lower triangular. The matrix $\mathbf{L} = \mathbf{E}_1^{-1}\mathbf{E}_2^{-1}\mathbf{E}_3^{-1}$ is a product of unit lower triangular matrices and so is unit lower triangular itself. Hence, $\mathbf{A}$ can be factored as follows:

$$\mathbf{A} = \mathbf{LU}, \quad \text{where} \quad \mathbf{L} = \begin{bmatrix} 1 & 0 & 0 \\ -\frac{3}{2} & 1 & 0 \\ 2 & -1 & 1 \end{bmatrix}, \quad \mathbf{U} = \begin{bmatrix} \textcircled{2} & 4 & 2 \\ 0 & \textcircled{1} & 8 \\ 0 & 0 & \textcircled{6} \end{bmatrix}. \tag{2.41}$$

We call $\mathbf{A} = \mathbf{LU}$ an LU-*factorization* of $\mathbf{A}$. Notice that each multiplier m_{ij} appears in the (i, j)-position of $\mathbf{L}$ in (2.41) and the pivots used in reduction lie on the main diagonal of **U**. The factorization (2.41) may be stored in a single matrix

$$\begin{bmatrix} 2 & 4 & 2 \\ m_{21} & 1 & 8 \\ m_{31} & m_{32} & 6 \end{bmatrix} = \begin{bmatrix} 2 & 4 & 2 \\ -\frac{3}{4} & 1 & 8 \\ 2 & -1 & 6 \end{bmatrix},$$

keeping in mind that **L** has a unit diagonal. ■

The theory in Illustration 2.5 is now stated in the general case.

Theorem 2.13 **LU-Factorization**

Let **A** *be an invertible* $n \times n$ *matrix. If forward reduction on* **A** *can be performed using* only *replacements, then* **A** *can be factored in the form* $\mathbf{A} = \mathbf{LU}$, *where* **L** *is a unit lower triangular matrix of multipliers and* **U** *is upper triangular.*

LDU-Factorization

Sometimes we wish to create a symmetric factorization of **A** in which both **L** and **U** are unit triangular. Define a diagonal matrix **D** having the pivots from **U** on its main diagonal. Then

$$\mathbf{LU} = \mathbf{LIU} = \mathbf{LDD}^{-1}\mathbf{U} = \mathbf{LD}(\mathbf{D}^{-1}\mathbf{U}),$$

and $\mathbf{D}^{-1}\mathbf{U}$ is unit upper triangular. Using (2.41) for illustration, we have

$$\mathbf{A} = \mathbf{LD}(\mathbf{D}^{-1}\mathbf{U}) = \begin{bmatrix} 1 & 0 & 0 \\ -\frac{3}{2} & 1 & 0 \\ 2 & -1 & 1 \end{bmatrix} \begin{bmatrix} \textcircled{2} & 0 & 0 \\ 0 & \textcircled{1} & 0 \\ 0 & 0 & \textcircled{6} \end{bmatrix} \begin{bmatrix} 1 & 2 & 1 \\ 0 & 1 & 8 \\ 0 & 0 & 1 \end{bmatrix}. \tag{2.42}$$

We call (2.42) an LDU-factorization of **A**.

Forward Reduction with Interchanges

Forward reduction on an invertible $n \times n$ matrix **A** will generally involve a mix of replacements and interchanges. We cite two reasons.

First, forward reduction on **A** reduces each entry below the main diagonal in column 1 through column $n-1$ to zero using a nonzero pivot p_{ii} on the main diagonal. If the entry on the main diagonal is zero at any stage, an interchange is necessary to bring a nonzero pivot onto the main diagonal. For example, an interchange is required immediately for column 1 in the case

$$\mathbf{A} = \begin{bmatrix} 0 & -3 & 5 \\ 4 & 19 & 17 \\ 2 & 8 & 9 \end{bmatrix}, \quad \text{because the } (1, 1)\text{-entry is zero,} \tag{2.43}$$

and for column 2 in the case

$$\mathbf{A} = \begin{bmatrix} 1 & 2 & 2 \\ 2 & 4 & 5 \\ -3 & -4 & -2 \end{bmatrix} \underset{\substack{\mathbf{r}_2 - 2\mathbf{r}_1 \to \mathbf{r}_2 \\ \mathbf{r}_3 + 3\mathbf{r}_1 \to \mathbf{r}_3}}{\sim} \begin{bmatrix} 1 & 2 & 2 \\ 0 & 0 & 1 \\ 0 & 2 & 4 \end{bmatrix} = \mathbf{B}, \tag{2.44}$$

where $p_{11} = 1$, $m_{21} = 2$, $m_{31} = -3$, but the (2, 2)-entry in **B** is zero.

Second, machine algorithms will select the entry of largest absolute value to be the pivot in the current pivot column. The largest entry may need to be brought onto the main diagonal using an interchange—this strategy is called *partial pivoting*.

The next illustration shows how interchanges are accommodated.

■ ILLUSTRATION 2.6

LU-Factorization, with Interchanges

We will forward reduce the matrix $\mathbf{A}$, where

$$\mathbf{A} = \begin{bmatrix} 0 & 6 & -9 & 8 \\ 4 & 8 & 16 & 12 \\ 2 & 6 & 4 & 6 \\ 1 & 2 & 4 & 2 \end{bmatrix}.$$

Convenient pivots have been chosen for purposes of illustration.

$$\begin{bmatrix} 0 & 6 & -9 & 8 \\ 4 & 8 & 16 & 12 \\ 2 & 6 & 4 & 6 \\ \textcircled{1} & 2 & 4 & 2 \end{bmatrix} \underset{\substack{\mathbf{r}_1 \leftrightarrow \mathbf{r}_4 \\ \text{eliminate} \\ \text{col 1}}}{\sim} \begin{bmatrix} \textcircled{1} & 2 & 4 & 2 \\ 0 & 0 & 0 & 4 \\ 0 & \textcircled{2} & -4 & 2 \\ 0 & 6 & -9 & 8 \end{bmatrix} \underset{\substack{\mathbf{r}_2 \leftrightarrow \mathbf{r}_3 \\ \text{eliminate} \\ \text{col 2}}}{\sim} \begin{bmatrix} \textcircled{1} & 2 & 4 & 2 \\ 0 & \textcircled{2} & -4 & 2 \\ 0 & 0 & 0 & 4 \\ 0 & 0 & \textcircled{3} & 2 \end{bmatrix}$$

$$\underset{\mathbf{r}_3 \leftrightarrow \mathbf{r}_4}{\sim} \begin{bmatrix} \textcircled{1} & 2 & 4 & 2 \\ 0 & \textcircled{2} & -4 & 2 \\ 0 & 0 & \textcircled{3} & 2 \\ 0 & 0 & 0 & \textcircled{4} \end{bmatrix} = \mathbf{U}$$

To see how the three interchanges above reorder rows, apply each in turn, beginning with the identity permutation $(1, 2, 3, 4)$. We have

$$(1, 2, 3, 4) \underset{\mathbf{r}_1 \leftrightarrow \mathbf{r}_4}{\rightarrow} (4, 2, 3, 1) \underset{\mathbf{r}_2 \leftrightarrow \mathbf{r}_3}{\rightarrow} (4, 3, 2, 1) \underset{\mathbf{r}_3 \leftrightarrow \mathbf{r}_4}{\rightarrow} (4, 3, 1, 2).$$

The permutation $(4, 3, 1, 2)$ defines a permutation matrix

$$\mathbf{P} = \begin{bmatrix} 0 & 0 & 0 & 1 \\ 0 & 0 & 1 & 0 \\ 1 & 0 & 0 & 0 \\ 0 & 1 & 0 & 0 \end{bmatrix}$$

and the product $\mathbf{PA}$ reorders the rows of $\mathbf{A}$ according to the permutation $(4, 3, 1, 2)$. Forward reduction can now be performed on $\mathbf{PA}$ (using the same replacements as before and no interchanges) to obtain an LU-factorization of

PA, namely

$$\mathbf{PA} = \begin{bmatrix} 1 & 2 & 4 & 2 \\ 2 & 6 & 4 & 6 \\ 0 & 6 & -9 & 8 \\ 4 & 8 & 19 & 12 \end{bmatrix} = \begin{bmatrix} 1 & 0 & 0 & 0 \\ 2 & 1 & 0 & 0 \\ 0 & 3 & 1 & 0 \\ 4 & 0 & 0 & 1 \end{bmatrix} \begin{bmatrix} 1 & 2 & 4 & 2 \\ 0 & 2 & -4 & 2 \\ 0 & 0 & 3 & 2 \\ 0 & 0 & 0 & 4 \end{bmatrix} = \mathbf{LU}.$$

Using Theorem 2.12, we have

$$\mathbf{PA} = \mathbf{LU} \quad \Rightarrow \quad \mathbf{A} = \mathbf{P}^{-1}\mathbf{LU} = \mathbf{P}^{\mathsf{T}}\mathbf{LU},$$

where $\mathbf{P}^{\mathsf{T}}$ is a permutation matrix. Thus, **A** can be written in the form $\mathbf{A} = \mathbf{L}_1\mathbf{U}$, where $\mathbf{L}_1 = \mathbf{P}^{\mathsf{T}}\mathbf{L}$ is **L** with its rows (possibly) permuted. ■

The theory in Illustration 2.6 is now stated in the general case.

Theorem 2.14 **PLU-Factorization**

Let **A** *be an invertible* $n \times n$ *matrix. The rows of* **A** *can be reordered using an* $n \times n$ *permutation matrix* **P** *in such a way that* $\mathbf{PA} = \mathbf{LU}$, *where* **L** *is a unit lower triangular matrix of multipliers and* **U** *is upper triangular. Furthermore,* $\mathbf{A} = \mathbf{P}^{\mathsf{T}}\mathbf{LU}$.

2.3.3 Solving Square Linear Systems

We can now apply LU-factorization to solve the square linear system (S) given by $\mathbf{Ax} = \mathbf{b}$ in the case when **A** is an invertible $n \times n$ matrix. A factorization $\mathbf{A} = \mathbf{LU}$ is used to *decompose* (S) into two triangular systems that are solved separately. Introduce a new column vector **y** of unknowns, where $\mathbf{y} = \mathbf{Ux}$. Then, using matrix associativity, we have

$$\mathbf{Ax} = \mathbf{b} \quad \Rightarrow \quad (\mathbf{LU})\mathbf{x} = \mathbf{L}(\mathbf{Ux}) = \mathbf{Ly} = \mathbf{b} \quad \Rightarrow \quad \mathbf{Ly} = \mathbf{b}.$$

Because **L** is lower triangular, the system $\mathbf{Ly} = \mathbf{b}$ can be solved for **y** by forward-substitution and because **U** is upper triangular, the system $\mathbf{Ux} = \mathbf{y}$ can then be solved for **x** by back-substitution.

EXAMPLE 5 **Solving by LU-Factorization**

We will solve the 3×3 system $\mathbf{Ax} = \mathbf{b}$ given by

$$(S) \quad \begin{cases} 2x_1 + 4x_2 + 2x_3 = 2 \\ -3x_1 - 5x_2 + 5x_3 = 0 \\ 4x_2 + 7x_2 + 2x_3 = 7 \end{cases}$$

The LU-factorization of the coefficient matrix **A** for (S) has already been found in Illustration 2.5. The system $\mathbf{Ly} = \mathbf{b}$ is solved by forward substitution:

$$\mathbf{Ly} = \mathbf{b} \quad \begin{cases} y_1 \qquad\qquad = 2 & \Rightarrow \quad y_1 = 2 \\ -\frac{3}{2}y_1 + y_2 \quad = 0 & \Rightarrow \quad y_2 = 3 \\ 2y_1 - y_2 + y_3 = 7 & \Rightarrow \quad y_3 = 6 \end{cases}$$

and then $\mathbf{Ux} = \mathbf{y}$ is solved by back-substitution:

$$\mathbf{Ux} = \mathbf{y} \qquad \begin{cases} 2x_1 + 4x_2 + 2x_3 = 2 & \Rightarrow \quad x_1 = 10 \\ \qquad\;\; x_2 + 8x_3 = 3 & \Rightarrow \quad x_2 = -5 \\ \qquad\qquad\;\; 6x_3 = 6 & \Rightarrow \quad x_3 = 1 \end{cases}$$

and thus (S) is consistent with unique solution $(x_1, x_2, x_3) = (10, -5, 1)$. ■

The next example illustrates *Doolittle's method* for finding an LU-factorization of $\mathbf{A}$ without Gauss elimination. The method works whenever all entries (pivots) on the main diagonal of $\mathbf{U}$ are nonzero.

EXAMPLE 6

Doolittle Factorization

We apply Doolittle's method to the given matrix $\mathbf{A}$ shown. Let $\mathbf{A} = [a_{ij}]$ and write $\mathbf{A} = \mathbf{LU}$, where

$$\underset{\mathbf{A}}{\begin{bmatrix} 2 & 8 & 9 \\ 4 & 19 & 17 \\ 0 & -3 & 5 \end{bmatrix}} = \underset{\mathbf{L}}{\begin{bmatrix} 1 & 0 & 0 \\ m_{21} & 1 & 0 \\ m_{31} & m_{32} & 1 \end{bmatrix}} \underset{\mathbf{U}}{\begin{bmatrix} u_{11} & u_{12} & u_{13} \\ 0 & u_{22} & u_{23} \\ 0 & 0 & u_{33} \end{bmatrix}} \tag{2.45}$$

Equate each entry in $\mathbf{A}$ to the corresponding row-by-column multiplication on the left of (2.45) as follows:

$$\text{column 1 from } \mathbf{U} \begin{cases} a_{11} = 2 = u_{11} \\ a_{21} = 4 = u_{11}m_{21} \\ a_{31} = 0 = u_{11}m_{31} \end{cases} \Rightarrow \begin{cases} u_{11} = 2 \\ m_{21} = 2 \\ m_{31} = 0 \end{cases}$$

$$\text{column 2 from } \mathbf{U} \begin{cases} a_{12} = 8 = u_{12} \\ a_{22} = 19 = u_{12}m_{21} + u_{22} \\ a_{32} = -3 = u_{12}m_{31} + u_{22}m_{32} \end{cases} \Rightarrow \begin{cases} u_{12} = 8 \\ u_{22} = 3 \\ m_{32} = -1 \end{cases}$$

$$\text{column 3 from } \mathbf{U} \begin{cases} a_{13} = 9 = u_{13} \\ a_{23} = 17 = u_{13}m_{21} + u_{23} \\ a_{33} = 5 = u_{13}m_{31} + u_{23}m_{32} + u_{33} \end{cases} \Rightarrow \begin{cases} u_{13} = 9 \\ u_{32} = -1 \\ u_{33} = 4 \end{cases}$$

Hence

$$\mathbf{A} = \begin{bmatrix} 1 & 0 & 0 \\ 2 & 1 & 0 \\ 0 & -1 & 1 \end{bmatrix} \begin{bmatrix} 2 & 8 & 9 \\ 0 & 3 & -1 \\ 0 & 0 & 4 \end{bmatrix}.$$ ■

A similar method can be used to obtain a factorization $\mathbf{A} = \mathbf{LU}$ with $\mathbf{U}$ having unit main diagonal instead of $\mathbf{L}$. This is called *Crout factorization.*

Computational Note

Forward elimination on an $n \times n$ matrix $\mathbf{A}$ takes about $2n^3/3$ flops and this is the cost of computing $\mathbf{A} = \mathbf{LU}$. A further $2n^2$ flops are needed to solve $\mathbf{Ax} = \mathbf{b}$ by forward- and back-substitution. Hence each additional solution of the system for a different $\mathbf{b}$ requires about $2n^2$ flops, which is small compared to n^3 for large n. Some machine algorithms to find $\mathbf{A}^{-1}$ for invertible $\mathbf{A}$ first factor $\mathbf{A}$ as $\mathbf{A} = \mathbf{LU}$, where $\mathbf{L}$ is lower triangular with (possibly) permuted rows. Then $\mathbf{A}^{-1} = \mathbf{U}^{-1}\mathbf{L}^{-1}$.

Note that forward elimination can be performed on any matrix (not just square or invertible ones) and so the theory outlined in this section applies more generally to nonsquare matrices (Exercises 2.3).

EXERCISES 2.3

Exercises 1–4. Find the inverse of the given triangular matrix $\mathbf{A}$ in the standard way by performing row reduction on $[\mathbf{A}|\mathbf{I}]$. Check that $\mathbf{AA}^{-1} = \mathbf{I}$. Note the form of the inverse.

1. $\begin{bmatrix} 1 & 2 \\ 0 & 2 \end{bmatrix}$ **2.** $\begin{bmatrix} 1 & 0 \\ 4 & 3 \end{bmatrix}$

3. $\begin{bmatrix} 2 & 1 & -1 \\ 0 & 3 & -2 \\ 0 & 0 & 5 \end{bmatrix}$ **4.** $\begin{bmatrix} 2 & 0 & 0 \\ 0 & 3 & 0 \\ 1 & 1 & 4 \end{bmatrix}$

Exercises 5–6. Find the inverse of each unit triangular matrix.

5. $\begin{bmatrix} 1 & 2 & -1 \\ 0 & 1 & 1 \\ 0 & 0 & 1 \end{bmatrix}$ **6.** $\begin{bmatrix} 1 & 0 & 0 \\ 2 & 1 & 0 \\ -3 & 2 & 1 \end{bmatrix}$

7. Explain why the inverse of a unit upper (lower) triangular matrix is unit upper (lower) triangular.

Exercises 8–10. Concerning permutation matrices.

8. Write down the permutation matrix $\mathbf{P}$ that corresponds to the permutation $(3, 4, 2, 1)$ on the rows of $\mathbf{I}_4$. Give the permutation that produces $\mathbf{P}$ by reordering the columns of $\mathbf{I}_4$.

9. Let $\mathbf{P}$ be a permutation matrix of order n. Show that $\mathbf{PP}^{\mathsf{T}} = \mathbf{I}_n$. Explain why $\mathbf{P}^{-1} = \mathbf{P}^{\mathsf{T}}$. Explain why $\mathbf{P}^{\mathsf{T}}$ is a permutation matrix.

10. Every permutation matrix $\mathbf{P}$ of order n is a product permutation (elementary) matrices $\mathbf{P}_{ij}$ defined by interchanging rows i and j in $\mathbf{I}_n$. Illustrate this fact by showing that the permutation matrix $\mathbf{P}$ defined by the permutation $(3, 4, 2, 1)$ on the rows of $\mathbf{I}_4$ can be expressed as the product $\mathbf{P} = \mathbf{P}_{14}\mathbf{P}_{23}\mathbf{P}_{34}$. Find $\mathbf{P}^{-1}$ in terms of $\mathbf{P}_{ij}$ and confirm that $\mathbf{P}^{-1} = \mathbf{P}^{\mathsf{T}}$ in this case.

Exercises 11–16. Find an LU-factorization of the given matrix $\mathbf{A}$. Check that $\mathbf{A} = \mathbf{LU}$.

11. $\begin{bmatrix} 3 & -4 \\ 9 & -7 \end{bmatrix}$ **12.** $\begin{bmatrix} 2 & 6 \\ -3 & 2 \end{bmatrix}$

13. $\begin{bmatrix} 1 & 5 & 1 \\ 2 & 4 & -6 \\ 4 & 14 & -13 \end{bmatrix}$ **14.** $\begin{bmatrix} 1 & 2 & 3 \\ -1 & 2 & 0 \\ 2 & 0 & 0 \end{bmatrix}$

15. $\begin{bmatrix} 1 & 2 & 3 \\ 0 & -1 & 4 \\ 0 & 0 & 8 \end{bmatrix}$ **16.** $\begin{bmatrix} 1 & 0 & 0 & 2 \\ 0 & 2 & 2 & 0 \\ 0 & 8 & 9 & 0 \\ -4 & 0 & 0 & 6 \end{bmatrix}$

17. Does the matrix equation $\mathbf{A} = \begin{bmatrix} 0 & 3 \\ 0 & 4 \end{bmatrix}$ have an LU-factorization? Explain.

Exercises 18–20. Use LU-factorization to solve the linear system $\mathbf{Ax} = \mathbf{b}$ for each given vector of constant terms $\mathbf{b}$. Cross-check your answers by substitution.

18. $\mathbf{A} = \begin{bmatrix} 1 & -2 \\ -3 & 0 \end{bmatrix}$, $\mathbf{b} = \begin{bmatrix} 1 \\ -3 \end{bmatrix}, \begin{bmatrix} 2 \\ 3 \end{bmatrix}, \begin{bmatrix} -4 \\ 5 \end{bmatrix}$

19. $\mathbf{A} = \begin{bmatrix} 1 & -2 & 2 \\ -3 & 12 & -6 \\ 0 & 1 & 1 \end{bmatrix}$, $\mathbf{b} = \begin{bmatrix} 7 \\ -33 \\ -1 \end{bmatrix}, \begin{bmatrix} 0 \\ 3 \\ 4 \end{bmatrix}, \begin{bmatrix} 1 \\ 1 \\ 1 \end{bmatrix}$

20. $\mathbf{A} = \begin{bmatrix} 1 & -1 & -1 & 1 \\ 0 & -2 & 2 & 0 \\ -1 & 5 & 2 & -1 \\ 1 & 3 & -13 & 2 \end{bmatrix}$,

$\mathbf{b} = \begin{bmatrix} -2 \\ -2 \\ 6 \\ -6 \end{bmatrix}, \begin{bmatrix} 0 \\ 0 \\ 1 \\ -1 \end{bmatrix}, \begin{bmatrix} 1 \\ 0 \\ 3 \\ 2 \end{bmatrix}$

Exercises 21–24. Use Doolittle's method to find an LU-factorization of the given matrix **A**. *Check that* **A** = **LU**.

21. $\begin{bmatrix} 2 & 0 \\ 3 & 4 \end{bmatrix}$ **22.** $\begin{bmatrix} 1 & -3 \\ 3 & 4 \end{bmatrix}$

23. $\begin{bmatrix} 1 & 2 & 3 \\ 4 & 5 & 6 \\ 7 & 8 & 1 \end{bmatrix}$ **24.** $\begin{bmatrix} 2 & 6 & 5 \\ -1 & 4 & -2 \\ 1 & 2 & 3 \end{bmatrix}$

Exercises 25–26. Find a Crout factorization of the given matrix **A**.

25. $\begin{bmatrix} 4 & -2 \\ 10 & 6 \end{bmatrix}$ **26.** $\begin{bmatrix} 1 & 2 & 0 \\ 2 & 3 & 4 \\ 0 & -2 & 1 \end{bmatrix}$

Exercises 27–28. Use LU-factorization to find $\mathbf{A}^{-1}$ *for the given matrix* **A**.

27. $\begin{bmatrix} 1 & 2 \\ 3 & 4 \end{bmatrix}$ **28.** $\begin{bmatrix} 2 & -2 & 4 \\ 1 & -3 & 1 \\ 3 & 7 & 5 \end{bmatrix}$

Exercises 29–30. Use forward reduction to find an LU-factorization **A** = **LU** *for the given (nonsquare) matrix* **A**.

29. $\begin{bmatrix} 1 & 2 \\ -1 & 7 \\ 1 & 5 \end{bmatrix}$ **30.** $\begin{bmatrix} 1 & -1 & 3 & -2 \\ 0 & 2 & -1 & 2 \\ 3 & 1 & -2 & 1 \end{bmatrix}$

USING MATLAB

Consult online help for the command **lu**, and related commands.

31. The code **[L,U] = lu(A)** computes an LU-factorization of **A**, where **L** is a product of a lower triangular and permutation matrices. Row reduction in MATLAB uses partial pivoting to minimize round-off error and so we expect that some row exchanges may occur.

Consider the matrix

$$\mathbf{A} = \begin{bmatrix} 1 & 2 & 3 \\ 2 & 6 & 5 \\ -1 & 4 & -2 \end{bmatrix}.$$

(a) Use hand calculation to reduce **A** to an upper triangular matrix **U** using partial pivoting, taking note of the interchanges that need to be made during reduction.

(b) Find a permutation matrix **P** that reorders the rows of **A** so that no interchanges need to be made during reduction.

(c) Use hand calculation to find an LU-factorization **PA** = **LU**.

(d) Show that $\mathbf{P}^{-1} = \mathbf{P}^{\mathsf{T}}$ and check that $\mathbf{A} = \mathbf{P}^{\mathsf{T}}\mathbf{L}\mathbf{U}$.

(e) Find the MATLAB LU-factorization **A**, and compare your results.

32. Use MATLAB to check hand calculations in this exercise set.

2.4 Applications

The applications in this section all involve either matrix powers or matrix-column vector multiplication.

2.4.1 Graph Theory

Graph theory is a branch of mathematics that experienced tremendous growth since the latter part of the 20th century. The development has been driven to some extent by practical applications of the theory, particularly to engineering (electrical and telecommunications networks, scheduling, optimization), computer science (algorithms, data structures), chemistry

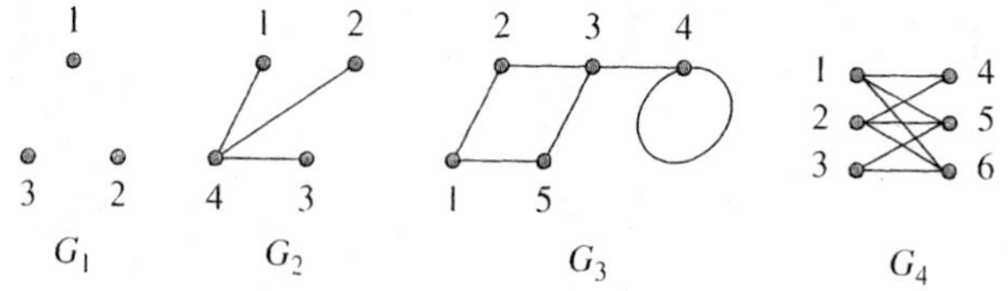

Figure 2.1 Undirected graphs.

(molecular structures), and economics. We will outline some rudimentary concepts[5] in the theory of graphs and see how matrices and matrix powers are applied in this field.

A *graph* G consists of a set of points, denoted by V, which are called *vertices* or *nodes*, together with a set of connecting lines, denoted by E, which are called *edges*. Each edge connects two (possibly coincident) vertices. A graph is denoted by $G = (V, E)$.

Refer to the examples of graphs in Figure 2.1 and make some general observations. The length or shape of an edge is unimportant in this theory. Although a graph must have at least one vertex, the set E may be empty (there may be no edges!), as in G_1. Some pairs of vertices may not be connected by an edge, as in G_2. An edge in G_3 connects vertex 4 to itself—such an edge is called a *loop*. A vertex not connected to any edge is called an *isolated* vertex. All three vertices in G_1 are isolated. Keep in mind that the graphs arising in some practical applications are *large-scale*, meaning that they contain large numbers of vertices and edges.

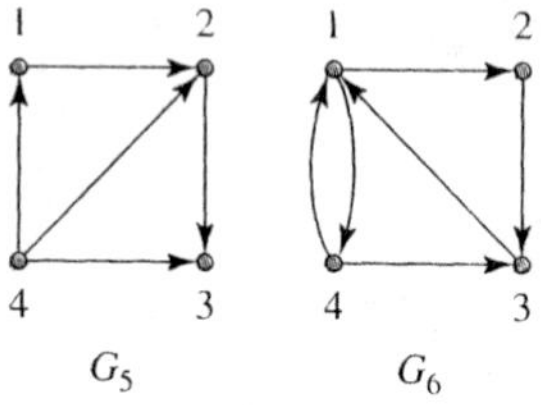

Figure 2.2 Digraphs.

In some applications a direction is assigned to each edge in the graph (as in a one-way system of roads). Such a graph is called a *directed graph*, or *digraph* for short. Two examples of digraphs are shown in Figure 2.2. The graphs in Figure 2.1 are called *undirected* graphs because no directions are assigned to the edges. A graph is called *simple* if there are no loops, and there is at most one edge between any pair of vertices. G_1, G_2, G_4, and G_5 are examples of simple graphs.

Adjacency

Suppose that $G = (V, E)$ is a graph (undirected or directed) with vertices labeled $v_1, v_2, \ldots, v_n$. We say that vertex v_j is *adjacent* to vertex v_i if there is an edge *from* v_i *to* v_j, and we use the symbol $v_i \to v_j$ to indicate this fact. If G is undirected, then $v_i \to v_j$ if and only if $v_j \to v_i$, for all i, j. Thus, for example, in G_2 we have $4 \to 2$ and $2 \to 4$. However, in the digraph G_5 we have $4 \to 2$, but vertex 4 is not adjacent to vertex 2.

Properties of graphs are often studied with the aid of a computer. Matrices provide the means by which graphs can be stored in a computer and manipulated by machine. The next definition gives a standard *mathematical model* for a graph G, meaning that the schematic diagram representing G is replaced by a matrix which carries all the relevant information about G.

[5] Caution! It should be pointed out that the terminology in this subject lacks uniformity and may vary from author to author.

Definition 2.12

Adjacency matrix

Suppose that $G = (V, E)$ is a graph (undirected or directed) with vertices labeled $v_1, v_2, \ldots, v_n$. The $n \times n$ *adjacency matrix* $\mathbf{A} = [a_{ij}]$ for G is defined by the assignment

$$a_{ij} = \begin{cases} 1 & \text{if there is an edge } \textit{from} \text{ vertex } v_i \textit{ to} \text{ vertex } v_j, \\ 0 & \text{otherwise.} \end{cases} \tag{2.46}$$

If G is undirected, then $a_{ij} = 1$ if and only if $a_{ji} = 1$ and so the adjacency matrix $\mathbf{A}$ is necessarily symmetric. Furthermore, the sum of entries in row i (or column i) of $\mathbf{A}$ indicates the number of edges *incident with* (connected to) vertex v_i.

EXAMPLE 1

Adjacency Matrix, Undirected Graph

The undirected graph G_3 in Figure 2.1 is modeled by the 5×5 adjacency matrix $\mathbf{A} = [a_{ij}]$ in (2.47).

$$\mathbf{A} = \begin{bmatrix} 0 & 1 & 0 & 0 & 1 \\ 1 & 0 & 1 & 0 & 0 \\ 0 & 1 & 0 & 1 & 1 \\ 0 & 0 & 1 & 1 & 0 \\ 1 & 0 & 1 & 0 & 0 \end{bmatrix}. \tag{2.47}$$

We have, for example, $a_{23} = a_{32} = 1$ because there is an undirected edge between vertex 2 and vertex 3 and $a_{24} = a_{42} = 0$ because no edge exists between the vertices 2 and 4. The entry $a_{44} = 1$ indicates that there is an edge joining vertex 4 to itself (a loop at vertex 4). The row sum for row 4 in $\mathbf{A}$ is 2, accounting for one edge and one loop incident with vertex 4. ■

Walks, Paths, and Cycles

An important theoretical question in graph theory is whether it is possible to travel along the edges of a graph from some vertex v_i to some vertex v_j (the two vertices may be coincident). Here is some terminology in this respect. A *walk* in a graph $G = (V, E)$ is a sequence of vertices $v_1, v_2, \ldots, v_k, v_{k+1}$ in G, such that v_{i+1} is adjacent to v_i, for $1 \le i \le k$, and we write

$$v_1 \to v_2 \to \ldots \to v_k \to v_{k+1}\ . \tag{2.48}$$

We say that the walk (2.48) has *length* k because there are k edges in it ($k + 1$ vertices). If the vertices in a walk are all distinct, we call the walk a *path*. The walk (2.48) with $k \ge 3$ and whose vertices are all distinct except that $v_1 = v_{k+1}$ is called a *cycle*. A cycle that contains every vertex in G is called a *Hamiltonian cycle*.[6] Note that all the preceding concepts apply also to digraphs, and we speak of a *directed walk*, *directed path*, and so on.

[6] Such cycles were studied by the Irish mathematician William Rowan Hamilton (1805–1865).

EXAMPLE 2

Walks, Paths, Cycles, Hamiltonian Cycle

The sequence $1 \to 2 \to 3 \to 2$ defines a walk in G_3 of length 3 from vertex 1 to vertex 2. The walk is not a path because the vertices are not distinct. The sequence $1 \to 2 \to 3 \to 4$ defines a path in G_3 of length 3 from vertex 1 to vertex 4. This path is not a cycle because the initial and final vertices are not identical. The sequence $1 \to 2 \to 3 \to 5 \to 1$ is a cycle G_3. There is no Hamiltonian cycle in either G_2 or G_3 (explain). There is a Hamiltonian cycle in G_4. In the digraph G_6, the sequence $1 \to 4 \to 1 \to 2 \to 3$ defines a directed walk (but not a path) of length 4 from vertex 1 to vertex 3. Although there are directed cycles in G_6, such as $1 \to 2 \to 3 \to 1$, there is no Hamiltonian directed cycle (explain). ■

Application: Matrix Powers

Let $\mathbf{A}$ be the adjacency matrix for a graph $G = (V, E)$ with vertices $v_1, v_2, \ldots, v_n$, and consider the product $\mathbf{A}^2 = \mathbf{B} = [b_{ij}]$. The entry b_{ij} in $\mathbf{B}$ is computed using the row-by-column product of row i with column j in $\mathbf{A}$, namely

$$b_{ij} = a_{i1}a_{1j} + a_{i2}a_{2j} + \cdots + a_{in}a_{nj} = p. \tag{2.49}$$

If $a_{i1} = a_{1j} = 1$, then there is an edge from v_i to v_1 and an edge from v_1 to v_j; that is, a walk $v_i \to v_1 \to v_j$ of length 2. If either a_{i1} or a_{1j} (or both) is zero, no walk of length 2 from v_i to v_j exists. Interpreting the other terms in (2.49) similarly shows that there are exactly p walks $(0 \le p \le n)$ in G of length 2 from vertex v_i to vertex v_j.

More generally, the (i, j)-entry in the power $\mathbf{A}^k$ counts the number of walks in G of length k from vertex v_i to vertex v_j.

EXAMPLE 3

Powers of the Adjacency Matrix, Undirected Graphs

Consider the undirected graph G_3 in Figure 2.1. Using $\mathbf{A}$ in (2.47) we have

$$\mathbf{A}^2 = \begin{bmatrix} 2 & 0 & 2 & 0 & 0 \\ 0 & 2 & 0 & 1 & 2 \\ 2 & 0 & 3 & 1 & 0 \\ 0 & 1 & 1 & 2 & 1 \\ 0 & 2 & 0 & 1 & 2 \end{bmatrix} = \mathbf{B}, \quad \mathbf{A}^3 = \begin{bmatrix} 0 & 4 & 0 & 2 & 4 \\ 4 & 0 & 5 & 1 & 0 \\ 0 & 5 & 1 & 4 & 5 \\ 2 & 1 & 4 & 3 & 1 \\ 4 & 0 & 5 & 1 & 0 \end{bmatrix} = \mathbf{C}.$$

The entry $b_{31} = 2$ in $\mathbf{B}$ indicates two walks, $3 \to 2 \to 1$ and $3 \to 5 \to 1$, of length 2 from vertex 3 to vertex 1. Also, $b_{14} = 0$ indicates that no walks of length 2 exist from vertex 1 to vertex 4. However, the entry $c_{14} = 2$ in $\mathbf{C}$ indicates that two walks of length 3 exist from vertex 1 to vertex 4. The diagonal entries in each power indicate the number of walks that are *based* (begin and end) at a vertex. For example, $b_{44} = 2$ indicates the walks $4 \to 3 \to 4$ and $4 \to 4 \to 4$ (the loop traversed twice) of length 2 based at vertex 4, and $c_{11} = 0$ indicates no walks of length 3 exist based at vertex 1. Note that the powers of $\mathbf{A}$ are symmetric matrices. ■

The adjacency matrix **A** for a digraph G will be nonsymmetric whenever there is a directed edge $v_i \to v_j$, but no directed edge $v_j \to v_i$, for some vertices v_i, v_j. The row sum for row i in **A** indicates the number of edges directed *from* vertex v_i and the column sum for column j indicates the number of edges directed *toward* vertex v_j. The (i, j)-entry in $\mathbf{A}^k$ indicates the number of directed walks in G from vertex v_i to vertex v_j.

EXAMPLE 4

Adjacency Matrix, Directed Graphs

Consider the adjacency matrix **A** and the power $\mathbf{A}^2 = \mathbf{B}$ for the graph G_6 in Figure 2.2.

$$\mathbf{A} = \begin{bmatrix} 0 & 1 & 0 & 1 \\ 0 & 0 & 1 & 0 \\ 1 & 0 & 0 & 0 \\ 1 & 0 & 1 & 0 \end{bmatrix}, \quad \mathbf{A}^2 = \begin{bmatrix} 1 & 0 & 2 & 0 \\ 1 & 0 & 0 & 0 \\ 0 & 1 & 0 & 1 \\ 1 & 1 & 0 & 1 \end{bmatrix} = \mathbf{B} \tag{2.50}$$

The edge $1 \to 2$ in G_6 defines $a_{12} = 1$ while $a_{21} = 0$. In (2.50) we have $b_{13} = 2$ indicating two directed walks $1 \to 2 \to 3$ and $1 \to 4 \to 3$ of length 2 from vertex 1 to vertex 3. The row sum for row 3 in **A** is 1, indicating one edge directed from vertex 3, and the column sum for column 3 in **A** is 2, indicating two edges directed towards vertex 3. ▪

Application: Simple Digraphs

Consider a simple digraph G with vertices $v_1, v_2, \ldots, v_n$. Broadly speaking, an edge in G directed from vertex v_i to vertex v_j might designate *dominance* or *influence* of v_i over v_j. For example, if G describes the outcome of matches played in a tennis competition, where the players (represented by positive integers) are the vertices of G, then $i \to j$ indicates that player i *beat* player j. In psychology (or in the world of business), if the vertices of G represent people in a therapy group (employees in a company), then $i \to j$ might indicate that person i has influence over (is superior to) person j.

▪ ILLUSTRATION 2.7

Analyzing a Tennis Competition

Consider a tennis competition that has been canceled due to rain after some matches have been played. Refer to Figure 2.3, which shows the simple digraph graph G_7, with six vertices representing the players and directed edges representing the outcome of the matches played.

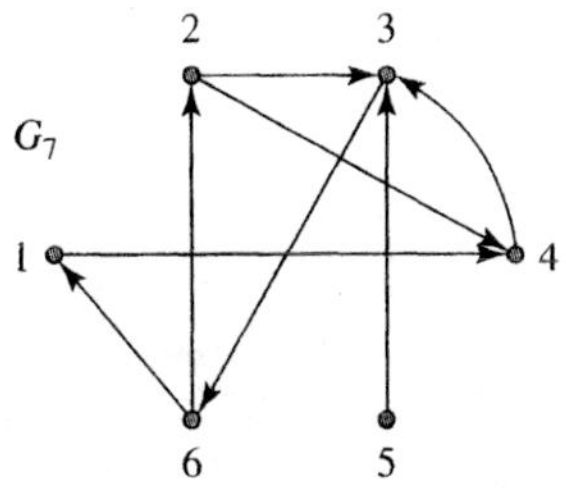

Figure 2.3 Tennis competition digraph.

The adjacency matrix **A** for G_7 is as follows:

$$\mathbf{A} = \begin{bmatrix} 0 & 0 & 0 & 1 & 0 & 0 \\ 0 & 0 & 1 & 1 & 0 & 0 \\ 0 & 0 & 0 & 0 & 0 & 1 \\ 0 & 0 & 1 & 0 & 0 & 0 \\ 0 & 0 & 1 & 0 & 0 & 0 \\ 1 & 1 & 0 & 0 & 0 & 0 \end{bmatrix}, \quad \text{row sums} \quad \begin{bmatrix} 1 \\ 2 \\ 1 \\ 1 \\ 1 \\ 2 \end{bmatrix}.$$

The row sums in **A** count the number of wins for each player (*first-order dominance*). Players 2 and 6 are tied, both having won two matches each. In order to resolve the problem of a tie, one might consider the idea of *second-order dominance*. Player i has second-order dominance over player j if there is an intermediate player h such that $i \to h \to j$; that is, player i beat player h who beat player j. In some sense, i is a stronger player than h with respect to j. The entries in the matrix $\mathbf{B} = \mathbf{A}^2$ record the number of directed paths of length 2 from vertex i to vertex j.

$$\mathbf{B} = [b_{ij}] = \begin{bmatrix} 0 & 0 & 1 & 0 & 0 & 0 \\ 0 & 0 & 1 & 0 & 0 & 1 \\ 1 & 1 & 0 & 0 & 0 & 0 \\ 0 & 0 & 0 & 0 & 0 & 1 \\ 0 & 0 & 0 & 0 & 0 & 1 \\ 0 & 0 & 1 & 2 & 0 & 0 \end{bmatrix}$$

For example, $b_{13} = 1$ corresponds to the path $1 \to 4 \to 3$ and the entry $b_{64} = 2$ corresponds to two paths $6 \to 1 \to 4$ and $6 \to 2 \to 4$. The entry c_{ij} in the matrix $\mathbf{C} = \mathbf{A} + \mathbf{A}^2$ indicates the number of paths of length 1 and of length 2 from vertex i to vertex j; that is, first- and second-order dominance of i over j. We have

$$\mathbf{C} = [c_{ij}] = \begin{bmatrix} 0 & 0 & 1 & 1 & 0 & 0 \\ 0 & 0 & 2 & 1 & 0 & 1 \\ 1 & 1 & 0 & 0 & 0 & 1 \\ 0 & 0 & 1 & 0 & 0 & 1 \\ 0 & 0 & 1 & 0 & 0 & 1 \\ 1 & 1 & 1 & 2 & 0 & 0 \end{bmatrix}, \quad \text{row sums} \quad \begin{bmatrix} 2 \\ 4 \\ 3 \\ 2 \\ 2 \\ 5 \end{bmatrix}.$$

Using this analysis, one might declare player 6 the winner and player 2 runner-up. ■

One idea from Illustration 2.7 is worth stating more generally. Suppose G is a graph with adjacency matrix **A**. Then the (i, j)-entry in **C**, where

$$\mathbf{C} = \mathbf{A} + \mathbf{A}^2 + \cdots + \mathbf{A}^k, \quad \text{and } k \text{ is a positive integer,} \tag{2.51}$$

records the number of walks in G of length $1, 2, \ldots, k$ from vertex v_i to vertex v_j. If the entries in **C** are all positive for some k, then a walk exists from any vertex v_i to any vertex v_j.

Eulerian Walks

Refer to Figure 2.4. In the 18th century, people of the town of Königsberg, East Prussia, wondered if it was possible to walk through the town in such a way that every one of the town's seven bridges would be crossed exactly once. The solution to the puzzle, which became famous, was published by Leonhard Euler (1707–1783) in 1736. Euler represented the land areas by four vertices, and the bridges by seven edges drawn between vertices, thus defining the graph G_8 in Figure 2.4. The problem then was to find a walk in G_8 that began at one

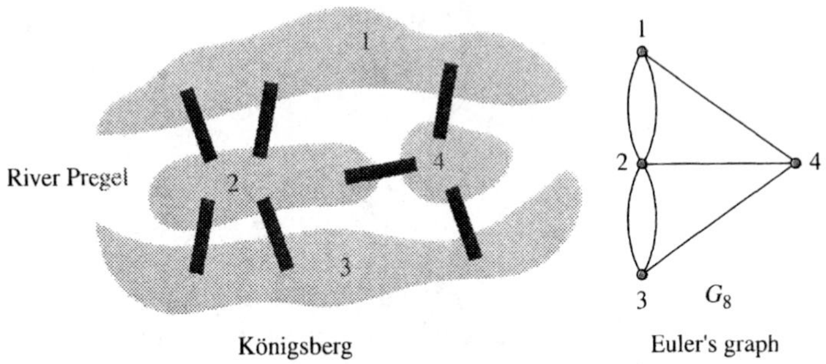

Figure 2.4 Euler's graph.

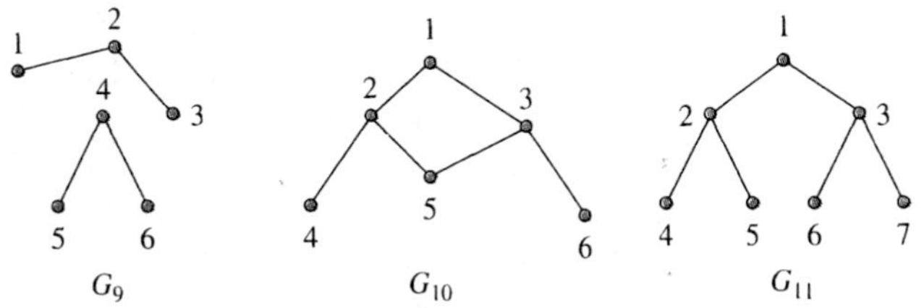

Figure 2.5 Conditions for a tree.

vertex, traversed each and every edge exactly once, and terminated either at the same vertex or some other vertex. Today, such a walk is called an *Eulerian walk.* By considering the number of edges incident with every vertex, Euler was able to show that no such walk exists (see Exercises 2.4).

A graph, such as G_8, in which some pairs of vertices are connected by more than one edge, is called a *multigraph.* Note that some modification of the definition of adjacency matrix for such graphs is required to take into account multiple edges.

Trees

Consider an undirected graph G and refer to Figure 2.5. We say that G is *connected* if there is a walk between any pair of vertices in G. The graph G_9 is clearly *disconnected*, being made up of two separate connected pieces, called *components.* A *tree* is a connected graph that has no cycles. The graph G_{10} is connected, but contains the cycle $1 \to 2 \to 5 \to 3 \to 1$. The graph G_{11} is an example of a tree.

Trees are used in many applications, particularly operations research and computer science (data structures, sorting and searching, optimization).

Weighted Graphs

If each edge in a graph G is assigned a real-value, called its *weight*, as in Figure 2.6, we obtain a *weighted graph.* The weight of an edge might represent distance, cost, time, and so on.

Remarks

Graph theory is notorious for problems that are easy to state yet extremely difficult to solve. We mention two of the most famous.

The Traveling Salesman Problem. Suppose the vertices of a graph G denote cities and edges denote connections (rail, air, road, other) between pairs of cities. Suppose also that each and every city must be visited, to deliver goods or maintain machinery, for example. Ideally, one would like to find a Hamiltonian cycle in G (visiting each city exactly once). If G is a weighted graph, with weights denoting costs, for example, an additional goal might be to find a particular Hamiltonian cycle that minimizes the total cost of the journey. To date, no complete solution to these problems has been found that applies to any graph, no matter how large-scale, although there are partial solutions that are satisfactory in particular cases. Research into this problem continues today.

The Four-Color Conjecture. Since the 19th century it was believed that four colors would suffice to color a map drawn in the plane in such a way that regions sharing a common boundary do not share the same color. The conjecture was first stated by F. Guthrie in 1853, and Arthur Cayley wrote the first research paper on the subject. Although many people tried, no correct proof was found for this conjecture until 1976 when Kenneth Appel and Wolfgang Haken at the University of Illinois gave a proof using graph-theoretic methods in conjunction with computer analysis of 1936 configurations.

Historical Notes

Leonhard Euler (1707–1783) Swiss mathematician and physicist (see profile for Chapter 8). Although graph theory was born in the year 1736 with Euler's (pronounced "Oiler") publication, the subject developed slowly over the next 200 years. The first systematic study was undertaken by D. König in the 1930s.

Gustav Robert Kirchhoff (1824–188) German physicist. Kirchhoff used trees in his study of electrical networks. Trees were later named by Arthur Cayley in 1875 and used by him in classifying saturated hydrocarbons[7] C_nH_{2n+2}, where n is the number of carbon atoms in the molecule.

2.4.2 Discrete Dynamical Systems

A *discrete dynamical system* (DDS, for short) is a time-dependent process that is monitored at discrete and equal intervals of time. Recording the number of cars at the various locations of a car rental company at the beginning of each and every month is one such process. Monitoring is done at equal intervals of time, and we call the process *discrete* (as opposed to *continuous*) because we choose not to take into account how the numbers of cars fluctuate within the time interval.

The following illustration will serve to introduce the basic idea of a DDS before a formal definition is given.

[7] A family of organic compounds composed entirely of carbon and hydrogen atoms. These include methane CH_4, ethane C_2H_6, propane C_3H_8, and butane C_4H_{10}.

■ ILLUSTRATION 2.8

Car Rentals

A car rental agency has three locations: Airport (A), Downtown (D), and Suburbs (S). At the beginning of month 0 (initial month) there are 600, 500, 400 cars at the locations A, D, S, respectively. The redistribution of cars during each successive month is assumed to be constant from month to month and is described by the digraph in Figure 2.6.

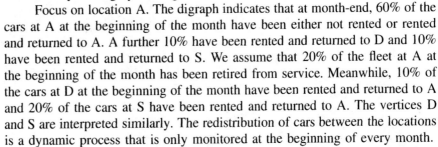

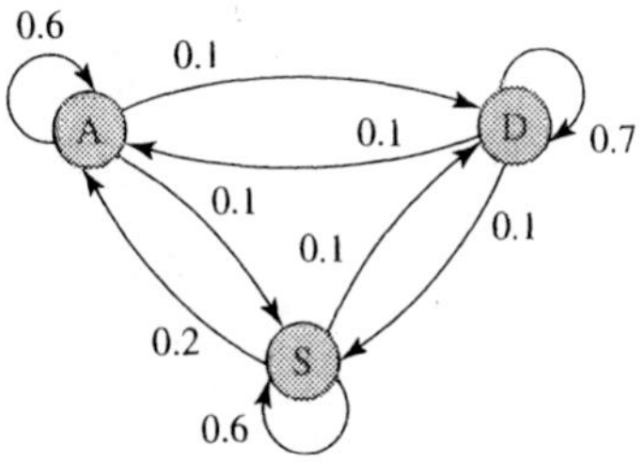

Figure 2.6 Redistribution of the car fleet.

Focus on location A. The digraph indicates that at month-end, 60% of the cars at A at the beginning of the month have been either not rented or rented and returned to A. A further 10% have been rented and returned to D and 10% have been rented and returned to S. We assume that 20% of the fleet at A at the beginning of the month has been retired from service. Meanwhile, 10% of the cars at D at the beginning of the month have been rented and returned to A and 20% of the cars at S have been rented and returned to A. The vertices D and S are interpreted similarly. The redistribution of cars between the locations is a dynamic process that is only monitored at the beginning of every month.

Let a_k, d_k, s_k denote the number cars at locations A, D, S, respectively, at the beginning of month k. We will refer to month k as *step* k. The initial *step vector* $\mathbf{u}_0$ and the kth step vector $\mathbf{u}_k$ are defined by

$$\mathbf{u}_0 = \begin{bmatrix} a_0 \\ d_0 \\ s_0 \end{bmatrix} = \begin{bmatrix} 600 \\ 500 \\ 400 \end{bmatrix}, \quad \mathbf{u}_k = \begin{bmatrix} a_k \\ d_k \\ s_k \end{bmatrix}.$$

Each step vector indicates the distribution of cars at that step. If we know the distribution of cars at step k, the distribution at step $k+1$ is given by a *system of difference equations* shown on the left side of (2.52), which can be rewritten in the matrix form shown on the right side. For $k = 0, 1, 2, \ldots$, we have

$$\begin{aligned} a_{k+1} &= 0.6a_k + 0.1d_k + 0.2s_k \\ d_{k+1} &= 0.1a_k + 0.7d_k + 0.1s_k \\ s_{k+1} &= 0.1a_k + 0.1d_k + 0.6s_k \end{aligned} \quad \Rightarrow \quad \begin{bmatrix} a_{k+1} \\ b_{k+1} \\ s_{k+1} \end{bmatrix} = \begin{bmatrix} 0.6 & 0.1 & 0.2 \\ 0.1 & 0.7 & 0.1 \\ 0.1 & 0.1 & 0.6 \end{bmatrix} \begin{bmatrix} a_k \\ d_k \\ s_k \end{bmatrix}. \tag{2.52}$$

The first equation in the system on the left side of (2.52) says that the number of cars at A at step $k+1$ is the sum of 60% of the cars at A at step k, plus 10% of the cars at D at step k, plus 20% of the number of cars at S at step k.

The matrix equation on the right side of (2.52) has the form

$$\mathbf{u}_{k+1} = \mathbf{A}\mathbf{u}_k, \quad k = 0, 1, 2, \ldots, \tag{2.53}$$

where the *step matrix* $\mathbf{A}$ is given by

$$\mathbf{A} = \begin{bmatrix} 0.6 & 0.1 & 0.2 \\ 0.1 & 0.7 & 0.1 \\ 0.1 & 0.1 & 0.6 \end{bmatrix}. \tag{2.54}$$

The recursive formula (2.53) defines a sequence of step vectors

$$\mathbf{u}_0, \quad \mathbf{u}_1\ (= \mathbf{A}\mathbf{u}_0), \quad \mathbf{u}_2\ (= \mathbf{A}\mathbf{u}_1), \quad \mathbf{u}_3\ (= \mathbf{A}\mathbf{u}_2), \quad \ldots \tag{2.55}$$

and in fact we can compute any step vector using the initial step vector $\mathbf{u}_0$ and a power of the step matrix $\mathbf{A}$. Recalling that $\mathbf{A}^0 = \mathbf{I}_3$, we have

$$\begin{aligned} & & \mathbf{u}_0 &= \mathbf{A}^0\mathbf{u}_0 \\ & & \mathbf{u}_1 &= \mathbf{A}^1\mathbf{u}_0 \\ \mathbf{u}_2 = \mathbf{A}\mathbf{u}_1 = \mathbf{A}(\mathbf{A}\mathbf{u}_0) &\Rightarrow & \mathbf{u}_2 &= \mathbf{A}^2\mathbf{u}_0 \\ \mathbf{u}_3 = \mathbf{A}\mathbf{u}_2 = \mathbf{A}(\mathbf{A}^2\mathbf{u}_0) &\Rightarrow & \mathbf{u}_3 &= \mathbf{A}^3\mathbf{u}_0 \\ & & &\vdots \end{aligned} \tag{2.56}$$

The pattern in (2.56) leads to the formula

$$\mathbf{u}_k = \mathbf{A}^k\mathbf{u}_0, \quad k = 0, 1, 2, \ldots . \tag{2.57}$$

Using (2.57), the distribution of cars at step 2 is computed as follows:

$$\mathbf{u}_2 = \mathbf{A}^2\mathbf{u}_0 = (0.1)^2 \begin{bmatrix} 6 & 1 & 2 \\ 1 & 7 & 1 \\ 1 & 1 & 6 \end{bmatrix}^2 \begin{bmatrix} 6 \\ 5 \\ 4 \end{bmatrix}$$

$$= 0.01 \begin{bmatrix} 39 & 15 & 25 \\ 14 & 51 & 15 \\ 13 & 14 & 39 \end{bmatrix} \begin{bmatrix} 600 \\ 500 \\ 400 \end{bmatrix} = \begin{bmatrix} 409 \\ 399 \\ 304 \end{bmatrix}.$$

Computing subsequent step vectors by hand is too laborious when machines are available. Using MATLAB we find that the distributions at steps 42 and 44 are

$$\mathbf{u}_{42} \simeq \begin{bmatrix} 1.17 \\ 1.26 \\ 0.92 \end{bmatrix}, \quad \mathbf{u}_{44} \simeq \begin{bmatrix} 0.87 \\ 0.95 \\ 0.69 \end{bmatrix}.$$

Rounding down, the third component in $\mathbf{u}_{42}$ indicates that C has no cars at the start of month 42 and $\mathbf{u}_{44}$ shows that the entire rental fleet will be depleted after 44 months. ■

The essential features in Illustration 2.8 provide the next general definition, which will be used in many other applications throughout the text.

Definition 2.13

Discrete dynamical system

A *discrete dynamical system* is a sequence of $n \times 1$ matrices (n-vectors)

$$\mathbf{u}_0, \mathbf{u}_1, \mathbf{u}_2, \ldots, \mathbf{u}_k, \ldots \tag{2.58}$$

that describe the state of the system at discrete and equal intervals of time. The sequence (2.58) is defined recursively by the formula

$$\mathbf{u}_{k+1} = \mathbf{A}\mathbf{u}_k, \quad \text{for } k = 0, 1, 2 \ldots, \tag{2.59}$$

where the *step matrix* $\mathbf{A}$ is $n \times n$. The kth step vector $\mathbf{u}_k$ can be found directly from the *initial step vector* $\mathbf{u}_0$ using the formula

$$\mathbf{u}_k = \mathbf{A}^k\mathbf{u}_0, \quad k = 0, 1, 2, \ldots$$

Two Questions

The following basic questions can be asked of any discrete dynamical system as given in Definition 2.13.

(a) What properties of the step matrix determine the long-term behavior of the system?
(b) What effect, if any, does the initial step vector have on the long-term behavior of the system?

Complete answers to these and related questions require the theory of *eigenvalues*, which is the subject of Chapter 6.

Open Systems, Closed Systems

The discrete dynamical system in Illustration 2.8 is called an *open system* because at each step some cars leave the system altogether. In general, an open system is indicated if at least one column sum in a step matrix **A**, such as (2.54), is less than 1. If all column sums equal 1, we called the system *closed*. The *transition matrices* which define Markov chains (Chapter 6) have column sums equal to 1 and are therefore *closed* discrete dynamical systems.

2.4.3 Age-Structured Population Models

Ecology is the study of the interaction between living organisms and their environment. Ecologists are interested in how environmental factors, such as death by predation, pollution, and so on, may affect the future population of a species. By studying population trends, action can be taken to protect endangered species.

Discrete dynamical systems provide a suitable mathematical model to study *population dynamics*. A discrete dynamical system is defined using actual population data, and assuming current trends remain the same, the system can then be used to make predictions about future increase or decline in population.

■ ILLUSTRATION 2.9

Modeling Bird Population

The following hypothetical study follows the ideas in [2] and illustrates the main features of age-structured population models.

White-naped Cranes *(Grus vipio)*

Consider a species of bird that has a maximum lifespan of six years. In this study, as in many studies of this type, only the female population is considered. We will assume that each female breeds once a year, beginning at age 3 years and any female reaching age 5 will breed before dying.

The lifespan of the species is divided into six age classes as shown in the following table.

Age	0	1	2	3	4	5
Age class (years)	(0–1)	(1–2)	(2–3)	(3–4)	(4–5)	(5–6)
Population in year t	n_0	n_1	n_2	n_3	n_4	n_5
Fecundity	0	0	0	f_3	f_4	f_5
Survival	s_0	s_1	s_2	s_3	s_4	—

In this study we assume that population is monitored just prior to breeding at the beginning of year t. Minor adjustments are required if population is monitored just after breeding (see [7] and Exercises 2.4). Refer to the preceding table. The number $n_k(t)$, $0 \le k \le 5$, denotes the female population of age k in year t. Hence, just prior to breeding at the beginning of year t, the populations $n_1(t)$ through $n_5(t)$ are known and there are no juveniles of age 0 at this time. The *fecundity* (reproduction) and *survival* parameters for each age class, which are assumed to be independent of population density and constant from year to year, are now explained.

The fecundity parameter f_k, $0 \le k \le 5$, represents the average number of female offspring produced in year t by each female in each age class. In this study, $f_0 = f_1 = f_2 = 0$. After breeding takes place in year t, the total number of female birds of age 0 is given by

$$n_0(t) = f_3 n_3(t) + f_4 n_4(t) + f_5 n_5(t). \tag{2.60}$$

The survival parameter s_k, $0 \le k \le 4$, represents the fraction of females of age k in year t which will survive to age $k+1$ in year $t+1$. Necessarily, $0 \le s_k \le 1$ and we may therefore view s_k as the probability[8] that a random female of age k will survive to age $k+1$. In particular,

$$n_1(t+1) = s_0 n_0(t) \tag{2.61}$$

and substituting (2.60) into (2.61) gives the population of females aged 1 at the beginning of year $t+1$, namely

$$n_1(t+1) = s_0 f_3 n_3(t) + s_0 f_4 n_4(t) + s_0 f_5 n_5(t). \tag{2.62}$$

The population of females in the remaining age classes at the beginning of year $(t+1)$ is given by the equations

$$\begin{aligned} n_2(t+1) &= s_1 n_1(t), & n_3(t+1) &= s_2 n_2(t) \\ n_4(t+1) &= s_3 n_3(t), & n_5(t+1) &= s_4 n_4(t) \end{aligned} . \tag{2.63}$$

Equations (2.62) and (2.63) are collectively called a *system of difference equations*. This system, which gives the population of females of ages 1 through 5 in year $(t+1)$ in terms of population in year t, can be rewritten as a single

[8] A *probability* is a real number p in the range $0 \le p \le 1$ that reflects the likelihood that a given event will occur. The value 0 represent impossibility and value 1 represents certainty.

matrix equation using matrix-column vector multiplication, namely

$$\underbrace{\begin{bmatrix} 0 & 0 & s_0 f_3 & s_0 f_4 & s_0 f_5 \\ s_1 & 0 & 0 & 0 & 0 \\ 0 & s_2 & 0 & 0 & 0 \\ 0 & 0 & s_3 & 0 & 0 \\ 0 & 0 & 0 & s_4 & 0 \end{bmatrix}}_{\mathbf{P}} \underbrace{\begin{bmatrix} n_1 \\ n_2 \\ n_3 \\ n_4 \\ n_5 \end{bmatrix}(t)}_{\mathbf{u}(t)} = \underbrace{\begin{bmatrix} n_1 \\ n_2 \\ n_3 \\ n_4 \\ n_5 \end{bmatrix}(t+1)}_{\mathbf{u}(t+1)} \tag{2.64}$$

The matrix **P** in (2.64) is referred to as the *population projection matrix* for this study and more generally, matrices of this type are called *Leslie matrices* (see Historical Notes). The vector $\mathbf{u}(t)$ is called the *age distribution vector* or *population vector* for year t.

Suppose $\mathbf{u}(0)$ is an *initial* age distribution vector that gives the population in year 0. Then (2.64) shows that $\mathbf{u}(t+1) = \mathbf{P}\mathbf{u}(t)$, for each nonnegative integer t. We obtain a discrete dynamical system

$$\mathbf{u}(1) = \mathbf{P}\mathbf{u}(0), \quad \mathbf{u}(2) = \mathbf{P}\mathbf{u}(1), \quad \mathbf{u}(3) = \mathbf{P}\mathbf{u}(2), \ldots \tag{2.65}$$

and as in (2.57), the age distribution vector $\mathbf{u}(t)$ can be calculated directly from $\mathbf{u}(0)$ using the formula

$$\mathbf{u}(t) = \mathbf{P}^t\mathbf{u}(0), \quad t = 1, 2, \ldots . \tag{2.66}$$

Case Study

Suppose the fecundity and survival parameters for our species of bird are estimated from field studies as follows:

Fecundity $f_3 = 2, \quad f_4 = 3, \quad f_5 = 2$

Survival $s_0 = 0.5, \quad s_1 = 0.4, \quad s_2 = 0.6, \quad s_3 = 0.7, \quad s_4 = 0.5.$

The population projection matrix **P** is shown on the left in (2.67).

$$\mathbf{P} = \begin{bmatrix} 0 & 0 & 1 & 1.5 & 1 \\ 0.4 & 0 & 0 & 0 & 0 \\ 0 & 0.6 & 0 & 0 & 0 \\ 0 & 0 & 0.7 & 0 & 0 \\ 0 & 0 & 0 & 0.5 & 0 \end{bmatrix}, \qquad \mathbf{u}(0) = \begin{bmatrix} 80 \\ 100 \\ 150 \\ 200 \\ 80 \end{bmatrix} \tag{2.67}$$

Current population data are used to define the initial population distribution vector $\mathbf{u}(0)$ for year $t = 0$ shown on the right in (2.67). Using (2.66) with $t = 10, 20, 30, 40$, we obtain projections of population distribution in subsequent 10-year periods. The results are shown in the following table.

Year t	0	10	20	30	40
Age 1	80	82	19	4	1
Age 2	100	43	9	2	0
Age 3	150	25	6	1	0
Age 4	200	17	5	1	0
Age 5	80	13	3	1	0

(2.68)

Table (2.68) shows that the species will become extinct in year 41. A full explanation for the decline in population requires an understanding of the *eigenvalues* of the matrix **P**. We will see later that the *dominant eigenvalue* (largest in absolute value) for **P** is $\lambda = 0.8637$ and because $\lambda < 1$, the age distribution vectors approach 0 and t increases, *for any* initial distribution vector $\mathbf{u}_0$.

Varying the fecundity and survival parameters will change the eigenvalues of **P** and thus affect the long-term behavior of the model. Some projects involving experimentation with fecundity and survival parameters are outlined in Exercises 2.4, and further analysis of these models is continued in Chapter 6. ■

Stable Population

Let **P** be the population projection matrix for an age-structured population model and let $\mathbf{u}(t)$ be the population distribution vector for stage t. If there is an integer t such that

$$\mathbf{u}(t+1) = \mathbf{P}\mathbf{u}(t) = \mathbf{u}(t),$$

then $\mathbf{u}(t)$ is called a *stationary* vector. If $\mathbf{u}(t)$ is stationary, then $\mathbf{u}(t+r) = \mathbf{u}(t)$, for all integers $r \geq 1$, and so the population distribution remains constant after stage t. We say that the population is *stable* or has reached *a steady state* after stage t. Properties of the population projection matrix **P** determine whether or not the population reaches a steady state. The theory of eigenvalues shows that if $\lambda = 1$ is an eigenvalue for the matrix **P**, then the population distribution approaches a steady state for *any* initial population distribution vector $\mathbf{u}(0)$.

EXAMPLE 5

Determining Stationary Vectors

Consider a population model defined by the Leslie matrix **P** in (2.69), where the fecundity and survival parameters are $f_0 = 0$, $f_1 = 1$, $f_3 = 3$, $s_0 = \frac{1}{2}$, and $s_1 = \frac{1}{3}$. Let $\mathbf{u}(t)$ be the age distribution vector for stage t.

$$\mathbf{P} = \begin{bmatrix} 0 & 1 & 3 \\ \frac{1}{2} & 0 & 0 \\ 0 & \frac{1}{3} & 0 \end{bmatrix}, \quad \mathbf{u}(t) = \begin{bmatrix} u_1 \\ u_2 \\ u_3 \end{bmatrix}(t) \tag{2.69}$$

Solving the matrix equation $\mathbf{P}\mathbf{u} = \mathbf{u}$ will determine whether or not a stationary vector $\mathbf{u}$ exists. Introducing the identity matrix $\mathbf{I}_3$, we have $\mathbf{P}\mathbf{u} = \mathbf{u}\mathbf{I}_3$ and so

$$(\mathbf{P} - \mathbf{I}_3)\mathbf{u} = \mathbf{0}. \tag{2.70}$$

Equation (2.70) is a homogeneous linear system that is solved by finding the reduced form $\mathbf{A}^*$ of the coefficient matrix $\mathbf{A} = \mathbf{P} - \mathbf{I}_3$. In this case we have

$$\mathbf{A} = \mathbf{P} - \mathbf{I}_3 = \begin{bmatrix} -1 & 1 & 3 \\ \frac{1}{2} & -1 & 0 \\ 0 & \frac{1}{3} & -1 \end{bmatrix} \sim \begin{bmatrix} 1 & 0 & -6 \\ 0 & 1 & -3 \\ 0 & 0 & 0 \end{bmatrix} = \mathbf{A}^*.$$

The solution to (2.70) is given by $(u_1, u_2, u_3) = (6\alpha, 3\alpha, \alpha)$, where α is a real parameter. The value $\alpha = 0$ gives the zero solution that is expected for any homogeneous linear system.

The population is stable if the ratio of population in the three age classes is 6 : 3 : 1. It will be seen in due course that $\lambda = 1$ is an eigenvalue for **P** in (2.69). ■

Historical Notes

The study of age-structured population models began with Bernardelli ([1]) and Lewis ([5]) and was then developed by Leslie ([3],[4]). The most general form for a population projection matrix is shown in (2.71), where f_k $(0 \leq k \leq m)$ are the fecundity parameters and s_k $(0 \leq k \leq m-1)$ is the survival parameter from age k to age $k+1$. The equation $\mathbf{Pu}(t) = \mathbf{u}(t+1)$ takes the form

$$\underbrace{\begin{bmatrix} f_0 & f_1 & f_2 & \cdots & f_{m-1} & f_m \\ s_0 & 0 & 0 & \cdots & 0 & 0 \\ 0 & s_1 & 0 & \cdots & 0 & 0 \\ \vdots & & & \vdots & & \vdots \\ 0 & 0 & & \cdots & s_{m-1} & 0 \end{bmatrix}}_{\mathbf{P}} \underbrace{\begin{bmatrix} n_0 \\ n_1 \\ n_2 \\ \vdots \\ n_m \end{bmatrix}(t)}_{\mathbf{u}(t)} = \underbrace{\begin{bmatrix} n_0 \\ n_1 \\ n_2 \\ \vdots \\ n_m \end{bmatrix}(t+1)}_{\mathbf{u}(t+1)}. \qquad (2.71)$$

We call **P** a *Leslie matrix*. Note that **P** is *sparse* because it has a large percentage of zero entries.

2.4.4 Input–Output Models

The 1973 Nobel Prize in Economics was awarded to Wassily W. Leontief *for the development and application of input–output models to economic problems*. Leontief was the sole and unchallenged creator of *input–output analysis*, which became the standard economic projection tool for countries and corporations around the world. This section introduces the basic ideas of input–output analysis.

■ ILLUSTRATION 2.10

A Closed Economic Model

Consider a simple economy based on three sectors: Machinery, Food, Farming. For simplicity, assume that each sector produces a single commodity (goods or services) in a given year and that *output* (income) is generated from the sale of this commodity. Furthermore, each commodity is produced by a single sector. The sectors are *interdependent*, meaning that a sector may need to purchase commodities from other sectors, including itself, in order to generate its output. The interdependency between sectors is shown in the following *input–output table*.

		Output (Income)		
		Machinery	Food	Farming
Input (Consumption)	Machinery	0.2	0.1	0.2
	Food	0.4	0.5	0.6
	Farming	0.4	0.4	0.2

Column 1 of the table shows that the Machinery sector consumes 20% of its own commodity, Food consumes another 40%, and Farming a further 40%. The other columns are interpreted similarly. The sum of each column in the table is 1, which indicates that the annual output from each sector is totally consumed within the economy and consequently the economy is called *closed*.

Let x_1, x_2, x_3 (million \$) denote, respectively, the annual output (income) produced by Machinery, Food, Farming from the sale of its commodity. Then the annual expenditure e by Machinery is (from row 1 of the table)

$$e = 0.2x_1 + 0.1x_2 + 0.2x_3.$$

Machinery spends $0.2x_1$ to buy 20% of its own commodity and spends a further $0.1x_2$ on Food and $0.2x_3$ on Farming. The annual income for Machinery is x_1. To be economically viable, we require that $e \le x_1$; that is, expenditure for the Machinery sector is not greater than its income. When $e = x_1$, the expenditure and income balance and we say that the Machinery sector is in *equilibrium*.

Assuming all three sectors are in equilibrium, the rows of the input–output table are used to define a linear system, namely

$$\begin{aligned} 0.2x_1 + 0.1x_2 + 0.2x_3 &= x_1 \\ 0.4x_1 + 0.5x_2 + 0.6x_3 &= x_2 \\ 0.4x_1 + 0.4x_2 + 0.2x_3 &= x_3, \end{aligned}$$

which can be rewritten in the matrix form

$$\mathbf{Cx} = \mathbf{x}, \tag{2.72}$$

where

$$\mathbf{C} = \begin{bmatrix} 0.2 & 0.1 & 0.2 \\ 0.4 & 0.5 & 0.6 \\ 0.4 & 0.4 & 0.2 \end{bmatrix}, \quad \text{and} \quad \mathbf{x} = \begin{bmatrix} x_1 \\ x_2 \\ x_3 \end{bmatrix}. \tag{2.73}$$

We call $\mathbf{C}$ the *consumption matrix* and $\mathbf{x}$ the *production vector* for the economy. Note that the entries in $\mathbf{x}$ are nonnegative real numbers (income). Introducing the identity matrix $\mathbf{I}_3$ into the right side of (2.72), we obtain a homogeneous linear system in matrix form

$$\mathbf{Ax} = \mathbf{0}, \quad \text{where} \quad \mathbf{A} = \mathbf{I}_3 - \mathbf{C} \tag{2.74}$$

From (2.73), we have

$$\mathbf{A} = \mathbf{I}_3 - \mathbf{C} = \begin{bmatrix} 0.8 & -0.1 & -0.2 \\ -0.4 & 0.5 & -0.6 \\ -0.4 & -0.4 & 0.8 \end{bmatrix}.$$

Notice that the column sums in $\mathbf{A}$ are all zero and so $\mathbf{Ax} = \mathbf{0}$ has infinitely many solutions (Exercises 2.4). Solving $\mathbf{Ax} = \mathbf{0}$, we have $(x_1, x_2, x_3) = (\frac{4}{9}t, \frac{14}{9}t, t)$, where t is a real parameter. Setting $t = 9$ (a convenient value) shows that when the output from Machinery, Food, Farming is 4, 14, 1 (million \$), respectively, the economy will be in equilibrium. ■

The analysis in Illustration 2.10 remains the same when there are n sectors (or industries) in the economy. In this case, the consumption matrix $\mathbf{C} = [c_{ij}]$

is $n \times n$ and entry c_{ij}, $1 \leq i, j \leq n$, represents the fraction of output of sector j that is consumed by (is an input to) sector i in some fixed time period. The nonnegative value x_i is the price for the total output from sector i. When the economy is closed, all column sums in $\mathbf{C}$ are equal to 1 and we sometimes call $\mathbf{C}$ an *exchange matrix* in this context. In the terminology of Chapter 6, $\mathbf{C}$ is *column stochastic* and $\lambda = 1$ is an *eigenvalue* for $\mathbf{C}$. As mentioned in the illustration, it is always possible to find a price vector $\mathbf{x}$ satisfying $\mathbf{Ax} = \mathbf{0}$, where $\mathbf{A} = \mathbf{I}_n - \mathbf{C}$, and such that $\mathbf{x} \geq \mathbf{0}$, with at least one positive component

We now modify the model in Illustration 2.10 to accommodate an *open economy.*

■ ILLUSTRATION 2.11

An Open Economic Model

Consider a simple economy based of three sectors: Hydro, Water, Gas. The input–output table and consumption matrix for this economy are as shown.

	Hydro	Water	Gas
Hydro	0.2	0.5	0.1
Water	0.4	0.2	0.2
Gas	0.1	0.3	0.3

$$\Rightarrow \quad \mathbf{C} = \begin{bmatrix} 0.2 & 0.5 & 0.1 \\ 0.4 & 0.2 & 0.2 \\ 0.1 & 0.3 & 0.3 \end{bmatrix}$$

The column sum for column 1 of $\mathbf{C}$ is less than 1, indicating that Hydro has an excess of output (income) over input (consumption). We say that Hydro is *productive*. Gas is also productive, while Water is said to be *nonproductive* because column 2 in $\mathbf{C}$ has sum 1. Whenever there is excess of output over input, we have an *open* economy and the excess income may be applied to satisfy an *external demand.* For this illustration, suppose there is an external demand (for goods or services) of 10, 20, 25 (million \$), respectively, from Hydro, Water, and Gas. Equating expenditures (internal consumption and external demand) to income, we have

$$\begin{array}{llcll} & \textbf{output} & & \textbf{internal consumption} & \textbf{external demand} \\ \text{Hydro} & x_1 & = & 0.2x_1 + 0.5x_2 + 0.1x_3 & +\ 10 \\ \text{Water} & x_2 & = & 0.4x_1 + 0.2x_2 + 0.2x_3 & +\ 20 \\ \text{Gas} & x_3 & = & 0.1x_1 + 0.3x_2 + 0.3x_3 & +\ 25. \end{array} \tag{2.75}$$

Taking terms from the right side to the left side in (2.75), we obtain a 3×3 linear system

$$(S) \quad \begin{cases} 0.8x_1 - 0.5x_2 - 0.1x_3 = 10 \\ -0.4x_1 + 0.8x_2 - 0.2x_3 = 20 \\ -0.1x_1 - 0.3x_2 + 0.7x_3 = 25 \end{cases} . \tag{2.76}$$

Multiplying each equation in (2.76) throughout by 10 to obtain integer coefficients and reducing the augmented matrix to reduced form, we obtain the unique solution

$$(x_1, x_2, x_3) = (74.35, 82.61, 81.74).$$

The solution indicates that Hydro, for example, must have an output of \$74.35 million in order to meet all demands from other sectors, plus an external demand of \$10 million. ■

The features of Illustration 2.11 will now be stated in a general context. Consider an economy based on n sectors. The $n \times n$ consumption matrix $\mathbf{C}$ for the economy has nonnegative entries and suppose at least one column sum less than 1. Then the economy is open and an excess of output over input for any productive sector can be used to satisfy an external demand. The price vector and the external demand vector are given by

$$\mathbf{x} = \begin{bmatrix} x_1 \\ \vdots \\ x_n \end{bmatrix}, \quad \mathbf{d} = \begin{bmatrix} d_1 \\ \vdots \\ d_n \end{bmatrix},$$

where the entries in $\mathbf{x}$ and $\mathbf{d}$ are necessarily nonnegative. Then, generalizing (2.75), we have

$$\mathbf{x} = \mathbf{Cx} + \mathbf{d}. \tag{2.77}$$

Introducing the identity matrix $\mathbf{I}$ on the left side of (2.77) gives

$$\mathbf{x} = \mathbf{Cx} + \mathbf{d} \quad \Rightarrow \quad \mathbf{Ix} = \mathbf{Cx} + \mathbf{d} \quad \Rightarrow \quad (\mathbf{I} - \mathbf{C})\mathbf{x} = \mathbf{d}.$$

Hence an n-sector open economy is modeled by an $n \times n$ linear system

$$(\mathbf{I} - \mathbf{C})\mathbf{x} = \mathbf{d}, \tag{2.78}$$

and this is the form of (2.76) when $n = 3$.

We require a solution to (2.78) such that $\mathbf{x} \geq \mathbf{0}$, with at least one component positive. Not all consumption matrices $\mathbf{C}$ will provide such a solution. If it happens that $(\mathbf{I} - \mathbf{C})^{-1}$ exists and has nonnegative entries, then (2.78) has a unique solution

$$\mathbf{x} = (\mathbf{I} - \mathbf{C})^{-1}\mathbf{d}. \tag{2.79}$$

Definition 2.14

Productive consumption matrices

An $n \times n$ consumption matrix $\mathbf{C}$ is called *productive* if $(\mathbf{I} - \mathbf{C})^{-1}$ exists and has nonnegative entries. When this is the case, we call $(\mathbf{I} - \mathbf{C})^{-1}$ a *Leontief inverse*.

Leontief Inverses, Approximation

Let r be a real number such that $|r| < 1$ and let k be a positive integer. The geometric series $1 + r + r^2 + \cdots$ converges to the value $1/1 - r$; that is,

$$1 + r + r^2 + \cdots + r^{k-1} \simeq (1 - r)^{-1}, \quad \text{for large } k. \tag{2.80}$$

Under certain conditions, this formula is valid if r is replaced by an $n \times n$ matrix $\mathbf{C}$, as we now show.

Multiplying out the product on the left side of (2.81) and canceling terms, we have

$$(\mathbf{I} - \mathbf{C})(\mathbf{I} + \mathbf{C} + \mathbf{C}^2 + \cdots + \mathbf{C}^{k-1}) = \mathbf{I} - \mathbf{C}^k. \tag{2.81}$$

If the powers $\mathbf{C}^k$ on the right side of (2.81) tend to the zero matrix $\mathbf{O}$ as k becomes large—that is, all entries in $\mathbf{C}^k$ approach zero, then

$$(\mathbf{I} - \mathbf{C})(\mathbf{I} + \mathbf{C} + \mathbf{C}^2 + \cdots + \mathbf{C}^{k-1}) \simeq \mathbf{I}$$

and so the sum $\mathbf{I} + \mathbf{C} + \mathbf{C}^2 + \cdots + \mathbf{C}^{k-1}$ is an approximate right-inverse for $\mathbf{I} - \mathbf{C}$. By Theorem 2.9 in Section 2.2, the matrix $\mathbf{I} - \mathbf{C}$ is invertible and

$$\mathbf{I} + \mathbf{C} + \mathbf{C}^2 + \cdots + \mathbf{C}^{k-1} \simeq (\mathbf{I} - \mathbf{C})^{-1}, \quad \text{as } k \text{ becomes large} \tag{2.82}$$

and (2.82) is the analogue of (2.80) for square matrices. We now state and prove a condition that will ensure that $\mathbf{C}^k$ tends to $\mathbf{O}$ as k becomes large.

Theorem 2.15 **A Sufficient Condition for Convergence**

Let $\mathbf{C} = [c_{ij}]$ be an $n \times n$ matrix in which all entries are nonnegative. If either all column sums in $\mathbf{C}$ are (strictly) less than 1 or all row sums are (strictly) less than 1, then $\mathbf{C}^k$ approaches $\mathbf{O}$ as k becomes large.

Proof Suppose all column sums are less than 1 and r is the maximum column sum in $\mathbf{C}$. Then $0 \le r < 1$. Let p be the maximum value of all entries in $\mathbf{C}$. Then $0 \le p < 1$ because each column entry must be less than r (< 1).

The (i, j)-entry c in $\mathbf{C}^2$ is the row-by-column product of row i in $\mathbf{C}$ with column j in $\mathbf{C}$. Because each entry in row i is less than or equal to p and the sum of column j is less than or equal to r, we have

$$\begin{aligned} c &= c_{i1}c_{1j} + c_{i2}c_{2j} + \cdots + c_{in}c_{nj} \\ &\le pc_{1j} + pc_{2j} + \cdots + pc_{nj} \\ &= p(c_{1j} + c_{2j} + \cdots + c_{nj}) \le pr, \end{aligned}$$

showing that $c \le pr$ for every entry c in $\mathbf{C}^2$. The same argument applies to $\mathbf{C}^3 = \mathbf{C}^2\mathbf{C}$, showing that each entry c in $\mathbf{C}^3$ satisfies $c \le pr^2$ and in general $c \le pr^{k-1}$ for each entry in c in $\mathbf{C}^k$. But the sequence r^k tends to 0 as k becomes large and so pr^k tends to 0 also. The proof for rows is similar. ■

EXAMPLE 6 **A Leontief Inverse, Visualizing Convergence**

Consider the 2×2 matrix $\mathbf{C} = \begin{bmatrix} 0.1 & 0.2 \\ 0.3 & 0.4 \end{bmatrix}$. We have $\text{rank}(\mathbf{I}_2 - \mathbf{C}) = 2$ and so $(\mathbf{I}_2 - \mathbf{C})$ is invertible. Hence

$$\mathbf{I}_2 - \mathbf{C} = \begin{bmatrix} 0.9 & -0.2 \\ -0.3 & 0.6 \end{bmatrix} \quad \Rightarrow \quad (\mathbf{I}_2 - \mathbf{C})^{-1} = \begin{bmatrix} 1.2500 & 0.4167 \\ 0.6250 & 1.8750 \end{bmatrix},$$

showing that $(\mathbf{I}_2 - \mathbf{C})^{-1}$ is a Leontief inverse (all entries are nonnegative). The matrix $\mathbf{C}$ is productive. The columns (and rows) of $\mathbf{C}$ satisfy the condition in Theorem 2.15. The convergence process

$$S(k) = \mathbf{I}_2 + \mathbf{C} + \mathbf{C}^2 + \cdots + \mathbf{C}^k \quad \to \quad (\mathbf{I}_2 - \mathbf{C})^{-1}$$

can be visualized by writing a MATLAB M-file to compute the sequence $S(1)$, $S(2), \ldots, S(k), \ldots$. For values $k = 14, 15, 16$, we have, respectively,

$$\begin{bmatrix} 1.2499 & 0.4165 \\ 0.6248 & 1.8747 \end{bmatrix}, \quad \begin{bmatrix} 1.2500 & 0.4166 \\ 0.6249 & 1.8749 \end{bmatrix}, \quad \begin{bmatrix} 1.2500 & 0.4167 \\ 0.6250 & 1.8750 \end{bmatrix}.$$ ▩

The matrix $\mathbf{I}_2 - \mathbf{C}$ in Example 6, which has positive diagonal entries and nonpositive off-diagonal entries, belongs to a class of so-called *M-matrices* that have many interesting properties.

Historical Notes

Wassily W. Leontief (1906–1999) was born in St. Petersburg, Russia, attended the university there, and went on to earn a Ph.D. from the University of Berlin in 1928. He emigrated to the United States in 1931, taught at Harvard University from 1931 to 1975, and was later appointed director of the Institute for Economic Analysis at New York University. In the 1940s Leontief led a two-year study of the economy of the United States. A 500×500 consumption matrix $\mathbf{C}$ was compiled showing input–output (in billions of dollars) of 500 industries in the United States. The linear systems arising from the analysis were painfully solved using the primitive computer technology of the day, using punched cards to input the data and commands. Leontief proved that it was possible to assign a price to the output of each industry so that the whole economy would be in equilibrium.

2.4.5 Pin-Jointed Planar Trusses

The following application of matrix multiplication is adapted from [6]. Figure 2.7 shows a *pin-jointed truss* in the vertical plane.

The framework has three rigid members and four pin-joints labeled A, B, C, D. The members are free to rotate at the pin-joints and are fastened to the wall at joints A, B, C. A horizontal force f_1 and a vertical force f_2 are applied at joint D. The weight of each member is considered negligible compared to the two forces.

We will assume that the forces f_1, f_2 applied at D cause a *compression force* within each member. Let t_i, $i = 1, 2, 3$ be the magnitude of the force in member i, with the understanding that t_i is negative if member i is in *extension*.

The forces at D cause the pin-joint to move d_1 units to the left and d_2 units down. The magnitudes of the forces and the displacements are recorded in 2-vectors, as follows:

$$\mathbf{f} = \begin{bmatrix} f_1 \\ f_2 \end{bmatrix}, \qquad \mathbf{d} = \begin{bmatrix} d_1 \\ d_2 \end{bmatrix}.$$

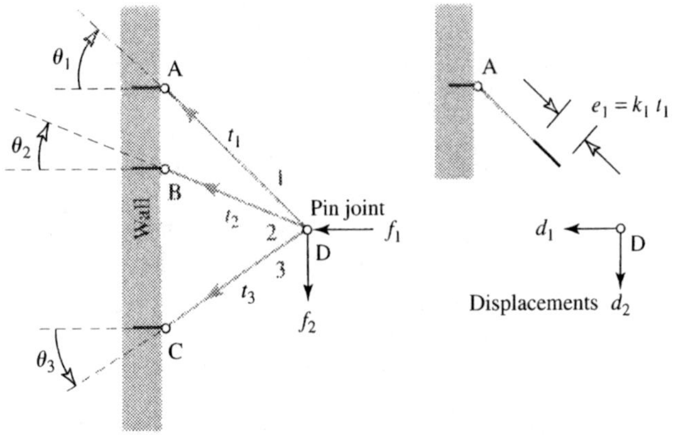

Figure 2.7 A pin-joined truss.

The problem is to find **d** in terms of **f** and we will outline briefly one possible solution.

Let θ_i be the angles shown in Figure 2.7, where θ_1 and θ_2 are negative angles, while θ_3 is positive. We assume that the angles remain constant before and after the displacement at D. Resolving forces horizontally and vertically at each pin-joint, and then summing gives the following 3×2 linear system

$$\left\{\begin{aligned} -t_1\cos\theta_1 - t_2\cos\theta_2 - t_3\cos\theta_3 &= f_1 \quad \text{(Horizontally)} \\ t_1\sin\theta_1 + t_2\sin\theta_2 + t_3\sin\theta_3 &= f_2 \quad \text{(Vertically)} \end{aligned}\right. \tag{2.83}$$

and (2.83) has matrix form

$$\mathbf{A} = \begin{bmatrix} -\cos\theta_1 & -\cos\theta_2 & -\cos\theta_3 \\ \sin\theta_1 & \sin\theta_2 & \sin\theta_3 \end{bmatrix}, \quad \mathbf{t} = \begin{bmatrix} t_1 \\ t_2 \\ t_3 \end{bmatrix} \quad \Rightarrow \quad \mathbf{At} = \mathbf{f} \tag{2.84}$$

We next relate the forces within the members to the extensions (compressions) e_i that they cause—e_i is negative for a compression. Assuming that Hooke's law applies, we have $e_i = k_i t_i$, where k_i is the coefficient of flexibility for each member $i = 1, 2, 3$. These equations are expressed in matrix form; thus

$$\mathbf{K} = \begin{bmatrix} k_1 & 0 & 0 \\ 0 & k_2 & 0 \\ 0 & 0 & k_3 \end{bmatrix}, \quad \mathbf{e} = \begin{bmatrix} e_1 \\ e_2 \\ e_3 \end{bmatrix} \quad \Rightarrow \quad \mathbf{e} = \mathbf{Kt}, \tag{2.85}$$

and assuming the diagonal entries in **K** are nonzero, we have $\mathbf{t} = \mathbf{K}^{-1}\mathbf{e}$.

The simplest way of relating the extensions **e** to the displacements **d** is to assume that d_1 and d_2 are small so that a linear relationship holds in the form

$$e_i = -d_1\cos\theta_i + d_2\sin\theta_i, \quad i = 1, 2, 3, \quad \Rightarrow \quad \mathbf{e} = \mathbf{A}^{\mathsf{T}}\mathbf{d}, \tag{2.86}$$

which is expressed in the matrix form on the right in (2.86). The equation relating **f** to **d** is then

$$\mathbf{f} = \mathbf{At} = \mathbf{A}(\mathbf{K}^{-1}\mathbf{e}) = (\mathbf{AK}^{-1})(\mathbf{A}^{\mathsf{T}}\mathbf{d}) = (\mathbf{AK}^{-1}\mathbf{A}^{\mathsf{T}})\mathbf{d}. \tag{2.87}$$

If the 2×2 matrix $\mathbf{B} = \mathbf{A}\mathbf{K}^{-1}\mathbf{A}^{\mathsf{T}}$ is invertible, then (2.87) implies that $\mathbf{d} = \mathbf{B}^{-1}\mathbf{f}$. The next step in the analysis would therefore be to investigate conditions under which $\mathbf{B}$ is invertible.

EXAMPLE 7

Case study: Pin-jointed planar truss

Refer to Figure 2.7. Assume that $\theta_1 = -30°$, $\theta_2 = -20°$, $\theta_3 = 40°$ and that the coefficient to flexibility $k_i = 0.02$ is constant for all members $i = 1, 2, 3$. Then, using MATLAB, we have

$$\mathbf{A} = \begin{bmatrix} -0.8660 & -0.9397 & -0.7660 \\ -0.5000 & -0.3420 & 0.6428 \end{bmatrix}, \quad \mathbf{B} = \mathbf{A}\mathbf{K}^{-1}\mathbf{A}^{\mathsf{T}} = \begin{bmatrix} 110.9923 & 13.1001 \\ 13.1001 & 39.0077 \end{bmatrix}$$

and $\mathbf{B}$ is invertible because $\det(\mathbf{B}) = 4157.9$ is nonzero. Assuming forces $(\mathrm{f}_1, \mathrm{f}_2) = (10, 20)$ units, we have

$$\mathbf{d} = \mathbf{B}^{-1}\mathbf{f} = \begin{bmatrix} 0.0094 & -0.0032 \\ -0.0032 & 0.0267 \end{bmatrix}\begin{bmatrix} 10 \\ 20 \end{bmatrix} = \begin{bmatrix} 0.0308 \\ 0.5024 \end{bmatrix} \text{ units.}$$ ■

REFERENCES

1. Bernardelli, H., "Population waves," *Journal of the Burma Research Society*, 31, 1–18 (1941).
2. Gillman, M. and Hails, R., *An Introduction to Ecological Modeling: Putting Practice into Theory* (eds J.H. Lawton and G.E. Likens) Methods in Ecology Series, Blackwell Scientific Publications, Oxford 1997.
3. Leslie, P.H., "On the uses of matrices in certain population mathematics," *Biometrika*, 33, 182–212 (1945).
4. Leslie, P.H.,"Some further notes on the use of matrices in population mathematics," *Biometrika*, 35, 213–45 (1948).
5. Lewis, E.G., "On the generation and growth of a population," *Sankhya*, 6, 93–96 (1942).
6. Noble, B. and Daniel, J.W., *Applied Linear Algebra*, 3rd. ed. Prentice Hall, Englewood Cliffs, NJ 1988.
7. Swatzman, G.L. and Kaluzny, S.P., *Ecological Simulation Primer*, Macmillan, New York 1987.

EXERCISES 2.4

SECTION 2.4.1 GRAPH THEORY

Exercises 1–4. Refer to Figures 2.1 and 2.2. (a) Write down the adjacency matrix for the given graph. (b) Use the adjacency matrix to find the number of walks of length 3 that begin at vertex 2 and list them all.

1. G_1 **2.** G_2 **3.** G_4 **4.** G_5

Exercises 5–8. Draw the undirected graph G corresponding to the given adjacency matrix. List any cycles in G of length 3.

5. $\begin{bmatrix} 0 & 0 \\ 0 & 0 \end{bmatrix}$ **6.** $\begin{bmatrix} 1 & 1 \\ 1 & 1 \end{bmatrix}$

7. $\begin{bmatrix} 1 & 0 & 1 \\ 0 & 0 & 1 \\ 1 & 1 & 1 \end{bmatrix}$ **8.** $\begin{bmatrix} 0 & 1 & 1 & 0 \\ 1 & 0 & 1 & 1 \\ 1 & 1 & 0 & 0 \\ 0 & 1 & 0 & 0 \end{bmatrix}$

9. Refer to Example 3. List the walks in G_3 corresponding to the entry $c_{23} = 5$ in $\mathbf{C} = \mathbf{A}^3$. How many paths are there in this list? List the walks in G_3 corresponding to $c_{44} = 3$. Are there any cycles based at vertex 4?

10. Find the adjacency matrix $\mathbf{A}$ for the digraph G_6 in Figure 2.2. Compute $\mathbf{D} = [d_{ij}] = \mathbf{A}^5$. Interpret the entry d_{44} in $\mathbf{D}$ by listing all the walks of length 5 based at vertex 4. Give an argument which justifies the value of the entry d_{23} in $\mathbf{D}$.

11. Use mathematical induction on k to show that the (i, j)-entry in $\mathbf{A}^k$ counts the number of walks of length k from vertex i to vertex j.

12. The *complete graph* on n vertices, denoted by K_n, is a simple undirected graph in which every pair of vertices is joined by an edge. Draw K_5 and find its adjacency matrix. Show that K_n has a Hamiltonian cycle and draw it in the case of K_5.

13. Suppose that each edge in the complete graph K_n is assigned a direction. The resulting digraph T_n is called a *tournament*. If the vertices in T_n represent players, then every player plays every other player (round-robin), and every match has a definite outcome. It can be shown that every tournament contains a directed path through all its vertices (a Hamiltonian path). Verify this fact in the case $n = 3, 4, 5, 6$.

14. Determine the simple undirected graph G on n vertices such that $b_{ij} = n - 2$ for all i, j, where $\mathbf{B} = [b_{ij}] = \mathbf{A}^2$ and $\mathbf{A}$ is the adjacency matrix for G.

Exercises 15–16. Refer to Figure 2.8. In each case, denote the adjacency matrix of the digraph by $\mathbf{A}$. *For* G_{12}, *explain each entry equal to 1 in* $\mathbf{B} = \mathbf{A}^3$. *For* G_{13}, *explain the nonzero entries in* $\mathbf{A}^2$, $\mathbf{A}^3$ *and the meaning of* $\mathbf{A}^4$.

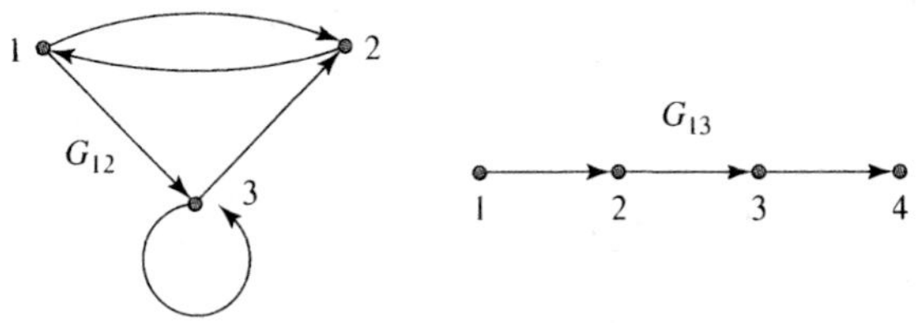

Figure 2.8

15. G_{12}

16. G_{13}

17. A coin is tossed in three successive trials. Use a tree to list all possible outcomes of the trials.

18. Six players, denoted by 1, 2, 3, 4, 5, 6, take part in a chess competition that is canceled after a certain number of matches have been played. The table lists the players along the top row and the player(s) they beat down each column.

1	2	3	4	5	6
3	1	2	2	6	1
4			3		
5			5		

Draw the digraph that models the competition. Use second-order dominance to decide who should win. Is it possible to declare a runner-up?

19. A simple graph $G = (V, E)$ is called *bipartite* if $V = V_1 \cup V_2$, where V_1 and V_2 have no vertex in common, and any edge joins a vertex in V_1 to a vertex in V_2. The graph G_4 in Figure 2.1 is bipartite. Write out the adjacency matrix for G_4. Show that the adjacency matrix for any bipartite graph G has the block form

$$\mathbf{A} = \begin{bmatrix} \mathbf{O} & \mathbf{B} \\ \mathbf{B}^{\mathsf{T}} & \mathbf{O} \end{bmatrix}$$

for some matrix $\mathbf{B}$. Show that G has no cycles of odd length.

20. Let G be an undirected graph or multigraph. The *degree* of a vertex v in G, denoted by $\delta(v)$ (Greek letter delta), is defined to be the number of edges incident with v (with loops counted twice).

(a) Find the degrees of all vertices in each of the graphs G_1, G_2, G_3, G_4 in Figure 2.1 and G_8 in Figure 2.4.

(b) For each graph in (a), relate the sum of the degrees to the number of edges in the graph.

(c) Conjecture (guess), and then prove a result regarding the sum of the degrees of all vertices that applies to any G.

(d) Prove that the number of vertices of odd degree must be an even integer.

21. Let G be an undirected graph or multigraph and let u and v be distinct vertices in G. Suppose an Eulerian walk in G begins at u and ends at v.

(a) Show that $\delta(u)$ and $\delta(v)$ are odd integers and the degrees of all other vertices are even integers.

(b) Conclude there is no Eulerian walk in G_8.

(c) If an Eulerian walk begins and ends at the same vertex in G, what can be said about the degrees of all vertices in G?

22. For what values of n does the complete graph K_n have an Eulerian walk?

SECTION 2.4.2 DISCRETE DYNAMICAL SYSTEMS

23. A video store has two locations: Downtown (D) and Uptown (U). The company has 2000 videos that can be rented and returned to either location. Any videos that are rented are returned the next day without fail. A video rented at D has a 90% chance of being returned to location D the morning after. A video rented at U has an 80% chance of being returned to location U the morning after.

(a) Find the step matrix $\mathbf{A}$ for the discrete dynamical system that models the distribution of videos from day to day.

(b) Assuming the videos are split equally between the locations on day 0, find the distribution of videos between the locations after 1, 2, and 3 days.

(c) Find the stationary distribution vector $\mathbf{u} = (u_1, u_2)$ which describes the distribution of videos in the long run.

24. A country has three parties: Democrat (D), Republican (R), and Green (G). There are 60 million people in the country, and 50% of these are guaranteed to vote in each election that is held every four years. Records show that 10% of those voting D in this election will vote R and 5% will vote G in the next election; 5% of those voting R in this election will vote D and 5% will vote G in the next election; 10% of those voting G in this election will vote D and 10% will vote R in the next election.

(a) Find the step matrix **A** for the discrete dynamical system that models the distribution of votes from election to election.

(b) Assuming that the ratio of votes between parties D, R, G is 5 : 4 : 1 after election 0, find the distribution of votes between parties after the next 3 elections.

(c) Find the stationary distribution vector $\mathbf{u} = (u_1, u_2, u_3)$ that describes the distribution of votes in the long run.

25. A community classifies its land in one of three ways: urban (U), agricultural (A), forested (F). A recent survey showed that 10% was classified U, 40% classified A, and 50% classified F. In a survey carried out five years later it was found that 80% of the urban land remained urban, 15% had become agricultural, and 5% had become forested. The remaining survey data are shown in the following table.

	U	A	F
U	80	5	5
A	15	90	5
F	5	5	90

Surveys are conducted every five years.

(a) Find the step matrix **A** for the discrete dynamical system that models the classification of land use from survey to survey.

(b) Assuming trends remain the same, predict the classification of land use for the next 3 surveys.

(c) Find the stationary distribution vector $\mathbf{u} = (u_1, u_2, u_3)$ that describes the classification of land use in the long run.

26. Consider the following step matrix

$$\mathbf{A} = 0.25\begin{bmatrix} 1 & 1 & 1 \\ 0 & 1 & 1 \\ 0 & 0 & 1 \end{bmatrix}.$$

(a) Compute $\mathbf{A}^k$ for $n = 2, 3, 4$.

(b) Find a formula for $\mathbf{A}^k$ for any positive integer k.

(c) If the initial distribution vector is $\mathbf{u}_0 = (100, 100, 100)$, find the smallest value of k for which each component in $\mathbf{u}_k$ is less than 1.

Exercise 27–28. Use mathematical induction to verify the formula for $\mathbf{A}^k$. *Beginning with an initial distribution vector* $\mathbf{u}_0 = (100, 200, 300)$, *determine the long term behavior of the discrete dynamical system with step matrix* **A**.

27. $\mathbf{A} = \begin{bmatrix} 1 & \frac{1}{2} & 0 \\ 0 & \frac{1}{2} & 1 \\ 0 & 0 & 0 \end{bmatrix}, \quad \mathbf{A}^k = \begin{bmatrix} 1 & 1-(\frac{1}{2})^k & 1-(\frac{1}{2})^{k-1} \\ 0 & (\frac{1}{2})^k & (\frac{1}{2})^{k-1} \\ 0 & 0 & 0 \end{bmatrix}$

28. $\mathbf{A} = \begin{bmatrix} 1 & \frac{1}{4} & 0 \\ 0 & \frac{1}{2} & 0 \\ 0 & \frac{1}{4} & 1 \end{bmatrix}, \quad \mathbf{A}^k = \begin{bmatrix} 1 & \frac{1}{2}-(\frac{1}{2})^{k+1} & 0 \\ 0 & (\frac{1}{2})^k & 0 \\ 0 & \frac{1}{2}-(\frac{1}{2})^{k+1} & 1 \end{bmatrix}$

SECTION 2.4.3 AGE-STRUCTURED POPULATION MODELS

Exercises 29–32. For each population projection matrix **P**, *find the age distribution vector that gives a stable population.*

29. $\begin{bmatrix} \frac{1}{2} & 3 \\ \frac{1}{6} & 0 \end{bmatrix}$

30. $\begin{bmatrix} 0.9 & 1 \\ 0.1 & 0 \end{bmatrix}$

31. $\begin{bmatrix} 0 & 3 & 5 \\ 0.25 & 0 & 0 \\ 0 & 0.20 & 0 \end{bmatrix}$

32. $\begin{bmatrix} \frac{1}{2} & 2 & 10 \\ \frac{1}{6} & 0 & 0 \\ 0 & \frac{1}{10} & 0 \end{bmatrix}$

33. The lifespan of a bird species is divided into three age classes 0, 1, 2. Each female produces on average two offspring in year 1 and a sample of 10 females produced a total of 14 offspring in year 2. Breeding ends after year 2. Only 40% of the females of age 0 survive to age 1 and 25% of those age 1 survive to year 2. Write down the population projection matrix for this model. Consider an initial population distribution vector $\mathbf{u}_0 = [200\ \ 100\ \ 50]^{\mathrm{T}}$. Describe the population evolution in the first few years. Will the population reach a steady state in the long run? Will the species prosper or die out?

34. The reproductive lifespan of a certain species of turtle is divided into four age classes 0, 1, 2, 3, each class having 3 year's duration. The population is modeled by the projection matrix

$$\mathbf{P} = \begin{bmatrix} \frac{1}{10} & \frac{1}{2} & \frac{3}{4} & \frac{1}{4} \\ \frac{4}{5} & 0 & 0 & 0 \\ 0 & \frac{3}{4} & 0 & 0 \\ 0 & 0 & \frac{1}{3} & 0 \end{bmatrix}$$

The fecundity parameter $f_0 = \frac{1}{10}$ means that, on average, one offspring is produced for every group of 10 females in age class 0. The other parameters are interpreted similarly. Determine whether or not the population will reach a steady state, and if so, determine the stable population distribution.

35. The lifespan of a species of duck is divided into two age classes 0 and 1. The population projection matrix and the age distribution vector are given, respectively, by

$$\mathbf{P} = \begin{bmatrix} 1 & 1.5 \\ 0.5 & 0 \end{bmatrix} \quad \text{and} \quad \mathbf{u}(t) = \begin{bmatrix} u_1 \\ u_2 \end{bmatrix}.$$

Suppose, in each year, k_1 and k_2 is the fraction of birds in age classes 0 and 1, respectively, killed by hunters.

(a) Show that this population model is defined by the equation $\mathbf{u}(t+1) = \mathbf{B}\mathbf{u}(t)$, $t = 0, 1, 2, \ldots$, where

$$\mathbf{B} = (\mathbf{I}_2 - \mathbf{K})\mathbf{P} \quad \text{and} \quad \mathbf{K} = \begin{bmatrix} k_1 & 0 \\ 0 & k_2 \end{bmatrix}.$$

(b) Let $k_1 = k_2 = k$. Show that there is a value of k that will allow the population to reach a steady state and find the steady state.

36. Refer to Illustration 2.4.3. Assume that each female breeds in years 1 through 5 and that populations $n_k(t)$, $0 \leq k \leq 4$, are recorded just after breeding at the beginning of year t. Observe that $n_5(t) = s_4 n_4(t-1)$ is the number of birds aged 5 who have just bred and will die out during year t. Show that the bird population is modeled by a discrete dynamical system defined, for $t = 0, 1, 2, \ldots$, by the equation

$$\begin{bmatrix} s_0 f_1 & s_1 f_2 & s_2 f_3 & s_3 f_4 & s_4 f_5 \\ s_0 & 0 & 0 & 0 & 0 \\ 0 & s_1 & 0 & 0 & 0 \\ 0 & 0 & s_2 & 0 & 0 \\ 0 & 0 & 0 & s_3 & 0 \end{bmatrix} \begin{bmatrix} n_0 \\ n_1 \\ n_2 \\ n_3 \\ n_4 \end{bmatrix} (t) = \begin{bmatrix} n_0 \\ n_1 \\ n_2 \\ n_3 \\ n_4 \end{bmatrix} (t+1).$$

SECTION 2.4.4 INPUT-OUTPUT MODELS

Exercises 37–40. Use the exchange matrix $\mathbf{C}$ *to find a nonnegative production vector* $\mathbf{x}$ *satisfying* $\mathbf{Cx} = \mathbf{x}$.

37. $\begin{bmatrix} 0.75 & 0.5 \\ 0.25 & 0.5 \end{bmatrix}$ **38.** $\begin{bmatrix} 1 & 0.5 \\ 0 & 0.5 \end{bmatrix}$

39. $\begin{bmatrix} 0.50 & 0.70 & 0.35 \\ 0.25 & 0.30 & 0.25 \\ 0.25 & 0 & 0.40 \end{bmatrix}$ **40.** $\begin{bmatrix} 0 & 0 & 0.75 \\ 1 & 0.50 & 0.25 \\ 0 & 0.50 & 0 \end{bmatrix}$

41. Suppose that $\mathbf{C}$ is an $n \times n$ consumption matrix such that each column sum in $\mathbf{C}$ is 1. Show that the homogeneous linear system $\mathbf{Ax} = \mathbf{0}$, where $\mathbf{A} = \mathbf{I} - \mathbf{C}$ has infinitely many solutions. Is $\mathbf{x}$ always a production vector?

42. A city economy has three interdependent sectors labeled Government (G), Housing (H), Transport (T). Each \$1 produced by G requires \$ 0.72 of G, \$ 0.18 of H and \$ 0.1 of T. Each \$1 produced by H requires \$ 0.35 of G, \$ 0.45 of H and \$ 0.2 of T. Each \$1 produced by T requires \$ 0.15 of G, \$ 0.25 of H and \$ 0.55 of T.

Construct the input-output table for this model. Is the model open or closed? Which sector uses the most transport?

43. Consider an economy based of three sectors that has the consumption matrix $\mathbf{C}$ and production vector $\mathbf{x}$ given by

$$\mathbf{C} = \begin{bmatrix} 0.2 & 0 & 0.2 \\ 0.4 & 0.5 & 0.2 \\ 0.4 & 0.5 & 0.6 \end{bmatrix}, \quad \mathbf{x} = \begin{bmatrix} x_1 \\ x_2 \\ x_3 \end{bmatrix}.$$

Solve the linear system $\mathbf{Cx} = \mathbf{x}$. Find $\mathbf{x}$ when the total output for the economy is 37 million dollars.

44. A closed economic model is based on Coal and Steel. The input-output table and production vector $\mathbf{x}$ are as follows:

	Coal	Steel
Coal	0.1	0.8
Steel	0.9	0.2

$$\mathbf{x} = \begin{bmatrix} x_1 \\ x_2 \end{bmatrix}$$

Compute $\mathbf{x}$ if the total output is 10 (million \$).

Exercises 45–48. In each case a consumption matrix $\mathbf{C}$ *and an external demand vector* $\mathbf{d}$ *(million \$) are given. Use hand calculation to find the production vector* $\mathbf{x}$.

45. $\begin{bmatrix} 0.10 & 0.40 \\ 0.30 & 0.20 \end{bmatrix}, \quad \begin{bmatrix} 1 \\ 2 \end{bmatrix}$

46. $\begin{bmatrix} 0.20 & 0.15 \\ 0.16 & 0.40 \end{bmatrix}, \quad \begin{bmatrix} 0.05 \\ 0.56 \end{bmatrix}$

47. $\begin{bmatrix} 0.50 & 0.20 & 0.10 \\ 0 & 0.40 & 0.20 \\ 0 & 0 & 0.50 \end{bmatrix}, \quad \begin{bmatrix} 0 \\ 1 \\ 5 \end{bmatrix}$

48. $\begin{bmatrix} 0.60 & 0.12 & 0 \\ 0 & 0.50 & 0.20 \\ 0.50 & 0.35 & 0.50 \end{bmatrix}, \quad \begin{bmatrix} 4 \\ 0 \\ 5 \end{bmatrix}$

49. A closed model for an economy with three industries (Heat, Light, Water) has the following consumption matrix C and production vector $\mathbf{x}$.

$$\mathbf{C} = \begin{bmatrix} 0.5 & 0.1 & 0.5 \\ 0.5 & 0.4 & 0.3 \\ 0 & 0.5 & 0.2 \end{bmatrix}, \quad \mathbf{x} = \begin{bmatrix} x_1 \\ x_2 \\ x_3 \end{bmatrix}$$

Write down the 3×3 linear system that models this economy. Solve for $\mathbf{x}$ given that the total production is \$3.92 (million).

USING MATLAB

50. *Experiment.* Write an M-file that accepts a positive integer n as input and generates a random $n \times n$ adjacency matrix $\mathbf{A}$ for an undirected graph G on n vertices. Draw the undirected graphs when the M-file is run for the cases $n = 3, 4, 5$. In each case, determine the number of paths of length 3 in G and list them all. Are there any cycles in G of length 3?

51. Denote the vertices of a graph by positive integers. To form the *adjacency list* for a digraph G we put i in column j of a table if there is a directed edge $j \to i$, as in the following example.

1	2	3	4	5	6
4	1	2	2	6	1
5			3		
			5		

Determine (and list) (a) the number of directed walks of length 6 from vertex 4 to vertex 4, (b) the number of directed paths of length 5 from vertex 3 to vertex 6, (c) the number of directed cycles of length 4 based at vertex 4.

52. Consult The Traveling Salesman Problem in MATLAB demos.

53. Consider the age-structured population model developed in Section 2.4.3. Assume that the fecundity and survival parameters are given by

$$f_0 = 0, \quad f_1 = 0, \quad f_2 = 0, \quad f_3 = 2, \quad f_4 = 4, \quad f_5 = 5$$

$$s_{0,1} = 0.4, \quad s_{1,2} = 0.6, \quad s_{2,3} = 0.7, \quad s_{3,4} = 0.7, \quad s_{4,5} = 0.5\,.$$

Investigate the population distributions in subsequent years given the initial population $\mathbf{u} = (300, 200, 200, 400, 500)$. What conclusions do you draw?

54. *Project.* Consider the Leslie matrix

$$\mathbf{P} = \begin{bmatrix} f_1 & f_2 & f_3 \\ s_0 & 0 & 0 \\ 0 & s_1 & 0 \end{bmatrix}$$

for an age-structured population study with three age classes. Choose survival parameters s_0 and s_1 and an initial age distribution vector $\mathbf{u}(0)$. Write an M-file that accepts the fecundity parameters as input and generates future population distribution vectors. Investigate how the fecundity parameters affect growth or decline of population.

55. *Project.* Assume that you are a biologist investigating the long term survival of a certain species classified vulnerable by the World Conservation Union. There are six age classes and the current population distribution is (220, 190, 180, 170, 150, 90). The survival parameters are estimated from recent surveys. The Leslie matrix has the form

$$\begin{bmatrix} 0 & f_2 & f_3 & f_4 & 0 & 0 \\ 0.94 & 0 & 0 & 0 & 0 & 0 \\ 0 & 0.96 & 0 & 0 & 0 & 0 \\ 0 & 0 & 0.96 & 0 & 0 & 0 \\ 0 & 0 & 0 & 0.8 & 0 & 0 \\ 0 & 0 & 0 & 0 & 0.5 & 0 \end{bmatrix}.$$

Vary the fecundity parameters in order to determine how the population might increase or decline. If possible, determine the fecundity parameters f_i, $2 \le i \le 4$, such that the population distribution will be stable.

56. The columns of the consumption matrix $\mathbf{C} = [c_{ij}]$ correspond, respectively, to an economy based on four industries A, B, C, D.

$$\mathbf{C} = \begin{bmatrix} 0.25 & 0.10 & 0.12 & 0.35 \\ 0.50 & 0.60 & 0 & 0.10 \\ 0.25 & 0 & 0.32 & 0 \\ 0 & 0 & 0 & 0.55 \end{bmatrix}$$

(a) Explain the entries $c_{41} = 0$ and c_{33}.

(b) Is industry A productive?

(c) Which industries are profitable?

(d) Is this an open or closed model?

(e) Which industries are independent of industry C?

Find the production vector $\mathbf{x} = (x_1, x_2, x_3, x_4)$ if there is an external demand on the economy of $(2, 4, 6, 2)$ million \$ for taxes.

57. Refer to the matrix **C** in the previous exercise. What is the smallest value of c such that the production vector $\mathbf{x} = (1, c, 1, 1)$ results in a demand vector **d** with all entries positive?

58. *Experiment.* Write an M-file to verify the results in Example 7. For which members is t_i $(i = 1, 2, 3)$ positive? Rework the example using the same angles $\theta_1 = -30°$, $\theta_2 = -20°$, $\theta_3 = 40°$ but with varying coefficients of flexibility k_i in the members $i = 1, 2, 3$. Report on your findings. Next, experiment with a variety of angles and coefficients of flexibility in order to conjecture if the matrix $\mathbf{B} = \mathbf{A}\mathbf{K}^{-1}\mathbf{A}^{\mathsf{T}}$ is invertible in every case for this particular truss.

CHAPTER 2 REVIEW

State whether each statement is true or false as stated. Provide a clear reason for your answer. If a false statement can be modified to make it true, then do so.

1. If $\begin{bmatrix} 1 & 2 \\ 4 & 6 \end{bmatrix}$, then $\mathbf{A}^2 - 7\mathbf{A} = 2\mathbf{I}_2$.

2. If $\mathbf{AD} = \mathbf{BD}$, for $n \times n$ matrices **A**, **B**, **D**, then $\mathbf{A} = \mathbf{B}$.

3. If **A** and **B** commute, then $\mathbf{A}^2\mathbf{B} = \mathbf{B}\mathbf{A}^2$.

4. If **A** is symmetric, then $-\mathbf{A}$ is skew-symmetric.

5. The product of two 2×2 skew-symmetric matrices is a scalar matrix.

6. The index of nilpotency of a nilpotent matrix can equal its order.

7. If **A** is idempotent, then $\mathbf{A}^k = \mathbf{A}$ for all positive integers k.

8. The matrix $\begin{bmatrix} 1 & -4 \\ 5 & 2 \end{bmatrix}$ is a product of elementary matrices.

9. The matrix $\begin{bmatrix} 2 \\ -1 \end{bmatrix} [2 \ \ -3]$ is a product of elementary matrices.

10. If **A** and **B** are $n \times n$ invertible matrices, then $\mathbf{A} + \mathbf{B}$ is invertible.

11. The equation $\mathbf{A}\begin{bmatrix} -1 \\ 3 \end{bmatrix} = \begin{bmatrix} 4.5 \\ -3.1 \end{bmatrix}$ is true for some matrix **A** with all nonzero entries.

12. The equation $\mathbf{A}\begin{bmatrix} 1 \\ 2 \end{bmatrix} = \begin{bmatrix} 5 \\ 11 \\ 17 \end{bmatrix}$ is valid for some matrix **A**.

13. If **A** is $n \times n$ and singular, then the linear system $\mathbf{Ax} = \mathbf{b}$ is inconsistent.

14. If **A** is $n \times n$ and not a product of elementary matrices, then $\mathbf{Ax} = \mathbf{0}$ has infinitely many solutions.

15. If $(\mathbf{A} - \mathbf{I})$ is invertible, then the equation $\mathbf{Ax} = \mathbf{x}$ has a nonzero solution **x**.

16. If $\mathbf{A} = \begin{bmatrix} 1 & 0 \\ 0 & 1 \\ 0 & 0 \end{bmatrix}$ and $\mathbf{B} = \begin{bmatrix} 1 & 0 & 1 \\ 0 & 1 & 0 \end{bmatrix}$, then one matrix is a right-inverse for the other.

17. The matrix $\begin{bmatrix} 1 & -1 \\ -3 & 3 \\ 7 & -7 \end{bmatrix}$ has a left-inverse.

18. If a square matrix has nonzero entries on the main diagonal, then **A** is invertible.

19. If **A** is a permutation matrix, then $\mathbf{A}^k$ is a permutation matrix for every $k \geq 2$.

20. The matrix $\begin{bmatrix} 1 & 2 & 3 \\ 1 & 3 & 4 \\ -1 & 0 & 1 \end{bmatrix}$ has an LU-factorization.

Josiah Willard Gibbs (1839–1903)

Josiah Gibbs graduated from Yale University in 1863 with the first doctorate of engineering ever awarded in the United States. In his thesis he used geometrical methods to study the design of gears. During the 1860s he studied in Europe, where he met and was influenced by the famous German physicists Hermann-Ludwig von Helmholtz and Gustav Kirchhoff. On returning to Yale in 1871 he was appointed professor of mathematical physics.

Gibbs's work was grounded in theoretical mathematics but had far-reaching applications to engineering, physics, chemistry, and astronomy. His pioneering work in *vector analysis* is considered a major contribution to pure and applied mathematics. Vector methods first appeared in printed notes to his students (1881–1884) but were not published until 1901. In 1880, Gibbs used vector analysis to compute the orbit of a comet using only three observations. His methods, applied in particular to Swift's comet, required less computation than those of Gauss (see Chapter 1, Profile).

A series of five papers (1882–1889) dealt with the electromagnetic theory of light. Gibbs's development of statistical mechanics provided a mathematical framework for quantum theory and for the work of the renowned Scottish mathematician and physicist James Clerk Maxwell (1831–1879).

Gibbs was a gifted engineer and scientist whose genius was not fully recognized in his lifetime. The American Mathematical Society named a lecture series in his honor in 1923 and a public lecture has been given by a distinguished mathematician in most years since that time.

Chapter 3

VECTORS

Introduction

In Chapter 1, a linear system was regarded as an *ordered list* of linear equations. In this chapter we look at linear systems in two other ways that come from writing a linear system in either its *vector form* or its *matrix form*. Each of these forms gives a different perspective on the geometric character of linear systems and their solutions. The 3×2 linear system (S) shown in (3.1) is used to develop the ideas.

$$(S)\quad \begin{cases} x_1 + 2x_2 = 3 \\ 3x_1 - x_2 = 2 \\ 4x_1 + 5x_2 = 9 \end{cases} \tag{3.1}$$

Linear Systems: Vector Form

Suppose we look at (S) not as an ordered list of three linear equations but as a single equation involving 3×1 matrices (3-vectors). Using the left side and right side of all equations in (S) gives the matrix equation

$$\begin{bmatrix} x_1 + 2x_2 \\ 3x_1 - x_2 \\ 4x_1 + 5x_2 \end{bmatrix} = \begin{bmatrix} 3 \\ 2 \\ 9 \end{bmatrix}, \tag{3.2}$$

because corresponding entries are equal. Using laws of matrix addition and scalar multiplication, the 3-vector on the left side of (3.2) can be rewritten as

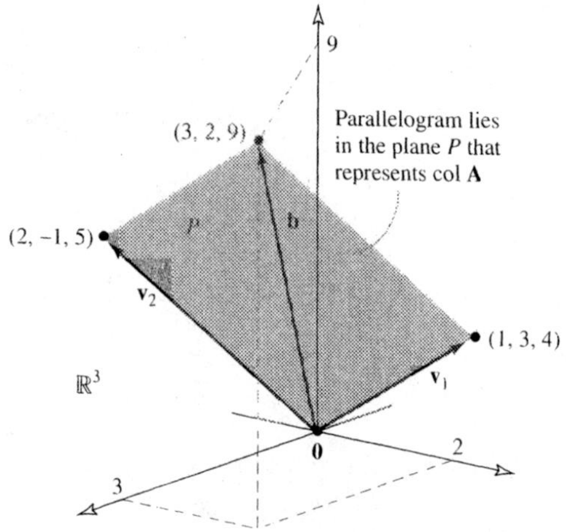

Figure 3.1 Column space.

follows:

$$\begin{bmatrix} x_1 + 2x_2 \\ 3x_1 - x_2 \\ 4x_1 + 5x_2 \end{bmatrix} = \begin{bmatrix} x_1 \\ 3x_1 \\ 4x_1 \end{bmatrix} + \begin{bmatrix} 2x_2 \\ -x_2 \\ 5x_2 \end{bmatrix} = x_1 \begin{bmatrix} 1 \\ 3 \\ 4 \end{bmatrix} + x_2 \begin{bmatrix} 2 \\ -1 \\ 5 \end{bmatrix}. \qquad (3.3)$$

Combining (3.2) and (3.3) gives a single *vector equation*

$$x_1 \begin{bmatrix} 1 \\ 3 \\ 4 \end{bmatrix} + x_2 \begin{bmatrix} 2 \\ -1 \\ 5 \end{bmatrix} = \begin{bmatrix} 3 \\ 2 \\ 9 \end{bmatrix} \qquad (3.4)$$

or in more compact form,

$$x_1\mathbf{v}_1 + x_2\mathbf{v}_2 = \mathbf{b}, \qquad (3.5)$$

where

$$\mathbf{v}_1 = \begin{bmatrix} 1 \\ 3 \\ 4 \end{bmatrix}, \qquad \mathbf{v}_2 = \begin{bmatrix} 2 \\ -1 \\ 5 \end{bmatrix}, \qquad \mathbf{b} = \begin{bmatrix} 3 \\ 2 \\ 9 \end{bmatrix}.$$

We call (3.4), or (3.5), the *vector form* of the linear system (S) in (3.1).

The expression $x_1\mathbf{v}_1 + x_2\mathbf{v}_2$ on the left side of (3.5) is called a *linear combination* of $\mathbf{v}_1$ and $\mathbf{v}_2$ using scalars x_1 and x_2. The vectors $\mathbf{v}_1$ and $\mathbf{v}_2$ are the columns of the coefficient matrix $\mathbf{A}$ for (3.1). We have

$$\mathbf{A} = [\,\mathbf{v}_1 \;\; \mathbf{v}_2\,] = \begin{bmatrix} 1 & 2 \\ 3 & -1 \\ 4 & 5 \end{bmatrix}$$

and writing the unknowns as a 2-vector $\mathbf{x} = \begin{bmatrix} x_1 \\ x_2 \end{bmatrix}$, we are able to express $x_1\mathbf{v}_1 + x_2\mathbf{v}_2$ as a matrix-column vector product, namely

$$\mathbf{A}\mathbf{x} = [\,\mathbf{v}_1 \;\; \mathbf{v}_2\,] \begin{bmatrix} x_1 \\ x_2 \end{bmatrix} = x_1\mathbf{v}_1 + x_2\mathbf{v}_2.$$

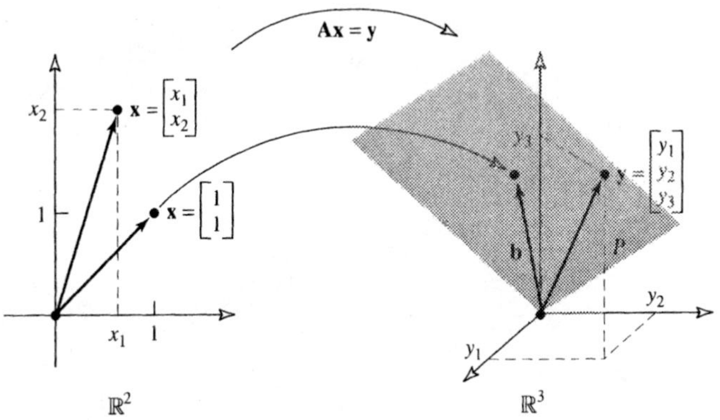

Figure 3.2 A transformation on $\mathbb{R}^2$.

The set of all linear combinations $x_1\mathbf{v}_1 + x_2\mathbf{v}_2$ as x_1 and x_2 run through all possible real values is called the *column space* of **A**, denoted by col **A**. Equivalently, col **A** is the set of all products **Ax** for all 2-vectors of unknowns **x**.

Refer to Figure 3.1. In this example, col **A** can be visualized geometrically as a plane P in $\mathbb{R}^3$ passing through the origin and containing the *arrows* that represent $\mathbf{v}_1$ and $\mathbf{v}_2$. If we view the system (S) in its vector form (3.5), then (S) is consistent if and only if **b** is a linear combination of the columns in **A**. In this example, setting $x_1 = x_2 = 1$ gives $\mathbf{v}_1 + \mathbf{v}_2 = \mathbf{b}$ and so (S) is consistent. Geometrically, the arrow that represents **b** lies in the plane P that represents col **A**.

Linear Systems: Matrix Form

We now take a different point of view. Recall from Section 2.1 that the linear system (S) in (3.1) has matrix form $\mathbf{Ax} = \mathbf{b}$, where $\mathbf{x} = \begin{bmatrix} x_1 \\ x_2 \end{bmatrix}$. Refer to Figure 3.2. The coefficient matrix **A** defines a *transformation* (function)

$$\mathbf{x} = \begin{bmatrix} x_1 \\ x_2 \end{bmatrix} \quad \xrightarrow{\text{multiply } \mathbf{A} \text{ by } \mathbf{x}} \quad \mathbf{y} = \begin{bmatrix} y_1 \\ y_2 \\ y_3 \end{bmatrix},$$

which maps each 2-vector **x** in $\mathbb{R}^2$ to a corresponding 3-vector $\mathbf{y}(= \mathbf{Ax})$ in $\mathbb{R}^3$. Forming the product **Ax**, we have

$$\mathbf{Ax} = \begin{bmatrix} 1 & 2 \\ 3 & -1 \\ 4 & 5 \end{bmatrix} \begin{bmatrix} x_1 \\ x_2 \end{bmatrix} = \begin{bmatrix} x_1 + 2x_2 \\ 3x_1 - x_2 \\ 4x_1 + 5x_2 \end{bmatrix} = \begin{bmatrix} y_1 \\ y_2 \\ y_3 \end{bmatrix} = \mathbf{y} \qquad (3.6)$$

and so the set of all vectors **y** in (3.6), as x_1 and x_2 run through all possible real values, is exactly col **A**. In other words, the transformation **Ax** maps all of $\mathbb{R}^2$ onto the plane P in $\mathbb{R}^3$ that represents col **A**. From this standpoint the system (S) in (3.1) is consistent if there is at least one 2-vector **x** in $\mathbb{R}^2$ that is mapped by **A** onto the vector **b** in $\mathbb{R}^3$. Figure 3.2 shows that this is indeed the case.

The vector and matrix forms of a linear system lead us in a natural way to study *spaces of vectors*, which is the subject of this chapter. Most of the concepts introduced in this chapter reappear again in Chapter 7, this time in the context of a general *vector space*. Chapter 3 should be viewed therefore as preparation for the more abstract theory that is to come.

3.1 Spaces of Vectors

Definition 3.1 ***The Spaces*** $\mathbb{R}^n$

The real number system is denoted by the symbol $\mathbb{R}$. For each positive integer $n = 1, 2, \ldots$, the symbol $\mathbb{R}^n$ (pronounced "R-n") denotes the set of all $n \times 1$ matrices with real entries. An object in $\mathbb{R}^n$ is called a *column vector* or *n-vector* and its entries are called *components*.

Vectors are denoted by boldfaced lowercase letters, as in the following examples

$$\mathbf{x} = \begin{bmatrix} x_1 \\ x_2 \end{bmatrix}, \quad \mathbf{b} = \begin{bmatrix} 3 \\ 2 \\ 9 \end{bmatrix}, \quad \ldots, \quad \mathbf{v} = \begin{bmatrix} v_1 \\ v_2 \\ \vdots \\ v_n \end{bmatrix}. \tag{3.7}$$

2-vector   3-vector n-vector

Visualizing $\mathbb{R}^n$

Refer to Figure 3.3(a). A two-dimensional space with a rectangular frame of reference (coordinate system) is called *the plane*. Each 2-vector $\mathbf{v} = \begin{bmatrix} v_1 \\ v_2 \end{bmatrix}$ in $\mathbb{R}^2$ corresponds to a unique point (v_1, v_2) in the plane and conversely, the coordinates of each point (v_1, v_2) in the plane define a unique 2-vector.[1] Each

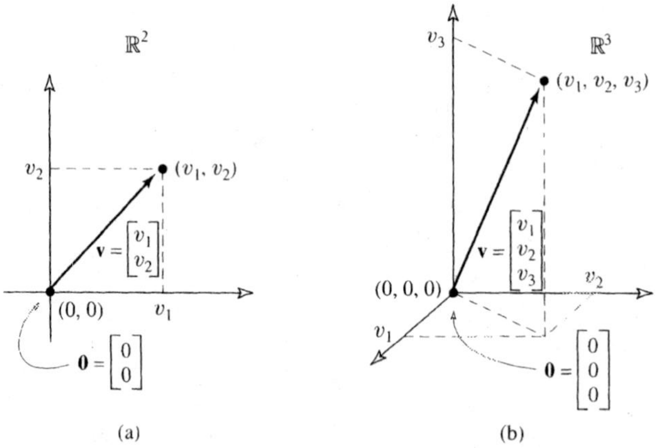

Figure 3.3 $\mathbb{R}^2$ and $\mathbb{R}^3$.

[1] In mathematical language, there is a *one-to-one* correspondence between the set of 2-vectors and the points in the plane.

2-vector is visualized as an *arrow* (*directed line segment*) with *initial point* $(0, 0)$ and *terminal point* (v_1, v_2), and $\mathbf{v}$ is called the *position vector* of the point (v_1, v_2). The origin $(0, 0)$ corresponds to the unique zero vector $\mathbf{0} = \begin{bmatrix} 0 \\ 0 \end{bmatrix}$ in $\mathbb{R}^2$. Thus, $\mathbb{R}^2$ is visualized geometrically as the set of all arrows in the plane with their initial points *anchored* at the origin $(0, 0)$.

Figure 3.3(b) shows three-dimensional space with a rectangular frame of reference. The components in each 3-vector $\mathbf{v} = \begin{bmatrix} v_1 \\ v_2 \\ v_3 \end{bmatrix}$ in $\mathbb{R}^3$ define a unique point (v_1, v_2, v_3) in three-dimensional space, and conversely. $\mathbb{R}^3$ is visualized geometrically as the set of all arrows anchored at the origin $(0, 0, 0)$.

In Figure 3.4 we see how the geometric representation of $\mathbb{R}^2$ and $\mathbb{R}^3$ is extended by analogy to give an abstract representation of $\mathbb{R}^n$. We think of the n-tuple $(v_1, v_2, \ldots, v_n)$ as the terminal point of the arrow that represents $\mathbf{v}$. Keep in mind that an n-vector and the associated n-tuple are both ordered lists of numbers that carry the same information in their components. However, they are distinct entities that cannot be equated.

3.1.1 Vector Algebra, Vector Geometry

An n-vector is an $n \times 1$ matrix and so $\mathbb{R}^n$ is exactly the set $\mathbb{R}^{n\times 1}$ of matrices defined in Section 2.1. The definitions of equality, addition, scalar multiplication, and subtraction given in Section 2.1 apply in particular to $\mathbb{R}^n$, and we will review these. Let $\mathbf{u}$ and $\mathbf{v}$ be n-vectors and let s be any real scalar.

- The vectors $\mathbf{u}$ and $\mathbf{v}$ are equal exactly when each of their corresponding components are equal, and we write $\mathbf{u} = \mathbf{v}$ when this is the case.
- The sum $\mathbf{u} + \mathbf{v}$ is the n-vector obtained by adding corresponding components in $\mathbf{u}$ and $\mathbf{v}$.
- The multiplication (scaling) of $\mathbf{u}$ by a scalar s is the n-vector $s\mathbf{u}$ obtained by multiplying all components in $\mathbf{u}$ by s.

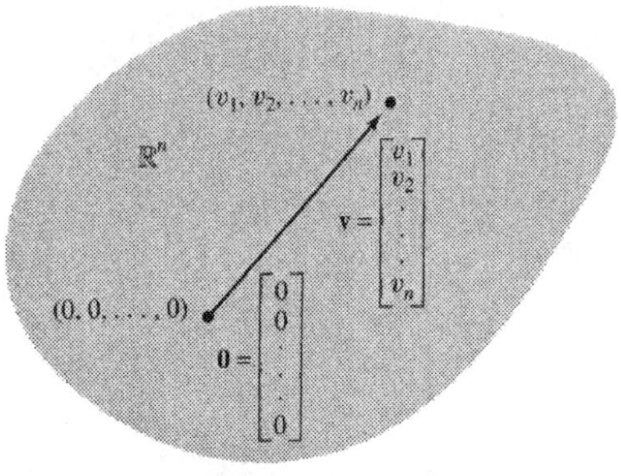

Figure 3.4 Visualizing $\mathbb{R}^n$.

- In each space $\mathbb{R}^n$ there is a unique zero vector $\mathbf{0}$ that has all components equal to zero.
- For each vector $\mathbf{v}$ in $\mathbb{R}^n$ there is a unique vector $-\mathbf{v}$ in $\mathbb{R}^n$, obtained by negating all the entries in $\mathbf{v}$, that satisfies the equation $\mathbf{v}+(-\mathbf{v})=\mathbf{0}$. The difference $\mathbf{u}-\mathbf{v}$ is defined by the equation $\mathbf{u}-\mathbf{v}=\mathbf{u}+(-\mathbf{v})$.

Refer to the remarks that precede Theorem 2.1 in Section 2.1. Theorem 2.1 states the eight rules for matrix addition and scalar multiplication on the set $\mathbb{R}^{m\times n}$ of all $m\times n$ matrices. The sets $\mathbb{R}^n$ and $\mathbb{R}^{n\times 1}$ are identical, and so the same rules, stated in Theorem 3.1, apply in particular to $\mathbb{R}^n$. The rules will appear in Section 7.1 as axioms which form part of the definition of an abstract vector space.

Theorem 3.1 **Algebraic Rules of Vector Algebra**

Let $\mathbf{u}$, $\mathbf{v}$, $\mathbf{w}$ *be vectors in* $\mathbb{R}^n$, $\mathbf{0}$ *the unique zero vector, and* s, t *any real scalars. Then the following rules apply.*

	Vector Addition		**Scalar Multiplication**
(A1)	$\mathbf{u}+\mathbf{v}=\mathbf{v}+\mathbf{u}$	**(A5)**	$s(\mathbf{u}+\mathbf{v})=s\mathbf{u}+s\mathbf{v}$
(A2)	$(\mathbf{u}+\mathbf{v})+\mathbf{w}=\mathbf{u}+(\mathbf{v}+\mathbf{w})$	**(A6)**	$(s+t)\mathbf{v}=s\mathbf{v}+t\mathbf{v}$
(A3)	$\mathbf{v}+\mathbf{0}=\mathbf{v}$	**(A7)**	$s(t\mathbf{v})=t(s\mathbf{v})=(st)\mathbf{v}$
(A4)	$\mathbf{v}+(-\mathbf{v})=\mathbf{0}$	**(A8)**	$1\mathbf{v}=\mathbf{v}$

Operations on n-vectors are performed on components, and so the rules **(A1)**–**(A8)** are generalizations of corresponding algebraic rules for real and complex numbers. For example, the *commutative law* $a+b=b+a$ for real and complex numbers a and b extends to 2-vectors $\mathbf{u}$ and $\mathbf{v}$ in the following way:

$$\mathbf{u}+\mathbf{v}=\begin{bmatrix} u_1+v_1 \\ u_2+v_2 \end{bmatrix}=\begin{bmatrix} v_1+u_1 \\ v_2+u_2 \end{bmatrix}=\mathbf{v}+\mathbf{u}.$$

EXAMPLE 1 **Vector Operations in $\mathbb{R}^2$**

$$\mathbf{u}=\begin{bmatrix}1\\3\end{bmatrix},\quad \mathbf{v}=\begin{bmatrix}2\\1\end{bmatrix}\quad\Rightarrow\quad \mathbf{u}+\mathbf{v}=\begin{bmatrix}1+2\\3+1\end{bmatrix}=\begin{bmatrix}3\\4\end{bmatrix}$$

$$0\mathbf{u}=0\begin{bmatrix}1\\3\end{bmatrix}=\begin{bmatrix}0\\0\end{bmatrix}\qquad -\mathbf{v}=(-1)\mathbf{v}=\begin{bmatrix}-2\\-1\end{bmatrix},$$

$$\mathbf{u}-\mathbf{v}=\mathbf{u}+(-1)\mathbf{v}=\begin{bmatrix}1\\3\end{bmatrix}+\begin{bmatrix}-2\\-1\end{bmatrix}=\begin{bmatrix}-1\\2\end{bmatrix}$$ ▪

The sum $\mathbf{u}+\mathbf{v}$ and difference $\mathbf{u}-\mathbf{v}$ of vectors $\mathbf{u}$ and $\mathbf{v}$ in $\mathbb{R}^n$ is visualized geometrically using the *parallelogram law*. Refer to Figure 3.5. The vectors

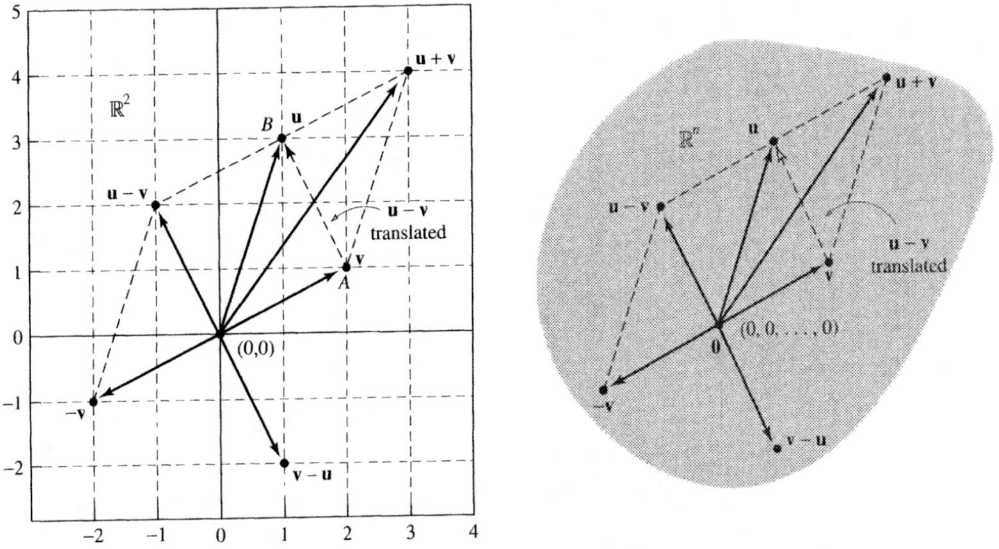

Figure 3.5 Parallelogram law in $\mathbb{R}^2$ and $\mathbb{R}^n$.

from Example 1 are plotted in $\mathbb{R}^2$. The sum $\mathbf{u} + \mathbf{v}$ is represented by the arrow with terminal point $(3, 4)$, the fourth vertex of the parallelogram formed by the arrows $\mathbf{u}$ and $\mathbf{v}$. The vector $\mathbf{u} - \mathbf{v}$ is represented by an arrow (anchored at the origin) with terminal point $(-1, 2)$, which is the fourth vertex of the parallelogram formed by the arrows $\mathbf{u}$ and $-\mathbf{v}$. Note that the directed line segment $\overrightarrow{AB}$ does not represent $\mathbf{u} - \mathbf{v}$ but does represent $\mathbf{u} - \mathbf{v}$ translated in the plane. Actually, $\overrightarrow{AB}$ is a called a *free vector* (not anchored at the origin) that has the same magnitude and direction as $\mathbf{u} - \mathbf{v}$. The parallelogram law in $\mathbb{R}^n$ is illustrated by the schematic picture on the right in Figure 3.5.

Free Vectors in Engineering and Science

Engineering, physics, and other applied fields define a *vector* quantity to be any physical quantity that has a *magnitude* and a *direction*.

In mechanics, for example, force, weight, velocity, and acceleration are examples of vector quantities. A physical quantity that has only magnitude is called a *scalar* and is represented by a single number using appropriate units. For example, mass, speed, temperature, and so on are scalar quantities.

In Figure 3.6 we have

$$\overrightarrow{AB} = (x_1, x_2) = (v_1 - u_1, v_2 - u_2)$$

and so the vectors in 2-space represented by the arrows labeled $\mathbf{f}$ are considered equal, having the same magnitude and direction. Similarly,

$$\overrightarrow{CD} = (x_1, x_2, x_3) = (v_1 - u_1, v_2 - u_2, v_3 - u_3)$$

implies that the vectors in 3-space labeled $\mathbf{f}$ are also equal.

In practical situations it is useful to translate an arrow that represents a force parallel to itself so that the force acts at a desired point. We call a vector

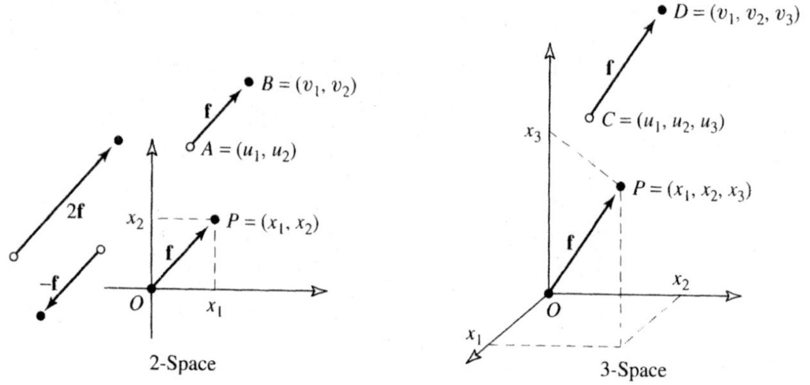

Figure 3.6 Free vectors in 2-space and 3-space.

free if it is allowed to be translated parallel to itself. However, in the context of linear algebra, keep in mind that $\mathbb{R}^n$ denotes the space of column vectors that are visualized as arrows with their initial points always anchored at the origin.

3.1.2 Subspaces

Certain subsets of vectors in $\mathbb{R}^n$ have special properties. The subsets in question, which are called *subspaces*, are defined after the following illustration.

■ ILLUSTRATION 3.1

Subspaces in $\mathbb{R}^3$

Refer to Figure 3.7. The equation $6x_1 - x_2 + 2x_3 = 0$ defines a plane P passing through the origin in $\mathbb{R}^3$. Solving the equation for x_2 shows that the set of all 3-vectors whose terminal points lie in P is given by

$$\begin{bmatrix} x_1 \\ 6x_1 + 2x_3 \\ x_3 \end{bmatrix}, \quad \text{for all real values } x_1, x_3. \tag{3.8}$$

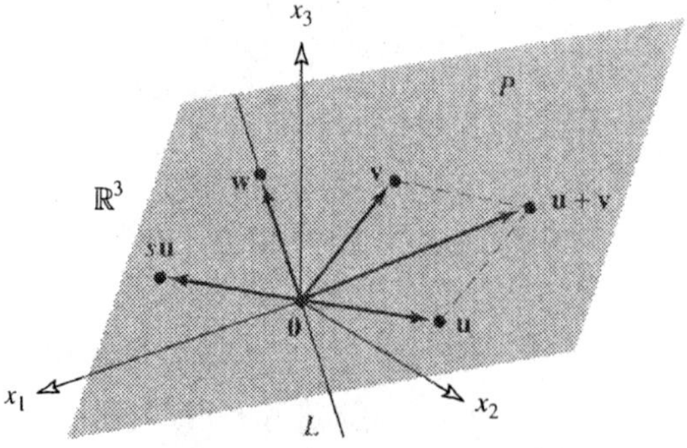

Figure 3.7 A plane in $\mathbb{R}^3$.

Let $\mathcal{U}$ be the subset of $\mathbb{R}^3$ consisting of all vectors defined in (3.8). Then $\mathcal{U}$ has the following two special properties:

(a) For any vectors $\mathbf{u}$ and $\mathbf{v}$ in $\mathcal{U}$, the sum $\mathbf{u}+\mathbf{v}$ is also in $\mathcal{U}$. For example, the vectors $\mathbf{u} = [0\ 2\ 1]^{\mathsf{T}}$ and $\mathbf{v} = [-1\ 0\ 3]^{\mathsf{T}}$ belongs to $\mathcal{U}$ (verify) and the sum $\mathbf{u}+\mathbf{v} = [-1\ 2\ 4]^{\mathsf{T}}$ also belongs to $\mathcal{U}$, as seen by setting $x_1 = -1$, $x_3 = 4$ in (3.8).

(b) If $\mathbf{u}$ belongs to $\mathcal{U}$ and s is *any* scalar, then the vector $s\mathbf{u}$ belongs to $\mathcal{U}$. For example, the vector $\mathbf{u} = [1\ 2\ -2]^{\mathsf{T}}$ belongs to $\mathcal{U}$ (verify) and setting $s = -0.5$, for example, we have $s\mathbf{u} = [-0.5\ -1\ 1]^{\mathsf{T}}$, which also belongs to $\mathcal{U}$, as seen by setting $x_1 = -0.5$ and $x_3 = 1$ in (3.8).

The subset $\mathcal{U}$ is a *linear space* of vectors in $\mathbb{R}^3$, in the sense that, if $\mathbf{u}$ and $\mathbf{v}$ are vectors in $\mathcal{U}$, then all vectors of the type $\alpha\mathbf{u}+\beta\mathbf{v}$ are in $\mathcal{U}$, for all scalars α, β. In geometric terms, the vectors $\alpha\mathbf{u}+\beta\mathbf{v}$ all lie in the plane P. The subset $\mathcal{U}$ is a *two-dimensional subspace* of $\mathbb{R}^3$ that is represented by P. In a similar fashion, a subset $\mathcal{W}$ of $\mathbb{R}^3$ consisting of all scalar multiplies $s\mathbf{w}$ of a fixed 3-vector $\mathbf{w}$ has properties (a) and (b). In this case $\mathcal{W}$ is a *one-dimensional subspace* of $\mathbb{R}^3$ that is represented by a line L passing through the origin in $\mathbb{R}^3$. ■

Definition 3.2

Subspace

Let $\mathcal{U}$ be a set containing one or more vectors in $\mathbb{R}^n$. Then $\mathcal{U}$ is a *subspace* of $\mathbb{R}^n$ if the following two conditions are satisfied:

(S1) If $\mathbf{u}$ and $\mathbf{v}$ belong to $\mathcal{U}$, then $\mathbf{u}+\mathbf{v}$ belongs to $\mathcal{U}$. We say that $\mathcal{U}$ is *closed with respect to vector addition*,

(S2) If $\mathbf{u}$ belongs to $\mathcal{U}$ and s is any scalar, then $s\mathbf{u}$ belongs to $\mathcal{U}$. We say that $\mathcal{U}$ is *closed with respect to scalar multiplication*.

One fundamental consequence of condition (**S2**) is this—suppose that $\mathbf{u}$ belongs to $\mathcal{U}$ and let $s = 0$, then $0\mathbf{u} = \mathbf{0}$ belongs to $\mathcal{U}$. In other words, every subspace of $\mathbb{R}^n$ contains the zero vector $\mathbf{0}$. Furthermore, any subset of $\mathbb{R}^n$ that *does not* contain the zero vector cannot be a subspace of $\mathbb{R}^n$.

It is important to be able to recognize which subsets of $\mathbb{R}^n$ are subspaces and which are not. In the next illustration we consider four subsets in $\mathbb{R}^2$, only one of which is a subspace.

■ ILLUSTRATION 3.2

Four Subsets in $\mathbb{R}^2$

Refer to Figure 3.8.

(a) Let $\mathcal{U}$ consist of all 2-vectors with terminal points (u_1, u_2), where $u_1 - u_2 = 1$. Then $\mathcal{U}$ is represented by a line L that does not pass through the origin. Clearly, the zero vector $\mathbf{0}$ is not in $\mathcal{U}$ and so $\mathcal{U}$ is not a subspace of $\mathbb{R}^2$.

(b) Let $\mathcal{U}$ be the set of all 2-vectors with terminal points (u_1, u_2) where $u_1 = 0$ or $u_2 = 0$. Taking $\mathbf{u} = [1\ 0]^{\mathsf{T}}$ and $\mathbf{v} = [0\ 1]^{\mathsf{T}}$, we have

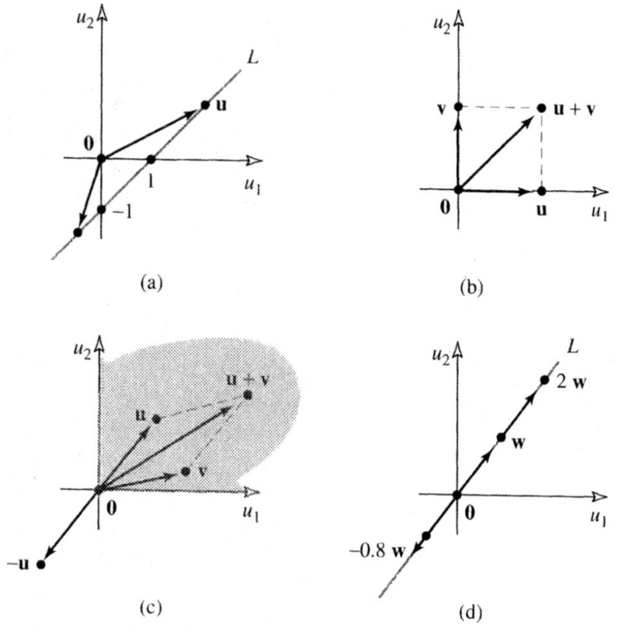

Figure 3.8 Subsets and subspaces in $\mathbb{R}^2$.

$\mathbf{u} + \mathbf{v} = [1\ \ 0]^{\mathsf{T}} + [0\ \ 1]^{\mathsf{T}} = [1\ \ 1]^{\mathsf{T}}$, which is not in $\mathcal{U}$. Thus (**S1**) fails and $\mathcal{U}$ is not a subspace of $\mathbb{R}^2$. However, as an exercise, verify that the condition (**S2**) is true.

(c) Let $\mathcal{U}$ be the set of all 2-vectors with terminal points (u_1, u_2), where $u_1 \geq 0$ and $u_2 \geq 0$. Then $\mathcal{U}$ consists of all vectors in $\mathbb{R}^2$ that have their terminal points in the first quadrant (including the axes). Applying the parallelogram law for vector addition shows that $\mathbf{u} + \mathbf{v}$ is again in $\mathcal{U}$ whenever $\mathbf{u}$ and $\mathbf{v}$ are in $\mathcal{U}$ and so (**S1**) is true. However, for any nonzero $\mathbf{u}$ in $\mathcal{U}$, take $s = -1$ and then $(-1)\mathbf{u} = -\mathbf{u}$ is not in $\mathcal{U}$. Hence (**S2**) fails and $\mathcal{U}$ is not a subspace of $\mathbb{R}^2$.

(d) Let $\mathcal{U}$ be the set of all 2-vectors whose terminal points lie on a line L passing through the origin O. Each vector $\mathbf{u}$ in $\mathcal{U}$ is a scalar multiple of a fixed nonzero vector $\mathbf{w}$ in the direction of L. Any two vectors in $\mathcal{U}$ have the form $\mathbf{u}_1 = \alpha_1\mathbf{w}$, $\mathbf{u}_2 = \alpha_2\mathbf{w}$, for some scalars α_1 and α_2. Hence,

$$\mathbf{u} + \mathbf{v} = \alpha_1\mathbf{w} + \alpha_2\mathbf{w} = (\alpha_1 + \alpha_2)\mathbf{w},$$

showing that $\mathbf{u} + \mathbf{v}$ is in $\mathcal{U}$ and so (**S1**) is true. If $\mathbf{u} = \alpha\mathbf{w}$ and s is any scalar, then $s\mathbf{u} = s(\alpha\mathbf{w}) = (s\alpha)\mathbf{w}$, showing that $s\mathbf{u}$ is in $\mathcal{U}$ and so (**S2**) is true. Hence $\mathcal{U}$ is a subspace of $\mathbb{R}^2$. Notice that the slope of the line L played no part in the discussion and so any line in $\mathbb{R}^2$ passing through the origin is the geometric representation of a subspace of $\mathbb{R}^2$. ▪

Special Subspaces

The space $\mathbb{R}^n$ contains two special subspaces that correspond to extreme possibilities. The subset $\mathcal{O} = \{\mathbf{0}\}$, containing only the zero vector $\mathbf{0}$, satisfies (**S1**) and (**S2**) and is therefore a subspace of $\mathbb{R}^n$. It is called the *zero* (or *trivial*) subspace. The subset $\mathcal{U} = \mathbb{R}^n$ satisfies (**S1**) and (**S2**) and so $\mathbb{R}^n$ is a subspace of $\mathbb{R}^n$. We call $\mathbb{R}^n$ the *whole space*. Between the two extremes, a subspace $\mathcal{U}$ of $\mathbb{R}^n$ is called a *proper subspace* of $\mathbb{R}^n$ if it is neither $\mathcal{O}$ nor $\mathbb{R}^n$. A proper subspace $\mathcal{U}$ must necessarily contain a nonzero vector and at the same time there is some vector in $\mathbb{R}^n$ that is not in $\mathcal{U}$.

Note that a proper subspace of $\mathbb{R}^2$ is represented geometrically by a line passing through the origin in $\mathbb{R}^2$ and a proper subspace of $\mathbb{R}^3$ is represented geometrically by either a line or a plane passing through the origin in $\mathbb{R}^3$.

3.1.3 Linear Combinations, Span

Definition 3.3

Linear Combination, Span

If $\mathcal{S} = \{\mathbf{v}_1, \mathbf{v}_2, \ldots, \mathbf{v}_k\}$ is a set of vectors in $\mathbb{R}^n$ and $x_1, x_2, \ldots, x_k$ is a set of scalars, then the expression

$$x_1\mathbf{v}_1 + x_2\mathbf{v}_2 + \cdots + x_k\mathbf{v}_k, \tag{3.9}$$

is called a *linear combination* of $\mathbf{v}_1, \mathbf{v}_2, \ldots, \mathbf{v}_k$ using the scalars $x_1, x_2, \ldots, x_k$. The set of linear combinations (3.9), as $x_1, x_2, \ldots, x_k$ range over all possible real values, is called the *span* of $\mathbf{v}_1, \mathbf{v}_2, \ldots, \mathbf{v}_k$ and is denoted by

$$\text{either} \quad \text{span}\{\mathbf{v}_1, \mathbf{v}_2, \ldots, \mathbf{v}_k\} \quad \text{or} \quad \text{span}\,\mathcal{S}.$$

Remarks

Refer to Definition 3.3 and make the following observations:

(a) The rules of vector algebra allow us to build up the linear combination (3.9) as follows: Form $x_1\mathbf{v}_1$, $x_2\mathbf{v}_2$, then $x_1\mathbf{v}_1 + x_2\mathbf{v}_2$, and $(x_1\mathbf{v}_1 + x_2\mathbf{v}_2) + x_3\mathbf{v}_3$, and so on, adding up to k summands.

(b) Setting $x_1 = 1$ and $x_i = 0$ for $2 \le i \le k$ in (3.9) shows that $\mathbf{v}_1$ belongs to span $\mathcal{S}$. A similar argument shows that all the other vectors in $\mathcal{S}$ are in span $\mathcal{S}$. Hence, $\mathcal{S}$ is a subset of span $\mathcal{S}$.

(c) $\mathcal{S}$ is a finite[2] set with k members. If $\mathcal{S}$ contains at least one nonzero vector, then span $\mathcal{S}$ is an infinite set.

■ ILLUSTRATION 3.3

Linear Combinations, Span

Define a set $\mathcal{S} = \{\mathbf{v}_1, \mathbf{v}_2\}$, where

$$\mathbf{v}_1 = \begin{bmatrix} 1 \\ 1 \end{bmatrix}, \qquad \mathbf{v}_2 = \begin{bmatrix} -1 \\ 2 \end{bmatrix}$$

[2] A set $\mathcal{S}$ is called *finite* if it contains a finite (nonnegative integer) number of members. For example, the set $\mathcal{S} = \{a, b, c\}$ is finite because it contains three members. A set is called *infinite* if it is not finite. For example, the set of all positive integers $\{1, 2, 3, \ldots\}$ is infinite.

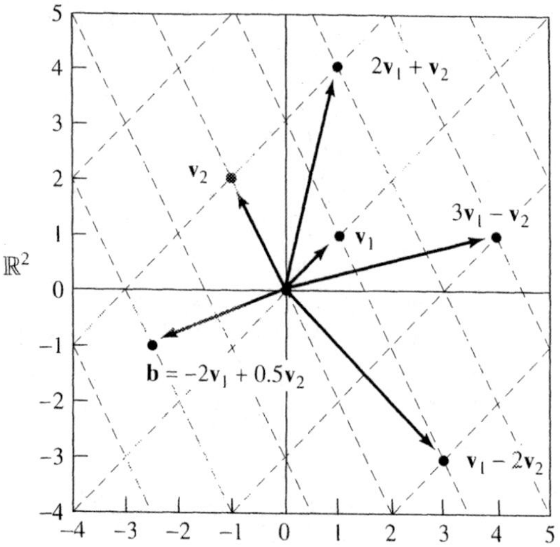

Figure 3.9 Linear combinations in $\mathbb{R}^2$.

Refer to Figure 3.9. The vectors $\mathbf{v}_1$ and $\mathbf{v}_2$ define a grid in $\mathbb{R}^2$ that may be used to locate any linear combination $x_1\mathbf{v}_1 + x_2\mathbf{v}_2$ of $\mathbf{v}_1$ and $\mathbf{v}_2$. Particular cases $2\mathbf{v}_1 + \mathbf{v}_2$, $3\mathbf{v}_1 - \mathbf{v}_2$, $\mathbf{v}_1 - 2\mathbf{v}_2$ are shown in the figure. Also, we have

$$-2\mathbf{v}_1 + 0.5\mathbf{v}_2 = -2\begin{bmatrix} 1 \\ 1 \end{bmatrix} + 0.5\begin{bmatrix} -1 \\ 2 \end{bmatrix} = \begin{bmatrix} -2.5 \\ -1 \end{bmatrix} = \mathbf{b}$$

so that $\mathbf{b}$ is a linear combination of $\mathbf{v}_1$, $\mathbf{v}_2$ using scalars $x_1 = -2$ and $x_2 = 0.5$, which means that $\mathbf{b}$ is *in* $\text{span}\{\mathbf{v}_1, \mathbf{v}_2\}$. By considering the grid in Figure 3.9 we can see geometrically that any 2-vector $\mathbf{b}$ is in $\text{span}\{\mathbf{v}_1, \mathbf{v}_2\}$; in other words, the set $\{\mathbf{v}_1, \mathbf{v}_2\}$ *generates* $\mathbb{R}^2$. This fact will be proved algebraically in Illustration 3.5. ■

Theorem 3.2 **The Span of a Set of Vectors is a Subspace**

If $\mathcal{S} = \{\mathbf{v}_1, \mathbf{v}_2, \ldots, \mathbf{v}_k\}$ *is a set of vectors in* $\mathbb{R}^n$, *then* $\text{span}\,\mathcal{S}$ *is a subspace of* $\mathbb{R}^n$.

Proof An illustration of the proof when $k = 2$ will suffice to show the general idea. Consider a subset $\mathcal{S} = \{\mathbf{v}_1, \mathbf{v}_2\}$ in $\mathbb{R}^n$. All vectors in $\text{span}\,\mathcal{S}$ have the form $\alpha\mathbf{v}_1 + \beta\mathbf{v}_2$ for some scalars α and β. Apply Definition 3.2.

(S1) Consider two vectors $\mathbf{u}_1$ and $\mathbf{u}_2$ in $\text{span}\,\mathcal{S}$. For some scalars x_1, x_2, y_1, y_2, we have $\mathbf{u}_1 = x_1\mathbf{v}_1 + x_2\mathbf{v}_2$ and $\mathbf{u}_2 = y_1\mathbf{v}_1 + y_2\mathbf{v}_2$. Then

$$\begin{aligned} \mathbf{u}_1 + \mathbf{u}_2 &= (x_1\mathbf{v}_1 + x_2\mathbf{v}_2) + (y_1\mathbf{v}_1 + y_2\mathbf{v}_2) \\ &= (x_1 + y_1)\mathbf{v}_1 + (x_2 + y_2)\mathbf{v}_2 \\ &= \alpha\mathbf{v}_1 + \beta\mathbf{v}_2, \quad \text{where } \alpha = x_1 + y_1,\ \beta = x_2 + y_2, \end{aligned}$$

which shows that $\mathbf{u}_1 + \mathbf{u}_2$ is a linear combination of $\mathbf{v}_1$ and $\mathbf{v}_2$; that is, $\mathbf{u}_1 + \mathbf{u}_2$ is in span $\mathcal{S}$.

(S2) Given the vector $\mathbf{u} = x_1\mathbf{v}_1 + x_2\mathbf{v}_2$ in span $\mathcal{S}$ and any scalar s, we have

$$s\mathbf{u} = s(x_1\mathbf{v}_1 + x_2\mathbf{v}_2) = (sx_1)\mathbf{v}_1 + (sx_2)\mathbf{v}_2 = \alpha\mathbf{v}_1 + \beta\mathbf{v}_2,$$

where $\alpha = sx_1$ and $\beta = sx_2$, showing that $s\mathbf{v}$ belongs to span $\mathcal{S}$.

The span of a finite set of vectors in $\mathbb{R}^n$ is a subspace of $\mathbb{R}^n$. We might wonder if there are subspaces of $\mathbb{R}^n$ that are not the span of a finite set of vectors. The answer is no! It will be seen in Section 3.2 that every subspace $\mathcal{U}$ of $\mathbb{R}^n$ is the span of a finite subset $\mathcal{B}$ in $\mathcal{U}$. The set $\mathcal{B}$ is called a *basis* for $\mathcal{U}$.

■ ILLUSTRATION 3.4

A Subspace of $\mathbb{R}^3$

Consider a linear system consisting of the single homogeneous linear equation

$$2x + 6y - 4z = 0. \tag{3.10}$$

Solving (3.10) in the standard way, we have $x_1 = -3s + 2t$, where $x_2 = s$ and $x_3 = t$ are real parameters, and writing the solution set in vector form, we have

$$\mathbf{x} = \begin{bmatrix} x \\ y \\ z \end{bmatrix} = s\mathbf{v}_1 + t\mathbf{v}_2, \quad \text{where } \mathbf{v}_1 = \begin{bmatrix} -3 \\ 1 \\ 0 \end{bmatrix}, \quad \mathbf{v}_2 = \begin{bmatrix} 2 \\ 0 \\ 1 \end{bmatrix}. \tag{3.11}$$

Let $\mathcal{U}$ be the set of all 3-vectors $\mathbf{x}$ in (3.11). Then $\mathcal{U} = \text{span}\{\mathbf{v}_1, \mathbf{v}_2\}$, and Theorem 3.2 tells us that $\mathcal{U}$ is a subspace of $\mathbb{R}^3$. Refer to Figure 3.10. The subspace $\mathcal{U}$ is represented geometrically by a plane P that passes through the origin $\mathbf{0}$ and whose orientation is defined by the vectors $\mathbf{v}_1$ and $\mathbf{v}_2$ that lie in P. The subspace $\mathcal{U}$ is called the *null space* of the 1×3 matrix $\mathbf{A} = [2 \;\; 6 \;\; -4]$ defined by the coefficients in (3.10). In Section 3.3 it will be shown that the null space of a general $m \times n$ matrix is a subspace of $\mathbb{R}^n$. ■

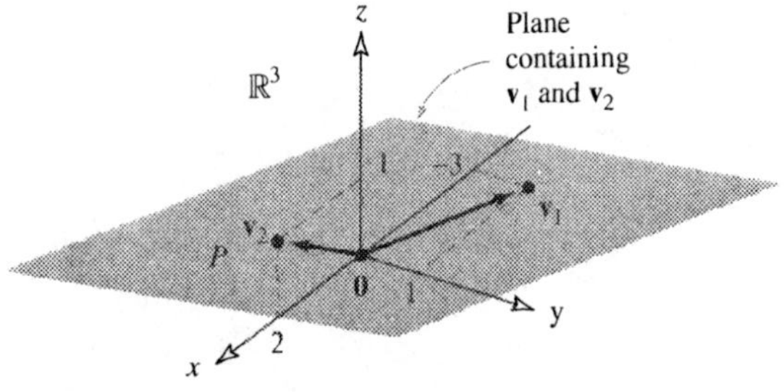

Figure 3.10 The plane $2x + 6y - 4z = 0$ in $\mathbb{R}^3$.

3.1.4 Vector Equations

Suppose we are given a finite set of vectors $\mathcal{S} = \{\mathbf{v}_1, \mathbf{v}_2, \ldots, \mathbf{v}_k\}$ in $\mathbb{R}^n$ together with a vector $\mathbf{b}$ in $\mathbb{R}^n$. How can we tell whether or, not $\mathbf{b}$ is a linear combination of the vectors in $\mathcal{S}$; that is, whether or not $\mathbf{b}$ is in $\mathcal{S}$? The answer to this question is obtained by solving the *vector equation*

$$x_1\mathbf{v}_1 + x_2\mathbf{v}_2 + \cdots + x_k\mathbf{v}_k = \mathbf{b} \tag{3.12}$$

for unknowns $x_1, x_2, \ldots, x_k$. Equation (3.12) can be rewritten as an $n \times k$ (note the size!) linear system (S), and as usual (S) can have zero, one, or infinitely many solutions. In terms of span, the cases are these.

(a) If (S) is inconsistent, then $\mathbf{b}$ is not in span $\mathcal{S}$.
(b) If (S) has a unique solution, then $\mathbf{b}$ is in span $\mathcal{S}$ and $\mathbf{b}$ is represented as a linear combination of vectors from $\mathcal{S}$ in only one way.
(c) If (S) has infinitely many solutions, then $\mathbf{b}$ is in span $\mathcal{S}$ and there are infinitely many ways of representing $\mathbf{b}$ as a linear combination of vectors from $\mathcal{S}$.

Note that cases (a), (b), and (c) can be rephrased (Exercises 3.1) in terms of the rank of the $n \times k$ coefficient matrix $\mathbf{A}$ for (S) and the rank of the $n \times (k+1)$ augmented matrix $\mathbf{M} = [\mathbf{A}|\,\mathbf{b}]$. Case (c) occurs in the next example.

EXAMPLE 2

Solving a Vector Equation

Let $\mathbf{v}_1$, $\mathbf{v}_2$, $\mathbf{v}_3$, and $\mathbf{b}$ be the following vectors in $\mathbb{R}^2$.

$$\mathbf{v}_1 = \begin{bmatrix} 1 \\ 2 \end{bmatrix}, \quad \mathbf{v}_2 = \begin{bmatrix} 3 \\ 4 \end{bmatrix}, \quad \mathbf{v}_3 = \begin{bmatrix} 5 \\ 6 \end{bmatrix}, \quad \mathbf{b} = \begin{bmatrix} 2 \\ 2 \end{bmatrix}$$

Then $\mathbf{b}$ is a linear combination of $\mathbf{v}_1$, $\mathbf{v}_2$, $\mathbf{v}_3$ if there are scalars x_1, x_2, x_3 such that $x_1\mathbf{v}_1 + x_2\mathbf{v}_2 + x_3\mathbf{v}_3 = \mathbf{b}$ [this is equation (3.12) with $k = 3$]. Using the laws of vector algebra, we have

$$x_1\begin{bmatrix} 1 \\ 2 \end{bmatrix} + x_2\begin{bmatrix} 3 \\ 4 \end{bmatrix} + x_3\begin{bmatrix} 5 \\ 6 \end{bmatrix} = \begin{bmatrix} x_1 + 3x_2 + 5x_3 \\ 2x_1 + 4x_2 + 6x_3 \end{bmatrix} = \begin{bmatrix} 2 \\ 2 \end{bmatrix} \tag{3.13}$$

From the last equality in (3.13) we see that $\mathbf{b}$ is a linear combination of $\mathbf{v}_1$, $\mathbf{v}_2$, $\mathbf{v}_3$ if and only if the 2×3 linear system

$$(S) \quad \begin{cases} x_1 + 3x_2 + 5x_3 = 2 \\ 2x_1 + 4x_2 + 6x_3 = 2 \end{cases}$$

is consistent. Reducing the augmented matrix $\mathbf{M}$ for (S) to its reduced form $\mathbf{M}^*$, we have

$$\mathbf{M} = \left[\begin{array}{ccc|c} 1 & 3 & 5 & 2 \\ 2 & 4 & 6 & 2 \end{array}\right] \sim \left[\begin{array}{ccc|c} 1 & 0 & -1 & -1 \\ 0 & 1 & 2 & 1 \end{array}\right] = \mathbf{M}^* .$$

Hence, $r = \operatorname{rank}\mathbf{A} = \operatorname{rank}\mathbf{M} = 2 < 3 = k$ and so (S) is consistent with $k - r = 3 - 2 = 1$ free variable in the solution set. Let $x_3 = t$, where t is a real parameter. The solution set for (S) is $(x_1, x_2, x_3) = (-1+t, 1-2t, t)$ and so, for any real t, we have

$$(-1+t)\mathbf{v}_1 + (1-2t)\mathbf{v}_2 + t\mathbf{v}_3 = \mathbf{b}.$$

There are infinitely many ways of writing $\mathbf{b}$ as a linear combination of $\mathbf{v}_1$, $\mathbf{v}_2$, $\mathbf{v}_3$. Choosing $t = 2$ and $t = 0$, for example, gives particular representations of $\mathbf{b}$, namely $\mathbf{v}_1 - 3\mathbf{v}_2 + 2\mathbf{v}_3 = \mathbf{b}$ and $-\mathbf{v}_1 + \mathbf{v}_2 + 0\mathbf{v}_3 = \mathbf{b}$ (or $\mathbf{v}_2 - \mathbf{v}_1 = \mathbf{b}$), respectively. Hence, $\mathbf{b}$ is in $\text{span}\{\mathbf{v}_1, \mathbf{v}_2, \mathbf{v}_3\}$ and in fact $\mathbf{b}$ is in $\text{span}\{\mathbf{v}_1, \mathbf{v}_2\}$, the point being that $\mathbf{b}$ can be represented in terms of $\mathbf{v}_1$ and $\mathbf{v}_2$ only. Can you see why $\mathbf{b}$ is also in $\text{span}\{\mathbf{v}_1, \mathbf{v}_3\}$ and in $\text{span}\{\mathbf{v}_2, \mathbf{v}_3\}$? ■

Definition 3.4

Spanning Set

Let $\mathcal{U}$ be a subspace of $\mathbb{R}^n$ and let $\mathcal{S}$ be a finite subset of $\mathcal{U}$. If $\text{span}\,\mathcal{S} = \mathcal{U}$, then $\mathcal{S}$ is called a *spanning set* for $\mathcal{U}$ and we say that $\mathcal{S}$ *generates* $\mathcal{U}$.

In Illustration 3.3 we have seen that $\mathcal{U} = \text{span}\{\mathbf{v}_1, \mathbf{v}_2\}$. The finite set $\mathcal{S} = \{\mathbf{v}_1, \mathbf{v}_2\}$ is a spanning set for the subspace $\mathcal{U}$, or in other words, $\mathcal{U}$ is generated by the vectors $\mathbf{v}_1$ and $\mathbf{v}_2$.

Finding spanning sets for the subspace $\mathcal{U} = \mathbb{R}^n$ (the whole space) is of particular interest. Here is an example when $n = 2$.

■ ILLUSTRATION 3.5

Generating $\mathbb{R}^2$

Refer to Illustration 3.3. Let $\mathcal{S} = \{\mathbf{v}_1, \mathbf{v}_2\}$, where

$$\mathbf{v}_1 = \begin{bmatrix} 1 \\ 1 \end{bmatrix}, \qquad \mathbf{v}_2 = \begin{bmatrix} -1 \\ 2 \end{bmatrix}$$

Our goal is to show that $\text{span}\,\mathcal{S} = \mathbb{R}^2$. Two set inclusions need to be proved.

(a) For any scalars x_1 and x_2, the linear combination $x_1\mathbf{v}_1 + x_2\mathbf{v}_2$ is a vector in $\mathbb{R}^2$ and so $\text{span}\,\mathcal{S}$ is a subset of $\mathbb{R}^2$.

(b) In order to show that $\mathbb{R}^2$ is a subset of $\text{span}\,\mathcal{S}$, we need to show that any 2-vector $\mathbf{b} = [\,p \;\; q\,]^{\mathsf{T}}$ in $\mathbb{R}^2$ can be expressed in the form $x_1\mathbf{v}_1 + x_2\mathbf{v}_2 = \mathbf{b}$, for some scalars x_1, x_2. Solving this vector equation amounts to solving a 2×2 linear system (S), with coefficient matrix $\mathbf{A}$, where

$$(S)\ \begin{cases} x_1 - \ \ x_1 = p \\ x_1 + 2x_2 = q \end{cases} \qquad \mathbf{A} = \begin{bmatrix} 1 & -1 \\ 1 & 2 \end{bmatrix}.$$

The matrix form of (S) is $\mathbf{Ax} = \mathbf{b}$, where $\mathbf{x} = [\,x_1 \;\; x_2\,]^{\mathsf{T}}$. Note that $\mathbf{A}$ is invertible (Section 2.2) because $\det(\mathbf{A}) = 1(2) - (-1)1 = 3 \neq 0$ and so

$\mathbf{x} = \mathbf{A}^{-1}\mathbf{b}$ gives the unique solution to (S). As an exercise, verify that $x_1 = (2p+q)/3$ and $x_2 = (q-p)/3$. The representation of $\mathbf{b}$ is

$$\frac{(2p+q)}{3}\begin{bmatrix}1\\1\end{bmatrix} + \frac{(q-p)}{3}\begin{bmatrix}-1\\2\end{bmatrix} = \begin{bmatrix}p\\q\end{bmatrix} \tag{3.14}$$

Hence, span $\mathcal{S} = \mathbb{R}^2$. The finite set $\mathcal{S} = \{\mathbf{v}_1, \mathbf{v}_2\}$ generates the infinite set $\mathbb{R}^2$. By the way, $\mathcal{S}$ is not unique: Any pair of nonzero, noncollinear vectors in $\mathbb{R}^2$ will generate $\mathbb{R}^2$ and any set $\mathcal{S} = \{\mathbf{v}_1, \mathbf{v}_2\}$ such that $\mathbf{v}_2 = s\mathbf{v}_1$, where s is a scalar, will not generate $\mathbb{R}^2$. Geometrically, two vectors lying on the same line through the origin in $\mathbb{R}^2$ cannot generate $\mathbb{R}^2$. ■

Application: Vector Equations

The introduction to this chapter illustrates how a linear system (S) can be viewed as a vector equation. The vector form of (S) treats the columns of the coefficient matrix for (S) as single entities. Writing a linear system in its vector form can often shed a different light on the problem at hand, as in the following application.

■ ILLUSTRATION 3.6

Scheduling a Work Force

An engineering company will hire new staff to develop new products using production teams, each team consisting of engineers (core personnel) plus support staff. The teams are labeled as follows:

D (Design), P (Production), A (Analysis), T (Testing)

and the numbers of core personnel in each team, together with the total number of engineers of each type to be hired, are as follows:

Type	Symbol	D	P	A	T	Total
Civil / Environmental	CE	2	3	1	0	16
Mechanical / Design	MD	4	0	3	1	19
Computer / Software	CS	2	6	3	2	35

with the understanding that each design team (D) consists of 2 CE, 4 MD, and 2 CS personnel; each production team (P) consists of 3 CE, 0 MD, and 6 CS personnel; and so on.

The Management's Problem

How does management schedule the new work force into teams, given the constraints: (a) All teams must be represented, (b) at least two teams of type T (testing) must be included? Can this be done so that each core employee appears in some team?

To solve the problem, let x_1, x_2, x_3, x_4 represent the number of teams of type D, P, A, and T, respectively, which are to be formed with the constraint that the unknowns must be nonnegative integers. Building a mathematical model in terms of numbers of engineers of each type (read the preceding table row by row), we arrive at the 3 × 4 linear system shown below.

$$(S)\quad \begin{cases} 2x_1 + 3x_2 + x_3 \phantom{{}+2x_4} = 16 & \text{CE} \\ 4x_1 \phantom{{}+3x_2} + 3x_3 + x_4 = 19 & \text{ME}\,. \\ 2x_1 + 6x_2 + 3x_3 + 2x_4 = 35 & \text{CE} \end{cases} \tag{3.15}$$

On the other hand, we can view (S) as the vector equation

$$\begin{array}{ccccccccccc} & \text{D} & & \text{P} & & \text{A} & & \text{T} & & \text{Total} & \\ x_1 & \begin{bmatrix} 2 \\ 4 \\ 2 \end{bmatrix} & + x_2 & \begin{bmatrix} 3 \\ 0 \\ 6 \end{bmatrix} & + x_3 & \begin{bmatrix} 1 \\ 3 \\ 3 \end{bmatrix} & + x_4 & \begin{bmatrix} 0 \\ 1 \\ 2 \end{bmatrix} & = & \begin{bmatrix} 16 \\ 19 \\ 35 \end{bmatrix} & \begin{matrix} \text{CE} \\ \text{MD} \\ \text{CS} \end{matrix} \end{array} \tag{3.16}$$

or, written more compactly (with the obvious of meaning of $\mathbf{v}_1, \mathbf{v}_2, \mathbf{v}_3, \mathbf{v}_4$),

$$x_1\mathbf{v}_1 + x_2\mathbf{v}_2 + x_3\mathbf{v}_3 + x_4\mathbf{v}_4 = \mathbf{b}.$$

Note how the vectors on the left side of (3.16) record the composition of each team, and these correspond to the columns of the coefficient matrix in (3.15). The management's problem can now be translated as follows: Is $\mathbf{b}$ a linear combination of the vectors $\{\mathbf{v}_1, \mathbf{v}_2, \mathbf{v}_3, \mathbf{v}_4\}$ in such a way that all scalars are nonzero, positive integers? We solve (3.15) as usual by reducing the augmented matrix $\mathbf{M}$ for (S). We have

$$\mathbf{M} = \left[\begin{array}{cccc|c} 2 & 3 & 1 & 0 & 16 \\ 4 & 0 & 3 & 1 & 19 \\ 2 & 6 & 3 & 2 & 35 \end{array}\right] \sim \left[\begin{array}{cccc|c} \textcircled{1} & 0 & 0 & -0.5 & 1 \\ 0 & \textcircled{1} & 0 & 0 & 3 \\ 0 & 0 & \textcircled{1} & 1 & 5 \end{array}\right] = \mathbf{M}^*$$

We see that x_2 is uniquely determined and x_4 is free. Let $x_4 = t$, where t is a real parameter and then the solution to (3.15) and (3.16) is

$$(x_1, x_2, x_3, x_4) = (1 + 0.5t, 3, 5 - t, t).$$

The constraint $x_3 = 5 - t \geq 0$ implies that $5 \geq t \geq 0$, where t is a nonnegative integer. However, we need t even in order to return an integer for x_1. There are therefore only three solutions, corresponding to $t = 0, 2, 4$, namely.

$$\mathbf{x} = (x_1, x_2, x_3, x_4) = (1, 3, 5, 0), \quad (2, 3, 3, 2), \quad (3, 3, 1, 4)$$

Only the last two solutions satisfy the management constraints. ■

Looking Ahead

Consider an $m \times n$ matrix $\mathbf{A}$ in terms of its columns (these are m-vectors) and its rows (considered as n-vectors).

$$\mathbf{A} = \begin{bmatrix} | & | & & | \\ \mathbf{v}_1 & \mathbf{v}_2 & \cdots & \mathbf{v}_n \\ | & | & & | \end{bmatrix} = \begin{bmatrix} — & \mathbf{r}_1 & — \\ — & \mathbf{r}_2 & — \\ & \vdots & \\ — & \mathbf{r}_m & — \end{bmatrix}$$

In Section 3.3 we consider the subspace of $\mathbb{R}^m$ spanned by the columns of **A** and the subspace of $\mathbb{R}^n$ spanned by the rows of **A**. These subspaces, called the *column space* and *row space* of **A**, respectively, play an important part in the theory of *linear transformations*, which is the subject of Section 3.4.

EXERCISES 3.1

Exercises 1–2. Find the values of r and s for which the equation is true.

1. $\begin{bmatrix} r-2 \\ s \end{bmatrix} = \begin{bmatrix} s \\ 4-r \end{bmatrix}$

2. $\begin{bmatrix} r+1 \\ 1 \end{bmatrix} = \begin{bmatrix} 2 \\ (s+1)^2 \end{bmatrix}$

Exercises 3–4. Plot the vectors **u** *and* **v** *accurately and construct the vectors* $-\mathbf{u}$, $-\mathbf{v}$, $\mathbf{u}+\mathbf{v}$, $\mathbf{u}-\mathbf{v}$, $2\mathbf{u}+3\mathbf{v}$.

3. $\mathbf{u} = \begin{bmatrix} -1 \\ 2 \end{bmatrix}$, $\mathbf{v} = \begin{bmatrix} 3 \\ -1 \end{bmatrix}$

4. $\mathbf{u} = \begin{bmatrix} 1 \\ 0 \end{bmatrix}$, $\mathbf{v} = \begin{bmatrix} 0 \\ 1 \end{bmatrix}$

Exercises 5–6. Each vector **b** *is a linear combination of 2-vectors. Compute* **b** *and draw an accurate diagram that shows how* **b** *is constructed from the linear combination that defines it.*

5. $-5\begin{bmatrix} 2 \\ -1 \end{bmatrix} + 7\begin{bmatrix} 3 \\ -1 \end{bmatrix} = \mathbf{b}$

6. $2\begin{bmatrix} 1 \\ -3 \end{bmatrix} + 3\begin{bmatrix} -6 \\ -1 \end{bmatrix} - 5\begin{bmatrix} 0 \\ 0 \end{bmatrix} = \mathbf{b}$

Exercises 7–8. Plot the vectors $\mathbf{v}_1$, $\mathbf{v}_2$, $\mathbf{v}_3$ *in the plane, where*

$$\mathbf{v}_1 = \begin{bmatrix} 1 \\ -1 \end{bmatrix}, \quad \mathbf{v}_2 = \begin{bmatrix} 0 \\ 1 \end{bmatrix}, \quad \mathbf{v}_3 = \begin{bmatrix} -1 \\ 3 \end{bmatrix}.$$

Compute the vector **b** *and construct* **b** *in your diagram.*

7. $\mathbf{b} = 2\mathbf{v}_1 - 3\mathbf{v}_2$

8. $\mathbf{b} = 3\mathbf{v}_1 + 5\mathbf{v}_2 - \mathbf{v}_3$

Exercises 9–10. Plot the vectors $\mathbf{v}_1$, $\mathbf{v}_2$, $\mathbf{v}_3$ *in* $\mathbb{R}^3$, *where*

$$\mathbf{v}_1 = \begin{bmatrix} 1 \\ 5 \\ 3 \end{bmatrix}, \quad \mathbf{v}_2 = \begin{bmatrix} 1 \\ 2 \\ 2 \end{bmatrix}, \quad \mathbf{v}_3 = \begin{bmatrix} 3 \\ 9 \\ 7 \end{bmatrix}.$$

Compute **b** *and draw an accurate diagram that shows* **b**.

9. $\mathbf{b} = 2\mathbf{v}_1 + 3\mathbf{v}_2 - \mathbf{v}_3$

10. $\mathbf{b} = 3\mathbf{v}_1 + 5\mathbf{v}_2 - 2\mathbf{v}_3$

Exercises 11–12. Solve the vector equation for **x**.

11. $3\mathbf{x}+4\mathbf{v}_1 = 4\mathbf{v}_2-2\mathbf{v}_3-\mathbf{x}$, where $\mathbf{v}_1 = \begin{bmatrix} 1 \\ 0 \end{bmatrix}$, $\mathbf{v}_2 = \begin{bmatrix} 1 \\ 1 \end{bmatrix}$, $\mathbf{v}_3 = \begin{bmatrix} -2 \\ -1 \end{bmatrix}$

12. $\mathbf{x}-2\mathbf{v}_1 = -3\mathbf{v}_2+\mathbf{v}_3-3\mathbf{x}$, where $\mathbf{v}_1 = \begin{bmatrix} 1 \\ 5 \\ 3 \end{bmatrix}$, $\mathbf{v}_2 = \begin{bmatrix} 1 \\ 2 \\ 2 \end{bmatrix}$, $\mathbf{v}_3 = \begin{bmatrix} 3 \\ 1 \\ 7 \end{bmatrix}$

Exercises 13–16. Determine whether the given subset $\mathcal{U}$ *of* $\mathbb{R}^2$ *is a subspace. Indicate (shade) the subset* $\mathcal{U}$ *in the plane. In case of failure, list the condition(s) that are false.*

13. $\{(x, y) \mid xy \geq 0\}$

14. $\{(x, y) \mid x = y\}$

15. $\{(x, y) \mid x^2 = y^2\}$

16. $\{(x, y) \mid x = y^2\}$

Exercises 17–20. Determine whether the given subset $\mathcal{U}$ of $\mathbb{R}^3$ is a subspace. Draw a diagram and indicate (shade) the subset $\mathcal{U}$. In case of failure, list the condition(s) that are false.

17. $\{(u_1, u_2, u_3) \mid u_3 = 0\}$

18. $\{(u_1, u_2, u_3) \mid u_1 = 0$ or $u_2 = 0\}$

19. $\{(u_1, u_2, u_3) \mid u_1^2 + u_2^2 \leq 200\}$

20. $\{(u_1, u_2, u_3) \mid u_1 = 2u_2 + 3u_3\}$

21. Consider the vectors

$$\mathbf{u} = \begin{bmatrix} 1 \\ 5 \end{bmatrix}, \quad \mathbf{v} = \begin{bmatrix} 2 \\ -1 \end{bmatrix}, \quad \mathbf{w} = \begin{bmatrix} 3 \\ 0 \end{bmatrix}$$

Plot the vectors $\mathbf{u}$, $\mathbf{v}$, $\mathbf{w}$, $\mathbf{u}+\mathbf{v}$, $(\mathbf{u}+\mathbf{v})+\mathbf{w}$, $\mathbf{v}+\mathbf{w}$, $\mathbf{u}+(\mathbf{v}+\mathbf{w})$ in the plane. Use these vectors to explain the associative law (**A2**) geometrically.

22. Prove that a subset $\mathcal{U}$ of $\mathbb{R}^n$ is a subspace if and only if the linear combination $s\mathbf{u}+t\mathbf{v}$ belongs to $\mathcal{U}$ for all vectors $\mathbf{u}$ and $\mathbf{v}$ in $\mathcal{U}$ and all real scalars s, t. *Hint*: There are two implications to prove.

Exercises 23–24. Let $\mathcal{U}$ be the set of all vectors $\mathbf{x}$ whose components satisfy the given equation. Write $\mathbf{x}$ as a linear combination of 3-vectors. Hence, show that $\mathcal{U}$ is a subspace of $\mathbb{R}^3$.

23. $2x_1 - x_2 + 4x_3 = 0$

24. $19x_1 + 3x_2 - 7x_3 = 0$

Exercises 25–30. Label each statement as either true (T) or false (F). If a false statement can be made true by modifying the wording, then do so.

25. The vector $\begin{bmatrix} -2 \\ 6 \end{bmatrix}$ is a linear combination of $\begin{bmatrix} 3 \\ -4 \end{bmatrix}$, $\begin{bmatrix} 1 \\ 2 \end{bmatrix}$.

26. If $2\mathbf{u}+2\mathbf{v} = 4\mathbf{w}-3\mathbf{v}+2\mathbf{u}$, then $\mathbf{w}$ is a linear combination of $\mathbf{u}$ and $\mathbf{u}+\mathbf{v}$.

27. If $\mathbf{u}$ is in $\mathbb{R}^3$, then $-\mathbf{u}$ is in span$\{\mathbf{u}\}$.

28. span$\{\mathbf{u}, \mathbf{u}-\mathbf{v}\}$ contains the vector $\mathbf{v}$.

29. The set span$\{\mathbf{u}, \mathbf{v}\}$ contains finitely many vectors.

30. Let $\mathbf{u}$, $\mathbf{v}$, $\mathbf{w}$ be distinct 2-vectors. Then span$\{\mathbf{u}, \mathbf{v}, \mathbf{w}\} = \mathbb{R}^2$.

Exercises 31–34. Write each vector equation $x_1\mathbf{v}_1 + x_2\mathbf{v}_2 + x_3\mathbf{v}_3 = \mathbf{b}$ as a linear system and hence find the ways in which the vector $\mathbf{b}$ can be expressed as a linear combination of $\mathbf{v}_1$, $\mathbf{v}_2$, $\mathbf{v}_3$.

31. $x_1 \begin{bmatrix} 1 \\ -3 \end{bmatrix} + x_2 \begin{bmatrix} -6 \\ -1 \end{bmatrix} - x_3 \begin{bmatrix} 0 \\ 0 \end{bmatrix} = \begin{bmatrix} 1 \\ 1 \end{bmatrix}$

32. $x_1 \begin{bmatrix} 2 \\ -1 \\ 0 \end{bmatrix} + x_2 \begin{bmatrix} -3 \\ 0 \\ -5 \end{bmatrix} + x_3 \begin{bmatrix} 1 \\ 1 \\ 0 \end{bmatrix} = \begin{bmatrix} 0 \\ 1 \\ 2 \end{bmatrix}$

33. $x_1 \begin{bmatrix} 1 \\ 0 \\ 0 \end{bmatrix} + x_2 \begin{bmatrix} 0 \\ 1 \\ 0 \end{bmatrix} + x_3 \begin{bmatrix} 0 \\ 0 \\ 1 \end{bmatrix} = \begin{bmatrix} a \\ b \\ c \end{bmatrix}$, for any real a, b, c

34. $x_1 \begin{bmatrix} 2 \\ 3 \\ 1 \\ 0 \end{bmatrix} + x_2 \begin{bmatrix} 1 \\ -1 \\ 0 \\ 1 \end{bmatrix} + x_3 \begin{bmatrix} 0 \\ 1 \\ 1 \\ 0 \end{bmatrix} = \begin{bmatrix} 0 \\ 1 \\ -1 \\ 2 \end{bmatrix}$

Exercises 35–38. Find all ways in which the last vector, denoted by **b**, *in the given list is a linear combination of previous vectors.*

35. $\begin{bmatrix} 1 \\ -3 \end{bmatrix}, \begin{bmatrix} -6 \\ 1 \end{bmatrix}, \begin{bmatrix} -34 \\ 0 \end{bmatrix}$

36. $\begin{bmatrix} 1 \\ 2 \end{bmatrix}, \begin{bmatrix} -1 \\ 1 \end{bmatrix}, \begin{bmatrix} 3 \\ 4 \end{bmatrix}, \begin{bmatrix} -1 \\ 2 \end{bmatrix}$

37. $\begin{bmatrix} 1 \\ 2 \\ 3 \end{bmatrix}, \begin{bmatrix} 4 \\ 5 \\ 6 \end{bmatrix}, \begin{bmatrix} 7 \\ 8 \\ 9 \end{bmatrix}, \begin{bmatrix} 0 \\ 0 \\ 0 \end{bmatrix}$

38. $\begin{bmatrix} 1 \\ 0 \\ 2 \end{bmatrix}, \begin{bmatrix} 0 \\ 3 \\ 6 \end{bmatrix}, \begin{bmatrix} 2 \\ -1 \\ 4 \end{bmatrix}, \begin{bmatrix} 2 \\ 2 \\ 6 \end{bmatrix}, \begin{bmatrix} 5 \\ -3 \\ 20 \end{bmatrix}$

Exercises 39–40. Write each linear system (S) as a vector equation and solve the vector equation.

39. $\begin{cases} 2x_1 + 2x_2 = 1 \\ x_1 - 4x_2 = 0 \\ 3x_1 - 2x_2 = 1 \end{cases}$

40. $\begin{cases} 2x_1 - 2x_2 + 3x_3 = 5 \\ 2x_1 - 4x_2 + 4x_3 = 2 \\ 3x_1 - 4x_2 + 5x_3 = 6 \end{cases}$

Exercises 41–48. In each case a set of vectors $\mathcal{U}$ is defined, where x, y, z are any real values. Determine whether $\mathcal{U}$ is a subspace. If so, find a spanning set for $\mathcal{U}$.

41. $\left\{ \begin{bmatrix} x \\ 0 \end{bmatrix} \right\}$

42. $\left\{ \begin{bmatrix} x \\ 2x \end{bmatrix} \right\}$

43. $\left\{\begin{bmatrix} x-y \\ y+2z \\ z \end{bmatrix}\right\}$ **44.** $\left\{\begin{bmatrix} x \\ y \end{bmatrix} : x+y \geq -1\right\}$

45. $\left\{\begin{bmatrix} x \\ y \\ z \end{bmatrix} : x+y+z=0\right\}$

46. The set of vectors whose terminal points lie on the line in $\mathbb{R}^2$ passing through the points $(1, 2)$ and $(-1, -0.5)$.

47. The set of vectors whose terminal points lie in the plane passing through the points $(1, 1, 3)$, $(0, 1, -2)$, $(-1, 1, 1)$.

48. The vectors in $\mathbb{R}^4$ with all components equal.

Exercises 49–50. Write each matrix equation $\mathbf{Ax} = \mathbf{b}$ *as a vector equation using the columns of* $\mathbf{A}$. *Solve the vector equation for* $\mathbf{x}$.

49. $\begin{bmatrix} 4 & 0 & 8 \\ -1 & 2 & 0 \end{bmatrix}\begin{bmatrix} x_1 \\ x_2 \\ x_3 \end{bmatrix} = \begin{bmatrix} 0 \\ 0 \end{bmatrix}$ **50.** $\begin{bmatrix} 1 & 2 \\ 3 & 4 \end{bmatrix}\begin{bmatrix} x_1 \\ x_2 \end{bmatrix} = \begin{bmatrix} 5 \\ 7 \end{bmatrix}$

51. Write the vector equation $x_1\mathbf{v}_1 + x_2\mathbf{v}_2 = \mathbf{b}$ as a linear system (S), where

$$\mathbf{v}_1 = \begin{bmatrix} 3 \\ 1 \\ 4 \end{bmatrix}, \quad \mathbf{v}_2 = \begin{bmatrix} 2 \\ -1 \\ 0 \end{bmatrix}, \quad \mathbf{b} = \begin{bmatrix} -1 \\ 0 \\ 2 \end{bmatrix}.$$

Determine if $\mathbf{b}$ is in $\operatorname{span}\{\mathbf{v}_1, \mathbf{v}_2\}$.

52. Consider the vectors

$$\mathbf{v}_1 = \begin{bmatrix} 4 \\ -1 \\ 2 \end{bmatrix}, \quad \mathbf{v}_2 = \begin{bmatrix} 0 \\ 2 \\ 4 \end{bmatrix}, \quad \mathbf{v}_3 = \begin{bmatrix} 1 \\ 0 \\ 1 \end{bmatrix},$$

$$\mathbf{u} = \begin{bmatrix} 1 \\ 1 \\ 3 \end{bmatrix}, \quad \mathbf{w} = \begin{bmatrix} 1 \\ 1 \\ 4 \end{bmatrix}.$$

(a) Find all solutions $\mathbf{x} = (x_1, x_2, x_3)$ to the vector equation $x_1\mathbf{v}_1 + x_2\mathbf{v}_2 + x_3\mathbf{v}_3 = \mathbf{0}$.

(b) Explain the solution in terms of $\operatorname{rank}\mathbf{A} = [\,\mathbf{v}_1\ \mathbf{v}_2\ \mathbf{v}_3\,]$.

(c) Show that $\mathbf{u}$ is in $\mathcal{U} = \operatorname{span}\{\mathbf{v}_1, \mathbf{v}_2, \mathbf{v}_3\}$ but that $\mathbf{w}$ is not in $\mathcal{U}$.

Exercises 53–54. Find solutions $\mathbf{x} = (x_1, x_2, x_3)$ *to the vector equations*

$$(a)\quad x_1\mathbf{v}_1 + x_2\mathbf{v}_2 + x_3\mathbf{v}_3 = \mathbf{0},$$

$$\text{and}\quad (b)\quad x_1\mathbf{v}_1 + x_2\mathbf{v}_2 + x_3\mathbf{v}_3 = \mathbf{b}.$$

Explain the connection between the two solution sets.

53. $\mathbf{v}_1 = \begin{bmatrix} 0 \\ 1 \\ 1 \end{bmatrix}, \quad \mathbf{v}_2 = \begin{bmatrix} 1 \\ 0 \\ 1 \end{bmatrix},$

$\mathbf{v}_3 = \begin{bmatrix} 1 \\ 1 \\ 0 \end{bmatrix}, \quad \mathbf{b} = \begin{bmatrix} 5 \\ -3 \\ 2 \end{bmatrix}.$

54. $\mathbf{v}_1 = \begin{bmatrix} 1 \\ 0 \\ 0 \end{bmatrix}, \quad \mathbf{v}_2 = \begin{bmatrix} 1 \\ 1 \\ 0 \end{bmatrix}, \quad \mathbf{v}_3 = \begin{bmatrix} 1 \\ 1 \\ 1 \end{bmatrix},$

$\mathbf{b} = \begin{bmatrix} 5 \\ -3 \\ 2 \end{bmatrix}.$

Exercises 55–57. Linear combinations and rank. Let $\mathbf{v}_1, \mathbf{v}_2, \ldots, \mathbf{v}_k$, $\mathbf{b}$ *be vectors in* $\mathbb{R}^n$ *and let* $\mathbf{M} = [\mathbf{A}\,|\,\mathbf{b}]$, *where* $\mathbf{A} = [\,\mathbf{v}_1\ \mathbf{v}_2\ \cdots\ \mathbf{v}_k\,]$. *Prove each statement.*

55. If $\operatorname{rank}\mathbf{A} < \operatorname{rank}\mathbf{M}$, then $\mathbf{b}$ is not in $\operatorname{span}\{\mathbf{v}_1, \mathbf{v}_2, \ldots, \mathbf{v}_k\}$.

56. If $\operatorname{rank}\mathbf{A} = \operatorname{rank}\mathbf{M} = k$, then $\mathbf{b}$ is a linear combination of $\mathbf{v}_1, \mathbf{v}_2, \ldots, \mathbf{v}_k$ in only one way.

57. If $\operatorname{rank}\mathbf{A} = \operatorname{rank}\mathbf{M} < k$, then $\mathbf{b}$ is a linear combination of $\mathbf{v}_1, \mathbf{v}_2, \ldots, \mathbf{v}_k$ in infinitely many ways.

58. Determine whether or not any column in the matrix

$$\mathbf{A} = [\,\mathbf{v}_1\ \mathbf{v}_2\ \mathbf{v}_3\,] = \begin{bmatrix} 1 & -2 & 1 \\ -1 & 1 & 2 \\ 2 & 4 & 3 \end{bmatrix}$$

is a linear combination of other columns. Give a general method for answering the same question for any $n \times n$ matrix $\mathbf{A}$.

59. Let $\mathcal{S} = \{\mathbf{v}_1, \mathbf{v}_2, \ldots, \mathbf{v}_k\}$ be a given set of vectors in $\mathbb{R}^n$ with $n \leq k$. Form the $n \times k$ matrix $\mathbf{A} = [\,\mathbf{v}_1\ \mathbf{v}_2\ \cdots\ \mathbf{v}_k\,]$ and suppose that $\operatorname{rank}\mathbf{A} = n$. Prove that $\operatorname{span}\{\mathbf{v}_1, \mathbf{v}_2, \ldots, \mathbf{v}_k\} = \mathbb{R}^n$.

60. Consider the columns of a square upper triangular matrix of order n. Give a condition that will guarantee that the columns will generate $\mathbb{R}^n$. Test your result in the case of the matrices

$$\mathbf{A} = \begin{bmatrix} 1 & 1 & 1 \\ 0 & 2 & 3 \\ 0 & 0 & 4 \end{bmatrix}, \qquad \mathbf{B} = \begin{bmatrix} 1 & 1 & 1 \\ 0 & 1 & 1 \\ 0 & 0 & 2 \end{bmatrix}.$$

USING MATLAB

61. If objects with masses m_1, m_2, m_3 measured in grams (g) are placed at the terminal points of the vectors $\mathbf{v}_1$, $\mathbf{v}_2$, $\mathbf{v}_3$, then the *center of mass* of the *system of point masses* is at the terminal point of the vector $\mathbf{v} = \frac{1}{m}(m_1\mathbf{v}_1 + m_2\mathbf{v}_2 + m_3\mathbf{v}_3)$, where $m = m_1 + m_2 + m_3$ is the *total mass* of the system. In many calculations, the system of point masses can then be replaced by an object of mass m at the terminal point of $\mathbf{v}$. Find the center of mass of the following systems. In each case, draw an accurate diagram showing the position of the point masses and the center of mass.

(a) $m_1 = 2\text{g}$ $\quad \mathbf{v}_1 = [1 \;\; 2]^{\mathsf{T}}$
$m_2 = 3\text{g}$ $\quad \mathbf{v}_2 = [-1 \;\; 4]^{\mathsf{T}}$
$m_3 = 5\text{g}$ $\quad \mathbf{v}_3 = [-3 \;\; -2]^{\mathsf{T}}$

(b) $m_1 = 2\text{g}$ $\quad \mathbf{v}_1 = [1 \;\; 0 \;\; 0]^{\mathsf{T}}$
$m_2 = 3\text{g}$ $\quad \mathbf{v}_2 = [0 \;\; 1 \;\; 0]^{\mathsf{T}}$
$m_3 = 4\text{g}$ $\quad \mathbf{v}_3 = [0 \;\; 0 \;\; 1]^{\mathsf{T}}$

Write an M-file to find the center of mass of any system of point masses in $\mathbb{R}^n$.

62. Two coins (C1 and C2) are alloys of silver (Ag), copper (Cu), and gold (Au). C1 weighs 1.2 oz and is made from 92% silver and 8% copper. C2 weighs 0.7 oz and is made from 4% silver, 6% copper, and 90% gold. If the vector $\mathbf{b} = (225.0, 25.5, 94.5)$ oz records the respective quantities of silver, copper, and gold available, find how many coins of each type can be struck.

63. The ordered triple (a, b, c) denotes a team containing a people of type A, b people of type B and c people of type C. Teams are indivisible units. Determine if it is possible to distribute 58 people of type A, 18 people of type B and 17 people of type C into teams of four types, namely (12,3,3), (1,1,0), (3,1,1), (1,0,1). Determine if this can be done so that all teams are represented.

Exercises 64–69. For each matrix $\mathbf{M} = [\mathbf{v}_1 \; \mathbf{v}_2 \; \mathbf{v}_3 \; \mathbf{b}]$ *determine if column* $\mathbf{b}$ *is a linear combination of earlier columns and whether* $\mathbf{v}_3$ *is a linear combination of the two previous columns.*

64. $\begin{bmatrix} 4 & 0 & -2 & 5 \\ -1 & -2 & 1 & 3 \\ 2 & 4 & 3 & 3 \end{bmatrix}$

65. $\begin{bmatrix} 1 & 3 & -1 & 4 \\ 1 & 2 & 1 & 4 \\ 1 & 1 & -1 & 4 \end{bmatrix}$

66. $\begin{bmatrix} 1 & 2 & 4 & 3 \\ 2 & 4 & 8 & 6 \\ -1 & 6 & 8 & -15 \\ 4 & 2 & 7 & 21 \end{bmatrix}$

67. $\begin{bmatrix} 2 & -3 & 1 & 7 \\ 5 & 2 & 4 & 8 \\ 1 & 1 & -1 & 1 \\ 3 & -2 & 2 & 8 \\ -2 & 7 & 0 & -11 \end{bmatrix}$

68. $\begin{bmatrix} 1.1 & 2.5 & 4.2 & 3.5 \\ 2.6 & 4.8 & 8.0 & 6.1 \\ 1 & 6.2 & 8.6 & -5.5 \\ 4.0 & 2.2 & 7.6 & 1.0 \end{bmatrix}$

69. The matrix **magic**(4)

70. *Project*. Let

$$\mathbf{v}_1 = \begin{bmatrix} 2 \\ -3 \end{bmatrix}, \qquad \mathbf{v}_2 = \begin{bmatrix} 3 \\ 1 \end{bmatrix}, \qquad \mathbf{v}_3 = \begin{bmatrix} -1 \\ 3 \end{bmatrix}$$

Your goal is to plot the vectors $\mathbf{v}_1$, $\mathbf{v}_2$, $\mathbf{v}_3$, $2\mathbf{v}_1 + 3\mathbf{v}_2$, $2\mathbf{v}_1 + 3\mathbf{v}_2 + 4\mathbf{v}_3$ in the same figure window. *Hint*: Define a 2-vector of x-values using the first components in $\mathbf{v}_1$ and $\mathbf{0}$ and a 2-vector of y-values using the second components, namely $\mathbf{x1} = [0, 2]$, $\mathbf{y1} = [0, -3]$ and then the 2-D command **plot**($\mathbf{x1}, \mathbf{y1}$) draws a line segment from (0,0) to (2,−3) representing the vector $\mathbf{v}_1$. Alternatively, consult online help for the command arrow.

3.2 Linear Independence, Basis, Dimension

We now come to some important concepts that lie at the core of linear algebra. The concepts developed here in the context of $\mathbb{R}^n$ appear also in the general setting of vector spaces (Chapter 7).

3.2.1 Linear Independence

Consider a set of vectors $\mathcal{S} = \{\mathbf{v}_1, \mathbf{v}_2, \ldots, \mathbf{v}_k\}$ in $\mathbb{R}^n$. If no vector in $\mathcal{S}$ is a linear combination of other vectors in $\mathcal{S}$, then $\mathcal{S}$ is *linearly independent*. Conversely, if at least one vector in $\mathcal{S}$ is a linear combination of other vectors in $\mathcal{S}$, then $\mathcal{S}$ is *linearly dependent*.

Our immediate goals are to be able to recognize which sets of vectors are linearly independent and which are linearly dependent. Even more, we wish to know which vectors in a linearly dependent set are linear combinations of other vectors in the set. The following definition holds the key to meeting all these goals.

Definition 3.5 ***Linear independence, linear dependence***

A set of vectors $\mathcal{S} = \{\mathbf{v}_1, \mathbf{v}_2, \ldots, \mathbf{v}_k\}$ in $\mathbb{R}^n$ is *linearly independent* if the vector equation

$$x_1\mathbf{v}_1 + x_2\mathbf{v}_2 + \cdots + x_k\mathbf{v}_k = \mathbf{0} \tag{3.17}$$

has *only* the zero (trivial) solution

$$(x_1, x_2, \ldots, x_k) = (0, 0, \ldots, 0).$$

$\mathcal{S}$ is *linearly dependent* if equation (3.17) has a nonzero solution $(x_1, x_2, \ldots, x_k)$, that is, a solution in which *at least one* of the scalars $x_1, x_2, \ldots, x_k$ is nonzero.

To be grammatically correct, we say either that the set $\mathcal{S} = \{\mathbf{v}_1, \mathbf{v}_2, \ldots, \mathbf{v}_k\}$ *is* linearly independent (dependent) or that the vectors $\mathbf{v}_1, \mathbf{v}_2, \ldots, \mathbf{v}_k$ *are* linearly independent (dependent).

The next theorem justifies the remarks that opened this subsection. The theorem and its proof applies more generally in the context of vector spaces (Section 7.2).

Theorem 3.3 **Characterization of Linear Dependence**

A set of vectors $\mathcal{S} = \{\mathbf{v}_1, \mathbf{v}_2, \ldots, \mathbf{v}_k\}$ in $\mathbb{R}^n$ is linearly dependent if and only if at least one vector in $\mathcal{S}$ is a linear combination of other vectors in $\mathcal{S}$.

Proof There are two implications to prove.

(a) Suppose $\mathcal{S}$ is linearly dependent. Then equation (3.17) has a nonzero solution $(x_1, x_2, \ldots, x_k)$. Hence, at least one of the scalars is nonzero. Suppose, for example, that $x_1 \neq 0$. Then, rearranging equation (3.17), we have

$$\mathbf{v}_1 = -\left(\frac{x_2}{x_1}\right)\mathbf{v}_2 - \left(\frac{x_3}{x_1}\right)\mathbf{v}_3 - \cdots - \left(\frac{x_k}{x_1}\right)\mathbf{v}_k,$$

showing that $\mathbf{v}_1$ is a linear combination of other vectors in $\mathcal{S}$. Using the same argument, any vector in $\mathcal{S}$ whose coefficient in (3.17) is nonzero can be represented as a linear combination of other vectors in $\mathcal{S}$.

(b) Suppose that some vector, denoted by $\mathbf{v}_1$, is a linear combination of other vectors in $\mathcal{S}$. Then

$$\mathbf{v}_1 = y_2\mathbf{v}_2 + \cdots + y_k\mathbf{v}_k$$

for some scalars $y_2, \ldots, y_k$ and rearranging, we have

$$(-1)\mathbf{v}_1 + y_2\mathbf{v}_2 + \cdots + y_k\mathbf{v}_k = \mathbf{0},$$

showing that (3.17) has the nonzero solution $x_1 = -1, x_2 = y_2, \ldots, x_k = y_k$. Hence, $\mathcal{S}$ is linearly dependent.

Caution! If a set of vectors $\mathcal{S}$ is linearly dependent, it may or may not be the case that *every* vector in $\mathcal{S}$ is a linear combination of other vectors in $\mathcal{S}$. Examples 1 and 2 illustrate the two possibilities.

Concerning Equation (3.17)

Keep the following points in mind:

(a) Solving equation (3.17) is the *standard test* for linear independence. The equation *always* has the zero solution and so the only important question is whether or not a nonzero solution exists.

(b) The vector equation (3.17) can be written as a homogeneous linear system (S) whose matrix form is $\mathbf{Ax} = \mathbf{0}$, where $\mathbf{A} = [\,\mathbf{v}_1\ \mathbf{v}_2\ \cdots\ \mathbf{v}_k\,]$ is the $n \times k$ coefficient matrix for (S). Hence, from this point of view, we have the statement: The set $\{\mathbf{v}_1, \mathbf{v}_2, \ldots, \mathbf{v}_k\}$ is linearly independent (dependent) if and only if $\mathbf{Ax} = \mathbf{0}$ has only the zero solution (has a nonzero solution). Note that the solution set $(x_1, x_2, \ldots, x_k)$ for $\mathbf{Ax} = \mathbf{0}$ can be found by computing the reduced form $\mathbf{A}^*$ of $\mathbf{A}$.

EXAMPLE 1

Three Vectors in $\mathbb{R}^3$

We wish to know if the subset $\mathcal{S} = \{\mathbf{v}_1, \mathbf{v}_2, \mathbf{v}_3\}$ of $\mathbb{R}^3$ is linearly independent or linearly dependent, where

$$\mathbf{v}_1 = \begin{bmatrix} 1 \\ 1 \\ 0 \end{bmatrix}, \quad \mathbf{v}_2 = \begin{bmatrix} 1 \\ -1 \\ 2 \end{bmatrix}, \quad \mathbf{v}_3 = \begin{bmatrix} 3 \\ -1 \\ 4 \end{bmatrix}.$$

Using the preceding remark (b), we solve the linear system defined by $\mathbf{Ax} = \mathbf{0}$, where $\mathbf{A} = [\,\mathbf{v}_1\ \mathbf{v}_2\ \mathbf{v}_3\,]$. Reducing $\mathbf{A}$ to its reduced form $\mathbf{A}^*$, we have

$$\mathbf{A} = \begin{bmatrix} 1 & 1 & 3 \\ 1 & -1 & -1 \\ 0 & 2 & 4 \end{bmatrix} \sim \begin{bmatrix} 1 & 0 & 1 \\ 0 & 1 & 2 \\ 0 & 0 & 0 \end{bmatrix} = \mathbf{A}^*.$$

Then $r = \operatorname{rank} \mathbf{A} = 2 < 3 = k$ (the number of unknowns) and so there is $k - r = 1$ free variable. Let $x_3 = t$ be a real parameter. The solution set is $(x_1, x_2, x_3) = (-t, -2t, t)$ and (3.17) (with $k = 3$) takes the form

$$-t\mathbf{v}_1 - 2t\mathbf{v}_2 + t\mathbf{v}_3 = \mathbf{0}. \tag{3.18}$$

A nonzero solution for (3.18) is obtained by taking any nonzero value for t. Thus $\mathcal{S}$ is a linearly dependent set. Taking $t = -1$ (a convenient nonzero value), we obtain $\mathbf{v}_1 + 2\mathbf{v}_2 - \mathbf{v}_3 = \mathbf{0}$, which shows that every vector in $\mathcal{S} = \{\mathbf{v}_1, \mathbf{v}_2, \mathbf{v}_3\}$ is a linear combination of the other vectors in $\mathcal{S}$. In this case, $\mathbf{v}_1$ is in span$\{\mathbf{v}_2, \mathbf{v}_3\}$, $\mathbf{v}_2$ is in span$\{\mathbf{v}_1, \mathbf{v}_3\}$, and $\mathbf{v}_3$ is in span$\{\mathbf{v}_1, \mathbf{v}_2\}$. ▩

The Linear Dependence Relation

Equation (3.17) is called the *linear dependence relation* for $\mathcal{S}$ because it tells us exactly which vectors in $\mathcal{S}$ are linear combinations of (or dependent on) other vectors in $\mathcal{S}$. Specifically, for each nonzero scalar x_p in the solution set $(x_1, x_2, \ldots, x_k)$ for (3.17), the equation can be solved for the corresponding vector $\mathbf{v}_p$, expressing $\mathbf{v}_p$ as a linear combination of other vectors in $\mathcal{S}$. Note that there are no nonzero scalars x_p in the solution set for (3.17) when $\mathcal{S}$ is linearly independent.

EXAMPLE 2

The Linear Dependence Relation

Suppose $\mathcal{S} = \{\mathbf{v}_1, \mathbf{v}_2, \mathbf{v}_3, \mathbf{v}_4\}$ is a set of vectors in $\mathbb{R}^n$ and that equation (3.17) is solved for these vectors giving the general solution

$$2t\mathbf{v}_1 + 0\mathbf{v}_2 - 4t\mathbf{v}_3 + t\mathbf{v}_4 = \mathbf{0}, \tag{3.19}$$

where t is a real parameter. Choosing $t = 1$ (a convenient nonzero value), we see that only $\mathbf{v}_1$, $\mathbf{v}_3$, and $\mathbf{v}_4$ are linear combinations of other vectors from $\mathcal{S}$. Solving (3.19), we have

$$\mathbf{v}_1 = 2\mathbf{v}_3 - 0.5\mathbf{v}_4, \qquad \mathbf{v}_3 = 0.5\mathbf{v}_1 + 0.25\mathbf{v}_4, \qquad \mathbf{v}_4 = -2\mathbf{v}_1 + 4\mathbf{v}_3.$$

Thus, $\mathbf{v}_1$ is in span$\{\mathbf{v}_3, \mathbf{v}_4\}$, $\mathbf{v}_3$ is in span$\{\mathbf{v}_1, \mathbf{v}_4\}$, $\mathbf{v}_4$ is in span$\{\mathbf{v}_1, \mathbf{v}_3\}$. However, the vector $\mathbf{v}_2$ whose coefficient is zero is not in span$\{\mathbf{v}_1, \mathbf{v}_3, \mathbf{v}_4\}$ because if $\mathbf{v}_2 = \alpha\mathbf{v}_1 + \beta\mathbf{v}_3 + \gamma\mathbf{v}_4$ for some scalars α, β, γ, then $\alpha\mathbf{v}_1 - \mathbf{v}_2 + \beta\mathbf{v}_3 + \gamma\mathbf{v}_4 = \mathbf{0}$, but this equation does appear in the solution (3.19) for any value t. ▩

Theorem 3.4

Linearly Independent Columns, Rank

Let $\mathbf{A} = [\mathbf{v}_1 \ \mathbf{v}_2 \ \cdots \ \mathbf{v}_k]$ *be an* $n \times k$ *matrix. The following statements are equivalent:*

(a) The columns of $\mathbf{A}$ are linear independent.
(b) rank $\mathbf{A} = k$.

Proof The columns of $\mathbf{A}$ are linearly independent if and only if $\mathbf{Ax} = \mathbf{0}$ has only the zero solution and this is the case if and only if rank $\mathbf{A} = k$ (Section 1.2, Theorem 1.6(a)).

EXAMPLE 3

Linearly Independent Columns

Consider a 4×3 matrix $\mathbf{A} = [\mathbf{v}_1 \ \mathbf{v}_2 \ \mathbf{v}_3]$, where

$$\mathbf{v}_1 = \begin{bmatrix} 1 \\ 0 \\ 0 \\ 0 \end{bmatrix}, \quad \mathbf{v}_2 = \begin{bmatrix} 1 \\ 2 \\ 0 \\ 0 \end{bmatrix}, \quad \mathbf{v}_3 = \begin{bmatrix} 1 \\ 2 \\ 3 \\ 0 \end{bmatrix} \quad \Rightarrow \quad \mathbf{A} = \begin{bmatrix} \textcircled{1} & 1 & 1 \\ 0 & \textcircled{2} & 2 \\ 0 & 0 & \textcircled{3} \\ 0 & 0 & 0 \end{bmatrix}.$$

Note that rank $\mathbf{A} = 3$ because $\mathbf{A}$ is already in echelon form. By Theorem 3.4, the columns of $\mathbf{A}$ form a linearly independent subset of $\mathbb{R}^4$. ■

We can now add one more condition for invertibility of a square matrix to those stated in Theorem 2.9 in Section 2.2.

Theorem 3.5 **Invertibility and Linear Independence**

An $n \times n$ matrix is invertible if and only if its columns (rows) are linearly independent.

Proof Let $\mathbf{A} = [\,\mathbf{v}_1\ \mathbf{v}_2\ \cdots\ \mathbf{v}_n\,]$ be an $n \times n$ matrix. By Theorem 2.9 in Section 2.2, $\mathbf{A}$ is invertible if and only if rank $\mathbf{A} = n$ and now apply Theorem 3.4 with $k = n$. By property (**P4**) in Section 2.2, $\mathbf{A}^\mathsf{T}$ is invertible if and only if its transpose $\mathbf{A}^\mathsf{T}$ is invertible, but the columns (rows) of $\mathbf{A}^\mathsf{T}$ are the rows (columns) of $\mathbf{A}$.

We now show that the number of vectors in a linearly independent subset of $\mathbb{R}^n$ cannot exceed n. We call n an *upper bound.*

Theorem 3.6 **The Number of Vectors in a Linearly Independent Set**

If the set $\mathcal{S} = \{\mathbf{v}_1, \mathbf{v}_2, \ldots, \mathbf{v}_k\}$ in $\mathbb{R}^n$ is linearly independent, then $k \leq n$.

Proof Form the matrix $\mathbf{A} = [\,\mathbf{v}_1\ \mathbf{v}_2\ \cdots\ \mathbf{v}_k\,]$ and apply Theorem 3.4. If $\mathcal{S}$ is linearly independent, then $k = \text{rank}\,\mathbf{A}$. But rank $\mathbf{A} \leq n$ and so $k \leq n$.

For example, any linearly independent subset of $\mathbb{R}^4$ contains 4 or fewer vectors and $n = 4$ is the upper bound in this case.

Caution! The implication in Theorem 3.6 goes only one way. A set of k vectors in $\mathbb{R}^n$, where $k \leq n$, is not necessarily linearly independent. You will need to check!

Note that the contrapositive (Appendix B: Toolbox) of the implication in Theorem 3.6 is this:

$$n < k \quad \Rightarrow \quad \mathcal{S} \text{ is linearly dependent.} \tag{3.20}$$

Hence, for example, 11, 20, or 300 vectors in $\mathbb{R}^{10}$ are linearly dependent.

Some Useful Facts

The proofs of the following facts are left for Exercises 3.2.

(L1) Any set $\mathcal{S}$ in $\mathbb{R}^n$ containing the zero vector $\mathbf{0}$ is linearly dependent.

(L2) A set $\mathcal{S} = \{\mathbf{v}\}$ containing a single nonzero vector $\mathbf{v}$ in $\mathbb{R}^n$ is linearly independent.

(L3) Two vectors in $\mathbb{R}^n$ form a linearly dependent set if and only if one vector is a scalar multiple of the other.

(L4) Any subset of a linearly independent set of vectors is linearly independent.

(L5) A (finite) set of vectors containing a linearly dependent subset is linearly dependent.

3.2.2 Bases

The next step in the theory is to define a frame of reference or *coordinate system* for a given subspace $\mathcal{U}$ of $\mathbb{R}^n$, and in order to do this we need to find a *basis* for $\mathcal{U}$.

Definition 3.6

Basis

Let $\mathcal{U}$ be a subspace of $\mathbb{R}^n$. A *basis* for $\mathcal{U}$ is a subset $\mathcal{B}$ of $\mathcal{U}$ such that (a) $\mathcal{B}$ is linearly independent, and (b) $\operatorname{span} \mathcal{B} = \mathcal{U}$.

Observe that if the set $\mathcal{B} = \{\mathbf{v}_1, \mathbf{v}_2, \ldots, \mathbf{v}_k\}$ in $\mathbb{R}^n$ is linearly independent, then $\mathcal{B}$ is a basis for the subspace $\mathcal{U} = \operatorname{span} \mathcal{B}$. Here is an example.

EXAMPLE 4 **Basis for a Subspace of $\mathbb{R}^3$**

Consider the vectors

$$\mathbf{v}_1 = \begin{bmatrix} 1 \\ 1 \\ 0 \end{bmatrix}, \qquad \mathbf{v}_2 = \begin{bmatrix} 2 \\ 2 \\ 2 \end{bmatrix}.$$

Using Fact (**L3**), the set $\mathcal{B} = \{\mathbf{v}_1, \mathbf{v}_2\}$ is linearly independent because neither vector is a scalar multiple of the other. Then $\mathcal{B}$ is a basis for the subspace $\mathcal{U} = \operatorname{span} \mathcal{B}$, which is visualized as a plane P passing through the origin in $\mathbb{R}^3$ containing the arrows that represent $\mathbf{v}_1$ and $\mathbf{v}_2$. ■

A subspace $\mathcal{U}$ of $\mathbb{R}^n$ has many possible *bases* (plural for basis) and we often need to choose one of these which has properties suitable for the application at hand. However, there are some convenient and natural choices, which we refer to as *standard*.

Standard Bases

Refer to Figure 3.11. The standard basis for $\mathbb{R}^2$ consists of two vectors, defined by

$$\mathbf{e}_1 = \begin{bmatrix} 1 \\ 0 \end{bmatrix}, \qquad \mathbf{e}_2 = \begin{bmatrix} 0 \\ 1 \end{bmatrix} \qquad \Rightarrow \qquad \mathcal{B} = \{\mathbf{e}_1, \mathbf{e}_2\}.$$

The standard basis for $\mathbb{R}^3$ consists of three vectors, defined by $\mathbf{e}_1, \mathbf{e}_2, \mathbf{e}_3$, where

$$\mathbf{e}_1 = \begin{bmatrix} 1 \\ 0 \\ 0 \end{bmatrix}, \quad \mathbf{e}_2 = \begin{bmatrix} 0 \\ 1 \\ 0 \end{bmatrix}, \quad \mathbf{e}_2 = \begin{bmatrix} 0 \\ 0 \\ 1 \end{bmatrix} \qquad \Rightarrow \qquad \mathcal{B} = \{\mathbf{e}_1, \mathbf{e}_2, \mathbf{e}_3\}$$

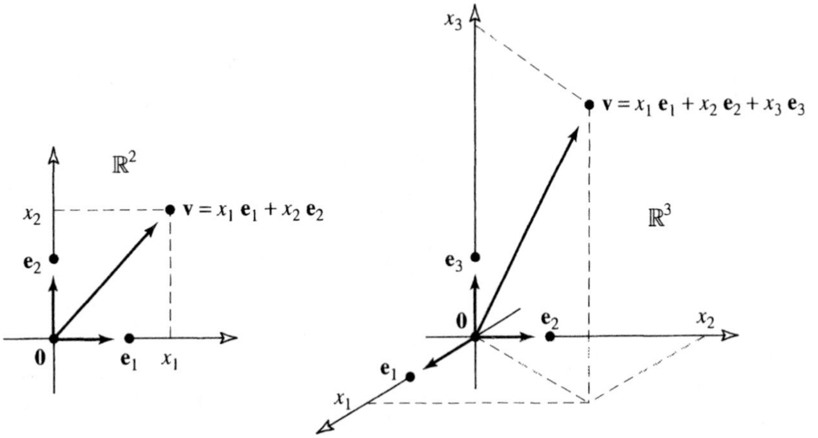

Figure 3.11 Standard bases in $\mathbb{R}^2$ and $\mathbb{R}^3$.

Note that the standard basis vectors for $\mathbb{R}^2$ and $\mathbb{R}^3$ are just the columns of the identity matrices $\mathbf{I}_2$ and $\mathbf{I}_3$, respectively.

$$\mathbf{I}_2 = \begin{bmatrix} 1 & 0 \\ 0 & 1 \end{bmatrix} = [\,\mathbf{e}_1 \;\; \mathbf{e}_2\,]$$

$$\mathbf{I}_3 = \begin{bmatrix} 1 & 0 & 0 \\ 0 & 1 & 0 \\ 0 & 0 & 1 \end{bmatrix} = [\,\mathbf{e}_1 \;\; \mathbf{e}_2 \;\; \mathbf{e}_3\,]$$

$$\mathbf{I}_n = \begin{bmatrix} 1 & 0 & \cdots & 0 \\ 0 & 1 & \cdots & 0 \\ \vdots & \vdots & \ddots & \vdots \\ 0 & 0 & \cdots & 1 \end{bmatrix} = [\,\mathbf{e}_1 \;\; \mathbf{e}_2 \;\; \mathbf{e}_3 \;\; \cdots \;\; \mathbf{e}_n\,]$$

The standard basis for $\mathbb{R}^n$ is the set of vectors $\mathcal{B} = \{\mathbf{e}_1,\ \mathbf{e}_2,\ \ldots\ ,\mathbf{e}_n\}$ that are the columns of the identity matrix $\mathbf{I}_n$. The set $\mathcal{B}$ is linearly independent because $\operatorname{rank} \mathbf{I}_n = n$ and any vector $\mathbf{v}$ in $\mathbb{R}^n$ is represented as a unique linear combination of vectors from $\mathcal{B}$ using the components in $\mathbf{v}$, namely

$$\mathbf{v} = \begin{bmatrix} x_1 \\ x_2 \\ \vdots \\ x_n \end{bmatrix} \quad \Rightarrow \quad \mathbf{v} = x_1\mathbf{e}_1 + x_2\mathbf{e}_2 + \cdots + x_n\mathbf{e}_n$$

The next result gives an approach to finding a basis for a given subspace $\mathcal{U}$ using a subset of vectors in $\mathcal{U}$. The result is true in a general vector and the proof appears in Section 7.2.

Theorem 3.7 **Constructing a Basis**

Suppose $\mathcal{U}$ is a subspace of $\mathbb{R}^n$ and let $\mathcal{S}$ be a finite subset of $\mathcal{U}$. There are two possibilities.

(a) Suppose $\mathcal{S}$ spans $\mathcal{U}$. If any vector $\mathbf{v}$ in $\mathcal{S}$ is a linear combination of other vectors in $\mathcal{S}$, then $\mathbf{v}$ can be deleted from $\mathcal{S}$ to form a smaller subset $\mathcal{S}'$ of

$\mathcal{S}$ which still spans $\mathcal{U}$. The process may be repeated (if necessary) until a linearly independent subset $\mathcal{B}$ of $\mathcal{S}$ is found such that span $\mathcal{B} = \mathcal{U}$.

(b) Suppose $\mathcal{S}$ is linearly independent. If $\mathcal{S}$ does not span $\mathcal{U}$, then there is a vector $\mathbf{v}$ in $\mathcal{U}$ which is not in span $\mathcal{S}$. The set $\mathcal{S}' = \mathcal{S} \cup \{\mathbf{v}\}$ obtained by adjoining $\mathbf{v}$ to $\mathcal{S}$ is linearly independent and span $\mathcal{S}$ is a subset of span $\mathcal{S}'$. The process may be repeated (if necessary) until a linearly independent set $\mathcal{B}$ is found that contains $\mathcal{S}$ and is such that span $\mathcal{B} = \mathcal{U}$.

Remarks

Consider Theorem 3.7(a). If $\mathbf{v}$ is any nonzero vector in $\mathcal{S}$, then the set $\{\mathbf{v}\}$ is linearly independent. Hence, deleting vectors (one at a time) from $\mathcal{S}$ will eventually result in a linearly independent spanning set for $\mathcal{U}$. Consider Theorem 3.7(b). There cannot be more than n vectors in any linearly independent subset of $\mathbb{R}^n$. Hence, the process of adding vectors to $\mathcal{S}$ (one at a time) will eventually result in a linearly independent set (containing n of fewer vectors) that spans $\mathcal{U}$.

EXAMPLE 5

Three Bases for $\mathbb{R}^2$

Consider the set $\mathcal{S} = \{\mathbf{v}_1, \mathbf{v}_2, \mathbf{v}_3\} \subset \mathbb{R}^2$, where

$$\mathbf{v}_1 = \begin{bmatrix} 1 \\ 1 \end{bmatrix}, \qquad \mathbf{v}_2 = \begin{bmatrix} 1 \\ 0 \end{bmatrix}, \qquad \mathbf{v}_3 = \begin{bmatrix} 2 \\ 1 \end{bmatrix}.$$

Using (3.20) on page 155 with $k = 3$ and $n = 2$ shows that $\mathcal{S}$ is linearly dependent. Moreover, the linear dependence relation for $\mathcal{S}$ is $\mathbf{v}_1 + \mathbf{v}_2 - \mathbf{v}_3 = \mathcal{O}$, which shows that any vector in $\mathcal{S}$ is a linear combination of the other two. Now let $\mathbf{A} = [\,\mathbf{v}_1 \;\; \mathbf{v}_2 \;\; \mathbf{v}_3\,]$ and let $\mathbf{b}$ be any 2-vector. The linear system $\mathbf{Ax} = \mathbf{b}$, which expresses $\mathbf{b}$ as a linear combination of the columns of $\mathbf{A}$, is consistent because rank $\mathbf{A} = 2 < 3 =$ number of unknowns. Hence, $\mathcal{S}$ spans $\mathbb{R}^2$. Using Fact (**L3**), any two vectors in $\mathcal{S}$ are linearly independent and so, using Theorem 3.7(a), deleting any vector from $\mathcal{S}$ results in a basis for $\mathbb{R}^2$. In this case, three different bases for $\mathbb{R}^2$ can be formed from $\mathcal{S}$. ▪

Constructing Bases for $\mathbb{R}^n$

Suppose we are given a set of vectors $\mathcal{S} = \{\mathbf{v}_1, \mathbf{v}_2, \ldots, \mathbf{v}_k\}$ in $\mathbb{R}^n$ and our goal is to find a basis for $\mathbb{R}^n$ that contains some vectors from $\mathcal{S}$. Proceed as follows: Form the $n \times (k+n)$ matrix $\mathbf{A} = [\,\mathbf{v}_1 \;\; \mathbf{v}_2 \;\; \cdots \;\; \mathbf{v}_k \;\; \mathbf{b}_1 \;\; \mathbf{b}_2 \;\; \cdots \;\; \mathbf{b}_n\,]$, where $\mathcal{B} = \{\mathbf{b}_1, \mathbf{b}_2, \ldots, \mathbf{b}_n\}$ is a basis for $\mathbb{R}^n$, and compute its reduced from $\mathbf{A}^*$. Then the pivot columns in $\mathbf{A}$ (these correspond to pivot columns in $\mathbf{A}^*$) are linearly independent and form a basis for $\mathbb{R}^n$.

▪ ILLUSTRATION 3.7

Constructing a Basis for $\mathbb{R}^4$

Let the set $\mathcal{S}$ consist of the first four columns in the matrix $\mathbf{A}$ shown below and let $\mathcal{B}$ be the standard basis for $\mathbb{R}^4$. Reducing $\mathbf{A}$ to $\mathbf{A}^*$, we have

$$\mathbf{A} = [\,\mathbf{v}_1\ \mathbf{v}_2\ \mathbf{v}_3\ \mathbf{v}_4\quad \mathbf{e}_1\ \mathbf{e}_2\ \mathbf{e}_3\ \mathbf{e}_4\,]$$

$$= \begin{bmatrix} 1 & 0 & 2 & 1 & 1 & 0 & 0 & 0 \\ -1 & -3 & 1 & 1 & 0 & 1 & 0 & 0 \\ 1 & 3 & -1 & 1 & 0 & 0 & 1 & 0 \\ 0 & -1 & 1 & 1 & 0 & 0 & 0 & 1 \end{bmatrix}$$

$$\sim \begin{bmatrix} \textcircled{1} & 0 & 2 & 0 & 0 & -2 & -1 & 3 \\ 0 & \textcircled{1} & -1 & 0 & 0 & 0.5 & 0.5 & -1 \\ 0 & 0 & 0 & \textcircled{1} & 0 & 0.5 & 0.5 & 0 \\ 0 & 0 & 0 & 0 & \textcircled{1} & 1.5 & 0.5 & -3 \end{bmatrix} = \mathbf{A}^*. \quad (3.21)$$

The pivots in $\mathbf{A}^*$ are circled. The matrix $\mathbf{B} = [\,\mathbf{v}_1\ \mathbf{v}_2\ \mathbf{v}_4\ \mathbf{e}_1\,]$ is row equivalent to $\mathbf{I}_4$ and so the columns of $\mathbf{B}$ are linearly independent. The columns of $\mathbf{B}$ also span $\mathbb{R}^4$. To see this, note that $\mathbf{B}$ is invertible and hence the matrix equation $\mathbf{Bx} = \mathbf{b}$ has the (unique) solution $\mathbf{x} = \mathbf{B}^{-1}\mathbf{b}$, for any 4-vector $\mathbf{b}$ and so any vector in $\mathbb{R}^4$ is a linear combination of the columns of $\mathbf{B}$. The argument just given can be eliminated if we look ahead to Theorem 3.10, which tells us that any set of four linearly independent vectors in $\mathbb{R}^4$ automatically spans the space $\mathbb{R}^4$. ■

3.2.3 Coordinate Systems

Every basis $\mathcal{B} = \{\mathbf{b}_1,\ \mathbf{b}_2,\ \dots\ ,\mathbf{b}_k\}$ for a subspace $\mathcal{U}$ of $\mathbb{R}^n$ imposes a *coordinate system* on $\mathcal{U}$ that depends on the *order* in which the basis vectors appear in the list $\{\mathbf{b}_1,\ \mathbf{b}_2,\ \dots\ ,\mathbf{b}_k\}$. Defining such a coordinate system depends on the next fundamental property of bases. The proof is cosmetically[3] the same for $\mathbb{R}^n$ and for vector spaces in general.

Theorem 3.8 **Unique Representation**

Let $\mathcal{U}$ be a subspace of $\mathbb{R}^n$ and let $\mathcal{B} = \{\mathbf{b}_1,\ \mathbf{b}_2,\ \dots\ ,\mathbf{b}_k\}$ be an ordered basis for $\mathcal{U}$. Then every vector $\mathbf{v}$ in $\mathcal{U}$ can be written in only one way as a linear combination of basis vectors, namely

$$\mathbf{v} = x_1\mathbf{b}_1 + x_2\mathbf{b}_2 + \cdots + x_k\mathbf{b}_k, \quad (3.22)$$

where the scalars $x_1, x_2, \dots, x_k$ are unique.

Proof Every vector $\mathbf{v}$ in $\mathcal{U}$ is a linear combination of basis vectors. Suppose $\mathbf{v}$ has two representations given by

$$x_1\mathbf{b}_1 + x_2\mathbf{b}_2 + \cdots + x_k\mathbf{b}_k = \mathbf{v} \quad \text{and} \quad y_1\mathbf{b}_1 + y_2\mathbf{b}_2 + \cdots + y_k\mathbf{b}_k = \mathbf{v}$$

for some scalars $x_1, x_2, \dots, x_k$ and $y_1, y_2, \dots, y_k$. Subtracting the two equations gives

$$(x_1 - y_1)\mathbf{b}_1 + (x_2 - y_2)\mathbf{b}_2 + \cdots + (x_k - y_k)\mathbf{b}_k = \mathbf{v} - \mathbf{v} = \mathbf{0}. \quad (3.23)$$

[3] For the general proof we regard $\mathbf{v}_1,\ \mathbf{v}_2,\ \dots\ ,\mathbf{v}_k$ as unspecified *objects* in a vector space.

But $\mathcal{B}$ is linearly independent and so (3.23) has only the zero solution. Hence, $x_1 = y_1, \ldots, x_k = y_k$, and the representation of $\mathbf{v}$ is unique.

The uniqueness property proved in Theorem 3.8 enables us to make the next definition. Keep in mind that the representation (3.22) depends on the order in which the basis vectors appear in the list $\{\mathbf{b}_1, \mathbf{b}_2, \ldots, \mathbf{b}_k\}$. Permuting the basis vectors will permute the scalars $x_1, x_2, \ldots, x_k$.

Definition 3.7

Coordinates

Let $\mathcal{U}$ be a subspace of $\mathbb{R}^n$ and let $\mathcal{B} = \{\mathbf{b}_1, \mathbf{b}_2, \ldots, \mathbf{b}_k\}$ be an *ordered* basis for $\mathcal{U}$. With every vector $\mathbf{v}$ in $\mathcal{U}$ is associated a unique set of scalars $x_1, x_2, \ldots, x_k$ given by (3.22) that are called the *coordinates* of $\mathbf{v}$ relative to $\mathcal{B}$ and are stored in the k-vector $[\mathbf{v}]_{\mathcal{B}}$, where

$$[\mathbf{v}]_{\mathcal{B}} = \begin{bmatrix} x_1 \\ x_2 \\ \vdots \\ x_k \end{bmatrix}$$

The coordinates of a vector $\mathbf{v}$ in $\mathbb{R}^n$ relative to the standard basis $\mathcal{B}$ for $\mathbb{R}^n$ are just the components of $\mathbf{v}$ and so $[\mathbf{v}]_{\mathcal{B}} = \mathbf{v}$. The standard base is therefore very easy to work with computationally because coordinates are self-evident. When using other bases, coordinates need to be computed.

EXAMPLE 6

Finding Coordinates in $\mathbb{R}^2$

Consider the problem of finding the coordinates of the vector $\mathbf{v}$ relative to the basis $\mathcal{B} = \{\mathbf{b}_1, \mathbf{b}_2\}$ for $\mathbb{R}^2$, where

$$\mathbf{b}_1 = \begin{bmatrix} 1 \\ 2 \end{bmatrix}, \quad \mathbf{b}_2 = \begin{bmatrix} 3 \\ 4 \end{bmatrix}, \quad \mathbf{v} = \begin{bmatrix} 5 \\ 6 \end{bmatrix}.$$

We solve the vector equation $x_1\mathbf{b}_1 + x_2\mathbf{b}_2 = \mathbf{b}$ by finding the reduced form $\mathbf{M}^*$ of the augmented matrix $\mathbf{M} = [\,\mathbf{b}_1 \;\; \mathbf{b}_2 \mid \mathbf{v}\,]$. We have

$$\mathbf{M} = \begin{bmatrix} 1 & 2 & 3 \\ 4 & 5 & 6 \end{bmatrix} \sim \begin{bmatrix} 1 & 0 & -1 \\ 0 & 1 & 2 \end{bmatrix} = \mathbf{M}^*$$

implies $x_1 = -1$, $x_2 = 2$. The coordinates of $\mathbf{v}$ relative to $\mathcal{B}$ are $[\mathbf{v}]_{\mathcal{B}} = \begin{bmatrix} -1 \\ 2 \end{bmatrix}$. ■

3.2.4 Dimension

We will now address the question of whether two different bases for a subspace can contain different numbers of vectors. The answer is no! For example, any basis for $\mathbb{R}^n$ contains n vectors. The justification for this fact is given in the next result which is true in a general vector space. The proof, which is given for $\mathbb{R}^n$, translates easily to the general setting.

Theorem 3.9 **Number of Basis Vectors**

Let $\mathcal{U}$ be a subspace of $\mathbb{R}^n$ and let $\mathcal{B} = \{\mathbf{b}_1, \mathbf{b}_2, \ldots, \mathbf{b}_p\}$ and $\mathcal{C} = \{\mathbf{c}_1, \mathbf{c}_2, \ldots, \mathbf{c}_q\}$ be two bases for $\mathcal{U}$. Then $p = q$.

Proof The proof is by contradiction (Appendix B: Toolbox). Assume that $p < q$. The matrix $\mathbf{B} = [\mathbf{b}_1\ \mathbf{b}_2\ \cdots\ \mathbf{b}_p]$ is $n \times p$ and $\mathbf{C} = [\mathbf{c}_1\ \mathbf{c}_2\ \cdots\ \mathbf{c}_q]$ is $n \times q$. The set $\mathcal{B}$ spans $\mathcal{U}$ and so each vector in $\mathcal{C}$ is a unique linear combination of vectors in $\mathcal{B}$, that is, $\mathbf{c}_1 = a_{11}\mathbf{b}_1 + \cdots + a_{p1}\mathbf{b}_p$ and $q-1$ similar equations hold for the other vectors in $\mathcal{C}$. We obtain q linear equations that can be expressed in the single matrix equation

$$\mathbf{C} = [\mathbf{c}_1\ \mathbf{c}_2\ \cdots\ \mathbf{c}_q] = [\mathbf{b}_1\ \mathbf{b}_2\ \cdots\ \mathbf{b}_p]\begin{bmatrix} a_{11} & \cdots & a_{1q} \\ \vdots & \ddots & \vdots \\ a_{p1} & \cdots & a_{pq} \end{bmatrix} = \mathbf{BA}, \quad (3.24)$$

where $\mathbf{A} = [a_{ij}]$ is the unique $p \times q$ matrix shown in (3.24). The columns of $\mathbf{A}$ record the coordinates of the vectors in $\mathcal{C}$ relative to $\mathbf{B}$. Now we argue as follows. The equation $\mathbf{Ax} = \mathbf{0}$ has a nonzero solution $\mathbf{x}$ because $p < q$. Multiplying both sides of the latter equation by $\mathbf{B}$ shows that $\mathbf{BAx} = \mathbf{0}$ has a nonzero solution $\mathbf{x}$. However, $\mathbf{C} = \mathbf{BA}$ from (3.24), which means that $\mathbf{Cx} = \mathbf{0}$ has a nonzero solution $\mathbf{x}$. But this is impossible because the columns of $\mathcal{C}$ are linearly independent. Our opening assumption must have been false and so $p \geq q$. Interchange $\mathcal{B}$ and $\mathcal{C}$, assume $p > q$, and apply the previous arguments to obtain the conclusion is $p \leq q$. Hence, $p = q$.

Theorem 3.9 provides the rationale for the next definition.

Definition 3.8 *Dimension*

Let $\mathcal{U}$ be a subspace of $\mathbb{R}^n$. The number of vectors k in a basis for $\mathcal{U}$ is called the *dimension* of $\mathcal{U}$. We say that $\mathcal{U}$ is *k-dimensional* and write $\dim \mathcal{U} = k$. Define $\dim \{\mathbf{0}\} = 0$.

Notice that the definition of dimension fits exactly with our intuition and notation in $\mathbb{R}^n$. Using the standard bases, we have

$$\dim \mathbb{R}^2 = 2, \quad \dim \mathbb{R}^3 = 3, \quad \ldots \quad , \dim \mathbb{R}^n = n.$$

A subspace of $\mathbb{R}^n$ spanned by a single nonzero vector is one-dimensional, a subspace of $\mathbb{R}^3$ spanned by two noncollinear, nonzero vectors (represented by a plane) is two-dimensional, and so on.

To verify that a finite set $\mathcal{B}$ is a basis for a subspace $\mathcal{U}$ requires two steps: Show that $\mathcal{B}$ is a linearly independent set and show that $\mathcal{B}$ spans $\mathcal{U}$. If we know the dimension of $\mathcal{U}$, then only one step is required! The result is stated here and proved in Chapter 7.

Theorem 3.10 **Constructing Bases Knowing the Dimension**

Let $\mathcal{U}$ be a subspace of $\mathbb{R}^n$ such that $\dim \mathcal{U} = k$.

(a) A subset $\mathcal{B}$ of $\mathcal{U}$ consisting of k linearly independent vectors automatically spans $\mathcal{U}$.

(b) A subset $\mathcal{B}$ of $\mathcal{U}$ consisting of k vectors that span $\mathcal{U}$ is automatically linearly independent.

3.2.5 Change of Basis

Consider a k-dimensional subspace $\mathcal{U}$ in $\mathbb{R}^n$ and let $\mathcal{B} = \{\mathbf{b}_1, \mathbf{b}_2, \ldots, \mathbf{b}_k\}$ and $\mathcal{C} = \{\mathbf{c}_1, \mathbf{c}_2, \ldots, \mathbf{c}_k\}$ be two different ordered bases for $\mathcal{U}$. Regard $\mathcal{C}$ as the current basis and $\mathcal{B}$ as the new basis. Each vector $\mathbf{v}$ in $\mathcal{U}$ is represented as a unique linear combination of basis vectors from $\mathcal{C}$. Our goal is to find the representation of $\mathbf{v}$ relative to the new basis $\mathcal{B}$. This problem is known as *change of basis*. Here is the theory.

Using the two sets of basis vectors, define $n \times k$ matrices $\mathbf{B} = [\mathbf{u}_1\ \mathbf{u}_2\ \cdots\ \mathbf{u}_k]$ and $\mathbf{C} = [\,\mathbf{v}_1\ \mathbf{v}_2\ \cdots\ \mathbf{v}_k\,]$. Taking $p = q = k$ in equation (3.24), we have

$$\mathbf{C} = [\,\mathbf{c}_1\ \mathbf{c}_2\ \cdots\ \mathbf{c}_k\,] = [\,\mathbf{b}_1\ \mathbf{b}_2\ \cdots\ \mathbf{b}_k\,] \begin{bmatrix} a_{11} & \cdots & a_{1k} \\ \vdots & \ddots & \vdots \\ a_{k1} & \cdots & a_{kk} \end{bmatrix} = \mathbf{BA}, \tag{3.25}$$

As noted earlier, the first column of $\mathbf{A}$ records the coordinates $[\,\mathbf{c}_1\,]_{\mathcal{B}}$ of $\mathbf{c}_1$ relative to $\mathcal{B}$, and similarly for the other columns in $\mathbf{A}$. Thus, the $k \times k$ matrix $\mathbf{A}$ is unique and has the form

$$\mathbf{A} = [\,[\,\mathbf{c}_1\,]_{\mathcal{B}}\ [\,\mathbf{c}_2\,]_{\mathcal{B}}\ \cdots\ [\,\mathbf{c}_k\,]_{\mathcal{B}}\,]. \tag{3.26}$$

Arguing as in proof of Theorem 3.9 shows that $\mathbf{Ax} = \mathbf{0}$ has only the zero solution, and this implies that $\mathbf{A}$ is invertible (Section 2.2, Theorem 2.9).

The coordinates $[\,\mathbf{v}\,]_{\mathcal{C}}$ of the vector $\mathbf{v}$ relative to the current basis $\mathcal{C}$ are obtained by solving the $n \times k$ linear system defined by the matrix equation

$$\mathbf{C}[\,\mathbf{v}\,]_{\mathcal{C}} = \mathbf{v}. \tag{3.27}$$

But $\mathbf{C} = \mathbf{BA}$ from (3.25) and so (3.27) becomes

$$\mathbf{BA}[\,\mathbf{v}\,]_{\mathcal{C}} = \mathbf{B}(\mathbf{A}[\,\mathbf{v}\,]_{\mathcal{C}}) = \mathbf{v} \tag{3.28}$$

The last equality in (3.28) represents $\mathbf{v}$ as a linear combination of the columns of $\mathbf{B}$ using scalars from the k-vector $\mathbf{A}[\,\mathbf{v}\,]_{\mathcal{C}}$. But the coordinates $[\,\mathbf{v}\,]_{\mathcal{B}}$ of $\mathbf{v}$ relative to $\mathcal{B}$ are unique, and so from (3.28) we can deduce the key relationship

$$[\,\mathbf{v}\,]_{\mathcal{B}} = \mathbf{A}[\,\mathbf{v}\,]_{\mathcal{C}}. \tag{3.29}$$

Using the fact that $\mathbf{A}$ is invertible, we also have

$$[\mathbf{v}]_{\mathcal{C}} = \mathbf{A}^{-1}[\mathbf{v}]_{\mathcal{B}}. \tag{3.30}$$

Equations (3.29) and (3.30) show exactly how to change coordinates between the two bases. The matrix $\mathbf{A}$ in (3.26) is called the *change of basis* or *transition matrix* from current basis $\mathcal{C}$ to new basis $\mathcal{B}$.

Consider the case when $\mathcal{U} = \mathbb{R}^n$ and $\mathcal{C}$ is the standard basis $\{\mathbf{e}_1, \mathbf{e}_2, \ldots, \mathbf{e}_n\}$. Writing $\mathbf{e}_1$ in terms of the new basis $\mathcal{B}$, we have

$$\mathbf{e}_1 = a_{11}\mathbf{u}_1 + a_{21}\mathbf{u}_2 + \cdots + a_{n1}\mathbf{u}_n = \mathbf{B}[\mathbf{e}_1]_{\mathcal{B}} \quad \Rightarrow \quad [\mathbf{e}_1]_{\mathcal{B}} = \mathbf{B}^{-1}\mathbf{e}_1,$$

and similarly for the other standard basis vectors $\mathbf{e}_2, \ldots, \mathbf{e}_n$. The change of basis matrix $\mathbf{A}$ in this case is

$$\mathbf{A} = [\,[\mathbf{e}_1]_{\mathcal{B}}\ [\mathbf{e}_2]_{\mathcal{B}}\ \cdots\ [\mathbf{e}_k]_{\mathcal{B}}\,] = \mathbf{B}^{-1}[\,\mathbf{e}_1\ \mathbf{e}_2\ \cdots\ \mathbf{e}_n\,] = \mathbf{B}^{-1}\mathbf{I}_n = \mathbf{B}^{-1}.$$

For any vector $\mathbf{v}$ in $\mathbb{R}^n$, we have $[\mathbf{v}]_{\mathcal{C}} = \mathbf{v}$, and so from (3.30) the coordinates $[\mathbf{v}]_{\mathcal{B}}$ of $\mathbf{v}$ relative to the new basis $\mathcal{B}$ are given by

$$[\mathbf{v}]_{\mathcal{B}} = \mathbf{B}^{-1}\mathbf{v}. \tag{3.31}$$

Here is an example.

EXAMPLE 7

Change of Basis in $\mathbb{R}^3$

Refer to Figure 3.12, which shows the following vectors:

$$\mathbf{b}_1 = \begin{bmatrix} 2 \\ 1 \\ 1 \end{bmatrix}, \quad \mathbf{b}_2 = \begin{bmatrix} 1 \\ -1 \\ 3 \end{bmatrix}, \quad \mathbf{b}_3 = \begin{bmatrix} 0 \\ 1 \\ 2 \end{bmatrix}, \quad \mathbf{v} = \begin{bmatrix} 2 \\ 3 \\ 4 \end{bmatrix}.$$

Let the current basis $\mathcal{C}$ be the standard basis and let $\mathcal{B} = \{\mathbf{b}_1, \mathbf{b}_2, \mathbf{b}_3\}$ be the new basis for $\mathbb{R}^3$. The coordinates of $\mathbf{v}$ relative to $\mathcal{C}$ are $[\mathbf{v}]_{\mathcal{C}} = \mathbf{v}$.

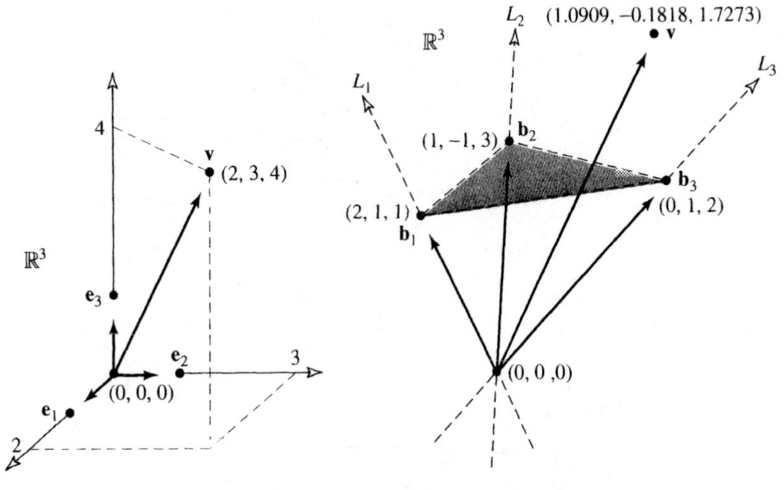

Figure 3.12 A coordinate system for $\mathbb{R}^3$.

We may think of the 3-tuple (2, 3, 4) as the terminal point of the arrow in $\mathbb{R}^3$ that represents $\mathbf{v}$. Form $\mathbf{B} = [\,\mathbf{b}_1\ \mathbf{b}_2\ \mathbf{b}_3\,]$ and check that rank $\mathbf{B} = 3$, which implies that $\mathcal{B}$ is a basis (Theorems 3.4 and 3.10(a)). Using (3.31), the MATLAB computation goes as follows:

$$[\mathbf{v}]_{\mathcal{B}} = \mathbf{B}^{-1}\mathbf{v} = \begin{bmatrix} 0.4545 & 0.1818 & -0.0909 \\ 0.0909 & -0.3636 & 0.1818 \\ -0.3636 & 0.4545 & 0.2727 \end{bmatrix} \begin{bmatrix} 2 \\ 3 \\ 4 \end{bmatrix}$$

$$= \begin{bmatrix} 1.0909 \\ -0.1818 \\ 1.7273 \end{bmatrix}.$$

Hence

$$\mathbf{v} = \begin{bmatrix} 2 \\ 3 \\ 4 \end{bmatrix} = 1.0909\mathbf{b}_1 - 0.1818\mathbf{b}_2 + 1.7273\mathbf{b}_3.$$

The ordered basis $\mathcal{B}$ defines a *coordinate system* for $\mathbb{R}^3$ with axes L_1, L_2, L_3 aligned along the basis vectors in $\mathcal{B}$. ▪

Ordered Bases: Engineering and Science

Engineers and physicists use alternative notation for the standard bases in $\mathbb{R}^2$ and $\mathbb{R}^3$ which is shown in Figure 3.13.

In $\mathbb{R}^3$, the ordered basis $\mathcal{B} = \{\mathbf{i}, \mathbf{j}, \mathbf{k}\}$ imposes an *orientation* on the standard rectangular coordinate system in $\mathbb{R}^3$ resulting in a *right-handed system*. The positive directions of the axes lie along the thumb (**i**), index finger (**j**), and middle finger (**k**) of the right hand, when these fingers are positioned so that they are mutually perpendicular, as in Figure 3.13. With this orientation, the *cross product* (Section 5.4.) of the vectors **i** and **j** is **k**; that is, $\mathbf{i} \times \mathbf{j} = \mathbf{k}$. Reversing the direction of **k** results in a *left-handed coordinate system* for $\mathbb{R}^3$.

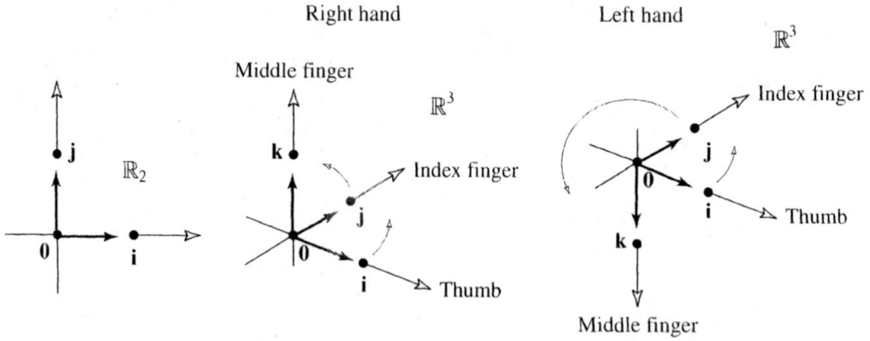

Figure 3.13 Ordered bases in $\mathbb{R}^2$ and $\mathbb{R}^3$.

EXERCISES 3.2

Exercises 1–10. Determine if the given set of vectors $\mathcal{S} = \{\mathbf{v}_1, \mathbf{v}_2 \ldots\}$ is linearly independent. When $\mathcal{S}$ is linearly dependent, give the linear dependence relation and state which vectors are linear combinations of other vectors in $\mathcal{S}$.

1. $\begin{bmatrix} 1 \\ 2 \end{bmatrix}, \begin{bmatrix} 3 \\ 4 \end{bmatrix}$

2. $\begin{bmatrix} 1 \\ -1 \end{bmatrix}, \begin{bmatrix} 1 \\ 0 \end{bmatrix}, \begin{bmatrix} 0 \\ -1 \end{bmatrix}$

3. $\begin{bmatrix} 1 \\ 2 \end{bmatrix}, \begin{bmatrix} 3 \\ 4 \end{bmatrix}, \begin{bmatrix} 5 \\ 6 \end{bmatrix}$

4. $\begin{bmatrix} -1 \\ 2 \end{bmatrix}, \begin{bmatrix} 3 \\ -6 \end{bmatrix}, \begin{bmatrix} 0 \\ 0 \end{bmatrix}, \begin{bmatrix} 1 \\ 1 \end{bmatrix}$

5. $\begin{bmatrix} 1 \\ 2 \\ 3 \end{bmatrix}, \begin{bmatrix} 4 \\ 5 \\ 6 \end{bmatrix}, \begin{bmatrix} 7 \\ 8 \\ 0 \end{bmatrix}$

6. $\begin{bmatrix} 0 \\ 1 \\ 1 \end{bmatrix}, \begin{bmatrix} 1 \\ 0 \\ 1 \end{bmatrix}, \begin{bmatrix} 1 \\ 1 \\ 1 \end{bmatrix}$

7. $\begin{bmatrix} 1 \\ 1 \\ 1 \end{bmatrix}, \begin{bmatrix} 2 \\ 2 \\ 2 \end{bmatrix}, \begin{bmatrix} 0 \\ 0 \\ 5 \end{bmatrix}, \begin{bmatrix} 1 \\ 2 \\ 3 \end{bmatrix}$

8. $\begin{bmatrix} 1 \\ 0 \\ 1 \end{bmatrix}, \begin{bmatrix} 0 \\ 2 \\ 0 \end{bmatrix}, \begin{bmatrix} 2 \\ 1 \\ 3 \end{bmatrix}, \begin{bmatrix} 3 \\ 2 \\ 5 \end{bmatrix}$

9. $\begin{bmatrix} 4 \\ -1 \\ 2 \end{bmatrix}, \begin{bmatrix} 0 \\ 2 \\ 4 \end{bmatrix}, \begin{bmatrix} 1 \\ 0 \\ 1 \end{bmatrix}$

10. $\begin{bmatrix} 2 \\ 1 \\ 3 \\ 1 \end{bmatrix}, \begin{bmatrix} 6 \\ 3 \\ 9 \\ 3 \end{bmatrix}, \begin{bmatrix} 1 \\ 0 \\ 0 \\ 1 \end{bmatrix}, \begin{bmatrix} 5 \\ 2 \\ 6 \\ 3 \end{bmatrix}$

Exercises 11–16. For what value(s) of k are the columns of the given matrix linearly dependent.

11. $\begin{bmatrix} 1 \\ k \end{bmatrix}$

12. $\begin{bmatrix} k & 0 & k \\ 0 & k & -k \end{bmatrix}$

13. $\begin{bmatrix} 2 & -1 \\ 3 & k \end{bmatrix}$

14. $\begin{bmatrix} 1 & -5 & 3 \\ 3 & -8 & -5 \\ -1 & 2 & k \end{bmatrix}$

15. $\begin{bmatrix} 1 & 2 & 2 \\ 1 & -k & 1 \\ -1 & 1 & k \end{bmatrix}$

16. $\begin{bmatrix} 1 & 0 & 1 \\ 2 & -1 & 1 \\ 1 & 3 & -k \\ 0 & k & -4 \end{bmatrix}$

17. Consider the set $\mathcal{S} = \{\mathbf{v}_1, \mathbf{v}_2, \mathbf{v}_3\}$, where

$$\mathbf{v}_1 = \begin{bmatrix} 1 \\ 2 \end{bmatrix}, \quad \mathbf{v}_2 = \begin{bmatrix} 3 \\ 1 \end{bmatrix}, \quad \mathbf{v}_3 = \begin{bmatrix} 2 \\ -1 \end{bmatrix}.$$

Explain why the sets of vectors $\{\mathbf{v}_1\}$, $\{\mathbf{v}_2\}$, $\{\mathbf{v}_3\}$, $\{\mathbf{v}_1, \mathbf{v}_2\}$, $\{\mathbf{v}_2, \mathbf{v}_3\}$, $\{\mathbf{v}_1, \mathbf{v}_3\}$ are linearly independent. Why is $\mathcal{S}$ linearly dependent? Find the linear dependence relation for $\mathcal{S}$. Which vectors in $\mathcal{S}$ are linear combinations of other vectors in $\mathcal{S}$?

18. The set $\mathcal{S} = \{\mathbf{v}_1, \mathbf{v}_2, \mathbf{v}_3\}$, where

$$\mathbf{v}_1 = \begin{bmatrix} 2 \\ -3 \end{bmatrix}, \quad \mathbf{v}_2 = \begin{bmatrix} -1 \\ 4 \end{bmatrix}, \quad \mathbf{v}_3 = \begin{bmatrix} 8 \\ -2 \end{bmatrix}$$

is linearly dependent. Express each vector in $\mathcal{S}$ as a linear combination of other vectors in $\mathcal{S}$.

Exercises 19–22. Use rank to determine if the columns of each matrix **A** *are linearly independent.*

19. $\begin{bmatrix} 1 & 2 & 1 \\ 0 & 2 & 3 \\ 1 & 5 & -1 \end{bmatrix}$

20. $\begin{bmatrix} 1 & 3 & -3 \\ 2 & 0 & 6 \\ -1 & -2 & 1 \end{bmatrix}$

21. $\begin{bmatrix} 1 & 1 & 2 \\ -1 & 3 & 2 \\ 0 & 4 & 8 \\ 2 & 3 & 6 \end{bmatrix}$

22. $\begin{bmatrix} 1 & 8 & 5 & -3 \\ 0 & 4 & 5 & 3 \\ 1 & 0 & -1 & 3 \\ -1 & 1 & 2 & -1 \end{bmatrix}$

Exercises 23–24. Show that the columns of the given matrix **A** *are linearly dependent and find the linear dependence relation between the columns. If possible, find a basis for $\mathbb{R}^3$ from the set of columns.* Hint: *The pivot columns in* **A** *are linearly independent.*

23. $\begin{bmatrix} 1 & 2 & -1 & 0 \\ 2 & 5 & -3 & 1 \\ -2 & 1 & 3 & 5 \end{bmatrix}$

24. $\begin{bmatrix} 0 & 1 & -1 & 3 \\ 2 & -2 & -5 & 9 \\ -2 & 3 & 4 & 7 \end{bmatrix}$

Exercises 25–26. The solution set to each linear system is a subspace $\mathcal{U} = \operatorname{span} \mathcal{S}$ in $\mathbb{R}^3$, for a some set of vectors $\mathcal{S}$. Find $\mathcal{S}$ and prove that $\mathcal{S}$ is linearly independent.

25. $x_1 - x_2 + x_3 = 0$

26. $\begin{cases} x_1 + x_2 + x_3 = 0 \\ 2x_1 + x_2 - x_3 = 0 \end{cases}$

27. Let $\mathcal{S}$ be the set of all binary vectors (entries are 0 or 1) in $\mathbb{R}^5$ that have exactly three entries equal to 1. Show that $\mathcal{S}$ is linearly dependent. If possible, find a basis $\mathcal{B}$ for $\mathbb{R}^5$ consisting of vectors from $\mathcal{S}$.

28. Consider the matrix

$$\mathbf{A} = \begin{bmatrix} 1 & 1 & -1 & 1 \\ 2 & 1 & -2 & 2 \\ 3 & 2 & -3 & 3 \end{bmatrix}.$$

The columns of $\mathbf{A}^\mathsf{T}$ are denoted by $\mathbf{v}_1$, $\mathbf{v}_2$, $\mathbf{v}_3$. Compute the reduced form for $\mathbf{A}$ and record the sequence of EROs used. Use this sequence to find scalars x_1, x_2, x_3 such that

$$x_1\mathbf{v}_1 + x_2\mathbf{v}_2 + x_3\mathbf{v}_3 = \mathbf{0}.$$

For an $m \times n$ matrix $\mathbf{A}$, how can you tell if its row vectors form a linearly dependent set?

Exercises 29–30. Let $\{\mathbf{u}, \mathbf{v}, \mathbf{w}\}$ be a linearly independent set of vectors in $\mathbb{R}^n$. Show that each given set is also linearly independent.

29. $\{\mathbf{u}, \mathbf{u}+\mathbf{v}, \mathbf{u}+\mathbf{v}+\mathbf{w}\}$

30. $\{\mathbf{u}+\mathbf{v}, \mathbf{v}+\mathbf{w}, \mathbf{w}+\mathbf{u}\}$

Exercises 31–32. For each set of vectors $\mathcal{S}$, find two linearly independent subsets of $\mathcal{S}$ which are as large as possible.

31. $\begin{bmatrix} 2 \\ -1 \\ 1 \end{bmatrix}, \begin{bmatrix} 10 \\ -5 \\ 3 \end{bmatrix}, \begin{bmatrix} -4 \\ 2 \\ 2 \end{bmatrix}$

32. $\begin{bmatrix} 0 \\ 1 \\ 2 \end{bmatrix}, \begin{bmatrix} 1 \\ -1 \\ 3 \end{bmatrix}, \begin{bmatrix} -1 \\ 2 \\ -1 \end{bmatrix}, \begin{bmatrix} 1 \\ 1 \\ 1 \end{bmatrix}$

Exercises 33–39. Use equation (3.17) to prove each statement.

33. The set $\mathcal{S} = \{\mathbf{0}\}$ consisting of the zero vector in $\mathbb{R}^n$ is linearly dependent.

34. A set $\mathcal{S} = \{\mathbf{v}\}$ containing a single nonzero vector $\mathbf{v}$ in $\mathbb{R}^n$ is linearly independent.

35. The set $\mathcal{S} = \{\mathbf{v}_1, \mathbf{v}_2\}$ in $\mathbb{R}^n$ is linearly dependent if and only if one vector is a scalar multiple of the other.

36. The set $\{\mathbf{0}, \mathbf{v}_1, \mathbf{v}_2, \ldots, \mathbf{v}_k\}$ is linearly dependent.

37. If $\{\mathbf{v}_1, \mathbf{v}_2, \ldots, \mathbf{v}_k\}$ is linearly dependent, then $\{\mathbf{v}_1, \mathbf{v}_2, \ldots, \mathbf{v}_k, \mathbf{u}_1, \mathbf{u}_2, \ldots, \mathbf{u}_p\}$ is linearly dependent.

38. If $\mathcal{S} = \{\mathbf{v}_1, \mathbf{v}_2, \ldots, \mathbf{v}_k\}$ is linearly independent, then any subset of $\mathcal{S}$ is linearly independent.

39. If $\{\mathbf{v}_1, \mathbf{v}_2\}$ is linearly independent and $\{\mathbf{v}_1, \mathbf{v}_2, \mathbf{v}_3\}$ is linearly dependent, then $\mathbf{v}_3$ is a linear combination of $\mathbf{v}_1$, $\mathbf{v}_2$. Furthermore, if $\mathbf{v}_4$ is a linear combination of $\mathbf{v}_1$, $\mathbf{v}_2$, $\mathbf{v}_3$, then $\mathbf{v}_4$ is a linear combination of $\mathbf{v}_1$, $\mathbf{v}_2$ only.

40. Suppose that $\mathcal{S} = \{\mathbf{v}_1, \mathbf{v}_2, \ldots, \mathbf{v}_k\}$ is a set of vectors in $\mathbb{R}^n$. Write out part (b) of the proof of Theorem 3.3 in the case when $\mathbf{v}_1 = \mathbf{0}$.

41. Find the linear dependence relation for the set of vectors $\mathcal{S} = \{\mathbf{v}_1, \mathbf{v}_2, \mathbf{v}_3, \mathbf{v}_4\}$ in $\mathbb{R}^4$, where

$$\mathbf{v}_1 = \begin{bmatrix} 1 \\ -1 \\ 2 \\ 1 \end{bmatrix}, \quad \mathbf{v}_2 = \begin{bmatrix} 2 \\ 1 \\ -1 \\ 0 \end{bmatrix},$$

$$\mathbf{v}_3 = \begin{bmatrix} -4 \\ -5 \\ 7 \\ 2 \end{bmatrix}, \quad \mathbf{v}_4 = \begin{bmatrix} 9 \\ 3 \\ -2 \\ 3 \end{bmatrix}.$$

Which vectors in $\mathcal{S}$ are linear combinations of other vectors in $\mathcal{S}$?

42. Prove that an ordered set of vectors $\mathcal{S} = \{\mathbf{v}_1, \mathbf{v}_2, \ldots, \mathbf{v}_k\}$ in $\mathbb{R}^n$ with $\mathbf{v}_1 \neq 0$ is linearly dependent if and only if one vector $\mathbf{v}_j$ is a linear combination of the preceding vectors in $\mathcal{S}$.

43. Let $\mathbf{A}$ be an $n \times n$ upper (lower) triangular matrix with nonzero entries on the main diagonal. Show that the columns in $\mathbf{A}$ are linearly independent. Are the row vectors in $\mathbf{A}$ linearly independent?

44. Find a basis for $\mathbb{R}^3$ that includes the vector $\mathbf{v} = \begin{bmatrix} 2 \\ 3 \\ 0 \end{bmatrix}$.

Hint: Consider Illustration 3.7.

45. Is it possible to find a subset $\mathcal{S}$ in $\mathbb{R}^3$ that contains four vectors and has the property that any pair of vectors in $\mathcal{S}$ are linearly independent and any subset of three vectors from $\mathcal{S}$ is linearly dependent?

Exercises 46–49. Find the dimension $d = \dim \mathcal{U}$ of the subspace $\mathcal{U}$ spanned by the columns of the given matrix **A**. *Find a basis for $\mathcal{U}$. Find $r = \operatorname{rank} \mathbf{A}$ and compare d with r.*

46. $\begin{bmatrix} 1 & 2 & 3 \\ 4 & 5 & 6 \end{bmatrix}$ **47.** $\begin{bmatrix} 1 & 2 & 3 \\ 4 & 8 & 12 \end{bmatrix}$

48. $\begin{bmatrix} 0 & 1 & 2 & 3 \\ 1 & 2 & 0 & 4 \\ 0 & 1 & 3 & 5 \end{bmatrix}$ **49.** $\begin{bmatrix} 1 & 2 & 3 \\ 4 & 5 & 6 \\ 7 & 8 & 9 \end{bmatrix}$

Exercises 50–56. Show that the columns of the given matrix form a basis $\mathcal{B}$ for a subspace of vectors. Find the coordinates of the given vector (a, b, c, d are real numbers) relative to $\mathcal{B}$.

50. $\begin{bmatrix} 1 & 0 \\ 0 & 1 \end{bmatrix}, \begin{bmatrix} a \\ b \end{bmatrix}$ **51.** $\begin{bmatrix} 1 & 2 \\ 2 & 1 \end{bmatrix}, \begin{bmatrix} a \\ b \end{bmatrix}$

52. $\begin{bmatrix} -1 & 1 \\ 2 & 3 \end{bmatrix}, \begin{bmatrix} a \\ b \end{bmatrix}$ **53.** $\begin{bmatrix} 1 & 2 \\ 0 & 4 \end{bmatrix}, \begin{bmatrix} a \\ b \end{bmatrix}$

54. $\begin{bmatrix} 1 & 2 & 3 \\ 0 & 2 & 3 \\ 0 & 0 & 3 \end{bmatrix}, \begin{bmatrix} a \\ b \\ c \end{bmatrix}$ **55.** $\begin{bmatrix} 1 & -1 & 1 \\ -1 & 0 & 1 \\ 0 & 1 & -1 \end{bmatrix}, \begin{bmatrix} a \\ b \\ c \end{bmatrix}$

56. $\begin{bmatrix} 1 & 0 & -2 & 1 \\ 1 & 1 & -1 & -1 \\ 1 & 2 & 2 & -2 \\ 2 & 1 & 3 & 0 \end{bmatrix}, \begin{bmatrix} a \\ b \\ c \\ d \end{bmatrix}$

57. Show that the set $\mathcal{B}$, where

$$\mathbf{b}_1 = \begin{bmatrix} 1 \\ 3 \end{bmatrix}, \quad \mathbf{b}_2 = \begin{bmatrix} -1 \\ 1 \end{bmatrix}$$

is a basis for $\mathbb{R}^2$. Find the coordinates $[\mathbf{v}]_{\mathcal{B}}$ of any vector

$\mathbf{v} = \begin{bmatrix} c \\ d \end{bmatrix}$ relative to $\mathcal{B}$.

Exercises 58–59. Show that the given set of vectors is a subspace $\mathcal{U}$ and find a basis for $\mathcal{U}$.

58. All vectors in $\mathbb{R}^5$ whose components sum to zero.

59. All vectors in $\mathbb{R}^3$ whose components are all equal.

60. Refer to Example 6. Plot the vectors $\mathbf{b}_1$, $\mathbf{b}_2$, $\mathbf{v}$ in the plane $\mathbb{R}^2$ and construct the vector $\mathbf{v}$ using the basis $\mathcal{B} = \{\mathbf{b}_1, \mathbf{b}_2\}$.

61. Find a basis for the subspace $\mathcal{U}$ of $\mathbb{R}^4$ that consists of all vectors lying in the hyperplane $x_1 - x_2 + x_3 - x_4 = 0$. Give the dimension of $\mathcal{U}$.

62. Find the set of vectors $\mathcal{U}$ in $\mathbb{R}^4$ that lie in the intersection of the hyperplanes $x_1 + x_2 + x_3 + x_4 = 0$ and $x_2 + 2x_3 - x_4 = 0$. Show that $\mathcal{U}$ is a subspace, find a basis for $\mathcal{U}$, and give its dimension.

63. Find the coordinates of any vector $\mathbf{v}$ in $\mathbb{R}^3$ relative to the basis formed from the columns of the matrix **A**, where

$$\mathbf{A} = \begin{bmatrix} 1 & -2 & 3 \\ 0 & 2 & 0 \\ 0 & 0 & -4 \end{bmatrix}, \quad \mathbf{v} = \begin{bmatrix} a \\ b \\ c \end{bmatrix}.$$

Exercises 64–65. Find the change of basis matrix **A** *that is used to change coordinates between the bases $\mathcal{C}$ and $\mathcal{B}$. Find the coordinates* **v** *relative to $\mathcal{C}$ and then use* **A** *to find the coordinates of* **v** *relative to $\mathcal{B}$. Cross-check by finding the coordinates of* **v** *relative to $\mathcal{B}$ directly.*

64. $\mathcal{C}: \begin{bmatrix} 1 \\ 2 \end{bmatrix}, \begin{bmatrix} 3 \\ 4 \end{bmatrix}, \quad \mathcal{B}: \begin{bmatrix} -1 \\ 1 \end{bmatrix}, \begin{bmatrix} -3 \\ -4 \end{bmatrix},$

$\mathbf{v} = \begin{bmatrix} 4 \\ -5 \end{bmatrix}$

65. $\mathcal{C}: \begin{bmatrix} 1 \\ 2 \\ 3 \end{bmatrix}, \begin{bmatrix} 1 \\ 0 \\ 1 \end{bmatrix}, \begin{bmatrix} 1 \\ 1 \\ 1 \end{bmatrix} \quad \mathcal{B}: \begin{bmatrix} 1 \\ 0 \\ 0 \end{bmatrix}, \begin{bmatrix} 0 \\ 1 \\ 1 \end{bmatrix}, \begin{bmatrix} -1 \\ 2 \\ 1 \end{bmatrix},$

$\mathbf{v} = \begin{bmatrix} -3 \\ 0 \\ 2 \end{bmatrix}$

66. Let **A** be an $n \times p$ matrix, let **B** be a $p \times q$ matrix, and let $\mathbf{C} = \mathbf{AB}$. Prove the following statements:

(a) If the columns of **B** are linearly dependent, then the columns of **C** are linearly dependent.

(b) If the rows of **A** are linearly dependent, then the rows of **C** are linearly dependent.

Exercises 67–70. Label each statement as either true (T) or false (F). If a false statement can be made true by modifying the wording, then do so.

67. Any set $\{\mathbf{v}_1, \mathbf{v}_2, \mathbf{v}_3, \mathbf{v}_4\}$ in $\mathbb{R}^n$ such that $\mathbf{v}_1 - \mathbf{v}_2 = \mathbf{v}_3 - \mathbf{v}_4$ is linearly dependent.

68. If a 6×4 matrix $\mathbf{A}$ has linearly independent columns, then the reduced form of $\mathbf{A}$ contains two zero rows.

69. If the rows of a matrix $\mathbf{A}$ are linearly independent, then the reduced form for $\mathbf{A}$ contains a zero row.

70. The coordinates of a vector $\mathbf{v}$ in $\mathbb{R}^n$ relative to a basis $\mathcal{B}$ define the terminal point of the arrow that represents $\mathbf{v}$.

USING MATLAB

Exercises 71–72. Find the linear dependence relation for the columns of the given matrix **M**. *Find two different bases for* $\mathbb{R}^3$ *from among the columns of* **M**.

71. $$\begin{bmatrix} 1 & -1 & 0 & 0 & 6 & 2 \\ 4 & 6 & -12 & -2 & -8 & -14 \\ 2 & 9 & 2 & 13 & -8 & -5 \end{bmatrix}$$

72. $$\begin{bmatrix} 9 & 4 & 4 & 1 & 30 & 33 & 9 \\ 2 & 8 & 0 & 26 & -4 & -6 & -6 \\ 6 & 7 & 8 & -13 & 37 & 55 & 7 \end{bmatrix}$$

73. Find as many different bases for $\mathbb{R}^4$ as possible from the columns of the matrix

$$\mathbf{A} = [\,\mathbf{v}_1\ \mathbf{v}_2\ \mathbf{v}_3\ \mathbf{v}_4\ \mathbf{v}_5\ \mathbf{v}_6\,] = \begin{bmatrix} 1 & 3 & -2 & 0 & 0 & 5 \\ 2 & 6 & -5 & -2 & -3 & 11 \\ 0 & 0 & 5 & 10 & 15 & -5 \\ 2 & 6 & 0 & 8 & 18 & 8 \end{bmatrix}.$$

74. Show that the columns of the matrix

$$\mathbf{A} = \begin{bmatrix} 2 & 2 & 8 & 0 \\ -2 & 0 & -2 & 3 \\ 0 & -4 & 2 & 1 \\ 2 & 0 & 6 & -1 \end{bmatrix}$$

do not form a basis for $\mathbb{R}^4$. Find a subspace $\mathcal{U}$ or $\mathbb{R}^4$ with a basis formed from among the columns of $\mathbf{A}$ and whose dimension is as large as possible.

75. Consider the matrix $\mathbf{M} = \mathbf{magic(8)}$. Find a subspace $\mathcal{U}$ of $\mathbb{R}^8$ with a basis formed from among the columns of $\mathbf{M}$ and whose dimension is as large as possible.

76. *Project.* Write an M-file that calls an $n \times k$ matrix $\mathbf{A}$ and returns all possible bases for $\mathbb{R}^n$ formed from the columns in $\mathbf{A}$ supplemented by the columns of the identity matrix $\mathbf{I}_n$.

77. *Experiment.* Estimate the probability of forming a basis for $\mathbb{R}^n$, $n = 2, 3, 4, 5$, from n randomly chosen n-vectors that have components chosen from the list $-1, 0, 1$.

3.3 Null Space, Column Space, Row Space

In this section we describe three fundamental subspaces that are associated with any given matrix **A**. They are called the *null space*, *column space*, and *row space* of **A**. These subspaces play a central role in the theory of *linear transformations*, which appears in Section 3.4.

3.3.1 Null Space

Definition 3.9 *The null space of a matrix*

Let $\mathbf{A}$ be an $m \times n$ matrix. The set of all vectors $\mathbf{x}$ in $\mathbb{R}^n$ that satisfy the matrix equation

$$\mathbf{Ax} = \mathbf{0} \tag{3.32}$$

is called the *null space* of **A** and is denoted by null **A**.

The set null **A** is a subspace of $\mathbb{R}^n$ (Exercises 3.3) and we describe null **A** by finding a basis for this subspace. The methods shown in the next illustration generalize to any $m \times n$ matrix. The matrix **A** and computations are used throughout this section.

■ ILLUSTRATION 3.8

Basis for the Null Space

Consider the 3×4 matrix

$$\mathbf{A} = \begin{bmatrix} 1 & 0 & 1 & 0 \\ 2 & 3 & 4 & 1 \\ 4 & 3 & 6 & 1 \end{bmatrix}.$$

The homogeneous linear system defined by $\mathbf{Ax} = \mathbf{0}$ is solved by computing the reduced form $\mathbf{A}^*$ of **A**. The sequence of EROs applied to **A** is recorded for later use.

$$\mathbf{r}_2 - 2\mathbf{r}_1 \to \mathbf{r}_2, \quad \mathbf{r}_3 - 4\mathbf{r}_1 \to \mathbf{r}_3, \quad \mathbf{r}_3 - \mathbf{r}_2 \to \mathbf{r}_3, \quad \frac{1}{3}\mathbf{r}_2 \to \mathbf{r}_2 \tag{3.33}$$

$$\mathbf{A} = \begin{bmatrix} 1 & 0 & 1 & 0 \\ 2 & 3 & 4 & 1 \\ 4 & 3 & 6 & 1 \end{bmatrix} \sim \begin{bmatrix} \textcircled{1} & 0 & 1 & 0 \\ 0 & \textcircled{1} & \frac{2}{3} & \frac{1}{3} \\ 0 & 0 & 0 & 0 \end{bmatrix} = \mathbf{A}^*$$

Observe that $\operatorname{rank} \mathbf{A} = 2 =$ number of pivots in $\mathbf{A}^*$ and in **A**. There are $n - r = 4 - 2 = 2$ free variables, and the general solution **x** to $\mathbf{Ax} = \mathbf{0}$ is given by

$$\mathbf{x} = (x_1, x_2, x_3, x_4) = \left(-t_1, -\frac{(2t_1 + t_2)}{3}, t_1, t_2\right), \tag{3.34}$$

where $x_3 = t_1$ and $x_4 = t_2$ are independent real parameters. Writing the solution set in vector form, we have

$$\mathbf{x} = \begin{bmatrix} x_1 \\ x_2 \\ x_3 \\ x_4 \end{bmatrix} = \begin{bmatrix} -t_1 \\ -(2t_1 + t_2)/3 \\ t_1 \\ t_2 \end{bmatrix} = t_1 \begin{bmatrix} -1 \\ -2/3 \\ 1 \\ 0 \end{bmatrix} + t_2 \begin{bmatrix} 0 \\ -1/3 \\ 0 \\ 1 \end{bmatrix}$$

$$= t_1\mathbf{v}_1 + t_2\mathbf{v}_2, \tag{3.35}$$

where

$$\mathbf{v}_1 = \begin{bmatrix} -1 \\ -2/3 \\ 1 \\ 0 \end{bmatrix}, \quad \mathbf{v}_2 = \begin{bmatrix} 0 \\ -1/3 \\ 0 \\ 1 \end{bmatrix}.$$

The set $\mathcal{B} = \{\mathbf{v}_1, \mathbf{v}_2\}$ is linearly independent, which can be seen by, forming the equation $y_1\mathbf{v}_1 + y_2\mathbf{v}_2 = \mathbf{0}$ and noting that the third and fourth components of $\mathbf{v}_1$ and $\mathbf{v}_2$ imply that the only solution is $y_1 = y_2 = 0$. From (3.35) we see that $\operatorname{null}\mathbf{A} = \operatorname{span}\mathcal{B}$ and so $\mathcal{B}$ is a basis for $\operatorname{null}\mathbf{A}$. ■

The underlying theory in Illustration 3.8 works for any matrix. The next result states the general principle.

Theorem 3.11 **Basis for Null Space**

Let $\mathbf{A}$ be an $m \times n$ matrix with $\operatorname{rank}\mathbf{A} = r$ and let (S) be the homogeneous linear system defined by the matrix equation $\mathbf{Ax} = \mathbf{0}$. The general solution $\mathbf{x}$ to (S) can be written as a linear combination of $n - r$ linearly independent vectors $\{\mathbf{v}_1, \mathbf{v}_2, \ldots, \mathbf{v}_{n-r}\}$, namely

$$\mathbf{x} = t_1\mathbf{v}_1 + t_2\mathbf{v}_2 + \cdots + t_{n-r}\mathbf{v}_{n-r},$$

where $t_1, t_2, \ldots, t_{n-r}$ are $n - r$ independent parameters that correspond to the free variables in the linear system (S). The set $\mathcal{B} = \{\mathbf{v}_1, \mathbf{v}_2, \ldots, \mathbf{v}_{n-r}\}$ is a basis for $\operatorname{null}\mathbf{A}$ and the dimension of $\operatorname{null}\mathbf{A}$ is $n - r$.

3.3.2 Solution Sets to Linear Systems

The solution set to a consistent $m \times n$ linear system (S) with coefficient matrix is $\mathbf{A}$ can be described in terms of $\operatorname{null}\mathbf{A}$ and a single *particular solution* for (S). This different perspective on solution sets, stated in the next result, augments the theory already presented in Sections 1.1 and 1.2.

Theorem 3.12 **Describing Solution Sets to Linear Systems**

Let a consistent $m \times n$ linear system (S) be written in matrix form $\mathbf{Ax} = \mathbf{b}$ and let $\mathcal{S}$ be the set of solutions to (S). Then $\mathcal{S}$ consists of all vectors in $\mathbb{R}^n$ of the type $\mathbf{x} = \mathbf{x}_p + \mathbf{x}_h$, where $\mathbf{x}_p$ is a particular solution for (S) and $\mathbf{x}_h$ ranges over all vectors in $\operatorname{null}\mathbf{A}$.

Proof We must show that the set $\mathcal{S}$ coincides with the set of all vectors of the type $\mathbf{x} = \mathbf{x}_p + \mathbf{x}_h$ and so there are two set inclusions to prove.

(a) Consider the vector $\mathbf{x} = \mathbf{x}_p + \mathbf{x}_h$ for any $\mathbf{x}_h$ in $\operatorname{null}\mathbf{A}$. Then

$$\mathbf{Ax} = \mathbf{A}(\mathbf{x}_p + \mathbf{x}_h) = \mathbf{Ax}_p + \mathbf{Ax}_h = \mathbf{b} + \mathbf{0} = \mathbf{b} \quad \Rightarrow \quad \mathbf{Ax} = \mathbf{b},$$

showing that $\mathbf{x}$ is in $\mathcal{S}$.

(b) Now suppose that $\mathbf{x}$ is any vector in $\mathcal{S}$ and consider $\mathbf{y} = \mathbf{x} - \mathbf{x}_p$. Then

$$\mathbf{Ay} = \mathbf{A}(\mathbf{x} - \mathbf{x}_p) = \mathbf{Ax} - \mathbf{Ax}_p = \mathbf{b} - \mathbf{b} = \mathbf{0},$$

showing that $\mathbf{y}$ is in $\operatorname{null}\mathbf{A}$ and so $\mathbf{x} = \mathbf{x}_p + \mathbf{y}$ has the required form.

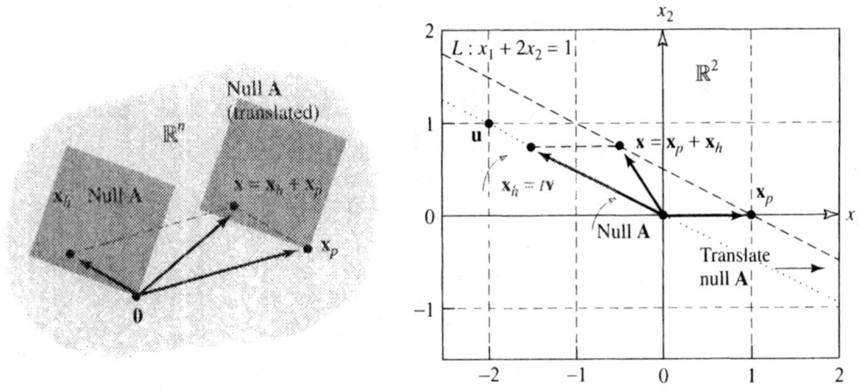

(a) (b)

Figure 3.14 Describing solution sets of linear systems.

Figure 3.14(a) shows the solution set $\mathcal{S}$ to the linear system defined by $\mathbf{Ax} = \mathbf{b}$. The terminal points of the arrows that represent vectors in $\mathcal{S}$ all lie in the subset of $\mathbb{R}^n$ formed by translating the subspace null $\mathbf{A}$ in $\mathbb{R}^n$ to the terminal point of the particular solution $\mathbf{x}_p$. Note that $\mathcal{S}$ is not in general a subspace of $\mathbb{R}^n$.

EXAMPLE 1

Visualizing Solution Sets

Consider the 2×2 linear system (S) with coefficient matrix $\mathbf{A}$.

$$(S) \quad \begin{cases} x_1 + 2x_2 = 1 \\ 2x_1 + 4x_2 = 2 \end{cases} \qquad \mathbf{A} = \begin{bmatrix} 1 & 2 \\ 2 & 4 \end{bmatrix}$$

Then (S) has solution set $\mathcal{U}$ is given by

$$\mathbf{x} = \begin{bmatrix} x_1 \\ x_2 \end{bmatrix} = \begin{bmatrix} 1 \\ 0 \end{bmatrix} + t\begin{bmatrix} -2 \\ 1 \end{bmatrix} = \mathbf{x}_p + t\mathbf{v}, \quad \text{where } \mathbf{x}_p = \begin{bmatrix} 1 \\ 0 \end{bmatrix}, \quad \mathbf{v} = \begin{bmatrix} -2 \\ 1 \end{bmatrix},$$

and t is a real parameter. The 2-vector $\mathbf{x}_p$ is a particular solution to $\mathbf{Ax} = \mathbf{b}$ and null $\mathbf{A} = \{t\mathbf{v}\}$. Hence, $\mathcal{S}$ consists of all vectors of the type $\mathbf{x} = \mathbf{x}_p + \mathbf{x}_h$, as stated in Theorem 3.12. Refer to Figure 3.14(b). Note that null $\mathbf{A}$ is a one-dimensional subspace of $\mathbb{R}^2$ with basis $\mathcal{B} = \{\mathbf{v}\}$. Each real value t locates a solution $\mathbf{x}_h = t\mathbf{v}$ to the equation $\mathbf{Ax} = \mathbf{0}$. Then $\mathbf{x} = \mathbf{x}_p + \mathbf{x}_h$ is a solution to (S), and all solutions to (S) are found in this way. ■

3.3.3 Column Space

Definition 3.10

Column space of a matrix

Let $\mathbf{A}$ be an $m \times n$ matrix. The subspace of $\mathbb{R}^m$ spanned by the columns of $\mathbf{A}$ is called the *column space* of $\mathbf{A}$ and is denoted by col $\mathbf{A}$.

Our goal is to describe col **A** by finding a basis for this subspace. The methods shown in the next illustration generalize to any $m \times n$ matrix.

■ ILLUSTRATION 3.9

Basis for column space

Using the matrix **A** in Illustration 3.8, we have

$$\mathbf{A} = [\,\mathbf{v}_1\ \mathbf{v}_2\ \mathbf{v}_3\ \mathbf{v}_4\,] = \begin{bmatrix} 1 & 0 & 1 & 0 \\ 2 & 3 & 4 & 1 \\ 4 & 3 & 6 & 1 \end{bmatrix} \sim \begin{bmatrix} \textcircled{1} & 0 & 1 & 0 \\ 0 & \textcircled{1} & \frac{2}{3} & \frac{1}{3} \\ 0 & 0 & 0 & 0 \end{bmatrix} = \mathbf{A}^*.$$

Recall from Section 3.2 that the linear dependence relation for the columns of **A** is found by solving the homogeneous linear system defined by $\mathbf{Ax} = \mathbf{0}$. The solution already appears in (3.34) and so the linear dependence relation for this matrix **A** is

$$-t_1 \begin{bmatrix} 1 \\ 2 \\ 4 \end{bmatrix} - \frac{(2t_1 + t_2)}{3} \begin{bmatrix} 0 \\ 3 \\ 3 \end{bmatrix} + t_1 \begin{bmatrix} 1 \\ 4 \\ 6 \end{bmatrix} + t_2 \begin{bmatrix} 0 \\ 1 \\ 1 \end{bmatrix} = \begin{bmatrix} 0 \\ 0 \\ 0 \end{bmatrix}, \tag{3.36}$$

where t_1 and t_2 are independent real parameters. Consider the set $\mathcal{B} = \{\mathbf{v}_1, \mathbf{v}_2\}$ of pivot columns in **A**. These columns correspond to the pivot columns (pivots are circled) in $\mathbf{A}^*$. To show that $\mathcal{B} = \{\mathbf{v}_1, \mathbf{v}_2\}$ is linearly independent, we solve the matrix equation $\mathbf{Bx} = \mathbf{0}$, where $\mathbf{B} = [\,\mathbf{v}_1\ \mathbf{v}_2\,]$. The EROs shown in (3.33) that reduce **A** to $\mathbf{A}^*$ also reduce **B** to $\mathbf{B}^*$, which is the submatrix of pivot columns in $\mathbf{A}^*$; that is,

$$\mathbf{B} = [\,\mathbf{v}_1\ \mathbf{v}_2\,] = \begin{bmatrix} 1 & 0 \\ 2 & 3 \\ 4 & 3 \end{bmatrix} \sim \begin{bmatrix} \textcircled{1} & 0 \\ 0 & \textcircled{1} \\ 0 & 0 \end{bmatrix} = \mathbf{B}^*.$$

Thus, $\mathbf{Bx} = \mathbf{0}$ has only the zero solution $\mathbf{x} = (0, 0)$. Now consider the linear dependence relation (3.36). The nonpivot columns in **A** are associated with the free variables in the solution set to $\mathbf{Ax} = \mathbf{0}$. Every nonpivot column in **A** can be expressed as a linear combination of the pivot columns. For this matrix, taking $t_1 = 1$ and $t_2 = 0$ in (3.36) expresses $\mathbf{v}_3$ in terms of $\mathbf{v}_1$, $\mathbf{v}_2$ and then taking $t_1 = 0$ and $t_2 = 1$ expresses $\mathbf{v}_4$ in terms of $\mathbf{v}_1$, $\mathbf{v}_2$. It follows that $\operatorname{col} \mathbf{A} = \operatorname{span}\{\mathbf{v}_1, \mathbf{v}_2, \mathbf{v}_3, \mathbf{v}_4\} = \operatorname{span}\{\mathbf{v}_1, \mathbf{v}_2\}$ and so $\mathcal{B}$ is a basis for col **A**.

For this matrix, col **A** can be visualized geometrically as the plane in $\mathbb{R}^3$ passing through the origin containing the arrows representing $\mathbf{v}_1$ and $\mathbf{v}_2$. ■

Remarks

For any $m \times n$ matrix **A**, the EROs that reduce **A** to $\mathbf{A}^*$ also reduce any submatrix of columns in **A** to the corresponding submatrix of columns in $\mathbf{A}^*$. If $\operatorname{rank} \mathbf{A} = r$, then **A** has an $m \times r$ submatrix **B** of pivot columns that reduce to an $m \times r$ binary submatrix $\mathbf{B}^*$ of pivot columns in $\mathbf{A}^*$ and the binary form of $\mathbf{B}^*$ guarantees the linear independence of the pivot columns in **A**.

Caution! A basis for col $\mathbf{A}$ is defined by taking the pivot columns from $\mathbf{A}$, not from $\mathbf{A}^*$. For example, in Illustration 3.9 note that the columns $\mathbf{v}_3$ and $\mathbf{v}_4$ in $\mathbf{A}$ are not linear combinations of pivot columns in $\mathbf{A}^*$. We will now state the general result.

Theorem 3.13 **Basis for Column Space**

Let $\mathbf{A}$ *be an* $m \times n$ *matrix with* rank $\mathbf{A} = r$. *The* r *pivot columns in* $\mathbf{A}$ *form a basis for* col $\mathbf{A}$ *and the dimension of* col $\mathbf{A}$ *is* r.

3.3.4 Row Space

The row vectors $\mathbf{r}_1, \mathbf{r}_2, \ldots, \mathbf{r}_m$ associated with an $m \times n$ matrix $\mathbf{A}$ are $1 \times n$ matrices. In order to express the row vectors as objects in $\mathbb{R}^n$, we can use the transpose operator as in the following example. In what follows we use the term rows loosely to mean row vectors expressed as objects in $\mathbb{R}^n$ and vice versa.

EXAMPLE 2 **Rows and Columns**

Consider the 2×3 matrix

$$\mathbf{A} = \begin{bmatrix} 1 & 0 & 5 \\ 2 & 3 & 4 \end{bmatrix},$$

where $m = 2$ and $n = 3$. Transposing the rows of $\mathbf{A}$ we have

$$\mathbf{A} = \begin{bmatrix} \mathbf{r}_1 \\ \mathbf{r}_2 \end{bmatrix}, \quad \begin{cases} \mathbf{r}_1 = [1 \;\; 0 \;\; 5] \\ \mathbf{r}_2 = [2 \;\; 3 \;\; 4] \end{cases} \quad \Rightarrow \quad \mathbf{r}_1^{\mathrm{T}} = \mathbf{u}_1 = \begin{bmatrix} 1 \\ 0 \\ 5 \end{bmatrix}, \quad \mathbf{r}_2^{\mathrm{T}} = \mathbf{u}_2 = \begin{bmatrix} 2 \\ 3 \\ 4 \end{bmatrix}.$$

The vectors $\mathbf{u}_1$ and $\mathbf{u}_2$ express rows of $\mathbf{A}$ as objects in $\mathbb{R}^3$. ■

Definition 3.11 *Row space of a matrix*

Let $\mathbf{A}$ be an $m \times n$ matrix. The subspace of $\mathbb{R}^n$ spanned by the rows of $\mathbf{A}$ is called the *row space* of $\mathbf{A}$ and is denoted by row $\mathbf{A}$.

We wish to describe row $\mathbf{A}$ by finding a basis for this subspace. The next illustration generalizes to any $m \times n$ matrix $\mathbf{A}$.

■ **ILLUSTRATION 3.10**

Basis for the row space

From Illustration 3.8 we have

$$\mathbf{A} = \begin{bmatrix} 1 & 0 & 1 & 0 \\ 2 & 3 & 4 & 1 \\ 4 & 3 & 6 & 1 \end{bmatrix} \sim \begin{bmatrix} \textcircled{1} & 0 & 1 & 0 \\ 0 & \textcircled{1} & \frac{2}{3} & \frac{1}{3} \\ 0 & 0 & 0 & 0 \end{bmatrix} = \begin{bmatrix} \mathbf{s}_1 \\ \mathbf{s}_2 \\ \mathbf{0} \end{bmatrix} = \mathbf{A}^*,$$

where $\mathbf{s}_1$ and $\mathbf{s}_2$ denote the nonzero rows in $\mathbf{A}^*$. A basis for row $\mathbf{A}$ is a set $\mathcal{B}$ in $\mathbb{R}^4$ that is linearly independent and such that span $\mathcal{B}$ = row $\mathbf{A}$. Let $\mathcal{B} = \{\mathbf{s}_1^\mathsf{T}, \mathbf{s}_2^\mathsf{T}\}$. Then $\mathcal{B}$ is linearly independent because the vector equation $y_1\mathbf{s}_1^\mathsf{T} + y_2\mathbf{s}_2^\mathsf{T} = \mathbf{0}$ has only the zero solution $y_1 = y_2 = 0$ (consider the first two components in $\mathbf{s}_1^\mathsf{T}$ and $\mathbf{s}_2^\mathsf{T}$). Each row in $\mathbf{A}$ is a linear combination of the nonzero rows in $\mathbf{A}^*$. The linear combinations are found by keeping track of the EROs that changed $\mathbf{A}$ into $\mathbf{A}^*$. Using (3.33), we have

$$\mathbf{r}_1 = \mathbf{s}_1, \qquad \mathbf{r}_2 = 2\mathbf{s}_1 + 3\mathbf{s}_2, \qquad \mathbf{r}_3 = 4\mathbf{s}_1 + 3\mathbf{s}_2.$$ ■

Theorem 3.14 **Basis for the Row Space**

If $\mathbf{A}$ *is an* $m \times n$ *matrix with* rank $\mathbf{A} = r > 0$, *then the set* $\mathcal{B}$ *of* r *nonzero row vectors (transposed) in the reduced form* $\mathbf{A}^*$ *of* $\mathbf{A}$ *is a basis for* row $\mathbf{A}$ *and* row $\mathbf{A}$ *has dimension* r. *If* rank $\mathbf{A} = 0$, *then row* $\mathbf{A} = \{0\}$.

Concerning the Transpose

Let $\mathbf{A}$ be an $m \times n$ matrix. The columns of $\mathbf{A}^\mathsf{T}$ are the rows of $\mathbf{A}$ and the rows of $\mathbf{A}^\mathsf{T}$ are the columns of $\mathbf{A}$. Hence, we obtain

$$\text{col}\,\mathbf{A}^\mathsf{T} = \text{row}\,\mathbf{A} \quad \text{and} \quad \text{row}\,\mathbf{A}^\mathsf{T} = \text{col}\,\mathbf{A}. \tag{3.37}$$

Note that a basis for row $\mathbf{A}$ uses rows from $\mathbf{A}^*$ and *not* $\mathbf{A}$. Appealing to (3.37), a basis for row $\mathbf{A}$ that uses rows from $\mathbf{A}$ itself is obtained by finding a basis for col $\mathbf{A}^\mathsf{T}$. The proof of the next result uses (3.37).

Theorem 3.15 **Rank of Transpose**

Let $\mathbf{A}$ *be an* $m \times n$ *matrix. Then* rank $\mathbf{A}^\mathsf{T}$ = rank $\mathbf{A}$.

Proof We have rank $\mathbf{A}^\mathsf{T}$ = dim (col $\mathbf{A}^\mathsf{T}$) = dim (row $\mathbf{A}$) = rank $\mathbf{A}$.

Let's reflect for a moment on one point—col $\mathbf{A}$ is a subspace of $\mathbb{R}^m$ of dimension r and row $\mathbf{A}$ is a subspace of $\mathbb{R}^n$ also of dimension r, where r = rank $\mathbf{A}$ is the number of pivots in $\mathbf{A}$. The rank of a matrix could therefore be defined as the common dimension of its column space and row space.

Rank and Nullity

Consider the solution $\mathbf{x} = (x_1, x_2, \ldots, x_n)$ to the linear system (S) defined by the matrix equation (3.32). There are r = rank $\mathbf{A}$ leading variables and $n - r$ free variables. The number of leading variables corresponds to the r linearly independent pivot columns in $\mathbf{A}$ that form a basis for col $\mathbf{A}$ (the pivot columns in $\mathbf{A}$ are identified using the pivot columns in $\mathbf{A}^*$).

The free variables provide $n - r$ independent parameters that enable the solution vector $\mathbf{x}$ to be written as a linear combination of $n - r$ linearly independent special vectors that form the basis for null $\mathbf{A}$. Thus, dim (null $\mathbf{A}$) = $n - r$. The integer $n - r$ is called the *nullity* of $\mathbf{A}$, denoted by ν or $\nu(\mathbf{A})$, where ν is the Greek letter *nu*. We obtain the fundamental relationship

$$n = \text{rank} + \text{nullity} \quad \text{or} \quad n = r + \nu. \tag{3.38}$$

Equation (3.38) is useful in determining the existence or nonexistence of matrices **A** with given properties. For example, there exist 2×2 matrices such that $\text{null}\,\mathbf{A} = \text{col}\,\mathbf{A}$ (Exercises 3.3), and (3.38) implies that the dimension of these subspaces must both be 1. If an $m \times n$ matrix **A** has rank n, then the dimension of its null space is necessarily zero, and so on.

EXERCISES 3.3

Exercises 1–10. For each matrix **A**, *find bases for* null **A**, col **A**, *and* row **A** *and state their dimension. Say how the relation* $n = \text{rank} + \text{nullity}$ *applies in each case. When possible, draw an accurate diagram that shows each subspace geometrically.*

1. $\begin{bmatrix} 0 & 0 & 0 \\ 0 & 0 & 0 \end{bmatrix}$

2. $\begin{bmatrix} 1 & 2 \\ 3 & 4 \end{bmatrix}$

3. $\begin{bmatrix} 1 & 2 \\ 3 & 4 \\ 5 & 6 \end{bmatrix}$

4. $\begin{bmatrix} -2 & 1 \\ -4 & 2 \\ 4 & -2 \end{bmatrix}$

5. $\begin{bmatrix} 0 & 1 & 0 \\ 0 & 0 & 0 \\ 1 & 0 & 0 \end{bmatrix}$

6. $\begin{bmatrix} 0 & 0 & 2 \\ 0 & 0 & 0 \\ 1 & 2 & 3 \end{bmatrix}$

7. $\begin{bmatrix} 1 & 0 & 1 & 0 \\ 0 & 1 & 0 & 1 \\ 1 & 0 & 1 & 0 \end{bmatrix}$

8. $\begin{bmatrix} 1 & -1 & 1 & -1 \\ -1 & 1 & -1 & 1 \\ 1 & -1 & 1 & -1 \end{bmatrix}$

9. $\begin{bmatrix} 1 & 2 & 1 & 3 & -1 \\ 1 & 2 & 2 & 4 & -1 \\ 2 & 4 & 1 & 3 & -2 \end{bmatrix}$

10. $\begin{bmatrix} 1 & 0 & 1 & 0 \\ 0 & 1 & 0 & 1 \\ 0 & 0 & 1 & 0 \\ 0 & 0 & 0 & 1 \end{bmatrix}$

11. Find row **A**, col **A**, null **A** for the matrix

$$\mathbf{A} = \begin{bmatrix} -1 & -7 & -9 \\ -4 & 1 & 1 \\ -6 & 6 & 0 \end{bmatrix}$$

Show that $\text{col}\,\mathbf{A} = \mathbb{R}^3$. Use the relation $n = \text{rank} + \text{nullity}$ to check that the dimension of null **A** is correct.

12. Show that null **A** is a subspace of $\mathbb{R}^n$, where n is the number of columns in **A**.

13. Let **A** be an $n \times n$ matrix. Prove that **A** is invertible if and only if null **A** is the zero subspace $\mathcal{O} = \{\mathbf{0}\}$ of $\mathbb{R}^n$.

14. Let **A** be an $m \times n$ matrix such that $\text{null}\,\mathbf{A} = \{\mathbf{0}\}$. Determine if the columns of **A** are linearly independent or dependent. Explain by considering the cases $m > n$, $m = n$, and $m < n$.

Exercises 15–16. For each matrix **A**, *determine if* $\text{col}\,\mathbf{A} = \mathbb{R}^3$.

15. $\begin{bmatrix} 1 & 2 & 3 \\ 0 & 4 & 5 \\ 0 & 0 & 6 \end{bmatrix}$

16. $\begin{bmatrix} 1 & -1 & 1 \\ -1 & 1 & -1 \\ 1 & -1 & 1 \end{bmatrix}$

17. Give a condition that will imply that an $n \times n$ upper (lower) triangular matrix **A** has $\text{col}\,\mathbf{A} = \mathbb{R}^n$.

18. Suppose you are given a 10×12 homogeneous linear system $\mathbf{Ax} = \mathbf{0}$. You find that all solutions can be generated from just two solutions, neither a scalar multiple of the other. Discuss whether or not the corresponding nonhomogeneous system $\mathbf{Ax} = \mathbf{b}$ has a solution for every vector **b**.

19. If **A** is a 6×7 matrix, what is the largest possible dimension of col **A**?

20. If **A** is a 7×6 matrix and $\dim\,\text{null}\,\mathbf{A} = 2$, what is the dimension of row **A**?

21. If **A** is a 6×8 matrix, what is the smallest possible dimension of null **A**?

22. If **A** is 6×4 what is the smallest possible dimension of null **A**?

Exercises 23–24. Find a basis $\mathcal{B}$ *for* row **A** *by finding* $\mathbf{A}^*$ *and then find a basis* $\mathcal{B}'$ *for* row **A** *using vectors from* **A** *itself. Write the basis vectors in* $\mathcal{B}'$ *in terms of vectors from* $\mathcal{B}$.

23. $\mathbf{A} = \begin{bmatrix} 4 & 1 & 3 & 2 \\ 2 & 4 & 5 & 1 \\ 2 & 5 & 8 & 3 \\ 2 & 3 & 2 & -1 \end{bmatrix}$

24. $\mathbf{A} = \begin{bmatrix} 1 & 3 & 0 & 1 & -1 \\ 2 & 6 & 3 & 1 & 0 \\ 1 & 1 & 0 & 1 & 1 \end{bmatrix}$

25. Find a basis $\mathcal{B}$ for col **A** and write the remaining columns as linear combinations of the basis vectors. Find a basis for null **A**.

$$\mathbf{A} = \begin{bmatrix} 1 & 2 & 1 & -2 & 1 \\ -1 & -2 & 0 & -1 & 1 \\ 2 & 4 & 2 & -4 & 2 \\ 3 & 6 & 3 & -6 & 3 \end{bmatrix}$$

Exercises 26–29. Find the null space of each given matrix **A**. *Solve the linear system (S) given by* $\mathbf{Ax} = \mathbf{b}$. *Draw an accurate diagram that shows* null **A** *and* null **A** *translated to form the solution set for (S).*

26. $[1 \;\; 1]$, $b = 1$

27. $[1 \;\; 0 \;\; -3]$, $b = 2$

28. $\begin{bmatrix} 1 & 2 \\ 3 & 4 \end{bmatrix}$, $\mathbf{b} = \begin{bmatrix} 1 \\ 1 \end{bmatrix}$

29. $\begin{bmatrix} 2 & -2 \\ -6 & 6 \end{bmatrix}$, $\mathbf{b} = \begin{bmatrix} -1 \\ 3 \end{bmatrix}$

30. Find a 2×2 matrix **A** for which null **A** = col **A**.

Exercises 31–42. Label each statement true (T) or false (F). If true, give an explanation, and if false, give a counterexample when possible.

31. There exists a 3×3 matrix **A** such that dim null **A** = dim col **A**.

32. The dimensions of the subspaces null **A** and row **A** are never equal.

33. If **A** is $m \times n$ with $m < n$, then null **A** is a subspace of $\mathbb{R}^m$ and col **A** is a subspace of $\mathbb{R}^n$.

34. If **A** is $m \times n$ with $m \geq n$, then null **A** is a subspace of $\mathbb{R}^n$ and col **A** is a subspace of $\mathbb{R}^m$.

35. A matrix **A** and its reduced echelon form $\mathbf{A}^*$ have the same column space.

36. A matrix **A** and an echelon form for **A** have the same row space.

37. If **A** is symmetric, then col **A** = row **A**.

38. If col **A** = row **A**, then **A** is symmetric.

39. If **A** is skew–symmetric ($\mathbf{A}^{\mathsf{T}} = -\mathbf{A}$), then col **A** = row **A**.

40. If null **A** is the zero subspace, then **A** is square.

41. If null **A** has a single basis vector $[1 \;\; 2 \;\; 0 \;\; 1]^{\mathsf{T}}$, then **A** has three pivot columns.

42. For an $m \times n$ matrix **A** with $m < n$, a basis for col **A** cannot include all the columns of **A**.

Exercises 43–46. If possible, find a matrix **A** *that satisfies the stated conditions.*

43. A basis for col **A** is $(1, 2, 0)$, $(0, 0, -1)$ and a basis for null **A** is $(1, 1, 2)$.

44. A basis for col **A** is $(1, 2, 3)$ and a basis for null **A** is $(1, 1, 0)$, $(1, 0, 1)$.

45. col **A** contains the vectors $(1, 0, 0, 2)$, $(0, 1, 1, 0)$ and a basis for null **A** is $(1, 2, 0, 1)$.

46. col **A** = row **A** and null **A** contains the vectors $(-1, -2, 3)$ and $(5, -2, -3)$.

47. Find the three fundamental subspaces in each part. **(a)** $\begin{bmatrix} 0 & 1 \\ 0 & 1 \end{bmatrix}$, **(b)** $\begin{bmatrix} 0 & 0 \\ 1 & 1 \end{bmatrix}$.

48. Use the row vectors in the matrix **A** and the vectors in the null space null **A** to find a basis for $\mathbb{R}^4$, where

$$\mathbf{A} = \begin{bmatrix} 1 & 1 & 1 & -1 \\ 1 & 1 & 3 & 5 \end{bmatrix}$$

USING MATLAB

Consult online help for the commands **null**, **rat** *and related commands.*

Exercises 49–52. Find a rational basis (entries are rational numbers) for the null **A**, col **A**, *and* row **A** *and the dimensions of these subspaces. Before you begin, try to predict what the dimensions are likely to be.*

49. $\begin{bmatrix} 2.1 & 3.5 & 4.6 \\ 1 & 2 & 4 \end{bmatrix}$

50. $\begin{bmatrix} 1 & 2 & 3 \\ 4 & 5 & 6 \\ 7 & 8 & 9 \end{bmatrix}$

51. **magic(3)**

52. **magic(4)**

Exercises 53–56. Find bases for the three fundamental subspaces of each matrix and state their dimensions.

53. $\begin{bmatrix} 3 & -6 & 1 & -1 \\ -1 & 2 & -2 & -2 \\ -2 & 4 & 5 & 8 \end{bmatrix}$

54. $\begin{bmatrix} 3 & -6 & 1 & -1 & 1 \\ -1 & 2 & -2 & -2 & 1 \\ -2 & 4 & 5 & 8 & 2 \end{bmatrix}$

55. $\begin{bmatrix} 1 & 0 & 1 & 0 \\ 0 & 1 & 0 & 1 \\ 1 & 0 & 1 & 0 \\ 0 & 1 & 0 & 1 \\ 1 & 0 & 1 & 0 \end{bmatrix}$

3.4 Linear Transformations on $\mathbb{R}^n$

Functions: A Review

Refer to Figure 3.15, which shows two (nonempty) sets $\mathcal{D}$ and $\mathcal{C}$.

A *function* F from $\mathcal{D}$ to $\mathcal{C}$ is a *rule* that assigns to every object x in $\mathcal{D}$ a corresponding object y in $\mathcal{C}$. We call y the *image* of x under F and write $F(x) = y$ to indicate that F *maps* x to y. The set $\mathcal{D}$ is called the *domain* of F and $\mathcal{C}$ is called the *codomain* of F. We write $F : \mathcal{D} \to \mathcal{C}$ to indicate that F is a function *on* $\mathcal{D}$.

It can happen that more than one object x in $\mathcal{D}$ is mapped by F to the same object y in $\mathcal{C}$. When this happens we say that F is *many-to-one*.

The set of all objects y in $\mathcal{C}$ that are the image of *some* x in $\mathcal{D}$ is called the *range* of F and is denoted by ran F. If at least one object y in $\mathcal{C}$ is not the image of any x in $\mathcal{D}$, then ran F is a proper subset of $\mathcal{C}$. This situation is illustrated in Figure 3.15, where ran F (the darker region) does not "fill out" all of $\mathcal{C}$. When every y in $\mathcal{C}$ is the image of some x in $\mathcal{D}$, we have ran $F = \mathcal{C}$ and we say that F is *onto* $\mathcal{C}$, or just simply that F is *onto*.

The preceding terminology will be reviewed using a familiar nonlinear function $F(x) = x^2$ defined on the closed interval $[-2, 2]$. In place of the usual graph of x^2 in $\mathbb{R}^2$ we show the domain and codomain of F as separate sets with arrows indicating the *mapping* of a point x in the domain to its image $F(x)$ in the range. Such a schematic picture may be useful in the work on linear transformations that begins after the example.

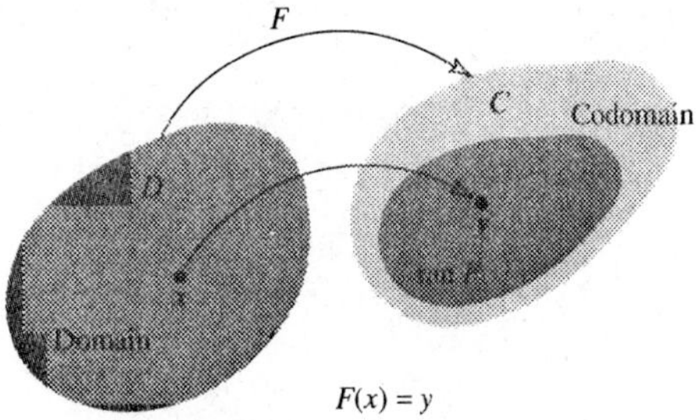

Figure 3.15 A schematic picture of a function F.

EXAMPLE 1

A Real-valued Function on $[-2, 2]$

Refer to Figure 3.16. The set of all real numbers on the real line between -2 and 2, end points included, is denoted by $[-2, 2]$. Take $\mathcal{D} = [-2, 2]$ to be the domain of the quadratic function $F(x) = x^2$ that maps each x in $[-2, 2]$ to the real value $y = x^2$. We write $F : [-2, 2] \to \mathbb{R}$ and identify the codomain $\mathcal{C}$ of F to be the set of all real numbers $\mathbb{R}$. The function F is not onto $\mathbb{R}$ because, for example, the point $y = 5$ is not the image of any x in the domain $[-2, 2]$, and we can see from the figure that ran $F = [0, 4]$. Note that $F(-2) = 4 = F(2)$, which shows that F is many-to-one. ■

In linear algebra, we are interested in *linear functions* $T : \mathcal{V} \to \mathcal{W}$, where the domain $\mathcal{V}$ and the codomain $\mathcal{W}$ are vector spaces (Chapter 7). It is usual to

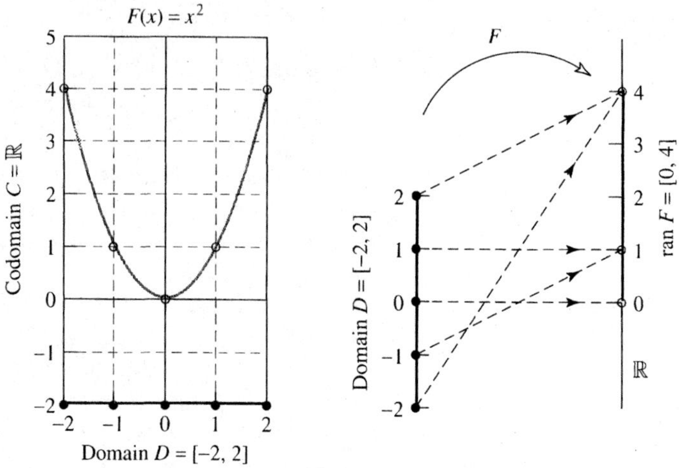

Figure 3.16 A real-valued function on $[-2, 2]$.

use the terms *transformation* or *mapping* in place of *function* in this context. In this section we restrict our attention to transformations on the particular vector spaces $\mathbb{R}^n$ and their subspaces.

Definition 3.12 *Linear transformation*

A transformation $T : \mathbb{R}^n \to \mathbb{R}^m$ with domain $\mathbb{R}^n$ and codomain $\mathbb{R}^m$ is called *linear* if two conditions are satisfied:

$$T(\mathbf{u} + \mathbf{v}) = T(\mathbf{u}) + T(\mathbf{v}), \tag{3.39}$$

$$T(c\mathbf{v}) = cT(\mathbf{v}), \tag{3.40}$$

for all vectors $\mathbf{u}$, $\mathbf{v}$ in $\mathbb{R}^n$ and scalars c.

Putting $c = 0$ in (3.40) gives $T(\mathbf{0}) = 0T(\mathbf{v}) = \mathbf{0}$, which shows that T maps the zero vector in $\mathbb{R}^n$ to the zero vector in $\mathbb{R}^m$.

Theorem 3.16 **Characterization of Linear Transformations**

A transformation T on $\mathbb{R}^n$ is linear if and only if the equation

$$T(\alpha\mathbf{u} + \beta\mathbf{v}) = \alpha T(\mathbf{u}) + \beta T(\mathbf{v}) \tag{3.41}$$

is true for all vectors $\mathbf{u}$, $\mathbf{v}$ in $\mathbb{R}^n$ and all real scalars α and β.

Proof There are two implications to prove.

(a) Suppose T is a linear transformation on $\mathbb{R}^n$. Replace $\mathbf{u}$ by $\alpha\mathbf{u}$ and $\mathbf{v}$ by $\beta\mathbf{v}$ in (3.39). Then, using (3.40), we have

$$T(\alpha\mathbf{u} + \beta\mathbf{v}) = T(\alpha\mathbf{u}) + T(\beta\mathbf{v}) = \alpha T(\mathbf{u}) + \beta T(\mathbf{v}).$$

(b) Now suppose equation (3.41) is satisfied for all vectors **u**, **v** and all scalars α, β. The assignment $\alpha = \beta = 1$ results in (3.39) and the assignment $\alpha = 0$, $\beta = c$ results in (3.40).

Equation (3.41) can be extended by mathematical induction to a finite collection of vectors $\mathbf{v}_1, \mathbf{v}_2, \ldots, \mathbf{v}_k$ in $\mathbb{R}^n$ and scalars $c_1, c_2, \ldots, c_k$. We have

$$T(c_1\mathbf{v}_1 + \cdots + c_k\mathbf{v}_k) = c_1T(\mathbf{v}_1) + \cdots + c_kT(\mathbf{v}_k), \tag{3.42}$$

showing that the image under T of a linear combination of vectors in $\mathbb{R}^n$ is a linear combination of images of the vectors using the *same* scalars $c_1, c_2, \ldots, c_k$.

A linear transformation T is a linear function from $\mathbb{R}^n$ to $\mathbb{R}^m$. Our goal is to show that *any* linear transformation T is represented by a matrix **A** using matrix-column vector multiplication. The next definition is a first step toward this goal.

Definition 3.13

Matrix transformations

Let **A** be an $m \times n$ matrix. The function $T : \mathbb{R}^n \to \mathbb{R}^m$ defined by $T(\mathbf{x}) = \mathbf{A}\mathbf{x}$ is called a *matrix transformation* from $\mathbb{R}^n$ to $\mathbb{R}^m$. The domain of T is $\mathbb{R}^n$ and the codomain of T is $\mathbb{R}^m$. The range of T is the set $\operatorname{ran} T$ consisting of all vectors **y** in $\mathbb{R}^m$ such that $T(\mathbf{x}) = \mathbf{A}\mathbf{x} = \mathbf{y}$ for at least one vector **x** in the domain $\mathbb{R}^n$.

A matrix transformation T on $\mathbb{R}^n$ is shown schematically in Figure 3.17. The image **y** of a vector **x** in $\mathbb{R}^n$ is found by computing $\mathbf{A}\mathbf{x} = \mathbf{y}$. Then **y** is the image of **x** under the mapping T.

When necessary, we will use the notation $T_{\mathbf{A}}$ to indicate that the matrix transformation T is *represented by* the matrix **A** via the rule $T(\mathbf{x}) = \mathbf{A}\mathbf{x}$.

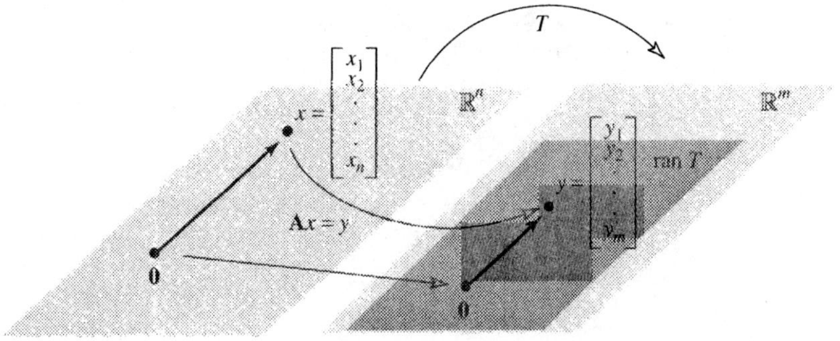

Figure 3.17 Schematic of a matrix transformation in $\mathbb{R}^n$.

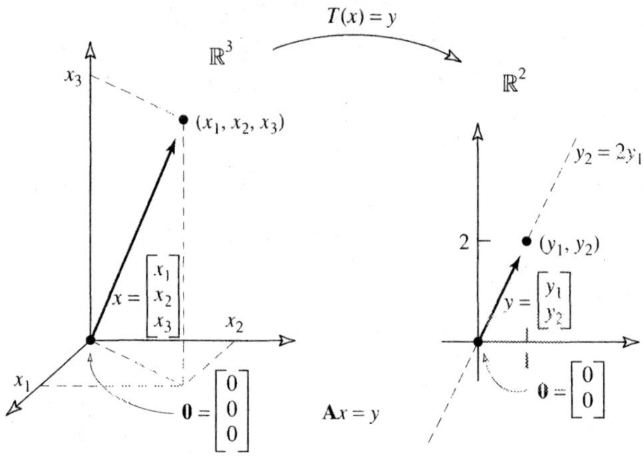

Figure 3.18 A matrix transformation from $\mathbb{R}^2$ to $\mathbb{R}^3$.

EXAMPLE 2

A Matrix Transformation on $\mathbb{R}^3$

Refer to Figure 3.18. The 2×3 matrix $\mathbf{A}$ shown on the left defines a matrix transformation $T : \mathbb{R}^3 \to \mathbb{R}^2$ given by $T(\mathbf{x}) = \mathbf{Ax}$. We have

$$\mathbf{A} = \begin{bmatrix} 1 & 0 & 1 \\ 2 & 1 & 1 \end{bmatrix}, \quad \mathbf{Ax} = \begin{bmatrix} 1 & 0 & 1 \\ 2 & 1 & 1 \end{bmatrix} \begin{bmatrix} x_1 \\ x_2 \\ x_3 \end{bmatrix} = \begin{bmatrix} x_1 + x_3 \\ 2x_1 + x_2 + x_3 \end{bmatrix} = \mathbf{y}. \tag{3.43}$$

The image of each 3-vector $\mathbf{x}$ in $\mathbb{R}^3$ is the 2-vector $\mathbf{y}$ in $\mathbb{R}^2$ shown on the right in (3.43). The components of $\mathbf{y}$ are linear expressions involving the components in $\mathbf{x}$. For example,

$$\mathbf{x} = \begin{bmatrix} 2 \\ -3 \\ 4 \end{bmatrix} \quad \text{is mapped to} \quad \begin{bmatrix} 2+4 \\ 4-3+4 \end{bmatrix} = \begin{bmatrix} 6 \\ 5 \end{bmatrix}.$$

Notice that T maps the zero vector $\mathbf{0}$ in $\mathbb{R}^3$ to the zero vector $\mathbf{0}$ in $\mathbb{R}^2$. ▪

Theorem 3.17 **Matrix Transformations are Linear Transformations**

Every matrix transformation on $\mathbb{R}^n$ is linear.

Proof Apply Theorem 3.16. Let T be a matrix transformation defined by an $m \times n$ matrix $\mathbf{A}$. Let $\mathbf{u}$ and $\mathbf{v}$ be vectors in $\mathbb{R}^n$. Let α and β be any scalars. Then, using the linearity of matrix-column vector multiplication, we have

$$\begin{aligned} T(\alpha\mathbf{u} + \beta\mathbf{v}) &= \mathbf{A}(\alpha\mathbf{u} + \beta\mathbf{v}) = \mathbf{A}(\alpha\mathbf{u}) + \mathbf{A}(\beta\mathbf{v}) \\ &= \alpha\mathbf{Au} + \beta\mathbf{Av} = \alpha T(\mathbf{u}) + \beta T(\mathbf{v}). \end{aligned}$$

—

Matrix transformations are linear transformations. But are there linear transformations on $\mathbb{R}^n$ that are not defined by matrix-column vector multiplication for some matrix **A**? The answer is no! The next result shows why.

Theorem 3.18 **Representation Theorem**

Let $T : \mathbb{R}^n \to \mathbb{R}^m$ be a linear transformation on $\mathbb{R}^n$. The $m \times n$ matrix

$$\mathbf{A} = [\, T(\mathbf{e}_1)\ \ T(\mathbf{e}_2)\ \ \cdots\ \ T(\mathbf{e}_n)\,], \tag{3.44}$$

*where $\{\mathbf{e}_1,\ \mathbf{e}_2,\ \ldots\ ,\mathbf{e}_n\}$ is the standard basis for $\mathbb{R}^n$, has the property that $T(\mathbf{x}) = \mathbf{A}\mathbf{x}$, for every vector $\mathbf{x}$ in $\mathbb{R}^n$. That is, T is represented by **A**. Moreover, **A** is unique.*

Proof Every vector $\mathbf{x}$ in $\mathbb{R}^n$ has a unique representation

$$\mathbf{x} = \begin{bmatrix} x_1 \\ \vdots \\ x_n \end{bmatrix} = x_1 \begin{bmatrix} 1 \\ 0 \\ \vdots \\ 0 \end{bmatrix} + \cdots + x_n \begin{bmatrix} 0 \\ 0 \\ \vdots \\ 1 \end{bmatrix} = x_1\mathbf{e}_1 + \cdots x_n\mathbf{e}_n.$$

Using the linearity property (3.42), we have

$$\begin{aligned} T(\mathbf{x}) &= T(x_1\mathbf{e}_1 + \cdots + x_n\mathbf{e}_n) = x_1T(\mathbf{e}_1) + \cdots + x_nT(\mathbf{e}_n) \\ &= [\, T(\mathbf{e}_1)\ \ \cdots\ \ T(\mathbf{e}_n)\,] \begin{bmatrix} x_1 \\ \vdots \\ x_n \end{bmatrix} = \mathbf{A}\mathbf{x}. \end{aligned}$$

To show that **A** is unique, suppose **B** is an $m \times n$ matrix such that $T(\mathbf{x}) = \mathbf{B}\mathbf{x}$ for all $\mathbf{x}$ in $\mathbb{R}^n$. But then $T(\mathbf{e}_j) = \mathbf{B}\mathbf{e}_j$ = column j of **B** and so column j of **A** equals column j of **B** for $1 \le j \le n$. Thus $\mathbf{A} = \mathbf{B}$.

—

Looking Ahead

We will mention at this point that the matrix **A** in (3.44) is only unique *relative to* the standard basis for $\mathbb{R}^n$. The same linear transformation T on $\mathbb{R}^n$ is represented, in a unique way, by an $m \times n$ matrix which is defined using another basis for $\mathbb{R}^n$. In particular, two matrices that represent the same linear transformation relative to two different bases are called *similar*. The topic of *similarity* is explained in Section 7.3 in the context of a general vector space.

For any linear transformation T, form **A** in (3.44) to find the matrix that defines T. The next examples use $\mathbb{R}^2$ and $\mathbb{R}^3$ to illustrate this process.

EXAMPLE 3 **Shear Transformations on $\mathbb{R}^2$**

Refer to Figure 3.19.

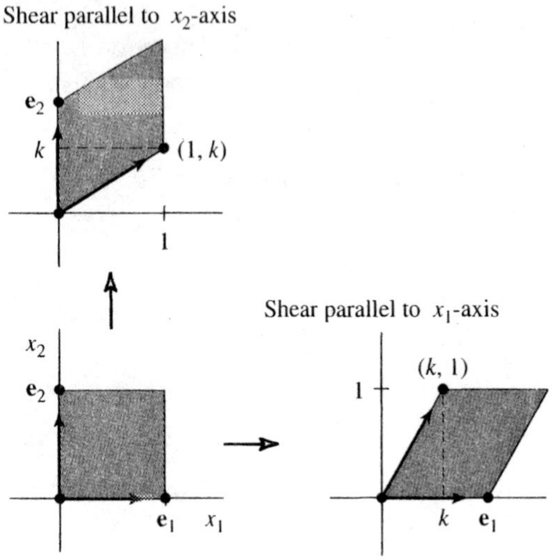

Figure 3.19 Shear.

Let k be a scalar. A *shear transformation* parallel to the x_1-axis in $\mathbb{R}^2$ transforms each point (x_1, x_2) to the point $(x_1 + kx_2, x_2)$. Using the standard basis $\{\mathbf{e}_1, \mathbf{e}_2\}$ in $\mathbb{R}^2$, we have

$$\mathbf{e}_1 = \begin{bmatrix} 1 \\ 0 \end{bmatrix}, \quad \mathbf{e}_2 = \begin{bmatrix} 0 \\ 1 \end{bmatrix}, \quad T(\mathbf{e}_1) = \mathbf{e}_1, \quad T(\mathbf{e}_2) = \begin{bmatrix} k \\ 1 \end{bmatrix} \quad \Rightarrow \quad \mathbf{A} = \begin{bmatrix} 1 & k \\ 0 & 1 \end{bmatrix}.$$

A shear transformation parallel to the x_2-axis transforms (x_1, x_2) to the point $(x_1, kx_1 + x_2)$ and we have

$$T(\mathbf{e}_1) = \begin{bmatrix} 1 \\ k \end{bmatrix}, \quad T(\mathbf{e}_2) = \mathbf{e}_2 \quad \Rightarrow \quad \mathbf{A} = \begin{bmatrix} 1 & 0 \\ k & 1 \end{bmatrix}.$$ ■

EXAMPLE 4

Rotation in $\mathbb{R}^2$ and $\mathbb{R}^3$

Refer to Figure 3.20.

Consider a transformation $T : \mathbb{R}^2 \to \mathbb{R}^2$ that rotates every point $P = (x_1, x_2)$ through a positive (counterclockwise) angle θ. We have

$$T(\mathbf{e}_1) = \begin{bmatrix} \cos\theta \\ \sin\theta \end{bmatrix}, \quad T(\mathbf{e}_2) = \begin{bmatrix} -\sin\theta \\ \cos\theta \end{bmatrix}, \quad \Rightarrow \quad \mathbf{A} = \begin{bmatrix} \cos\theta & -\sin\theta \\ \sin\theta & \cos\theta \end{bmatrix}.$$

Rotating a general point $\mathbf{x}$ in $\mathbb{R}^2$, we have

$$\mathbf{x} = \begin{bmatrix} x_1 \\ x_2 \end{bmatrix}, \quad T(\mathbf{x}) = \mathbf{A}\mathbf{x} = \begin{bmatrix} \cos\theta - \sin\theta \\ \sin\theta + \cos\theta \end{bmatrix} \begin{bmatrix} x_1 \\ x_2 \end{bmatrix} = \begin{bmatrix} x_1\cos\theta - x_2\sin\theta \\ x_1\sin\theta + x_2\cos\theta \end{bmatrix}.$$

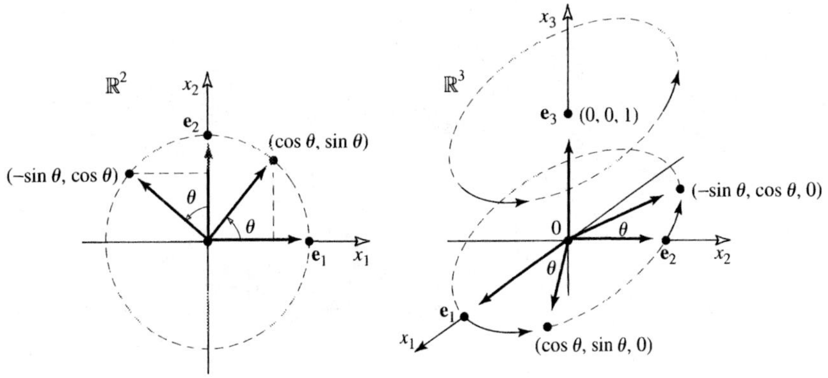

Figure 3.20 Rotation.

Rotation in $\mathbb{R}^3$ through a positive angle θ about the x_3-axis is accomplished via the matrix

$$\mathbf{A} = \begin{bmatrix} \cos\theta & -\sin\theta & 0 \\ \sin\theta & \cos\theta & 0 \\ 0 & 0 & 1 \end{bmatrix}.$$

Rotation matrices are *orthogonal* matrices (Section 4.2); they preserve *norms* (lengths) and angles. A linear transformation T represented by an orthogonal matrix is called an *isometry*. ■

Kernel and Range

The three fundamental subspaces, null **A**, col **A**, row **A**, associated with a matrix **A** were explained in Section 3.3. These subspaces play a fundamental role in describing the action of the linear transformation T defined by **A**.

Definition 3.14

Kernel and range

Let $T : \mathbb{R}^n \to \mathbb{R}^m$ be a linear transformation on $\mathbb{R}^n$ represented by a matrix **A**. The *kernel* of T, denoted by $\ker T$, is the set of all vectors **x** in $\mathbb{R}^n$ such that $T(\mathbf{x}) = \mathbf{0}$. The *range* of T, denoted by $\operatorname{ran} T$, is the set of all vectors **b** in $\mathbb{R}^m$ such that $T(\mathbf{x}) = \mathbf{b}$ for some vector **x** in $\mathbb{R}^n$.

Refer to Figure 3.21 which gives a schematic picture of kernel and range. In terms of the fundamental subspaces associated with **A**, we have

$$\ker T_{\mathbf{A}} = \operatorname{null} \mathbf{A}, \quad \operatorname{ran} T_{\mathbf{A}} = \operatorname{col} \mathbf{A}, \quad \operatorname{ran} T_{\mathbf{A}^{\mathsf{T}}} = \operatorname{row} \mathbf{A}.$$

Definition 3.15

Onto transformations

Let $T : \mathbb{R}^n \to \mathbb{R}^m$ be a linear transformation on $\mathbb{R}^n$ represented by an $m \times n$ matrix **A**. A transformation for which $\operatorname{ran} T = \mathbb{R}^m$ is called an *onto transformation* and we say that T maps $\mathbb{R}^n$ *onto* $\mathbb{R}^m$.

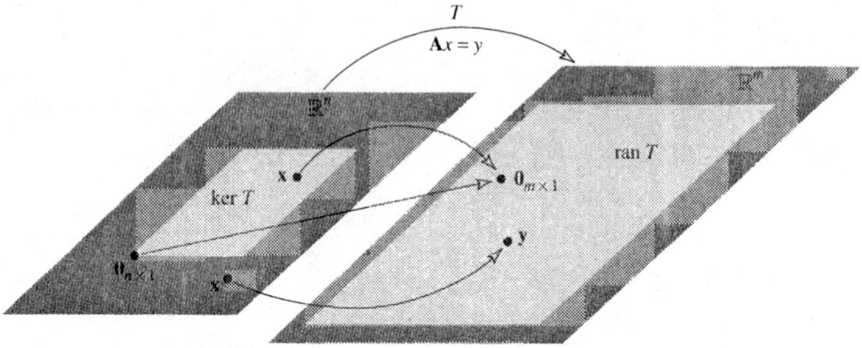

Figure 3.21 Schematic showing the kernel.

EXAMPLE 5

Kernel, Range, onto

The matrix

$$\mathbf{A} = \begin{bmatrix} 2 & -3 & 1 \\ 1 & 0 & 1 \end{bmatrix}$$

defines a linear transformation $T : \mathbb{R}^3 \to \mathbb{R}^2$. Referring to Section 3.3, the kernel of T is null $\mathbf{A}$ which is found by computing $\mathbf{A}^*$. We have

$$\mathbf{A} = [\,\mathbf{v}_1\ \mathbf{v}_2\ \mathbf{v}_3\,] = \begin{bmatrix} 2 & -3 & 1 \\ 1 & 0 & 1 \end{bmatrix} \sim \begin{bmatrix} \textcircled{1} & 0 & 1 \\ 0 & \textcircled{1} & 1/3 \end{bmatrix} = \mathbf{A}^*$$

and the solution set is $(x_1, x_2, x_3) = (-t, -\frac{t}{3}, t)$, where t is a real parameter.

To avoid fractions we write $t = 3s$ to obtain

$$\mathbf{x} = \begin{bmatrix} x_1 \\ x_2 \\ x_3 \end{bmatrix} = s \begin{bmatrix} -3 \\ 1 \\ 3 \end{bmatrix} = s\mathbf{v}, \quad \text{where } s \text{ is a real parameter,}$$

and so null $\mathbf{A}$ is a one-dimensional subspace of $\mathbb{R}^3$ with basis $\mathcal{B} = \{\mathbf{v}\}$. Refer to Figure 3.22, which represents ker T (null $\mathbf{A}$) as a line L in $\mathbb{R}^3$ passing through the origin in the direction of the basis vector $\mathbf{v}$.

The set $\mathcal{B} = \{\mathbf{v}_1, \mathbf{v}_2\}$ is a basis for col $\mathbf{A}$. The columns $\mathbf{v}_1$ and $\mathbf{v}_2$ in $\mathbf{A}$ are identified using the pivot columns in $\mathbf{A}^*$. We have ran T = col $\mathbf{A} = \mathbb{R}^2$, showing that T is an onto transformation. ■

The equation $\mathbf{Ax} = \mathbf{b}$ defines an $m \times n$ linear system (S). Finding the vectors $\mathbf{x}$ in $\mathbb{R}^n$ that are mapped to a vector $\mathbf{b}$ in $\mathbb{R}^m$ amounts to solving (S). Thus the linear transformation T represented by $\mathbf{A}$ is onto if and only if (S) is consistent for every $\mathbf{b}$ in $\mathbb{R}^m$.

Theorem 3.19

Onto Transformations and Rank

Let $T : \mathbb{R}^n \to \mathbb{R}^m$ be a linear transformation represented by an $m \times n$ matrix $\mathbf{A}$ and suppose $m \le n$. If rank $\mathbf{A} = m$*, then T is onto $\mathbb{R}^m$.*

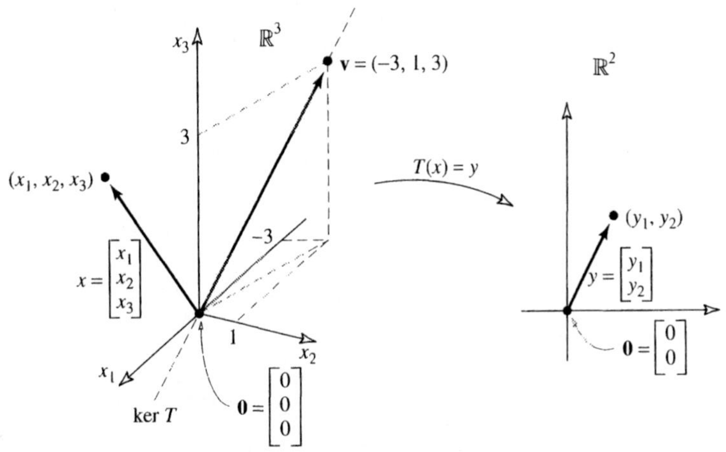

Figure 3.22 The kernel of T.

Proof We appeal to Theorem 1.5 in Section 1.2. In essence, rank $\mathbf{A} = m$ implies rank $\mathbf{A} \le n$ and so the linear system (S) defined by $\mathbf{Ax} = \mathbf{b}$ is consistent for every vector $\mathbf{b}$ in $\mathbb{R}^m$. When $m < n$, (S) has infinitely many solutions $\mathbf{x}$ and when $m = n$, (S) has a unique solution. ■

Use Figure 3.15 at the beginning of the section to visualize the following general property of functions.

Definition 3.16

One-to-one function

A function $F : \mathcal{D} \to \mathcal{C}$ is called a *one-to-one* function if

$$F(x_1) = y = F(x_2) \quad \text{implies} \quad x_1 = x_2 \tag{3.45}$$

for all pairs of objects x_1 and x_2 in $\mathcal{D}$.

If T is a linear transformation on $\mathbb{R}^n$ represented by a matrix $\mathbf{A}$, condition (3.45) says that T is one-to-one if $T(\mathbf{u}_1) = T(\mathbf{u}_2)$ implies $\mathbf{u}_1 = \mathbf{u}_2$, for each pair of vectors $\mathbf{u}_1$, $\mathbf{u}_2$ in $\mathbb{R}^n$ (no two vectors are mapped to the same vector). The property of being one-to-one may be established by looking at the null space of $\mathbf{A}$.

Theorem 3.20 **One-to-one Linear Transformations**

A linear transformation $T : \mathbb{R}^n \to \mathbb{R}^m$ represented by a matrix $\mathbf{A}$ is one-to-one if and only if ker T (= null $\mathbf{A}$) *is the zero subspace of* $\mathbb{R}^n$.

Proof There are two implications to prove.

(a) Suppose T is one-to-one. For any vector $\mathbf{u}$ in null $\mathbf{A}$ we have $\mathbf{Au} = \mathbf{0}$. Also $\mathbf{A0} = \mathbf{0}$ because $\mathbf{0}$ is in null $\mathbf{A}$. Using the fact that T is one-to-one, we have $T(\mathbf{u}) = T(\mathbf{0})$ implies $\mathbf{Au} = \mathbf{A0}$ implies $\mathbf{u} = \mathbf{0}$, showing that null $\mathbf{A}$ contains only the zero vector in $\mathbb{R}^n$.

(b) Suppose $\operatorname{null}\mathbf{A} = \{\mathbf{0}\}$ and there are vectors $\mathbf{u}_1$ and $\mathbf{u}_2$ in $\mathbb{R}^n$ such that $T(\mathbf{u}_1) = T(\mathbf{u}_2)$. Then

$$T(\mathbf{u}_1) = T(\mathbf{u}_2) \;\Rightarrow\; \mathbf{A}\mathbf{u}_1 = \mathbf{A}\mathbf{u}_2$$
$$\Rightarrow\; \mathbf{A}(\mathbf{u}_1 - \mathbf{u}_2) = \mathbf{0} \quad\Rightarrow\quad \mathbf{u}_1 - \mathbf{u}_2 = \mathbf{0},$$

so that $\mathbf{u}_1 = \mathbf{u}_2$, and T is one-to-one by (3.45).

Consider the linear transformation T represented by the $m \times n$ matrix $\mathbf{A}$. Then T is one-to-one if and only if the $m \times n$ homogeneous linear system (S) defined by $\mathbf{Ax} = \mathbf{0}$ has only the zero solution. Thus

$$T \text{ is one-to-one if and only if } n \le m \text{ and } \operatorname{rank}\mathbf{A} = n. \tag{3.46}$$

Note that when $m < n$ the system (S) has infinitely many solutions implying that $\operatorname{null}\mathbf{A}$ has a positive dimension and that T is not one-to-one.

Each matrix, shown next with its reduced form, defines a one-to-one linear transformation T. The symbols a, b, c are any nonzero real numbers and $\mathbf{O}$ is a 1×2 zero row vector.

$$\begin{bmatrix} 1 & 2 \\ 3 & 4 \end{bmatrix} \sim \mathbf{I}_2, \quad \begin{bmatrix} 1.1 & -2.4 \\ -0.6 & 4.1 \\ 0.5 & 3.7 \end{bmatrix} \sim \begin{bmatrix} 1 & 0 \\ 0 & 1 \\ 0 & 0 \end{bmatrix} = \begin{bmatrix} \mathbf{I}_2 \\ \mathbf{O} \end{bmatrix}, \quad \begin{bmatrix} a & 2 & 3 \\ 0 & b & 4 \\ 0 & 0 & c \end{bmatrix} \sim \mathbf{I}_3$$

Composition of Transformations

Refer to Figure 3.23, which shows a function F defined on a set $\mathcal{D}$ and a function G defined on a set $\mathcal{C}$. Then F maps an object x to y in $\operatorname{ran} F$ and G maps y to z. The *composition* of F and G, denoted by $G \circ F$ (note the order), maps x directly to z and is defined by the rule $(G \circ F)(x) = G(F(x)) = z$.

Consider linear transformations $T_{\mathbf{A}} : \mathbb{R}^n \to \mathbb{R}^m$ and $T_{\mathbf{B}} : \mathbb{R}^m \to \mathbb{R}^p$ represented, respectively, by the $m \times n$ matrix $\mathbf{A}$ and the $m \times p$ matrix $\mathbf{B}$. The composite transformation $(T_{\mathbf{B}} \circ T_{\mathbf{A}}) : \mathbb{R}^n \to \mathbb{R}^p$ is represented by the matrix $\mathbf{BA}$ because the equations $\mathbf{Ax} = \mathbf{y}$ and $\mathbf{By} = \mathbf{z}$ imply that $(\mathbf{BA})\mathbf{x} = \mathbf{B}(\mathbf{A}(\mathbf{x})) = \mathbf{By} = \mathbf{z}$, for every vector $\mathbf{x}$ in $\mathbb{R}^n$. Thus, $(T_{\mathbf{B}} \circ T_{\mathbf{A}}) = T_{\mathbf{C}}$, where $\mathbf{C} = \mathbf{BA}$.

EXAMPLE 6

Dilation and Rotation

Refer to Figure 3.24. Our goal is to find the linear transformation $T : \mathbb{R}^2 \to \mathbb{R}^2$ that maps the unit square in $\mathbb{R}^2$ onto the rectangle $OD'E'F'$. The transformation can be seen as a composition of two transformations $T_{\mathbf{A}}$ and $T_{\mathbf{B}}$ represented by

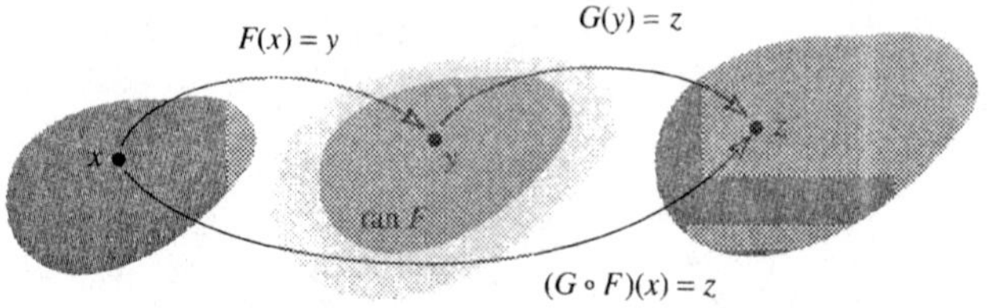

Figure 3.23 Schematic of composition.

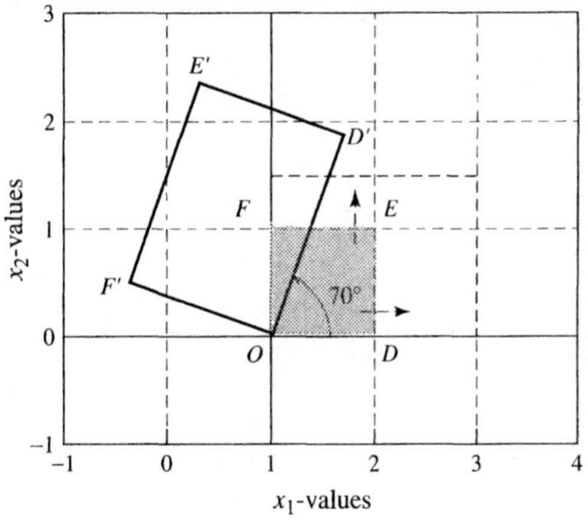

Figure 3.24 A composite transformation on $\mathbb{R}^2$.

matrices **A** and **B** shown in (3.47). The diagonal matrix **A** dilates the unit square in $\mathbb{R}^2$ by 200% in the x_1-direction and 150% in the x_2-direction. The matrix **B** defines rotation through a positive (counterclockwise) angle $70° \simeq 1.2217$ radians.

$$\mathbf{A} = \begin{bmatrix} 2 & 0 \\ 0 & 1.5 \end{bmatrix}, \quad \mathbf{B} = \begin{bmatrix} \cos 70° & -\sin 70° \\ \sin 70° & \cos 70° \end{bmatrix} = \begin{bmatrix} 0.3420 & -0.9397 \\ 0.9397 & 0.3420 \end{bmatrix} \tag{3.47}$$

We have

$$\mathbf{C} = \mathbf{BA} = \begin{bmatrix} 0.6840 & -1.4095 \\ 1.8794 & 0.5130 \end{bmatrix}.$$

If the vectors representing the vertices D, E, F in the unit square are stored as columns in the matrix **X**, the images of the vertices are D', E', F' given, respectively, by the columns of the matrix $\mathbf{Y} = \mathbf{CX}$. We have

$$\mathbf{X} = \begin{bmatrix} 1 & 1 & 0 \\ 0 & 1 & 1 \end{bmatrix} \quad \Rightarrow \quad \mathbf{CX} = \begin{bmatrix} 0.6840 & -0.7255 & -1.4095 \\ 1.8794 & 2.3924 & 0.5130 \end{bmatrix} = \mathbf{Y}. \quad ■$$

Inverse Transformations

Refer to Figure 3.25. Consider a one-to-one linear transformation $T : \mathbb{R}^n \to \mathbb{R}^m$. Each vector $\mathbf{u}$ in $\mathbb{R}^n$ is mapped by T onto a unique vector $\mathbf{v}$ in ran T. Hence, for each vector $\mathbf{v}$ in ran T there is a unique vector $\mathbf{u}$ in $\mathbb{R}^n$ such that $T(\mathbf{u}) = \mathbf{v}$. We are therefore able to define a transformation $T^{-1} : \text{ran}\, T \to \mathbb{R}^n$ as follows: For each $\mathbf{v}$ in ran T,

$$T^{-1}(\mathbf{v}) = \mathbf{u} \quad \text{if and only if} \quad T(\mathbf{u}) = \mathbf{v}.$$

We call T^{-1} an *inverse* for T. In terms of composite transformations, we have

$$(T^{-1} \circ T)(\mathbf{u}) = \mathbf{u}, \text{ for all } \mathbf{u} \text{ in } \mathbb{R}^n,$$
$$(T \circ T^{-1})(\mathbf{v}) = \mathbf{v}, \text{ for all } \mathbf{v} \text{ in ran } T.$$

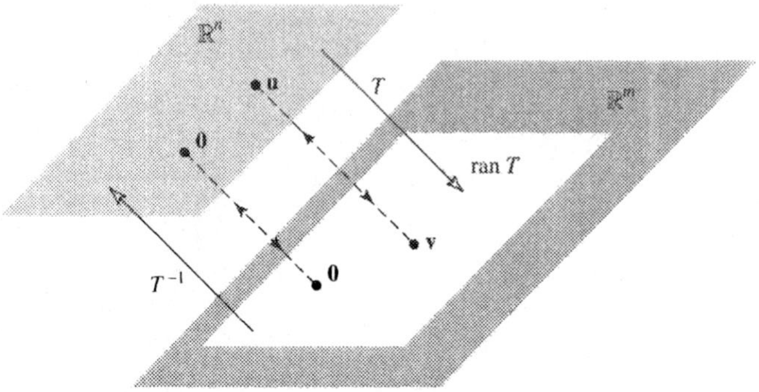

Figure 3.25 Inverse transformation on ran T.

Suppose T is a one-to-one linear transformation on $\mathbb{R}^n$ represented by an $m \times n$ matrix $\mathbf{A}$. Then an inverse T^{-1} exists and the linearity of matrix-column vector multiplication shows that T^{-1} is linear (Exercises 3.4). Our goal is to find a matrix $\mathbf{B}$ which defines T^{-1}. We argue as follows:

(a) Using fact (3.46), T is one-to-one if and only if $n \leq m$ and rank $\mathbf{A} = n$.

(b) Appealing to Theorem 2.9 in Section 2.2, the condition in (a) implies that $\mathbf{A}$ has a left-inverse $\mathbf{B}$, that is, $\mathbf{BA} = \mathbf{I}_n$. Note that $\mathbf{B}$ is not in general unique.

(c) Let $\mathbf{u}$ be in $\mathbb{R}^n$ and $\mathbf{v}$ be in $\mathbb{R}^m$ such that $\mathbf{Au} = \mathbf{v}$. Then $\mathbf{Bv} = \mathbf{B}(\mathbf{Au}) = (\mathbf{BA})\mathbf{u} = \mathbf{I}_n\mathbf{u} = \mathbf{u}$ and so T^{-1} is represented by the matrix $\mathbf{B}$. The composite transformation $T_\mathbf{B} \circ T_\mathbf{A}$ is the identity transformation on $\mathbb{R}^n$ represented by the identity matrix $\mathbf{I}_n$. The composite transformation $T_\mathbf{A} \circ T_\mathbf{B}$ is represented by a matrix $\mathbf{C}$, where $\mathbf{C} = \mathbf{AB}$, and $\mathbf{Cv} = \mathbf{v}$, for all $\mathbf{v}$ in col $\mathbf{A}$ ($=$ ran T).

■ ILLUSTRATION 3.11

A Linear Transformation on $\mathbb{R}^2$

At this point we return to the linear system (S) shown in (3.1) that was used in the introduction to motivate the ideas of this chapter. The coefficient matrix $\mathbf{A}$ for (S) is shown on the left in (3.48). Consider the linear transformation $T : \mathbb{R}^2 \to \mathbb{R}^3$ represented by $\mathbf{A}$. We have $n = 2$, $m = 3$ and rank $\mathbf{A} = 2$ in this case and so the conditions in item (a) above are met. A left-inverse $\mathbf{B}$ for $\mathbf{A}$ exists and $\mathbf{B}$ can be found, as in Section 2.2, by solving a 4×6 linear system. The matrices $\mathbf{B}$, shown on the right, where s and t are real parameters, are all left-inverses for $\mathbf{A}$.

$$\mathbf{A} = [\,\mathbf{v}_1\ \mathbf{v}_2\,] = \begin{bmatrix} 1 & 2 \\ 3 & -1 \\ 4 & 5 \end{bmatrix}, \quad \mathbf{B} = \frac{1}{7}\begin{bmatrix} (1 - 19s) & (2 - 3s) & 7s \\ (3 - 19t) & (-1 - 3t) & 7t \end{bmatrix} \tag{3.48}$$

Observe that ran T ($=$ ran $\mathbf{T}$) is a two-dimensional subspace of $\mathbb{R}^3$ spanned by the columns $\mathbf{v}_1$, $\mathbf{v}_2$ of $\mathbf{A}$. Each matrix $\mathbf{B}$ defines a one-to-one linear transformation $T_\mathbf{B}$ whose domain is ran T. As mentioned in the introduction, the vector $\mathbf{x} = (1, 1)$ is mapped by $T_\mathbf{A}$ to the vector $\mathbf{y} = (3, 2, 9)$ and $\mathbf{y} = (3, 2, 9)$ is mapped by $T_\mathbf{B}$ to $\mathbf{x} = (1, 1)$; that is, $\mathbf{Ax} = \mathbf{y}$ and $\mathbf{By} = \mathbf{x}$. ■

Linear Operators on $\mathbb{R}^n$

A linear transformation $T : \mathbb{R}^n \to \mathbb{R}^n$ represented by an $n \times n$ matrix $\mathbf{A}$ is called a *linear operator* on $\mathbb{R}^n$.

Suppose first that $\text{rank}\,\mathbf{A} = n$. Then T is one-to-one and the columns of $\mathbf{A}$ span $\mathbb{R}^n$. Thus, $\text{ran}\,T = \mathbb{R}^n$ and so T is onto. The matrix $\mathbf{A}$ is invertible and $\mathbf{A}^{-1}$ defines a unique inverse linear operator T^{-1} on $\mathbb{R}^n$. If $\text{rank}\,\mathbf{A} < n$, then $\mathbf{A}$ is singular and in this case the linear operator T represented by $\mathbf{A}$ is not one-to-one and not onto $\mathbb{R}^n$.

It may be interesting to note that each linear operator represented by the regular magic squares of order $2k+1$, $k = 1, 2, 3, \ldots,$ are one-to-one and onto because each matrix is invertible. The linear operators represented by magic squares of order $2k+2$ $k = 1, 2, 3, \ldots$ are not one-to-one and not onto because the matrices are all singular (see Exercises 3.4).

Looking Forward

Projection and *reflection* matrices are introduced in Chapter 4. These matrices define linear operators on $\mathbb{R}^n$ that, respectively, project a vector $\mathbf{v}$ onto a subspace of $\mathbb{R}^n$ and reflect $\mathbf{v}$ in a subspace of $\mathbb{R}^n$. *Orthogonal* matrices, introduced in Section 4.2, preserve lengths (norms) and angles—the linear operators defined on $\mathbb{R}^n$ using these matrices are called *isometries*.

EXERCISES 3.4

Exercises 1–4. The linear transformation T maps basis vectors as shown. Determine (a) the domain and codomain of T, (b) $\ker T$ *and* $\text{ran}\,T$*, (c) whether T is one-to-one, (d) whether T is onto.*

1. $T(1,0) = (1,1),\ T(0,1) = (2,3)$
2. $T(1,0) = (1,-1,2),\ T(0,1) = (2,2,4)$
3. $T(1,0,0) = (1,0),\ T(0,1,0) = (2,3),\ T(0,0,1) = (1,-1)$
4. $T(1,0,0) = (1,0,1),\ T(0,1,0) = (1,1,0),\ T(0,0,1) = (1,3,-2)$

Exercises 5–8. Find a matrix $\mathbf{A}$ that represents the given linear transformation T. Determine (a) the domain and codomain of T, (b) $\ker T$ *and* $\text{ran}\,T$*, (c) whether T is one-to-one, (d) whether T is onto.*

5. $T(x_1, x_2) = (x_1 - x_2, x_1 + x_2)$
6. $T(x_1, x_2) = (x_1, x_1 - x_2, x_1 + x_2)$
7. $T(x_1, x_2, x_3) = (2x_1 - 3x_2 + x_3, x_1 + x_3)$
8. $T(x_1, x_2, x_3) = (x_1 - x_2, 2x_1, x_2 + x_3, 2x_1 + 3x_3)$

Exercises 9–12. Consider the linear transformation represented by the given matrix. Use the concepts of rank and dimension to determine whether (a) T is one-to-one, (b) T is onto. Confirm your answers by computing a basis for $\text{null}\,\mathbf{A}$ *and* $\text{col}\,\mathbf{A}$.

9. $\begin{bmatrix} 1 & -1 \\ -2 & 1 \end{bmatrix}$ **10.** $\begin{bmatrix} 2 & -4 & 3 \\ -4 & 8 & -6 \end{bmatrix}$

11. $\begin{bmatrix} 1 & -1 & 2 \\ 2 & 1 & 0 \\ 3 & 0 & 2 \end{bmatrix}$ **12.** $\begin{bmatrix} 1 & 2 & 3 \\ 4 & 5 & 6 \\ 7 & 8 & 9 \end{bmatrix}$

Exercises 13–16. Consider the linear operator T on $\mathbb{R}^2$ represented by the given matrix $\mathbf{A}$. Find the vectors $\mathbf{x}$ such that $T(\mathbf{x}) = \mathbf{x}$. The vectors $\mathbf{x}$ are called invariant *under the action of the operator T.*

13. $\begin{bmatrix} 2 & -3 \\ -1 & 4 \end{bmatrix}$ **15.** $\begin{bmatrix} 1 & 0 \\ 0 & -1 \end{bmatrix}$

14. $\begin{bmatrix} 5 & -2 \\ 7 & 2 \end{bmatrix}$ **16.** $\begin{bmatrix} 4 & -1 \\ 3 & 2 \end{bmatrix}$

Exercises 17–22. Find a matrix $\mathbf{A}$ that represents the linear operator T on $\mathbb{R}^2$ that transforms points as described.

17. Projects each point (x, y) onto $(-x, 0)$

18. Rotates each point (x, y) through $-\pi/4$ radians
19. Rotates each point (x, y) through $\pi/2$ radians and then reflects the image in the origin
20. Reflects each point (x, y) in the line $y = x$
21. Projects each point (x, y) onto the x-axis and then reflects the image in the y–axis
22. Moves each point (x, y) to $(2x, 3y)$

Exercises 23–24. Find a matrix **A** *that represents a linear operator T on* $\mathbb{R}^2$ *satisfying the given conditions.*

23. T maps points $(4, -2)$ to $(7, -2)$ and $(-1, 2)$ to $(-8, 7)$
24. T maps points $(1, 0)$ to $(1, 1)$ and $(-1, 0)$ to $(2, 2)$
25. Consider the line L in $\mathbb{R}^2$ with equation $y = 3-2x$. Show that the image of points on L under the transformation T defined by $\mathbf{A} = \begin{bmatrix} 1 & -2 \\ -3 & -4 \end{bmatrix}$ is a line L'. Find the equation of L'. Draw an accurate diagram.
26. Consider the linear operator T on $\mathbb{R}^2$ represented by the matrix $\mathbf{A} = \begin{bmatrix} 1 & 3 \\ 2 & 6 \end{bmatrix}$. Find bases for (a) null **A**, (b) col **A**. Draw an accurate diagram showing $\ker T$ and $\operatorname{ran} T$. Interpret the fundamental equation n = rank + nullity in this case.
27. Consider the matrix $\mathbf{B} = \begin{bmatrix} 1 & 0 & 1 \\ 2 & 1 & 1 \end{bmatrix}$. Use rank to explain why there is a matrix **A** such that $\mathbf{BA} = \mathbf{I}_2$. Find all right-inverses **A** for **B**. Describe the composite transformation $T_{\mathbf{B}} \circ T_{\mathbf{A}}$. Is this transformation one-to-one? Is it onto? Find a matrix **C** that represents the composite transformation $T_{\mathbf{A}} \circ T_{\mathbf{B}}$. Find the domain and range of $T_{\mathbf{C}}$. Is $T_{\mathbf{C}}$ one-to-one? Is $T_{\mathbf{C}}$ onto?
28. The following matrices **A** and **B** represent linear transformations $T_{\mathbf{A}}$ and $T_{\mathbf{B}}$, respectively. Find $\ker(T_{\mathbf{B}} \circ T_{\mathbf{A}})$ and $\operatorname{ran}(T_{\mathbf{B}} \circ T_{\mathbf{A}})$.

$$\mathbf{A} = \begin{bmatrix} 1 & 2 \\ 0 & -1 \\ 4 & 6 \end{bmatrix}, \quad \mathbf{B} = \begin{bmatrix} 2 & -1 & 0 \\ 0 & -1 & 3 \\ 0 & 0 & 1 \end{bmatrix}$$

29. Let T be a one-to-one linear transformation on $\mathbb{R}^n$ represented by a matrix **A**. Show that the inverse transformation T^{-1} defined on $\operatorname{ran} T$ is linear; that is, show that
$$T^{-1}(\alpha\mathbf{v}_1 + \beta\mathbf{v}_2) = \alpha T^{-1}(\mathbf{v}_1) + \beta T^{-1}(\mathbf{v}_2),$$
for all vectors $\mathbf{v}_1$, $\mathbf{v}_2$ in $\operatorname{ran} T$ and all scalars α, β.
30. Let T be a linear operator on $\mathbb{R}^n$ represented by an $n \times n$ matrix **A** such that $\operatorname{rank} \mathbf{A} < n$. Show that T is not one–one and not onto $\mathbb{R}^n$.

Exercises 31–38. Label each statement as either true (T) or false (F). If a false statement can be made true by modifying the wording, then do so.

31. There is a linear operator T on $\mathbb{R}^n$ such that $\ker T = \operatorname{ran} T$.
32. There are linear operators on $\mathbb{R}^n$ that are one-to-one but not onto.
33. There are linear operators on $\mathbb{R}^n$ that are onto but not one-to-one.
34. $\ker(T_{\mathbf{B}} \circ T_{\mathbf{A}}) = \ker T_{\mathbf{A}}$.
35. $\operatorname{ran}(T_{\mathbf{B}} \circ T_{\mathbf{A}}) = \operatorname{ran} T_{\mathbf{B}}$.
36. The linear transformation represented by a square upper triangular matrix has an inverse.
37. There exist linear operators T such that $T^{-1} = T$.
38. There is a linear operator T on $\mathbb{R}^3$ represented by a matrix **A** with at least one nonzero entry such that $\ker(T \circ T \circ T) = \mathbb{R}^3$.

USING MATLAB

Exercises 39–40. Let T be the linear transformation represented by the given matrix, denoted by **A**. *Find a bases for* $\ker T$ *and* $\operatorname{ran} T$. *If possible, find the matrix* **B** *that represents the inverse transformation* T^{-1} *defined on* $\operatorname{ran} T$.

39. **magic(3)**
40. **magic(4)**
41. Use rank to explain why the matrix
$$\mathbf{A} = \begin{bmatrix} 1 & 2 & 1 \\ 0 & -1 & 0 \\ 1 & 0 & 2 \\ 0 & 0 & 1 \end{bmatrix}$$
has a left-inverse **B**. Write down, in block form, the augmented matrix of the linear system (S) that is solved to find **B**. Compute **B** and check that $\mathbf{BA} = \mathbf{I}_3$.
42. Complete the details of Illustration 3.11. Find the left-inverses **B** of the matrix $\mathbf{A} = \begin{bmatrix} 1 & 2 \\ 3 & -1 \\ 4 & 5 \end{bmatrix}$. If $\mathbf{AB} = \mathbf{C}$, show that $\mathbf{Cv} = \mathbf{v}$, for all vectors **v** in col **A**. This means that **C** defines the identity transformation on $\operatorname{ran} T$.

Exercises 43–44. Explain why the given matrix **A** *has a left–inverse* **B**. *Express* **A** *and* **B** *in block form and then compute* **B**. *Check that* $\mathbf{BA} = \mathbf{I}_3$.

43. $\begin{bmatrix} 1 & 2 & 1 \\ 0 & -1 & 0 \\ 1 & 0 & 2 \\ 0 & 0 & 0 \end{bmatrix}$

44. $\begin{bmatrix} 1 & 2 & 1 & 0 \\ 2 & 0 & 1 & 1 \\ -1 & 4 & 1 & 1 \\ 1 & 3 & 1 & 0 \\ 0 & 0 & 0 & 0 \end{bmatrix}$

45. Consult online help for the command **patch**, and related commands. Refer to Figure 3.26. Find a linear transformation T that maps $\triangle OAB$ onto $\triangle OA'B'$. Reproduce Figure 3.26 in MATLAB. Color the triangles using the **patch** command.

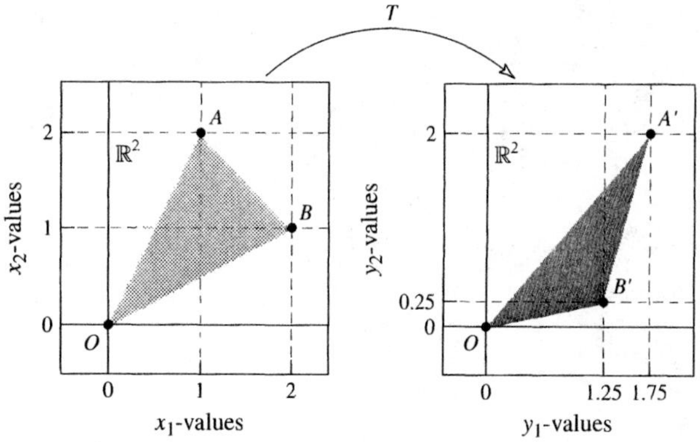

Figure 3.26 Mapping the plane.

CHAPTER 3 REVIEW

State whether each statement is true or false as stated. Provide a clear reason for your answer. If a false statement can be modified to make it true, then do so.

1. The vector $\begin{bmatrix} -2 \\ 6 \end{bmatrix}$ is a linear combination of $\begin{bmatrix} 3 \\ -4 \end{bmatrix}$, $\begin{bmatrix} 1 \\ 2 \end{bmatrix}$.

2. If $2\mathbf{u}+2\mathbf{v} = 4\mathbf{w}-3\mathbf{v}+2\mathbf{u}$, then $\mathbf{w}$ is a linear combination of $\mathbf{u}$ and $\mathbf{u}+\mathbf{v}$.

3. If $\mathbf{u}$ is in $\mathbb{R}^3$, then $-\mathbf{u}$ is in span$\{\mathbf{u}\}$.

4. span$\{\mathbf{u}, \mathbf{u}-\mathbf{v}\}$ does not contain the vector $\mathbf{v}$.

5. The set span$\{\mathbf{u}, \mathbf{v}\}$ contains finitely many vectors.

6. span $\mathcal{S}$ contains infinitely many vectors for any nonempty set $\mathcal{S}$.

7. The columns of the matrix $\mathbf{A} = \begin{bmatrix} 2 & 4 & 6 \\ 1 & 2 & 3 \end{bmatrix}$ span $\mathbb{R}^2$.

8. The columns of the matrix $\begin{bmatrix} 2 & 3 \\ 0 & 4 \end{bmatrix}$ form a basis for $\mathbb{R}^2$.

9. If $\mathbf{I}_n$ is the reduced form of a matrix $\mathbf{A}$, then $\mathbf{A}$ has linearly independent rows.

10. The null space of the matrix $\begin{bmatrix} 4 & 0 & 1 \\ -1 & 2 & 0 \\ 2 & 4 & 1 \end{bmatrix}$ has dimension 1.

11. There exists a 3×5 matrix whose column space has dimension 4.

12. If the columns of a 4×3 matrix $\mathbf{A}$ are linearly independent, then dim col $\mathbf{A} = 4$.

13. The span of a finite set of vectors in $\mathbb{R}^n$ is a subspace of $\mathbb{R}^n$.

14. Every linear transformation on $\mathbb{R}^2$ is represented by a 2×2 matrix $\mathbf{A}$.

15. A set of five vectors in $\mathbb{R}^3$ is linearly dependent only if the vectors are nonzero.

16. If the vectors $\mathbf{v}_1, \mathbf{v}_2, \ldots, \mathbf{v}_k$ in $\mathbb{R}^n$ are linearly independent and T is a linear transformation on $\mathbb{R}^n$, then the set $\{T(\mathbf{v}_1), T(\mathbf{v}_2), \ldots, T(\mathbf{v}_k)\}$ is linearly independent.

17. The dimensions of the column space and the rows space of a matrix can be unequal.

18. If the vectors $\mathbf{v}_1, \mathbf{v}_2, \ldots, \mathbf{v}_k$ in $\mathbb{R}^n$ are linearly independent and T is a linear transformation on $\mathbb{R}^n$, then the set $\{T(\mathbf{v}_1), T(\mathbf{v}_2), \ldots, T(\mathbf{v}_k)\}$ is linearly independent.

19. If null $\mathbf{A} = \{\mathbf{0}\}$, for some matrix $\mathbf{A}$, then the transformation T represented by $\mathbf{A}$ is onto.

20. The linear transformation T represented by the matrix $\begin{bmatrix} 1 & 2 \\ 3 & 4 \\ -1 & 0 \end{bmatrix}$ is onto.

Karl Pearson graduated from Cambridge University in 1879. His professional interests covered a wide spectrum—applied mathematics, biometrics, statistics, eugenics (the study of selective human breeding), and the philosophy of science. His other interests included poetry, law, history of religion, and the theory of elasticity.

At the age of 27, Pearson was appointed chair of applied mathematics at University College, London, England. Pearson's book *The Grammar of Science* (1892) extended the influence of science in a general way and was remarkable in that it anticipated some of the ideas of the theory of relativity. Following this publication, Pearson became interested in developing mathematical and statistical methods for studying the processes of heredity and evolution.

Pearson laid the foundation of much of 20th-century statistics. His 18 papers, entitled *Mathematical Contribution to the Theory of Evolution*, published during the period from 1893 to 1912, are considered his most important work. In these papers he defined *standard deviation* (1893), the *product-moment correlation coefficient r* (Section 4.4), *multiple and partial correlation, normal curve, product moment*, and the *chi-square test of statistical significance* (1900). We remember Pearson here particularly for his work on *regression analysis*.

Pearson was the first Galton professor of eugenics at University College. Francis Galton (1822–1911), who was a cousin of Charles Darwin (1809–1882), is famous for his pioneering studies in evolution. Galton and Pearson cofounded the statistical journal *Biometrika*.

Chapter 4

ORTHOGONALITY

Introduction

An algebraic condition, called the *dot product*, that tells us when two vectors in $\mathbb{R}^n$ are *orthogonal* (meaning perpendicular or at right angles) is the key idea in this chapter. The properties of dot products and Euclidean norm (length) are developed in Section 4.1.

In Section 4.2 we consider orthogonal sets of vectors, orthogonal matrices, and orthogonal transformations (*isometries*). Section 4.3 begins with the basic theory of orthogonal subspaces and goes on to consider the orthogonal projection of a vector onto a subspace of $\mathbb{R}^n$, extending the ideas from the end of Section 4.1. Orthonormal bases (the vectors have unit norm and each pair of distinct vectors are orthogonal) are computationally pleasant to work with. Section 4.3 ends with the theory of the *Gram-Schmidt process* for constructing orthonormal bases and the *QR factorization* of a matrix that arises naturally from this process.

The work of engineers and scientists often involves making predictions or estimates based on observed or measured experimental data. A *data point* (t, c) in $\mathbb{R}^2$, for example, might record a temperature reading c (degrees Celsius) of an object at time t (minutes). Some experimental data from 50 readings are plotted in Figure 4.1.

The configuration of a set of data points, or the nature of the experiment, may suggest that a *functional relationship* exists between the data points. The data shown in Figure 4.1 suggests that the relationship is *linear* and we assume that experimental conditions or error in observation have caused the data points to be noncollinear. The *method of least squares* can be applied to find the *line of regression* L that lies among a set of data points, and that is in a certain sense the best possible *fit* for the data. When least squares is applied to the

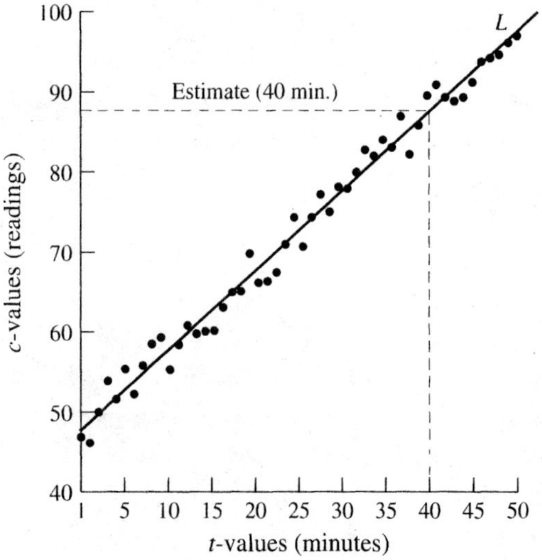

Figure 4.1 50 data points and line of regression.

data shown in Figure 4.1, we find that the equation of L is $c = 1.01t + 46.69$. The equation of L can now be used to estimate the value of c for any given value of t.

The method of least squares is widely used in mathematical modeling and is fundamental to statistical analysis. The method is explained in Section 4.4 together with an application to statistics.

4.1 Dot Product, Norm

We will begin by establishing a simple algebraic condition that tells us when a pair of vectors in $\mathbb{R}^2$ or $\mathbb{R}^3$ are orthogonal. Refer to Figure 4.2(a), which shows nonzero, noncollinear vectors[1] $\mathbf{u} = (u_1, u_2)$ and $\mathbf{v} = (v_1, v_2)$.

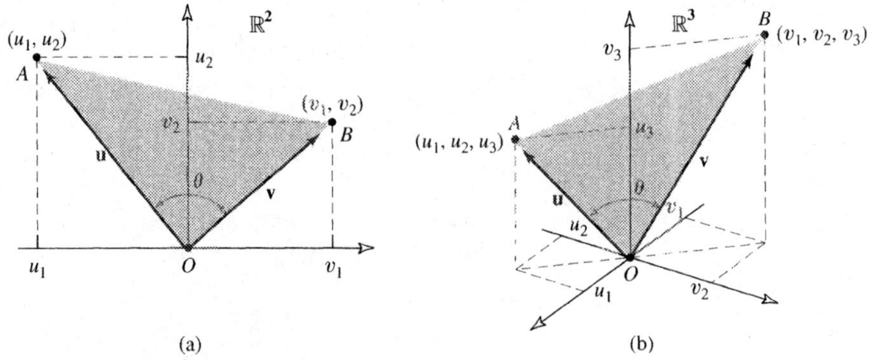

Figure 4.2 Triangles in $\mathbb{R}^2$ and $\mathbb{R}^3$ defined by vectors **u** and **v**.

[1] It will be convenient in this section to use n-tuples to denote n-vectors.

The arrows that represent **u** and **v** define a triangle ΔOAB in the plane. Applying the Pythagorean theorem, the lengths of the arrows representing **u** and **v** are given by

$$|OA| = \sqrt{u_1^2 + u_2^2}, \quad \text{and} \quad |OB| = \sqrt{v_1^2 + v_2^2}. \quad \text{respectively.} \tag{4.1}$$

The Euclidean distance $|AB|$ between the terminal points $A = (u_1, u_2)$ and $B = (v_1, v_2)$ is given by

$$|AB| = \sqrt{(u_1 - v_1)^2 + (u_2 - v_2)^2}. \tag{4.2}$$

Assuming the angle $\theta = 90°$ ($= \pi/2$ radians), the Pythagorean theorem applies in ΔOAB, namely

$$|OA|^2 + |OB|^2 = |AB|^2 \tag{4.3}$$

and using (4.1) and (4.2), we have

$$(u_1^2 + u_2^2) + (v_1^2 + v_2^2) = (u_1 - v_1)^2 + (u_2 - v_2)^2. \tag{4.4}$$

Simplifying the right side of (4.4), canceling squared terms on both sides, and then a common factor -2 gives the equation

$$u_1 v_1 + u_2 v_2 = 0. \tag{4.5}$$

Hence, if the arrows representing **u** and **v** are orthogonal, then condition (4.5) is satisfied. We will see on page 200 that the converse is true, namely, that if condition (4.5) is satisfied, then **u** and **v** are orthogonal. We say that (4.5) is both a *necessary and sufficient condition* for the vectors **u** and **v** to be orthogonal.

Condition (4.5) generalizes to $\mathbb{R}^3$ in a natural way. Refer to Figure 4.2(b) which shows nonzero, noncollinear vectors $\mathbf{u} = (u_1, u_2, u_3)$ and $\mathbf{v} = (v_1, v_2, v_3)$ in $\mathbb{R}^3$. The arrows that represent **u** and **v** define a triangle ΔOAB lying in a plane passing through the origin in $\mathbb{R}^3$. The lengths of the arrows representing **u** and **v** are given, respectively, by

$$|OA| = \sqrt{u_1^2 + u_2^2 + u_3^2}, \quad \text{and} \quad |OB| = \sqrt{v_1^2 + v_2^2 + v_3^2}. \tag{4.6}$$

The Euclidean distance $|AB|$ between the terminal points $A = (u_1, u_2, u_3)$ and $B = (v_1, v_2, v_3)$ is given by

$$|AB| = \sqrt{(u_1 - v_1)^2 + (u_2 - v_2)^2 + (u_3 - v_3)^2}. \tag{4.7}$$

Assuming the angle $\theta = 90°$, the Pythagorean theorem can be applied in ΔOAB as before and we obtain a condition analogous to (4.5), namely

$$u_1 v_1 + u_2 v_2 + u_3 v_3 = 0. \tag{4.8}$$

Once again, condition (4.8) is both necessary and sufficient for **u** and **v** to be orthogonal in $\mathbb{R}^3$.

The algebraic expressions on the left side of equations (4.5) and (4.8) hold the key to much of the theory of this chapter. These expressions are generalized to $\mathbb{R}^n$ in the next definition.

Definition 4.1 *Dot product*

Let $\mathbf{u} = (u_1, u_2, \dots, u_n)$ and $\mathbf{v} = (v_1, v_2, \dots, v_n)$ be vectors in $\mathbb{R}^n$. The number $\mathbf{u}\cdot\mathbf{v}$ (read "**u** dot **v**") defined by the expression

$$\mathbf{u}\cdot\mathbf{v} = u_1v_1 + u_2v_2 + \cdots + u_nv_n \tag{4.9}$$

is called the *dot product*[2], or *scalar product*, of **u** and **v**.

Caution! The dot product of two n-vectors whose components are complex numbers is defined differently (Section 8.5). However, the definition in the complex case agrees with (4.9) when all components are real.

Note that **u** and **v** are (column) vectors in $\mathbb{R}^n$. Using matrix multiplication and omitting square brackets on 1×1 matrices, we have

$$\mathbf{u}\cdot\mathbf{v} = [\,u_1 \;\; u_2 \;\; \cdots \;\; u_n\,]\begin{bmatrix} v_1 \\ v_2 \\ \vdots \\ v_n \end{bmatrix} = [\,v_1 \;\; v_2 \;\; \cdots \;\; v_n\,]\begin{bmatrix} u_1 \\ u_2 \\ \vdots \\ u_n \end{bmatrix},$$

showing that $\mathbf{u}\cdot\mathbf{v} = \mathbf{u}^{\mathsf{T}}\mathbf{v} = \mathbf{v}^{\mathsf{T}}\mathbf{u}$. Notice also that if **A** and **B** are real matrices with **A** conformable to **B** for multiplication, then the (i, j)-entry in **AB** is just the dot product of row i in **A** with column j in **B**.

The proofs of the following fundamental properties of dot product are straightforward and left to Exercises 4.1.

Properties of Dot Product

Let **u**, **v**, **w** be vectors in $\mathbb{R}^n$ and let c be a real scalar.

(**DP1**) $\mathbf{u}\cdot\mathbf{v} = \mathbf{v}\cdot\mathbf{u}$ (Commutativity)
(**DP2**) $\mathbf{u}\cdot(\mathbf{v}+\mathbf{w}) = \mathbf{u}\cdot\mathbf{v} + \mathbf{u}\cdot\mathbf{w}$ (Distributivity)
(**DP3**) $c(\mathbf{u}\cdot\mathbf{v}) = (c\mathbf{u})\cdot\mathbf{v} = \mathbf{u}\cdot(c\mathbf{v})$
(**DP4**) $\mathbf{v}\cdot\mathbf{v} \geq 0$ and $\mathbf{v}\cdot\mathbf{v} = 0$ if and only if $\mathbf{v} = \mathbf{0}$.

Inner Product Spaces

Properties (**DP1**)–(**DP4**) are used as *axioms* to define an abstract concept called an *inner product*. A vector space (Chapter 7) on which an inner product can be defined is called an *inner product space*. The vector space $\mathbb{R}^n$ together with the dot product is an example of an inner product space.

EXAMPLE 1

Using Properties of Dot Product

We wish to expand the expression $(x\mathbf{u}+\mathbf{v})\cdot(x\mathbf{u}+\mathbf{v})$, where **u** and **v** are nonzero vectors, and x is a real scalar. Using the properties noted on the right side of

[2] The term *dot product* agrees with the MATLAB command **dot(u,v)**.

each step, we have

$$\begin{aligned}(x\mathbf{u}+\mathbf{v})\cdot(x\mathbf{u}+\mathbf{v}) &= (x\mathbf{u}+\mathbf{v})\cdot(x\mathbf{u}) + (x\mathbf{u}+\mathbf{v})\cdot\mathbf{v} && \textbf{(DP2)}\\ &= (x\mathbf{u})\cdot(x\mathbf{u}+\mathbf{v}) + \mathbf{v}\cdot(x\mathbf{u}+\mathbf{v}) && \textbf{(DP1)}\\ &= (x\mathbf{u})\cdot(x\mathbf{u}) + (x\mathbf{u})\cdot\mathbf{v} + \mathbf{v}\cdot(x\mathbf{u}) + \mathbf{v}\cdot\mathbf{v} && \textbf{(DP2)}\\ &= (\mathbf{u}\cdot\mathbf{u})x^2 + 2(\mathbf{u}\cdot\mathbf{v})x + \mathbf{v}\cdot\mathbf{v} && \textbf{(DP3)}, \textbf{(DP1)}\end{aligned}$$

Note that the resulting quadratic expression could have been achieved quickly by expanding the product $(x\mathbf{u}+\mathbf{v})\cdot(x\mathbf{u}+\mathbf{v})$ using the laws of algebra. ■

4.1.1 Norm

In engineering and science, a vector is an entity that has both magnitude (length) and direction. In mathematics, the magnitude of an n-vector $\mathbf{v}$, and more generally a matrix $\mathbf{A}$, can be defined in a number of ways, each of which is pertinent to specific needs. In mathematics, we often use the more general term *norm* instead of *magnitude*. For example, the norm of a vector $\mathbf{v}$ in $\mathbb{R}^2$ or $\mathbb{R}^3$ may be defined to be the Euclidean length of the arrow that represents $\mathbf{v}$, as given in (4.1) and (4.6). Other possibilities for defining the norm of a vector are mentioned in Exercises 4.1.

Definition 4.2

Euclidean norm

Let $\mathbf{v} = (v_1, v_2, \ldots, v_n)$ be a vector in $\mathbb{R}^n$. The *Euclidean norm* or 2-norm of $\mathbf{v}$ is the nonnegative real number $\|\mathbf{v}\|$ given by

$$\|\mathbf{v}\| = \sqrt{v_1^2 + v_2^2 + \cdots + v_n^2}. \tag{4.10}$$

When necessary, the symbol $\|\mathbf{v}\|_2$ will be used to denote the Euclidean norm.

Observe that $\|\mathbf{v}\| = \sqrt{\mathbf{v}\cdot\mathbf{v}}$ and $\|\mathbf{v}\|^2 = \mathbf{v}\cdot\mathbf{v}$.

EXAMPLE 2

Euclidean Norms in $\mathbb{R}^4$

For the 4-vector $\mathbf{v} = (3, -1, 2, 0)$ in $\mathbb{R}^4$, we have

$$\|\mathbf{v}\| = \sqrt{3^2 + (-1)^2 + 2^2 + 0^2} = \sqrt{14} \simeq 3.7417 \text{ units}.$$

The standard basis for $\mathbb{R}^4$ is set $\{\mathbf{e}_1, \mathbf{e}_2, \mathbf{e}_3, \mathbf{e}_4\}$ of columns of the identity matrix $\mathbf{I}_4$. We have $\|\mathbf{e}_k\| = 1$, for $1 \le k \le 4$, and we say that the basis vectors have *unit norm*. ■

The proofs of the following fundamental properties of norm are straightforward and left to Exercises 4.1. A third property will be added shortly.

Properties of Euclidean Norm

Let **u** and **v** be vectors in $\mathbb{R}^n$ and let c be a real scalar.

(**N1**) $\|\mathbf{v}\| \geq 0$ and $\|\mathbf{v}\| = 0$ if and only if $\mathbf{v} = \mathbf{0}$.
(**N2**) $\|c\mathbf{v}\| = |c|\ \|\mathbf{v}\|$.

For example, $\|-\mathbf{v}\| = |-1|\ \|\mathbf{v}\| = \|\mathbf{v}\|$, using (**N2**).

The next theorem is a classic result that connects the concepts of dot product and norm, and gives an *upper bound* on the absolute value of the dot product.

Theorem 4.1 **Cauchy–Schwarz Inequality**

Let **u** *and* **v** *be vectors in* $\mathbb{R}^n$. *Then*

$$|\mathbf{u}\cdot\mathbf{v}| \leq \|\mathbf{u}\|\ \|\mathbf{v}\|, \tag{4.11}$$

where $|\mathbf{u}\cdot\mathbf{v}|$ *denotes absolute value of the dot product.*

Proof There are two cases to consider.

(a) If $\mathbf{u} = \mathbf{0}$, then for any **v** we have

$$|\mathbf{0}\cdot\mathbf{v}| = 0 = \|\mathbf{0}\|\ \|\mathbf{v}\|,$$

showing that (4.11) is true in this case.

(b) If **u** is nonzero, consider the vector $\mathbf{w} = x\mathbf{u} + \mathbf{v}$, where x is a real scalar. Note that $\|\mathbf{w}\|^2 = \mathbf{w}\cdot\mathbf{w} \geq 0$ for all x. Using Example 1, we have

$$\begin{aligned} \|\mathbf{w}\|^2 = \|x\mathbf{u}+\mathbf{v}\|^2 &= (x\mathbf{u}+\mathbf{v})\cdot(x\mathbf{u}+\mathbf{v}) \\ &= \|\mathbf{u}\|^2x^2 + 2(\mathbf{u}\cdot\mathbf{v})x + \|\mathbf{v}\|^2 \geq 0. \end{aligned} \tag{4.12}$$

Let $a = \|\mathbf{u}\|^2$, $b = 2\mathbf{u}\cdot\mathbf{v}$ and $c = \|\mathbf{v}\|^2$. Now $a \neq 0$ and so the left side of (4.12) is the quadratic $ax^2 + bx + c$ that cannot have two distinct real roots because it is nonnegative for all x. Thus, the discriminant of $ax^2 + bx + c$, namely $b^2 - 4ac$, satisfies $b^2 - 4ac \leq 0$. Hence

$$b^2 - 4ac = 4(\mathbf{u}\cdot\mathbf{v})^2 - 4\|\mathbf{u}\|^2\|\mathbf{v}\|^2 \leq 0 \quad \Rightarrow \quad |\mathbf{u}\cdot\mathbf{v}|^2 \leq \|\mathbf{u}\|^2\ \|\mathbf{v}\|^2,$$

and we take positive square roots in the last inequality to obtain the result. ■

Refer to the figure opposite. A plane triangle OAC in $\mathbb{R}^2$ or $\mathbb{R}^3$ is defined by nonzero, noncollinear vectors **u** and **v**. Using geometrical arguments, we have

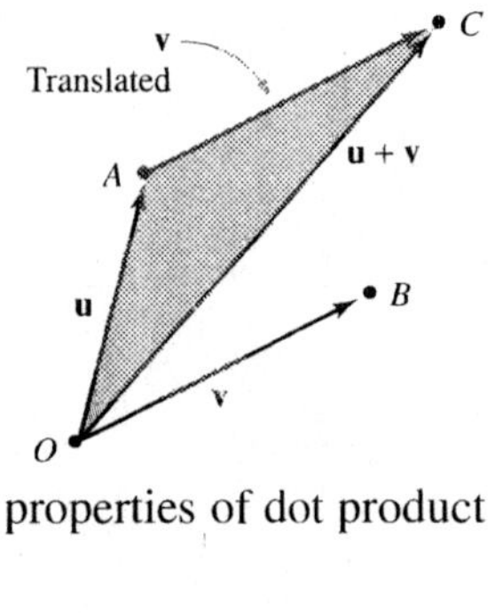

$$|OC| \leq |OA| + |AC|,$$

and so

$$\|\mathbf{u}+\mathbf{v}\| \leq \|\mathbf{u}\| + \|\mathbf{v}\|.$$

The latter inequality is also true in $\mathbb{R}^n$. The proof uses properties of dot product and norm.

(**N3**) For all vectors $\mathbf{u}$ and $\mathbf{v}$ in $\mathbb{R}^n$, we have

$$\|\mathbf{u}+\mathbf{v}\| \leq \|\mathbf{u}\| + \|\mathbf{v}\| \qquad \text{(The triangle inequality)}$$

Proof Setting $x = 1$ in Example 1, we have

$$\begin{aligned}
\|\mathbf{u}+\mathbf{v}\|^2 &= (\mathbf{u}+\mathbf{v})\cdot(\mathbf{u}+\mathbf{v}) \\
&= \mathbf{u}\cdot\mathbf{u} + 2(\mathbf{u}\cdot\mathbf{v}) + \mathbf{v}\cdot\mathbf{v} \\
&\leq \|\mathbf{u}\|^2 + 2|\mathbf{u}\cdot\mathbf{v}| + \|\mathbf{v}\|^2 \quad [\text{because } \mathbf{u}\cdot\mathbf{v} \leq |\mathbf{u}\cdot\mathbf{v}|] \\
&\leq \|\mathbf{u}\|^2 + 2\|\mathbf{u}\|\,\|\mathbf{v}\| + \mathbf{v}\cdot\mathbf{v} \quad [\text{using Cauchy–Schwarz}] \\
&= (\|\mathbf{u}\| + \|\mathbf{v}\|)^2,
\end{aligned}$$

and the result follows by taking positive square roots.

—

Normed Linear Spaces

A *norm* on a vector space $\mathcal{V}$ is a function that assigns to every object $\mathbf{v}$ in $\mathcal{V}$ a real number $\|\mathbf{v}\|$, called the *norm* of $\mathbf{v}$, such that properties (**N1**), (**N2**), (**N3**) are satisfied. We then call $\mathcal{V}$ a *normed linear space.* $\mathbb{R}^n$ with the Euclidean norm is an example of a normed linear space.

Distance between Two Vectors

The distance $d(p, q)$ between two points p and q on the real axis $\mathbb{R}$ is given by the absolute value $d(p, q) = |p - q|$. When $p = -2$ and $q = 3$, for example, we have $d(-2, 3) = |-2-3| = |-5| = 5$ units. By analogy, we define the distance $d(\mathbf{u}, \mathbf{v})$ between vectors $\mathbf{u}$ and $\mathbf{v}$ in $\mathbb{R}^2$ to be the Euclidean distance between the terminal points of the arrows that represent $\mathbf{u}$ and $\mathbf{v}$, as in (4.2). Equation (4.7) is the analogous statement for $\mathbb{R}^3$. Both cases can be combined into the single expression $|AB| = d(\mathbf{u}, \mathbf{v}) = \|\mathbf{u}-\mathbf{v}\|$, which is now generalized to $\mathbb{R}^n$.

Definition 4.3

Euclidean distance between two vectors in $\mathbb{R}^n$

Let $\mathbf{u} = (u_1, u_2, \ldots, u_n)$ and $\mathbf{v} = (v_1, v_2, \ldots, v_n)$ be vectors in $\mathbb{R}^n$. The Euclidean *distance* between $\mathbf{u}$ and $\mathbf{v}$ is the scalar $d(\mathbf{u}, \mathbf{v})$ given by

$$d(\mathbf{u}, \mathbf{v}) = \|\mathbf{u}-\mathbf{v}\| = \sqrt{(u_1 - v_1)^2 + (u_2 - v_2)^2 + \cdots + (u_n - v_n)^2}. \tag{4.13}$$

Angle between Two Vectors

Refer to Figures 4.2(a) and (b). In both cases, the nonzero, noncollinear vectors $\mathbf{u}$ and $\mathbf{v}$ define a plane triangle $\triangle OAB$ in which the included angle $\angle AOB = \theta$.

The *law of cosines* in $\triangle OAB$, which is proved using the Pythagorean theorem, states that

$$|AB|^2 = |OA|^2 + |OB|^2 - 2|OA||OB|\cos\theta. \tag{4.14}$$

Using the Euclidean norm, (4.14) becomes

$$\| \mathbf{u} - \mathbf{v} \|^2 = \| \mathbf{u} \|^2 + \| \mathbf{v} \|^2 - 2\| \mathbf{u} \| \; \| \mathbf{v} \| \cos\theta,$$

and expanding $\| \mathbf{u} - \mathbf{v} \|^2$ as in Example 1, we have

$$\| \mathbf{u} \|^2 - 2(\mathbf{u} \cdot \mathbf{v}) + \| \mathbf{v} \|^2 = \| \mathbf{u} \|^2 + \| \mathbf{v} \|^2 - 2\| \mathbf{u} \| \; \| \mathbf{v} \| \cos\theta.$$

Canceling squared terms and a factor -2 gives a formula for $\cos\theta$ in terms of dot product and norms, namely

$$\cos\theta = \frac{\mathbf{u} \cdot \mathbf{v}}{\| \mathbf{u} \| \; \| \mathbf{v} \|}, \qquad (4.15)$$

noting that the denominator in (4.15) is nonzero. The necessary and sufficient condition for orthogonality of two nonzero vectors in $\mathbb{R}^2$ or $\mathbb{R}^3$ that was mentioned at the start of the section now follows. First, if $\theta = 90°$, then $\cos\theta = 0$ and (4.15) implies that $\mathbf{u} \cdot \mathbf{v} = 0$, as in (4.5) and (4.8). Second, if $\mathbf{u} \cdot \mathbf{v} = 0$, then (4.15) implies that $\cos\theta = 0$, and so $\theta = 90°$.

Our next goal is to extend formula (4.15) so that it applies to *any* vectors $\mathbf{u}$ and $\mathbf{v}$ in $\mathbb{R}^n$. Recall that if p and $q > 0$ are real numbers such that $|p| \leq q$, then

$$|p| \leq q \quad \Leftrightarrow \quad -q \leq p \leq q \quad \Leftrightarrow -1 \leq \frac{p}{q} \leq 1. \qquad (4.16)$$

Suppose $\mathbf{u}$ and $\mathbf{v}$ are nonzero vectors in $\mathbb{R}^n$ as shown in Figure 4.3. Let $p = \mathbf{u} \cdot \mathbf{v}$ and $q = \| \mathbf{u} \| \; \| \mathbf{v} \|$. Using (4.16), the Cauchy–Schwarz inequality (4.11) becomes

$$-1 \leq \frac{\mathbf{u} \cdot \mathbf{v}}{\| \mathbf{u} \| \; \| \mathbf{v} \|} \leq 1.$$

Refer to Figure 4.4, which shows the graph $y = \cos\theta$ for values of the angle[3] θ in the interval $[0, \pi]$ (endpoints included).

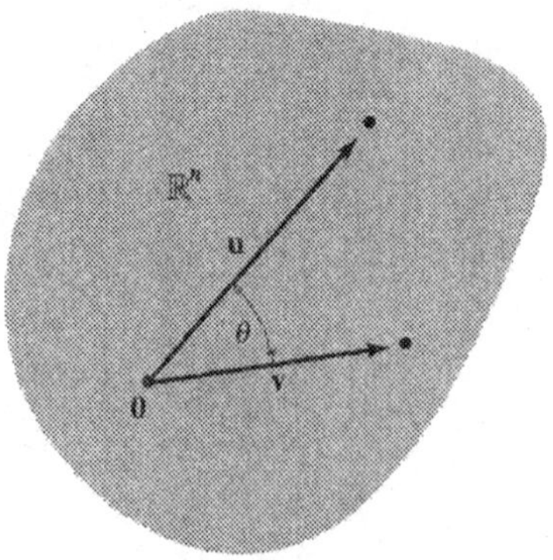

Figure 4.3 The angle θ between $\mathbf{u}$ and $\mathbf{v}$ in $\mathbb{R}^n$.

[3] Recall that an angle can be measured in either radians (*circular measure*) or degrees (*sexagesimal measure*). The conversion between radians and degrees is: π radians $= 180°$; that is, 1 radian $\simeq 57.2958°$ and $1° \simeq 0.0175$ radians.

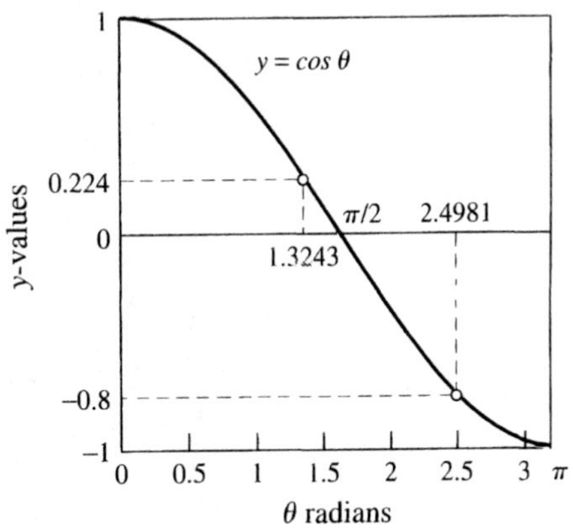

Figure 4.4 Graph of $y = \cos\theta$ on the interval $[0, \pi]$.

For each value $y = \dfrac{\mathbf{u}\cdot\mathbf{v}}{\|\mathbf{u}\|\,\|\mathbf{v}\|}$ in the interval $[-1, 1]$ (endpoints included) on the y-axis there is a *unique* angle θ in the interval $[0, \pi]$ such that $y = \cos\theta$. For example, $\cos(1.3243) \simeq 0.224$ and $\cos(2.4981) \simeq -0.8$. Note that $0 < y < 1$ implies that θ is acute, and $-1 < y < 0$ implies that θ is obtuse. At the extremes, $\cos\theta = 1$ implies $\theta = 0$, and so $\mathbf{u}$ and $\mathbf{v}$ are parallel (in the same direction), and $\cos\theta = -1$ implies $\theta = \pi$ radians, and so $\mathbf{u}$ and $\mathbf{v}$ are parallel (in opposite directions).

Definition 4.4

Angle between two vectors

Let $\mathbf{u}$ and $\mathbf{v}$ be nonzero vectors in $\mathbb{R}^n$. The *angle* between $\mathbf{u}$ and $\mathbf{v}$ is the angle θ, $0 \le \theta \le \pi$, which satisfies the equation

$$\cos\theta = \frac{\mathbf{u}\cdot\mathbf{v}}{\|\mathbf{u}\|\,\|\mathbf{v}\|}. \tag{4.17}$$

Thus (4.17) extends formula (4.15) to $\mathbb{R}^n$. Rearranging (4.17) gives the *trigonometric representation* of dot product, namely

$$\mathbf{u}\cdot\mathbf{v} = \|\mathbf{u}\|\,\|\mathbf{v}\|\cos\theta. \tag{4.18}$$

EXAMPLE 3

Angle between Two Vectors in $\mathbb{R}^4$

Let $\mathbf{u} = (-1, 2, 3, 0)$ and $\mathbf{v} = (2, -1, 3, 4)$. Using (4.17), we have

$$\cos\theta = \frac{\mathbf{u}\cdot\mathbf{v}}{\|\mathbf{u}\|\,\|\mathbf{v}\|} = \frac{5}{\sqrt{14}\sqrt{30}} \simeq 0.2440$$

so that $\theta = \arccos(0.2440) \simeq 1.3243$ radians, or equivalently, $\simeq 75.8787^\circ$. ■

Orthogonality

From equation (4.17), we have

$$\theta = \pi/2\ (= 90^\circ) \quad \Leftrightarrow \quad \cos\theta = 0 \quad \Leftrightarrow \quad \mathbf{u}\cdot\mathbf{v} = 0,$$

and this observation leads to a definition of orthogonality in $\mathbb{R}^n$.

Definition 4.5

Orthogonality

The vectors $\mathbf{u}$ and $\mathbf{v}$ in $\mathbb{R}^n$ are called *orthogonal* if $\mathbf{u}\cdot\mathbf{v} = 0$ and we write $\mathbf{u} \perp \mathbf{v}$ when this is the case.

Note that if at least one of $\mathbf{u}$, $\mathbf{v}$ is the zero vector $\mathbf{0}$, then $\mathbf{u}\cdot\mathbf{v} = 0$. Hence, $\mathbf{0} \perp \mathbf{v}$ for every $\mathbf{v}$ in $\mathbb{R}^n$.

Unit Vectors

A vector $\mathbf{u}$ is called a *unit vector* if $\|\mathbf{u}\| = 1$. If $\mathbf{v}$ is any nonzero vector then the vector $\mathbf{u}$ given by

$$\mathbf{u} = \frac{\mathbf{v}}{\|\mathbf{v}\|} \tag{4.19}$$

is a unit vector in the direction of $\mathbf{v}$. To *normalize* a nonzero vector $\mathbf{v}$ means to compute $\mathbf{u}$ as in (4.19).

EXAMPLE 4

Orthogonal Vectors, Unit Vectors

Consider the vectors $\mathbf{v} = (-1, 2, 3, 1)$ and $\mathbf{w} = (4, -5, 6, -4)$ in $\mathbb{R}^4$. Then $\mathbf{v}\cdot\mathbf{w} = 0$ and so $\mathbf{v} \perp \mathbf{w}$. Normalizing $\mathbf{v}$ will provide a unit vector $\mathbf{u}$ in the direction of $\mathbf{v}$. We have $\|\mathbf{v}\| = \sqrt{15}$, and so

$$\mathbf{u} = \frac{\mathbf{v}}{\sqrt{15}} = \frac{1}{\sqrt{15}}(-1, 2, 3, 1) \simeq (-0.26, 0.52, 0.77, 0.26).$$

The standard basis $\mathcal{S} = \{\mathbf{e}_1, \mathbf{e}_2, \mathbf{e}_3, \mathbf{e}_4\}$ for $\mathbb{R}^4$ has the property that $\mathbf{e}_i \perp \mathbf{e}_j$, for each pair of distinct integers i and j with $1 \leq i, j \leq 4$. The vectors in $\mathcal{S}$ are said to be *mutually orthogonal*. Also, $\|\mathbf{e}_i\| = 1$ for each i, showing that all the vectors in $\mathcal{S}$ have unit norm. ■

The classical Pythagorean theorem applies to a right-angled plane triangle OAB. See Figures 4.2(a) and (b) with $\theta = 90^\circ$. The same result applies more generally in $\mathbb{R}^n$.

Theorem 4.2

The Pythagorean Theorem

Let $\mathbf{u}$ *and* $\mathbf{v}$ *be vectors in* $\mathbb{R}^n$. *Then*

$$\|\mathbf{u} + \mathbf{v}\|^2 = \|\mathbf{u}\|^2 + \|\mathbf{v}\|^2 \tag{4.20}$$

if and only if $\mathbf{u}$ *and* $\mathbf{v}$ *are orthogonal.*

Proof Assume **u** and **v** are nonzero and set $x = 1$ in Example 1. We have

$$\|\mathbf{u}+\mathbf{v}\|^2 = \|\mathbf{u}\|^2 + 2(\mathbf{u}\cdot\mathbf{v}) + \|\mathbf{v}\|^2.$$

Hence, (4.20) if and only if $\mathbf{u}\cdot\mathbf{v} = 0$. The result is clearly true when **u** or **v** (or both) are the zero vector.

■

Equations of Lines and Planes, Normal Form

The dot product provides the means to express the equations of lines and planes in an economical way. Refer to Figure 4.5(a). A line L in $\mathbb{R}^2$ is defined uniquely by specifying a point (p_1, p_2) on L and a vector $\mathbf{n} = (a, b)$ orthogonal to L. We call **n** a *normal vector*. Let $\mathbf{x} = (x_1, x_2)$ be any vector whose terminal point (x_1, x_2) lies on L. The vector $\mathbf{x} - \mathbf{p} = (x_1 - p_1, x_2 - p_2)$ is parallel to L and orthogonal to **n**. Hence, the condition for (x_1, x_2) to lie on L is

$$\mathbf{n}\cdot(\mathbf{x}-\mathbf{p}) = 0 \quad \text{or} \quad (a, b)\cdot(x_1 - p_1, x_2 - p_2) = 0 \tag{4.21}$$

and either of these equations is called the *normal form* for the equation of L. Note from (4.21) that the Cartesian equation for L is $ax_1 + bx_2 = d$, where $d = ap_1 + bp_2$. For example, when $\mathbf{n} = (-1, 3)$ and $\mathbf{p} = (4, 5)$, the equation of L is $-x_1 + 3x_2 = 11$.

The equation $ax_1 + bx_2 = d$ defines a line L in $\mathbb{R}^2$. Such an equation can be written in the normal form (4.21) by taking $\mathbf{n} = (a, b)$ and, assuming $a \neq 0$, $\mathbf{p} = (d/a, 0)$.

Refer to Figure 4.5(b). A plane P in $\mathbb{R}^3$ is defined uniquely by specifying a point (p_1, p_2, p_3) on P and a vector $\mathbf{n} = (a, b, c)$ orthogonal to P. Let $\mathbf{x} = (x_1, x_2, x_3)$ be any vector whose terminal point (x_1, x_2, x_3) lies on P. Then the vector $\mathbf{x} - \mathbf{p} = (x_1 - p_1, x_2 - p_2, x_3 - p_3)$ is parallel to P and orthogonal to **n**. Hence, the condition for (x_1, x_2, x_3) to lie on P is

$$\mathbf{n}\cdot(\mathbf{x}-\mathbf{p}) = 0 \quad \text{or} \quad (a, b, c)\cdot(x_1 - p_1, x_2 - p_2, x_3 - p_3) = 0 \tag{4.22}$$

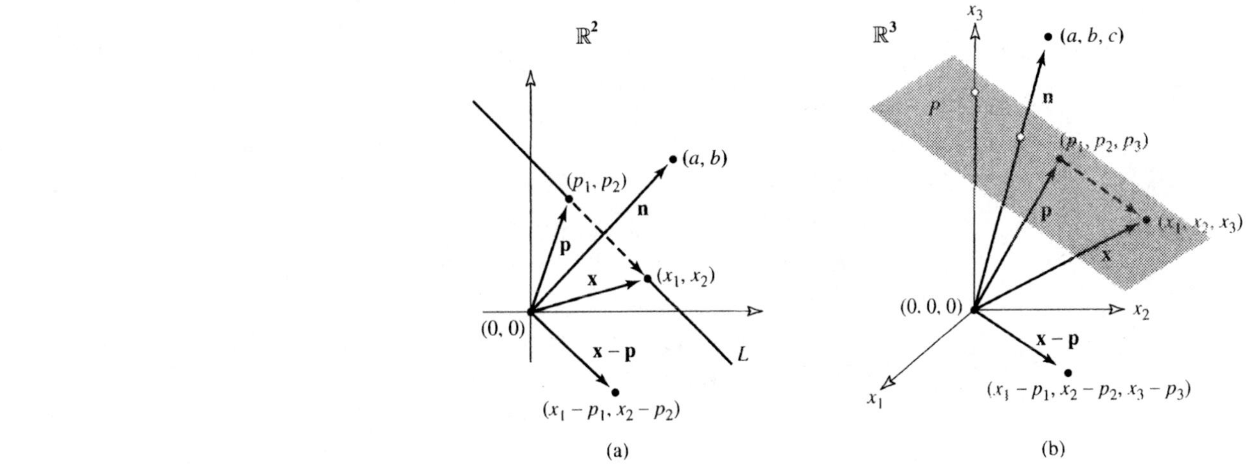

Figure 4.5 Lines in $\mathbb{R}^2$, planes in $\mathbb{R}^3$.

and either of these equations is called the *normal form* for the equation of P. From (4.22) the Cartesian equation of P is $ax_1 + bx_2 + cx_3 = d$, where $d = ap_1 + bp_2 + cp_3$. For example, when $\mathbf{n} = (2, -3, 4)$ and $\mathbf{p} = (1, 1, 2)$, the equation of P is $2x_1 - 3x_2 + 4x_3 = 7$.

The equation $ax_1 + bx_2 + cx_3 = d$ defines a plane P in $\mathbb{R}^3$. Such an equation can be written in the normal form (4.22) by taking $\mathbf{n} = (a, b, c)$ and, assuming $a \neq 0$, $\mathbf{p} = (d/a, 0, 0)$.

EXAMPLE 5

Finding the Equation of a Plane

We wish to find the equation of the plane P passing through the (noncollinear) points in $\mathbb{R}^3$ that are the terminal points of the vectors

$$\mathbf{p}_1 = (3, 0, -1), \quad \mathbf{p}_2 = (1, 1, -1), \quad \mathbf{p}_3 = (2, 2, 2).$$

Then

$$\mathbf{u} = \mathbf{p}_1 - \mathbf{p}_2 = (3, 0, -1) - (1, 1, -1) = (2, -1, 0)$$

$$\mathbf{v} = \mathbf{p}_1 - \mathbf{p}_3 = (3, 0, -1) - (2, 2, 2) = (1, -2, -3),$$

and $\mathbf{u}$, $\mathbf{v}$ are parallel to P. Let $\mathbf{n} = (a, b, c)$ be a normal to P. Then $\mathbf{n}\cdot\mathbf{u} = 0$ and $\mathbf{n}\cdot\mathbf{v} = 0$ and we obtain a 2×3 homogeneous linear system (S) that can be solved for a, b, c.

$$(S) \quad \begin{cases} 2a - b = 0 \\ a - 2b - 3c = 0 \end{cases} \quad \Rightarrow \quad (a, b, c) = t(-1, -2, 1),$$
where t is a real parameter.

Choosing $t = -1$ gives $\mathbf{n} = (1, 2, -1)$. Letting $\mathbf{x} = (x_1, x_2, x_3)$ and using $\mathbf{p} = (3, 0, -1)$, the equation of P in normal form, as given by (4.22), is

$$(1, 2, -1)\cdot(x_1 - 3, x_2, x_3 + 1) = 0,$$

and the Cartesian form is $x_1 + 2x_2 - x_3 = -4$. ■

Projection and Reflection

Consider a nonzero vector $\mathbf{v}$ in $\mathbb{R}^n$ and a one-dimensional subspace $\mathcal{U}$ (a line) with basis $\{\mathbf{b}\}$, where $\mathbf{b}$ is a unit vector not in the direction of $\mathbf{v}$. Figure 4.6 shows the case of $\mathbb{R}^2$ in which $\mathcal{U}$ is represented by a line L passing through the origin. We see (geometrically) that there is a value of s such that the vectors $\mathbf{v} - s\mathbf{b}$ and $\mathbf{b}$ are orthogonal. The terminal point P of $s\mathbf{b}$ is the point on L closest to A, lying at the foot of the perpendicular from A to L. The value of s is determined as follows:

$$(\mathbf{v} - s\mathbf{b})\cdot\mathbf{b} = \mathbf{v}\cdot\mathbf{b} - (s\mathbf{b})\cdot\mathbf{b} = \mathbf{v}\cdot\mathbf{b} - s(\mathbf{b}\cdot\mathbf{b}) = \mathbf{v}\cdot\mathbf{b} - s\|\,\mathbf{b}\,\|^2 = \mathbf{v}\cdot\mathbf{b} - s = 0,$$

because $\mathbf{b}$ is a unit vector, and so $s = \mathbf{v}\cdot\mathbf{b}$.

The vector $\text{proj}_{\mathcal{U}}(\mathbf{v}) = (\mathbf{v}\cdot\mathbf{b})\mathbf{b}$ is called the *orthogonal projection* of $\mathbf{v}$ onto $\mathcal{U}$ and $\text{comp}_{\mathcal{U}}(\mathbf{v}) = \mathbf{v} - (\mathbf{v}\cdot\mathbf{b})\mathbf{b}$ is called the *component of* $\mathbf{v}$ *orthogonal to* $\mathcal{U}$.

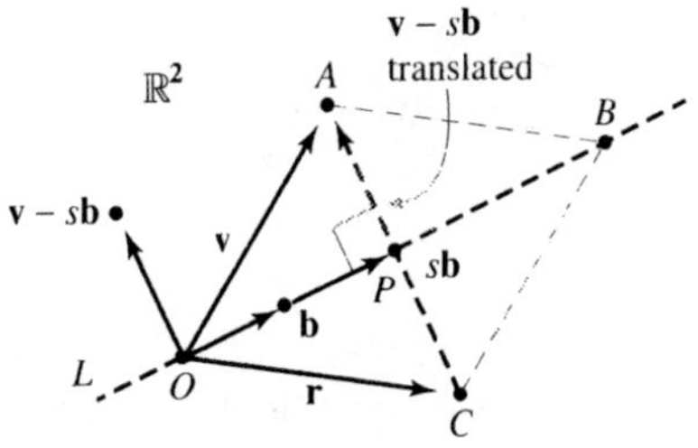

Figure 4.6 Projection and reflection.

The point C in Figure 4.6 is the reflection of A in the line L. In parallelogram $OABC$, we have $\overrightarrow{OC} = \overrightarrow{OB} + \overrightarrow{BC} = 2\overrightarrow{OP} - \overrightarrow{OA}$ and the vector $\mathbf{r}$ with terminal point C is therefore given by $\mathbf{r} = \text{refl}_{\mathcal{U}}(\mathbf{v}) = 2(\mathbf{v} \cdot \mathbf{b})\mathbf{b} - \mathbf{v}$. We call $\text{refl}_{\mathcal{U}}(\mathbf{v})$ the *reflection of* $\mathbf{v}$ *in* $\mathcal{U}$.

The development just given for $\mathbb{R}^2$, using geometric arguments, forms the basis of the next definition which applies to any one-dimensional subspace $\mathcal{U}$ in $\mathbb{R}^n$. We will modify the preceding theory slightly to gain greater generality—if $\mathbf{u}$ is any nonzero vector in $\mathcal{U}$, then $\mathbf{b} = \mathbf{u}/\|\mathbf{u}\|^2$ is a unit basis vector in $\mathcal{U}$ in the direction of $\mathbf{u}$.

Definition 4.6

Orthogonal projection, component, reflection

Let $\mathbf{v}$ be any nonzero vector in $\mathbb{R}^n$ and let $\mathcal{U}$ be a one-dimensional subspace of $\mathbb{R}^n$ with basis $\{\mathbf{u}\}$. The *projection*, *component*, and *reflection* of $\mathbf{v}$ with respect to $\mathcal{U}$ are given by

Projection of $\mathbf{v}$ onto $\mathcal{U}$
$$\text{proj}_{\mathcal{U}}(\mathbf{v}) = \frac{\mathbf{v} \cdot \mathbf{u}}{\|\mathbf{u}\|^2}\mathbf{u}, \tag{4.23}$$

Component of $\mathbf{v}$ orthogonal to $\mathcal{U}$
$$\text{comp}_{\mathcal{U}}(\mathbf{v}) = \mathbf{v} - \frac{\mathbf{v} \cdot \mathbf{u}}{\|\mathbf{u}\|^2}\mathbf{u}, \tag{4.24}$$

Reflection of $\mathbf{v}$ in $\mathcal{U}$
$$\text{refl}_{\mathcal{U}}(\mathbf{v}) = 2\frac{(\mathbf{v} \cdot \mathbf{u})}{\|\mathbf{u}\|^2}\mathbf{u} - \mathbf{v}. \tag{4.25}$$

EXAMPLE 6

Projection, Reflection in $\mathbb{R}^4$

Let $\mathbf{v} = (1, 2, 0, -1)$ and let $\mathcal{U} = \text{span}\{\mathbf{u}\}$, where $\mathbf{u} = (1, 0, 1, -1)$. Then $\mathbf{u} \cdot \mathbf{v} = 2$ and $\|\mathbf{u}\| = \sqrt{3}$. Using (4.23), we have

$$\text{proj}_{\mathcal{U}}(\mathbf{v}) = \frac{\mathbf{v} \cdot \mathbf{u}}{\|\mathbf{u}\|^2}\,\mathbf{u} = \frac{2}{(\sqrt{3})^2}(1, 0, 1, -1) = \frac{2}{3}(1, 0, 1, -1).$$

Also,

$$\text{comp}_{\mathcal{U}}(\mathbf{v}) = (1, 2, 0, -1) - \frac{2}{3}(1, 0, 1, -1) = \frac{1}{3}(1, 6, -2, -1)$$

and

$$\text{refl}_{\mathcal{U}}(\mathbf{v}) = 2\frac{(\mathbf{v}\cdot\mathbf{u})}{\|\mathbf{u}\|^2}\mathbf{u} - \mathbf{v} = \frac{4}{3}(1,0,1,-1) - (1,2,0,-1) = \frac{1}{3}(1,-6,4,-1)$$

▪

Projection and Reflection Matrices

Each of the equations (4.23), (4.24), and (4.25) defines a transformation on $\mathbb{R}^n$. We will now show that there exist $n \times n$ matrices **P**, **C**, and **R** such that $\text{proj}_{\mathcal{U}}(\mathbf{v}) = \mathbf{Pv}$, $\text{comp}_{\mathcal{U}}(\mathbf{v}) = \mathbf{Cv}$, and $\text{refl}_{\mathcal{U}}(\mathbf{v}) = \mathbf{Rv}$, and consequently each of these transformations is linear (Section 3.4), being defined by matrix-column vector multiplication. Consider (4.23) first. It is easy to show (Exercises 4.1) that $(\mathbf{v}\cdot\mathbf{u})\mathbf{u} = \mathbf{u}\mathbf{u}^{\mathsf{T}}\mathbf{v}$, where $\mathbf{u}\mathbf{u}^{\mathsf{T}}$ is $n \times n$ and symmetric. Noting that $\|\mathbf{u}\|^2 = \mathbf{u}^{\mathsf{T}}\mathbf{u}$, equation (4.23) becomes

$$\text{proj}_{\mathcal{U}}(\mathbf{v}) = \frac{\mathbf{u}\mathbf{u}^{\mathsf{T}}}{\mathbf{u}^{\mathsf{T}}\mathbf{u}}\mathbf{v}, \quad \Rightarrow \quad \mathbf{P} = \frac{\mathbf{u}\mathbf{u}^{\mathsf{T}}}{\mathbf{u}^{\mathsf{T}}\mathbf{u}}. \tag{4.26}$$

The matrix **P** on the right side of (4.26) is called a *projection matrix*. Using the fact that $\mathbf{v} = \mathbf{I}_n\mathbf{v}$, the two remaining transformations are defined by the matrices **C** and **R**, as follows:

$$\text{comp}_{\mathcal{U}}(\mathbf{v}) = \left(\mathbf{I}_n - \frac{\mathbf{u}\mathbf{u}^{\mathsf{T}}}{\mathbf{u}^{\mathsf{T}}\mathbf{u}}\right)\mathbf{v}, \quad \Rightarrow \quad \mathbf{C} = \mathbf{I}_n - \frac{\mathbf{u}\mathbf{u}^{\mathsf{T}}}{\mathbf{u}^{\mathsf{T}}\mathbf{u}} \tag{4.27}$$

$$\text{refl}_{\mathcal{U}}(\mathbf{v}) = \left(2\frac{\mathbf{u}\mathbf{u}^{\mathsf{T}}}{\mathbf{u}^{\mathsf{T}}\mathbf{u}} - \mathbf{I}_n\right)\mathbf{v} \quad \Rightarrow \quad \mathbf{R} = 2\frac{\mathbf{u}\mathbf{u}^{\mathsf{T}}}{\mathbf{u}^{\mathsf{T}}\mathbf{u}} - \mathbf{I}_n \tag{4.28}$$

The matrix **R** given on the right side of (4.28) is called a *reflection matrix*.

EXAMPLE 7

Projection and Reflection Matrices

We wish to find the orthogonal projection of $\mathbf{v} = (5, 3)$ onto the one-dimensional subspace $\mathcal{U}$ of $\mathbb{R}^2$ with basis $\mathcal{B} = \{\mathbf{u}\}$, where $\mathbf{u} = (2, -1)$. Using (4.26), we have

$$\begin{aligned}\text{proj}_{\mathcal{U}}(\mathbf{v}) &= \frac{\mathbf{u}\mathbf{u}^{\mathsf{T}}}{\mathbf{u}^{\mathsf{T}}\mathbf{u}}\mathbf{v} = 0.2\begin{bmatrix} 2 \\ -1 \end{bmatrix}[2 \ \ -1]\begin{bmatrix} 5 \\ 3 \end{bmatrix} \\ &= 0.2\begin{bmatrix} 4 & -2 \\ -2 & 1 \end{bmatrix}\begin{bmatrix} 5 \\ 3 \end{bmatrix} = \begin{bmatrix} 2.8 \\ -1.4 \end{bmatrix}\end{aligned}$$

and using (4.28), we have

$$\text{refl}_{\mathcal{U}}(\mathbf{v}) = \left(2\frac{\mathbf{u}\mathbf{u}^{\mathsf{T}}}{\mathbf{u}^{\mathsf{T}}\mathbf{u}} - \mathbf{I}_2\right)\mathbf{v} = 0.2\begin{bmatrix} 3 & -4 \\ -4 & -3 \end{bmatrix}\begin{bmatrix} 5 \\ 3 \end{bmatrix} = \begin{bmatrix} 0.6 \\ -5.8 \end{bmatrix}.$$

The component of $\mathbf{v}$ orthogonal to $\mathcal{U}$ is

$$\text{comp}_{\mathcal{U}}(\mathbf{v}) = \mathbf{v} - \text{proj}_{\mathcal{U}}(\mathbf{v}) = \begin{bmatrix} 5 \\ 3 \end{bmatrix} - \begin{bmatrix} 2.8 \\ -1.4 \end{bmatrix} = \begin{bmatrix} 2.2 \\ 4.4 \end{bmatrix}.$$ ■

Looking Ahead

It follows from (4.26) that the orthogonal projection of a vector $\mathbf{v}$ onto a subspace $\mathcal{U}$ with basis $\{\mathbf{u}\}$, where $\mathbf{u}$ is a unit vector is $\mathbf{Pv}$, where $\mathbf{P} = \mathbf{uu}^\mathsf{T}$. We will see in Section 4.3 that the projection of $\mathbf{v}$ onto an arbitrary subspace $\mathcal{U}$ of $\mathbb{R}^n$ is given by $\mathbf{Pv}$, where $\mathbf{P} = \mathbf{QQ}^\mathsf{T}$, where the columns of $\mathbf{Q}$ are formed using an orthonormal basis for $\mathcal{U}$.

Work

Refer to the diagram, in which a constant force $\mathbf{f}$ acts on a particle P as P moves along the vector $\mathbf{d}$, where $0 \le \theta \le 180°$. The *work done* by $\mathbf{f}$ on P during the motion is the scalar quantity W given by the product

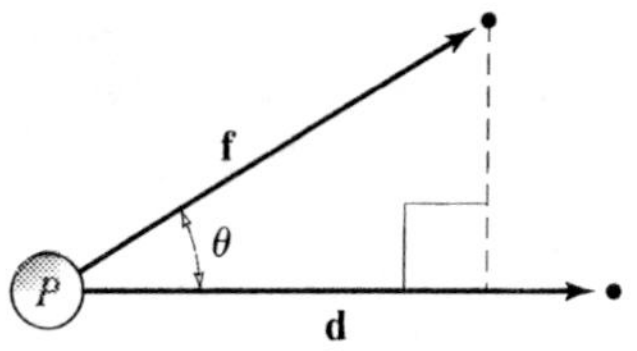

$$\begin{aligned} W &= (\text{magnitude of } \mathbf{f} \text{ in direction } \mathbf{d}) \\ &\quad \times (\text{distance traveled by } P) \\ &= (\|\mathbf{f}\| \cos\theta)\ (\|\mathbf{d}\|) = \mathbf{f} \cdot \mathbf{d} \end{aligned}$$

Then $0 < W$ when $0 \le \theta < 90°$ and $W < 0$ when $90 < \theta \le 180°$. If $\mathbf{f} \perp \mathbf{d}$ then $W = 0$.[4]

Historical Notes

Augustin-Louis Cauchy (1789–1857) (See historical profile for Chapter 5).

Hermann Amandus Schwarz (1843–1921) German mathematician. Schwarz was professor at the University of Berlin and is known for his work in complex analysis, calculus of variations and differential geometry.

Versions of the Cauchy–Schwarz inequality, which we attribute jointly to Cauchy and Schwarz, were discovered independently by several mathematicians in the late 19th century.

James Prescott Joule (1818–1889) English physicist. Joule's law states that the internal energy of a given mass of gas is independent of its volume or pressure and depends only on temperature.

[4] The standard unit of work is the Joule (J). One Joule (1 J) is the amount of work done when the point of application of a force of one newton (1 N) is displaced through a distance of one meter (1 m) in the direction of the force. 1 J $\simeq 10^7$ ergs $\simeq$ 0.74 foot-pounds.

EXERCISES 4.1

Exercises 1–10. Consider the vectors

$$\mathbf{u} = (2, 1, -2), \quad \mathbf{v} = (3, -4, 0), \quad \mathbf{w} = (1, 0, 1).$$

Compute the expressions and compare results. When appropriate, comment on what property of dot product or norm is illustrated by the results.

1. $\mathbf{u}\cdot\mathbf{v}$, $\mathbf{v}\cdot\mathbf{u}$
2. $\mathbf{w}\cdot(\mathbf{u}+\mathbf{v})$, $\mathbf{w}\cdot\mathbf{u}+\mathbf{w}\cdot\mathbf{v}$
3. $(\mathbf{u}-\mathbf{v})\cdot\mathbf{w}$, $\mathbf{u}\cdot\mathbf{w}-\mathbf{v}\cdot\mathbf{w}$
4. $\|2\mathbf{u}-\mathbf{v}\|$, $\|\mathbf{v}-2\mathbf{u}\|$
5. $3(\mathbf{u}\cdot\mathbf{v})$, $(3\mathbf{u})\cdot\mathbf{v}$, $\mathbf{u}\cdot(3\mathbf{v})$
6. $\|\mathbf{u}\|$, $\|\mathbf{v}\|$, $\|\mathbf{w}\|$
7. $\|\mathbf{u}+\mathbf{v}\|$, $\|\mathbf{u}\|+\|\mathbf{v}\|$
8. $\|\mathbf{v}-\mathbf{w}\|$, $\|\mathbf{v}\|-\|\mathbf{w}\|$
9. $\|(\mathbf{u}\cdot\mathbf{v})\mathbf{w}\|$, $\|(\mathbf{u}\cdot\mathbf{w})\mathbf{v}\|$
10. $\|\mathbf{u}+\mathbf{v}+\mathbf{w}\|$, $\|\mathbf{u}\|+\|\mathbf{v}\|+\|\mathbf{w}\|$

Exercises 11–14. Find the angle (to the nearest degree) between each pair of vectors.

11. $(2, -3), (5, 2)$
12. $(3, -4), (4, 3)$
13. $(2, -3, 1), (4, -2, -3)$
14. $(2, 0, 5), (-5, -4, 2)$
15. Find (a) all vectors orthogonal to $\mathbf{u} = (-3, 4)$, (b) all unit vectors orthogonal to $\mathbf{u}$.
16. Consider the vectors
$$\mathbf{u} = (-1, 3, 2), \quad \mathbf{v} = (5, 1, 1), \quad \mathbf{w} = (-2, 6, 4).$$
 (a) Find the cosine of the angle between each pair of vectors. Which pairs are orthogonal?
 (b) Verify the Cauchy–Schwarz inequality for each pair of vectors.

Exercises 17–18. Find the value of k such that the vectors are orthogonal.

17. $(2, -3, k), (-2, 5, -1)$
18. $(3, -k, 0), (5, 1, 7)$
19. If $\mathbf{u}$ and $\mathbf{v}$ are adjacent sides of an equilateral triangle, find $\mathbf{u}\cdot\mathbf{v}$.
20. Find (a) the set S of all vectors orthogonal to both $(-2, 5, 1)$ and $(-3, 4, 1)$, (b) all unit vectors in S.
21. Given three mutually orthogonal vectors $\mathbf{u}$, $\mathbf{v}$, $\mathbf{w}$ such that $\|\mathbf{u}\| = \|\mathbf{v}\| = 2$, and $\|\mathbf{w}\| = 3$, compute $\|\mathbf{u}+\mathbf{v}+\mathbf{w}\|^2$.
22. Suppose that $\mathbf{u} = (6, 3, -2)$, $\mathbf{v} = (-2, k, -4)$, and that the angle between $\mathbf{u}$ and $\mathbf{v}$ is $\text{Cos}^{-1}(4/21)$. Find k.
23. Let A, B, C be points in $\mathbb{R}^3$ such that $\overrightarrow{AB} = (3, 2, -3)$ and $\overrightarrow{AC} = (8, -1, 1)$. Find the smallest interior angle in the triangle $\triangle ABC$.
24. Find the area of the triangle in $\mathbb{R}^3$ defined by the terminal points of the vectors $\mathbf{u} = (-3, -1, 5)$, $\mathbf{v} = (2, 0, -4)$, $\mathbf{w} = (5, -2, 0)$.
25. Use vector projection to find the area of the parallelogram defined by the points $(2, -3)$, $(1, 1)$, $(5, -6)$, $(4, -2)$.
26. Find the value of k so that the vectors $\mathbf{u} = (2, -1, 1)$, $\mathbf{v} = (1, 2, -3)$, $\mathbf{w} = (3, k, 5)$ are coplanar.

Exercises 27–28. Consider nonzero vectors **a**, **b**, **c**, *and suppose that $\triangle ABC$ is defined by the vector formula $\mathbf{a} = \mathbf{b}+\mathbf{c}$, where vertex A is opposite the side defined by* **a**, *and so on. Give a trigonometric formula that corresponds to each identity.*

27. $\mathbf{a}\cdot\mathbf{a} = \mathbf{a}\cdot(\mathbf{b}+\mathbf{c})$
28. $\mathbf{a}\cdot\mathbf{a} = (\mathbf{b}+\mathbf{c})\cdot(\mathbf{b}+\mathbf{c})$
29. A median is a line drawn from a vertex of a triangle to the midpoint of the opposite side. Use vector arguments to prove that the median to the base of an isosceles triangle is orthogonal to the base.
30. Use dot product to prove that the angle subtended at a point P on the circumference of a circle by a diameter is $90°$.

Exercises 31–32. Prove each identity for all vectors **u**, **v** *in $\mathbb{R}^n$.*

31. $(\mathbf{u}+\mathbf{v})\cdot(\mathbf{u}-\mathbf{v}) = \|\mathbf{u}\|^2 - \|\mathbf{v}\|^2$
32. $\|\mathbf{u}+\mathbf{v}\|^2 + \|\mathbf{u}-\mathbf{v}\|^2 = 2\|\mathbf{u}\|^2 + 2\|\mathbf{v}\|^2$
33. Express $\|\mathbf{u}+\mathbf{v}\|^2 - \|\mathbf{u}-\mathbf{v}\|^2$ as a dot product.
34. Prove that $(\mathbf{u}+\mathbf{v})$ is orthogonal to $(\mathbf{u}-\mathbf{v})$ if and only if $\|\mathbf{u}\| = \|\mathbf{v}\|$. What geometric property is illustrated by this result?
35. Find the distance between the vectors $\mathbf{u} = (-1, 3, 0)$, $\mathbf{v} = (1, -2, 3)$.
36. Let $\mathbf{u}$ and $\mathbf{v}$ be orthogonal unit vectors in $\mathbb{R}^n$. Simplify $(3\mathbf{u}-\mathbf{v})\cdot(\mathbf{u}+\mathbf{v})$.
37. $S = \{\mathbf{u}, \mathbf{v}, \mathbf{w}\}$ is a set of unit vectors in $\mathbb{R}^n$ and the angle between each pair is $\pi/4$ radians. Find the value of the expression $(2\mathbf{u}+3\mathbf{v}-\mathbf{w})\cdot(\mathbf{u}-\mathbf{v}+\mathbf{w})$.

Exercises 38–39. Verify the Pythagorean theorem in each case.

38. $\mathbf{u} = (4, 3)$, $\mathbf{v} = (-4.5, 6)$
39. $\mathbf{u} = (1, 2, 3, 4)$, $\mathbf{v} = (2, 1, 0, -1)$
40. Let $S = \{\mathbf{u}, \mathbf{v}, \mathbf{w}\}$ be a set of vectors in $\mathbb{R}^n$. If $\mathbf{x}$ is orthogonal to each of the vectors in S, prove that $\mathbf{x}$ is orthogonal to every vector in span S.
41. Let $\mathbf{u} = (2, 3)$ and $\mathbf{v} = (-1, 4)$. Find the vector projection of $\mathbf{u}$ onto the line $\mathcal{L}$ defined by $\mathbf{v}$ and of $\mathbf{v}$ onto the line $\mathcal{L}$ defined by $\mathbf{u}$. Find the component of $\mathbf{v}$ orthogonal to $\mathbf{u}$ and the component of $\mathbf{u}$ orthogonal to $\mathbf{v}$. Draw an accurate diagram.
42. If $\mathbf{u}$ and $\mathbf{v}$ are nonzero vectors in $\mathbb{R}^n$ with $a = \|\mathbf{u}\|$ and $b = \|\mathbf{v}\|$, show that the vector $\mathbf{w} = \dfrac{1}{a+b}(b\mathbf{u} + a\mathbf{v})$ bisects the angle between $\mathbf{u}$ and $\mathbf{v}$.

Exercises 43–46. Prove the properties of dot product.

43. (**DP1**) 44. (**DP2**) 45. (**DP3**) 46. (**DP4**).

Exercises 47–48. Prove the properties of Euclidean norm.

47. (**N1**) 48. (**N2**).

49. Verify property (**N2**) in the case $\mathbf{u} = (1, 2, -1)$ and $c = -2$.

Exercises 50–51. Norms. *In each case, a real-valued function is defined on $\mathbb{R}^n$, where $\mathbf{v} = (v_1, v_2, \ldots, v_n)$ is any vector. Show that properties (**N1**), (**N2**), (**N3**) are satisfied. Hence, the function defines a norm on $\mathbb{R}^n$.*

50. The 1-norm on $\mathbb{R}^n$ is defined by
$$\|\mathbf{v}\|_1 = |v_1| + \cdots + |v_n|, \text{ for each } \mathbf{v} \text{ in } \mathbb{R}^n.$$
51. The ∞-norm on $\mathbb{R}^n$ is defined by
$$\|\mathbf{v}\|_\infty = \max\{|v_1|, \ldots, |v_n|\}, \text{ for each } \mathbf{v} \text{ in } \mathbb{R}^n.$$

Exercises 52–53. Find the equation of the plane in $\mathbb{R}^3$ passing through the given points.

52. $(1, 2, 3)$, $(-1, 2, 0)$, $(2, -3, 4)$
53. $(0, 1, 2)$, $(-2, 1, 1)$, $(3, -5, 1)$
54. Confirm the calculations in Example 7 by accurately plotting the line L in the plane and the vectors $\text{proj}_L(\mathbf{v})$ and $\text{refl}_L(\mathbf{v})$.

Exercises 55–56. Find the orthogonal projection of **v** *onto the line L defined by* **u**, *and the reflection of* **v** *in L. Interchange* **u** *and* **v**, *and repeat the exercise.*

55. $\mathbf{u} = (4, 1)$, $\mathbf{v} = (1, 2)$
56. $\mathbf{u} = (1, -1, 1)$, $\mathbf{v} = (4, 3, 2)$
57. If **v** and **u** are vectors in $\mathbb{R}^n$, show that $(\mathbf{v}\cdot\mathbf{u})\mathbf{u} = \mathbf{u}\mathbf{u}^{\mathsf{T}}\mathbf{v}$.

Hint: Begin with the case $\mathbf{v} = \begin{bmatrix} v_1 \\ v_2 \end{bmatrix}$, $\mathbf{u} = \begin{bmatrix} u_1 \\ u_2 \end{bmatrix}$ and then generalize to $\mathbb{R}^n$.

58. In order to see their form, write out the matrices **P**, **C**, **R** relating to projection and reflection relative to a subspace of $\mathbb{R}^3$, where $\mathbf{u} = (u_1, u_2, u_3)$.
59. Rework Example 6. Use projection and reflection matrices to find the orthogonal projection of the vector $\mathbf{v} = (1, 2, 0, -1)$ onto the subspace $\mathcal{U}$ of $\mathbb{R}^4$ spanned by the vector $\mathbf{u} = (1, 0, 1, -1)$. Find also the component of **v** orthogonal to $\mathcal{U}$ and the reflection of **v** in $\mathcal{U}$.
60. Let L be a line through the origin in $\mathbb{R}^2$ inclined at an angle θ to the positive x-axis, where $0 \le \theta < \pi$. Let Q be a point in $\mathbb{R}^2$. The orthogonal projection of Q on L can be accomplished be a composition of three linear transformations on $\mathbb{R}^2$. (a) Rotate L through an angle $-\theta$, (b) project onto the x-axis, (c) rotate through an angle θ. Find the matrices that define the three transformations, and take their product (in correct order) to find the projection matrix **P** in trigonometric form. Draw an accurate diagram of this process.
61. Find the projection and reflection matrices **P** and **R** for the subspace $\mathcal{U}$ of $\mathbb{R}^2$ represented by a line L through the origin inclined at $\pi/3$ radians to the positive x–axis. (a) Find the image of the vector $\mathbf{v} = (4, 2)$ under the projection and reflection transformations. (b) Find the null spaces of **P** and **R**.
62. Find the projection and reflection matrices **P** and **R** for the subspace $\mathcal{U}$ of $\mathbb{R}^2$ with basis $\mathbf{u} = (1, 1, 1)$. (a) Find the image of the vector $\mathbf{v} = (1, -1, 4)$ under the projection and reflection transformations. (b) Find the null spaces of **P** and **R**.
63. Prove that a projection matrix **P** is idempotent; that is, $\mathbf{P}^2 = \mathbf{P}$. Interpret this property geometrically.
64. Show that a reflection matrix **R** is a square root of the identity $\mathbf{I}_n$. Interpret this property geometrically.
65. Let **w** be a nonzero vector in $\mathbb{R}^n$. The $n \times n$ matrix
$$\mathbf{H} = \mathbf{I}_n - 2\frac{\mathbf{w}\mathbf{w}^{\mathsf{T}}}{\mathbf{w}^{\mathsf{T}}\mathbf{w}}$$
is called a *Householder matrix*. Show that the linear transformation **Hv** on $\mathbb{R}^n$ defines a reflection of **v** in a line orthogonal to **w**.
66. Find the projection of the vector $\mathbf{v} = (1, 2, -1)$ onto the plane π in $\mathbb{R}^3$ that is orthogonal to the line $(x, y, z) = t(1, 1, 1)$, where t is a real parameter.
67. Determine, with reasons, which of the matrices (if any) define an orthogonal projection or a reflection. When possible, determine the line L of projection or reflection.
$$\mathbf{A} = \frac{1}{2}\begin{bmatrix} 1 & 1 \\ 1 & 1 \end{bmatrix} \quad \mathbf{B} = \frac{1}{17}\begin{bmatrix} 0 & -4 \\ -4 & 1 \end{bmatrix}$$
$$\mathbf{C} = \begin{bmatrix} 0 & 0.5 \\ 0.5 & 1 \end{bmatrix} \quad \mathbf{D} = \begin{bmatrix} -1 & 0 \\ 0 & -1 \end{bmatrix}$$
$$\mathbf{E} = \frac{1}{5}\begin{bmatrix} 1 & 2 \\ 3 & 1 \end{bmatrix} \quad \mathbf{F} = \begin{bmatrix} 1/\sqrt{2} & -1/\sqrt{6} \\ -1/\sqrt{6} & 1/\sqrt{3} \end{bmatrix}$$

Exercises 68–69. Find (a) the work done by the given force **F** *over the given displacement, (b) the component of the given force in the given direction.*

68. $\mathbf{F} = (2, 3)$, $\vec{d} = (4, 1)$
69. $\mathbf{F} = (-3, 5, 1)$, $\vec{d} = (-4, -2, 3)$
70. Consider points $A = (1, 3, 5)$, $B = (2, 1, 3)$ and $C = (3, 4, 7)$ in $\mathbb{R}^3$. Find the work done by a force of 15 N (Newtons) acting in the direction A to B over a displacement $\overrightarrow{AC}$.

71. Refer to Figure 4.7. A body B weighing 200 lb is in equilibrium on a frictionless surface inclined at an angle θ to the horizontal. Introducing the xy-coordinate system shown, the weight of B is represented by the vector $\mathbf{w} = (0, -200)$. Find the components of $\mathbf{w}$ parallel to and orthogonal to the surface. What force is required to keep B at rest on the surface when $\theta = 0.45$ radians?

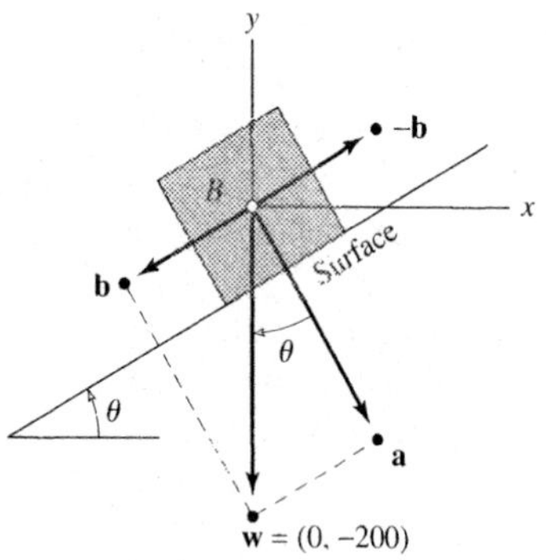

Figure 4.7

USING MATLAB

Consult online help for the commands **norm**, **acos**, and related commands.

72. $(-2.3, 0.4)$

73. $(1.2, -0.3, 1.5)$

74. $(1, 2, 3.4, -1.5, 6.7)$

75. Write a function M-file **iangle.m** that calls a pair of vectors $\mathbf{u}$ and $\mathbf{v}$ in $\mathbb{R}^n$ and returns the angle between $\mathbf{u}$ and $\mathbf{v}$ in radians and in degrees. Use the command **acos** (inverse cosine function) and test your program on the following pairs $\mathbf{u}$, $\mathbf{v}$:

$\mathbf{u}$	$\mathbf{v}$
$(1, 2, 3)$,	$(-1, 0, 4)$
$(1.5, -2.2, 0)$,	$(-1, -1, 3.5)$
$(1, -2, 3, -4, 5)$,	$(-1, 0, 4, 6, -2)$

76. *Project.* Investigate the infinity and Frobenius norms for matrices. Are the properties (**N1**), (**N2**), (**N3**) true in these cases?

77. Use MATLAB to cross-check hand calculations in this exercise set.

4.2 Orthogonal Sets, Orthogonal Matrices

Recall from Section 4.1 that two (column) vectors $\mathbf{u}$ and $\mathbf{v}$ in $\mathbb{R}^n$ are orthogonal if $\mathbf{u}\cdot\mathbf{v} = 0$, and we write $\mathbf{u} \perp \mathbf{v}$ when this is the case. Using matrix multiplication, the condition $\mathbf{u}\cdot\mathbf{v} = 0$ can be rewritten in either of the forms $\mathbf{u}^{\mathsf{T}}\mathbf{v} = 0$ or $\mathbf{v}^{\mathsf{T}}\mathbf{u} = 0$. Our first step is to generalize the concept of orthogonality from two vectors to a *set* of vectors in $\mathbb{R}^n$.

Definition 4.7 *Orthogonal set, orthonormal set*

A set of vectors $\mathcal{S} = \{\mathbf{u}_1, \mathbf{u}_2, \ldots, \mathbf{u}_k\}$ in $\mathbb{R}^n$ is called *orthogonal* if $\mathbf{u}_i \cdot \mathbf{u}_j = 0$ for all pairs of subscripts i, j with $i \neq j$. The vectors in $\mathcal{S}$ are said to be *pairwise* or *mutually* orthogonal. An orthogonal set is called *orthonormal* if each vector in $\mathcal{S}$ has unit norm.

EXAMPLE 1 **Orthogonal Sets, Orthonormal Sets**

Consider the set $\mathcal{S} = \{\mathbf{u}_1, \mathbf{u}_2, \mathbf{u}_3\}$ in $\mathbb{R}^3$, where

$$\mathbf{u}_1 = \begin{bmatrix} 2 \\ 1 \\ 0 \end{bmatrix}, \quad \mathbf{u}_2 = \begin{bmatrix} -1 \\ 2 \\ -1 \end{bmatrix}, \quad \mathbf{u}_3 = \begin{bmatrix} -1 \\ 2 \\ 5 \end{bmatrix}.$$

Checking orthogonality for the three pairs of distinct vectors in $\mathcal{S}$, we have

$$\mathbf{u}_1^{\mathsf{T}}\mathbf{u}_2 = \mathbf{u}_1^{\mathsf{T}}\mathbf{u}_3 = \mathbf{u}_2^{\mathsf{T}}\mathbf{u}_3 = 0,$$

and so $\mathcal{S}$ is an orthogonal set but not orthonormal. Normalizing the vectors in $\mathcal{S}$ results in an orthonormal set $\mathcal{S}' = \{\mathbf{u}_1', \mathbf{u}_2', \mathbf{u}_3'\}$, where

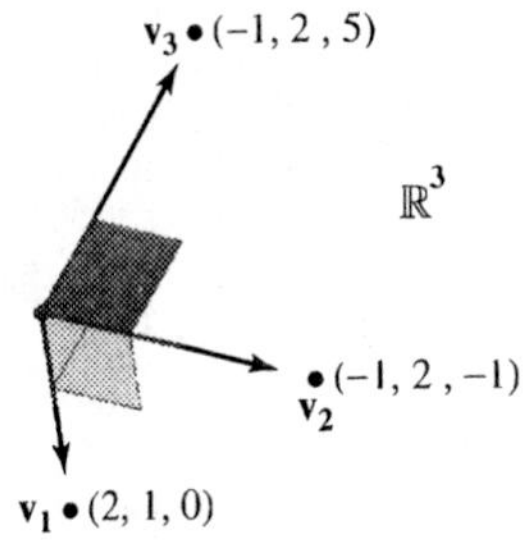

$$\mathbf{u}_1' = \frac{1}{\sqrt{5}}\begin{bmatrix} 2 \\ 1 \\ 0 \end{bmatrix}, \quad \mathbf{u}_2' = \frac{1}{\sqrt{6}}\begin{bmatrix} -1 \\ 2 \\ -1 \end{bmatrix}, \quad \mathbf{u}_3' = \frac{1}{\sqrt{30}}\begin{bmatrix} -1 \\ 2 \\ 5 \end{bmatrix}.$$

A geometric argument shows that an orthogonal set of nonzero vectors in $\mathbb{R}^3$ contains at most three vectors.

The standard basis $\{\mathbf{e}_1, \mathbf{e}_2, \ldots, \mathbf{e}_n\}$ for $\mathbb{R}^n$ is an orthonormal set because $\mathbf{e}_i \cdot \mathbf{e}_j = 0$, for all pairs of distinct subscripts i and j and each vector has unit norm. ■

The connection between orthogonality and linear independence is given next. The condition that the vectors are nonzero is essential—a set of vectors containing the zero vector is linearly dependent.

Theorem 4.3 **Orthogonality and Linear Independence**

Let $\mathcal{S} = \{\mathbf{u}_1, \mathbf{u}_2, \ldots, \mathbf{u}_k\}$ be a set of nonzero vectors in $\mathbb{R}^n$. If $\mathcal{S}$ is orthogonal, then $\mathcal{S}$ is linearly independent.

Proof We test linear independence in the standard way by solving the vector equation

$$x_1\mathbf{u}_1 + x_2\mathbf{u}_2 + \cdots + x_k\mathbf{u}_k = \mathbf{0} \tag{4.29}$$

for the scalars $x_1, x_2, \ldots, x_k$. Taking the dot product of $\mathbf{u}_1$ with left and right sides of (4.29), and using linear properties of dot product, we obtain

$$x_1(\mathbf{u}_1 \cdot \mathbf{u}_1) + x_2(\mathbf{u}_1 \cdot \mathbf{u}_2) + \cdots + x_k(\mathbf{u}_1 \cdot \mathbf{u}_k) = \mathbf{u}_1 \cdot \mathbf{0} = 0.$$

However, $\mathcal{S}$ is orthogonal, and so dot products of pairs of distinct vectors are zero. Equation (4.29) reduces to $x_1(\mathbf{u}_1 \cdot \mathbf{u}_1) = x_1\|\mathbf{u}_1\|^2 = 0$. But $\|\mathbf{u}_1\| > 0$ because $\mathbf{u}_1$ is nonzero and so $x_1 = 0$. Repeating the argument for each vector in the list $\mathbf{u}_2, \mathbf{u}_3, \ldots, \mathbf{u}_k$ shows that $x_2 = 0, x_3 = 0, \ldots, x_k = 0$. Hence, (4.29) has only the zero solution and consequently $\mathcal{S}$ is linearly independent. ■

Remarks

Refer to Theorem 4.3.

(a) The converse implication in the theorem is false in general. That is, a linearly independent set of nonzero vectors in $\mathbb{R}^n$ may not be orthogonal (Exercises 4.2).

(b) If a set of nonzero vectors $\mathcal{S}$ in a subspace of $\mathbb{R}^n$ is orthogonal, then $\mathcal{S}$ is linearly independent. If dim $\mathcal{U} = k$ and $\mathcal{S}$ contains k vectors, then the results in Section 3.2 show that span $\mathcal{S} = \mathcal{U}$. Hence, $\mathcal{S}$ is a basis for $\mathcal{U}$. For example, the set $\mathcal{S}$ in Example 1 is a basis for $\mathbb{R}^3$.

(c) Recall from Section 3.2 that any linearly independent subset of $\mathbb{R}^n$ contains at most n vectors. Thus, any orthogonal set of nonzero vectors in $\mathbb{R}^n$ contains at most n vectors.

Definition 4.8 ***Orthogonal basis, orthonormal basis***

A basis $\mathcal{B}$ for a subspace $\mathcal{U}$ of $\mathbb{R}^n$ is called *orthogonal* (respectively, *orthonormal*) if $\mathcal{B}$ is an orthogonal (respectively, orthonormal).

In Example 1, the set $\mathcal{S}$ is an orthogonal basis for $\mathbb{R}^3$ and $\mathcal{S}'$ is an orthonormal basis. For each n, the standard basis $\{\mathbf{e}_1, \mathbf{e}_2, \cdots, \mathbf{e}_n\}$ for $\mathbb{R}^n$ is orthonormal.

EXAMPLE 2 **Finding an Orthonormal Basis for a Subspace of $\mathbb{R}^3$**

Refer to Figure 4.8. Consider the subspace $\mathcal{U}$ of $\mathbb{R}^3$ defined by the plane with equation $x - 2y + z = 0$. Setting $y = s$ and $z = t$, where s and t are parameters, we have $x = 2s - t$ and so all vectors (x, y, z) in $\mathcal{U}$ (written as columns) are given by

$$\begin{bmatrix} x \\ y \\ z \end{bmatrix} = s\begin{bmatrix} 2 \\ 1 \\ 0 \end{bmatrix} + t\begin{bmatrix} -1 \\ 0 \\ 1 \end{bmatrix} = s\mathbf{u} + t\mathbf{v}, \quad \text{where } \mathbf{u} = \begin{bmatrix} 2 \\ 1 \\ 0 \end{bmatrix}, \ \mathbf{v} = \begin{bmatrix} -1 \\ 0 \\ 1 \end{bmatrix}.$$

The set $\mathcal{B} = \{\mathbf{u}, \mathbf{v}\}$ is a basis for $\mathcal{U}$ that is not orthogonal.

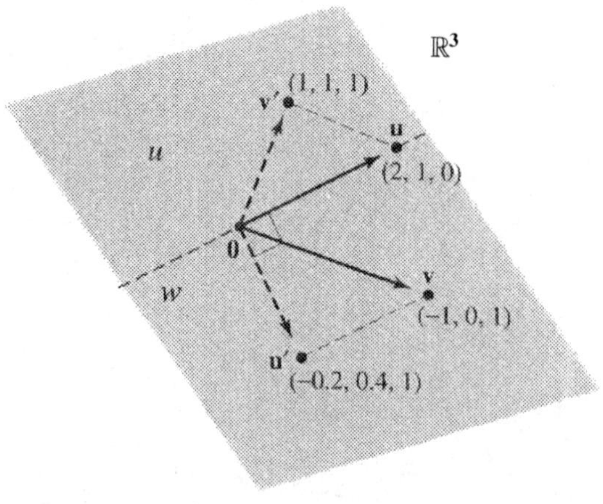

Figure 4.8 A subspace of $\mathbb{R}^3$.

Our goal is to construct an orthogonal basis $\mathcal{B}'$ for $\mathcal{U}$ using vectors from $\mathcal{B}$. Let $\mathcal{W}$ be the subspace of $\mathbb{R}^3$ with basis $\{\mathbf{u}\}$. Then, from Section 4.1, $\mathbf{u}' = \mathbf{v} - \text{proj}_{\mathcal{W}}(\mathbf{v})$ is the component of $\mathbf{v}$ orthogonal to $\mathbf{u}$, and we have

$$\mathbf{u}' = \mathbf{v} - \text{proj}_{\mathcal{W}}(\mathbf{v}) = \mathbf{v} - \frac{\mathbf{u}\mathbf{u}^{\mathsf{T}}}{\|\mathbf{u}\|^2}\mathbf{v}$$

$$= \left(\mathbf{I}_3 - \frac{1}{5}\begin{bmatrix} 4 & 2 & 0 \\ 2 & 1 & 0 \\ 0 & 0 & 0 \end{bmatrix}\right)\begin{bmatrix} -1 \\ 0 \\ 1 \end{bmatrix} = \begin{bmatrix} -0.2 \\ 0.4 \\ 1 \end{bmatrix}.$$

Hence, $\mathcal{B}' = \{\mathbf{u}, \mathbf{u}'\}$ is an orthogonal basis for $\mathcal{U}$ (explain). A similar calculation constructs an orthogonal basis $\mathcal{B}' = \{\mathbf{v}, \mathbf{v}'\}$ for $\mathcal{U}$. Normalizing the vectors in either set $\mathcal{B}'$ results in an orthonormal basis for $\mathcal{U}$. ■

Suppose $\mathcal{B}$ is a basis for a subspace $\mathcal{U}$ of $\mathbb{R}^n$. Then the coordinates of a vector $\mathbf{u}$ in $\mathcal{U}$ relative to $\mathcal{B}$ are easily computed if the basis happens to be orthogonal.

Theorem 4.4 **Coordinates Relative to an Orthogonal Basis**

Let $\mathcal{U}$ be a subspace of $\mathbb{R}^n$ and let $\mathcal{B} = \{\mathbf{u}_1, \mathbf{u}_2, \ldots, \mathbf{u}_k\}$ be an orthogonal basis for $\mathcal{U}$. Then, any vector $\mathbf{u}$ in $\mathcal{U}$ has the unique representation

$$\mathbf{u} = \frac{\mathbf{u}\cdot\mathbf{u}_1}{\|\mathbf{u}_1\|^2}\mathbf{u}_1 + \frac{\mathbf{u}\cdot\mathbf{u}_2}{\|\mathbf{u}_2\|^2}\mathbf{u}_2 + \cdots + \frac{\mathbf{u}\cdot\mathbf{u}_k}{\|\mathbf{u}_k\|^2}\mathbf{u}_k. \tag{4.30}$$

Proof Note that the vectors in $\mathcal{B}$ are necessarily nonzero because $\mathcal{B}$ is linearly independent. By the unique representation theorem in Section 3.2, any vector $\mathbf{u}$ in $\mathcal{U}$ has a representation

$$\mathbf{u} = x_1\mathbf{u}_1 + x_2\mathbf{u}_2 + \cdots + x_k\mathbf{u}_k, \tag{4.31}$$

where the scalars (coordinates) $x_1, x_2, \ldots, x_k$ are unique. Taking the dot product of $\mathbf{u}_1$ with left and right sides of (4.31) gives

$$\mathbf{u}\cdot\mathbf{u}_1 = x_1(\mathbf{u}_1\cdot\mathbf{u}_1) + x_2(\mathbf{u}_2\cdot\mathbf{u}_1) + \cdots + x_k(\mathbf{u}_k\cdot\mathbf{u}_1) = x_1\|\mathbf{u}_1\|^2,$$

because $\mathbf{u}_i \perp \mathbf{u}_j$ for $i \neq j$. However, $\|\mathbf{u}_1\| > 0$ and so $x_1 = (\mathbf{u}\cdot\mathbf{u}_1)/\|\mathbf{u}_1\|^2$. The other coordinates are computed similarly. ■

EXAMPLE 3 **Finding Coordinates Relative to an Orthogonal Basis for $\mathbb{R}^3$**

You may verify that the set $\mathcal{B} = \{\mathbf{u}_1, \mathbf{u}_2, \mathbf{u}_3\}$ is orthogonal, where

$$\mathbf{u}_1 = \begin{bmatrix} -1 \\ 1 \\ 2 \end{bmatrix}, \quad \mathbf{u}_2 = \begin{bmatrix} 1 \\ -1 \\ 1 \end{bmatrix}, \quad \mathbf{u}_3 = \begin{bmatrix} 1 \\ 1 \\ 0 \end{bmatrix}, \qquad \mathbf{u} = \begin{bmatrix} p \\ q \\ r \end{bmatrix} \tag{4.32}$$

By Theorem 4.3 and Remark (b) that follows it, $\mathcal{B}$ is an orthogonal basis for $\mathbb{R}^3$. The coordinates of the vector $\mathbf{u}$ in (4.32) relative to the standard basis for

$\mathbb{R}^3$ are (p, q, r). Using Theorem 4.4, the coordinates of $\mathbf{u}$ relative to $\mathcal{B}$ are (x_1, x_2, x_3), where

$$x_1 = \frac{\mathbf{u}\cdot\mathbf{u}_1}{\|\mathbf{u}_1\|^2} = \frac{-p+q+2r}{6}, \qquad x_2 = \frac{\mathbf{u}\cdot\mathbf{u}_2}{\|\mathbf{u}_2\|^2} = \frac{p-q+r}{3},$$

$$x_3 = \frac{\mathbf{u}\cdot\mathbf{u}_3}{\|\mathbf{u}_3\|^2} = \frac{p+q}{2}.$$

For example, if $\mathbf{u} = \begin{bmatrix} 1 \\ 1 \\ 1 \end{bmatrix}$, then $\mathbf{u} = \frac{1}{3}\,\mathbf{u}_1 + \frac{1}{3}\,\mathbf{u}_2 + \mathbf{u}_3$. ■

The next result, which follows directly from Theorem 4.3, shows that calculation of coordinates relative to an orthonormal basis is even more efficient.

Theorem 4.5 **Coordinates Relative to an Orthonormal Basis**

Let $\mathcal{U}$ be a subspace of $\mathbb{R}^n$ and let $\mathcal{B} = \{\mathbf{u}_1, \mathbf{u}_2, \ldots, \mathbf{u}_k\}$ be a orthonormal basis for $\mathcal{U}$. Then any vector $\mathbf{u}$ in $\mathcal{U}$ has the unique representation

$$\mathbf{u} = (\mathbf{u}\cdot\mathbf{u}_1)\,\mathbf{u}_1 + (\mathbf{u}\cdot\mathbf{u}_2)\,\mathbf{u}_2 + \cdots + (\mathbf{u}\cdot\mathbf{u}_k)\,\mathbf{u}_k.$$

Proof In equation (4.30), we have $\|\mathbf{u}_i\|^2 = 1$, for $1 \le i \le k$.

Notice that each coefficient of $\mathbf{u}_i$, $1 \le i \le k$, in Theorem 4.5 is $\mathbf{u}\cdot\mathbf{u}_i = \|\mathbf{u}\|\cos\theta$, where θ is the angle between $\mathbf{u}$ and $\mathbf{u}_i$. The scalar $\mathbf{u}\cdot\mathbf{u}_i$ is called the *scalar projection* of $\mathbf{u}$ onto $\mathbf{u}_i$.

Orthogonal Matrices

Consider an $n \times k$ matrix $\mathbf{A} = [\,\mathbf{v}_1\ \ \mathbf{v}_2\ \cdots\ \mathbf{v}_k\,]$ whose columns $\{\mathbf{v}_1, \mathbf{v}_2, \ldots, \mathbf{v}_k\}$ form an orthogonal set in $\mathbb{R}^n$.

Caution! Strangely enough, we do not call $\mathbf{A}$ an orthogonal matrix!

Noting that the rows of $\mathbf{A}^{\mathsf{T}}$ are $\mathbf{v}_1^{\mathsf{T}}, \mathbf{v}_2^{\mathsf{T}}, \ldots, \mathbf{v}_k^{\mathsf{T}}$, we have

$$\mathbf{A}^{\mathsf{T}}\mathbf{A} = \begin{bmatrix} \mathbf{v}_1^{\mathsf{T}} \\ \mathbf{v}_2^{\mathsf{T}} \\ \vdots \\ \mathbf{v}_k \end{bmatrix} [\,\mathbf{v}_1\ \ \mathbf{v}_2\ \cdots \mathbf{v}_k\,] = \begin{bmatrix} \|\mathbf{v}_1\|^2 & \mathbf{v}_1^{\mathsf{T}}\mathbf{v}_2 & \cdots & \mathbf{v}_1^{\mathsf{T}}\mathbf{v}_k \\ \mathbf{v}_2^{\mathsf{T}}\mathbf{v}_1 & \|\mathbf{v}_2\|^2 & \cdots & \mathbf{v}_2^{\mathsf{T}}\mathbf{v}_k \\ \vdots & \vdots & \ddots & \vdots \\ \mathbf{v}_k^{\mathsf{T}}\mathbf{v}_1 & \mathbf{v}_k^{\mathsf{T}}\mathbf{v}_2 & \cdots & \|\mathbf{v}_k\|^2 \end{bmatrix}$$

$$= \begin{bmatrix} \|\mathbf{v}_1\|^2 & 0 & \cdots & 0 \\ 0 & \|\mathbf{v}_2\|^2 & \cdots & 0 \\ \vdots & \vdots & \ddots & \vdots \\ 0 & 0 & \cdots & \|\mathbf{v}_k\|^2 \end{bmatrix}, \tag{4.33}$$

which shows that $\mathbf{A}^{\mathsf{T}}\mathbf{A}$ is a $k \times k$ diagonal matrix. However, note that $\mathbf{A}\mathbf{A}^{\mathsf{T}}$ is an $n \times n$ symmetric matrix that is not generally diagonal.

EXAMPLE 4

Comparing $\mathbf{A}^{\mathsf{T}}\mathbf{A}$ and $\mathbf{A}\mathbf{A}^{\mathsf{T}}$

The 3×2 matrix $\mathbf{A}$ shown has orthogonal columns.

$$\mathbf{A} = [\,\mathbf{v}_1 \;\; \mathbf{v}_2\,] = \begin{bmatrix} 1 & -1 \\ 1 & 1 \\ 1 & 0 \end{bmatrix}, \qquad \mathbf{A}^{\mathsf{T}} = \begin{bmatrix} \mathbf{v}_1^{\mathsf{T}} \\ \mathbf{v}_2^{\mathsf{T}} \end{bmatrix} = \begin{bmatrix} 1 & 1 & 1 \\ -1 & 1 & 0 \end{bmatrix}.$$

Computing $\mathbf{A}^{\mathsf{T}}\mathbf{A}$ using (4.33), we have

$$\mathbf{A}^{\mathsf{T}}\mathbf{A} = \begin{bmatrix} \|\mathbf{v}_1\|^2 & 0 \\ 0 & \|\mathbf{v}_2\|^2 \end{bmatrix} = \begin{bmatrix} 3 & 0 \\ 0 & 2 \end{bmatrix}, \quad \text{while} \quad \mathbf{A}\mathbf{A}^{\mathsf{T}} = \begin{bmatrix} 2 & 0 & 1 \\ 0 & 2 & 1 \\ 1 & 1 & 1 \end{bmatrix}.$$

■

Equation (4.33) leads to a condition for $\mathbf{A}^{\mathsf{T}}$ to be a left-inverse (Section 2.2) for $\mathbf{A}$ that is both necessary and sufficient. The proof is straightforward.

Theorem 4.6

A Left-Inverse for A

Let $\mathbf{A}$ be an $n \times k$ matrix. Then $\mathbf{A}^{\mathsf{T}}\mathbf{A} = \mathbf{I}_k$ if and only if the columns of $\mathbf{A}$ form an orthonormal set in $\mathbb{R}^n$.

The case when $n = k$ in Theorem 4.6 is of particular interest. It is customary in this case to denote the $n \times n$ matrix $\mathbf{A}$ by the symbol $\mathbf{Q}$.

Definition 4.9

Orthogonal matrices, orthogonal transformations

An $n \times n$ matrix $\mathbf{Q}$ is called *orthogonal* if $\mathbf{Q}^{\mathsf{T}}\mathbf{Q} = \mathbf{I}_n$. A linear transformation T on $\mathbb{R}^n$ defined by an orthogonal matrix $\mathbf{Q}$ is called an *orthogonal transformation.*

Theorem 4.6 tells us that $\mathbf{Q}$ is orthogonal if and only if $\mathbf{Q}$ has orthonormal columns.

Caution! An orthogonal matrix is square and has columns with unit norms.

Remarks

The following facts are deduced from Definition 4.9.

(a) $\mathbf{Q}^{\mathsf{T}}$ is a left-inverse for $\mathbf{Q}$. Hence, by Theorem 2.9 in Section 2.2, every orthogonal matrix is invertible and its inverse $\mathbf{Q}^{-1} = \mathbf{Q}^{\mathsf{T}}$ is easy to compute.

(b) Because a left-inverse for a square matrix is also a right-inverse, we have $\mathbf{Q}\mathbf{Q}^{\mathsf{T}} = \mathbf{I}_n$, which shows that the set of row vectors in $\mathbf{Q}$ is orthonormal. Hence the rows (and columns) in an orthogonal matrix are orthonormal.

(c) Looking ahead to Section 5.2, we will see that the *determinant* of an orthogonal matrix is either $+1$ or -1.

The next result uses dot product to give a necessary and sufficient condition for a square matrix to be orthogonal.

Theorem 4.7 **Characterization of Orthogonal Matrices**

Let $\mathbf{Q}$ *be an* $n \times n$ *matrix. The following statements are equivalent.*

(a) $\mathbf{Q}$ is orthogonal.
(b) $\mathbf{Qu} \cdot \mathbf{Qv} = \mathbf{u} \cdot \mathbf{v}$, for any vectors $\mathbf{u}$ and $\mathbf{v}$ in $\mathbb{R}^n$.

Proof There are two implications to prove.

Assume condition (a) is true. Then $\mathbf{Q}^{\mathsf{T}}\mathbf{Q} = \mathbf{I}_n$ and for any vectors $\mathbf{u}$, $\mathbf{v}$ in $\mathbb{R}^n$, we have

$$\mathbf{Qu} \cdot \mathbf{Qv} = (\mathbf{Qu})^{\mathsf{T}}\mathbf{Qv} = \mathbf{u}^{\mathsf{T}}\mathbf{Q}^{\mathsf{T}}\mathbf{Qv} = \mathbf{u}^{\mathsf{T}}\mathbf{v} = \mathbf{u} \cdot \mathbf{v}.$$

Assume condition (b) is true. Consider an $n \times n$ matrix $\mathbf{Q} = [\,\mathbf{v}_1 \;\; \mathbf{v}_2 \;\; \cdots \;\; \mathbf{v}_n\,]$. Let $\mathbf{v}_i$ and $\mathbf{v}_j$ be any two columns (possibly identical) in $\mathbf{Q}$. Then $\mathbf{v}_i = \mathbf{Qe}_i$ and $\mathbf{v}_j = \mathbf{Qe}_j$, where $\mathbf{e}_i$, $\mathbf{e}_j$ are standard basis vectors in $\mathbb{R}^n$. Then

$$\mathbf{v}_i \cdot \mathbf{v}_j = \mathbf{Qe}_i \cdot \mathbf{Qe}_j = \mathbf{e}_i \cdot \mathbf{e}_j = \begin{cases} 1 & \text{if } i = j \\ 0 & \text{if } i \neq j \end{cases}.$$

■

Setting $\mathbf{u} = \mathbf{v}$ in condition (b) of Theorem 4.7 shows that $\|\,\mathbf{Qv}\,\|^2 = \|\,\mathbf{v}\,\|^2$ and by taking positive roots, we have

$$\|\,\mathbf{Qv}\,\| = \|\,\mathbf{v}\,\|. \tag{4.34}$$

Equation (4.34) says that an orthogonal transformation (linear operator) T on $\mathbb{R}^n$ represented by $\mathbf{Q}$ preserves norms (lengths). We call T an *isometry*. In fact, the only linear isometries on $\mathbb{R}^n$ are those represented by orthogonal matrices. Isometries also preserve angles between pairs of vectors (see Exercises 4.2).

■ ILLUSTRATION 4.1

The Orthogonal Transformations on $\mathbb{R}^2$

Our goal is to describe all orthogonal 2×2 matrices $\mathbf{Q}$. These matrices represent the orthogonal transformations on $\mathbb{R}^2$. The columns of $\mathbf{Q}$ are unit orthogonal vectors. Refer to Figure 4.9, which shows the unit circle in $\mathbb{R}^2$. Let the first column of $\mathbf{Q}$ be a unit vector $(\cos\theta, \sin\theta)$ obtained by rotating the x-axis through a positive (counterclockwise) angle θ. The two possible choices for the second column of $\mathbf{Q}$, are $(-\sin\theta, \cos\theta)$ or $(\sin\theta, -\cos\theta)$. Hence, all orthogonal 2×2 matrices $\mathbf{Q}$ have one of two forms, namely

$$(\text{Type I}) \quad \begin{bmatrix} \cos\theta & -\sin\theta \\ \sin\theta & \cos\theta \end{bmatrix} \quad \text{or} \quad (\text{Type II}) \quad \begin{bmatrix} \cos\theta & \sin\theta \\ \sin\theta & -\cos\theta \end{bmatrix}.$$

Type I matrices represent rotations of the plane through a positive angle θ. Type II matrices represent reflections in a line L passing through the origin in $\mathbb{R}^2$ determined by rotating the x-axis through an angle $\theta/2$. The details left for Exercises 4.2. ■

Other important classes of orthogonal transformations are defined by permutation matrices (Sections 2.3) and reflection matrices (Sections 2.3).

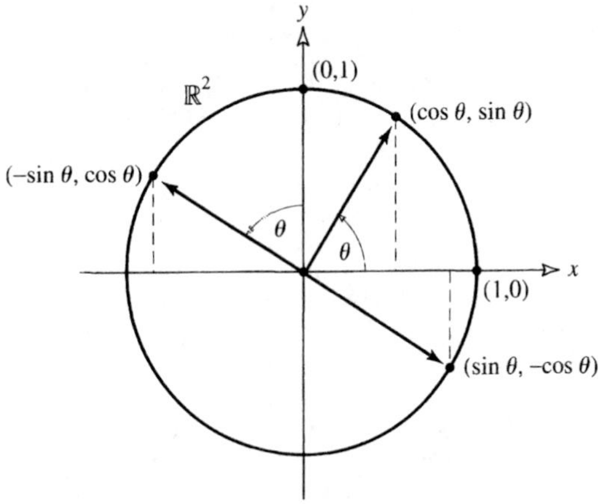

Figure 4.9 Orthogonal unit vectors in $\mathbb{R}^2$.

EXERCISES 4.2

1. Give a geometric argument that shows that an orthogonal set of nonzero vectors in $\mathbb{R}^3$ contains at most 3 vectors.

2. Find a set of three vectors in $\mathbb{R}^3$ that are linearly independent but not orthogonal.

3. Find a set of three vectors in $\mathbb{R}^n$ that are orthogonal and linearly dependent.

4. Suppose $\mathcal{V}$ is a subspace of $\mathbb{R}^n$ of dimension k. Explain why any orthogonal subset $\mathcal{B}$ in $\mathcal{V}$ containing exactly k nonzero vectors is a basis for $\mathcal{V}$.

Exercises 5–6. Find an orthonormal basis for the subspace of $\mathbb{R}^3$ defined by the given equation.

5. $4x_1 + 3x_3 = 0.$

6. $x_1 - x_2 - x_3 = 0.$

7. The vector $\mathbf{u}$ lies in the subspace of $\mathbb{R}^3$ spanned by the set $\mathcal{B} = \{\mathbf{u}_1, \mathbf{u}_2\}$. Verify that the conditions in Theorem 4.4 are met and then find the coordinates of $\mathbf{u}$ relative to $\mathcal{B}$, where

$$\mathbf{u} = \begin{bmatrix} 1 \\ 1 \\ 3 \end{bmatrix}, \quad \mathcal{B}: \mathbf{u}_1 = \begin{bmatrix} -2 \\ 1 \\ 0 \end{bmatrix}, \quad \mathbf{u}_2 = \begin{bmatrix} 2 \\ 4 \\ 10 \end{bmatrix}.$$

*Exercises 8–9. Verify that the set $\mathcal{B}$ is orthogonal. Check that the conditions in Theorem 4.4 are met and then find the coordinates of **u** relative to $\mathcal{B}$.*

8. $\mathbf{u} = \begin{bmatrix} 4 \\ 5 \end{bmatrix}, \quad \mathcal{B}: \mathbf{u}_1 = \begin{bmatrix} 2 \\ 1 \end{bmatrix}, \quad \mathbf{u}_2 = \begin{bmatrix} -2 \\ 4 \end{bmatrix}$

9. $\mathbf{u} = \begin{bmatrix} 1 \\ -1 \\ 2 \end{bmatrix}, \quad \mathcal{B}: \mathbf{u}_1 = \begin{bmatrix} 1 \\ 1 \\ 2 \end{bmatrix}, \quad \mathbf{u}_2 = \begin{bmatrix} -\frac{1}{2} \\ \frac{1}{2} \\ 0 \end{bmatrix},$

$$\mathbf{u}_3 = \begin{bmatrix} \frac{2}{3} \\ \frac{2}{3} \\ -\frac{2}{3} \end{bmatrix}$$

10. Verify that the set $\mathcal{B}$ is orthonormal. Check that the conditions in Theorem 4.5 are met and then find the coordinates of $\mathbf{u}$ relative to $\mathcal{B}$, where

$$\mathcal{B}: \mathbf{u}_1 = \begin{bmatrix} 0 \\ \frac{1}{\sqrt{2}} \\ \frac{1}{\sqrt{2}} \end{bmatrix}, \quad \mathbf{u}_2 = \begin{bmatrix} \frac{2}{\sqrt{6}} \\ \frac{1}{\sqrt{6}} \\ -\frac{1}{\sqrt{6}} \end{bmatrix},$$

$$\mathbf{u}_3 = \begin{bmatrix} \frac{1}{\sqrt{3}} \\ -\frac{1}{\sqrt{3}} \\ \frac{1}{\sqrt{3}} \end{bmatrix}, \quad \mathbf{u} = \begin{bmatrix} 1 \\ -2 \\ 3 \end{bmatrix}.$$

Exercises 11–16. Determine if the matrix is orthogonal.

11. $\begin{bmatrix} 0 & -1 \\ 1 & 0 \end{bmatrix}$

12. $\begin{bmatrix} 1/\sqrt{2} & 1/\sqrt{2} \\ 1/\sqrt{2} & -1/\sqrt{2} \end{bmatrix}$

13. $\begin{bmatrix} 1/\sqrt{5} & 2/\sqrt{5} \\ -2/\sqrt{5} & 1/\sqrt{5} \end{bmatrix}$

14. The 4×4 matrix $\mathbf{A} = [a_{ij}]$, where $a_{ij} = -0.5$ when $i = j$ and $a_{ij} = 0.5$ when $i \neq j$.

15. $\begin{bmatrix} 1 & -1 & 1 \\ -1 & 1 & -1 \\ 1 & 1 & 1 \end{bmatrix}$ **16.** $\begin{bmatrix} 0 & 1 & 0 \\ 1 & 0 & 0 \\ 0 & 0 & 1 \end{bmatrix}$

Exercises 17–20. Verify that the given matrix is orthogonal.

17. $\dfrac{1}{\sqrt{2}}\begin{bmatrix} 1 & 1 \\ -1 & 1 \end{bmatrix}$

18. $\begin{bmatrix} a & b \\ b & -a \end{bmatrix}$, where $a^2 + b^2 = 1$

19. $\begin{bmatrix} 1/3 & 2\sqrt{2}/3 & 0 \\ 2/3 & -\sqrt{2}/6 & \sqrt{2}/2 \\ -2/3 & \sqrt{2}/6 & \sqrt{2}/2 \end{bmatrix}$

20. $\dfrac{1}{4}\begin{bmatrix} 2\sqrt{3} & -\sqrt{3} & 1 \\ 2 & 3 & -\sqrt{3} \\ 0 & 2 & 2\sqrt{3} \end{bmatrix}$

21. Find the values of x, y, and z for which the matrix $\begin{bmatrix} 0 & 2y & z \\ x & y & -z \\ x & -y & z \end{bmatrix}$ is orthogonal.

Exercises 22–28. Prove each statement. When required, use previous statements in the list.

22. If $\mathbf{A}$ is an $n \times n$ orthogonal matrix, then $\mathbf{A}^{\mathsf{T}}$ is orthogonal.

23. The inverse $\mathbf{A}^{-1}$ of an $n \times n$ orthogonal matrix is orthogonal.

24. If $\mathbf{A}$ and $\mathbf{B}$ are orthogonal, then $\mathbf{AB}$ and $\mathbf{BA}$ are orthogonal.

25. If $\mathbf{A}$ and $\mathbf{B}$ are orthogonal, then $\mathbf{A}(\mathbf{A}^{\mathsf{T}} + \mathbf{B}^{\mathsf{T}})\mathbf{B} = \mathbf{A} + \mathbf{B}$.

26. Every $n \times n$ permutation matrix $\mathbf{P}$ is orthogonal (Section 2.3).

27. A matrix $\mathbf{B}$ obtained by rearranging the rows (columns) of an $n \times n$ orthogonal matrix $\mathbf{A}$ is orthogonal.

28. If $\mathbf{A}$ and $\mathbf{B}$ are orthogonal matrices, then $\mathbf{A} = \mathbf{BP}$, for some orthogonal matrix $\mathbf{P}$.

29. Justify the claim in Illustration 4.1. Let $\mathbf{u} = \begin{bmatrix} u_1 \\ u_2 \end{bmatrix}$ be a unit vector in the direction of a line L passing through the origin in $\mathbb{R}^2$. Recall from Section 4.1 that the reflection transformation which reflects each vector $\mathbf{v}$ in the line L is defined by the reflection matrix

$$\begin{bmatrix} 2u_1^2 - 1 & 2u_1u_2 \\ 2u_2u_1 & 2u_2^2 - 1 \end{bmatrix}.$$

Let $\mathbf{u} = \left[\cos\frac{\theta}{2}, \sin\frac{\theta}{2}\right]^{\mathsf{T}}$ and hence prove that every 2×2 orthogonal matrix $\mathbf{Q}$ of Type II defines a reflection transformation in the line L.

30. Let $\mathcal{S} = \{\mathbf{u}_1, \mathbf{u}_2, \ldots, \mathbf{u}_k\}$ be an orthogonal set of vectors in $\mathbb{R}^n$. Show by mathematical induction on k that

$$\|\mathbf{u}_1 + \mathbf{u}_2 + \cdots + \mathbf{u}_k\|^2 = \|\mathbf{u}_1\|^2 + \|\mathbf{u}_2\|^2 + \cdots + \|\mathbf{u}_k\|^2.$$

31. Let $\mathcal{B} = \{\mathbf{b}_1, \ldots, \mathbf{b}_k\}$ be an orthogonal basis for a subspace $\mathcal{U}$ of $\mathbb{R}^n$ and let L_i, $1 \leq i \leq k$, be the line in $\mathbb{R}^n$ defined by all scalar multiples $\mathbf{b}_i$. Show that any vector $\mathbf{v}$ in $\mathbb{R}^n$ can be written in the form

$$\mathbf{v} = \text{proj}_{L_1}\mathbf{v} + \text{proj}_{L_2}\mathbf{v} + \cdots + \text{proj}_{L_k}\mathbf{v}.$$

32. Consider the previous exercise in the case of $\mathbb{R}^2$. Let $\mathbf{b}_1 = (2, 1)$, $\mathbf{b}_2 = (-2, 4)$ and $\mathbf{v} = (3, 4)$. Find proj_{L_1} and proj_{L_2} and show that $\mathbf{v} = \text{proj}_{L_1}\mathbf{v} + \text{proj}_{L_2}\mathbf{v}$. Draw an accurate diagram showing L_1 and L_2 and the projections of $\mathbf{v}$ onto these lines.

33. The matrix

$$\mathbf{A} = \begin{bmatrix} 1 & -1 \\ 1 & 1 \\ 1 & 0 \end{bmatrix}$$

in Example 4 has orthogonal columns and defines a linear transformation T on $\mathbb{R}^2$. Find $\ker T$ and $\text{ran}\, T$. Show that T is one-to-one and find the matrix that defines the inverse transformation on $\text{ran}\, T$.

34. Show that the linear transformation T on $\mathbb{R}^4$ defined by the matrix

$$\mathbf{Q} = \frac{1}{2}\begin{bmatrix} 1 & -1 & -1 & -1 \\ 1 & -1 & 1 & 1 \\ 1 & 1 & -1 & 1 \\ 1 & 1 & 1 & -1 \end{bmatrix}$$

is orthogonal. Find bases for $\ker T$, $\text{ran}\, T$.

35. Let $\mathbf{A}$ and $\mathbf{B}$ be square matrices. If the block matrix

$$\begin{bmatrix} \mathbf{A} & \mathbf{C} \\ \mathbf{O} & \mathbf{B} \end{bmatrix}$$

is orthogonal, show that $\mathbf{A}$ and $\mathbf{B}$ are orthogonal and that $\mathbf{C} = \mathbf{O}$.

USING MATLAB

36. Write an M-file that calls a pair of nonzero n-vectors $\{\mathbf{v}_1, \mathbf{v}_2\}$ and finds an orthonormal basis for the subspace of $\mathbb{R}^n$ spanned by $\mathbf{v}_1$ and $\mathbf{v}_2$. Test the M-file using the following pairs

$$\begin{bmatrix} 1 \\ 0 \end{bmatrix}, \begin{bmatrix} 1 \\ -1 \end{bmatrix} \quad \text{and} \quad \begin{bmatrix} 1 \\ -2 \\ 1 \end{bmatrix}, \begin{bmatrix} -2 \\ 3 \\ -1 \end{bmatrix}.$$

37. Consider a given unit vector $\mathbf{u} = [\, x_1 \; y_2 \; y_3 \; \cdots \; y_n \,]^{\mathsf{T}}$ in $\mathbb{R}^n$, where $|x_1| < 1$. Let $d = 1/(1 - x_1)$ and $\mathbf{v} = [\, y_2 \; y_3 \; \cdots \; y_n \,]^{\mathsf{T}}$. Verify that the matrix

$$\mathbf{Q} = \left[\begin{array}{c|c} x_1 & \mathbf{v}^{\mathsf{T}} \\ \hline \mathbf{v} & \mathbf{I} - d\mathbf{v}\mathbf{v}^{\mathsf{T}} \end{array}\right]$$

is orthogonal. Note that the value $x_1 = 1$ gives the orthogonal matrix $\mathbf{I}_n$. Write an M-file that calls a vector $\mathbf{u}$ as input and outputs the matrix $\mathbf{Q}$. Test your program on the initial vectors $[\, \frac{3}{5} \; \frac{4}{5} \,]^{\mathsf{T}}$ and $[\, \frac{1}{9} \; -\frac{8}{9} \; -\frac{4}{9} \,]^{\mathsf{T}}$. This process is used in applications to construct an orthonormal basis for $\mathbb{R}^n$ a given initial vector $\mathbf{u}$.

38. Use MATLAB to check hand calculations in this exercise set.

4.3 Orthogonal Subspaces, Projections, Bases

In Section 4.2 we considered orthogonal sets of vectors—each pair of vectors in the set is orthogonal. We begin this section by extending the idea of orthogonality to pairs of subspaces in $\mathbb{R}^n$.

Definition 4.10

Orthogonal subspaces

The subspaces $\mathcal{U}$ and $\mathcal{V}$ in $\mathbb{R}^n$ are called *orthogonal* if every vector $\mathbf{u}$ in $\mathcal{U}$ is orthogonal to every vector $\mathbf{v}$ in $\mathcal{V}$. We write $\mathcal{U} \perp \mathcal{V}$ when this is the case.

Theorem 4.8

Condition for Two Subspaces to be Orthogonal

Let $\mathcal{U} = \text{span}\{\mathbf{u}_1, \mathbf{u}_2, \ldots, \mathbf{u}_p\}$ *and* $\mathcal{V} = \text{span}\{\mathbf{v}_1, \mathbf{v}_2, \ldots, \mathbf{v}_q\}$ *be subspaces of* $\mathbb{R}^n$. *The following statements are equivalent.*

(a) $\mathcal{U} \perp \mathcal{V}$

(b) $\mathbf{u}_i \perp \mathbf{v}_j$ for every pairs of subscripts (i, j), with $1 \le i \le p$, $1 \le j \le q$

Proof There are two implications to prove.

If (a) is true, then (b) follows by Definition 4.10. Suppose (b) is true. A particular instance with $p = q = 2$ will suffice to illustrate the general proof. Let $\mathbf{u}$ and $\mathbf{v}$ be in $\mathcal{U}$ and $\mathcal{V}$, respectively. Then $\mathbf{u} = x_1\mathbf{u}_1 + x_2\mathbf{u}_2$ and $\mathbf{v} = y_1\mathbf{v}_1 + y_2\mathbf{v}_2$, for some scalars x_1, x_2, y_1, y_2. Using properties of dot product, we have

$$\begin{aligned}\mathbf{u}\cdot\mathbf{v} &= (x_1\mathbf{u}_1 + x_2\mathbf{u}_2)\cdot(y_1\mathbf{v}_1 + y_2\mathbf{v}_2) \\ &= x_1y_1\mathbf{u}_1\cdot\mathbf{v}_1 + x_1y_2\mathbf{u}_1\cdot\mathbf{v}_2 + x_2y_1\mathbf{u}_1\cdot\mathbf{v}_2 + x_2y_2\mathbf{u}_2\cdot\mathbf{v}_2 = 0\end{aligned}$$

because all dot products in the sum are zero by assumption (b). The proof for general p, q follows the same pattern.

EXAMPLE 1

Orthogonal Subspaces in $\mathbb{R}^3$

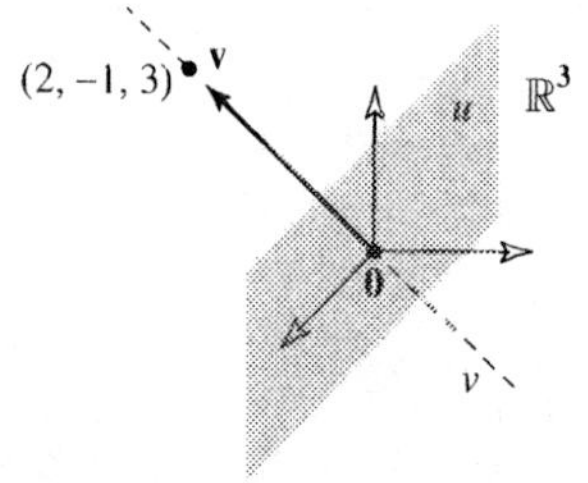

Figure 4.10 Orthogonal subspaces in $\mathbb{R}^3$.

Refer to Figure 4.10. The subspace $\mathcal{U}$ of $\mathbb{R}^3$ shown is defined by a plane with equation $2x_1 - x_2 + 3x_3 = 0$.

Letting x_1 and x_3 be free variables and solving for x_2, we have $x_2 = 2x_1 + x_3$, and the vectors $\mathbf{x}$ lying in $\mathcal{U}$ are given by

$$\mathbf{x} = \begin{bmatrix} x_1 \\ x_2 \\ x_3 \end{bmatrix} = s_1 \begin{bmatrix} 1 \\ 2 \\ 0 \end{bmatrix} + s_2 \begin{bmatrix} 0 \\ 3 \\ 1 \end{bmatrix} = s_1\mathbf{u}_1 + s_2\mathbf{u}_2, \quad \text{where}$$

$$\mathbf{u}_1 = \begin{bmatrix} 1 \\ 2 \\ 0 \end{bmatrix}, \quad \mathbf{u}_2 = \begin{bmatrix} 0 \\ 3 \\ 1 \end{bmatrix},$$

and s_1, s_2 are real parameters. The subspace $\mathcal{U}$ has dimension 2 and basis $\mathcal{B} = \{\mathbf{u}_1, \mathbf{u}_2\}$.

Let $\mathcal{V}$ be the subspace of $\mathbb{R}^3$ spanned by $\mathbf{v} = (2, -1, 3)$. Then $\mathbf{v}\cdot\mathbf{u}_1 = \mathbf{v}\cdot\mathbf{u}_2 = 0$, and so by Theorem 4.8 we have $\mathcal{V} \perp \mathcal{U}$. You may have noticed that $\mathcal{U} = \text{null}\,\mathbf{A}$, where $\mathbf{A} = [\,2 \;\; -1 \;\; 3\,]$ is a 1×3 matrix and that $\mathcal{V} = \text{row}\,\mathbf{A}$. Thus $\text{row}\,\mathbf{A} \perp \text{null}\,\mathbf{A}$ in this case and we will see in due course that this relationship holds for any $m \times n$ matrix $\mathbf{A}$. ■

Caution! Refer to Figure 4.11, which shows two planes in $\mathbb{R}^3$ that intersect at right angles. Let $\mathcal{U} = \text{span}\{\mathbf{e}_1, \mathbf{e}_2\}$ and $\mathcal{V} = \text{span}\{\mathbf{e}_2, \mathbf{e}_3\}$, where $\{\mathbf{e}_1, \mathbf{e}_2, \mathbf{e}_3\}$ is the standard basis for $\mathbb{R}^3$. Theorem 4.8 shows that $\mathcal{U}$ and $\mathcal{V}$ are not orthogonal because $\mathbf{e}_2\cdot\mathbf{e}_2 \neq 0$.

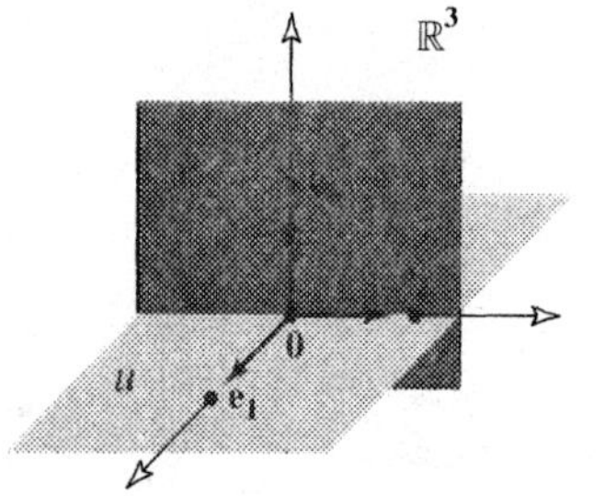

Figure 4.11 Subspaces in $\mathbb{R}^3$ defined by standard basis vectors.

Fundamental Subspaces

Consider anorthogonality $m \times n$ matrix $\mathbf{A}$. We will now look a little more closely at the orthogonality relationships between the fundamental subspaces (Section 3.3) of $\mathbf{A}$ and its transpose $\mathbf{A}^\mathsf{T}$. At first sight there appear to be six fundamental subspaces to consider:

$$\begin{array}{lll} \text{null}\,\mathbf{A} & \text{col}\,\mathbf{A} & \text{row}\,\mathbf{A} \\ \text{null}\,\mathbf{A}^\mathsf{T} & \text{row}\,\mathbf{A}^\mathsf{T} & \text{col}\,\mathbf{A}^\mathsf{T} \end{array}$$

However, $\text{row}\,\mathbf{A}^\mathsf{T} = \text{col}\,\mathbf{A}$ and $\text{col}\,\mathbf{A}^\mathsf{T} = \text{row}\,\mathbf{A}$, which leaves the four basic subspaces $\text{null}\,\mathbf{A}$, $\text{null}\,\mathbf{A}^\mathsf{T}$, $\text{col}\,\mathbf{A}$, and $\text{row}\,\mathbf{A}$ to be considered.

Recall that $\text{null}\,\mathbf{A}$ is the subspace of $\mathbb{R}^n$ consisting of all column vectors $\mathbf{x}$ such that $\mathbf{Ax} = \mathbf{0}$ and $\text{row}\,\mathbf{A}$ is the subspace of $\mathbb{R}^n$ spanned by the rows of $\mathbf{A}$. The equation $\mathbf{Ax} = \mathbf{0}$ implies that $\mathbf{x}$ is orthogonal to every row vector $\mathbf{r}$ in $\mathbf{A}$, and consequently $\mathbf{x}$ is orthogonal to every linear combination of rows of $\mathbf{A}$. Hence,

$$\text{row}\,\mathbf{A} \perp \text{null}\,\mathbf{A}. \tag{4.35}$$

■ ILLUSTRATION 4.2

The Row Space and Null Space are Orthogonal

We will illustrate (4.35) by considering the 2×3 matrix

$$\mathbf{A} = \begin{bmatrix} 1 & -2 & 1 \\ -1 & 1 & 2 \end{bmatrix} = \begin{bmatrix} \mathbf{r}_1 \\ \mathbf{r}_2 \end{bmatrix},$$

where $\mathbf{r}_1 = [\,1 \;\; -2 \;\; 1\,]$ and $\mathbf{r}_2 = [\,-1 \;\; 1 \;\; 2\,]$ are the row vectors in $\mathbf{A}$. Solving the linear system defined by $\mathbf{Ax} = \mathbf{0}$, we have $\mathbf{x} = (x_1, x_2, x_3) = (5t, 3t, t)$, where t is a real parameter, and so $\operatorname{null} \mathbf{A} = \operatorname{span}\{\mathbf{u}\}$, where $\mathbf{u} = [\,5 \;\; 3 \;\; 1\,]^{\mathsf{T}}$. The vectors $\mathbf{r}_1$, $\mathbf{r}_2$ of $\mathbf{A}$ are linearly independent and form a basis for $\operatorname{row} \mathbf{A}$, which is visualized as a plane passing through the origin in $\mathbb{R}^3$ containing the arrows that represent the vectors $\mathbf{r}_1^{\mathsf{T}}$ and $\mathbf{r}_2^{\mathsf{T}}$ in $\mathbb{R}^3$. Recall the fundamental equation $n = \text{rank} + \text{nullity}$. We have $n = 3$ in this case and $\operatorname{rank} \mathbf{A} = \operatorname{rank} \mathbf{A}^{\mathsf{T}} = 2$ so that $\operatorname{row} \mathbf{A}$ has dimension 2 while $\operatorname{null} \mathbf{A}$ has dimension 1. Verify that $\mathbf{r}_1 \perp \mathbf{u}$ and $\mathbf{r}_2 \perp \mathbf{u}$. Theorem 4.8 shows that $\operatorname{row} \mathbf{A} \perp \operatorname{null} \mathbf{A}$. ■

We now note that $\mathbf{A}^{\mathsf{T}}$ is $n \times m$ so that $\operatorname{col} \mathbf{A}^{\mathsf{T}}$ and $\operatorname{null} \mathbf{A}$ are subspaces of $\mathbb{R}^n$. Analogously, $\operatorname{null} \mathbf{A}^{\mathsf{T}}$ and $\operatorname{col} \mathbf{A}$ are subspaces of $\mathbb{R}^m$. Using the relationship (4.35) we arrive at fundamental connections between null space and column space in $\mathbb{R}^n$ and $\mathbb{R}^m$, respectively, namely

$$\operatorname{col} \mathbf{A}^{\mathsf{T}} \perp \operatorname{null} \mathbf{A}, \quad \text{and} \quad \operatorname{col} \mathbf{A} \perp \operatorname{null} \mathbf{A}^{\mathsf{T}}. \tag{4.36}$$

Definition 4.11

Orthogonal complement

Let $\mathcal{U}$ be a subspace of $\mathbb{R}^n$. The *orthogonal complement* of $\mathcal{U}$, denoted by $\mathcal{U}^{\perp}$, is the set of all vectors $\mathbf{v}$ in $\mathbb{R}^n$ such that $\mathbf{v} \perp \mathbf{u}$ for every vector $\mathbf{u}$ in $\mathcal{U}$.

Refer to Example 1 and note that $\mathcal{U}^{\perp} = \mathcal{V}$ and $\mathcal{V}^{\perp} = \mathcal{U}$. Hence $(\mathcal{U}^{\perp})^{\perp} = \mathcal{U}$ and $(\mathcal{V}^{\perp})^{\perp} = \mathcal{V}$.

Theorem 4.9

Properties of Orthogonal Complements

Let $\mathcal{U}$ be a subspace of $\mathbb{R}^n$. Then

(a) $\mathcal{U}^{\perp}$ is a subspace of $\mathbb{R}^n$.

(b) The intersection of the subspaces $\mathcal{U}$ and $\mathcal{U}^{\perp}$ is the zero subspace; that is, $\mathcal{U}^{\perp} \cap \mathcal{U} = \{\mathbf{0}\}$.

(c) $(\mathcal{U}^{\perp})^{\perp} = \mathcal{U}$.

Proof

(a) Let $\mathbf{v}_1$, $\mathbf{v}_2$ be any vectors in $\mathcal{U}^{\perp}$ and let $\mathbf{u}$ be any vector on $\mathcal{U}$. For any scalars c_1 and c_2 we have $(c_1\mathbf{v}_1 + c_2\mathbf{v}_2)\cdot\mathbf{u} = c_1\mathbf{v}_1\cdot\mathbf{u} + c_2\mathbf{v}_2\cdot\mathbf{u} = 0$, because $\mathbf{u} \perp \mathbf{v}_1$ and $\mathbf{u} \perp \mathbf{v}_2$, showing that $\mathcal{U}^{\perp}$ is a subspace of $\mathbb{R}^n$.

(b) Note that if a vector $\mathbf{w}$ is in both sets $\mathcal{U}$ and $\mathcal{U}^{\perp}$ then $\mathbf{w}\cdot\mathbf{w} = \|\,\mathbf{w}\,\|^2 = 0$, and so necessarily $\mathbf{w} = \mathbf{0}$.

(c) See Exercises 4.3.

—

EXAMPLE 2

Orthogonal Subspaces and Orthogonal Complements

Refer to Figure 4.11 and let $\mathcal{U} = \text{span}\{\mathbf{e}_1\}$ and $\mathcal{V} = \text{span}\{\mathbf{e}_2\}$. Then $\mathcal{U} \perp \mathcal{V}$ and also $\mathcal{U}^\perp = \text{span}\{\mathbf{e}_2, \mathbf{e}_3\}$ and $\mathcal{V}^\perp = \text{span}\{\mathbf{e}_1, \mathbf{e}_3\}$. In this case $\mathcal{V}$ is a proper subspace of $\mathcal{U}^\perp = \text{span}\{\mathbf{e}_2, \mathbf{e}_3\}$, and $\mathcal{U}$ is a proper subspace of $\mathcal{V}^\perp = \text{span}\{\mathbf{e}_1, \mathbf{e}_3\}$. ■

For the next result, refer back to the fundamental relationships in (4.36).

Theorem 4.10 **Orthogonal Complements of Fundamental Subspaces**

Let $\mathbf{A}$ *be an* $m \times n$ *matrix. Then*

$$(a)\quad (\text{col}\,\mathbf{A})^\perp = \text{null}\,\mathbf{A}^\mathsf{T} \quad (b)\quad (\text{row}\,\mathbf{A})^\perp = \text{null}\,\mathbf{A}.$$

Proof (a) The subspace col $\mathbf{A}$ consists of all linear combinations of columns of $\mathbf{A}$. Thus col $\mathbf{A}$ is the set $\mathbf{Ax}$, for all vectors $\mathbf{x}$ in $\mathbb{R}^n$. A vector $\mathbf{v}$ is in $(\text{col}\,\mathbf{A})^\perp$ if and only if $\mathbf{v}\cdot\mathbf{Ax} = \mathbf{v}^\mathsf{T}\mathbf{Ax} = (\mathbf{A}^\mathsf{T}\mathbf{v})^\mathsf{T}\mathbf{x} = 0$, for all $\mathbf{x}$ in $\mathbb{R}^n$. Using Exercises 4.3, 13, we have

$$(\mathbf{A}^\mathsf{T}\mathbf{v})^\mathsf{T}\mathbf{x} = 0 \text{ for all } \mathbf{x} \text{ in } \mathbb{R}^n \quad \text{if and only if} \quad (\mathbf{A}^\mathsf{T})\mathbf{v} = 0, \tag{4.37}$$

and so (4.3) implies that $\mathbf{v}$ is in $(\text{col}\,\mathbf{A})^\mathsf{T}$ if and only if $\mathbf{v}$ is in null $\mathbf{A}^\mathsf{T}$. (b) Substitute $\mathbf{A}^\mathsf{T}$ for $\mathbf{A}$ in (a).

The fundamental equation $n = \text{rank} + \text{nullity}$ is used next to determine the dimension $\mathcal{U}^\perp$ in terms of n and the dimension of $\mathcal{U}$.

Theorem 4.11 **Orthogonal Complement, Dimension**

Let $\mathcal{U}$ *be a subspace of* $\mathbb{R}^n$ *with* $\dim\mathcal{U} = k$, $k \le n$. *Then,* $\dim\mathcal{U}^\perp = n - k$; *that is*

$$\dim\mathcal{U} + \dim\mathcal{U}^\perp = n. \tag{4.38}$$

If $\mathcal{B} = \{\mathbf{u}_1, \mathbf{u}_2, \ldots, \mathbf{u}_k\}$ *is a basis for* $\mathcal{U}$ *and* $\mathcal{B}^\perp = \{\mathbf{v}_{k+1}, \mathbf{v}_{k+2}, \cdots, \mathbf{v}_n\}$ *is a basis for* $\mathcal{U}^\perp$, *then the union* $\mathcal{B} \cup \mathcal{B}^\perp = \{\mathbf{u}_1, \ldots, \mathbf{u}_k, \mathbf{v}_{k+1}, \ldots, \mathbf{v}_n\}$ *is a basis for* $\mathbb{R}^n$.

Proof

(a) If $\dim\mathcal{U} = 0$ then $\mathcal{U} = \{\mathbf{0}\}$ and $\mathcal{U}^\perp = \mathbb{R}^n$ and (4.38) is true.

(b) Suppose that $\dim\mathcal{U} = k > 0$. Use the basis for $\mathcal{U}$ to form a $k \times n$ matrix $\mathbf{A} = [\,\mathbf{v}_1\ \mathbf{v}_2\ \cdots\ \mathbf{v}_k\,]^\mathsf{T}$ and note that rank $\mathbf{A} = k$ because $\mathcal{B}$ is linearly independent. The vectors in null $\mathbf{A}$ are orthogonal to vectors in $\mathcal{B}$ and consequently, the vectors in null $\mathbf{A}$ are orthogonal to the vectors $\mathcal{U}$. Hence, null $\mathbf{A} = \mathcal{U}^\perp$. Thus, $\dim\mathcal{U}^\perp = \text{nullity}\,\mathbf{A} = n - k$ because rank $\mathbf{A}$ + nullity $\mathbf{A} = n$, for any $\mathbf{A}$.

We may assume that the bases $\mathcal{B} = \{\mathbf{u}_1, \mathbf{u}_2, \ldots, \mathbf{u}_k\}$ and $\mathcal{B}' = \{\mathbf{u}_{k+1}, \mathbf{u}_{k+2}, \cdots, \mathbf{u}_n\}$ are orthogonal (their existence is guaranteed by the Gram–Schmidt process later in the section). Then the set $\mathcal{B} \cup \mathcal{B}'$ is orthogonal and is linearly independent by Theorem 4.3. However, $\mathcal{B}$ contains n vectors and so is a basis for $\mathbb{R}^n$.

Orthogonal Projections

Recall from the end of Section 4.1 that the projection of a vector $\mathbf{v}$ onto a one-dimensional subspace of $\mathbb{R}^n$ with basis $\{\mathbf{u}\}$ is given by

$$\text{proj}_{\mathcal{U}}(\mathbf{v}) = \frac{\mathbf{v} \cdot \mathbf{u}}{\| \mathbf{u} \|^2}\mathbf{u} = \frac{\mathbf{v}^{\mathsf{T}}\mathbf{u}}{\mathbf{u}^{\mathsf{T}}\mathbf{u}}\mathbf{u}.$$

This result is now generalized to an arbitrary subspace $\mathcal{U}$ of $\mathbb{R}^n$.

Definition 4.12

Projection onto a subspace of $\mathbb{R}^n$

Let $\mathcal{U}$ be a subspace of $\mathbb{R}^n$ with *orthogonal* basis $\{\mathbf{u}_1, \mathbf{u}_2, \ldots, \mathbf{u}_k\}$ and let $\mathbf{v}$ be a vector in $\mathbb{R}^n$. The *orthogonal projection* of $\mathbf{v}$ onto $\mathcal{U}$, denoted by either $\text{proj}_{\mathcal{U}}(\mathbf{v})$ or $\overline{\mathbf{v}}$, is defined by

$$\overline{\mathbf{v}} = \text{proj}_{\mathcal{U}}(\mathbf{v}) = p_1\mathbf{u}_1 + p_2\mathbf{u}_2 + \cdots + p_k\mathbf{u}_k, \tag{4.39}$$

where

$$p_i = \frac{\mathbf{v} \cdot \mathbf{u}_i}{\| \mathbf{u}_i \|^2} = \frac{\mathbf{v}^{\mathsf{T}}\mathbf{u}_i}{\mathbf{u}_i^{\mathsf{T}}\mathbf{u}_i}, \quad 1 \le i \le k.$$

The *component of* $\mathbf{v}$ *orthogonal to* $\mathcal{U}$ is defined to be $\text{comp}_{\mathcal{U}}(\mathbf{v}) = \mathbf{v} - \overline{\mathbf{v}}$.

Thus $\overline{\mathbf{v}}$ is the sum of the projections of $\mathbf{v}$ onto each basis vector in $\{\mathbf{u}_1, \mathbf{u}_2, \ldots, \mathbf{u}_k\}$.

The calculation in (4.39) is simplified if the basis $\mathcal{U}$ happens to be orthonormal. Then $\| \mathbf{u}_i \| = 1$, for $1 \le i \le k$, and we have

$$\text{proj}_{\mathcal{U}}(\mathbf{v}) = (\mathbf{v} \cdot \mathbf{u}_1)\, \mathbf{u}_1 + (\mathbf{v} \cdot \mathbf{u}_2)\, \mathbf{u}_2 + \cdots + (\mathbf{v} \cdot \mathbf{u}_k)\, \mathbf{u}_k. \tag{4.40}$$

Projection Matrices

Form an $n \times k$ matrix $[\,\mathbf{u}_1\ \mathbf{u}_2\ \cdots\ \mathbf{u}_k\,]$ using the basis vectors in Definition 4.12. Then (4.39) can be written as the matrix-column vector product

$$\text{proj}_{\mathcal{U}}(\mathbf{v}) = [\,\mathbf{u}_1\ \mathbf{u}_2\ \cdots\ \mathbf{u}_k\,] \begin{bmatrix} p_1 \\ p_2 \\ \vdots \\ p_k \end{bmatrix}. \tag{4.41}$$

If the basis $\{\mathbf{u}_1, \mathbf{u}_2, \ldots, \mathbf{u}_k\}$ is orthonormal, then $p_i = \mathbf{u}_i^{\mathsf{T}}\mathbf{v}$, $1 \le i \le k$. Letting $\mathbf{Q} = [\,\mathbf{u}_1 \;\; \mathbf{u}_2 \;\; \cdots \;\; \mathbf{u}_k\,]$, (4.41) takes the form

$$\operatorname{proj}_{\mathcal{U}}(\mathbf{v}) = [\,\mathbf{u}_1 \;\; \mathbf{u}_2 \;\; \cdots \;\; \mathbf{u}_k\,] \begin{bmatrix} \mathbf{u}_1^{\mathsf{T}} \\ \mathbf{u}_2^{\mathsf{T}} \\ \vdots \\ \mathbf{u}_k^{\mathsf{T}} \end{bmatrix} \mathbf{v} = \mathbf{Q}\mathbf{Q}^{\mathsf{T}}\mathbf{v} = \mathbf{P}\mathbf{v}, \tag{4.42}$$

where $\mathbf{P} = \mathbf{Q}\mathbf{Q}^{\mathsf{T}}$ is called the *projection matrix* of $\mathbf{v}$ onto $\mathcal{U}$. Notice that this generalizes the formula $\mathbf{u}\mathbf{u}^{\mathsf{T}}\mathbf{v}$ obtained in Section 4.1 for the projection of $\mathbf{v}$ onto the subspace spanned by the unit vector $\mathbf{u}$.

Caution! $\mathbf{P} = \mathbf{Q}\mathbf{Q}^{\mathsf{T}}$ in (4.42) is $n \times n$, while $\mathbf{Q}^{\mathsf{T}}\mathbf{Q} = \mathbf{I}_k$.

The next result is fundamental and will be required in later work. It describes how a vector $\mathbf{v}$ can be *decomposed* into a sum of two special vectors.

Theorem 4.12 **Orthogonal Decomposition**

Let $\mathcal{U}$ be a subspace of $\mathbb{R}^n$ having an orthogonal basis $\{\mathbf{u}_1, \mathbf{u}_2, \ldots, \mathbf{u}_k\}$. Every vector $\mathbf{v}$ in $\mathbb{R}^n$ has a unique decomposition *as the sum*

$$\mathbf{v} = \overline{\mathbf{v}} + \mathbf{w},$$

where $\overline{\mathbf{v}} = \operatorname{proj}_{\mathcal{U}}(\mathbf{v})$ is in $\mathcal{U}$ and $\mathbf{w} = \operatorname{comp}_{\mathcal{U}}(\mathbf{v}) = \mathbf{v} - \overline{\mathbf{v}}$ is in $\mathcal{U}^{\perp}$.

Proof Note that $\overline{\mathbf{v}}$ given in (4.39) is a linear combination of basis vectors and so $\overline{\mathbf{v}}$ lies in $\mathcal{U}$. Theorem 4.12 in Section 4.2 tells us that $\mathbf{w}$ will lie in $\mathcal{U}^{\perp}$ if $\mathbf{w}$ is orthogonal to each basis vector in $\{\mathbf{u}_1, \mathbf{u}_2, \ldots, \mathbf{u}_k\}$. For example, taking the dot product of $\mathbf{w}$ with $\mathbf{u}_1$ and using the orthogonality of the basis, we have

$$\begin{aligned}
\mathbf{w}\cdot\mathbf{u}_1 &= (\mathbf{v} - \overline{\mathbf{v}})\cdot\mathbf{u}_1 = \mathbf{v}\cdot\mathbf{u}_1 - \overline{\mathbf{v}}\cdot\mathbf{u}_1 \\
&= \mathbf{v}\cdot\mathbf{u}_1 - \frac{\mathbf{v}\cdot\mathbf{u}_1}{\|\mathbf{u}_1\|^2}(\mathbf{u}_1\cdot\mathbf{u}_1) - \frac{\mathbf{v}\cdot\mathbf{u}_2}{\|\mathbf{u}_2\|^2}(\mathbf{u}_1\cdot\mathbf{u}_2) - \cdots - \frac{\mathbf{v}\cdot\mathbf{u}_k}{\|\mathbf{u}_k\|^2}(\mathbf{u}_1\cdot\mathbf{u}_k) \\
&= \mathbf{v}\cdot\mathbf{u}_1 - \mathbf{v}\cdot\mathbf{u}_1 = 0,
\end{aligned}$$

and a similar computation applies to the rest of the basis vectors. Hence $\mathbf{w}$ is in $\mathcal{U}^{\perp}$.

In order to show that there can be only one such decomposition, suppose $\mathbf{v}$ has a second decomposition $\mathbf{v} = \mathbf{x} + \mathbf{y}$, for some vectors $\mathbf{x}$ in $\mathcal{U}$ and $\mathbf{y}$ in $\mathcal{U}^{\perp}$. Then $\mathbf{v} - \mathbf{v} = (\overline{\mathbf{v}} + \mathbf{w}) - (\mathbf{x} + \mathbf{y}) = \mathbf{0}$, and so $\overline{\mathbf{v}} + \mathbf{w} = \mathbf{x} + \mathbf{y}$. Define $\mathbf{z} = \overline{\mathbf{v}} - \mathbf{x}$, then

$$\mathbf{z} = \overline{\mathbf{v}} - \mathbf{x} = \mathbf{y} - \mathbf{w}.$$

However, $\mathcal{U}$ and $\mathcal{U}^{\perp}$ are subspaces of $\mathbb{R}^n$ and so $\mathbf{z} = \overline{\mathbf{v}} - \mathbf{x}$ is in $\mathcal{U}$ and $\mathbf{z} = \mathbf{y} - \mathbf{w}$ is in $\mathcal{U}^{\perp}$. By Theorem 4.9, $\mathbf{z} = \mathbf{0}$; that is, $\overline{\mathbf{v}} = \mathbf{x}$ and $\mathbf{w} = \mathbf{y}$.

EXAMPLE 3

Orthogonal Decomposition

To illustrate the proof of Theorem 4.12, consider the subspace $\mathcal{U} = \text{span}\{\mathbf{u}_1, \mathbf{u}_2\}$ in $\mathbb{R}^3$ and the vector $\mathbf{v}$, where

$$\mathbf{u}_1 = [1\ \ -1\ \ 2]^{\mathsf{T}}, \quad \mathbf{u}_2 = [1\ \ 1\ \ 0]^{\mathsf{T}}, \quad \mathbf{v} = [1\ \ 1\ \ 1]^{\mathsf{T}}.$$

From (4.39) we have

$$\bar{\mathbf{v}} = \text{proj}_{\mathcal{U}}\mathbf{v} = \frac{\mathbf{v}\cdot\mathbf{b}_1}{\|\mathbf{b}_1\|^2}\,\mathbf{b}_1 + \frac{\mathbf{v}\cdot\mathbf{b}_2}{\|\mathbf{b}_2\|^2}\,\mathbf{b}_2 = \frac{2}{6}\begin{bmatrix}1\\-1\\2\end{bmatrix} + \frac{2}{2}\begin{bmatrix}1\\1\\0\end{bmatrix} = \begin{bmatrix}4/3\\2/3\\2/3\end{bmatrix}$$

and

$$\mathbf{w} = \mathbf{v} - \bar{\mathbf{v}} = \begin{bmatrix}1\\1\\1\end{bmatrix} - \begin{bmatrix}4/3\\2/3\\2/3\end{bmatrix} = \begin{bmatrix}-1/3\\1/3\\1/3\end{bmatrix}.$$

Then $\mathbf{w}$ is orthogonal to both $\mathbf{u}_1$ and $\mathbf{u}_2$ and $\mathbf{v} = \bar{\mathbf{v}} + \mathbf{w}$. ■

Best Approximation

Refer to Figure 4.12. The orthogonal projection $\bar{\mathbf{v}} = \text{proj}_{\mathcal{U}}(\mathbf{v})$ has a special property—it happens to be the vector in $\mathcal{U}$ that is *closest to* $\mathbf{v}$, in the sense that the Euclidean distance $d(\mathbf{v}, \mathbf{x}) = \|\mathbf{v} - \mathbf{x}\|$, computed for all vectors $\mathbf{x}$ in $\mathcal{U}$, is a minimum when $\mathbf{x} = \bar{\mathbf{v}}$. We say that $\bar{\mathbf{v}}$ is the *best approximation* to $\mathbf{v}$ relative to $\mathcal{U}$.

Theorem 4.13 **Best Approximation**

Let $\mathcal{U}$ be a subspace of $\mathbb{R}^n$ and let $\mathbf{v}$ be any vector in $\mathbb{R}^n$. The orthogonal projection vector $\bar{\mathbf{v}} = \text{proj}_{\mathcal{U}}\mathbf{v}$ satisfies

$$\|\mathbf{v} - \mathbf{x}\| > \|\mathbf{v} - \bar{\mathbf{v}}\|, \tag{4.43}$$

for all vectors $\mathbf{x}$ in $\mathcal{U}$ distinct from $\bar{\mathbf{v}}$.

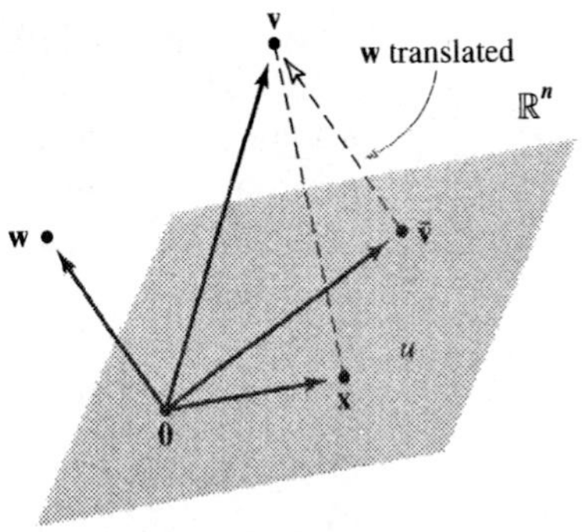

Figure 4.12 Best approximation.

Proof Let $\mathbf{x}$ be any vector in $\mathcal{U}$ distinct from $\overline{\mathbf{v}}$. Then $\overline{\mathbf{v}} - \mathbf{x}$ is in $\mathcal{U}$. By the orthogonal decomposition theorem, $\mathbf{v} - \overline{\mathbf{v}}$ is in $\mathcal{U}^{\perp}$ and so $\mathbf{v} - \overline{\mathbf{v}}$ is orthogonal to $\overline{\mathbf{v}} - \mathbf{x}$. Applying the Pythagorean theorem (Section 4.1), we have

$$\| \mathbf{v} - \mathbf{x} \|^2 = \| (\mathbf{v} - \overline{\mathbf{v}}) + (\overline{\mathbf{v}} - \mathbf{x}) \|^2 = \| (\mathbf{v} - \overline{\mathbf{v}}) \|^2 + \| (\overline{\mathbf{v}} - \mathbf{x}) \|^2 \qquad (4.44)$$

But $\| \overline{\mathbf{v}} - \mathbf{x} \|^2 > 0$ because $\overline{\mathbf{v}}$ and $\mathbf{x}$ are distinct. Hence, (4.44) gives the inequality $\| \mathbf{v} - \mathbf{x} \|^2 > \| \mathbf{v} - \overline{\mathbf{v}} \|^2$ and taking positive square roots gives (4.43).

EXAMPLE 4

Best Approximation

The equation $x_1 + 2x_2 - x_3 = 0$ determines a subspace $\mathcal{U}$ of $\mathbb{R}^3$. It can be verified that $\mathcal{B} = \{\mathbf{u}_1, \mathbf{u}_2\}$ is an orthonormal basis for $\mathcal{U}$, where

$$\mathbf{u}_1 = \frac{1}{\sqrt{2}} \begin{bmatrix} 1 \\ 0 \\ 1 \end{bmatrix}, \quad \mathbf{u}_2 = \frac{1}{\sqrt{3}} \begin{bmatrix} -1 \\ 1 \\ 1 \end{bmatrix}.$$

Consider the vector $\mathbf{v} = (1, 4, 3)$. Using (4.42), the orthogonal projection of $\mathbf{v}$ onto $\mathcal{U}$ is given by $\overline{\mathbf{v}} = \mathbf{Q}\mathbf{Q}^{\mathsf{T}}\mathbf{v}$, where $\mathbf{Q} = [\,\mathbf{u}_1 \;\; \mathbf{u}_2\,]$. We have

$$\overline{\mathbf{v}} = \mathbf{Q}\mathbf{Q}^{\mathsf{T}}\mathbf{v} = \begin{bmatrix} 1/\sqrt{2} & -1/\sqrt{3} \\ 0 & 1/\sqrt{3} \\ 1/\sqrt{2} & 1/\sqrt{3} \end{bmatrix} \begin{bmatrix} 1/\sqrt{2} & 0 & 1/\sqrt{2} \\ -1/\sqrt{3} & 1/\sqrt{3} & 1/\sqrt{3} \end{bmatrix} \begin{bmatrix} 1 \\ 4 \\ 3 \end{bmatrix} = \begin{bmatrix} 0 \\ 2 \\ 4 \end{bmatrix},$$

and so $\mathbf{w} = \operatorname{comp}_{\mathcal{U}}(\mathbf{v}) = \mathbf{v} - \overline{\mathbf{v}} = (1, 2, -1)$. The minimum distance of $\mathbf{v}$ to the plane representing $\mathcal{U}$ is $d(\mathbf{v}, \overline{\mathbf{v}}) = \| \mathbf{w} \| = \sqrt{6} \simeq 2.45$ units. ▩

The Gram–Schmidt Process, QR-factorization

In many applications, the choice of a convenient basis for a given subspace is an important consideration. Well-chosen bases make for ease of computation. An orthonormal basis, such as the standard basis for $\mathbb{R}^n$, has numerical advantages. In fact, any subspace $\mathcal{U}$ of $\mathbb{R}^n$, and more generally, any finite-dimensional *inner product space* has an orthonormal basis. The standard method for constructing orthonormal bases is called the *Gram–Schmidt process* and this will now be explained. As a byproduct we obtain the QR-*factorization* of an $m \times n$ matrix. This factorization has important applications, including the computation of *eigenvalues* of a square matrix. QR factorization will be employed in Section 4.4 to improve the accuracy of least squares calculations.

We will begin with an illustration before going on to make more general statements.

■ ILLUSTRATION 4.3

Gram–Schmidt Process and QR-Factorization

The columns $\{\mathbf{u}_1, \mathbf{u}_2, \mathbf{u}_3\}$ of the matrix $\mathbf{A}$ shown in (4.45) are linearly independent and form a basis $\mathcal{B}$ for the subspace $\operatorname{col}\mathbf{A}$ which is equal to $\mathbb{R}^3$ in this case.

$$\mathbf{A} = \begin{bmatrix} 1 & -2 & -3 \\ 2 & 0 & -3 \\ 2 & 4 & 3 \end{bmatrix} \quad \Rightarrow \quad \mathbf{u}_1 = \begin{bmatrix} 1 \\ 2 \\ 2 \end{bmatrix}, \quad \mathbf{u}_2 = \begin{bmatrix} -2 \\ 0 \\ 4 \end{bmatrix}, \quad \mathbf{u}_3 = \begin{bmatrix} -3 \\ -3 \\ 3 \end{bmatrix} \tag{4.45}$$

The basis $\mathcal{B}$ is not orthogonal and our goal is to construct an orthonormal (normalized) basis $\mathcal{Q} = \{\mathbf{q}_1, \mathbf{q}_2, \mathbf{q}_3\}$ for col $\mathbf{A}$ using the vectors $\{\mathbf{u}_1, \mathbf{u}_2, \mathbf{u}_3\}$. The significance of the scalars r_{ij} defined in the steps below will be explained in due course. Refer to Figure 4.13.

Step 1 Let $r_{11} = \| \mathbf{u}_1 \|$. We have $r_{11} = 3$ in this case and we define

$$\mathbf{q}_1 = \frac{\mathbf{u}_1}{r_{11}} = \frac{1}{3}\begin{bmatrix} 1 \\ 2 \\ 2 \end{bmatrix} = \begin{bmatrix} \frac{1}{3} \\ \frac{2}{3} \\ \frac{2}{3} \end{bmatrix}. \tag{4.46}$$

Then $\mathbf{q}_1$ is a unit vector and span$\{\mathbf{u}_1\}$ = span$\{\mathbf{q}_1\}$.

Step 2 Take the component of $\mathbf{u}_2$ orthogonal to span$\{\mathbf{q}_1\}$ and normalize it to obtain $\mathbf{q}_2$. Note that the orthogonal projection of $\mathbf{u}_2$ onto $\mathbf{q}_1$ is $(\mathbf{q}_1 \cdot \mathbf{u}_2)\mathbf{q}_1$, because $\mathbf{q}_1$ is a unit vector. Letting $r_{12} = \mathbf{q}_1 \cdot \mathbf{u}_2 = \mathbf{q}_1^{\mathsf{T}}\mathbf{u}_2$, we have

$$\mathbf{q}_2 = \frac{\mathbf{u}_2 - r_{12}\mathbf{q}_1}{\| \mathbf{u}_2 - r_{12}\mathbf{q}_1 \|}. \tag{4.47}$$

Then $\mathbf{q}_2$ is a unit vector orthogonal to $\mathbf{q}_1$ and span$\{\mathbf{u}_1, \mathbf{u}_2\}$ = span$\{\mathbf{q}_1, \mathbf{q}_2\}$ (explain). We have $r_{12} = 2$ in this case, and so the numerator in (4.47) is

$$\mathbf{u}_2 - r_{12}\mathbf{q}_1 = \begin{bmatrix} -2 \\ 0 \\ 4 \end{bmatrix} - 2\begin{bmatrix} \frac{1}{3} \\ \frac{2}{3} \\ \frac{2}{3} \end{bmatrix} = \begin{bmatrix} -\frac{8}{3} \\ -\frac{4}{3} \\ \frac{8}{3} \end{bmatrix}.$$

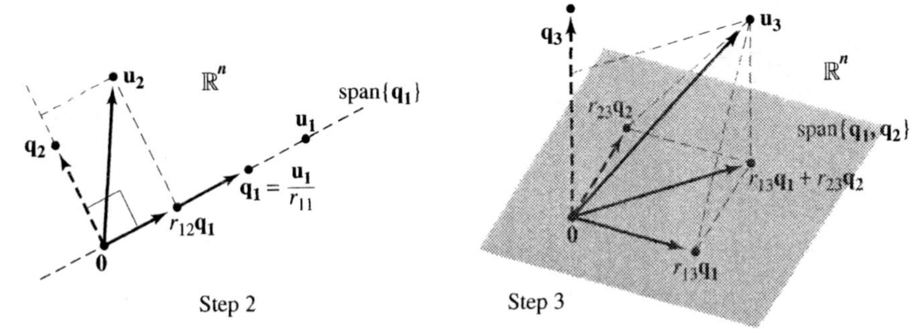

Figure 4.13 The Gram–Schmidt process: Steps 2 and 3.

Let $r_{22} = \| \mathbf{u}_2 - r_{12}\mathbf{q}_1 \|$. Then $r_{22} = 4$ in this case and (4.47) becomes

$$\mathbf{q}_2 = \frac{\mathbf{u}_2 - r_{12}\mathbf{q}_1}{r_{22}} = \frac{1}{4}\begin{bmatrix} -\frac{8}{3} \\ -\frac{4}{3} \\ \frac{8}{3} \end{bmatrix} = \begin{bmatrix} -\frac{2}{3} \\ -\frac{1}{3} \\ \frac{2}{3} \end{bmatrix}. \tag{4.48}$$

Note that $\mathbf{q}_2$ in (4.48) must be nonzero—if $\mathbf{q}_2 = \mathbf{0}$, then $\mathbf{u}_2 = r_{12}\mathbf{q}_1$, which is false because $\mathbf{q}_1$ and $\mathbf{u}_2$ are linearly independent.

Step 3 Take the component of $\mathbf{u}_3$ orthogonal to span$\{\mathbf{q}_1, \mathbf{q}_2\}$ and normalize it to obtain $\mathbf{q}_3$. The orthogonal projection of $\mathbf{u}_3$ onto $\mathbf{q}_1$ is $(\mathbf{q}_1 \cdot \mathbf{u}_3)\mathbf{q}_1$ and the orthogonal projection of $\mathbf{u}_3$ onto $\mathbf{q}_2$ is $(\mathbf{q}_2 \cdot \mathbf{u}_3)\mathbf{q}_2$. Letting $r_{13} = \mathbf{q}_1^T\mathbf{u}_3$ and $r_{23} = \mathbf{q}_2^T\mathbf{u}_3$, we have

$$\mathbf{q}_3 = \frac{\mathbf{u}_3 - r_{13}\mathbf{q}_1 - r_{23}\mathbf{q}_2}{\| \mathbf{u}_3 - r_{13}\mathbf{q}_1 - r_{23}\mathbf{q}_2 \|}. \tag{4.49}$$

Then $\mathbf{q}_3$ is a unit vector orthogonal to $\mathbf{q}_1$ and $\mathbf{q}_2$ and span$\{\mathbf{u}_1, \mathbf{u}_2, \mathbf{u}_3\}$ = span$\{\mathbf{q}_1, \mathbf{q}_2, \mathbf{q}_3\}$ (explain). We have $r_{13} = -1$ and $r_{23} = 5$ in this case and so the numerator in (4.49) is

$$\mathbf{u}_3 - r_{13}\mathbf{q}_1 - r_{23}\mathbf{q}_2 = \begin{bmatrix} -3 \\ -3 \\ 3 \end{bmatrix} + \begin{bmatrix} \frac{1}{3} \\ \frac{2}{3} \\ \frac{2}{3} \end{bmatrix} - 5\begin{bmatrix} -\frac{2}{3} \\ -\frac{1}{3} \\ \frac{2}{3} \end{bmatrix} = \begin{bmatrix} \frac{2}{3} \\ -\frac{2}{3} \\ \frac{1}{3} \end{bmatrix}. \tag{4.50}$$

Let $r_{33} = \| \mathbf{u}_3 - r_{13}\mathbf{q}_1 - r_{23}\mathbf{q}_2 \|$. Then $r_{33} = 1$ in this case and (4.49) becomes

$$\mathbf{q}_3 = \frac{\mathbf{u}_3 - r_{13}\mathbf{q}_1 - r_{23}\mathbf{q}_2}{r_{33}} = \begin{bmatrix} \frac{2}{3} \\ -\frac{2}{3} \\ \frac{1}{3} \end{bmatrix}.$$

As in Step 2, $\mathbf{q}_3$ is nonzero—if $\mathbf{q}_3 = \mathbf{0}$, then $\mathbf{u}_3 = r_{13}\mathbf{q}_1 + r_{23}\mathbf{q}_2$, which is false because $\mathbf{u}_3, \mathbf{q}_1, \mathbf{q}_2$ are linearly independent (explain).

The Gram–Schmidt process stops here—we have found an orthonormal basis $\mathcal{Q} = \{\mathbf{q}_1, \mathbf{q}_2, \mathbf{q}_3\}$ for col $\mathbf{A}$. At every step, the $\mathbf{q}$'s that have been constructed so far form an orthogonal set and are therefore linearly independent.

We are now ready to explain how the Gram–Schmidt process gives a QR-factorization of $\mathbf{A}$. Use equations (4.46), (4.48), and (4.50) to write the following vector equations:

$$\begin{aligned} \mathbf{u}_1 &= r_{11}\mathbf{q}_1 \\ \mathbf{u}_2 &= r_{12}\mathbf{q}_1 + r_{22}\mathbf{q}_2 \\ \mathbf{u}_3 &= r_{13}\mathbf{q}_1 + r_{23}\mathbf{q}_2 + r_{33}\mathbf{q}_3. \end{aligned} \tag{4.51}$$

It is now clear that each vector in the original basis $\mathcal{B}$ is a linear combination of vectors from the orthonormal basis $\mathcal{Q}$, and conversely. Let

$\mathbf{Q} = [\,\mathbf{q}_1\ \mathbf{q}_2\ \mathbf{q}_3\,]$. All the equations in (4.51) can be written compactly as one matrix equation, namely

$$\mathbf{A} = [\,\mathbf{u}_1\ \mathbf{u}_2\ \mathbf{u}_3\,] = [\,\mathbf{q}_1\ \mathbf{q}_2\ \mathbf{q}_3\,]\begin{bmatrix} r_{11} & r_{12} & r_{13} \\ 0 & r_{22} & r_{23} \\ 0 & 0 & r_{33} \end{bmatrix} = \mathbf{QR},$$

and the significance of the scalars r_{ij} defined in the steps above is now clear. In the specific example, we have

$$\mathbf{A} = \begin{bmatrix} 1 & -2 & -3 \\ 2 & 0 & -3 \\ 2 & 4 & 3 \end{bmatrix} = \begin{bmatrix} \frac{1}{3} & -\frac{2}{3} & \frac{2}{3} \\ \frac{2}{3} & -\frac{1}{3} & -\frac{2}{3} \\ \frac{2}{3} & \frac{2}{3} & \frac{1}{3} \end{bmatrix}\begin{bmatrix} 3 & 2 & -1 \\ 0 & 4 & 5 \\ 0 & 0 & 1 \end{bmatrix} = \mathbf{QR}$$

Note that $\mathbf{Q}$ has orthonormal columns, and $\mathbf{R}$ is invertible and upper triangular. ■

Gram–Schmidt Process

The condition in Illustration 4.3 that the columns of the matrix $\mathbf{A}$ be linearly independent will be relaxed in order to formulate the process for any matrix, even with dependent columns.

Consider an $m \times n$ matrix $\mathbf{A} = [\,\mathbf{u}_1\ \mathbf{u}_2\ \cdots\ \mathbf{u}_n\,]$. The vectors $\mathcal{S} = \{\mathbf{u}_1, \mathbf{u}_2, \ldots, \mathbf{u}_n\}$ span col $\mathbf{A}$ and if dim col $\mathbf{A} = p$, the work of Section 3.3 ensures that there exists a set of p vectors in $\mathcal{S}$ that are linearly independent and form a basis $\mathcal{B}$ for col $\mathbf{A}$. The Gram–Schmidt process can be applied to $\mathcal{S}$ to find an orthonormal basis $\mathcal{Q} = \{\mathbf{q}_1, \mathbf{q}_2, \ldots, \mathbf{q}_p\}$ for col $\mathbf{A}$ using the algorithm (4.52). Note that if any column $\mathbf{u}_j$ is a linear combination of the $\mathbf{q}$'s constructed so far, we go on to consider $\mathbf{u}_{j+1}$.

```
For j = 1 to n
   q_j = u_j
   For i = 1 to j − 1
       r_ij = q_i^T u_j
       q_j = q_j − r_ij q_i
   end
   If q_j = 0, next j
   r_jj = ‖ q_j ‖
   q_j = q_j / r_jj
end
```
(4.52)

The process ensures that span$\{\mathbf{u}_1, \mathbf{u}_2, \ldots, \mathbf{u}_n\}$ = span$\{\mathbf{q}_1, \mathbf{q}_2, \ldots, \mathbf{q}_p\}$ = col $\mathbf{A}$.

The Gram–Schmidt process yields a factorization $\mathbf{A} = \mathbf{QR}$ for any $m \times n$ matrix $\mathbf{A}$, not only square matrices, as is the case in Illustration 4.3.

Theorem 4.14 **QR-Factorization**

Let $\mathbf{A}$ *be an* $m \times n$ *matrix such that* rank $\mathbf{A} = n$. *Then* $\mathbf{A}$ *has a factorization* $\mathbf{A} = \mathbf{QR}$, *where* $\mathbf{Q} = [\mathbf{q}_1 \ \mathbf{q}_2 \ \cdots \ \mathbf{q}_n]$ *is* $m \times n$ *with orthonormal columns* $\mathbf{q}_1, \mathbf{q}_2, \ldots, \mathbf{q}_p$ *and* $\mathbf{R}$ *is an* $n \times n$ *invertible upper triangular matrix*

$$\mathbf{R} = \begin{bmatrix} r_{11} & r_{12} & \cdots & r_{1n} \\ 0 & r_{22} & \cdots & r_{2n} \\ \vdots & \vdots & \ddots & \vdots \\ 0 & 0 & \cdots & r_{nn} \end{bmatrix}.$$

Proof There are various methods for constructing $\mathbf{Q}$ and $\mathbf{R}$, one being the Gram–Schmidt process outlined in (4.52).

Computational Note

The algorithm (4.52) is fast compared to other methods, such as Householder's method, for example. However, (4.52) suffers from *numerical instability*—the numerator in the quotient that computes $\mathbf{q}_j$ [see (4.47) and (4.49), for example] may be the difference of almost equal vectors, and the error from such differences is potentially large. It is customary to use a modified form of Gram–Schmidt to overcome some of these concerns (see [1] for a more complete treatment).

REFERENCES

1. Hager, W. W., *Applied Numerical Linear Algebra*, Prentice Hall, Englewood Cliffs, N.J., 1988.

Historical Notes

The algorithm for constructing orthogonal sets was discovered jointly by Jörgen Pederson Gram and Erhardt Schmidt.

Jörgen Pederson Gram (1850–1916) Danish actuary. While working for the Hafnia Life Insurance Company, Gram developed the mathematical theory of accident insurance. His research work toward a Ph.D. in mathematics formulated the orthogonalization process.

Erhardt Schmidt (1876–1959) German mathematician. Schmidt graduated from the University of Göttingen and later taught at Berlin University. He made important contributions to many fields of mathematics and first formulated mathematically the orthogonalization process in a paper dated 1907.

EXERCISES 4.3

Exercises 1–4. Determine if the subspaces $\mathcal{U}$ and $\mathcal{V}$ spanned by the given sets of vectors are orthogonal.

1. $\mathcal{U}: \begin{bmatrix} 1 \\ -1 \end{bmatrix}, \quad \mathcal{V}: \begin{bmatrix} 5 \\ 5 \end{bmatrix}$

2. $\mathcal{U}: \begin{bmatrix} 1 \\ 2 \end{bmatrix}, \begin{bmatrix} -2 \\ 1 \end{bmatrix} \quad \mathcal{V}: \begin{bmatrix} 6 \\ -3 \end{bmatrix}$

3. $\mathcal{U}: \begin{bmatrix} 0 \\ 0 \\ 0 \end{bmatrix}, \begin{bmatrix} 2 \\ -3 \\ 3 \end{bmatrix}, \begin{bmatrix} 1 \\ 1 \\ 1 \end{bmatrix}, \quad \mathcal{V}: \begin{bmatrix} -1 \\ 1 \\ 0 \end{bmatrix}$

4. $\mathcal{U}: \begin{bmatrix} 1 \\ 2 \\ 3 \end{bmatrix} \quad \mathcal{V}: \begin{bmatrix} -2 \\ 1 \\ 0 \end{bmatrix}, \begin{bmatrix} -3 \\ 0 \\ 1 \end{bmatrix}$

Exercises 5–7. Find the projection of the given vector **v** *onto the subspace spanned by the vectors $\mathcal{B}$.*

5. $\mathbf{v} = \begin{bmatrix} 3 \\ -1 \\ 7 \end{bmatrix}, \quad \mathcal{B}: \begin{bmatrix} 1 \\ 1 \\ 0 \end{bmatrix}, \begin{bmatrix} 1 \\ -1 \\ 2 \end{bmatrix}$

6. $\mathbf{v} = \begin{bmatrix} 1 \\ 2 \\ 3 \end{bmatrix}, \quad \mathcal{B}: \begin{bmatrix} 1 \\ 1 \\ 0 \end{bmatrix}, \begin{bmatrix} 1 \\ -1 \\ 2 \end{bmatrix}, \begin{bmatrix} -1 \\ 1 \\ 1 \end{bmatrix}$

7. $\mathbf{v} = \begin{bmatrix} 1 \\ 0 \\ -1 \\ 4 \end{bmatrix}, \quad \mathcal{B}: \begin{bmatrix} 1 \\ 1 \\ 1 \\ 1 \end{bmatrix}, \begin{bmatrix} 1 \\ -1 \\ 1 \\ -1 \end{bmatrix}$

8. Let $\mathbf{u} = [0\ 0\ 3]^{\mathsf{T}}$, $\mathbf{v} = [1\ 1\ 0]^{\mathsf{T}}$, $\mathbf{w} = [-1\ 2\ 0]^{\mathsf{T}}$ be vectors in $\mathbb{R}^3$. Define subspaces $\mathcal{U} = \text{span}\{\mathbf{u}\}$ and $\mathcal{V} = \text{span}\{\mathbf{v}, \mathbf{w}\}$. Show that $\mathcal{U} \perp \mathcal{V}$.

9. Show that $(\mathcal{U}^{\perp})^{\perp} = \mathcal{U}$ for any subspace $\mathcal{U}$ of $\mathbb{R}^n$.

Exercises 10–11. For the given matrix **A**, *verify that* $(\text{col}\,\mathbf{A})^{\perp} = \text{null}\,\mathbf{A}^{\mathsf{T}}$.

10. $\begin{bmatrix} 1 & -1 \\ 2 & 4 \\ 0 & 1 \end{bmatrix}$

11. $\begin{bmatrix} 1 & 2 & 3 \\ 0 & 1 & 1 \\ -1 & 3 & 2 \end{bmatrix}$

12. Find the orthogonal complement to the subspace of $\mathbb{R}^3$ spanned by the row vectors from the matrix

$$\mathbf{A} = \begin{bmatrix} 1 & 1 & -1 \\ -1 & 0 & -1 \end{bmatrix}.$$

13. Refer to Theorem 4.10. Prove that if $(\mathbf{A}^{\mathsf{T}}\mathbf{v})^{\mathsf{T}}\mathbf{x} = 0$, for all vectors **x** in $\mathbb{R}^n$, then $(\mathbf{A}^{\mathsf{T}})\mathbf{v} = \mathbf{0}$, and conversely.

14. To illustrate Theorem 4.11, let $\mathcal{U} = \text{span}\{\mathbf{u}_1, \mathbf{u}_2\}$, where $\mathbf{u}_1 = [1\ -1\ 0\ 1]^{\mathsf{T}}$, $\mathbf{u}_2 = [2\ 0\ 1\ -1]^{\mathsf{T}}$. Find bases $\mathcal{B}$ and $\mathcal{B}^{\perp}$ for $\mathcal{U}$ and $\mathcal{U}^{\perp}$, respectively, and confirm that $\dim \mathcal{U} + \dim \mathcal{U}^{\perp} = 4$.

15. Find the orthogonal projection of the vector $\mathbf{v} = [-1\ 3\ 2]^{\mathsf{T}}$ onto the subspace $\mathcal{U} = \text{span}\{\mathbf{u}_1, \mathbf{u}_2\}$, where $\mathbf{u}_1 = [1\ -2\ -1]^{\mathsf{T}}$ and $\mathbf{u}_1 = [3\ -1\ 0]^{\mathsf{T}}$. Find the component of **v** orthogonal to $\mathcal{U}$.

16. Find the orthogonal projection of the vector $\mathbf{v} = [1\ -4\ 3]^{\mathsf{T}}$ onto the subspace $\mathcal{U}$ of $\mathbb{R}^3$ defined by the equation $2x_1 - 2x_2 + x_3 = 0$. Find the component of **v** orthogonal to $\mathcal{U}$.

17. Find the shortest distance from the point $P(-2, 3, 1)$ to the subspace $\mathcal{U} = \text{span}\{\mathbf{u}_1, \mathbf{u}_2\}$, where $\mathbf{u}_1 = [2\ -1\ 0]^{\mathsf{T}}$ and $\mathbf{u}_2 = [-1\ 4\ 4]^{\mathsf{T}}$.

18. Find the shortest distance from the point $Q(-1, 4, -2, 2)$ to the subspace $\mathcal{U}$ of $\mathbb{R}^4$ having vectors whose terminal points (x_1, x_2, x_3, x_4) satisfy the equation $2x_1 - 3x_3 + 2x_4 = 0$.

19. Decompose the given vector **v** into the form $\mathbf{v} = \overline{\mathbf{v}} + \mathbf{w}$, where $\overline{\mathbf{v}}$ is in $\mathcal{U}$ and **w** is in $\mathcal{U}^{\perp}$.

(a) $\mathcal{U} = \text{span}\{[3\ -1\ 2]^{\mathsf{T}},\ [2\ 0\ -3]^{\mathsf{T}}\}$, $\mathbf{v} = [2\ 1\ 6]^{\mathsf{T}}$.

(b) $\mathcal{U} = \text{span}\{[1\ 1\ 1\ 1]^{\mathsf{T}}, [1\ 1\ -1\ -1]^{\mathsf{T}}, [1\ -1\ 1\ -1]^{\mathsf{T}}\}$, $\mathbf{v} = [2\ 0\ 1\ 6]^{\mathsf{T}}$.

Exercises 20–21. Consider the projection matrix **P** *as defined in (4.42). Prove each property of* **P**.

20. **P** is symmetric.

21. **P** is idempotent; that is, $\mathbf{P}^2 = \mathbf{P}$.

22. Find the orthogonal projection of the vector $\mathbf{v} = [1\ 0\ -1\ 2]^{\mathsf{T}}$ onto the subspace of $\mathbb{R}^4$ spanned by the vectors $\mathbf{u}_1 = 0.5[1\ 1\ -1\ -1]^{\mathsf{T}}$, $\mathbf{u}_2 = 0.5[1\ -1\ 1\ -1]^{\mathsf{T}}$.

23. Use the Gram–Schmidt process to construct an orthonormal basis for $\mathbb{R}^2$ using the vectors $\mathbf{u}_1 = [-1\ 3]^{\mathsf{T}}$ and $\mathbf{u}_2 = [4\ 3]^{\mathsf{T}}$. Begin with $\mathbf{u}_1$. Rework the problem beginning with $\mathbf{u}_2$. Draw an accurate diagram that illustrates the process in $\mathbb{R}^2$ in both cases.

Exercises 24–25. Show that the set $\mathcal{S}$ is linearly independent. Use the Gram–Schmidt process to find an orthonormal basis for $\mathbb{R}^3$.

24. $\mathcal{S}: \mathbf{u}_1 = [1\ 1\ 2]^\mathsf{T},\ \mathbf{u}_2 = [2\ 1\ 1]^\mathsf{T},\ \mathbf{u}_3 = [1\ 2\ 1]^\mathsf{T}$

25. $\mathcal{S}: \mathbf{u}_1 = [3\ 0\ 0\ 0]^\mathsf{T},\ \mathbf{u}_2 = [0\ 1\ 2\ 1]^\mathsf{T},\ \mathbf{u}_3 = [0\ -1\ 3\ 2]^\mathsf{T}$

26. Find an orthonormal basis for the subspace $\mathcal{U}$ of $\mathbb{R}^3$ whose vectors have terminal points lying in the plane with equation $2x - 6y + 8z = 0$.

Exercises 27–28. For each given matrix **A**, *find orthonormal bases for* col **A**, col $\mathbf{A}^\mathsf{T}$, null **A**.

27. $\begin{bmatrix} 2 & 1 \\ 1 & 1 \\ 2 & 1 \end{bmatrix}$

28. $\begin{bmatrix} 2 & 1 & -1 \\ 1 & -1 & 0 \\ -1 & 1 & 0 \end{bmatrix}$

USING MATLAB

29. *Project.* Consult online help for the commands **colspace** and **null**. Write an M-file that calls an $n \times p$ matrix $\mathbf{A} = [\,\mathbf{v}\ \mathbf{v}_1\ \mathbf{v}_2\ \cdots\ \mathbf{v}_p\,]$ and finds the best approximation to $\mathbf{v}$ relative to the subspace $\mathcal{U}$ of $\mathbb{R}^n$ spanned by the columns $\{\mathbf{v}_1, \mathbf{v}_2, \ldots, \mathbf{v}_p\}$.

30. *Project.* Write an M-file to implement the Gram–Schmidt process (4.52). Test your program using the columns of the matrices

$$\begin{bmatrix} 1 & 2 & 3 \\ 4 & 5 & 6 \end{bmatrix}, \quad \begin{bmatrix} 1 & 3 \\ 1 & 5 \\ 1 & 0 \end{bmatrix}, \quad \begin{bmatrix} 1 & 2 & 3 & 2 \\ 2 & 1 & 3 & 1 \\ -1 & 0 & -1 & 2 \\ 3 & 4 & 7 & 1 \end{bmatrix}.$$

31. Consult online help for the command **QR**, and related commands. Use hand calculation to find the QR-factorization of the matrix

$$\mathbf{A} = \begin{bmatrix} 3 & -1 & 2 \\ 4 & 4 & -1 \end{bmatrix}$$

and compare with the machine output.

4.4 Applications

4.4.1 Method of Least Squares

The method of least squares has many important applications. In particular, it is used to find the equation of a line, curve or plane that *best fits* a given set of data points (see the Introduction to this chapter). The theory, explained next, uses the idea of orthogonal projection of a vector onto a subspace that was developed in Section 4.3.

Consider an $m \times n$ linear system (S) defined by the matrix equation

$$\mathbf{A}\mathbf{x} = \mathbf{b}, \tag{4.53}$$

where $\mathbf{A} = [\,\mathbf{v}_1\ \mathbf{v}_2\ \cdots\ \mathbf{v}_n\,]$ is the $m \times n$ coefficient matrix for (S) with columns $\{\mathbf{v}_1, \mathbf{v}_2, \ldots, \mathbf{v}_n\}$ in $\mathbb{R}^m$. The vector $\mathbf{b}$ of constant terms is in $\mathbb{R}^m$ and the vector $\mathbf{x}$ of unknowns is in $\mathbb{R}^n$.

Note that the product $\mathbf{A}\mathbf{x}$ is a linear combination of columns in $\mathbf{A}$ using scalars from the vector $\mathbf{x}$. In fact, the set of all products $\mathbf{A}\mathbf{x}$ as $\mathbf{x}$ ranges over all vectors in $\mathbb{R}^n$ is exactly col $\mathbf{A}$, the column space of $\mathbf{A}$. Recall that col $\mathbf{A}$ is a subspace of $\mathbb{R}^m$ and (S) is consistent if and only if $\mathbf{b}$ lies in col $\mathbf{A}$.

Refer to Figure 4.14. When (S) is inconsistent, $\mathbf{A}\mathbf{x}$ and $\mathbf{b}$ are not equal for any $\mathbf{x}$ in $\mathbb{R}^n$ and consequently $\mathbf{b}$ does not lie in col $\mathbf{A}$. Given any vector $\mathbf{x}$ in $\mathbb{R}^n$, the m-vector

$$\mathbf{r}(\mathbf{x}) = \mathbf{b} - \mathbf{A}\mathbf{x}$$

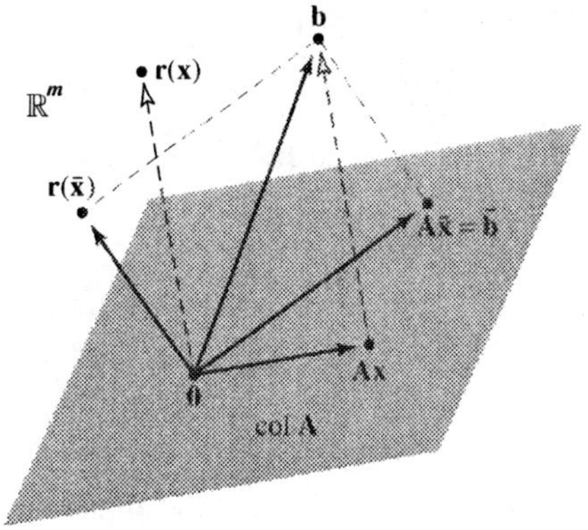

Figure 4.14 Solving an inconsistent linear system in least squares.

is called the *residual* or *error* at $\mathbf{x}$. The system (S) is consistent if and only if $\mathbf{r}(\mathbf{x}) = \mathbf{0}$ for some solution vector $\mathbf{x}$ in $\mathbb{R}^n$. When (S) is inconsistent we have $\| \mathbf{r}(\mathbf{x}) \| > 0$ for all $\mathbf{x}$ because $\mathbf{r}(\mathbf{x})$ is nonzero. The best we can do to solve (4.53) in some way is to find, if possible, a vector $\mathbf{x}$ in $\mathbb{R}^n$ that minimizes the distance $\| \mathbf{r}(\mathbf{x}) \|$ as $\mathbf{x}$ ranges over all of $\mathbb{R}^n$.

Let $\overline{\mathbf{b}} = \text{proj}_{\mathcal{U}}\mathbf{b}$ be the orthogonal projection of $\mathbf{b}$ onto col $\mathbf{A}$. Then $\overline{\mathbf{b}}$ lies in col $\mathbf{A}$ and so there is at least one solution vector $\overline{\mathbf{x}}$ in $\mathbb{R}^n$ with $\mathbf{A}\overline{\mathbf{x}} = \overline{\mathbf{b}}$. Refer to Theorem 4.13 (best approximation) in Section 4.3. In the theorem, replace n with m and let $\mathcal{U} = \text{col}\,\mathbf{A}$ be the given subspace $\mathbb{R}^m$. Let $\mathbf{v} = \mathbf{b}$ be the given vector in $\mathbb{R}^m$. Then

$$\| \mathbf{r}(\mathbf{x}) \| = \| \mathbf{b} - \mathbf{A}\mathbf{x} \| > \| \mathbf{b} - \mathbf{A}\overline{\mathbf{x}} \| = \| \mathbf{r}(\overline{\mathbf{x}}) \|$$

for all other vectors $\mathbf{A}\mathbf{x}$ in col $\mathbf{A}$ distinct from $\overline{\mathbf{b}}$ and $\mathbf{r}(\overline{\mathbf{x}})$ is the minimum we require.

There are other methods of finding $\overline{\mathbf{x}}$ (using calculus, for example) that use the fact that $\| \mathbf{r}(\overline{\mathbf{x}}) \|$ is the minimum if and only if $\| \mathbf{r}(\overline{\mathbf{x}}) \|^2 = r_1^2 + r_2^2 + \cdots + r_m^2$ is minimum, where $\mathbf{r}(\overline{\mathbf{x}}) = (r_1, r_2, \ldots, r_m)$, and this gives rise to the terminology *least squares solution*.

Normal Equations

The vector $\mathbf{w} = \mathbf{b} - \overline{\mathbf{b}} = \mathbf{b} - \mathbf{A}\overline{\mathbf{x}}$ is in $(\text{col}\,\mathbf{A})^\perp$ and $(\text{col}\,\mathbf{A})^\perp = \text{null}\,\mathbf{A}^\mathsf{T}$ so that $\mathbf{b} - \mathbf{A}\overline{\mathbf{x}}$ is in null $\mathbf{A}^\mathsf{T}$. Thus, $\mathbf{A}^\mathsf{T}(\mathbf{b} - \mathbf{A}\overline{\mathbf{x}}) = \mathbf{0}$ and rearranging, we have

$$\mathbf{A}^\mathsf{T}\mathbf{A}\overline{\mathbf{x}} = \mathbf{A}^\mathsf{T}\mathbf{b}. \tag{4.54}$$

Equation (4.54) is called the *normal equation* for $\mathbf{A}$, or alternatively, the *system of normal equations* (the linear system is $n \times n$). The solution $\overline{\mathbf{x}}$ to (4.54) is the *least squares solution* to (4.53).

The theory guarantees that the system of normal equations is always consistent and, as usual, there are two possibilities— either (4.54) has a unique solution, or (4.54) has infinitely many solutions. Which of the two possibilities occurs depends on the rank of the $n \times n$ symmetric matrix $\mathbf{A}^\mathsf{T}\mathbf{A}$. In order to classify the cases, we will require the following result.

Theorem 4.15 **Rank of $\mathbf{A}^{\mathsf{T}}\mathbf{A}$**

For any $m \times n$ matrix $\mathbf{A}$, *we have* $\operatorname{rank}\mathbf{A} = \operatorname{rank}\mathbf{A}^{\mathsf{T}}\mathbf{A}$.

Proof We first prove that $\operatorname{null}\mathbf{A} = \operatorname{null}\mathbf{A}^{\mathsf{T}}\mathbf{A}$ by showing the set inclusions:

(a) $\operatorname{null}\mathbf{A} \subseteq \operatorname{null}\mathbf{A}^{\mathsf{T}}\mathbf{A}$ and (b) $\operatorname{null}\mathbf{A}^{\mathsf{T}}\mathbf{A} \subseteq \operatorname{null}\mathbf{A}$.

(a) If $\mathbf{x}$ is in $\operatorname{null}\mathbf{A}$, then $\mathbf{Ax} = \mathbf{0}$. Hence, $\mathbf{A}^{\mathsf{T}}(\mathbf{Ax}) = \mathbf{A}^{\mathsf{T}}\mathbf{0} = \mathbf{0}$, that is, $(\mathbf{A}^{\mathsf{T}}\mathbf{A})\mathbf{x} = \mathbf{0}$, showing that $\mathbf{x}$ is in $\operatorname{null}(\mathbf{A}^{\mathsf{T}}\mathbf{A})$.

(b) If $\mathbf{x}$ is in $\operatorname{null}(\mathbf{A}^{\mathsf{T}}\mathbf{A})$, then $\mathbf{A}^{\mathsf{T}}\mathbf{Ax} = \mathbf{0}$. Premultiply the latter equation by $\mathbf{x}^{\mathsf{T}}$ to obtain $\mathbf{x}^{\mathsf{T}}\mathbf{A}^{\mathsf{T}}\mathbf{Ax} = \mathbf{0}$ that gives $(\mathbf{Ax})^{\mathsf{T}}(\mathbf{Ax}) = (\mathbf{Ax})\cdot(\mathbf{Ax}) = \|\mathbf{Ax}\|^2 = 0$. Hence, $\mathbf{Ax} = \mathbf{0}$, and so $\mathbf{x}$ is in $\operatorname{null}\mathbf{A}$.

To complete the proof, note that $\mathbf{A}$ and $\mathbf{A}^{\mathsf{T}}\mathbf{A}$ both have n columns and so the fundamental equation connecting rank and nullity gives

$$\operatorname{rank}\mathbf{A} + \operatorname{nullity}\mathbf{A} = n = \operatorname{rank}(\mathbf{A}^{\mathsf{T}}\mathbf{A}) + \operatorname{nullity}(\mathbf{A}^{\mathsf{T}}\mathbf{A})$$

However the nullity of $\mathbf{A}$ and $\mathbf{A}^{\mathsf{T}}\mathbf{A}$ are equal, and we are done. ■

Theorem 4.16 **Classifying Solutions to Normal Equations**

(a) If $\operatorname{rank}\mathbf{A} = n$, then classifying solutions $\mathbf{A}^{\mathsf{T}}\mathbf{A}$ is invertible and the linear system (4.54) has the unique solution

$$\bar{\mathbf{x}} = (\mathbf{A}^{\mathsf{T}}\mathbf{A})^{-1}\mathbf{A}^{\mathsf{T}}\mathbf{b}. \tag{4.55}$$

(b) If $\operatorname{rank}\mathbf{A} < n$, then $\mathbf{A}^{\mathsf{T}}\mathbf{A}$ is singular and the linear system (4.54) has infinitely many solutions.

Projection Matrix

Consider the case when $\operatorname{rank}\mathbf{A} = n$. We have $\bar{\mathbf{b}} = \mathbf{A}\bar{\mathbf{x}}$ and substituting $\bar{\mathbf{x}}$ from (4.55) gives

$$\bar{\mathbf{b}} = \mathbf{A}(\mathbf{A}^{\mathsf{T}}\mathbf{A})^{-1}\mathbf{A}^{\mathsf{T}}\mathbf{b} = \mathbf{Pb},$$

where $\mathbf{P} = \mathbf{A}(\mathbf{A}^{\mathsf{T}}\mathbf{A})^{-1}\mathbf{A}^{\mathsf{T}}$ is the projection matrix that projects $\mathbf{b}$ onto $\bar{\mathbf{b}} = \operatorname{proj}_{\operatorname{col}\mathbf{A}}(\mathbf{b})$. When the columns of $\mathbf{A}$ form an orthonormal basis for $\operatorname{col}\mathbf{A}$, then $\mathbf{A}^{\mathsf{T}}\mathbf{A} = \mathbf{I}_n$ and $\mathbf{P} = \mathbf{AA}^{\mathsf{T}}$, as seen in Section 4.3. Note that $\mathbf{P}^2 = \mathbf{P}$ and that $\mathbf{Pb} = \mathbf{b}$ if $\mathbf{b}$ is in $\operatorname{col}\mathbf{A}$— in other words, $\mathbf{P}$ acts as the identity relative to the subspace $\operatorname{col}\mathbf{A}$.

EXAMPLE 1 **Normal Equations, Unique Solution**

Consider the 3×2 linear system (S) written in the form $\mathbf{Ax} = \mathbf{b}$, where

$$(S)\quad \begin{cases} x_1 + x_2 = 1 & \mathrm{L}_1 \\ \quad\;\; x_2 = 1 & \mathrm{L}_2 \\ x_1 - x_2 = 0 & \mathrm{L}_3 \end{cases} \qquad \mathbf{A} = \begin{bmatrix} 1 & 1 \\ 0 & 1 \\ 1 & -1 \end{bmatrix}, \quad \mathbf{x} = \begin{bmatrix} x_1 \\ x_2 \end{bmatrix}, \quad \mathbf{b} = \begin{bmatrix} 1 \\ 1 \\ 0 \end{bmatrix}.$$

Note that (S) is inconsistent because $\operatorname{rank}\mathbf{A} = 2 < 3 = \operatorname{rank}[\mathbf{A}\,|\,\mathbf{b}]$. We expect the normal equations to have a unique solution because $\operatorname{rank}\mathbf{A} = 2 =$ number of columns in $\mathbf{A}$.

We have $\mathbf{A}^{\mathsf{T}}\mathbf{A} = \begin{bmatrix} 2 & 0 \\ 0 & 3 \end{bmatrix}$ and $(\mathbf{A}^{\mathsf{T}}\mathbf{A})^{-1} = \begin{bmatrix} \frac{1}{2} & 0 \\ 0 & \frac{1}{3} \end{bmatrix}$. Then

$$\overline{\mathbf{x}} = \begin{bmatrix} \overline{x}_1 \\ \overline{x}_2 \end{bmatrix} = (\mathbf{A}^{\mathsf{T}}\mathbf{A})^{-1}\mathbf{A}^{\mathsf{T}}\mathbf{b} = \begin{bmatrix} \frac{1}{2} & 0 \\ 0 & \frac{1}{3} \end{bmatrix} \begin{bmatrix} 1 & 0 & 1 \\ 1 & 1 & -1 \end{bmatrix} \begin{bmatrix} 1 \\ 1 \\ 0 \end{bmatrix} = \begin{bmatrix} \frac{1}{2} \\ \frac{2}{3} \end{bmatrix},$$

showing that $(\overline{x}_1, \overline{x}_2) = (\frac{1}{2}, \frac{2}{3})$ is the best possible solution. Figure 4.15 shows the three lines and solution. ■

EXAMPLE 2

Normal Equations, Infinitely Many Solutions

Consider the linear system (S) defined by $\mathbf{Ax} = \mathbf{b}$, where

$$\mathbf{A} = \begin{bmatrix} 1 & -2 \\ -1 & 2 \\ 1 & -2 \end{bmatrix}, \quad \mathbf{x} = \begin{bmatrix} x_1 \\ x_2 \end{bmatrix}, \quad \mathbf{b} = \begin{bmatrix} 1 \\ 1 \\ 0 \end{bmatrix}.$$

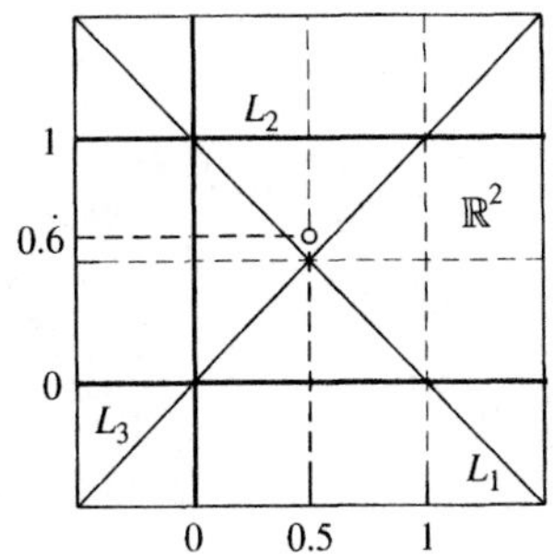

Figure 4.15 Solving a linear system in least squares.

Then (S) is inconsistent because $\operatorname{rank}\mathbf{A} = 1 < 2 = \operatorname{rank}[\mathbf{A}\,|\,\mathbf{b}]$. We expect the normal equations to have infinitely many solutions because $\operatorname{rank}\mathbf{A} = 1 <$ number of columns in $\mathbf{A}$. We have

$$\mathbf{A}^{\mathsf{T}}\mathbf{A} = \begin{bmatrix} 3 & -6 \\ -6 & 12 \end{bmatrix} \quad \text{and} \quad \mathbf{A}^{\mathsf{T}}\mathbf{b} = \begin{bmatrix} 0 \\ 0 \end{bmatrix}.$$

Let $\overline{\mathbf{x}} = \begin{bmatrix} \overline{x}_1 \\ \overline{x}_2 \end{bmatrix}$. Then the system of normal equations $\mathbf{A}^{\mathsf{T}}\mathbf{A}\overline{\mathbf{x}} = \mathbf{A}^{\mathsf{T}}\mathbf{b}$ in this case becomes

$$\begin{bmatrix} 3 & -6 \\ -6 & 12 \end{bmatrix} \begin{bmatrix} \overline{x}_1 \\ \overline{x}_2 \end{bmatrix} = \begin{bmatrix} 0 \\ 0 \end{bmatrix}$$

and the solutions are $(\overline{x}_1, \overline{x}_2) = (2t, t)$, where t is a real parameter. ■

Computational Note

There are computational difficulties that arise in practice when using the normal equation to solve a least squares problem. Special examples illustrate that computing $\mathbf{A}^{\mathsf{T}}\mathbf{A}$ and then solving $\mathbf{A}^{\mathsf{T}}\mathbf{A}\overline{\mathbf{x}} = \mathbf{A}^{\mathsf{T}}\mathbf{b}$ can give poor accuracy (see [1], for example). The use of QR-factorization or full Householder methods provides ways to circumvent these difficulties and gain greater accuracy.

For example, suppose $\mathbf{A}$ is an $m \times n$ matrix with $\operatorname{rank}\mathbf{A} = n$ (the columns of $\mathbf{A}$ are linearly independent). By Theorem 4.14, we have $\mathbf{A} = \mathbf{QR}$, where $\mathbf{Q}$ is $m \times n$ with orthonormal columns and $\mathbf{R}$ is an invertible upper triangular matrix. In practice, $\mathbf{Q}$ and $\mathbf{R}$ can be found relatively accurately and quickly. Substituting for $\mathbf{A}$ in the normal equation gives

$$\mathbf{R}\overline{\mathbf{x}} = \mathbf{Q}^{\mathsf{T}}\mathbf{b}, \tag{4.56}$$

because $\mathbf{Q}^{\mathsf{T}}\mathbf{Q} = \mathbf{I}_n$ and $\mathbf{R}^{\mathsf{T}}$ is invertible.

EXAMPLE 3 **Least Squares Solution Using QR-Factorization**

Consider the matrix equation $\mathbf{Ax} = \mathbf{b}$, where

$$\mathbf{A} = \begin{bmatrix} 2 & -1 \\ 1 & 4 \\ -2 & 10 \end{bmatrix}, \quad \mathbf{x} = \begin{bmatrix} x_1 \\ x_2 \end{bmatrix}, \quad \mathbf{b} = \begin{bmatrix} 1 \\ 2 \\ -1 \end{bmatrix}.$$

The 3×2 linear system defined by $\mathbf{Ax} = \mathbf{b}$ is inconsistent (verify). The columns of $\mathbf{A}$ are linearly independent and so rank $\mathbf{A} = 2$ and $\mathbf{A}$ has a QR-factorization

$$\mathbf{A} = \begin{bmatrix} \frac{2}{3} & \frac{1}{3} \\ \frac{1}{3} & \frac{2}{3} \\ -\frac{2}{3} & \frac{2}{3} \end{bmatrix} \begin{bmatrix} 3 & -6 \\ 0 & 9 \end{bmatrix} = \mathbf{QR}.$$

In this case, (4.56) becomes

$$\begin{bmatrix} 3 & -6 \\ 0 & 9 \end{bmatrix} \begin{bmatrix} \overline{x_1} \\ \overline{x_2} \end{bmatrix} = \begin{bmatrix} \frac{2}{3} & \frac{1}{3} & -\frac{2}{3} \\ \frac{1}{3} & \frac{2}{3} & \frac{2}{3} \end{bmatrix} \begin{bmatrix} 1 \\ 2 \\ -1 \end{bmatrix} = \begin{bmatrix} 2 \\ 1 \end{bmatrix}. \tag{4.57}$$

Back-substitution in (4.57) gives the least squares solution $\overline{\mathbf{x}} = \begin{bmatrix} \frac{8}{9} \\ \frac{1}{9} \end{bmatrix}$. ▪

EXAMPLE 4 **Linear Fitting**

A scientific experiment has produced four data points

$$\mathrm{A} = (0, 1), \quad \mathrm{B} = (1, 2), \quad \mathrm{C} = (2, 1), \quad \mathrm{D} = (3, 5). \tag{4.58}$$

We conjecture that a linear relationship $y = mx + c$ exits between the data points and we will find m (slope of line) and c (y–intercept) using least squares. It is convenient to write

$$c + mx = y. \tag{4.59}$$

Substituting the values from (4.58) into (4.59) gives a 4×2 linear system

$$(S) \quad \begin{cases} c + 0m = 1 \\ c + m = 2 \\ c + 2m = 1 \\ c + 3m = 5 \end{cases}.$$

The system (S) is inconsistent (explain). However, a least squares solution to (S) can be found. Let $\mathbf{Ax} = \mathbf{b}$ be the matrix form of (S), where

$$\mathbf{A} = \begin{bmatrix} 1 & 0 \\ 1 & 1 \\ 1 & 2 \\ 1 & 3 \end{bmatrix}, \quad \mathbf{x} = \begin{bmatrix} c \\ m \end{bmatrix}, \quad \mathbf{b} = \begin{bmatrix} 1 \\ 2 \\ 1 \\ 5 \end{bmatrix}.$$

Note that rank **A** = 2 and so the normal equations have the unique solution

$$\overline{\mathbf{x}} = \begin{bmatrix} \overline{c} \\ \overline{m} \end{bmatrix} = (\mathbf{A}^{\mathsf{T}}\mathbf{A})^{-1}\mathbf{A}^{\mathsf{T}}\mathbf{b}.$$

Computing by hand, we have

$$\mathbf{A}^{\mathsf{T}}\mathbf{A} = \begin{bmatrix} 4 & 6 \\ 6 & 14 \end{bmatrix}, \quad (\mathbf{A}^{\mathsf{T}}\mathbf{A})^{-1} = \begin{bmatrix} 0.7 & -0.3 \\ -0.3 & 0.2 \end{bmatrix}, \quad \mathbf{A}^{\mathsf{T}}\mathbf{b} = \begin{bmatrix} 9 \\ 19 \end{bmatrix}$$

and so

$$(\mathbf{A}^{\mathsf{T}}\mathbf{A})^{-1}\mathbf{A}^{\mathsf{T}}\mathbf{b} = \begin{bmatrix} 0.7 & -0.3 \\ -0.3 & 0.2 \end{bmatrix}\begin{bmatrix} 9 \\ 19 \end{bmatrix} = \begin{bmatrix} 0.6 \\ 1.1 \end{bmatrix},$$

giving $\overline{c} = 0.6$ and $\overline{m} = 1.1$ and $y = 1.1x + 0.6$. MATLAB was used to plot the data points and graph in Figure 4.16.

Computing the residual at $\overline{\mathbf{x}}$ we have

$$\mathbf{r}(\overline{\mathbf{x}}) = \mathbf{b} - \mathbf{A}\overline{\mathbf{x}} = \begin{bmatrix} 1 \\ 2 \\ 1 \\ 5 \end{bmatrix} - \begin{bmatrix} 1 & 0 \\ 1 & 1 \\ 1 & 2 \\ 1 & 3 \end{bmatrix}\begin{bmatrix} 0.6 \\ 1.1 \end{bmatrix} = \begin{bmatrix} 1 \\ 2 \\ 1 \\ 5 \end{bmatrix} - \begin{bmatrix} 0.6 \\ 1.7 \\ 2.8 \\ 3.9 \end{bmatrix} = \begin{bmatrix} 0.4 \\ 0.3 \\ -1.8 \\ 1.1 \end{bmatrix}.$$

The residual at $\overline{\mathbf{x}}$ gives the vertical distances from the data points (positive for points above, negative for points below) to the line $y = 1.1x + 0.6$. Least squares minimizes the function $\| \mathbf{r}(\mathbf{x}) \|^2$ and in this case $\| \mathbf{r}(\overline{\mathbf{x}}) \|^2 = 4.7$ is the sum of the squares of the vertical distances between the data points and the graph.

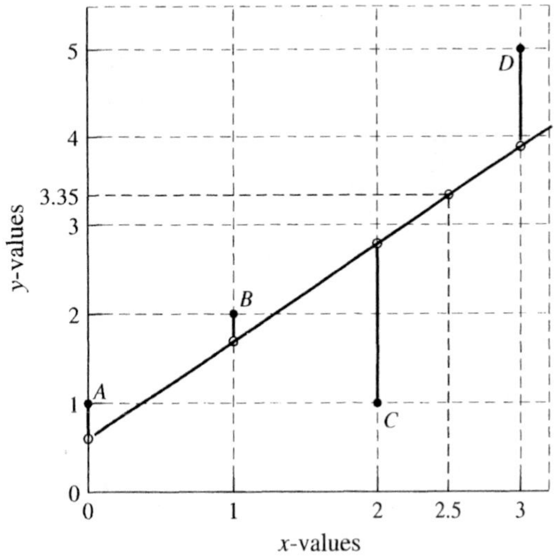

Figure 4.16 Linear fitting with least squares.

The equation of the line can now be used to estimate any value of y corresponding to any given value of x. For example, given $x = 2.5$, we have $y = (1.1)(2.5) + 0.6 = 3.35$. ■

4.4.2 Statistical Correlation

Statistics is a branch of the mathematical sciences that deals with the classification and interpretation of data in accordance with the theory of probability and the methods of hypothesis testing. Statisticians analyze empirical data and investigate the characteristics of the data itself. Statistics is called *descriptive* if it summarizes or classifies data and *inferential* when it uses a sample of data to draw conclusions or make predictions about a larger population that includes the sample itself. Inferential statistics are used in engineering, the sciences, and other applied fields.

A *correlation coefficient* is a real number s, where $|s| \leq 1$, that measures the extent to which two variables x and y are related. There is a *positive correlation* if one variable tends to increase (decrease) as the other does and there is a *negative correlation* if one variable tends to increase as the other decreases. Values of s close to zero indicate that there is little correlation between the variables. We call s a *statistic*.

Of the many correlation statistics, one of the most common is the *product moment correlation coefficient* defined by Karl Pearson and denoted by r. This statistic, often referred to as Pearson's r, satisfies $1 \leq r \leq 1$ and measures the degree of linear relationship (correlation) between the variables. We interpret the value of r as follows:

$-1.0 \leq r < -0.8$	or	$0.8 < r \leq 1.0$	High correlation
$-0.8 \leq r < -0.2$	or	$0.2 < r \leq 0.8$	Moderate correlation
$-0.2 \leq r < 0$	or	$0 < r \leq 0.2$	Low correlation

and $r = 0$ indicates no correlation. Plots of two data samples, displayed next, and the value of Pearson's r for each sample is shown in Figure 4.17.

	Sample 1	Sample 2
x-values	$\mathbf{u} = (1, 2, 3, 4, 5, 6, 7, 8, 9, 10)$	$\mathbf{u} = (1, 2, 3, 4, 5, 6, 7, 8, 9, 10)$
y-values	$\mathbf{v} = (2, 3, 2, 5, 8, 4, 6, 8, 7, 10)$	$\mathbf{v} = (9, 4, 6, 3, 6, 5, 4, 1, 5, 3)$

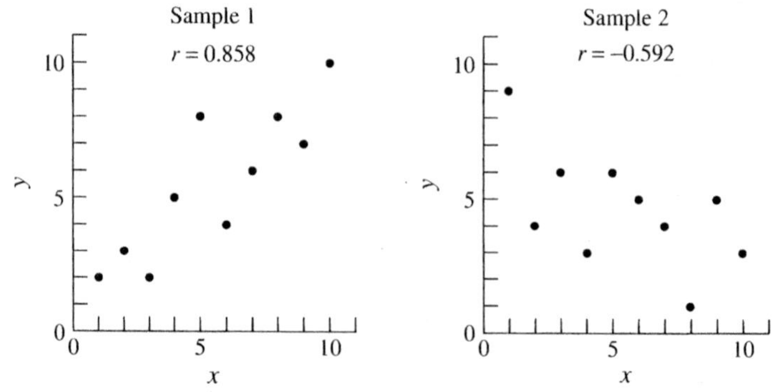

Figure 4.17 Positive and negative correlation.

There is a high positive correlation between the variables for Sample 1, while for Sample 2 there is a moderate negative correlation.

We will now define Pearson's r using dot product of vectors. For any vector $\mathbf{u}$ in $\mathbb{R}^n$, the scalar $\overline{u}$ denotes the *arithmetic mean* (or average) of the entries in $\mathbf{u}$. If $\mathbf{c} = (1, 1, \ldots, 1)$ is in $\mathbb{R}^n$ then using the dot product

$$\overline{u} = \frac{\mathbf{u} \cdot \mathbf{c}}{n} \quad \text{or} \quad n\overline{u} = \mathbf{u} \cdot \mathbf{c},$$

and $\overline{u}\mathbf{c}$ is the vector in $\mathbb{R}^n$ that has all components equal to the mean $\overline{u}$ of $\mathbf{u}$.

Definition 4.13

Pearson's product moment correlation

Let $\mathbf{u}$ and $\mathbf{v}$ be vectors in $\mathbb{R}^n$ such that the components of $\mathbf{u}$ are not all equal, and similarly for $\mathbf{v}$. The scalar r defined by

$$r = \frac{(\mathbf{u} - \overline{u}\mathbf{c}) \cdot (\mathbf{v} - \overline{v}\mathbf{c})}{\|\mathbf{u} - \overline{u}\mathbf{c}\| \, \|\mathbf{v} - \overline{v}\mathbf{c}\|} \tag{4.60}$$

is Pearson's correlation coefficient for the vectors $\mathbf{u}$ and $\mathbf{v}$.

Expressing equation (4.60) differently, we have $r = \cos\theta$, where θ is the angle between the vectors $\mathbf{u} - \overline{u}\mathbf{c}$ and $\mathbf{v} - \overline{v}\mathbf{c}$ representing the deviations of $\mathbf{u}$ and $\mathbf{v}$ from their average $\overline{u}$ and $\overline{v}$, respectively.

Suppose the vectors $\mathbf{u}$ and $\mathbf{v}$ in $\mathbb{R}^n$ each represent a set of data. For example, $\mathbf{u}$ records IQ scores for n people and $\mathbf{v}$ records their respective test results in mathematics. How are the two sets of data are related to one another? Such problems are called *correlation problems*. We will illustrate with an example.

EXAMPLE 5

Correlation of Sets of Data

A Children's Hospital recorded the weights and lengths of 10 babies born on 18 July this year, and the data are recorded in the vectors $\mathbf{w}$ and $\mathbf{h}$, where

$$\begin{aligned} \text{Weight/lb} \quad \mathbf{w} &= (8.0, 4.5, 7.5, 7.0, 6.5, 5.0, 5.5, 9.0, 7.5, 9.5), \\ \text{Lengths/in} \quad \mathbf{h} &= (22, 16, 20, 21, 18, 19, 18, 22, 17, 21). \end{aligned}$$

We will compute Pearson's r for $\mathbf{w}$ and $\mathbf{h}$. We have

$$\begin{aligned} \overline{w} &= (8 + 4.5 + 7.5 + 7.0 + 6.5 + 5.0 + 5.5 + 9.0 + 7.5 + 9.5)/10 \\ &= 70/10 = 7.0 \\ \overline{h} &= (22 + 16 + 20 + 21 + 18 + 19 + 18 + 22 + 17 + 21)/10 \\ &= 194/10 = 19.4 \end{aligned}$$

$$\mathbf{w} - \overline{w}\mathbf{c} = (1,\ -2.5,\ 0.5,\ 0,\ -0.5,\ -2,\ -1.5,\ 2,\ 0.5,\ 2.5)$$

$$\mathbf{h} - \overline{h}\mathbf{c} = (2.6,\ -3.4,\ 0.6,\ 1.6,\ -1.4,\ -0.4,\ -1.4,\ 2.6,\ -2.4,\ 1.6).$$

Then

$$(\mathbf{w}-\overline{w}\mathbf{c})\cdot(\mathbf{h}-\overline{h}\mathbf{c}) = 2.6+8.5+0.3+0+0.7+0.8+2.1+5.2-1.2+4.0 = 23.0$$

$$\begin{aligned} \|\mathbf{w} - \overline{w}\mathbf{c}\| &= (1 + 6.25 + 0.25 + 0 + 0.25 + 4 + 2.25 + 4 + 0.25 + 6.25)^{1/2} \\ &= \sqrt{24.5} \sim 4.9497 \\ \|\mathbf{h} - \overline{h}\mathbf{c}\| &= (6.76 + 11.56 + 0.36 + 2.56 + 1.96 + 0.16 \\ &\quad + 1.96 + 6.76 + 5.76 + 2.56)^{1/2} \\ &= \sqrt{40.40} \sim 6.3561. \end{aligned}$$

Pearson's r for the data of weights and lengths is

$$r = \frac{23.0}{(4.9497)(6.3561)} = \frac{23.0}{31.4608} \sim 0.73.$$

There is a strong positive correlation between **w** and **h**. ▪

Statisticians use the correlation r to define the *coefficient of determination* r^2. This statistic provides a slightly different interpretation of the data.

REFERENCES

1. Noble, B. and Daniel, J. W., *Applied Linear Algebra*, 3rd ed. Prentice Hall, Englewood Cliffs, N.J., 1988.

Historical Notes

See profile of **Karl Pearson** that opens this chapter.

EXERCISES 4.4

4.4.1 Least Squares Approximation

1. Let $\mathbf{A}$ be an $m \times n$ matrix and consider the equation $\mathbf{Ax} = \mathbf{b}$. Record the size of each matrix on the left and right sides of the equation and check conformability for multiplication (Section 2.1). Do the same for the equation $\mathbf{A}^\mathsf{T}\mathbf{Ax} = \mathbf{A}^\mathsf{T}\mathbf{b}$.
2. Let $\mathbf{A}$ be an $m \times n$ matrix with linearly independent columns. The projection of a vector $\mathbf{b}$ onto col $\mathbf{A}$ is given by $\overline{\mathbf{x}} = \mathbf{A}(\mathbf{A}^\mathsf{T}\mathbf{A})^{-1}\mathbf{A}^\mathsf{T}\mathbf{b}$. Explain the fallacy in writing

$$\begin{aligned} \mathbf{P} &= \mathbf{A}(\mathbf{A}^\mathsf{T}\mathbf{A})^{-1}\mathbf{A}^\mathsf{T} = \mathbf{A}(\mathbf{A}^{-1}(\mathbf{A}^\mathsf{T})^{-1})\mathbf{A}^\mathsf{T} \\ &= (\mathbf{A}\mathbf{A}^{-1})(\mathbf{A}^\mathsf{T})^{-1}\mathbf{A}^\mathsf{T} = \mathbf{I}_n \end{aligned}$$

so that $\overline{\mathbf{x}} = \mathbf{I}_n\mathbf{b} = \mathbf{b}$.

3. Refer to Exercise 2. Suppose $\mathbf{A}$ is $n \times n$ and invertible. Prove that $\mathbf{A}(\mathbf{A}^\mathsf{T}\mathbf{A})^{-1}\mathbf{A}^\mathsf{T} = \mathbf{I}_n$, and explain the meaning of this result.
4. Show that the projection matrix $\mathbf{P} = \mathbf{A}(\mathbf{A}^\mathsf{T}\mathbf{A})^{-1}\mathbf{A}^\mathsf{T}$ symmetric and idempotent (see Exercises 4.1 for the one-dimensional case).

Exercises 5–6. Find the system of normal equations and the least squares solution to $\mathbf{Ax} = \mathbf{b}$. *Give a geometric interpretation of the least squares solution, as in Figure 4.16, for example.*

5. $\mathbf{A} = \begin{bmatrix} 1 \\ -1 \\ 1 \end{bmatrix}, \quad \mathbf{b} = \begin{bmatrix} 1 \\ 2 \\ 3 \end{bmatrix}$

6. $\mathbf{A} = \begin{bmatrix} 1 & 0 \\ 0 & 1 \\ 1 & 1 \end{bmatrix}, \quad \mathbf{b} = \begin{bmatrix} 1 \\ 1 \\ 0 \end{bmatrix}$

Exercises 7–8. Compute the residual at the given vector $\mathbf{x}_0$ *for the matrix* **A** *and column vector* **b**. *Compare this value with the minimum residual obtained by least squares approximation. Find the least squares solution to* $\mathbf{Ax} = \mathbf{b}$.

7. $\mathbf{A} = \begin{bmatrix} 1 & -2 \\ 3 & 2 \\ -1 & 2 \end{bmatrix}$, $\mathbf{b} = \begin{bmatrix} 1 \\ 0 \\ 2 \end{bmatrix}$, $\mathbf{x}_0 = \begin{bmatrix} 1 \\ 1 \end{bmatrix}$

8. $\mathbf{A} = \begin{bmatrix} 1 & -2 \\ 1 & 1 \\ 1 & 5 \\ 1 & -2 \end{bmatrix}$, $\mathbf{b} = \begin{bmatrix} 1 \\ 2 \\ 3 \\ 4 \end{bmatrix}$, $\mathbf{x}_0 = \begin{bmatrix} 0 \\ 1 \end{bmatrix}$

Exercises 9–10. Find the orthogonal projection of **b** *onto* col **A** *for each matrix* **A** *and column vector* **b** *in the stated exercise.*

9. Exercise 7 **10.** Exercise 8

11. Find all least squares solutions to the matrix equation $\mathbf{Ax} = \mathbf{b}$, where

$$\mathbf{A} = \begin{bmatrix} 1 & 2 & 3 \\ -1 & 1 & 3 \\ 2 & 2 & 2 \\ 3 & 1 & -1 \end{bmatrix}, \quad \mathbf{b} = \begin{bmatrix} -1 \\ 0 \\ 2 \\ 1 \end{bmatrix}.$$

4.4.2 Statistical Correlation

12. As a doctor working in a children's hospital, how would you use the results from Example 5 as an indicator of possible abnormalities in your patients?

13. If $\mathbf{u} = k\mathbf{v}$ for some $k \neq 0$ and $\mathbf{u}$ is not a multiple of $\mathbf{c}$; that is, $\mathbf{u}$ contains at least two different data, show that the correlation coefficient of $\mathbf{u}$ and $\mathbf{v}$ is ± 1.

14. Show that the coefficient r^2 of correlation between $\mathbf{u}$ and $\mathbf{v}$ can be written

$$r^2 = \frac{(\mathbf{u} \cdot \mathbf{v} - n\bar{u}\bar{v})^2}{(\| \mathbf{u} \|^2 - n\bar{u}^2)(\| \mathbf{v} \|^2 - n\bar{v}^2)}.$$

Give a condition on $\mathbf{u}$ and $\mathbf{v}$ such that (a) $r^2 = 0$, (b) $r^2 = 1$.

15. Suppose $\mathbf{u}$, $\mathbf{v}$, $\mathbf{w}$ are three n-vectors such that there is a high correlation between $\mathbf{u}$, $\mathbf{v}$ and between $\mathbf{v}$, $\mathbf{w}$. Is there a high correlation between $\mathbf{u}$, $\mathbf{w}$?

16. A construction company provided the following data on the size of work crew and the level of bonus pay for eight work crews assigned to a project.

Crew Size $\mathbf{s} = (4, 4, 4, 4, 6, 6, 6, 6)$
Bonus Pay $\mathbf{b} = (2, 2, 3, 3, 2, 2, 3, 3)$

Show that the correlation coefficient of the vectors $\mathbf{s}$ and $\mathbf{b}$ is zero. What do you conclude about the linear relationship between the vectors $\mathbf{c}$ and $\mathbf{b}$?

USING MATLAB

Given an $m \times n$ matrix **A** and an $m \times 1$ column vector **b**, the MATLAB command $\mathbf{x} = \mathbf{A}\backslash\mathbf{b}$ computes the least squares solution to $\mathbf{Ax} = \mathbf{b}$. Where possible, work the following exercises from first principals, and then cross-check using the computer.

Exercises 17–18. Use least squares approximation to find a linear function $y = mx + c$ *that best fits the given data points. Estimate the dependent value* y_0 *at the given independent value* x_0. *Draw an accurate diagram.*

17. $(1, 2)$, $(-2, 3)$, $(5, 1)$, $x_0 = 6$

18. $(-2, 4)$, $(-1, 1)$, $(2, -6)$, $x_0 = 1.4$

19. Using least squares, find the cubic equation $y = ax^3 + bx + c$ that best fits the following data points:

$$(-3, -5), \quad (-1, -1), \quad (0, 1), \quad (1, 3)$$

20. An experiment measures a quantity y given values of two parameters x_1 and x_2. The nature of the experiment suggests that a functional relationship of the form $y = \alpha_1 x_1 + \alpha_2 x_2$ exists, for certain constants α_1 and α_2. Find the function y in least squares given the data points (x_1, x_2, y) shown next. Estimate the value of y when $x_1 = 3$ and $x_2 = 4$.

$$(1, 0, 3), \quad (0, 1, 4), \quad (1, 1, 1), \quad (-1, 1, 0), \quad (1, -1, -1)$$

21. Rearrange the equation of the circle $(x - a)^2 + (y - b)^2 = r^2$ into the form

$$2xa + 2yb + c = x^2 + y^2, \quad \text{where } c = r^2 - a^2 - b^2. \tag{4.61}$$

Substituting data points (x, y) into (4.61) gives a linear system that can be solved in least squares for a, b, c and then $r = \sqrt{c + a^2 + b^2}$. Find the circle that best fits the data points

$$(1, 0), \quad (1, -1), \quad (-1, -2), \quad (-1, 3), \quad (1, 1),$$

22. *Cost-Demand Curves* A *cost-demand curve* $d = f(c)$ shows the relationship between the cost of a commodity c (independent variable) and the demand d (dependent variable) for the commodity. When cost is low, there is usually greater demand, and when cost is high there is usually less demand. A cost-demand curve is invariably a *decreasing* function of c. If the cost-demand data points (c, d) give the demand d corresponding to the cost c, where

$$(c, d) = (1, 7), \quad (3, 3), \quad (6, 1),$$

find a cost-demand curve in the form $d = \dfrac{\alpha}{c} + \beta$ that best fits the data. Estimate the demand d when the cost is $c_0 = 7.2$.

Exercises 23–25. In each exercise, it is believed that a quadratic relationship $y = a + bx + cx^2$ (c nonzero) exists between the data points. Find the approximating curve that best fits the given data and use the command **plot(x,y)** *to plot the data points and the graph of the approximating function y on the same axes.*

23. The following readings were obtained from an experiment.

$$(x, y): \ (0, 1.4), \ (1, 1.2), \ (2, 3.9), \ (3, 9.8), \ (4, 17.2)$$

Interpolate the value of y when $x = 3.1$.

24. The drying time y (minutes) for a house paint depends on the amount of drying agent x (ounces) added to the paint. Use the given data points (x, y) to interpolate the value of y when $x = 2.3$.

$$(x, y): \ (0, 9), \ (1, 8.5), \ (2, 7.4), \ (3, 7.8)$$

25. Given the set of data points

$$(x, y): \ (-2, 1.4), \ (-1, 0.8), \ (0, 0.6), \ (1, 1.2), \ (2, 1.6),$$

find the approximating parabola y. Find the vertex of the parabola and plot the vertex on the graph.

26. Consider the experimental data shown.

$$\mathbf{D} = \begin{bmatrix} 1 & 2 & \textcircled{3} & 4 & \textcircled{5} & 6 & 7 & \textcircled{8} & 9 & 10 \\ 10 & 11 & 20 & 12 & 10 & 14 & 14 & 12 & 17 & 18 \end{bmatrix} \begin{matrix} \text{(Minutes)} \\ \text{(Temperature)} \end{matrix}$$

Find Pearson's r for these data. Consult **help subplot**. Plot the data points and the line of regression which passes through them in **subplot(1,2,1)**. Suppose the readings at 3, 5, 8 min are subject to error. The corresponding data points are therefore considered to be *outliers* that skew the data. Replace the three temperature values with the mean (integer) value for the rest of the data points. Plot the data points again in **subplot(1,2,2)** and draw the new line of regression. What is Pearson's r now?

27. Generate a set of 50 random data points using the commands

```
r = rand(1, 50);
  x = 1 : 50;
  y = 44 + x + 6 * r;
save mydata1.mat x y;
```

Write an M-file to plot the data points in $\mathbb{R}^2$ and to find the equation of a line of best fit through the data points. Plot the line on the same axes.

28. Write an function M-file called `pearson.m` that will compute Pearson's **r** statistic using the command **r = pearson(u, v)** Create data sets **u** and **v** with just two or three components and use these to test your program, then test your program using the data from Example 5.

29. The grades that 10 students obtained in their engineering statics and dynamics courses are given.

$$\begin{aligned} \mathbf{s} &= \text{grade in statics} \\ &= (71, 83, 56, 71, 45, 99, 85, 88, 91, 49) \\ \mathbf{b} &= \text{grade in dynamics} \\ &= (69, 61, 59, 86, 53, 98, 81, 88, 90, 36) \end{aligned}$$

Determine the correlation coefficient of the vectors **s** and **b** and interpret the result.

30. The weights of 10 children at birth, and at 10 years of age are given.

Weight in lb at birth $\mathbf{b} = (6, 5, 8, 9, 4, 7, 6, 7, 8, 5)$

Weight in lb at 10 years $\mathbf{t} = (63, 71, 70, 58, 68, 70, 55, 62, 49, 65)$

Determine the correlation coefficient of the vectors **b** and **t** and interpret the result.

31. The term marks and final examination mark in a linear algebra course for ten students are recorded.

Term marks $\mathbf{t} = (45, 22, 36, 41, 49, 18, 17, 28, 35, 12)$

Final marks $\mathbf{f} = (48, 31, 42, 50, 42, 31, 8, 24, 78, 17)$

Determine the correlation coefficient of the vectors **t** and **f**.

CHAPTER 4 REVIEW

State whether each statement is true or false as stated. Provide a clear reason for your answer. If a false statement can be modified to make it true, then do so.

1. If **a**, **b**, **c** be vectors with **a** orthogonal to **b**, then $(\mathbf{a} \cdot \mathbf{b})\mathbf{c} = \mathbf{a}(\mathbf{b} \cdot \mathbf{c})$, for any **c**.
2. $\mathbf{a} \cdot (\mathbf{b} + \mathbf{c}) = \mathbf{c} \cdot \mathbf{a} + \mathbf{b} \cdot \mathbf{a}$, for all vectors **a**, **b**, **c** in $\mathbb{R}^n$.
3. The angle θ between the vectors $\mathbf{u} = (-5, -1, 1)$ and $\mathbf{v} = (2, 4, -4)$ is $\simeq 125°$.
4. For all real t, the vector $\mathbf{c} = (10t, 9t, 17t)$ is orthogonal to the plane defined by the vectors $\mathbf{a} = (1, -3, 1)$ and $\mathbf{b} = (5, 2, -4)$.
5. A linearly independent set in $\mathbb{R}^n$ that contains p vectors is orthogonal if $p < n$.

6. If a set of vectors in $\mathbb{R}^n$ is orthogonal, then the set is linearly independent.
7. If $\mathbf{A}$ has orthogonal rows, then $\mathbf{A}^\mathsf{T}\mathbf{A} = \mathbf{I}$.
8. An orthogonal matrix is necessarily square.
9. There exists an orthogonal matrix $\mathbf{Q}$ of order 4 with entries ± 0.5.
10. The vector $(-2, 2, 2)$ lies in the orthogonal complement of $\mathcal{U}$, where $\mathcal{U}$ is the subspace of $\mathbb{R}^3$ defined by the equation $x_1 - x_2 - x_3 = 0$.
11. If $\mathcal{U}$, $\mathcal{V}$, and $\mathcal{W}$ are subspaces of $\mathbb{R}^n$ such that $\mathcal{U} \perp \mathcal{V}$ and $\mathcal{V} \perp \mathcal{W}$, then $\mathcal{U} \perp \mathcal{W}$.
12. If $\mathbf{A}$ is $n \times k$, then ${}^\mathsf{T}\mathbf{A}$ is a left-inverse for $\mathbf{A}$.
13. The orthogonal matrix $\begin{bmatrix} \cos\theta & \sin\theta \\ \sin\theta & -\cos\theta \end{bmatrix}$ defines a rotation in the plane.
14. Every vector in null $\mathbf{A}^\mathsf{T}$ is orthogonal to row $\mathbf{A}$.
15. Every vector orthogonal to null $\mathbf{A}^\perp$ lies in null $\mathbf{A}$.
16. The projection matrix $\mathbf{P} = \mathbf{Q}\mathbf{Q}^\mathsf{T}$ is symmetric.
17. The system of normal equations is either inconsistent or has infinitely many solutions.
18. Every matrix equation $\mathbf{Ax} = \mathbf{b}$ has a least squares solution.
19. The Gram–Schmidt process can only be applied to a linearly independent set of vectors.
20. Every matrix has a QR-factorization.

Augustin Louis Cauchy (1789–1857)

Cauchy was born in Paris just after the start of the French Revolution. His interest in mathematics began around 1802 with encouragement from the famous mathematicians Pierre Simon Laplace (1749–1827), and Joseph Louis Lagrange (1736–1813) who were friends of the family. Following Lagrange's advice, Cauchy obtained a good grounding in languages at the École Centrale du Parthénon before studying mathematics. Cauchy finished his formal mathematical training at the École Polytechnique in 1807 and then entered the engineering school École des Ponts et Chaussées, where he became an outstanding student.

Cauchy started his professional career as a civil engineer, working in the port of Cherbourg. At the same time he read the works of Laplace and Lagrange and started his own research.

After 1815 Cauchy held positions at École Polytechnique, the Sorbonne, and the Collège de France. His lectures often combined brilliance with confusion, and he was known for his undiplomatic dealings with colleagues. On occasion, his strong religious views clouded his scientific judgment.

Cauchy was a partisan of the Bourbons and after the revolution of 1830 was forced to resign his position in Paris and was subsequently refused public employment for the next 18 years. During this time, he taught at the University of Turin, Northern Italy, where a special chair of theoretical physics was created for him. On leaving Turin in 1848, he was allowed to return to his old position in Paris without taking an oath of allegiance to the new regime.

Cauchy's work had a great impact on virtually all branches of mathematics. He laid the foundations of modern analysis in terms of limits, continuity, convergent and divergent series, and functions of a complex variable and made notable contributions to group theory, geometry, and number theory. We remember him here particularly for his pioneering work on *determinants*.

Cauchy was an indefatigable worker, publishing 789 works. After Leonhard Euler (see Chapter 8, Profile), he ranks as one of the most prolific writers of mathematics of all time.

Chapter 5

DETERMINANTS

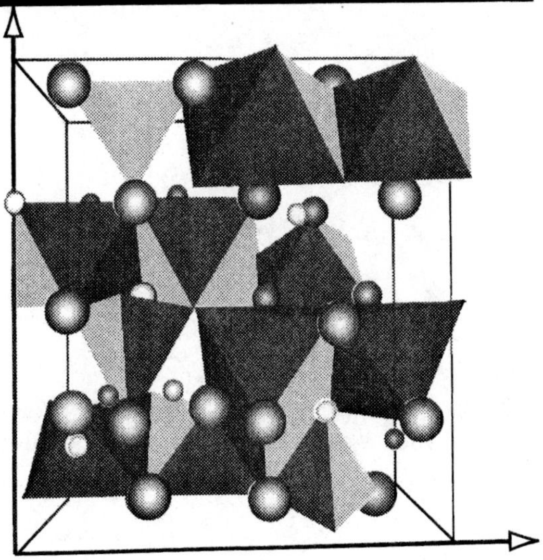

Introduction

With every square matrix $\mathbf{A}$ we associate a unique number, denoted by either $\det(\mathbf{A})$ or $|\mathbf{A}|$, called the *determinant* of $\mathbf{A}$. Just as the rank of a matrix reflects certain properties of the matrix, so does $\det(\mathbf{A})$. In particular, $\mathbf{A}$ is invertible if and only if $\det(\mathbf{A})$ is nonzero (Section 5.2, Theorem 5.6), and one consequence of this fact is that the square linear system (S) defined by $\mathbf{Ax} = \mathbf{b}$ has a unique solution for every vector $\mathbf{b}$ if and only if $\det(\mathbf{A})$ is nonzero. If $\det(\mathbf{A}) = 0$, then (S) is either inconsistent or has infinitely many solutions. In this way $\det(\mathbf{A})$ *determines* the *character* of the solution to (S).

The idea of a determinant appeared simultaneously in Japan and Europe in the year 1683, and for the next 250 years or so mathematicians developed their properties and applications. Maclaurin (1698–1746), Cramer (1704–1752), Bézout (1730–1783), and Vandermonde (1735–1796) all contributed to the theory. Laplace (1749–1827) invented the method for evaluating determinants that today bears his name. The 19th century brought forth an even greater rise in research activity in this area of algebra through the work of Sylvester, Cayley, Cauchy and others. By 1906, the available known facts concerning determinants were assembled in the four-volume work of Thomas Muir called *The Theory of Determinants in the Historical Order of Development*, but by the mid-20th century interest in determinants as a topic of research or as a computational tool had diminished considerably.

Fashions come and fashions go, even in mathematics. Today, determinants are of little interest from a computational point of view, though they do enter into the theory of linear algebra (eigenvalue problems) and mathematical analysis (Jacobians), and an understanding of the basic theory is therefore important. Section 5.1 gives the definition, main properties, and standard algorithm for computing a determinant. Section 5.2 is mainly devoted to inverses and consequences of the *product rule*. The section closes with some geometric properties of the *cross product*.

Historical Notes

Takakazu Seki Kōwa (1642–1708) Japanese mathematician. In 1683 he gave general methods for calculating "determinants" of order 2, 3, 4, 5. Independently in the same year, a letter from **Baron Gottfried Wilhelm von Leibniz** (1646–1716) to **Guillaume François Antoine Marquis de L'Hôpital** (1661–1704) mentioned the idea of the "determinant" of a 3 × 3 matrix. However, it was Cauchy who first coined the term *determinant*.

5.1 Definition and Computation

Consider the $n \times n$ matrix

$$\mathbf{A} = \begin{bmatrix} a_{11} & \cdots & a_{1j} & \cdots & a_{1n} \\ \vdots & & \vdots & & \vdots \\ a_{i1} & \cdots & a_{ij} & \cdots & a_{in} \\ \vdots & & \vdots & & \vdots \\ a_{n1} & \cdots & a_{nj} & \cdots & a_{nn} \end{bmatrix}. \tag{5.1}$$

Square Submatrices

Let a_{ij} be any entry in (5.1). Deleting row i and column j in $\mathbf{A}$, we obtain a square submatrix of $\mathbf{A}$ of order $n-1$ that will be denoted by $\mathbf{A}_{ij}$. Because there are n^2 entries in $\mathbf{A}$, there are therefore n^2 possible submatrices $\mathbf{A}_{ij}$.

EXAMPLE 1

Square Submatrices

Consider the 3 × 3 matrix

$$\mathbf{A} = \begin{bmatrix} 1 & 2 & 3 \\ 4 & 5 & 6 \\ 7 & 8 & 9 \end{bmatrix}.$$

We will form the submatrices $\mathbf{A}_{11}$ and $\mathbf{A}_{23}$.

$$\text{Delete row } i = 1 \text{ and column } j = 1 \text{ in } \mathbf{A} \quad \Rightarrow \quad \mathbf{A}_{11} = \begin{bmatrix} 5 & 6 \\ 8 & 9 \end{bmatrix}.$$

$$\text{Delete row } i = 2 \text{ and column } j = 3 \text{ in } \mathbf{A} \quad \Rightarrow \quad \mathbf{A}_{23} = \begin{bmatrix} 1 & 2 \\ 7 & 8 \end{bmatrix}.$$

There are seven other submatrices $\mathbf{A}_{ij}$ that can be formed from $\mathbf{A}$. ■

The determinant of a square matrix $\mathbf{A}$ of order n is defined in terms of the determinants of certain square submatrices $\mathbf{A}_{ij}$ of $\mathbf{A}$ of order $(n-1)$. To prepare the way for Definition 5.2 we introduce the ideas of *minor* and *cofactor*.

Definition 5.1

Minor and cofactor

The (i, j)-*minor* of $\mathbf{A}$ is the number

$$m_{ij} = \det(\mathbf{A}_{ij}).$$

The (i, j)-*cofactor* of $\mathbf{A}$ is the number

$$c_{ij} = (-1)^{i+j} m_{ij} = (-1)^{i+j} \det(\mathbf{A}_{ij}).$$

It should be mentioned that the (i, j)-cofactor of $\mathbf{A}$ is traditionally denoted by C_{ij}, with a capital letter. However, we have chosen to break with tradition and write c_{ij} in order to emphasize the fact that a cofactor is a scalar.

Notice that the cofactor c_{ij} is just m_{ij} multiplied by either $+1$ or -1 according to the following pattern, which replaces the entries in $\mathbf{A}$ by alternate $+$ and $-$ signs.

$$\begin{bmatrix} + & - & + & \cdots \\ - & + & - & \cdots \\ + & - & + & \cdots \\ \vdots & \vdots & \vdots & \ddots \end{bmatrix}$$

Hence

$$\begin{aligned} c_{11} &= (-1)^{1+1} m_{11} = +m_{11}, \\ c_{12} &= (-1)^{1+2} m_{12} = -m_{12}, \\ c_{23} &= (-1)^{2+3} m_{23} = -m_{23} \end{aligned}$$

and so on. A square matrix of order n has n^2 minors and n^2 cofactors.

Definition 5.2

Determinant

(a) The determinant of the 1×1 matrix $[a]$ is given by $\det([a]) = a$.
(b) The determinant of the $n \times n$ matrix $\mathbf{A} = [a_{ij}]$ is given by

$$\det(\mathbf{A}) = \sum_{j=1}^{n} a_{1j} c_{1j} = a_{11} c_{11} + a_{12} c_{12} + \cdots + a_{1n} c_{1n}. \tag{5.2}$$

The expression on the left side of (5.2) is called the *Laplace expansion* of $\det(\mathbf{A})$ on row 1 of $\mathbf{A}$. Each entry in row 1 of $\mathbf{A}$ is multiplied by the corresponding cofactor, and the resulting products are summed. Each cofactor c_{ij} in (5.2) involves a determinant of order $(n - 1)$, and so the definition is *recursive*. This means that, for example, a determinant of order 3 is computed using 3 determinants of order 2, and a determinant of order 2 is computed using 2 determinants of order 1, as we see next.

Determinants of Order 2

Consider the general 2 × 2 matrix

$$\mathbf{A} = \begin{bmatrix} a_{11} & a_{12} \\ a_{21} & a_{22} \end{bmatrix}.$$

Expanding on row 1 of **A**, as in (5.2), we have

$$\begin{aligned} \det(\mathbf{A}) &= a_{11}c_{11} + a_{12}c_{12} = a_{11}\det([a_{22}]) + a_{12}(-\det([a_{21}])) \\ &= a_{11}a_{22} - a_{12}a_{21}. \end{aligned} \tag{5.3}$$

The following diagonal scheme helps to visualize formula (5.3).

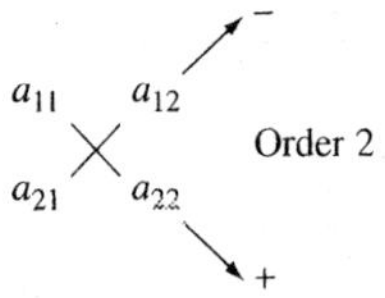

EXAMPLE 2

Finding a Specified Minor and Cofactor

With **A** as in Example 1, the minor and cofactor corresponding to the entry a_{11} in **A** is:

$$\mathbf{A} = \begin{bmatrix} 1 & 2 & 3 \\ 4 & 5 & 6 \\ 7 & 8 & 9 \end{bmatrix} \Rightarrow \mathbf{A}_{11} = \begin{bmatrix} 5 & 6 \\ 8 & 9 \end{bmatrix} \Rightarrow m_{11} = \begin{vmatrix} 5 & 6 \\ 8 & 9 \end{vmatrix}$$

$$= 5(9) - 6(8) = -3,$$

and so $c_{11} = (-1)^{1+1}m_{11} = -3$. For the entry a_{32} in **A**, we have

$$\mathbf{A} = \begin{bmatrix} 1 & 2 & 3 \\ 4 & 5 & 6 \\ 7 & 8 & 9 \end{bmatrix} \Rightarrow \mathbf{A}_{32} = \begin{bmatrix} 1 & 3 \\ 4 & 6 \end{bmatrix} \Rightarrow m_{32} = \begin{vmatrix} 1 & 3 \\ 4 & 6 \end{vmatrix}$$

$$= 1(6) - 3(4) = -6,$$

and so $c_{32} = (-1)^{3+2}m_{32} = (-1)(-6) = 6$. ▩

Determinants of Order 3

Consider the general 3 × 3 matrix

$$\mathbf{A} = \begin{bmatrix} a_{11} & a_{12} & a_{13} \\ a_{21} & a_{22} & a_{23} \\ a_{31} & a_{32} & a_{33} \end{bmatrix}.$$

Expanding on row 1 of **A** and using (5.3), we have

$$\det(\mathbf{A}) = a_{11}c_{11} + a_{12}c_{12} + a_{13}c_{13}$$

$$= a_{11}\begin{vmatrix} a_{22} & a_{23} \\ a_{32} & a_{33} \end{vmatrix} - a_{12}\begin{vmatrix} a_{21} & a_{23} \\ a_{31} & a_{33} \end{vmatrix} + a_{13}\begin{vmatrix} a_{21} & a_{22} \\ a_{31} & a_{32} \end{vmatrix}$$

$$= a_{11}a_{22}a_{33} + a_{12}a_{23}a_{31} + a_{13}a_{21}a_{32} \qquad (5.4)$$
$$-a_{11}a_{23}a_{32} - a_{12}a_{21}a_{33} - a_{13}a_{22}a_{31}.$$

The expansion (5.4) can be formed by writing down columns 1 and 2 from **A** (in order) on the right side of the matrix **A**.

$$\begin{bmatrix} a_{11} & a_{12} & a_{13} \\ a_{21} & a_{22} & a_{23} \\ a_{31} & a_{32} & a_{33} \end{bmatrix} \begin{matrix} a_{11} & a_{12} \\ a_{21} & a_{23} \\ a_{31} & a_{32} \end{matrix}$$

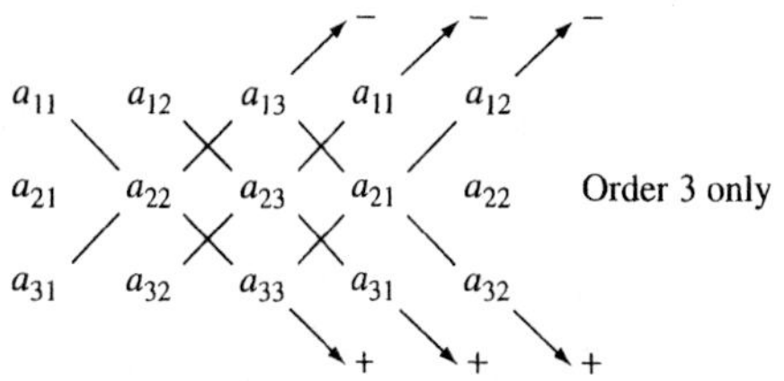

Multiply entries on diagonals as shown, each product with a sign + or − as indicated in the scheme and sum the six products.

Caution! No such scheme exists for square matrices of order $n \geq 4$.

Computational Note

The expansions (5.3) and (5.4) demonstrate a property of determinants that is true for any order n: det(**A**) is a sum of $n!$ *signed* (either + or −) products. There are n entries from **A** in each product—one entry from each row and one from each column; no two entries from any row or any column (you may check this for $n = 2, 3$). When **A** has order $n = 4$, there are $4! = 4 \cdot 3 \cdot 2 \cdot 1 = 24$ signed products in the total expansion of det(**A**) and when $n = 10$, there are $10! = 36288 \times 10^2 \simeq 3.6$ million signed products. Using the Laplace expansion to find det(**A**) is therefore not computationally viable. In practice, determinants are computed by performing forward reduction on the matrix.

The definition of det(**A**) given by (5.2) uses row 1 of **A**. In fact, expansion on *any* row or *any* column of **A** results in value of det(**A**) given by (5.2). This fact is recorded next, without proof.

Theorem 5.1 **Row and Column Expansions**

Let $\mathbf{A} = [a_{ij}]$ *be an* $n \times n$ *matrix and let* $1 \leq i, j \leq n$.

Expanding on row i *of* **A**, *we have*

$$\det(\mathbf{A}) = \sum_{j=1}^{n} a_{ij}c_{ij} = a_{i1}c_{i1} + a_{i2}c_{i2} + \cdots + a_{in}c_{in}. \qquad (5.5)$$

Expanding on column j of **A**, *we have*

$$\det(\mathbf{A}) = \sum_{i=1}^{n} a_{ij}c_{ij} = a_{1j}c_{1j} + a_{2j}c_{2j} + \cdots + a_{nj}c_{nj}. \tag{5.6}$$

The expressions on the right side of (5.5) and (5.6) are called *Laplace expansions*.

EXAMPLE 3

Laplace Expansion, a Determinant of Order 4

When using Theorem 5.1, expanding on a row or column containing zero entries will shorten the computation. Consider the matrix

$$\mathbf{A} = \begin{bmatrix} 1 & 2 & 5 & 4 \\ 3 & 1 & 4 & 2 \\ 0 & 3 & 0 & 4 \\ -1 & 2 & 2 & 3 \end{bmatrix}.$$

The Laplace expansion on row 3 only requires computation of c_{32} and c_{34}. Each minor is evaluated by expanding on row 1 of the corresponding submatrix. Thus

$$m_{32} = \begin{vmatrix} 1 & 5 & 4 \\ 3 & 4 & 2 \\ -1 & 2 & 3 \end{vmatrix} = (1)\begin{vmatrix} 4 & 2 \\ 2 & 3 \end{vmatrix} - 5\begin{vmatrix} 3 & 2 \\ -1 & 3 \end{vmatrix} + 4\begin{vmatrix} 3 & 4 \\ -1 & 2 \end{vmatrix}$$
$$= (12-4) - 5(9-(-2)) + 4(6-(-4)) = -7$$

and

$$m_{34} = \begin{vmatrix} 1 & 2 & 5 \\ 3 & 1 & 4 \\ -1 & 2 & 2 \end{vmatrix} = (1)\begin{vmatrix} 1 & 4 \\ 2 & 2 \end{vmatrix} - 2\begin{vmatrix} 3 & 4 \\ -1 & 2 \end{vmatrix} + 5\begin{vmatrix} 3 & 1 \\ -1 & 2 \end{vmatrix}$$
$$= (2-8) - 2(6-(-4)) + 5(6-(-1)) = 9.$$

Then $c_{32} = -m_{32} = 7$ and $c_{34} = -m_{34} = -9$. Hence

$$\det(\mathbf{A}) = a_{31}c_{31} + a_{32}c_{32} + a_{33}c_{33} + a_{34}c_{34} = 0 + 3(7) + 0 + 4(-9) = -15.$$

▩

Some of the most basic properties of determinants are stated next. Proofs are included here in order to reinforce the understanding of the definition of determinant.

Theorem 5.2

Properties of Determinants

(a) $\det(\mathbf{A}^{\mathsf{T}}) = \det(\mathbf{A})$. Hence any property of $\det(\mathbf{A})$ stated in terms of rows is also true when stated for columns.
(b) If **A** has a zero row (column), then $\det(\mathbf{A}) = 0$.
(c) If **A** has two rows (columns) equal, then $\det(\mathbf{A}) = 0$.
(d) If α is any number, then $\det(\alpha\mathbf{A}) = \alpha^n \det(\mathbf{A})$.

Proof

(a) The rows of $\mathbf{A}$ are the columns of $\mathbf{A}^{\mathrm{T}}$, and conversely. Appeal to Theorem 5.1.

(b) Let row i in $\mathbf{A}$ be a zero row. Expanding on row i, we have $\det(\mathbf{A}) = 0c_{i1} + 0c_{i2} + \cdots + 0c_{in} = 0$.

(c) Apply mathematical induction on the order n of $\mathbf{A}$. The property is true for order $n = 2$. Suppose the property is true for order $n - 1 \geq 2$ and consider a matrix $\mathbf{A}$ of order n that has two rows equal. Expand $\det(\mathbf{A})$ on any row i that is not one of the two equal rows. The cofactors in (5.5) involve determinants of order $n - 1$, each with two rows equal—these determinants are zero by the inductive hypothesis. Hence

$$\det(\mathbf{A}) = a_{i1}c_{i1} + \cdots + a_{in}c_{in} = a_{i1}0 + \cdots + a_{in}0 = 0.$$

(d) Apply mathematical induction on the order n of $\mathbf{A}$. The property is true for order $n = 1$. Suppose the property is true for order $n - 1 \geq 1$ and consider a matrix $\mathbf{A}$ of order n. Expanding $\det(\alpha\mathbf{A})$ on any row i, we have

$$\begin{aligned}\det(\alpha\mathbf{A}) &= \alpha a_{i1}c'_{i1} + \cdots + \alpha a_{in}c'_{in} \quad (c' \text{ denotes a cofactor of } \alpha\mathbf{A}) \\ &= \alpha a_{i1}\alpha^{n-1}c_{i1} + \cdots + \alpha a_{in}\alpha^{n-1}c_{in} \quad \text{(inductive hypothesis)} \\ &= \alpha^n \det(\mathbf{A}).\end{aligned}$$

The next result tells us how the value of $\det(\mathbf{A})$ changes when a single elementary row operation (ERO) is performed on $\mathbf{A}$.

Theorem 5.3 **Elementary Row Operations and the Determinant**

Let $\mathbf{A}$ *be an* $n \times n$ *matrix and let* m *and* α *be nonzero numbers.*

(a) Performing the replacement $\mathbf{r}_i - m\mathbf{r}_j \to \mathbf{r}_i$ on row i of $\mathbf{A}$ results in a matrix $\mathbf{A}_1$ and

$$\det(\mathbf{A}_1) = \det(\mathbf{A}).$$

(b) Performing the interchange $\mathbf{r}_i \leftrightarrow \mathbf{r}_j$ on $\mathbf{A}$, where $i \neq j$, results in a matrix $\mathbf{A}_1$ and

$$\det(\mathbf{A}_1) = -\det(\mathbf{A}).$$

(c) Performing the scaling $\alpha\mathbf{r}_i \to \mathbf{r}_i$ on row i of $\mathbf{A}$ results in a matrix $\mathbf{A}_1$ and

$$\det(\mathbf{A}_1) = \alpha \det(\mathbf{A}).$$

Proofs of properties (a), (b), (c) are left for Exercises 5.1.

Consider each ERO defined in cases (a), (b), (c) of Theorem 5.3. Recall from Section 2.2 that each ERO defines an elementary matrix $\mathbf{E}$ that is obtained by performing the same ERO on the identity matrix $\mathbf{I}$, and in every case we have $\mathbf{A}_1 = \mathbf{E}\mathbf{A}$. It is easy to prove (Exercises 5.1) that

$$\det(\mathbf{E}\mathbf{A}) = \det(\mathbf{E})\det(\mathbf{A}), \tag{5.7}$$

and, in particular, (5.7) is true when $\mathbf{A}$ is an elementary matrix.

Linearity

Note that the equation

$$\det(\mathbf{A} + \mathbf{B}) = \det(\mathbf{A}) + \det(\mathbf{B}) \tag{5.8}$$

is *false* in general. For example, we have

$$\mathbf{A} = \begin{bmatrix} 1 & 2 \\ 3 & 4 \end{bmatrix}, \quad \mathbf{B} = \begin{bmatrix} 5 & 6 \\ 7 & 8 \end{bmatrix}, \quad \mathbf{A} + \mathbf{B} = \begin{bmatrix} 6 & 8 \\ 10 & 12 \end{bmatrix},$$

and $\det(\mathbf{A}) = \det(\mathbf{B}) = -2$, while $\det(\mathbf{A} + \mathbf{B}) = -8 \neq \det(\mathbf{A}) + \det(\mathbf{B})$.

The linearity property for determinants is expressed in the next result.

Theorem 5.4 Linearity

Suppose that $n \times n$ matrices $\mathbf{A}$, $\mathbf{B}$, $\mathbf{C}$ are identical except for row (column) i. Let $\mathbf{a} = (a_1, a_2, \ldots, a_n)$ and $\mathbf{b} = (b_1, b_2, \ldots, b_n)$ denote row (column) i in $\mathbf{A}$ and $\mathbf{B}$, respectively. Let row (column) i in $\mathbf{C}$ be $s\mathbf{a} + t\mathbf{b}$, where s and t are scalars. Then

$$\det(\mathbf{C}) = s \det(\mathbf{A}) + t \det(\mathbf{B}). \tag{5.9}$$

Proof Expanding $\det(\mathbf{C})$ on row i, remembering that the cofactors of $\mathbf{A}$, $\mathbf{B}$, $\mathbf{C}$ for row i are identical, we have

$$\begin{aligned} \det(\mathbf{C}) &= (sa_1 + tb_1)c_{i1} + \cdots + (sa_n + tb_n)c_{in} \\ &= s(a_1 c_{i1} + \cdots + a_n c_{in}) + t(b_1 c_{i1} + \cdots + b_n c_{in}) \\ &= s \det(\mathbf{A}) + t \det(\mathbf{B}). \end{aligned}$$

Illustrating Theorem 5.4 when $n = 2$, using row $i = 2$, we have

$$\mathbf{A} = \begin{bmatrix} p_1 & p_2 \\ a_1 & a_2 \end{bmatrix}, \quad \mathbf{B} = \begin{bmatrix} p_1 & p_2 \\ b_1 & b_2 \end{bmatrix},$$

$$|\mathbf{C}| = \begin{vmatrix} p_1 & p_2 \\ sa_1 + ta_1 & sb_1 + tb_2 \end{vmatrix} = s \begin{vmatrix} p_1 & p_2 \\ a_1 & a_2 \end{vmatrix} + t \begin{vmatrix} p_1 & p_2 \\ b_1 & b_2 \end{vmatrix} = s|\mathbf{A}| + t|\mathbf{B}|.$$

Computation

We will now explain how determinants are computed in practice. The first step is to consider a triangular matrix.

Theorem 5.5 Triangular Matrices

Let $\mathbf{U}$ be an $n \times n$ upper (lower) triangular matrix. Then $\det(\mathbf{U})$ *is the product of the entries on the main diagonal of* $\mathbf{U}$.

Proof An illustration of the proof when $n = 4$ shows the general idea. Consider an upper triangular matrix

$$\mathbf{U} = \begin{bmatrix} a_{11} & * & * & * \\ 0 & a_{22} & * & * \\ 0 & 0 & a_{33} & * \\ 0 & 0 & 0 & a_{44} \end{bmatrix},$$

where each $*$ represents a number (0 accepted). The Laplace expansion on column 1 of $\mathbf{U}$ gives $\det(\mathbf{U}) = a_{11}c_{11} = a_{11}m_{11}$. The minor $m_{11} = \det(\mathbf{U}_{11})$ is computed using the Laplace expansion on col 1 of $\mathbf{U}_{11}$, and so on. Hence

$$\begin{vmatrix} a_{11} & * & * & * \\ 0 & a_{22} & * & * \\ 0 & 0 & a_{33} & * \\ 0 & 0 & 0 & a_{44} \end{vmatrix} = a_{11} \begin{vmatrix} a_{22} & * & * \\ 0 & a_{33} & * \\ 0 & 0 & a_{44} \end{vmatrix} = a_{11}a_{22} \begin{vmatrix} a_{33} & * \\ 0 & a_{44} \end{vmatrix}$$
$$= a_{11}a_{22}a_{33}a_{44}.$$

Loosely speaking, each entry on the main diagonal of $\mathbf{U}$ is "pulled out" of the determinant at each step of the calculation. Note that the transpose of a lower triangular matrix is upper triangular and Theorem 5.2(a) can be applied.

Applying Theorem 5.5, we have

$$\begin{vmatrix} 2 & 10 & 0 \\ 0 & -3 & 6 \\ 0 & 0 & 5 \end{vmatrix} = (2)(-3)(5) = -30 \quad \text{and} \quad \begin{vmatrix} 2 & 10 & 0 \\ 0 & 0 & 6 \\ 0 & 0 & 5 \end{vmatrix} = (2)(0)(5) = 0.$$

Clearly, $\det(\mathbf{T}) = 0$ for any triangular matrix $\mathbf{T}$ that has a zero entry on its main diagonal.

We now make an important observation. Theorem 5.3 shows that performing a single ERO on $\mathbf{A}$ results in a matrix $\mathbf{A}_1$, and

$$\det(\mathbf{A}) = p\det(\mathbf{A}_1), \quad \text{where } p = +1, -1 \text{ or } 1/\alpha \ (\alpha \text{ nonzero}). \tag{5.10}$$

We have seen in Section 1.2 that reduction of $\mathbf{A}$ to an echelon form $\mathbf{U}$ is accomplished using a sequence of elementary row operations ERO_1, ERO_2, $\cdots$, ERO_k, and $\mathbf{U}$ is necessarily upper triangular. Applying each ERO in turn and computing the determinant at each step using (5.10), we have

$$\mathbf{A} \underset{\text{ERO}_1}{\sim} \mathbf{A}_1 \underset{\text{ERO}_2}{\sim} \mathbf{A}_2 \underset{\text{ERO}_3}{\sim} \cdots \underset{\text{ERO}_k}{\sim} \mathbf{A}_k = \mathbf{U} \tag{5.11}$$

$$|\mathbf{A}| = p_1\,|\mathbf{A}_1| = p_1\,p_2\,|\mathbf{A}_2| = \cdots = p_1\,p_2 \cdots p_k\,|\mathbf{A}_k| = q\,|\mathbf{U}|, \tag{5.12}$$

where $p_1, p_2, \ldots, p_k$ are the scalars corresponding to the EROs and $q = p_1 p_2 \cdots p_k$. Note that q is nonzero. If $u_1, u_2, \ldots, u_n$ are the entries on the main diagonal of $\mathbf{U}$, then by Theorem 5.5, we have

$$\det(\mathbf{A}) = q\,u_1\,u_2 \cdots u_n, \tag{5.13}$$

and $\det(\mathbf{A}) = 0$ whenever $\mathbf{U}$ has a zero entry on its main diagonal.

The formula (5.13) is the key to the efficient computation of a determinant.

EXAMPLE 4

Evaluating a Determinant, Forward Reduction

We will compute $|\mathbf{A}|$ by applying forward reduction on $\mathbf{A}$. We choose a pivot (circled) in column 1 that is convenient for hand calculation.

$$\begin{vmatrix} 2 & 2 & 3 \\ \textcircled{1} & -2 & 3 \\ -1 & 4 & 1 \end{vmatrix} \underset{\mathbf{r}_1 \leftrightarrow \mathbf{r}_2}{=} - \begin{vmatrix} \textcircled{1} & -2 & 3 \\ 2 & 2 & 3 \\ -1 & 4 & 1 \end{vmatrix} \underset{\substack{\mathbf{r}_2 - 2\mathbf{r}_1 \to \mathbf{r}_2 \\ \mathbf{r}_3 + \mathbf{r}_1 \to \mathbf{r}_3}}{=} - \begin{vmatrix} \textcircled{1} & -2 & 3 \\ 0 & \textcircled{6} & -3 \\ 0 & 2 & 4 \end{vmatrix}$$

$$\underset{\mathbf{r}_3 - \frac{1}{3}\mathbf{r}_2 \to \mathbf{r}_3}{=} - \begin{vmatrix} \textcircled{1} & 2 & 3 \\ 0 & \textcircled{6} & -3 \\ 0 & 0 & \textcircled{5} \end{vmatrix} = (-1)(1)(6)(5) = -30.$$

Referring to (5.11), there are $k = 4$ EROs in this case and $p_1 = -1$, $p_2 = p_3 = p_4 = 1$. Comparing with (5.13), we see that $q = -1$, $u_1 = 1$, $u_2 = 6$, $u_3 = 5$ and $|\mathbf{A}| = -30$. ■

Determinant Equations

The determinants of matrices having entries that are variables or functions can be computed by forward reduction in the usual way.

EXAMPLE 5

Solving a Determinant Equation, EROs

We wish to solve the following determinant equation for x.

$$\begin{vmatrix} x & x-2 & 2x+2 \\ x+2 & 2x+2 & x \\ 2x-2 & x & x-2 \end{vmatrix} = 0 \tag{5.14}$$

Applying the sequence of EROs $\mathbf{r}_1 + \mathbf{r}_2 \to \mathbf{r}_1$, $\mathbf{r}_1 + \mathbf{r}_3 \to \mathbf{r}_1$, $\mathbf{r}_3 - 2\mathbf{r}_2 \to \mathbf{r}_3$ gives

$$\begin{vmatrix} x & x-2 & 2x+2 \\ x+2 & 2x+2 & x \\ 2x-2 & x & x-2 \end{vmatrix} = \begin{vmatrix} 4x & 4x & 4x \\ x+2 & 2x+2 & x \\ -6 & -3x-4 & -x-2 \end{vmatrix}$$

and we see that $x = 0$ is a solution to (5.14). Pull out a factor of $4x$ from row 1 and -1 from row 3 to give

$$-(4x) \begin{vmatrix} 1 & 1 & 1 \\ x+2 & 2x+2 & x \\ 6 & 3x+4 & x+2 \end{vmatrix} \underset{\substack{\mathbf{r}_2 - (x+2)\mathbf{r}_1 \to \mathbf{r}_2 \\ \mathbf{r}_1 - 6\mathbf{r}_2 \to \mathbf{r}_1}}{=} -4x \begin{vmatrix} 1 & 1 & 1 \\ 0 & x & -2 \\ 0 & 3x-2 & x-4 \end{vmatrix}.$$

Using a final replacement, we have, assuming x is nonzero

$$-4x \begin{vmatrix} 1 & 1 & 1 \\ 0 & x & -2 \\ 0 & 3x-2 & x-4 \end{vmatrix} \underset{\mathbf{r}_3 - \frac{(3x-2)}{x}\mathbf{r}_2 \to \mathbf{r}_3}{=} -4x \begin{vmatrix} 1 & 1 & 1 \\ 0 & x & -2 \\ 0 & 0 & \dfrac{(x^2+2x-4)}{x} \end{vmatrix}$$

and the value of the last determinant is $x^2 + 2x - 4$. Thus the solutions x to (5.14) are the roots of the polynomial equation $-4x(x^2 + 2x - 4) = 0$ which has factors $4x$ and $x^2 + 2x - 4$. The roots are $x = 0$ and (using the quadratic formula) $x = -1 \pm \sqrt{5}$. ■

Historical Notes

Pierre Simon Marquis de Laplace (1749–1827) French mathematician (analysis, probability) and physicist. Laplace is regarded by many as the greatest theorist of celestial mechanics since Sir Isaac Newton (1643–1727). Laplace held a number of high-profile posts in France, survived the fall of successive regimes in that country and was ennobled both under the Empire and by Louis XVIII.

EXERCISES 5.1

Exercises 1–4. Evaluate the determinants of order 2.

1. $\begin{vmatrix} 1.2 & -4 \\ 0 & 3.4 \end{vmatrix}$

2. $\begin{vmatrix} -7 & -8 \\ 1 & -4 \end{vmatrix}$

3. $\begin{vmatrix} a & b \\ b & a \end{vmatrix}$

4. $\begin{vmatrix} x & x+2 \\ x+1 & x+3 \end{vmatrix}$

Exercises 5–6. Let **A** *be the given matrix. Write down the square submatrices* $\mathbf{A}_{11}$, $\mathbf{A}_{12}$, $\mathbf{A}_{13}$. *Compute the minors* m_{11}, m_{12}, m_{13} *and then find* $\det(\mathbf{A})$ *using Laplace expansion on row 1 of* **A**.

5. $\begin{bmatrix} 1 & 2 & 3 \\ 4 & 5 & 6 \\ 7 & 8 & 9 \end{bmatrix}$

6. $\begin{bmatrix} 1 & -2 & -3 & 0 \\ -1 & 0 & 0 & 3 \\ 0 & 1 & 3 & 0 \\ 1 & 2 & 0 & 2 \end{bmatrix}$

Exercises 7–12. Evaluate each determinant using the most computationally efficient manner.

7. $\begin{vmatrix} 6 & 2 & 5 & 4 \\ 7 & 3 & 6 & 0 \\ 1 & 1 & 0 & 0 \\ -2 & 0 & 0 & 0 \end{vmatrix}$

8. $\begin{vmatrix} 2 & 0 & 0 & 4 \\ 0 & 0 & 6 & 4 \\ 0 & -1 & 0 & 4 \\ 0 & 0 & 0 & 4 \end{vmatrix}$

9. $\begin{vmatrix} 4 & 0 & 0 & 1 \\ 1 & 2 & 0 & 0 \\ 0 & -1 & 1 & 0 \\ 0 & 0 & 1 & -2 \end{vmatrix}$

10. $\begin{vmatrix} -3 & 2 & 6 & 9 \\ 0 & 1 & -9 & 2 \\ 0 & 0 & 5 & 4 \\ 0 & 0 & 2 & 1 \end{vmatrix}$

11. $\begin{vmatrix} -3 & 2 & 9 & 1 \\ 4 & 1 & -7 & 6 \\ 0 & 0 & 5 & 1 \\ 0 & 0 & -6 & -2 \end{vmatrix}$

12. $\begin{vmatrix} a_1 & 0 & 0 & b_1 \\ 0 & a_2 & 0 & b_2 \\ 0 & 0 & a_3 & b_3 \\ c_1 & c_2 & c_3 & 0 \end{vmatrix}$

Exercises 13–14. Evaluate, using the Laplace expansion on the column that requires least computation. Cross-check using Laplace expansion on row 1 of **A**.

13. $\begin{vmatrix} 3 & 2 & -4 \\ 0 & 4 & 7 \\ 0 & 1 & 5 \end{vmatrix}$

14. $\begin{vmatrix} 1 & 3 & 2 \\ 8 & 4 & 0 \\ 2 & 1 & 2 \end{vmatrix}$

Exercises 15–16. Evaluate, using Laplace expansion on row 2. Cross-check using the mnemonic scheme that is valid only for determinants of order 3.

15. $\begin{vmatrix} 5 & -2 & 1 \\ 3 & 2 & 7 \\ 1 & 1 & 3 \end{vmatrix}$

16. $\begin{vmatrix} 1 & 2 & 5 \\ 1 & 1 & 1 \\ 1 & 4 & 3 \end{vmatrix}$

17. Find det(**A**) by expanding on a row of your choice. Cross-check by expanding on any column of your choice, where

$$\mathbf{A} = \begin{bmatrix} 1 & 2 & 3 \\ -3 & 4 & -5 \\ 6 & 7 & 8 \end{bmatrix}.$$

18. Show that $\begin{vmatrix} a & 0 & d & c \\ b & 0 & -c & d \\ 0 & c & -b & a \\ 0 & d & a & b \end{vmatrix} = 0.$

Exercises 19–20. Verify each identity using a row or column expansion on the left side of each identity.

19. $\begin{vmatrix} 0 & x & y & 0 \\ x & 0 & 0 & a \\ y & 0 & 0 & b \\ 0 & a & b & 0 \end{vmatrix} = \begin{vmatrix} x & y \\ a & b \end{vmatrix}^2$

20. $\begin{vmatrix} a & b & x \\ c & d & y \\ 0 & 0 & 1 \end{vmatrix} = \begin{vmatrix} a & b \\ c & d \end{vmatrix}$

Exercises 21–22. Find the values of a and b for which the determinant is zero.

21. $\begin{vmatrix} a & b & 0 \\ b & a & b \\ 0 & b & a \end{vmatrix}$

22. $\begin{vmatrix} a & b & 0 \\ b & a & b \\ a & b & a \end{vmatrix}$

Exercises 23–24. Evaluate each determinant using the fact that $\begin{vmatrix} 1 & 2 & 5 \\ 1 & 1 & 1 \\ 1 & 4 & 3 \end{vmatrix} = 10.$

23. $\begin{vmatrix} 10 & 20 & 50 \\ 10 & 10 & 10 \\ 1 & 4 & 3 \end{vmatrix}$

24. $\begin{vmatrix} 1 & 1 & 1 \\ 1 & 4 & 3 \\ 1 & 2 & 5 \end{vmatrix}$

Exercises 25–28. Use elementary row operations (EROs) to evaluate each determinant given that $\begin{vmatrix} 3 & 2 & 5 \\ 1 & 4 & 6 \\ 2 & 3 & 7 \end{vmatrix} = 15.$

25. $\begin{vmatrix} 9 & 6 & 15 \\ 2 & 8 & 12 \\ -2 & -3 & -7 \end{vmatrix}$

26. $\begin{vmatrix} 1 & 4 & 6 \\ 2 & 3 & 7 \\ 3 & 2 & 5 \end{vmatrix}$

27. $\begin{vmatrix} 4 & 6 & 11 \\ 1 & 4 & 6 \\ 2 & 3 & 7 \end{vmatrix}$

28. $\begin{vmatrix} 1 & 4 & 6 \\ 5 & 5 & 12 \\ 2 & 3 & 7 \end{vmatrix}$

Exercises 29–32. Use forward reduction to evaluate each determinant. Hint: *Use replacements in order to obtain a convenient pivot.*

29. $\begin{vmatrix} 6 & -6 & 6 \\ 2 & 4 & -6 \\ 10 & -5 & 5 \end{vmatrix}$

30. $\begin{vmatrix} 17 & 7 & 5 \\ 47 & 21 & 16 \\ 29 & 14 & 11 \end{vmatrix}$

31. $\begin{vmatrix} 35 & 36 & 23 \\ 20 & 25 & 16 \\ 38 & 39 & 25 \end{vmatrix}$

32. $\begin{vmatrix} 27 & 8 & 61 \\ 27 & 8 & 61 \\ 35 & -9 & 10 \end{vmatrix}$

Exercises 33–34. Verify that each determinant is zero.

33. $\begin{vmatrix} a+b & c & 1 \\ b+c & a & 1 \\ c+a & b & 1 \end{vmatrix}$

34. $\begin{vmatrix} \sin^2\theta & 1 & \cos^2\theta \\ \sin^2\phi & 1 & \cos^2\phi \\ \sin^2 x & 1 & \cos^2 x \end{vmatrix}$

Exercises 35–38. Solve the determinant equation for x.

35. $\begin{vmatrix} 1 & 2+x & 3 \\ 2 & 1 & 3+x \\ 3 & 2+x & 1 \end{vmatrix} = 0$

36. $\begin{vmatrix} 2x+1 & x-7 \\ x-8 & 2x+2 \end{vmatrix} = 0$

37. $\begin{vmatrix} 1 & 3 & 9 \\ 1 & 2 & 4 \\ 1 & x & x^2 \end{vmatrix} = 0$

38. $\begin{vmatrix} x+3 & 3x+3 & x+1 \\ x-5 & x-7 & x-2 \\ x & x+3 & x+6 \end{vmatrix} = 0$

39. Solve the determinant equation

$$\begin{vmatrix} x & 2x & 3x \\ 5 & 1 & 2 \\ 7 & 0 & 5 \end{vmatrix} + \begin{vmatrix} x & 0 \\ 0 & x \end{vmatrix} = 0.$$

40. Prove that $\begin{vmatrix} 1 & x & x^2 \\ 1 & y & y^2 \\ 1 & z & z^2 \end{vmatrix} = (x-y)(y-z)(z-x).$

41. Prove that $\begin{vmatrix} 4 & x+1 & x+1 \\ x+1 & (x+2)^2 & 1 \\ x+1 & 1 & (x+2)^2 \end{vmatrix}$
$= 2(x+1)(x+3)^3$.

42. Prove properties (a), (b), (c) in Theorem 5.3.

43. Consider the cases (a), (b), (c) in Theorem 5.3. In each case let **E** be the elementary matrix that corresponds to the stated ERO. Prove that $\det(\mathbf{EA}) = \det(\mathbf{E})\det(\mathbf{A})$ in each case.

44. Suppose row i and row j in **A** are equal, where $i \neq j$. Use a row interchange to show that $\det(\mathbf{A}) = 0$.

45. A square matrix **A** of order n is skew-symmetric if $\mathbf{A}^\mathsf{T} = -\mathbf{A}$. Show that **A** is singular if n is odd. Which nonzero skew-symmetric matrices of order 2 are singular? Find two sparse 4×4 skew-symmetric matrices **A** and **B** with entries $0, \pm 1$ and such that **A** is singular and **B** is invertible.

46. Show that

$$\begin{vmatrix} x & y & 1 \\ x_1 & y_1 & 1 \\ x_2 & y_2 & 1 \end{vmatrix} = 0$$

gives the equation of the line through distinct points (x_1, y_1) and (x_2, y_2) in the plane. Find the equation of the line through $(1, 2)$ and $(3, -1)$.

USING MATLAB

Consult online help for **det**, **syms**, **determ**, **solve**, and related commands.

Exercises 47–48. Use the Symbolic Math Toolbox to solve each determinant equation for x.

47. $\begin{vmatrix} 1 & 2 & 3 & 4 \\ 1 & x+1 & 3 & 4 \\ 1 & 2 & x+1 & 4 \\ 1 & 2 & 3 & x+1 \end{vmatrix} = 0$

48. $\begin{vmatrix} x & a & b \\ a & x & b \\ a & b & x \end{vmatrix} = 0$

Exercises 49–50. The equation of a plane in 3-space passing through the points (x_1, y_1, z_1), (x_2, y_2, z_2), (x_3, y_3, z_3) is given by $ax + by + cz + d = 0$, which can be expressed as the determinant equation $\det(\mathbf{A}) = 0$, *where*

$$\mathbf{A} = \begin{bmatrix} x & y & z & 1 \\ x_1 & y_1 & z_1 & 1 \\ x_2 & y_2 & z_2 & 1 \\ x_3 & y_3 & z_3 & 1 \end{bmatrix}.$$

Use the Symbolic Math Toolbox to find the equation of the plane passing through the given points.

49. $(1, 1, 1)$, $(-1, 2, 3)$, $(1, 2, 3)$

50. $(0, 1, 1)$, $(-1, -2, 0)$, $(3, 4, 5)$

51. *Project.* Prove that $\mathbf{A}(\epsilon) = \begin{vmatrix} 1+\epsilon & 2 & 3 \\ 4 & 5 & 6 \\ 7 & 8 & 9 \end{vmatrix} = -3\epsilon$. Write an M-file to investigate whether or not $\mathbf{A}(\epsilon)$ is singular for various small values of ϵ.

52. Use MATLAB to check hand calculations in this exercise set.

5.2 Inverses and Products

The first result of this section adds one more condition for the invertibility of a square matrix to those already stated in Sections 2.2 and 3.2.

Theorem 5.6 **Determinant Condition for Invertibility**

The $n \times n$ *matrix* $\mathbf{A}$ *is invertible if and only if* $\det(\mathbf{A}) \neq 0$.

Proof Refer to equations (5.12) and (5.13) in Section 5.1. There are two implications to prove.

(a) If $\mathbf{A}$ is invertible, then $\mathbf{A}$ is row equivalent to $\mathbf{I}_n$; that is, $\mathbf{U} = \mathbf{I}_n$ in (5.12) and so $\det(\mathbf{A}) = q$ in (5.13), where q is nonzero.

(b) If $\mathbf{A}$ is singular, then $\operatorname{rank} \mathbf{A} < n$ and $\mathbf{A}$ is row equivalent to an echelon form $\mathbf{U}$ that has a zero row at the bottom of the matrix. Equation (5.13) then gives $\det(\mathbf{A}) = q \cdot 0 = 0$. Taking the contrapositive of the implication just proved, it follows that $\det(\mathbf{A}) \neq 0$ implies $\mathbf{A}$ is invertible, and this completes the proof.

■

The equivalent form of Theorem 5.6, namely

$$\mathbf{A} \text{ is singular if and only if } \det(\mathbf{A}) = 0, \tag{5.16}$$

is of particular interest as it provides the key condition for the computation of eigenvalues in Chapter 6.

Computational Note

As mentioned in the introduction to this chapter, the value (zero or nonzero) of $\det(\mathbf{A})$ determines the character of the solution to the $n \times n$ linear system defined by $\mathbf{Ax} = \mathbf{b}$. When using the determinant in this context keep in mind that machine computation may result in a value for $\det(\mathbf{A})$ that is close to zero, when the true value should actually be zero.

Theorem 5.7 **Product Rule**

If $\mathbf{A}$ *and* $\mathbf{B}$ *are* $n \times n$ *matrices, then* $\det(\mathbf{AB}) = \det(\mathbf{A})\det(\mathbf{B})$.

Proof There are two cases to consider here.

(a) Suppose that $\mathbf{A}$ is invertible. Then, by Theorem 2.9 in Section 2.2, $\mathbf{A} = \mathbf{E}_1\mathbf{E}_2\cdots\mathbf{E}_k$, for elementary matrices $\mathbf{E}_1, \mathbf{E}_2, \ldots, \mathbf{E}_k$. Multiplying by $\mathbf{B}$, we have $\mathbf{AB} = \mathbf{E}_1\mathbf{E}_2\cdots\mathbf{E}_k\mathbf{B}$, and using (5.7) from Section 5.1 repeatedly, we obtain

$$\begin{aligned}
\det(\mathbf{AB}) &= \det(\mathbf{E}_1\mathbf{E}_2\cdots\mathbf{E}_k\mathbf{B}) \\
&= \det(\mathbf{E}_1)\det(\mathbf{E}_2\cdots\mathbf{E}_k\mathbf{B}) \\
&\ \ \vdots \\
&= \det(\mathbf{E}_1)\det(\mathbf{E}_2)\cdots\det(\mathbf{E}_k)\det(\mathbf{B}) \\
&= \det(\mathbf{E}_1\mathbf{E}_2\cdots\mathbf{E}_k)\det(\mathbf{B}) = \det(\mathbf{A})\det(\mathbf{B})
\end{aligned}$$

(b) Suppose that $\mathbf{A}$ is singular. Then $\det(\mathbf{A}) = 0$ by (5.16). From Section 2.2, there is an invertible matrix $\mathbf{F}$ such that $\mathbf{FA} = \mathbf{U}$, where $\mathbf{U}$ is an echelon form for $\mathbf{A}$, and because $\operatorname{rank} \mathbf{A} < n$, $\mathbf{U}$ has a zero row at the bottom of the matrix. Consider the equation **FAB** = **UB**. The matrix **UB** has a zero row at the bottom of the matrix. Hence det(**UB**) = 0, and so **UB** (= **FAB**) is singular by (5.16). Now note that **AB** must be singular, for otherwise **F**(**AB**) would be invertible. Thus, using (5.16) again, we have

$$\det(\mathbf{AB}) = 0 = 0\det(\mathbf{B}) = \det(\mathbf{A})\det(\mathbf{B}).$$

Some consequences of the product rule are stated next. In the interests of clarity, we use the alternative notation for determinant here.

(a) If **A** and **B** are $n \times n$ matrices, then

$$|\mathbf{AB}| = |\mathbf{A}||\mathbf{B}| = |\mathbf{B}||\mathbf{A}| = |\mathbf{BA}|,$$

and so, surprisingly, $|\mathbf{AB}| = |\mathbf{BA}|$, whether **A** and **B** commute or not.

(b) Let $\mathbf{A}_1, \mathbf{A}_2, \ldots, \mathbf{A}_k$ be $n \times n$ matrices. Using mathematical induction, we have

$$|\mathbf{A}_1\mathbf{A}_2\cdots\mathbf{A}_k| = |\mathbf{A}_1||\mathbf{A}_2|\ \cdots\ |\mathbf{A}_k|, \quad k = 1, 2, \ldots\ . \tag{5.17}$$

(c) Setting $\mathbf{A}_1 = \mathbf{A}_2 = \cdots = \mathbf{A}_k = \mathbf{A}$ in (5.17), we obtain

$$|\mathbf{A}^k| = |\mathbf{A}|^k, \quad k = 1, 2, \ldots \tag{5.18}$$

(d) Suppose the $n \times n$ matrix **A** is invertible. Then there is an $n \times n$ matrix $\mathbf{A}^{-1}$ such that $\mathbf{AA}^{-1} = \mathbf{I}_n$. Using the product rule, we have

$$|\mathbf{AA}^{-1}| = |\mathbf{I}_n| \quad \Rightarrow \quad |\mathbf{A}||\mathbf{A}^{-1}| = 1, \tag{5.19}$$

and, by Theorem 5.6, $|\mathbf{A}| \neq 0$, so that

$$|\mathbf{A}^{-1}| = \frac{1}{|\mathbf{A}|}. \tag{5.20}$$

Recall from Section 2.2 that $(\mathbf{A}^{-1})^k = (\mathbf{A}^k)^{-1} = \mathbf{A}^{-k}$ for any positive integer k. Equation (5.20) for k factors becomes

$$|\mathbf{A}^{-k}| = |\mathbf{A}|^{-k}, \quad k = 1, 2, \ldots\ , \tag{5.21}$$

and combining (5.21) with (5.18) shows that (5.18) holds for any integer k (positive, negative or zero).

EXAMPLE 1

Computations Using the Product Rule

Suppose **A** and **B** are square matrices of order 6 such that

$$|\mathbf{AB}^2| = 72 \quad \text{and} \quad |\mathbf{A}^2\mathbf{B}^2| = 144.$$

Our goal is to find $|\mathbf{A}|$, $|2\mathbf{A}|$, and $|\mathbf{AB}|^6$.

Let $|\mathbf{A}| = a$ and $|\mathbf{B}| = b$. Then $ab^2 = 72$ and $a^2b^2 = 144$ so that $a = a^2b^2/ab^2 = 2$ and $b^2 = 36$ implies $b = \pm 6$. Then $|2\mathbf{A}| = 2^6|\mathbf{A}| = 128$, and $|\mathbf{AB}^6| = |\mathbf{A}||\mathbf{B}|^6 = 2(36)^3 = 93312$. ■

The Adjoint

Let $\mathbf{A}$ be an $n \times n$ matrix and let c_{ij}, $1 \le i, j \le n$, denote the n^2 cofactors of $\mathbf{A}$. The *adjoint* of $\mathbf{A}$ is the $n \times n$ matrix $\operatorname{adj} \mathbf{A} = [c_{ij}]^\mathsf{T}$.

EXAMPLE 2

Finding the Adjoint

Taking $n = 3$, the $n^2 = 9$ cofactors for the 3×3 matrix $\mathbf{A}$ shown on the left side are listed on the right side.

$$\mathbf{A} = \begin{bmatrix} 2 & -1 & 3 \\ 1 & 2 & -1 \\ 3 & 2 & 2 \end{bmatrix} \qquad \begin{matrix} c_{11} = 6 & c_{12} = -5 & c_{13} = -4 \\ c_{21} = 8 & c_{22} = -5 & c_{23} = -7 \\ c_{31} = -5 & c_{32} = 5 & c_{33} = 5 \end{matrix}$$

Then

$$\operatorname{adj} \mathbf{A} = \begin{bmatrix} 6 & -5 & -4 \\ 8 & -5 & -7 \\ -5 & 5 & 5 \end{bmatrix}^\mathsf{T} = \begin{bmatrix} 6 & 8 & -5 \\ -5 & -5 & 5 \\ -4 & -7 & 5 \end{bmatrix}.$$

Caution! Remember to transpose the matrix of cofactors. ▪

Let $\mathbf{A}$ be the matrix in Example 2 and consider the product $\mathbf{A}(\operatorname{adj} \mathbf{A})$. We have

$$\mathbf{A}(\operatorname{adj} \mathbf{A}) = \begin{bmatrix} 2 & -1 & 3 \\ 1 & 2 & -1 \\ 3 & 2 & 2 \end{bmatrix} \begin{bmatrix} 6 & 8 & -5 \\ -5 & -5 & 5 \\ -4 & -7 & 5 \end{bmatrix} = \begin{bmatrix} 5 & 0 & 0 \\ 0 & 5 & 0 \\ 0 & 0 & 5 \end{bmatrix} = \begin{bmatrix} |\mathbf{A}| & 0 & 0 \\ 0 & |\mathbf{A}| & 0 \\ 0 & 0 & |\mathbf{A}| \end{bmatrix}.$$

Make the following observations. The $(1, 1)$-entry in $\mathbf{A}(\operatorname{adj} \mathbf{A})$ is the row-by-column product p of row 1 in $\mathbf{A}$ with column 1 in $\operatorname{adj} \mathbf{A}$ (these are the cofactors from row 1 in $\mathbf{A}$), and p is just $|\mathbf{A}|$. The other diagonal entries in $\mathbf{A}(\operatorname{adj} \mathbf{A})$ are explained similarly. The $(1, 2)$-entry in $\mathbf{A}(\operatorname{adj} \mathbf{A})$ is the row-by-column product p of row $i = 1$ in $\mathbf{A}$ and column $j = 2$ in $\operatorname{adj} A$ (these are the cofactors from row 2 in $\mathbf{A}$). We have

$$p = a_{11}c_{21} + a_{12}c_{22} + a_{13}c_{23}.$$

Form a new matrix $\mathbf{B}$ from $\mathbf{A}$ by replacing row $j = 2$ in $\mathbf{A}$ with row $i = 1$, as follows:

$$\mathbf{B} = \begin{bmatrix} 2 & -1 & 3 \\ 2 & -1 & 3 \\ 3 & 2 & 2 \end{bmatrix}.$$

Note that c_{21}, c_{22}, c_{23} are also the cofactors for $\mathbf{B}$ from row $j = 2$ and so the expansion of $\det(\mathbf{B})$ on row $j = 2$ is p. But $p = \det(\mathbf{B}) = 0$ because $\mathbf{B}$ has two equal rows. The same arguments apply to the other (i, j)-entries in $\mathbf{A}(\operatorname{adj} \mathbf{A})$ with $i \neq j$. More generally, for any $n \times n$ matrix $\mathbf{A}$, we have

$$\mathbf{A}(\operatorname{adj} \mathbf{A}) = \det(\mathbf{A})\mathbf{I}_n \tag{5.22}$$

Modifying the preceding arguments, working with columns instead of rows, shows that $(\text{adj}\,\mathbf{A})\mathbf{A} = \det(\mathbf{A})\mathbf{I}_n$.

If $\mathbf{A}$ is invertible, then $\det(\mathbf{A}) \neq 0$ by Theorem 5.6, and rearranging (5.22) gives

$$\mathbf{A}^{-1} = \frac{\text{adj}\,\mathbf{A}}{\det(\mathbf{A})}. \tag{5.23}$$

We call (5.23) the *adjoint formula* for the inverse of $\mathbf{A}$.

EXAMPLE 3

Using the Adjoint Formula

Consider the matrix $\mathbf{A}$ in Example 2. You may verify that $\det(\mathbf{A}) = 5 \neq 0$. Hence $\mathbf{A}$ is invertible, and applying (5.23), we have

$$\mathbf{A}^{-1} = \frac{1}{5}\begin{bmatrix} 6 & 8 & -5 \\ -5 & -5 & 5 \\ -4 & -7 & 5 \end{bmatrix}.$$

■

The adjoint formula (5.23) provides an easy way to compute the inverse of a general 2×2 matrix $\mathbf{A} = \begin{bmatrix} a & b \\ c & d \end{bmatrix}$. Assuming $\det(\mathbf{A}) = ad - bc \neq 0$, we have

$$\mathbf{A}^{-1} = \frac{1}{(ad - bc)}\begin{bmatrix} d & -b \\ -c & a \end{bmatrix},$$

and this agrees with the formula given in Section 2.2.

Theorem 5.8

Cramer's Rule

Let (S) be an $n \times n$ linear system defined by $\mathbf{Ax} = \mathbf{b}$, where the $n \times n$ matrix $\mathbf{A}$ is invertible. The n unknowns in $\mathbf{x}$ are uniquely determined by the expressions

$$x_j = \frac{|\mathbf{A}_j|}{|\mathbf{A}|}, \quad j = 1, 2, \ldots, n,$$

where $\mathbf{A}_j$ is the $n \times n$ matrix obtained by replacing column j in $\mathbf{A}$ by $\mathbf{b}$.

Proof Exercises 5.2.

EXAMPLE 4

Applying Cramer's Rule

We will solve the linear system (S) shown for x_1.

$$(S) \quad \begin{cases} 2x_1 - x_2 + 3x_3 = 4 \\ x_1 + 2x_2 - x_3 = 2 \\ 3x_1 + 2x_2 + 2x_3 = 6 \end{cases} \qquad \mathbf{A} = \begin{bmatrix} 2 & -1 & 3 \\ 1 & 2 & -1 \\ 3 & 2 & 2 \end{bmatrix}, \quad \mathbf{b} = \begin{bmatrix} 4 \\ 2 \\ 7 \end{bmatrix}.$$

Forward reduction on **A** gives $\det(\mathbf{A}) = 5 \neq 0$, showing that (S) has a unique solution. To find x_1 we form the matrix $\mathbf{A}_1$, replacing the first column of **A** by **b**. We have

$$\mathbf{A}_1 = \begin{bmatrix} 4 & -1 & 3 \\ 2 & 2 & -1 \\ 6 & 2 & 2 \end{bmatrix}$$

and forward reduction on $\mathbf{A}_1$ shows that $\det(\mathbf{A}_1) = 10$, and Cramer's rule gives $x_1 = \dfrac{\det(\mathbf{A}_1)}{\det(\mathbf{A})} = \dfrac{10}{5} = 2.$ ■

Computational Note

Cramer's rule is of theoretical interest but of limited use computationally. It only applies to square linear systems with a unique solution, and is only useful for finding specific unknowns in small linear systems.

Application: Cryptology, Unimodular Matrices

Secure transference of information (banking, company, Internet commerce) via global telecommunications requires *encryption* and *decryption* of messages or data. The discipline devoted to secret systems is called *cryptology*, and the part of cryptology dealing with the design and implementation of secret systems is called *cryptography*. *Cryptoanalysis* deals with *breaking* (decoding) these systems.

A *cipher* is a method of transforming a message from plaintext into a coded message called the *cipher text*. The transformation used to *encode* the plaintext message is called the *key*.

One of the earliest *ciphering* devices dates from the time Julius Caesar sent secret messages to his troops. These simple translation methods became known as *Caesar ciphers*. They depend on translating each character in the plaintext message a certain fixed number of places along the alphabet string—translating by three places, for example, A $\rightarrow$ D and with wrap-around, Z $\rightarrow$ C). More interesting ciphers are build using *modular arithmetic* and prime numbers (an integer p is *prime* if it has no factors other than 1 and itself ($p = 2, 3, 5, 7 \ldots$ are prime numbers).

Matrix cyphers use matrix multiplication in the following way. Using the one-to-one correspondence between the letters A, B, C, ... , Z and the positive integer $1, 2, 3, \ldots, 26$, we can *encode* any plain text message into a sequence of integers. Zeros or other integers might be included to indicate blank spaces and punctuation. The plaintext message MATH, for example, corresponds to the unique 4-tuple $\mathbf{v} = (13, 1, 20, 8)$ and can then be stored in the columns of a 2×2 matrix $\mathbf{M} = \begin{bmatrix} 13 & 20 \\ 1 & 8 \end{bmatrix}$. Our message could then be encoded using a 2×2 encoding matrix **E** with integer entries, which we are at liberty to choose. For example,

$$\mathbf{E} = \begin{bmatrix} 1 & 2 \\ 3 & 4 \end{bmatrix} \qquad \mathbf{EM} = \begin{bmatrix} 1 & 2 \\ 3 & 4 \end{bmatrix}\begin{bmatrix} 13 & 20 \\ 1 & 8 \end{bmatrix} = \begin{bmatrix} 15 & 36 \\ 43 & 92 \end{bmatrix} = \mathbf{N}$$

and **N** can be interpreted as the 4-tuple $\mathbf{u} = (15, 43, 36, 92)$ (encoded message). If the receiver knows **E** and **E** is invertible, the receiver can *decode* the matrix **N** that has been received using the decoding matrix $\mathbf{D} = \mathbf{E}^{-1}$ because $\mathbf{DN} = \mathbf{D}(\mathbf{EM}) = (\mathbf{DE})\mathbf{M} = \mathbf{M}$, where $\mathbf{DE} = \mathbf{I}$.

The inverse of an invertible matrix with integer entries may not have integer entries. There may be inaccuracies due to round-off error, and through this the decryption would be prone to error. One possibility is to insist that **E** and **D** both have integer entries.

Unimodular Matrices

An $n \times n$ matrix **A** is called *unimodular* if $\det(\mathbf{A}) = -1$ or $+1$. A unimodular matrix with integer entries is necessarily invertible and its inverse also has integer entries because $\mathbf{A}^{-1} = \pm\operatorname{adj}\mathbf{A}$.

An $n \times n$ unimodular matrix can be constructed by performing a sequence of replacements or interchanges (no scaling allowed) on the identity matrix $\mathbf{I}_n$. A pair of row interchanges will change the determinant by $(-1)^2 = 1$, and row replacements leave a determinant unchanged.

Cross Product

We end this section by using a determinant to define the *cross product*[1] of a pair of vectors **u** and **v** in 3-space $\mathbb{R}^3$. We will see how the elementary operations on determinants are employed in developing the properties of this vector product. In this theory it will be convenient to write a 3-vector **u** as the 3-tuple (u_1, u_2, u_3) that locates the terminal point of the arrow representing **u**.

Definition 5.3

Cross product

Let $\mathbf{u} = (u_1, u_2, u_3)$ and $\mathbf{v} = (v_1, v_2, v_3)$ be vectors in $\mathbb{R}^3$. The *cross product* of **u** and **v**, denoted by $\mathbf{u} \times \mathbf{v}$, is defined by the rule

$$\mathbf{u} \times \mathbf{v} = (u_2v_3 - u_3v_2,\ u_3v_1 - u_1v_3,\ u_1v_2 - u_2v_1) \tag{5.23}$$

Appealing to the alternative notation mentioned at the end of Section 3.1 for the standard basis in $\mathbb{R}^3$, we have $\mathbf{i} = (1, 0, 0)$, $\mathbf{j} = (0, 1, 0)$, $\mathbf{k} = (0, 0, 1)$ and so any 3-vector (a, b, c) can be expressed in the form $a\mathbf{i} + b\mathbf{j} + c\mathbf{k}$. Thus (5.23) becomes

$$\mathbf{u} \times \mathbf{v} = (u_2v_3 - u_3v_2)\,\mathbf{i} + (u_3v_1 - u_1v_3)\,\mathbf{j} + (u_1v_2 - u_2v_1)\,\mathbf{k}, \tag{5.24}$$

and the right side of (5.24) is exactly the Laplace expansion on row 1 of the determinant[2] on the right side of (5.25). Hence, we have

$$\mathbf{u} \times \mathbf{v} = \begin{vmatrix} \mathbf{i} & \mathbf{j} & \mathbf{k} \\ u_1 & u_2 & u_3 \\ v_1 & v_2 & v_3 \end{vmatrix}. \tag{5.25}$$

[1] Also called the *vector product*. The name cross product agrees with the MATLAB command **cross (u,v)**.

[2] This expression is sometimes called a *pseudo determinant*, being a mixture of vectors and numbers. However, we shall operate on it using the usual rules for determinants.

EXAMPLE 5

Cross Product

Let $\mathbf{u} = 2\mathbf{i} - 3\mathbf{j} + \mathbf{k}$ and $\mathbf{v} = 3\mathbf{i} + 2\mathbf{j} - 3\mathbf{k}$. The Laplace expansion on row 1 of (5.26) gives

$$\mathbf{u} \times \mathbf{v} = \begin{vmatrix} \mathbf{i} & \mathbf{j} & \mathbf{k} \\ 2 & -3 & 1 \\ 3 & 2 & -3 \end{vmatrix} = (9-2)\mathbf{i} - (-6-3)\mathbf{j} + (4+9)\mathbf{k} = 7\mathbf{i} + 9\mathbf{j} + 13\mathbf{k}.$$

■

Properties of Products

Let $\mathbf{u} = (u_1, u_2, u_3)$, $\mathbf{v} = (v_1, v_2, v_3)$, $\mathbf{w} = (w_1, w_2, w_3)$. The following properties of cross product are proved using basic determinant operations.

(P1) $\mathbf{u} \times \mathbf{u} = \mathbf{0}$.
Two rows in (5.26) are equal, hence the determinant is zero.

(P2) $\mathbf{u} \times \mathbf{v} = -\mathbf{v} \times \mathbf{u}$ (cross product is noncommutative)
Interchanging rows 2 and 3 in (5.26) multiplies the determinant by -1.

Caution! The order of terms $\mathbf{u}$ and $\mathbf{v}$ in $\mathbf{u} \times \mathbf{v}$ matters!

(P3) If $\mathbf{v} = s\mathbf{u}$ for some scalar s, then $\mathbf{u} \times \mathbf{v} = \mathbf{0}$.
Suppose $\mathbf{v} = s\mathbf{u}$. Then the ERO $\mathbf{r}_3 - s\mathbf{r}_2 \to \mathbf{r}_3$ reduces row 3 in (5.26) to zero and leaves the determinant value unchanged.

(P4) $(\mathbf{u} \times \mathbf{v}) \cdot \mathbf{u} = (\mathbf{u} \times \mathbf{v}) \cdot \mathbf{v} = 0$ ($\mathbf{u} \times \mathbf{v}$ is orthogonal to both $\mathbf{u}$ and $\mathbf{v}$)

$$(\mathbf{u} \times \mathbf{v}) \cdot \mathbf{u} = \begin{vmatrix} u_1 & u_2 & u_3 \\ u_1 & u_2 & u_3 \\ v_1 & v_2 & v_3 \end{vmatrix}, \quad (\mathbf{u} \times \mathbf{v}) \cdot \mathbf{v} = \begin{vmatrix} v_1 & v_2 & v_3 \\ u_1 & u_2 & u_3 \\ v_1 & v_2 & v_3 \end{vmatrix}$$

and a determinant is zero when two rows are equal.

(P5) $\mathbf{w} \times (\mathbf{u} + \mathbf{v}) = (\mathbf{w} \times \mathbf{u}) + (\mathbf{w} \times \mathbf{v})$ (cross product is distributive over addition)
Using the linearity property of determinants we have

$$\mathbf{w} \times (\mathbf{u} + \mathbf{v}) = \begin{vmatrix} \mathbf{i} & \mathbf{j} & \mathbf{k} \\ a_1 & a_2 & a_3 \\ b_1 + c_1 & b_2 + c_2 & b_3 + c_3 \end{vmatrix}$$

$$= \begin{vmatrix} \mathbf{i} & \mathbf{j} & \mathbf{k} \\ a_1 & a_2 & a_3 \\ b_1 & b_2 & b_3 \end{vmatrix} + \begin{vmatrix} \mathbf{i} & \mathbf{j} & \mathbf{k} \\ a_1 & a_2 & a_3 \\ c_1 & c_2 & c_3 \end{vmatrix} = (\mathbf{w} \times \mathbf{u}) + (\mathbf{w} \times \mathbf{v}).$$

(P6) $\mathbf{i} \times \mathbf{j} = \mathbf{k}$, $\mathbf{j} \times \mathbf{k} = \mathbf{i}$, $\mathbf{k} \times \mathbf{i} = \mathbf{j}$ and $\mathbf{i} \times \mathbf{i} = 0$, $\mathbf{j} \times \mathbf{j} = 0$, $\mathbf{k} \times \mathbf{k} = 0$.
These properties follow easily from (5.26).

EXAMPLE 6

Taking a Cross Product

Because cross product is noncommutative, computing a cross product using properties (**P5**) and (**P6**) requires strict observance of order of operations, as is seen next.

$$\mathbf{u}\times\mathbf{v} = (2\mathbf{i}-3\mathbf{j}+\mathbf{k})\times(3\mathbf{i}+2\mathbf{j}-3\mathbf{k})$$
$$= 6(\mathbf{i}\times\mathbf{i})+4(\mathbf{i}\times\mathbf{j})-6(\mathbf{i}\times\mathbf{k})-9(\mathbf{j}\times\mathbf{i})-6(\mathbf{j}\times\mathbf{j})+9(\mathbf{j}\times\mathbf{k})$$
$$+3(\mathbf{k}\times\mathbf{i})+2(\mathbf{k}\times\mathbf{j})-3(\mathbf{k}\times\mathbf{k})$$
$$= 4\mathbf{k}+6\mathbf{j}+9\mathbf{k}+9\mathbf{i}+3\mathbf{j}-2\mathbf{i} = 7\mathbf{i}+9\mathbf{j}+13\mathbf{k}$$ ■

(P7) $(\mathbf{u}\times\mathbf{v})\cdot\mathbf{w} = (\mathbf{v}\times\mathbf{w})\cdot\mathbf{u} = (\mathbf{w}\times\mathbf{u})\cdot\mathbf{v}$ (scalar triple product)
There are three equalities in (**P7**), all proved similarly. For example, noting that two row interchanges multiply a determinant by $(-1)^2$, for the first equality, we have

$$(\mathbf{u}\times\mathbf{v})\cdot\mathbf{w} = \begin{vmatrix} w_1 & w_2 & w_3 \\ u_1 & u_2 & u_3 \\ v_1 & v_2 & v_3 \end{vmatrix} \underset{\substack{\mathbf{r}_2\leftrightarrow\mathbf{r}_3 \\ \mathbf{r}_1\leftrightarrow\mathbf{r}_3}}{=} (-1)^2 \begin{vmatrix} u_1 & u_2 & u_3 \\ v_1 & v_2 & v_3 \\ w_1 & w_2 & w_3 \end{vmatrix}$$
$$= (\mathbf{v}\times\mathbf{w})\cdot\mathbf{u}.$$

Taking the transpose of the determinant equal to $(\mathbf{v}\times\mathbf{w})\cdot\mathbf{u}$, we have

$$(\mathbf{v}\times\mathbf{w})\cdot\mathbf{u} = \det([\,\mathbf{u}\ \ \mathbf{v}\ \ \mathbf{w}\,]), \tag{5.27}$$

where $\mathbf{u}$, $\mathbf{v}$, $\mathbf{w}$ are now interpreted as column vectors in the 3×3 matrix $[\,\mathbf{u}\ \ \mathbf{v}\ \ \mathbf{w}\,]$.

(P8) $(\mathbf{u}\times\mathbf{v})\times\mathbf{w} = (\mathbf{u}\cdot\mathbf{w})\mathbf{v} - (\mathbf{v}\cdot\mathbf{w})\mathbf{u}$
The vector $(\mathbf{u}\times\mathbf{v})\times\mathbf{w}$ is called the *vector triple product* of $\mathbf{u}$, $\mathbf{v}$, $\mathbf{w}$.

Proof Assuming $\mathbf{u}$ and $\mathbf{v}$ are noncollinear, the vector $\mathbf{u}\times\mathbf{v}$ is orthogonal to the plane determined by $\mathbf{u}$ and $\mathbf{v}$ and $(\mathbf{u}\times\mathbf{v})\times\mathbf{w}$ is orthogonal to $\mathbf{u}\times\mathbf{v}$ (and to $\mathbf{w}$). Hence, the vectors $\mathbf{u}$, $\mathbf{v}$ and $(\mathbf{u}\times\mathbf{v})\times\mathbf{w}$ are coplanar, in which case $(\mathbf{u}\times\mathbf{v})\times\mathbf{w}$ must be a linear combination of $\mathbf{u}$ and $\mathbf{v}$.

Assume that $(\mathbf{u}\times\mathbf{v})\times\mathbf{w} = s\mathbf{u}+t\mathbf{v}$, where s and t are unique scalars. Then $\mathbf{w}\cdot(\mathbf{u}\times\mathbf{v})\times\mathbf{w} = 0 = s(\mathbf{w}\cdot\mathbf{u})+t(\mathbf{w}\cdot\mathbf{v})$. A solution is $s = -r(\mathbf{w}\cdot\mathbf{v})$ and $t = r(\mathbf{w}\cdot\mathbf{u})$, where r is independent of $\mathbf{u}$ and $\mathbf{v}$ and this must be the only solution. The choice $\mathbf{u}=\mathbf{i}$, $\mathbf{v}=\mathbf{j}$, $\mathbf{w}=\mathbf{i}$, shows that $r=1$. If $\mathbf{u}$ and $\mathbf{v}$ are collinear, then both sides of (**P8**) are the zero vector.

▬

For the next property, we assume that $\mathbf{u}$ and $\mathbf{v}$ are nonzero and that θ is the angle between $\mathbf{u}$ and $\mathbf{v}$, where $0\le\theta\le\pi$.

(P9) $\|\mathbf{u}\times\mathbf{v}\| = \|\mathbf{u}\|\,\|\mathbf{v}\|\sin\theta$

Proof We have

$$\|\mathbf{u}\times\mathbf{v}\|^2. = (u_2v_3-u_3v_2)^2+(u_3v_1-u_1v_3)^2+(u_1v_2-u_2v_1)^2$$
$$= (u_1^2+u_2^2+u_3^2)(v_1^2+v_2^2+v_3^2)-(u_1v_2+u_2v_2+u_3v_3)^2$$
$$= \|\mathbf{u}\|^2\|\mathbf{v}\|^2-(\mathbf{u}\cdot\mathbf{v})^2 \tag{5.28}$$

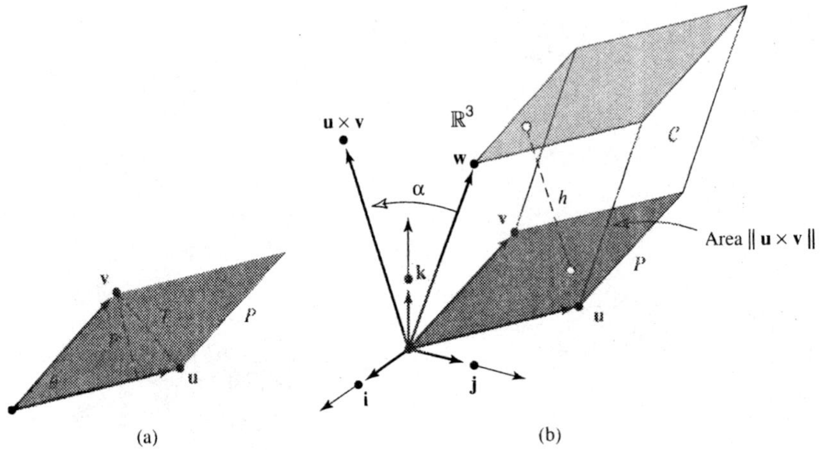

Figure 5.1 Cross product in $\mathbb{R}^3$.

Recalling from Section 4.1 that $(\mathbf{u} \cdot \mathbf{v}) = \| \mathbf{u} \| \| \mathbf{v} \| \cos\theta$ and that $\sin^2\theta + \cos^2\theta = 1$, equation (5.28) becomes

$$\| \mathbf{u} \times \mathbf{v} \|^2 = \| \mathbf{u} \|^2 \| \mathbf{v} \|^2 (1 - \cos^2\theta) = \| \mathbf{u} \|^2 \| \mathbf{v} \|^2 \sin^2\theta,$$

and we take the square root of both sides, noting that $\sin\theta$ is nonnegative because $0 \le \theta \le \pi$.

Geometric Interpretation of Properties (P8) and (P9)

Figure 5.1 (a) shows a plane parallelogram P with side length $\| \mathbf{u} \|$ and height $p = \| \mathbf{v} \| \sin\theta$.

The area (base × height) of P is

$$a_P = \| \mathbf{u} \| \| \mathbf{v} \| \sin\theta = \| \mathbf{u} \times \mathbf{v} \| \text{ units}^2$$

The area of the triangle $\mathcal{T}$ with sides $\mathbf{u}$ and $\mathbf{v}$ is $a_T = \frac{1}{2} \| \mathbf{u} \times \mathbf{v} \|$ units^2.

The parallelepiped (crystal shape) C shown in Figure 5.1(b) has height h and the plane parallelogram P for its base. The volume (base area × height) of C is

$$v_C = \| \mathbf{u} \times \mathbf{v} \| \| \mathbf{w} \| \cos\alpha = (\mathbf{u} \times \mathbf{v}) \cdot \mathbf{w} \quad \text{(scalar triple product)}$$

and the volume is given by the absolute value of the determinant in (5.27).

EXAMPLE 7

Area and Volume

Refer again to Figure 5.1 and let $\mathbf{u} = (-1, 2, 1)$ and $\mathbf{v} = (2, -1, 2)$.

$$\mathbf{u} \times \mathbf{v} = \begin{vmatrix} \mathbf{i} & \mathbf{j} & \mathbf{k} \\ -1 & 2 & 1 \\ 2 & -1 & 2 \end{vmatrix} = 5\mathbf{i} + 4\mathbf{j} - 3\mathbf{k} \quad \Rightarrow$$

$$a_P = \| \mathbf{u} \times \mathbf{v} \| = \sqrt{25 + 16 + 9}$$

which gives $a_P = \sqrt{50} \simeq 7.0711$ units ^{2} and $a_T \simeq 3.5355$ units^2.

Set $\mathbf{w} = (2.5, -4.3, 4.5)$ and compute the volume of the parallelepiped C determined by **u**, **v**, **w**. Viewing **u**, **v**, **w** as column vectors, we have

$$[\mathbf{u}\ \mathbf{v}\ \mathbf{w}] = \begin{vmatrix} -1 & 2 & 2.5 \\ 2 & -1 & -4.3 \\ 1 & 2 & 4.5 \end{vmatrix} = -18.2 \quad \Rightarrow \quad v_C = 18.2 \text{ units}^3$$

▪

Postscript

Have you ever wondered about the beautiful geometric shapes of natural crystals, such as quartz? These shapes are produced by the regular stacking of atoms at the microscopic level, as in the highly magnified image of the mineral topaz that appears on page 245. Other high magnification images of minerals and gems may be seen at the French web site:

http://www.ill.fr/dif/3D-crystals/minerals.html

Historical Notes

The product rule for determinants was proved by Cauchy in 1812. The French mathematician **Jacques Binet** (1786–1856) gave a less satisfactory proof in the same year. Binet wrote over 50 research papers in mathematics and his early contributions to matrix theory prepared the way for the work of Arthur Cayley and others.

Gabriel Cramer (1704–1752) Swiss mathematician. Cramer is remembered as a compiler and disseminator of mathematical ideas. Cramer's rule appeared in the appendix of his work *Introduction à l'analyse des lignes courbes algébriques* (1750) although the result was probably known to **Colin Maclaurin** (1698–1746), and others as early as 1729. By way of compensation, the famous infinite series (expansion of a function) credited to Maclaurin's name should probably be attributed to **Brook Taylor** (1685–1731) and **James Stirling** (1692–1770).

EXERCISES 5.2

Exercises 1–4. Find det(**A**) *and hence determine if the matrix* **A** *is invertible.*

1. $\begin{bmatrix} 1 & -2 & 3 & -4 \\ 0 & 1 & -1 & 1 \\ 1 & 3 & 0 & -3 \\ 0 & -7 & 3 & 1 \end{bmatrix}$

2. $\begin{bmatrix} 2 & -3 & 1 & -4 \\ 5 & 2 & 4 & 2 \\ 1 & 1 & -1 & 1 \\ 3 & -2 & 2 & -3 \end{bmatrix}$

3. $\begin{bmatrix} 0 & 2 & -5 & 1 & -6 \\ 1 & 0 & 2 & 2 & 3 \\ 1 & 1 & -4 & 0 & -3 \\ 0 & 1 & 3 & 1 & -2 \\ 1 & 2 & -1 & 6 & -2 \end{bmatrix}$

4. $\begin{bmatrix} 3 & 4 & -5 & 7 \\ 2 & -3 & 3 & -2 \\ 4 & 11 & -13 & 16 \\ 7 & -2 & 1 & 3 \end{bmatrix}$

5. Let **A** and **B** be square matrices of order 4 such that $|\mathbf{A}| = 3$ and $|\mathbf{B}| = 5$. Find $|\mathbf{A}^4|$, $|\mathbf{A}^{-4}|$, $|\mathbf{A}^2\mathbf{B}^3|$, $|\mathbf{A}^3\mathbf{B}^{-5}|$.

6. Let **A** and **B** be square matrices of order 3 such that $|\mathbf{A}^2\mathbf{B}^3| = 576$ and $|\mathbf{A}^3\mathbf{B}^2| = 432$. Find $|\mathbf{A}^2\mathbf{B}^2|$, $|\mathbf{A}^{-1}\mathbf{B}|$ and $|\mathbf{A}^5\mathbf{B}^{-7}|$.

7. Let **A** and **B** be square matrices of order 6 such that $|\mathbf{A}^3\mathbf{B}^5| = 1944$ and $|\mathbf{A}^5\mathbf{B}^2| = 288$. Find $|2\mathbf{A}|$, $|\mathbf{A}^2/4|$, $|\mathbf{A}^3\mathbf{B}^{-4}|$, $|\mathbf{A}\mathbf{B}^4|$, $|\mathbf{A}^4\mathbf{B}^{-4}|$.

8. Given $\mathbf{A} = \begin{bmatrix} -2 & 3 & 1.2 \\ 0 & 1 & 3.1 \\ 2 & -3 & 4.2 \end{bmatrix}$, compute $|\mathbf{A}^{-1}|$.

Exercises 9–12. Find the adjoint of each matrix.

9. $\begin{bmatrix} 1 & 2 \\ 3 & 4 \end{bmatrix}$

10. $\begin{bmatrix} a & b \\ c & d \end{bmatrix}$

11. $\begin{bmatrix} 1 & 1 & 1 \\ 0 & 1 & 1 \\ 0 & 0 & 1 \end{bmatrix}$

12. $\begin{bmatrix} 0 & a & 0 \\ b & 0 & b \\ 0 & a & 0 \end{bmatrix}$

Exercises 13–16. Use the adjoint formula to find the inverse of the given matrix. Verify that $|\mathbf{A}^{-1}| = |\mathbf{A}|^{-1}$.

13. $\begin{bmatrix} 1 & 2 \\ 3 & 4 \end{bmatrix}$

14. $\begin{bmatrix} 2 & -1 \\ 5 & 4 \end{bmatrix}$

15. $\begin{bmatrix} 2 & 2 & 0 \\ 0 & 4 & -1 \\ 4 & 0 & 3 \end{bmatrix}$

16. $\begin{bmatrix} 1 & -1 & 1 \\ 0 & 2 & 1 \\ 0 & 0 & 1 \end{bmatrix}$

17. Show that $|\mathbf{A}\mathbf{A}^{\mathsf{T}}|$ is nonnegative for any $n \times n$ matrix $\mathbf{A}$.

18. Refer to Section 1.3. The Vandermonde matrix $\mathbf{V} = \mathbf{V}(x_1, x_2, \ldots, x_n)$ is given by

$$\mathbf{V} = \begin{bmatrix} x_1^{n-1} & x_1^{n-2} & \cdots & x_1^2 & x_1 & 1 \\ x_2^{n-1} & x_2^{n-2} & \cdots & x_2^2 & x_2 & 1 \\ \vdots & \vdots & \ddots & \vdots & \vdots & \vdots \\ x_n^{n-1} & x_2^{n-2} & \cdots & x_n^2 & x_n & 1 \end{bmatrix}$$

Prove by mathematical induction that $\mathbf{V}$ is invertible precisely when all the values of x_i are distinct. *Hint*: Consider $\mathbf{V}^{\mathsf{T}}$, take x_1 times row 2 from row 1, x_1 times row 3 from row 2, and so on. Remove a factor from each column in the resulting determinant and apply induction.

19. Solve the determinant equation

$$\begin{vmatrix} x^3 & x^2 & x & 1 \\ 2^3 & 2^2 & 2 & 1 \\ 3^3 & 3^2 & 3 & 1 \\ 4^3 & 4^2 & 4 & 1 \end{vmatrix} = 0$$

20. Let $\mathbf{A}$ be any $n \times n$ matrix. Show that $\det \mathbf{A} = \pm\sqrt{\det(\mathbf{B})}$ for some symmetric matrix $\mathbf{B}$. The result is due to Lagrange.

21. Let $\mathbf{A}$ be an $n \times n$ matrix such that $|\mathbf{A}^k| = 1$, for some positive integer k. Prove that $|\mathbf{A}^2| = 1$.

22. Let $\mathbf{A}$ be an $n \times n$ matrix and let $\mathbf{P}$ be an $n \times n$ invertible matrix. Prove that $\det(\mathbf{P}^{-1}\mathbf{A}\mathbf{P}) = \det(\mathbf{A})$.

23. Show that if an $n \times n$ matrix $\mathbf{A}$ is orthogonal (Section 4.2), then $\det(\mathbf{A}) = \pm 1$.

24. Let $\mathbf{A} = \begin{bmatrix} a & b & c \\ c & a & b \\ b & c & a \end{bmatrix}$ and $\mathbf{B} = \begin{bmatrix} -a & b & c \\ -b & c & a \\ -c & a & b \end{bmatrix}$. Consider the product $\mathbf{B}\mathbf{A}$ and show that

$$\begin{vmatrix} 2bc - a^2 & c^2 & b^2 \\ c^2 & 2ac - b^2 & a^2 \\ b^2 & a^2 & 2ab - c^2 \end{vmatrix} = (a^3 + b^3 + c^3 - 3abc)^2.$$

Exercises 25–32. Find the inverse of each matrix using the adjoint formula.

25. $\begin{bmatrix} 2 & 5 \\ 3 & 4 \end{bmatrix}$

26. $\begin{bmatrix} 1.5 & -0.5 \\ 0.5 & \frac{\sqrt{3}}{2} \end{bmatrix}$

27. $\begin{bmatrix} 15 & 14 \\ 16 & 15 \end{bmatrix}$

28. $\begin{bmatrix} -18 & 19 \\ 19 & 22 \end{bmatrix}$

29. $\begin{bmatrix} \sin\theta & \cos\theta \\ -\cos\theta & \sin\theta \end{bmatrix}$

30. $\begin{bmatrix} -a & a & a \\ c & c & -c \\ b & -b & b \end{bmatrix}$

31. $\begin{bmatrix} 3 & 5 & 4 \\ 4 & 9 & 5 \\ 6 & 14 & 6 \end{bmatrix}$

32. $\begin{bmatrix} 6 & -5 & 1 \\ 7 & -6 & 1 \\ 10 & -11 & 3 \end{bmatrix}$

33. Let n and d be determinants as shown. Express the quotient n/d as a determinant.

$$n = \begin{vmatrix} 2bc - a^2 & a^2 & a^2 \\ b^2 & 2ca - b^2 & b^2 \\ c^2 & c^2 & 2ab - c^2 \end{vmatrix}$$

$$d = \begin{vmatrix} -a & a & a \\ c & c & -c \\ b & -b & b \end{vmatrix}.$$

34. Show that the adjoint of a triangular matrix is triangular and deduce that the inverse of an triangular invertible matrix is triangular.

35. Prove that $\operatorname{adj}(\mathbf{A}^{\mathsf{T}}) = (\operatorname{adj} \mathbf{A})^{\mathsf{T}}$.

36. Prove that $\text{adj}(\alpha\mathbf{A}) = \alpha^{n-1}\text{adj}\,\mathbf{A}$, where α is any scalar.

37. $\det(\text{adj}\,\mathbf{A}) = \det(\mathbf{A})^{n-1}$. This is *Cauchy's formula. Hint*: Consider cases: $\mathbf{A}$ invertible and $\mathbf{A}$ singular.

38. Let $\mathbf{A}$ be invertible. Show that $\text{adj}\,\mathbf{A}^{-1} = (\text{adj}\,\mathbf{A})^{-1}$. When $n \geq 2$ show that $\text{adj}(\text{adj}\,\mathbf{A}) = |\mathbf{A}|^{n-2}\mathbf{A}$. Illustrate both properties of the adjoint using sparse 3×3 matrices.

39. $\mathbf{A}$ and $\mathbf{B}$ are $n \times n$ invertible matrices. Show that $\text{adj}(\mathbf{AB}) = (\text{adj}\,\mathbf{B})(\text{adj}\,\mathbf{A})$.

40. Prove that the adjoint of an invertible skew-symmetric matrix of order n is skew-symmetric.

Exercises 41–42. Solve each square system by Cramer's Rule.

41. $\begin{cases} x_1 + 5x_2 = 2 \\ -3x_1 - 9x_2 = 6 \end{cases}$ **42.** $\begin{cases} 1.5x_1 - x_2 = 0.5 \\ 3x_1 + 2x_2 = 1.5 \end{cases}$

Exercises 43–44. Use Cramer's rule to find the values of x_2 and x_3.

43. $\begin{cases} x_1 + x_2 + 2x_3 = -1 \\ 2x_1 - x_2 + 2x_3 = -4 \\ 4x_1 + x_2 + 4x_3 = -2 \end{cases}$

44. $\begin{cases} x_1 + 2x_2 + 4x_3 = 31 \\ 5x_1 + x_2 + 2x_3 = 29 \\ 3x_1 - x_2 + x_3 = 10 \end{cases}$

Exercises 45–46. Each linear system can be described as sparse, *having a relatively large number of zero coefficients. Use Cramer's rule to find x_4.*

45. $\begin{cases} x_1 + 3x_3 - 5x_4 = 1 \\ x_2 + 2x_4 = 0 \\ 2x_3 - x_4 = -1 \\ x_1 + x_2 + 3x_3 = 2 \end{cases}$

46. $\begin{cases} -x_1 \quad x_4 - x_5 = 0 \\ 2x_2 - x_5 = 1 \\ x_3 + x_4 = 1 \\ x_4 + x_5 = 1 \\ x_4 - x_5 = 2 \end{cases}$

47. Suppose that $\mathbf{Ax} = \mathbf{b}$ represents a square linear system, where $\mathbf{A} = [\mathbf{a}_1\ \mathbf{a}_2\ \cdots\ \mathbf{a}_n]$, $\det(\mathbf{A}) \neq 0$, and $\mathbf{x} = (x_1, x_2, \ldots, x_n)$. Use Cramer's rule to explain the following facts. Illustrate each fact using a simple 3×3 system of your choice.

(a) If $\mathbf{b} = 3\mathbf{a}_1$, then $x_1 = 3$ and all other unknowns are zero.

(b) If $\mathbf{b} = 2\mathbf{a}_1 + 3\mathbf{a}_2$, then $x_1 = 2$, $x_2 = 3$ and all other variables are zero.

48. Use the adjoint formula for the inverse to prove Cramer's rule.

Exercises 49–52. Find $\det(\mathbf{A})$ in each case and hence determine if the given matrix is unimodular. In case $\mathbf{A}$ is unimodular, verify that $\mathbf{A}^{-1}$ has integer entries.

49. $\begin{bmatrix} 1 & -1 \\ 1 & 1 \end{bmatrix}$ **50.** $\begin{bmatrix} 31 & 6 \\ 36 & 7 \end{bmatrix}$

51. $\begin{bmatrix} 8 & 0 & 3 \\ 2 & -1 & 0 \\ -7 & 2 & -1 \end{bmatrix}$

52. $\begin{bmatrix} 8 & 0 & 3 \\ 2 & -1 & 0 \\ 7 & 2 & -1 \end{bmatrix}$

Exercises 53–58. Let $\mathbf{u} = (-3, 2, 4)$, $\mathbf{v} = (-2, 1, -1)$, $\mathbf{w} = (0, 5, -1)$. Compute each expression.

53. $\mathbf{u} \times \mathbf{v}$

54. $\mathbf{v} \times \mathbf{w}$

55. $\mathbf{w} \times \mathbf{u}$

56. $\mathbf{u} \cdot (\mathbf{v} \times \mathbf{w})$

57. $\mathbf{u} \times (\mathbf{v} \times \mathbf{w})$

58. $(\mathbf{u} \times \mathbf{v}) \times \mathbf{w}$

59. Use cross product to find the equation of the plane in $\mathbb{R}^3$ that passes through the points $P_1 = (1, 1, 1)$, $P_2 = (-1, 0, 1)$, $P_3 = (0, 1, -1)$ and find the area of the triangle formed by the three points.

60. Find the area of the ΔABC in $\mathbb{R}^3$ with vertices $A = (2, 3, 4)$, $B = (-3, 1, 0)$, $C = (-1, -1, -1.5)$.

Exercises 61–62. Evaluate each expression.

61. $(\mathbf{i} + \mathbf{j}) \times (\mathbf{i} - \mathbf{j} + \mathbf{k}) \cdot (\mathbf{i} + \mathbf{j} + \mathbf{k})$

62. $(\mathbf{i} - 2\mathbf{j} + \mathbf{k}) \times (\mathbf{i} + \mathbf{k}) \times (\mathbf{j} + \mathbf{k})$

63. Find the volume v_C of the parallelepiped shown in Figure 5.1 in the case when $\mathbf{u} = (1, -1, 2)$, $\mathbf{v} = (2, 0, 3)$, $\mathbf{w} = (0, 4, -1)$.

USING MATLAB

64. *Project.* Write an M-file that defines a function $\mathbf{C} = \mathbf{cof(A)}$ that calls a square matrix $\mathbf{A}$ and returns adj $\mathbf{A}$. Test your program on the matrices

$$\begin{bmatrix} 1 & 0 \\ 0 & 1 \end{bmatrix}, \quad \begin{bmatrix} 1 & 2 & 3 \\ 4 & 5 & 6 \\ 7 & 8 & 9 \end{bmatrix}, \quad \begin{bmatrix} 1 & 1 & 1 \\ 0 & 1 & 1 \\ 0 & 0 & 1 \end{bmatrix}.$$

65. *Project.* Write an M-file that will apply Cramer's rule to solve a square linear system for a specified unknown. The input should be the coefficient matrix $\mathbf{A}$, the column of constants $\mathbf{b}$ and the index of the unknown.

Exercises 66–67. Determine if each matrix is singular.

66. $$\begin{bmatrix} 1 & -2 & 1 & 1 \\ -2 & 2 & 2 & 7 \\ 5 & -6 & 1 & 2 \\ -7 & 8 & 1 & 5 \end{bmatrix}$$

67. $$\begin{bmatrix} 1 & -2 & -3 & 4 \\ -2 & 3 & 4 & -5 \\ 3 & -4 & -5 & 6 \\ -4 & 5 & 6 & -7 \end{bmatrix}$$

68. Write an M-file that will encode the plain text message ET IN ARCADIA EGO (let 0 denote a space) using the encoding matrix

$$\mathbf{E} = \begin{bmatrix} 2 & 2 & 1 \\ 1 & 1 & 0 \\ 1 & 2 & 1 \end{bmatrix}.$$

Decode the encoded message using the decoding matrix $\mathbf{D}$. Extend your program to encode any message using an appropriate $n \times n$ matrix $\mathbf{E}$.

Exercises 69–71. Show that the equation of the plane passing through points $P_1 = (x_1, y_1, z_1)$, $P_2 = (x_2, y_2, z_2)$, $P_3 = (x_3, y_3, z_3)$ is given by

$$\begin{vmatrix} x & y & z & 1 \\ x_1 & y_1 & z_1 & 1 \\ x_2 & y_2 & z_1 & 1 \\ x_3 & y_3 & z_3 & 1 \end{vmatrix} = 0$$

Define symbolic variables x, y, z and use the formula to find the equation of the plane in $\mathbb{R}^3$ passing through the points P_1, P_2, P_3, respectively, given in each exercise.

69. $(1, 1, 6)$, $(2, -1, -1)$, $(5, -2, 3)$

70. $(-3, 7, -3)$, $(-2, 6, -2)$, $(4, 6, -2)$

71. $(1, 4, 1)$, $(2, 11, 0)$, $(3, 6, 5)$

72. If one of the points (x_1, y_1, z_1), (x_2, y_2, z_2), (x_3, y_3, z_3), (x_4, y_4, z_4) is not in the plane determined by the other three, we obtain the tetrahedron C shown.

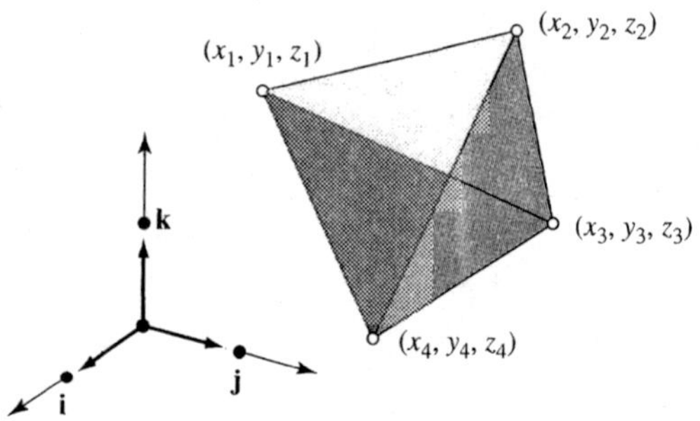

The volume of C is given by the absolute value of the determinant

$$v_C = \frac{1}{6} \begin{vmatrix} x_1 & y_1 & z_1 & 1 \\ x_2 & y_2 & z_2 & 1 \\ x_3 & y_3 & z_3 & 1 \\ x_4 & y_4 & z_4 & 1 \end{vmatrix}$$

Test the formula by calculating by hand the volume of the tetrahedron determined by each set of points, then cross-check using machine calculation.

(a) (0,0,0), (1,0,0), (0,1,0), (0,0,1)

(b) (1,0,0), (0,1,0), (0,0,1), (1,1,1)

73. Use MATLAB to check hand calculations in this exercise set.

CHAPTER 5 REVIEW

State whether each statement is true or false as stated. Provide a clear reason for your answer. If a false statement can be modified to make it true, then do so.

1. The minor m_{21} in $\begin{bmatrix} 3 & 1 & -5 & 0 \\ 1 & 0 & 2 & 3 \\ 0 & 1 & 3 & -2 \\ 1 & 2 & -1 & -2 \end{bmatrix}$ is -2.

2. If $\begin{vmatrix} a & b & c \\ d & e & f \\ g & h & i \end{vmatrix} = 6$, then $\begin{vmatrix} a-d & b-e & c-f \\ 5a & 5b & 5c \\ -g & -h & -i \end{vmatrix} = 30$.

3. If **B** is formed by scaling row i in **A** by 2, then $\det(\mathbf{A}) = 0.5\det(\mathbf{B})$.

4. If **B** is formed by adding a linear combination of two rows of **A** to a different row of **A**, then $\det(\mathbf{B}) = \det(\mathbf{A})$.

5. If $\mathbf{AB} = \begin{bmatrix} 3 & 4 \\ 5 & 6 \end{bmatrix}$ and $\det(\mathbf{B}) = 2$, then $\det(\mathbf{A}) = 1$.

6. $\det(-2\mathbf{A}) = -2\det(\mathbf{A})$.

7. The determinant of any square triangular matrix is nonzero.

8. $\det(\mathbf{A} + \mathbf{B}) = \det(\mathbf{A}) + \det(\mathbf{B})$ for no matrices **A** and **B**.

9. The equation $\begin{vmatrix} x-1 & -2 \\ 1 & x-1 \end{vmatrix} = 0$ is true for no real x.

10. If **A** has two columns equal, then $\det(\mathbf{A}) = 0$.

11. If $\det(\mathbf{A}^4) = 0$, then $\det(\mathbf{A}) = 0$.

12. If **A** is symmetric, then $\det(\mathbf{A}) = 0$.

13. The symmetric matrix $\begin{bmatrix} 0 & 1 \\ 1 & 0 \end{bmatrix}$ cannot be written in the form $\mathbf{AA}^\mathrm{T}$ for any **A**.

14. If **A** is nilpotent, then **A** is singular.

15. For some square matrices **A**, $\det(-\mathbf{A}) = -\det(\mathbf{A})$.

16. For any invertible matrix **A**, $\det(\mathbf{A}) = \det(\mathbf{A}^{-1})^{-1}$.

17. If **A** and **B** are square matrices and **AB** is invertible, then both **A** and **B** have nonzero determinant.

18. If **A** is 3×3 and $\det(\mathbf{A}) = 0$, then $\operatorname{rank} \mathbf{A} = 2$.

19. If **A** is unimodular, then $\det(\mathbf{A}^{-1}) = \det(\mathbf{A})$.

20. Any consistent square linear system can be solved by Cramer's rule.

Andrei Andreyevich Markov (1856–1922)

Andrei Markov was born into the upper classes of St. Petersburg, Russia. At that time, the city was the capital of czarist Russia and a vibrant center for commerce and the arts. Markov earned undergraduate and graduate degrees from St. Petersburg University. He received the university's gold medal and was appointed to the faculty. Markov was elected to the St. Petersburg Academy of Science, nominated by his former professor P. L. Chebychev.

Markov was known for his rebellious spirit, and this characteristic may account for problems he had later in life with the government and his colleagues, but it may have also contributed to his drive to develop new ideas in mathematics. His interest in poetry and literature led him to apply probability to analyze the distribution of vowels and consonants in the masterpiece Eugene Onegin (1833) by Aleksandr Sergeëvich Pushkin (1799–1837). This research helped lay the foundation of a discipline called *mathematical linguistics*.

Markov taught at the university until 1905 and made significant contributions in the fields of number theory, analysis, and infinite series. He is best known, however, for his work in probability, particularly his development of Markov chains (Section 6.4) as a model for studying random variables. This research opened up the study of stochastics (which includes Markov chains), which today is a vast and growing subfield of mathematics with important applications.

Markov wrote an excellent textbook on probability and statistics and influenced many future and famous mathematicians. He wrote little about applications, other than linguistic analysis, but today Markov chains have significant use and are applied, in particular, in the theory of information retrieval and search engines of the Internet.

Chapter **6**

EIGENVALUE PROBLEMS

Introduction

Many problems arising in engineering, science, and other applied fields depend for their solution on the theory of *eigenvalues* and *eigenvectors*. We refer to these problems generally as *eigenvalue problems*. Some of the applications introduced in Section 2.4 are eigenvalue problems, and a deeper understanding of those models cannot be attained without the theory developed in this chapter.

As a first step, we will introduce the idea of eigenvalues and eigenvectors geometrically in $\mathbb{R}^2$. Consider the matrix $\mathbf{A}$ in (6.1) and any vector $\mathbf{u}$ in $\mathbb{R}^2$.

$$\mathbf{A} = \begin{bmatrix} 1 & 1 \\ -2 & 4 \end{bmatrix} \tag{6.1}$$

Recall from Section 3.4 that the matrix $\mathbf{A}$ defines a linear transformation T on $\mathbb{R}^2$ that maps each vector $\mathbf{u}$ to its image $T(\mathbf{u}) = \mathbf{Au} = \mathbf{v}$ in $\mathbb{R}^2$. In general, we can expect that the directions of the arrows that represent $\mathbf{u}$ and $\mathbf{v}$ will be different, as the following two cases show.

$$\mathbf{u}_1 = \begin{bmatrix} 1 \\ 0 \end{bmatrix} \quad \Rightarrow T(\mathbf{u}_1) = \mathbf{Au}_1 = \begin{bmatrix} 1 & 1 \\ -2 & 4 \end{bmatrix}\begin{bmatrix} 1 \\ 0 \end{bmatrix} = \begin{bmatrix} 1 \\ -2 \end{bmatrix} = \mathbf{v}_1$$

$$\mathbf{u}_2 = \begin{bmatrix} -1 \\ 1 \end{bmatrix} \quad \Rightarrow T(\mathbf{u}_2) = \mathbf{Au}_2 = \begin{bmatrix} 1 & 1 \\ -2 & 4 \end{bmatrix}\begin{bmatrix} -1 \\ 1 \end{bmatrix} = \begin{bmatrix} 0 \\ 6 \end{bmatrix} = \mathbf{v}_2$$

The vectors $\mathbf{u}_1$, $\mathbf{u}_2$ and their respective images $\mathbf{v}_1$, $\mathbf{v}_2$ are shown in Figure 6.1.

However, there are some vectors $\mathbf{u}$ in $\mathbb{R}^2$ that have the property that the image $\mathbf{v}$ of $\mathbf{u}$ is a scalar multiple of $\mathbf{u}$. In such cases the arrows representing $\mathbf{u}$ and $\mathbf{v}$ lie on the same line L through the origin in $\mathbb{R}^2$. It is these particular vectors $\mathbf{u}$ that are of interest, and we will now give a brief preview of the theory that is to come.

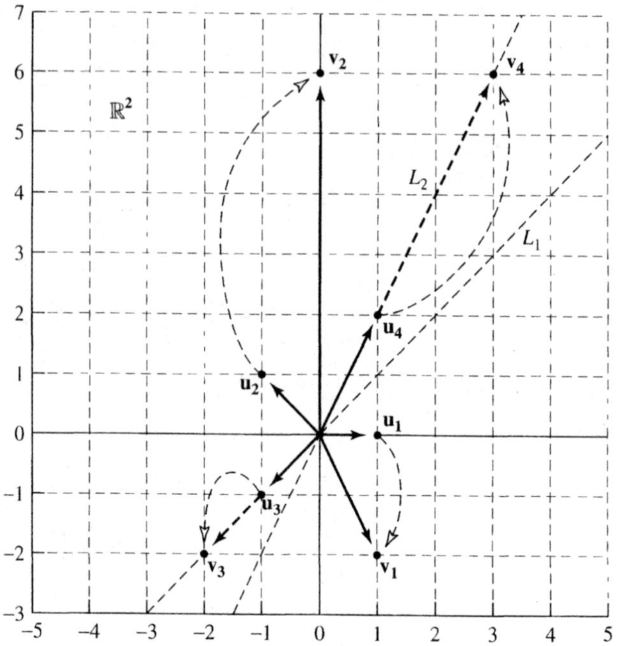

Figure 6.1 The action of the matrix **A**.

Refer to Figure 6.1. The matrix **A** in (6.1) determines two one-dimensional subspaces $\mathcal{S}_1$ and $\mathcal{S}_2$ of $\mathbb{R}^2$ that are represented, respectively, by lines L_1 and L_2 passing through the origin. The action of **A** transforms any vector with terminal point on L_1 to its image **v** whose terminal point also lies on L_1, and similarly for L_2. For example,

$$\mathbf{u}_3 = \begin{bmatrix} -1 \\ -1 \end{bmatrix} \Rightarrow T(\mathbf{u}_3) = \mathbf{A}\mathbf{u}_3 = \begin{bmatrix} 1 & 1 \\ -2 & 4 \end{bmatrix} \begin{bmatrix} -1 \\ -1 \end{bmatrix} = \begin{bmatrix} -2 \\ -2 \end{bmatrix} = 2 \begin{bmatrix} -1 \\ -1 \end{bmatrix} = \mathbf{v}_3,$$

$$\mathbf{u}_4 = \begin{bmatrix} 1 \\ 2 \end{bmatrix} \Rightarrow T(\mathbf{u}_4) = \mathbf{A}\mathbf{u}_4 = \begin{bmatrix} 1 & 1 \\ -2 & 4 \end{bmatrix} \begin{bmatrix} 1 \\ 2 \end{bmatrix} = \begin{bmatrix} 3 \\ 6 \end{bmatrix} = 3 \begin{bmatrix} 1 \\ 2 \end{bmatrix} = \mathbf{v}_4,$$

showing that $\mathbf{v}_3 = 2\mathbf{u}_3$ and $\mathbf{v}_4 = 3\mathbf{u}_4$. The vectors $\mathbf{u}_3$ and $\mathbf{u}_4$ are defined by the equations

$$\mathbf{A}\mathbf{u}_3 = \lambda\mathbf{u}_3, \quad \text{where } \lambda = 2, \tag{6.2}$$

$$\mathbf{A}\mathbf{u}_4 = \lambda\mathbf{u}_4, \quad \text{where } \lambda = 3. \tag{6.3}$$

The scalar $\lambda = 2$ in (6.2) is called an *eigenvalue* of **A** and the *nonzero* vector $\mathbf{u}_3$ is called an *eigenvector* of **A** *associated with* the eigenvalue $\lambda = 2$. Equation (6.3) shows that the nonzero vector $\mathbf{u}_4$ is an eigenvector of **A** associated with the eigenvalue $\lambda = 3$. Eigenvectors are (by definition) nonzero vectors and are never unique—if c is a nonzero real scalar, then $c\mathbf{u}_3$ is an eigenvector of **A** associated with $\lambda = 2$. It will be shown in Section 6.1 that an $n \times n$ matrix has n (not necessarily distinct) real or complex eigenvalues. Hence, $\lambda = 2, 3$

are the only eigenvalues of **A** in (6.1). The subspaces $\mathcal{S}_1$ and $\mathcal{S}_2$ in $\mathbb{R}^2$ are called *eigenspaces* of **A** associated, respectively, with the eigenvalues $\lambda = 2$ and $\lambda = 3$.

Some care is needed in extending the preceding ideas to other real 2×2 matrices. For example, the matrix **B** in (6.4) has two equal real eigenvalues $\lambda = 3, 3$ (we record *repeated* eigenvalues).

$$\mathbf{B} = \begin{bmatrix} 4 & 1 \\ -1 & 2 \end{bmatrix}, \qquad \mathbf{u} = c\begin{bmatrix} 1 \\ -1 \end{bmatrix}, \qquad c \text{ a nonzero real scalar.} \tag{6.4}$$

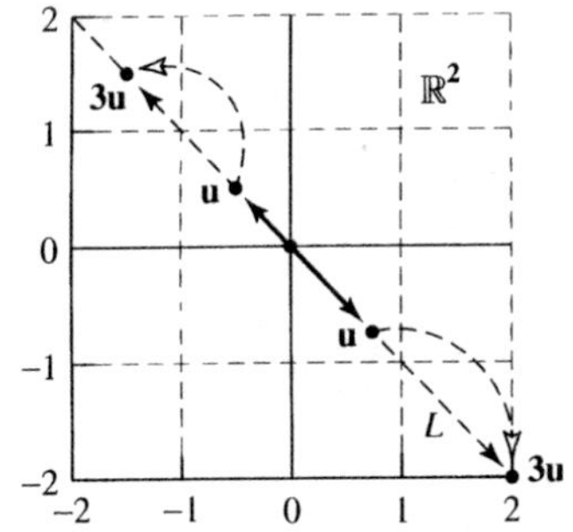

Figure 6.2 The eigenspace of the matrix **B**.

The nonzero vectors **u**, shown on the right in (6.4), satisfy $\mathbf{Bu} = 3\mathbf{u}$ and are the eigenvectors of **B** associated with the eigenvalue $\lambda = 3$. The matrix **B** defines just one eigenspace $\mathcal{S}$ represented by a line L passing through the origin in $\mathbb{R}^2$. Figure 6.2 shows L and the action of **B** on the eigenvectors

$$\mathbf{u} = \begin{bmatrix} 2/3 \\ -2/3 \end{bmatrix}, \quad \begin{bmatrix} -1/2 \\ 1/2 \end{bmatrix}.$$

Complex numbers enter into the theory of eigenvalues in a natural way. For example, the real matrix **C** shown in (6.5) has two distinct complex eigenvalues $\lambda = 2 + i$, $\lambda = 2 - i$ (complex conjugates), where $i^2 = -1$.

$$\mathbf{C} = \begin{bmatrix} 2 & 1 \\ -1 & 2 \end{bmatrix} \qquad \begin{matrix} \mathbf{u} = c\begin{bmatrix} 1 \\ i \end{bmatrix} & (\lambda = 2 + i) \\ \mathbf{u} = c\begin{bmatrix} 1 \\ -i \end{bmatrix} & (\lambda = 2 - i) \end{matrix} \tag{6.5}$$

The set of complex eigenvectors **u** associated with each eigenvalue of **C** is shown on the right in (6.5), where c is a nonzero complex scalar. In this case the eigenvectors of **C** cannot be visualized in terms of subspaces of $\mathbb{R}^2$. Although some real square matrices have complex eigenvalues, a real symmetric matrix has only real eigenvalues—the proof given in Section 8.6 uses complex numbers.

The basic theory of eigenvalues and eigenvectors appears in Section 6.1. The theory is applied to the *diagonalization* of square matrices in Section 6.2. Later sections continue the analysis of the eigenvalue problems that were introduced in Section 2.4 and discuss other applications, including Markov chains and systems of differential equations.

6.1 Eigenvalues and Eigenvectors

The theory of eigenvalues and eigenvectors applies only to square matrices.

Definition 6.1

Eigenvalues, eigenvectors

Let **A** be an $n \times n$ matrix. A real or complex number λ is called an *eigenvalue* of **A** if the matrix equation

$$\mathbf{Ax} = \lambda\mathbf{x} \tag{6.6}$$

is satisfied for some *nonzero* vector $\mathbf{x}$ in $\mathbb{R}^n$. The vector $\mathbf{x}$ is called an *eigenvector* of $\mathbf{A}$ *associated with* the eigenvalue λ.

Note that the German word *eigen* is pronounced "i-gen," with a hard g. Eigenvalues are also called *characteristic roots* or *latent roots.*

Remarks

(a) Suppose that λ is an eigenvalue of $\mathbf{A}$ associated with the eigenvector $\mathbf{x}$. Then, for any nonzero scalar c, we have

$$\mathbf{A}(c\mathbf{x}) = c\mathbf{A}\mathbf{x} = c\lambda\mathbf{x} = \lambda(c\mathbf{x}),$$

showing that $c\mathbf{x}$ is also an eigenvector of $\mathbf{A}$ associated with the eigenvalue λ.

(b) Putting $\mathbf{x} = \mathbf{0}$ in (6.6) gives $\mathbf{A0} = \mathbf{0} = \lambda\mathbf{0}$, for any λ. However, the zero vector $\mathbf{0}$ is not considered to be an eigenvector of $\mathbf{A}$ associated with any eigenvalue λ. Note that *some* square matrices have an eigenvalue $\lambda = 0$.

Finding Eigenvalues

If λ is an eigenvalue of an $n \times n$ matrix $\mathbf{A}$, then equation (6.6) is true for some nonzero vector $\mathbf{x}$ in $\mathbb{R}^n$. Write $\mathbf{x} = \mathbf{I}_n\mathbf{x}$ and rearrange (6.6) as follows:

$$\begin{aligned} \mathbf{A}\mathbf{x} = \lambda\mathbf{x} \quad &\text{if and only if} \quad \mathbf{A}\mathbf{x} = \lambda\,\mathbf{I}_n\,\mathbf{x} \\ &\text{if and only if} \quad (\mathbf{A} - \lambda\mathbf{I}_n)\mathbf{x} = \mathbf{0}, \end{aligned} \tag{6.7}$$

and (6.7) is the matrix form of a homogeneous $n \times n$ linear system (S). We are only interested in the nonzero solutions $\mathbf{x}$ for (S). Using the theory of Sections 2.2 and 5.1, we have

$$\begin{aligned} (S) \text{ has a nonzero solution } \mathbf{x} \quad &\text{if and only if} \quad (\mathbf{A} - \lambda\mathbf{I}_n) \text{ is singular} \\ &\text{if and only if} \quad \det(\mathbf{A} - \lambda\mathbf{I}_n) = 0. \end{aligned} \tag{6.8}$$

Thus, the eigenvalues of an $n \times n$ matrix $\mathbf{A}$ are the solutions λ of the determinant equation

$$\det(\mathbf{A} - \lambda\mathbf{I}_n) = 0. \tag{6.9}$$

Note that (6.9) is equivalent (Exercises 6.1) to the equation

$$\det(\lambda\mathbf{I}_n - \mathbf{A}) = 0, \tag{6.10}$$

which will be convenient when we wish to express the *characteristic polynomial* of $\mathbf{A}$ in standard form.

Square Matrices of Order 2

Consider a general 2×2 matrix

$$\mathbf{A} = \begin{bmatrix} a & b \\ c & d \end{bmatrix}.$$

Computing the determinant in (6.10), we have

$$\det(\lambda \mathbf{I}_2 - \mathbf{A}) = \begin{vmatrix} \lambda - a & -b \\ -c & \lambda - d \end{vmatrix} = (\lambda - a)(\lambda - d) - bc$$

$$= \lambda^2 - \lambda(a + d) + (ad - bc),$$

and so equation (6.10) in this case is

$$\lambda^2 - \lambda(a + d) + (ad - bc) = 0. \tag{6.11}$$

The eigenvalues λ_1, λ_2 of **A** are the roots of the quadratic equation (6.11), which may be found using the quadratic formula.[1] Equation (6.11) can now be written in terms of its factors, namely $(\lambda - \lambda_1)(\lambda - \lambda_2) = 0$, and so

$$\lambda^2 - (\lambda_1 + \lambda_2) + \lambda_1\lambda_2 = 0. \tag{6.12}$$

Comparing (6.12) with (6.11), we have

$$\lambda_1 + \lambda_2 = a + d = \text{trace}\,\mathbf{A} \quad \text{and} \quad \lambda_1\lambda_2 = ad - bc = \det(\mathbf{A}),$$

where trace **A** is equal to the sum of entries on main diagonal of **A**.

EXAMPLE 1

Finding Eigenvalues in the 2 × 2 Case

Consider the matrix **A** in (6.1). Using (6.11), we have

$$\mathbf{A} = \begin{bmatrix} 1 & 1 \\ -2 & 4 \end{bmatrix} \quad \Rightarrow \quad \lambda^2 - 5\lambda + 6 = (\lambda - 2)(\lambda - 3) = 0,$$

giving two real and distinct eigenvalues of **A**, namely $\lambda_1 = 2$ and $\lambda_2 = 3$. Note that trace $\mathbf{A} = 5 = \lambda_1 + \lambda_2$ and $\det(\mathbf{A}) = 6 = \lambda_1\lambda_2$ as expected. ■

Characteristic Polynomial and Equation

Let $\mathbf{A} = [a_{ij}]$ be an $n \times n$ matrix. The determinant in (6.10) is

$$\det(\lambda \mathbf{I} - \mathbf{A}) = \begin{vmatrix} \lambda - a_{11} & -a_{12} & \cdots & -a_{1n} \\ -a_{21} & \lambda - a_{22} & \cdots & -a_{2n} \\ \vdots & \vdots & \ddots & \vdots \\ -a_{n1} & -a_{n2} & \cdots & \lambda - a_{nn} \end{vmatrix} \tag{6.13}$$

and the Laplace expansion of (6.13) results in a polynomial $p(\lambda)$ of degree n of the form

$$p(\lambda) = \lambda^n + c_{n-1}\lambda^{n-1} + c_{n-2}\lambda^{n-2} + \cdots + c_1\lambda + c_0, \tag{6.14}$$

where the coefficients $c_{n-1}, c_{n-2}, \ldots, c_1, c_0$ depend on the entries in **A**.

---

[1] Recall that the two roots of the quadratic $\lambda^2 + p\lambda + q = 0$ are determined by the discriminant $\Delta = p^2 - 4q$. The roots are real and unequal when $\Delta > 0$, real and equal when $\Delta = 0$, and a complex conjugate pair when $\Delta < 0$.

Definition 6.2

Characteristic polynomial, characteristic equation

The polynomial $p(\lambda)$ in (6.14) is called the *characteristic polynomial* of **A** and the equation $p(\lambda) = 0$ is called the *characteristic equation* of **A**.

For example, the characteristic polynomial of the matrix **A** in Example 1 is $p(\lambda) = \lambda^2 - 5\lambda + 6$ and $\lambda^2 - 5\lambda + 6 = 0$ is the characteristic equation of **A**.

The coefficients c_{n-1} and c_0 in (6.14) are of particular interest and can be used to cross-check accuracy when computing the characteristic polynomial. A careful analysis of the Laplace expansion that gives (6.14) shows that

$$c_{n-1} = -(a_{11} + a_{22} + \cdots + a_{nn}) = -\operatorname{trace}\mathbf{A}, \tag{6.15}$$

where $\operatorname{trace}\mathbf{A} = a_{11} + a_{22} + \cdots + a_{nn}$. Also, (6.14) is true for all λ, and so we have

$$c_0 = p(0) = \det(-\mathbf{A}) = (-1)^n \det(\mathbf{A}). \tag{6.16}$$

For example, when (6.11) is compared to (6.14) (with $n = 2$), we have $c_1 = -\operatorname{trace}\mathbf{A} = -(a + d)$ and $c_0 = (-1)^2 \det(\mathbf{A}) = \det(\mathbf{A})$.

By the fundamental theorem of algebra, the equation $p(\lambda) = 0$ has n real or complex roots $\lambda_1, \lambda_2, \ldots, \lambda_n$ (the eigenvalues of **A**). The characteristic polynomial $p(\lambda)$ therefore has a factorization in linear factors

$$\begin{aligned} p(\lambda) &= (\lambda - \lambda_1)(\lambda - \lambda_2)\cdots(\lambda - \lambda_n) \\ &= \lambda^n - (\lambda_1 + \lambda_2 + \cdots + \lambda_n)\lambda^{n-1} + \cdots + (-1)^n \lambda_1\lambda_2\cdots\lambda_n \end{aligned} \tag{6.17}$$

and comparing the expansion (6.17) with (6.16) and (6.15), we have

$$\operatorname{trace}\mathbf{A} = \text{sum of the eigenvalues,}$$

$$\det(\mathbf{A}) = \text{product of eigenvalues.}$$

If $p(\lambda)$ is a polynomial with real coefficients and $a+bi$ is a complex root of $p(\lambda) = 0$, then $a - bi$ is also a root (Chapter 8). That is, the complex roots of $p(\lambda) = 0$ appear in complex conjugate pairs, as is the case for **C** in (6.5). However, as mentioned in the Introduction to this chapter, the characteristic equation for a real symmetric matrix has only real roots.

The next two examples illustrate some basic techniques in hand calculation. Recall that $\det(\mathbf{A}) = \det(\mathbf{A}^T)$ and so row replacement or column replacement operations on a determinant will leave its value unchanged.

EXAMPLE 2

Distinct Eigenvalues

To find the eigenvalues of the matrix **A** shown next, we solve the determinant equation $\det(\lambda\mathbf{I} - \mathbf{A}) = 0$.

$$\mathbf{A} = \begin{bmatrix} 1 & -1 & -2 \\ 1 & -1 & -2 \\ -1 & -1 & 0 \end{bmatrix} \quad \Rightarrow \quad \det(\lambda\mathbf{I} - \mathbf{A}) = \begin{vmatrix} \lambda - 1 & 1 & 2 \\ -1 & \lambda + 1 & 2 \\ 1 & 1 & \lambda \end{vmatrix}.$$

Perform the replacement $\mathbf{r}_2+\mathbf{r}_3 \to \mathbf{r}_2$, factor out $\lambda+2$ from row 2 and expand the last determinant on column 1, as follows:

$$\begin{vmatrix} \lambda-1 & 1 & 2 \\ -1 & \lambda+1 & 2 \\ 1 & 1 & \lambda \end{vmatrix} = \begin{vmatrix} \lambda-1 & 1 & 2 \\ 0 & \lambda+2 & \lambda+2 \\ 1 & 1 & \lambda \end{vmatrix} = (\lambda+2)\begin{vmatrix} \lambda-1 & 1 & 2 \\ 0 & 1 & 1 \\ 1 & 1 & \lambda \end{vmatrix}$$

$$= (\lambda+2)\,[(\lambda-1)(\lambda-1)+(-1)] = (\lambda-2)(\lambda^2-2\lambda)$$

$$= \lambda(\lambda-2)(\lambda+2). \tag{6.18}$$

The eigenvalues of $\mathbf{A}$ are $\lambda = -2, 0, 2$, and multiplying out (6.18), the characteristic polynomial of $\mathbf{A}$ is

$$p(\lambda) = \lambda^3 - 4\lambda.$$

Comparing $p(\lambda)$ with (6.14) when $n = 3$, we have $c_2 = -\operatorname{trace}\mathbf{A} = 0$ (hence $\operatorname{trace}\mathbf{A} = 0 =$ sum of eigenvalues), $c_1 = -4$, and $c_0 = 0 = (-1)^3 \det\mathbf{A}$ (hence $\det\mathbf{A} = 0 =$ product of eigenvalues). ■

EXAMPLE 3

Repeated Eigenvalues

We will find the eigenvalues of the matrix $\mathbf{A}$ shown on the left.

$$\mathbf{A} = \begin{bmatrix} 1 & -2 & 4 \\ 3 & -4 & 4 \\ 3 & -2 & 2 \end{bmatrix} \quad \Rightarrow \quad \det(\lambda\mathbf{I}-\mathbf{A}) = \begin{vmatrix} \lambda-1 & 2 & -4 \\ -3 & \lambda+4 & -4 \\ -3 & 2 & \lambda-2 \end{vmatrix}$$

Performing the row replacement $\mathbf{r}_3 - \mathbf{r}_2 \to \mathbf{r}_3$, then the column replacement $\mathbf{c}_2 + \mathbf{c}_3 \to \mathbf{c}_2$, and expanding on row 3, we have

$$\det(\lambda\mathbf{I}-\mathbf{A}) = \begin{vmatrix} \lambda-1 & -2 & -4 \\ -3 & \lambda & -4 \\ 0 & 0 & \lambda+2 \end{vmatrix} = (\lambda+2)\begin{vmatrix} \lambda-1 & -2 \\ -3 & \lambda \end{vmatrix}$$

$$= (\lambda+2)(\lambda^2-\lambda-6) = (\lambda+2)^2(\lambda-3). \tag{6.19}$$

The eigenvalues of $\mathbf{A}$ are $\lambda = -2, -2, 3$, and multiplying out (6.19), the characteristic polynomial of $\mathbf{A}$ is

$$p(\lambda) = \lambda^3 + \lambda^2 - 8\lambda - 12.$$

Comparing $p(\lambda)$ with (6.14) when $n = 3$, we have $c_2 = -\operatorname{trace}\mathbf{A} = 1$ (hence $\operatorname{trace}\mathbf{A} = -1 =$ sum of eigenvalues), $c_1 = -8$, and $c_0 = (-1)^3\det\mathbf{A} = -12$ (hence $\det\mathbf{A} = 12 =$ product of eigenvalues). ■

Definition 6.3

Algebraic multiplicity

The *algebraic multiplicity* of an eigenvalue λ of a matrix $\mathbf{A}$, denoted by a_λ, is the number of times λ appears in the list of eigenvalues of $\mathbf{A}$.

In Example 3, we have $a_{-2} = 2$ and $a_3 = 1$.

Suppose **A** has distinct eigenvalues $\lambda_1, \ldots, \lambda_k$, where $1 \le k \le n$. Then the characteristic polynomial of **A**, given in (6.14), has a factorization

$$p(\lambda) = (\lambda - \lambda_1)^{a_{\lambda_1}}(\lambda - \lambda_2)^{a_{\lambda_2}} \cdots (\lambda - \lambda_r)^{a_{\lambda_k}}.$$

Eigenvectors, Eigenspaces

Let **A** be an $n \times n$ matrix and let λ be an eigenvalue of **A**. From (6.7) we see that the eigenvectors associated with λ are the nonzero solutions **x** to the $n \times n$ homogeneous linear system

$$(\mathbf{A} - \lambda \mathbf{I}_n)\mathbf{x} = \mathbf{0}, \tag{6.20}$$

or, in other words, the eigenvectors associated with λ are the nonzero vectors **x** in null$(\mathbf{A} - \lambda\mathbf{I}_n)$, the null space of $\mathbf{A} - \lambda\mathbf{I}_n$.

Definition 6.4

Eigenspace

Let **A** be an $n \times n$ matrix with eigenvalue λ. The *eigenspace* of **A** associated with λ, denoted by E_λ, is the set of nonzero vectors in null$(\mathbf{A} - \lambda\mathbf{I}_n)$.

Refer to Section 3.3 for the theory of null spaces.

EXAMPLE 4

Finding Eigenspaces

Return to Example 2, in which **A** has distinct eigenvalues $\lambda = -2, 0, 2$. For each λ, compute the reduced form of $(\mathbf{A} - \lambda\mathbf{I}_3)$. Throughout, t is a real parameter.

For $\lambda_2 = -2$, we have

$$(\mathbf{A} + 2\mathbf{I}_3) = \begin{bmatrix} 3 & -1 & -2 \\ 1 & 1 & -2 \\ -1 & -1 & 2 \end{bmatrix} \sim \begin{bmatrix} 1 & 0 & -1 \\ 0 & 1 & -1 \\ 0 & 0 & 0 \end{bmatrix} \quad \Rightarrow \quad E_{-2} = \left\{ t \begin{bmatrix} 1 \\ 1 \\ 1 \end{bmatrix} \right\}.$$

For $\lambda = 0$, we have

$$(\mathbf{A} - 0\mathbf{I}_3) = \begin{bmatrix} 1 & -1 & -2 \\ 1 & -1 & -2 \\ -1 & -1 & 0 \end{bmatrix} \sim \begin{bmatrix} 1 & 0 & -1 \\ 0 & 1 & 1 \\ 0 & 0 & 0 \end{bmatrix} \quad \Rightarrow \quad E_0 = \left\{ t \begin{bmatrix} 1 \\ -1 \\ 1 \end{bmatrix} \right\}.$$

For $\lambda_3 = 2$, we have

$$(\mathbf{A} - 2\mathbf{I}_3) = \begin{bmatrix} -1 & -1 & -2 \\ 1 & -3 & -2 \\ -1 & -1 & -2 \end{bmatrix} \sim \begin{bmatrix} 1 & 0 & 1 \\ 0 & 1 & 1 \\ 0 & 0 & 0 \end{bmatrix} \quad \Rightarrow \quad E_2 = \left\{ t \begin{bmatrix} -1 \\ -1 \\ 1 \end{bmatrix} \right\}.$$

■

It follows from Example 4 that each eigenspace E_0, E_{-2}, E_2 is a one-dimensional subspace of $\mathbb{R}^3$ spanned, respectively, by the vectors

$$\mathbf{b}_1 = \begin{bmatrix} 1 \\ -1 \\ 1 \end{bmatrix}, \quad \mathbf{b}_2 = \begin{bmatrix} 1 \\ 1 \\ 1 \end{bmatrix}, \quad \mathbf{b}_3 = \begin{bmatrix} -1 \\ -1 \\ 1 \end{bmatrix}.$$

In this case, the set $\mathcal{B} = \{\mathbf{b}_1, \mathbf{b}_2, \mathbf{b}_3\}$, which is formed by taking a basis vector from each eigenspace, is linearly independent and so forms a basis for $\mathbb{R}^3$. This situation does not extend to all 3×3 matrices.

Definition 6.5

Geometric multiplicity

The *geometric multiplicity* of an eigenvalue λ, denoted by g_λ, is dimension of the eigenspace associated with λ.

Refer to Example 4. Note that $g_{-2} = g_0 = g_2 = 1$, and the geometric multiplicity of each eigenvalue happens to coincide with its algebraic multiplicity in that case. However, more generally, if λ is an eigenvalue of a given matrix $\mathbf{A}$, it can be proved that

$$g_\lambda \leq a_\lambda.$$

and it may happen that $g_\lambda < a_\lambda$ for certain λ, as will be seen in Example 6.

EXAMPLE 5

Repeated Eigenvalues, Geometric Multiplicity

Return to Example 3 in which $\mathbf{A}$ has eigenvalues $\lambda = -2$ and $\lambda = 3$ with algebraic multiplicities $a_{-2} = 2$ and $a_3 = 1$, respectively. We compute the reduced form for $(\mathbf{A} - \lambda\mathbf{I}_3)$ to find the eigenvectors of $\mathbf{A}$ associated with each λ (s and t are real parameters throughout).

For $\lambda = -2$, we have

$$\begin{bmatrix} 3 & -2 & 4 \\ 3 & -2 & 4 \\ 3 & -2 & 4 \end{bmatrix} \sim \begin{bmatrix} 1 & -\frac{2}{3} & \frac{4}{3} \\ 0 & 0 & 0 \\ 0 & 0 & 0 \end{bmatrix} \Rightarrow$$

$$E_{-2} = \begin{bmatrix} \frac{2}{3}s - \frac{4}{3}t \\ s \\ t \end{bmatrix} = s\begin{bmatrix} \frac{2}{3} \\ 1 \\ 0 \end{bmatrix} + t\begin{bmatrix} -\frac{4}{3} \\ 0 \\ 1 \end{bmatrix}.$$

Setting $s = 3$, $t = 0$ and then $s = 0$, $t = 3$ (convenient values) gives a basis $\mathcal{B} = \{\mathbf{b}_1, \mathbf{b}_2\}$ for the eigenspace E_{-2}, where

$$\mathbf{b}_1 = \begin{bmatrix} 2 \\ 3 \\ 0 \end{bmatrix}, \quad \mathbf{b}_2 = \begin{bmatrix} -4 \\ 0 \\ 3 \end{bmatrix}.$$

Thus E_{-2} is a two-dimensional subspace of $\mathbb{R}^3$.

For $\lambda = 3$, we have

$$\begin{bmatrix} -2 & -2 & 4 \\ 3 & -7 & 4 \\ 3 & -2 & -1 \end{bmatrix} \sim \begin{bmatrix} 1 & 0 & -1 \\ 0 & 1 & -1 \\ 0 & 0 & 0 \end{bmatrix}$$

and so

$$E_3 = \{t\mathbf{b}\}, \quad \text{where} \quad \mathbf{b} = \begin{bmatrix} 1 \\ 1 \\ 1 \end{bmatrix}.$$

Hence, **b** is a basis for the eigenspace E_3.

Note that E_{-2} is represented by a plane in $\mathbb{R}^3$ passing through the origin and defined by the linearly independent vectors $\mathbf{b}_1$ and $\mathbf{b}_2$. The eigenspace E_3 is a one-dimensional subspace of $\mathbb{R}^3$. We have $g_{-2} = a_{-2} = 2$; that is, the geometric and algebraic multiplicities of $\lambda = -2$ are identical in this case. ■

EXAMPLE 6

Geometric Multiplicity Less than Algebraic Multiplicity

Consider the matrix

$$\mathbf{A} = \begin{bmatrix} 2 & 1 & 0 \\ 0 & 1 & -1 \\ 0 & 2 & 4 \end{bmatrix}.$$

The eigenvalues **A** are $\lambda_1 = \lambda_2 = 2$ and $\lambda_3 = 3$ (verify), and so $a_2 = 2$ and $a_3 = 1$.

For $\lambda = 2$, we have

$$(\mathbf{A} - 2\mathbf{I}_3) = \begin{bmatrix} 0 & 1 & 0 \\ 0 & -1 & -1 \\ 0 & 2 & 2 \end{bmatrix} \sim \begin{bmatrix} 0 & 1 & 0 \\ 0 & 0 & 1 \\ 0 & 0 & 0 \end{bmatrix} \Rightarrow E_2 = \left\{ t \begin{bmatrix} 1 \\ 0 \\ 0 \end{bmatrix} \right\},$$

where t is a real parameter. Thus, E_2 is a one-dimensional subspace of $\mathbb{R}^3$ and $g_2 = 1 < 2 = a_2$. You may verify that E_3 is also a one-dimensional subspace of $\mathbb{R}^3$ and so $g_3 = 1 = a_3$. For this particular matrix **A**, it is impossible to find a basis for $\mathbb{R}^3$ consisting of eigenvectors associated with the eigenvalues of **A** and in the language of Section 6.2, **A** is not diagonalizable. ■

The following famous theorem is due jointly to Arthur Cayley and William Rowan Hamilton.

Theorem 6.1 **Cayley–Hamilton**

*Every $n \times n$ matrix **A** satisfies its own characteristic equation.*

EXAMPLE 7 **Cayley–Hamilton in Action**

Returning to Example 1, we have

$$\mathbf{A} = \begin{bmatrix} 1 & 1 \\ -2 & 4 \end{bmatrix} \quad \Rightarrow \quad p(\lambda) = \lambda^2 - 5\lambda + 6.$$

Substituting **A** for λ in $p(\lambda)$ and introducing the identity $\mathbf{I}_2$, we have

$$\begin{bmatrix} 1 & 1 \\ -2 & 4 \end{bmatrix}^2 - 5\begin{bmatrix} 1 & 1 \\ -2 & 4 \end{bmatrix} + 6\begin{bmatrix} 1 & 0 \\ 0 & 1 \end{bmatrix} = \begin{bmatrix} -1 & 5 \\ -10 & 14 \end{bmatrix}$$
$$- \begin{bmatrix} 5 & 5 \\ -10 & 20 \end{bmatrix} + \begin{bmatrix} 6 & 0 \\ 0 & 6 \end{bmatrix} = \begin{bmatrix} 0 & 0 \\ 0 & 0 \end{bmatrix}.$$ ■

We close this section by adding one more condition for the invertibility of a square matrix to those already listed (Section 2.2, Section 3.2, Section 5.1). The proof is left to Exercises 6.1.

Theorem 6.2 **Invertible Matrices**

An $n \times n$ matrix **A** *is invertible if and only if $\lambda = 0$ is not an eigenvalue of* **A**.

Theorem 6.2 may be applied to show that the matrix **A** in Example 2 is singular and the matrix **A** in Example 3 is invertible.

Computational Note

A primary goal in any eigenvalue problem is to compute the eigenvalues of the square matrix in question. Although these values appear as the roots of the characteristic polynomial, it is not computationally effective to find such roots, except in simple cases. In practice, other techniques are used in finding eigenvalues. In some applications, only the *dominant eigenvalue* (the largest in absolute value) is of interest and a primary goal is to compute this value.

EXERCISES 6.1

Exercises 1–2. Determine whether **x** *is an eigenvector of the given matrix.*

1. $\mathbf{x} = \begin{bmatrix} -2 \\ 1 \end{bmatrix}, \quad \begin{bmatrix} 5 & 4 \\ -1 & 4 \end{bmatrix}$

2. $\mathbf{x} = \begin{bmatrix} 1 \\ -1 \\ 4 \end{bmatrix}, \quad \begin{bmatrix} 3 & -5 & -1 \\ 1 & 9 & 1 \\ 1 & 5 & 5 \end{bmatrix}$

Exercises 3–12. Find the eigenvalues and an associated eigenvector for each eigenvalue.

3. $\begin{bmatrix} -2 & 2 \\ 3 & 4 \end{bmatrix}$

4. $\begin{bmatrix} 5 & -4 \\ 6 & -5 \end{bmatrix}$

5. $\begin{bmatrix} 5 & 4 \\ 4 & 5 \end{bmatrix}$

6. $\begin{bmatrix} -3 & 15 \\ -2 & 8 \end{bmatrix}$

7. $\begin{bmatrix} 1 & 0 & 2 \\ 0 & 1 & 0 \\ 2 & 0 & 1 \end{bmatrix}$

8. $\begin{bmatrix} -1 & 2 & 0 \\ 0 & 1 & 0 \\ -2 & 2 & 1 \end{bmatrix}$

9. $\begin{bmatrix} 1 & 1 & 1 \\ 2 & 3 & 4 \\ -2 & -3 & -4 \end{bmatrix}$

10. $\begin{bmatrix} 3 & 0 & 4 \\ 0 & 5 & 0 \\ 4 & 0 & 3 \end{bmatrix}$

11. $\begin{bmatrix} -15 & 28 & -14 \\ -10 & 19 & -10 \\ -4 & 8 & -5 \end{bmatrix}$

12. $\begin{bmatrix} -2 & 4 & -4 \\ -3 & 5 & -4 \\ -3 & 3 & -2 \end{bmatrix}$

13. Find a matrix **A** that has eigenvalues $\lambda_1 = 1$, $\lambda_2 = 2$ and associated eigenvectors, respectively:

$$\mathbf{x}_1 = \begin{bmatrix} 2 \\ -5 \end{bmatrix}, \quad \mathbf{x}_2 = \begin{bmatrix} -1 \\ 3 \end{bmatrix}.$$

14. Use the quadratic formula to show that the eigenvalues of the symmetric matrix

$$\begin{bmatrix} a & b \\ b & d \end{bmatrix}$$

are real numbers.

15. Determine if $\lambda = 3, 4$ are eigenvalues of $\mathbf{A} = \begin{bmatrix} 3 & 2 \\ 3 & -2 \end{bmatrix}$.

Exercises 16–19. Find a 2×2 *matrix* **A** *with nonzero entries which satisfies the given conditions.*

16. **A** has real and unequal eigenvalues.

17. **A** has real and equal eigenvalues.

18. **A** has complex eigenvalues.

19. **A** has eigenvalues 0 and 1.

20. For what values of k does the matrix $\begin{bmatrix} 1 & k \\ 1 & 1 \end{bmatrix}$ have eigenvalues $\lambda = 0.5, 1.5$?

21. Let a and b be any scalars. The linear transformation represented by the matrix $\mathbf{A} = \begin{bmatrix} a & b \\ b & -a \end{bmatrix}$ is a reflection-dilation in $\mathbb{R}^2$. Find the eigenvalues of **A** and draw an accurate diagram that shows how the eigenvalues relate to the action of **A** in general.

22. Suppose $\mathbf{A} = [a_{ij}]$ is an $n \times n$ upper (lower) triangular matrix. Show that the characteristic polynomial of **A** is given by

$$p(\lambda) = (\lambda - a_{11})(\lambda - a_{22}) \cdots (\lambda - a_{nn}).$$

What are the eigenvalues of **A**?

23. Find the eigenvalues and associated eigenvectors of the zero matrix $\mathbf{O} = \begin{bmatrix} 0 & 0 \\ 0 & 0 \end{bmatrix}$. State the algebraic and geometric multiplicities of each eigenvalue. Generalize the results to the case of the $n \times n$ zero matrix **O**. Determine if it is possible to find a basis for $\mathbb{R}^n$ consisting of eigenvectors of **O**.

24. Find the eigenvalues and associated eigenvectors of the matrix $\mathbf{I}_2 = \begin{bmatrix} 1 & 0 \\ 0 & 1 \end{bmatrix}$. State the algebraic and geometric multiplicities of each eigenvalue. Generalize the results to the case of the $n \times n$ identity matrix $\mathbf{I}_n$. Determine if it is possible to find a basis for $\mathbb{R}^n$ consisting of eigenvectors of $\mathbf{I}_n$.

25. Without computing eigenvalues, show that the matrices all have the same eigenvalues.

$$\begin{bmatrix} 1 & 2 \\ 3 & 4 \end{bmatrix}, \quad \begin{bmatrix} 10 & 4 \\ -12 & -5 \end{bmatrix}, \quad \begin{bmatrix} 19 & -11 \\ 24 & -14 \end{bmatrix}$$

26. Suppose that $p(\lambda) = \lambda^3 - 7\lambda + 6$ is the characteristic polynomial of the matrix **A**. Find the eigenvalues of **A**. Is it possible to determine the order of the square matrix **A**? *Hint*: If r is a root of $p(\lambda)$, then $(\lambda - r)$ is a factor of $p(\lambda)$.

27. Find a symmetric 2×2 matrix with positive integer entries and eigenvalues $-1, 3$.

28. Show that $\det(\mathbf{A} - \lambda\mathbf{I}_n) = 0$ if and only if $\det(\mathbf{A} - \lambda\mathbf{I}_n) = 0$.

29. Show that $p(\lambda) = \lambda^2 + c_1\lambda + c_0$ is the characteristic polynomial of the matrix $\mathbf{C} = \begin{bmatrix} -c_1 & -c_0 \\ 1 & 0 \end{bmatrix}$. Hence find a matrix **C** with eigenvalues $\lambda = 3.2, -2.8$.

30. Generalize the previous exercise by showing that

$$p(\lambda) = \lambda^n + c_{n-1}\lambda^{n-1} + \cdots + c_2\lambda^2 + c_1\lambda + c_0$$

is the characteristic polynomial of the matrix

$$\mathbf{C} = \begin{bmatrix} -c_{n-1} & -c_{n-2} & \cdots & -c_2 & -c_1 & -c_0 \\ 1 & 0 & \cdots & 0 & 0 & 0 \\ 0 & 1 & \cdots & 0 & 0 & 0 \\ \vdots & \vdots & \ddots & \vdots & \vdots & \vdots \\ 0 & 0 & \cdots & 0 & 1 & 0 \end{bmatrix}.$$

We call $\mathbf{C}$ the *companion matrix* for $p(\lambda)$. *Hint*: Apply the column operations $\mathbf{c}_{k+1}+\lambda\mathbf{c}_k \to \mathbf{c}_{k+1}$, $k = 1, 2, \ldots, (n-1)$ to $\det(\lambda\mathbf{I} - \mathbf{C})$ and then use Laplace expansion on column n.

31. Find the eigenvalues and associated eigenspaces of the matrix $\mathbf{A}$ in (6.1). Hence confirm the position of lines L_1 and L_2 in Figure 6.1.

32. Prove Theorem 6.2.

33. Suppose the sum of the entries in each column of an $n \times n$ matrix $\mathbf{A}$ are the same scalar α. Show that α is an eigenvalue of $\mathbf{A}$. Hence, a *stochastic* matrix (Section 6.4) has an eigenvalue $\lambda = 1$.

Exercises 34–35. Verify that each matrix has an eigenvalue equal to the column (row) sum.

34. $\begin{bmatrix} 0.6 & 0.3 \\ 0.4 & 0.7 \end{bmatrix}$ **35.** $\begin{bmatrix} 1 & 1 & 1 \\ 1.4 & 0.6 & 1 \\ 0.5 & 2 & 0.5 \end{bmatrix}$

36. Show that the matrix $\mathbf{A} = \begin{bmatrix} 2 & 3 & -2 \\ 0 & 3 & -1 \\ 1 & 5 & -2 \end{bmatrix}$ has a repeated eigenvalue λ. State the algebraic multiplicity of λ. Find the eigenspace E_λ corresponding to λ and state the geometric multiplicity of λ.

37. In Section 7.3, two $n \times n$ matrices $\mathbf{A}$ and $\mathbf{B}$ are called *similar* if $\mathbf{B} = \mathbf{P}^{-1}\mathbf{A}\mathbf{P}$ for some invertible $n \times n$ matrix $\mathbf{P}$. Show that similar matrices have the same eigenvalues. How are the eigenvectors of similar matrices $\mathbf{A}$ and $\mathbf{B}$ related? Show that the matrices

$$\mathbf{A} = \begin{bmatrix} 2 & 1 \\ 1 & 2 \end{bmatrix} \quad \text{and} \quad \mathbf{B} = \begin{bmatrix} 4 & 1 \\ 1 & 1 \end{bmatrix}$$

are not similar.

Exercises 38–43. Let $\mathbf{A}$ *be an* $n \times n$ *matrix. Prove each result.*

38. If λ is an eigenvalue of $\mathbf{A}$, then λ^k is an eigenvalue of $\mathbf{A}^k$, where k is a positive integer. How are the eigenvectors of $\mathbf{A}$ and $\mathbf{A}^k$ related?

39. λ is an eigenvalue of the invertible matrix $\mathbf{A}$ if and only if λ^{-1} is an eigenvalue of $\mathbf{A}^{-1}$. How are the eigenvectors of $\mathbf{A}$ and $\mathbf{A}^{-1}$ related?

40. The matrix $\mathbf{A}$ and its transpose $\mathbf{A}^\mathsf{T}$ have the same eigenvalues. How are the eigenvectors of $\mathbf{A}$ and $\mathbf{A}^\mathsf{T}$ related?

41. The eigenvalues of a nilpotent matrix (Section 2.1) are all zero.

42. The eigenvalues of an idempotent matrix (Section 2.1) are 0 or 1.

43. The eigenvalues of an orthogonal matrix (Section 4.2) are either $+1$ or -1.

Exercises 44–47. Let $\mathbf{A}$ *and* $\mathbf{B}$ *be* $n \times n$ *matrices and* s *any scalar. Prove each property of trace.*

44. $\operatorname{trace}(\mathbf{A} \pm \mathbf{B}) = \operatorname{trace}\mathbf{A} \pm \operatorname{trace}\mathbf{B}$

45. $\operatorname{trace}(s\mathbf{A}) = s(\operatorname{trace}\mathbf{A})$

46. $\operatorname{trace}(\mathbf{AB}) = \operatorname{trace}(\mathbf{BA})$

47. If $\mathbf{A}$ is similar to $\mathbf{B}$ then $\operatorname{trace}\mathbf{A} = \operatorname{trace}\mathbf{B}$

Exercises 48–49. Verify the Cayley–Hamilton theorem for each given matrix.

48. $\begin{bmatrix} -1 & 1 \\ 0 & 2 \end{bmatrix}$ **49.** $\begin{bmatrix} 0 & 1 & 0 \\ 0 & 1 & -3 \\ 0 & 0 & 2 \end{bmatrix}$

USING MATLAB

Consult online help for the commands **eig**, **compan**, **poly**, **roots**, and related commands.

50. The commands
p = [1, −3, 2], **C = compan(p)**
define a polynomial $p(\lambda) = \lambda^2 - 3\lambda + 2$ and the companion matrix $\mathbf{C}$ for $p(\lambda)$ (Exercise 30). Use hand calculation to find the eigenspaces associated with the eigenvalues of $\mathbf{C}$.

51. Find the companion matrix $\mathbf{C}$ for the polynomial $\lambda^3 - 6\lambda^2 + 11\lambda - 6$. Find the eigenvalues of $\mathbf{C}$. For each eigenvalue λ show by hand calculation that $\mathbf{x} = \begin{bmatrix} \lambda^2, & \lambda, & 1 \end{bmatrix}^\mathsf{T}$ is an eigenvector of $\mathbf{C}$ associated with the eigenvalue λ. Enter the command **[X,D] = eig(C)** and explain the output in terms of the eigenvectors found by hand.

52. The command **p** = [1, 4, −6, 4, −1] defines a polynomial $p(\lambda)$. Find the companion matrix $\mathbf{C}$ of $p(\lambda)$. Find the eigenvalues of $\mathbf{C}$ and verify that these are the roots of $p(\lambda)$ using the command **roots(p)**. Use the fundamental theorem of algebra to explain the form in which the roots appear (real, complex).

53. Use MATLAB to check hand calculations in this exercise set.

6.2 Diagonalization

Diagonal matrices are particularly pleasant to work with computationally, as can be seen from the following examples:

$$\begin{bmatrix} 2.5 & 0 \\ 0 & -3.2 \end{bmatrix} \begin{bmatrix} x \\ y \end{bmatrix} = \begin{bmatrix} 2.5x \\ -3.2y \end{bmatrix}$$

and

$$\begin{bmatrix} 0.5 & 0 \\ 0 & 0.1 \end{bmatrix}^k = \begin{bmatrix} 0.5^k & 0 \\ 0 & 0.1^k \end{bmatrix}, \quad k = 1, 2, 3, \ldots .$$

Sometimes a square matrix **A** can be *transformed* into a diagonal matrix **D** in order to take advantage of the algebraic properties of **D**. The transformation of **A** into **D**, which is called *diagonalization*, is not always possible and depends on the eigenvectors of **A**. Diagonalization of matrices is used in many pure and applied applications, including matrix powers, linear transformations, discrete dynamical systems, and systems of differential equations.

Definition 6.6

Diagonalizable, diagonalizing matrix

An $n \times n$ matrix **A** is *diagonalizable* if there is an invertible $n \times n$ matrix **P** and an $n \times n$ diagonal matrix **D** such that $\mathbf{P}^{-1}\mathbf{AP} = \mathbf{D}$. We say that **P** *diagonalizes* **A** and we call **P** a *diagonalizing matrix* for **A**.

When **A** is diagonalizable with diagonalizing matrix **P**, we have

$$\begin{aligned} \mathbf{P}^{-1}\mathbf{AP} = \mathbf{D} &\Leftrightarrow \mathbf{PP}^{-1}\mathbf{AP} = \mathbf{PD} &\Leftrightarrow \mathbf{AP} = \mathbf{PD} \\ \mathbf{AP} = \mathbf{PD} &\Leftrightarrow \mathbf{APP}^{-1} = \mathbf{PDP}^{-1} &\Leftrightarrow \mathbf{A} = \mathbf{PDP}^{-1}. \end{aligned} \tag{6.21}$$

In previous chapters we have seen the importance of matrix factorizations, such as LU and QR. The equivalencies in (6.21) give another factorization—if **A** is diagonalizable, then $\mathbf{A} = \mathbf{PDP}^{-1}$ for suitable square matrices **P** and **D**.

We now address two fundamental questions:

(a) When is a square matrix diagonalizable?
(b) If **A** is diagonalizable, how can the matrices **P** and **D** be found?

Some clue to the answers to both questions lies in the next illustration.

■ ILLUSTRATION 6.1

Diagonalizing a Matrix of Order 2

You may verify that the matrix

$$\mathbf{A} = \begin{bmatrix} 3 & 2 \\ -1 & 0 \end{bmatrix}$$

has eigenvalues λ_1, λ_2, and that particular eigenvectors associated with each eigenvalue are as follows:

$$\lambda_1 = 1, \quad \mathbf{x}_1 = \begin{bmatrix} -1 \\ 1 \end{bmatrix}, \qquad \lambda_2 = 2, \quad \mathbf{x}_2 = \begin{bmatrix} -2 \\ 1 \end{bmatrix}.$$

Using $\mathbf{x}_1$ and $\mathbf{x}_2$, form a matrix

$$\mathbf{P} = [\mathbf{x}_1\ \mathbf{x}_2] = \begin{bmatrix} -1 & -2 \\ 1 & 1 \end{bmatrix}$$

and note that $\mathbf{P}$ is invertible because its columns are linearly independent. You may verify that

$$\mathbf{P}^{-1} = \begin{bmatrix} 1 & 2 \\ -1 & -1 \end{bmatrix}, \quad \text{and} \quad \mathbf{P}^{-1}\mathbf{A}\mathbf{P} = \begin{bmatrix} 1 & 0 \\ 0 & 2 \end{bmatrix} = \mathbf{D}.$$

Notice that the eigenvalues of $\mathbf{A}$ appear on the diagonal of $\mathbf{D}$ in the same order that the associated eigenvectors appear in $\mathbf{P}$. Permuting the columns of $\mathbf{P}$ will result in a corresponding permutation of the eigenvalues of $\mathbf{A}$ on the diagonal of $\mathbf{D}$. ■

The key to the diagonalization of an $n \times n$ matrix $\mathbf{A}$ depends on being able to construct the columns of a diagonalizing matrix $\mathbf{P}$ using n linearly independent eigenvectors of $\mathbf{A}$, as in Illustration 6.1, where $n = 2$.

Theorem 6.3 **Characterizing Diagonalizable Matrices**

An $n \times n$ matrix $\mathbf{A}$ is diagonalizable if and only if there is a set of n linearly independent eigenvectors of $\mathbf{A}$.

Proof Let $\mathbf{A}$ be an $n \times n$ matrix. There are two implications to prove.

(a) Suppose that $\mathbf{A}$ is diagonalizable. Then there exists a diagonalizing matrix $\mathbf{P}$ such that $\mathbf{P}^{-1}\mathbf{A}\mathbf{P} = \mathbf{D}$, where $\mathbf{D}$ is an $n \times n$ diagonal matrix. Let $\mathbf{P} = [\,\mathbf{x}_1\ \mathbf{x}_2\ \cdots\ \mathbf{x}_n\,]$, where $\mathbf{x}_1,\ \mathbf{x}_2,\ \ldots,\mathbf{x}_n$ are nonzero vectors in $\mathbb{R}^n$ and suppose that $\mathbf{D}$ has the form

$$\mathbf{D} = \begin{bmatrix} d_1 & 0 & \cdots & 0 \\ 0 & d_2 & \cdots & 0 \\ \vdots & \vdots & \ddots & \vdots \\ 0 & 0 & \cdots & d_n \end{bmatrix}.$$

Computing $\mathbf{AP}$ and $\mathbf{PD}$, we have

$$\mathbf{AP} = \mathbf{A}[\,\mathbf{x}_1\ \cdots\ \mathbf{x}_n\,] = [\,\mathbf{Ax}_1\ \cdots\ \mathbf{Ax}_n\,],$$

$$\mathbf{PD} = [\,\mathbf{x}_1\ \cdots\ \mathbf{x}_n\,]\begin{bmatrix} d_1 & 0 & \cdots & 0 \\ 0 & d_2 & \cdots & 0 \\ \vdots & \vdots & \ddots & \vdots \\ 0 & 0 & \cdots & d_n \end{bmatrix} = [\,d_1\mathbf{x}_1\ \cdots\ d_n\mathbf{x}_n\,].$$

But $\mathbf{AP} = \mathbf{PD}$ from (6.21) and so $[\,\mathbf{Ax}_1\ \cdots\ \mathbf{Ax}_n\,] = [\,d_1\mathbf{x}_1\ \cdots\ d_n\mathbf{x}_n\,]$. Equating corresponding columns in the latter equation, we have

$$\mathbf{Ax}_1 = d_1\mathbf{x}_1, \quad \mathbf{Ax}_2 = d_2\mathbf{x}_2, \quad \ldots, \quad \mathbf{Ax}_n = d_n\mathbf{x}_n, \tag{6.22}$$

which shows that $\mathbf{x}_i$ is an eigenvector of $\mathbf{A}$ associated with the eigenvalue $\lambda_i = d_i$, $1 \le i \le n$. Because $\mathbf{P}$ is invertible, its columns $\mathbf{x}_1,\ \mathbf{x}_2,\ \ldots,\mathbf{x}_n$ are linearly independent.

(b) Suppose that $\mathbf{A}$ has a set of linearly independent eigenvectors $\mathbf{x}_1, \mathbf{x}_2, \ldots, \mathbf{x}_n$. Then $\mathbf{A}\mathbf{x}_i = \lambda_i \mathbf{x}_i$, $1 \leq i \leq n$, where the associated eigenvalues $\lambda_1, \lambda_2, \ldots, \lambda_n$ are not necessarily distinct. Form the matrix $\mathbf{P} = [\,\mathbf{x}_1 \;\; \mathbf{x}_2 \;\; \cdots \;\; \mathbf{x}_n\,]$. Then

$$\begin{aligned}\mathbf{AP} = \mathbf{A}[\,\mathbf{x}_1 \;\; \mathbf{x}_2 \;\; \cdots \;\; \mathbf{x}_n\,] &= [\,\mathbf{A}\mathbf{x}_1 \;\; \mathbf{A}\mathbf{x}_2 \;\; \cdots \;\; \mathbf{A}\mathbf{x}_n\,] \\ &= [\,\lambda_1\mathbf{x}_1 \;\; \lambda_2\mathbf{x}_2 \;\; \cdots \;\; \lambda_n\mathbf{x}_n\,] \\ &= [\,\mathbf{x}_1 \;\; \mathbf{x}_2 \;\; \cdots \;\; \mathbf{x}_n\,]\begin{bmatrix}\lambda_1 & 0 & \cdots & 0 \\ 0 & \lambda_2 & \cdots & 0 \\ \vdots & \vdots & \ddots & \vdots \\ 0 & 0 & \cdots & \lambda_n\end{bmatrix} = \mathbf{PD},\end{aligned}$$

showing that $\mathbf{AP} = \mathbf{PD}$. The matrix $\mathbf{P}$ is invertible because $\mathbf{x}_1, \mathbf{x}_2, \ldots, \mathbf{x}_n$ are linearly independent, and from (6.21) it follows that $\mathbf{P}^{-1}\mathbf{AP} = \mathbf{D}$. Thus $\mathbf{A}$ is diagonalizable.

Remarks

Refer to Theorem 6.3.

(a) When $\mathbf{A}$ is diagonalizable, the diagonalizing matrix $\mathbf{P}$ must necessarily have eigenvectors of $\mathbf{A}$ for its columns and the diagonal entries in $\mathbf{D}$ are necessarily the eigenvalues of $\mathbf{A}$ associated with the corresponding eigenvectors.

(b) The matrices $\mathbf{P}$ and $\mathbf{D}$ are not unique—the columns of $\mathbf{P}$ may be scaled (in some applications the columns of $\mathbf{P}$ are normalized), and permuting the columns of $\mathbf{P}$ will permute the eigenvalues on the main diagonal of $\mathbf{D}$ according to the order in which the associated eigenvectors were permuted.

An $n \times n$ matrix $\mathbf{A}$ is not diagonalizable if $\mathbf{A}$ has fewer than n linearly independent eigenvectors.

EXAMPLE 1

Diagonalization fails

The matrix $\mathbf{A}$ shown in (6.23) is not diagonalizable.

$$\mathbf{A} = \begin{bmatrix} 1 & -1 \\ 0 & 1 \end{bmatrix}, \quad \mathrm{E}_1 = \left\{ t \begin{bmatrix} 1 \\ 0 \end{bmatrix} \right\}. \tag{6.23}$$

The eigenvalues of $\mathbf{A}$ are $\lambda = 1, 1$, and so $a_1 = 2$. The eigenspace for $\lambda = 1$ is shown in (6.23) and so $g_1 = 1$. Note that if a matrix $\mathbf{P}$ is formed using scalar multiples of the vector $\begin{bmatrix} 1 \\ 0 \end{bmatrix}$, then $\mathbf{AP} = \mathbf{PD}$. For example,

$$\mathbf{AP} = \begin{bmatrix} 1 & -1 \\ 0 & 1 \end{bmatrix}\begin{bmatrix} 1 & 2 \\ 0 & 0 \end{bmatrix} = \begin{bmatrix} 1 & 2 \\ 0 & 0 \end{bmatrix}\begin{bmatrix} 1 & 0 \\ 0 & 1 \end{bmatrix} = \mathbf{PD}.$$

However, $\mathbf{P}$ is singular and two linearly independent eigenvectors cannot be found in this case to form an invertible matrix $\mathbf{P}$. ■

Theorem 6.4 **Criterion for Diagonalization**

Suppose an $n \times n$ matrix $\mathbf{A}$ has distinct eigenvalues $\lambda_1, \lambda_2, \ldots \lambda_k$, where $k \le n$. Let $\mathcal{B}_i$ be a basis for the eigenspace E_{λ_i}, where $1 \le i \le k$, and let

$$\mathcal{B} = \mathcal{B}_1 \cup \mathcal{B}_2 \cup \cdots \cup \mathcal{B}_k.$$

Then $\mathcal{B}$ is a linearly independent subset of $\mathbb{R}^n$ and $\mathbf{A}$ is diagonalizable if and only if $\mathcal{B}$ contains n vectors.

▪ ILLUSTRATION 6.2

Theorem 6.4 and its Proof

Consider the matrix $\mathbf{A}$ shown in (6.24), which has eigenvalues $\lambda = 3, 2, 2$. The eigenspace E_3 is one-dimensional with basis $\mathcal{B}_1 = \{\mathbf{u}_1\}$, and the eigenspace E_2 is two-dimensional with basis $\mathcal{B}_2 = \{\mathbf{u}_2, \mathbf{u}_3\}$, where $\mathbf{u}_1$, $\mathbf{u}_2$, $\mathbf{u}_3$ are shown in (6.24).

$$\mathbf{A} = \begin{bmatrix} 3 & 1 & 1 \\ -1 & 1 & -1 \\ 1 & 1 & 3 \end{bmatrix}, \quad \mathbf{u}_1 = \begin{bmatrix} 1 \\ -1 \\ 1 \end{bmatrix}, \quad \mathbf{u}_2 = \begin{bmatrix} -1 \\ 1 \\ 0 \end{bmatrix}, \quad \mathbf{u}_3 = \begin{bmatrix} -1 \\ 0 \\ 1 \end{bmatrix} \tag{6.24}$$

Our goal is to show how the proof of Theorem 6.4 works for this particular matrix $\mathbf{A}$. We will show that the set $\mathcal{B} = \mathcal{B}_1 \cup \mathcal{B}_2$ is a linearly independent subset of $\mathbb{R}^3$ by showing that the vector equation

$$x_1\mathbf{u}_1 + x_2\mathbf{u}_2 + x_3\mathbf{u}_3 = \mathbf{0}$$

has only the zero solution $x_1 = x_2 = x_3 = 0$. Let $\lambda_1 = 3$ and $\lambda_2 = 2$ be the distinct eigenvalues of $\mathbf{A}$. Consider the vector equation

$$x_1\mathbf{u}_1 + \mathbf{w} = \mathbf{0}, \tag{6.25}$$

where $\mathbf{w} = x_2\mathbf{u}_2 + x_3\mathbf{u}_3$. Multiplying equation (6.25) by $\mathbf{A}$, we have

$$\mathbf{A}(x_1\mathbf{u}_1) + \mathbf{A}\mathbf{w} = \mathbf{A}\mathbf{0}, \quad \text{which implies that}$$

$$x_1\lambda_1\mathbf{u}_1 + \lambda_2\mathbf{w} = \mathbf{0}. \tag{6.26}$$

Multiplying (6.25) by $\lambda_2 (\neq 0)$ and subtracting from (6.26) gives $(\lambda_2 - \lambda_1)x_1\mathbf{u}_1 = \mathbf{0}$. But λ_1 and λ_2 are distinct eigenvalues and so $x_1 = 0$ because $\mathbf{u}_1$ is nonzero. Thus, $\mathbf{w} = \mathbf{0}$ from (6.25) and so $x_2 = x_3 = 0$, because $\mathbf{u}_2$, $\mathbf{u}_3$ are linearly independent.

You are asked in Exercises 6.2 to give a similar proof in a case when $\lambda = 0$ is an eigenvalue of the given matrix $\mathbf{A}$. ■

The next important result is a special case of Theorem 6.4.

Theorem 6.5 **Distinct Eigenvalues**

An $n \times n$ matrix $\mathbf{A}$ with n distinct eigenvalues $\lambda_1, \lambda_2, \ldots, \lambda_n$ is diagonalizable.

Proof Assume that $\mathbf{A}$ has n distinct eigenvalues $\lambda_1, \lambda_2, \ldots, \lambda_n$. Because $g_\lambda \le a_\lambda$, for any eigenvalue λ, each eigenspace is one-dimensional, and in Theorem 6.4, we have

$$\mathcal{B} = \mathcal{B}_1 \cup \mathcal{B}_2 \cup \cdots \cup \mathcal{B}_n,$$

where each set $\mathcal{B}_i$ contains one basis vector. Thus $\mathcal{B}$ contains n vectors and $\mathbf{A}$ is therefore diagonalizable by Theorem 6.4.

Theorem 6.6 **Real Symmetric Matrices**

Let $\mathbf{A}$ *be a real symmetric matrix of order* n. *Then*

(a) The eigenvalues for $\mathbf{A}$ are all real.
(b) $\mathbf{A}$ has n linearly independent eigenvectors.
(c) The dimension of each eigenspace is equal to the algebraic multiplicity of the eigenvalue with which it is associated.

Proof For (a), see Section 8.5.

It follows from Theorem 6.4 and Theorem 6.6(b) that a real symmetric matrix is diagonalizable.

Powers of a Square Matrix

There are applications that require computation of the powers $\mathbf{A}^k$ of an $n \times n$ matrix $\mathbf{A}$, where k is large. It is important in such cases to choose an efficient method of computation. We describe next a method that uses diagonalization—a second method that has general applicability is given in Appendix B: Toolbox.

Using Diagonalization

If $\mathbf{A}$ is diagonalizable, then $\mathbf{A} = \mathbf{PDP}^{-1}$ for some diagonalizing matrix for $\mathbf{P}$ and diagonal matrix $\mathbf{D}$. Computing $\mathbf{A}^2$, we have

$$\mathbf{A}^2 = (\mathbf{PDP}^{-1})(\mathbf{PDP}^{-1}) = (\mathbf{PD})(\mathbf{P}^{-1}\mathbf{P})(\mathbf{DP}^{-1}) = \mathbf{PD}^2\mathbf{P}^{-1}$$

and applying mathematical induction, we have

$$\mathbf{A}^k = \mathbf{PD}^k\mathbf{P}^{-1}, \quad k = 1, 2, \ldots . \tag{6.27}$$

Using (6.27) to compute $\mathbf{A}^k$ for any k is extremely efficient: Finding $\mathbf{D}^k$ is trivial and $\mathbf{P}^{-1}$, computed only once, is used for any k. The factorization (6.27) can also be used to obtain a *closed form expression* for $\mathbf{A}^k$ (a formula for $\mathbf{A}^k$ in terms of k).

EXAMPLE 2 **Powers of a Square Matrix, Closed Form Expressions**

The following matrix $\mathbf{A}$ is diagonalizable, and the associated matrices $\mathbf{P}$ and $\mathbf{D}$ have already been found.

$$\mathbf{A} = \begin{bmatrix} 1 & -1 & -2 \\ 1 & -1 & -2 \\ -1 & -1 & 0 \end{bmatrix}, \quad \mathbf{P} = \begin{bmatrix} 1 & 1 & 1 \\ -1 & 1 & 1 \\ 1 & 1 & -1 \end{bmatrix}, \quad \mathbf{D} = \begin{bmatrix} 0 & 0 & 0 \\ 0 & -2 & 0 \\ 0 & 0 & 2 \end{bmatrix}$$

Check that $\mathbf{P}^{-1} = \frac{1}{2}\begin{bmatrix} 1 & -1 & 0 \\ 0 & 1 & 1 \\ 1 & 0 & -1 \end{bmatrix}$. Hence, for example,

$$\mathbf{A}^6 = \frac{1}{2}\begin{bmatrix} 1 & 1 & 1 \\ -1 & 1 & 1 \\ 1 & 1 & -1 \end{bmatrix}\begin{bmatrix} 0 & 0 & 0 \\ 0 & 64 & 0 \\ 0 & 0 & 64 \end{bmatrix}\begin{bmatrix} 1 & -1 & 0 \\ 0 & 1 & 1 \\ 1 & 0 & -1 \end{bmatrix} = 32\begin{bmatrix} 1 & 1 & 0 \\ 1 & 1 & 0 \\ -1 & 1 & 2 \end{bmatrix} \tag{6.28}$$

and more generally for $k = 1, 2, \ldots$ we have

$$\mathbf{A}^k = \frac{1}{2}\begin{bmatrix} 1 & 1 & 1 \\ -1 & 1 & 1 \\ 1 & 1 & -1 \end{bmatrix}\begin{bmatrix} 0 & 0 & 0 \\ 0 & (-2)^k & 0 \\ 0 & 0 & 2^k \end{bmatrix}\begin{bmatrix} 1 & -1 & 0 \\ 0 & 1 & 1 \\ 1 & 0 & -1 \end{bmatrix}$$

$$= 2^{k-1}\begin{bmatrix} 0 & (-1)^k & 1 \\ 0 & (-1)^k & 1 \\ 0 & (-1)^k & -1 \end{bmatrix}\begin{bmatrix} 1 & -1 & 0 \\ 0 & 1 & 1 \\ 1 & 0 & -1 \end{bmatrix}$$

$$= 2^{k-1}\begin{bmatrix} 1 & (-1)^k & (-1)^k - 1 \\ 1 & (-1)^k & (-1)^k - 1 \\ -1 & (-1)^k & (-1)^k + 1 \end{bmatrix}. \tag{6.29}$$

Then (6.29) is a closed form expression for $\mathbf{A}^k$. Putting $k = 6$ in (6.29) gives (6.28). ■

Linear Operators on $\mathbb{R}^n$

The next comment connects with Sections 3.4 and 7.3. A diagonalizable $n \times n$ matrix $\mathbf{A}$ has a factorization $\mathbf{A} = \mathbf{PDP}^{-1}$, where $\mathbf{P}$ is $n \times n$ and invertible and $\mathbf{D}$ is diagonal. Consider the linear operator (transformation) $T : \mathbb{R}^n \to \mathbb{R}^n$ defined by $T(\mathbf{u}) = \mathbf{Au} = \mathbf{v}$, for all $\mathbf{u}$ in $\mathbb{R}^n$. Then T can be viewed as the *composition* of three linear operators—the vector $\mathbf{u}$ is mapped by $\mathbf{P}^{-1}$ to $\mathbf{v}_1 = \mathbf{P}^{-1}\mathbf{u}$, then $\mathbf{v}_1$ is mapped by $\mathbf{D}$ to $\mathbf{v}_2 = \mathbf{D}\mathbf{v}_1$, and finally, $\mathbf{v}_2$ is mapped by $\mathbf{P}$ to $\mathbf{v} = \mathbf{P}\mathbf{v}_2$. Using associativity of matrix multiplication, we have

$$T(\mathbf{u}) = \mathbf{Au} = (\mathbf{PDP}^{-1})\mathbf{u} = (\mathbf{PD})\mathbf{P}^{-1}\mathbf{u} = \mathbf{P}(\mathbf{D})\mathbf{v}_1 = \mathbf{P}\mathbf{v}_2 = \mathbf{v}.$$

EXERCISES 6.2

Exercises 1–2. *Construct a diagonalizable matrix* $\mathbf{A}$ *from the given matrices* $\mathbf{P}$ *and* $\mathbf{D}$. *What are the eigenvalues of* $\mathbf{A}$?

1. $\mathbf{P} = \begin{bmatrix} 1 & 1 \\ -1 & 1 \end{bmatrix}, \quad \mathbf{D} = \begin{bmatrix} 2 & 0 \\ 0 & 3 \end{bmatrix}$

2. $\mathbf{P} = \begin{bmatrix} 1 & 1 & 2 \\ 0 & 1 & 0 \\ 0 & 0 & 3 \end{bmatrix}, \quad \mathbf{D} = \begin{bmatrix} 1 & 0 & 0 \\ 0 & 3 & 0 \\ 0 & 0 & -1 \end{bmatrix}$

Exercises 3–4. Find a diagonalizable matrix **A** *given the eigenvalues and associated eigenvectors of* **A**.

3. $\lambda = -1, \begin{bmatrix} 2 \\ -5 \end{bmatrix}, \quad \lambda = 2, \begin{bmatrix} -1 \\ 3 \end{bmatrix}.$

4. $\lambda = 2, \begin{bmatrix} 2 \\ 7 \end{bmatrix}, \quad \lambda = 3, \begin{bmatrix} 1 \\ 3 \end{bmatrix}$

Exercises 5–10. Determine if the given matrix **A** *is diagonalizable. If so, find a diagonalizing matrix* **P** *for* **A** *and check that* $\mathbf{P}^{-1}\mathbf{AP} = \mathbf{D}$.

5. $\begin{bmatrix} 0 & 0 \\ 0 & 0 \end{bmatrix}$

6. $\begin{bmatrix} 1 & 1 \\ 1 & 1 \end{bmatrix}$

7. $\begin{bmatrix} 2 & -2 \\ -2 & 2 \end{bmatrix}$

8. $\begin{bmatrix} 15 & 6 \\ -6 & 3 \end{bmatrix}$

9. $\begin{bmatrix} 0 & 1 & 0 \\ 1 & 0 & 0 \\ 0 & 0 & 1 \end{bmatrix}$

10. $\begin{bmatrix} 9 & -27 & 27 \\ 1 & 0 & 0 \\ 0 & 1 & 0 \end{bmatrix}$

11. If $(a - d)^2 + 4bc > 0$, show that the real matrix $\mathbf{A} = \begin{bmatrix} a & b \\ c & d \end{bmatrix}$ is diagonalizable. Is the converse true?

12. Find examples of real 2×2 matrices **A**, **B**, **C**, **D** that are diagonalizable and such that $\mathbf{A} + \mathbf{B}$ and **CD** are not diagonalizable.

13. If an $n \times n$ matrix **A** is diagonalizable show that det(**A**) is equal to the product of the eigenvalues of **A**. *Hint:* Use determinants. Is this fact true for nondiagonalizable **A**?

14. Prove that the matrix $\mathbf{A} = \begin{bmatrix} 1 & 1 \\ 1 & 1 \end{bmatrix}$ is diagonalizable. What can you conclude about the invertibility of diagonalizable matrices?

15. Suppose **A** is invertible. Show that **A** is diagonalizable implies $\mathbf{A}^{-1}$ diagonalizable.

16. Suppose **A** and **B** are diagonalizable, with the same diagonalizing matrix **P**. Show that $\mathbf{AB} = \mathbf{BA}$.

17. Find all square matrices **A** of order n that are diagonalizable and have only one distinct eigenvalue λ.

18. Apply Theorem 6.4 to the matrix **A** shown. Find bases for the eigenspaces of **A**. Form the set $\mathcal{B}$, and prove that it is linearly independent using the technique shown in Illustration 6.2. Explain why **A** can be diagonalized.

$$\mathbf{A} = \begin{bmatrix} 1 & 1 & 1 \\ 1 & 1 & 1 \\ 1 & 1 & 1 \end{bmatrix}$$

19. Verify that the real symmetric matrix

$$\mathbf{A} = \begin{bmatrix} 0 & 0 & -1 \\ 0 & 0 & 1 \\ -1 & 1 & 1 \end{bmatrix}$$

is diagonalizable. Find a diagonalizing matrix **P** for **A** and check that $\mathbf{P}^{-1}\mathbf{AP} = \mathbf{D}$.

20. Suppose that **A** is $n \times n$ and diagonalizable. Which of the following matrices are also diagonalizable? (a) $\mathbf{A}^{\mathsf{T}}$ (b) $\mathbf{A}^k$, $k = 1, 2, \ldots$ (c) $s\mathbf{A}$ for any scalar s.

Exercises 21–24. Matrix Powers, Closed Form Expressions. Find a closed form expression for the kth power of each matrix and hence compute the give power.

21. $\begin{bmatrix} 3 & -2 \\ 4 & -3 \end{bmatrix}^{100}$

22. $\begin{bmatrix} 1 & 2 \\ 2 & 1 \end{bmatrix}^{10}$

23. $\begin{bmatrix} 1 & 0 & 2 \\ 0 & 1 & 0 \\ 2 & 0 & 1 \end{bmatrix}^{10}$

24. $\begin{bmatrix} -1 & 2 & 0 \\ 0 & 1 & 0 \\ -2 & 2 & 1 \end{bmatrix}^{10}$

Exercises 25–27. For the given matrix **A**, *find a closed form expression for* $\mathbf{A}^k$.

25. $\begin{bmatrix} 1 & 0.5 & 0 \\ 0 & 0.5 & 1 \\ 0 & 0 & 0 \end{bmatrix}$

26. $\begin{bmatrix} 1 & 0.25 & 0 \\ 0 & 0.5 & 0 \\ 0 & 0.25 & 1 \end{bmatrix}$

27. $\mathbf{T} = \frac{1}{5}\begin{bmatrix} 3 & 1 & 2 \\ 1 & 3 & 2 \\ 1 & 1 & 1 \end{bmatrix}$

USING MATLAB

28. Consider the matrix $\mathbf{A} = \begin{bmatrix} 5 & -2 & 2 \\ 4 & -3 & 4 \\ 4 & -6 & 7 \end{bmatrix}$.

(a) Find the eigenvalues of **A** and explain why is **A** diagonalizable.

(b) Find a matrix **P** and a diagonal matrix **D** such that $\mathbf{AP} = \mathbf{PD}$. Compute the eigenvectors of **A** by hand and explain your calculations in terms of the machine output.

(c) Compare det(**A**) and det(**D**) and prove a general result connecting these two numbers.

29. *Project.* Investigate whether the magic matrices are diagonalizable. Report on your findings.

30. Use MATLAB to check hand calculations in this exercise set.

6.3 Applied Eigenvalue Problems

The first part of this section gives further analysis of the discrete dynamical systems that were introduced in Section 2.4, and we begin by reviewing the general definition. A *discrete dynamical system* consists of a sequence of column vectors, called *step vectors,*

$$\mathbf{x}_0,\ \mathbf{x}_1,\ \mathbf{x}_2,\ \ldots\ ,\ \mathbf{x}_k,\ \ldots \tag{6.30}$$

together with an $n \times n$ matrix **A**, called the *step matrix* for system. The step vector $\mathbf{x}_{k+1}$ is computed from $\mathbf{x}_k$ and **A** by matrix-vector multiplication

$$\mathbf{x}_{k+1} = \mathbf{A}\mathbf{x}_k, \quad k = 0, 1, 2, \ldots\ . \tag{6.31}$$

Given an initial step vector $\mathbf{x}_0$, successive step vectors in (6.30) are found as follows:

$$\begin{array}{lrcll} \text{Step 1} & \mathbf{x}_1 & = & \mathbf{A}\mathbf{x}_0, & \\ \text{Step 2} & \mathbf{x}_2 & = & \mathbf{A}\mathbf{x}_1 = \mathbf{A}\ (\mathbf{A}\mathbf{x}_0) & = \mathbf{A}^2\mathbf{x}_0, \\ \text{Step 3} & \mathbf{x}_3 & = & \mathbf{A}\mathbf{x}_2 = \mathbf{A}\ (\mathbf{A}^2\mathbf{x}_0) & = \mathbf{A}^3\mathbf{x}_0, \\ & & \vdots & & \end{array} \tag{6.32}$$

and the pattern in (6.32) suggests that $\mathbf{x}_k$ can be computed from $\mathbf{x}_0$ directly, thus

$$\mathbf{x}_k = \mathbf{A}^k\mathbf{x}_0, \quad k = 0, 1, 2, 3, \ldots \tag{6.33}$$

stepping from $\mathbf{x}_0$ directly to $\mathbf{x}_k$ using the power $\mathbf{A}^k$.

The first illustration is typical of the dynamical systems that model the interaction between a pair of competing species. These are called *predator-prey models*. The specific model analyzed below serves to show how eigenvalues and eigenvectors enter into the analysis of dynamical systems in general.

■ ILLUSTRATION 6.3

Predator-prey Models

The stocks of Northern Cod in the Northwest Atlantic are a source of food for the harp seals that inhabit that region. One goal in studying the ecology of the two species is to understand the relationship between the population of seals (predators) and the population of cod (prey) over discrete and equal intervals of time, say from year to year. Too few cod and too many seals might suggest a future decline in the seal population, and the culling of too many seals might indicate that the cod population should rise.[2]

[2] The Northern Cod fishery was placed under a moratorium in 1992. The cause of the decimation of cod stocks during the previous decade was a complex issue that included overfishing and the dramatic increase in the seal population during the same period. No conclusive scientific evidence has been found that linked the decline of cod stocks to the increase in seal population.

Suppose that s_k and c_k denote, respectively, the population of seals (in thousands) and of cod (in tons) in year k. Assume that the change in population from year k to year $k+1$ can be modeled by the following *system of difference equations* (the numerical data are hypothetical, chosen for purpose of illustration).

$$\begin{cases} s_{k+1} = 0.6s_k + 0.5c_k \\ c_{k+1} = -0.24s_k + 1.4c_k \end{cases} \tag{6.34}$$

Interpret the coefficients in equations (6.34) as follows: Without any cod, the seals would decline by 40% per year ($s_{k+1} = 0.6s_k$), but the cod population (food source) in year k contributes $0.5c_k$ to s_{k+1}. In the absence of seals, the cod stocks would increase by 40% per year ($c_{k+1} = 1.4c_k$), but this number is reduced by $0.24s_k$, which represents the proportion of cod consumed by the seals.

The matrix form of (6.34) is

$$\begin{bmatrix} s_{k+1} \\ c_{k+1} \end{bmatrix} = \begin{bmatrix} 0.6 & 0.5 \\ -0.24 & 1.4 \end{bmatrix} \begin{bmatrix} s_k \\ c_k \end{bmatrix}. \tag{6.35}$$

The *population vector* (step vector)

$$\mathbf{x}_k = \begin{bmatrix} s_k \\ c_k \end{bmatrix}, \quad k = 0, 1, 2, \ldots$$

records the population in year k and equation (6.35) defines a discrete dynamical system with step matrix

$$\mathbf{A} = \begin{bmatrix} 0.6 & 0.5 \\ -0.24 & 1.4 \end{bmatrix}. \tag{6.36}$$

Note that $\mathbf{A}$ defines a linear transformation $T_{\mathbf{A}} : \mathbb{R}^2 \to \mathbb{R}^2$ and so we may view the dynamical system as a sequence of transformations on $\mathbb{R}^2$.

The following vector $\mathbf{x}_0$ gives the initial population distribution for seals and cod in year zero, and the distributions in two subsequent years are also shown.

$$\mathbf{x}_0 = \begin{bmatrix} 500 \\ 300 \end{bmatrix}, \quad \mathbf{x}_1 = \mathbf{A}\mathbf{x}_0 = \begin{bmatrix} 450 \\ 300 \end{bmatrix}, \quad \mathbf{x}_2 = \mathbf{A}\mathbf{x}_1 = \begin{bmatrix} 420 \\ 312 \end{bmatrix}, \quad \ldots \tag{6.37}$$

It appears from (6.37) that the seal population may be in decline, but a more thorough analysis of the model will show otherwise.

Using MATLAB, we have

$$\mathbf{x}_9 = \mathbf{A}^9\mathbf{x}_0 = \begin{bmatrix} 0.6 & 0.5 \\ -0.24 & 1.4 \end{bmatrix}^9 \begin{bmatrix} 500 \\ 300 \end{bmatrix} \simeq \begin{bmatrix} 695 \\ 794 \end{bmatrix},$$

and it now appears that by year nine both seal and cod populations have increased from their initial levels in year zero. Similar computations can be carried out, starting from any initial population vector, but our goal is a deeper understanding of the underlying structure of the model.

Recall the basic questions we may ask about discrete dynamical systems:

(a) What is the long-term behavior of the system?

(b) What effect does the initial step vector have on the long-term behavior?

We answer both questions in general by considering the eigenvalues and eigenvectors of the step matrix **A**. With **A** as in (6.36), proceed as follows.

The MATLAB command

[P,D] = eig(A)

returns 2×2 matrices **P** and **D** such that **AP** = **PD**. In this case

$$\mathbf{P} = \begin{bmatrix} -0.9285 & -0.6402 \\ -0.3714 & -0.7682 \end{bmatrix}, \quad \mathbf{D} = \begin{bmatrix} 0.8 & 0 \\ 0 & 1.2 \end{bmatrix}.$$

The eigenvalues of **A** are $\lambda_1 = 0.8$ and $\lambda_2 = 1.2$ and these appear on the diagonal of **D**. They are distinct and so **A** is diagonalizable, **P** is nonsingular and **A** has the factorization $\mathbf{A} = \mathbf{PDP}^{-1}$. The columns of **P** are normalized (unit) eigenvectors

$$\mathbf{u} = \begin{bmatrix} -0.9285 \\ -0.3714 \end{bmatrix}, \quad \mathbf{v} = \begin{bmatrix} -0.6402 \\ -0.7682 \end{bmatrix} \tag{6.38}$$

(computed by the machine algorithm) associated with λ_1 and λ_2, respectively. Note that the command **poly(A)** returns the row vector $\mathbf{r} = [\,1 \;\; -2 \;\; 0.96\,]$ that records the coefficients the polynomial $p(\lambda) = \lambda^2 - 2\lambda + 0.96$, which is the characteristic polynomial for **A**. Recall that $-\mathbf{r}(2) = \operatorname{trace} \mathbf{A} = 2 = \lambda_1 + \lambda_2$ and $\mathbf{r}(3) = 0.96 = \det(\mathbf{A}) = \lambda_1\lambda_2$.

The long-term behavior of the system can now be considered, starting from any given initial population vector $\mathbf{x}_0$.

Refer to Figure 6.3 and observe that the eigenvectors **u**, **v** define lines L_1, L_2 in $\mathbb{R}^2$ with slopes $m_1 = 2/5$ and $m_2 = 6/5$, respectively.

If λ is an eigenvalue of **A** and $\mathbf{x}_0$ is an eigenvector associated with λ, then we have

$$\begin{aligned} \mathbf{x}_1 &= \mathbf{A}\mathbf{x}_0 = \lambda\mathbf{x}_0, & &\Rightarrow \mathbf{x}_1 = \lambda\mathbf{x}_0 \\ \mathbf{x}_2 &= \mathbf{A}\mathbf{x}_1 = \lambda\mathbf{x}_1 = \lambda^2\mathbf{x}_0 & &\Rightarrow \mathbf{x}_2 = \lambda^2\mathbf{x}_0, \\ & & &\quad\vdots \end{aligned}$$

and the pattern gives the general rule

$$\mathbf{x}_k = \lambda^k \mathbf{x}_0, \quad k = 0, 1, 2 \ldots . \tag{6.39}$$

Consider eigenvectors[3] $\mathbf{u}_0$ and $\mathbf{v}_0$ defined by

$$\mathbf{u}_0 = \begin{bmatrix} 500 \\ 200 \end{bmatrix}, \quad \mathbf{v}_0 = \begin{bmatrix} 300 \\ 360 \end{bmatrix}$$

that are associated with λ_1 and λ_2, respectively. Note that $\mathbf{u}_0$ and $\mathbf{v}_0$ are scalar multiples of **u** and **v** in (6.38).

[3] Actually, any nonzero scalar multiples of $\mathbf{u}_0$ and $\mathbf{v}_0$ will work just as well here.

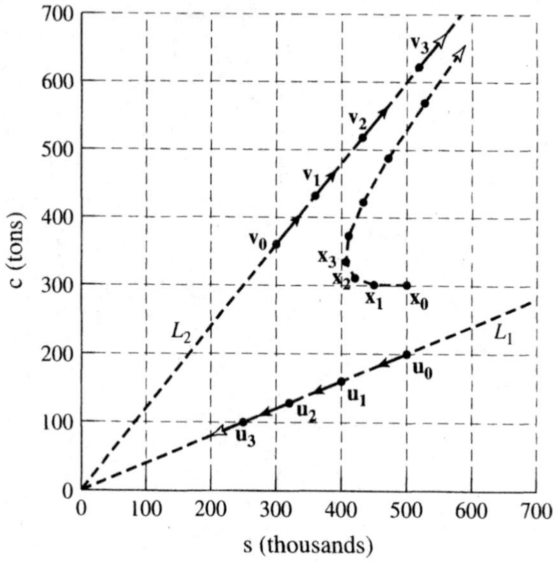

Figure 6.3 Phase portrait.

Observe that $\mathcal{B} = \{\mathbf{u}_0, \mathbf{v}_0\}$ is a basis for $\mathbb{R}^2$, being a linearly independent spanning set for $\mathbb{R}^2$. Thus, any initial population vector $\mathbf{x}_0$ can be written as a unique linear combination of $\mathbf{u}_0$ and $\mathbf{v}_0$, namely,

$$\mathbf{x}_0 = c_1\mathbf{u}_0 + c_2\mathbf{v}_0, \tag{6.40}$$

where c_1 and c_2 depend on $\mathbf{x}_0$. Multiplying (6.40) by $\mathbf{A}^k$ and using (6.33) and (6.39), we obtain

$$\begin{aligned}\mathbf{x}_k &= \mathbf{A}^k\mathbf{x}_0 = \mathbf{A}^k(c_1\mathbf{u}_0 + c_2\mathbf{v}_0)\\ &= c_1\mathbf{A}^k\mathbf{u}_0 + c_2\mathbf{A}^k\mathbf{v}_0\\ &= c_1\lambda_1^k\mathbf{u}_0 + c_2\lambda_2^k\mathbf{v}_0.\end{aligned} \tag{6.41}$$

Equation (6.41) gives a complete answer to our fundamental questions. There are three cases to consider.

(a) Using the initial population vector $\mathbf{x}_0$ in (6.37) and computing c_1, c_2 in this case, we find that

$$\mathbf{x}_0 = \begin{bmatrix} 500 \\ 300 \end{bmatrix} = \frac{3}{4}\mathbf{u}_0 + \frac{5}{12}\mathbf{v}_0;$$

that is, $c_1 = 3/4$, $c_2 = 5/12$, and applying equation (6.41) we have

$$\mathbf{x}_k = c_1(0.8)^k\mathbf{u}_0 + c_2(1.2)^k\mathbf{v}_0. \tag{6.42}$$

But[4] $(0.8)^k \to 0$ as $k \to \infty$, and so

$$\mathbf{x}_k \simeq c_2(1.2)^k\mathbf{v}_0, \quad \text{for large } k$$

4 See Appendix B: Toolbox.

The terminal points of the vectors $\mathbf{x}_0, \mathbf{x}_1, \mathbf{x}_2, \ldots$ lie on a curve in $\mathbb{R}^2$ that becomes *asymptotic* to the line L_2. The population of seals and cod both increase in the long-term such that $m_2 = c_k/s_k \simeq 360/300 = 6/5$, for large k that is the slope of L_2.

(b) For an initial population vector $\mathbf{u}_0 = \begin{bmatrix} 500 \\ 200 \end{bmatrix}$ (lying along L_1), we have

$$\mathbf{u}_{k+1} = \mathbf{A}\mathbf{u}_k = 0.8\mathbf{u}_k, \quad k = 0, 1, 2, \ldots$$

and so each application of the transformation $\mathbf{A}$ dilates (shrinks) $\mathbf{u}_k$ by a factor of 0.8 and so the population of seals and cod decrease by 20% per year and both species will die out in the long term.

$$\mathbf{u}_0 = \begin{bmatrix} 500 \\ 200 \end{bmatrix}, \quad \mathbf{u}_1 = \begin{bmatrix} 400 \\ 160 \end{bmatrix}, \quad \mathbf{u}_2 = \begin{bmatrix} 320 \\ 128 \end{bmatrix}, \quad \mathbf{u}_3 = \begin{bmatrix} 256.0 \\ 102.4 \end{bmatrix}, \quad \ldots$$

(c) For the initial population vector $\mathbf{v}_0 = \begin{bmatrix} 300 \\ 360 \end{bmatrix}$ (lying along L_2) we have

$$\mathbf{v}_{k+1} = \mathbf{A}\mathbf{v}_k = 1.2\mathbf{v}_k, \quad k = 0, 1, 2, \ldots$$

Each application of the transformation $\mathbf{A}$ stretches $\mathbf{v}_k$ by a factor of 1.2. The seal and cod population each increase by 20% per year and both species will thrive in the long term.

$$\mathbf{v}_0 = \begin{bmatrix} 300 \\ 600 \end{bmatrix}, \quad \mathbf{v}_1 = \begin{bmatrix} 360 \\ 432 \end{bmatrix}, \quad \mathbf{v}_2 = \begin{bmatrix} 432.0 \\ 518.4 \end{bmatrix}, \quad \mathbf{v}_3 = \begin{bmatrix} 518.4 \\ 622.1 \end{bmatrix}, \quad \ldots$$

The vectors $\mathbf{u}_k$ and $\mathbf{v}_k$ thus provide the simplest solutions to equation (6.31) in the sense that at each step the population next year is a scalar multiple of the population this year.

The key to the problem lies in the size of the eigenvalues, relative to each other and relative to 1. In this case, $0.8 = \lambda_1 < \lambda_2 = 1.2$, and we call λ_2 the *dominant* eigenvalue. Because $1 < \lambda_2$, initial population vectors along L_2 lead to growth and because $\lambda_1 < 1$, initial population vectors along L_1 lead to decline. In Exercises 6.3, 11 you are asked to investigate how the system evolves when starting from other initial population vectors. ■

The General Model

The illustration points the way to the of analysis discrete dynamical systems in general. The step matrix $\mathbf{A}$ plays the key role. Keep in mind that, in general, $\mathbf{A}$ may not be diagonalizable and some of its eigenvalues may be complex. We describe the most straightforward case.

Let $\mathbf{A}$ be a real $n \times n$ step matrix $\mathbf{A}$ that is diagonalizable. The n real or complex eigenvalues of $\mathbf{A}$ are labeled $\lambda_1, \lambda_2, \ldots, \lambda_n$ (repeated values are included), where

$$|\lambda_n| \le \cdots \le |\lambda_2| \le |\lambda_1|. \tag{6.43}$$

For complex eigenvalues, $|\lambda|$ is the modulus of λ. If $|\lambda_2| < |\lambda_1|$ in (6.43), we call λ_1 the *dominant eigenvalue* of $\mathbf{A}$.

There exists a set of linearly independent eigenvectors $\mathbf{v}_1, \mathbf{v}_2, \ldots, \mathbf{v}_n$ associated with the eigenvalues $\lambda_1, \lambda_2, \ldots, \lambda_n$. The set $\mathcal{B} = \{\mathbf{v}_1, \mathbf{v}_2, \ldots, \mathbf{v}_n\}$ spans $\mathbb{R}^n$ and so is a basis for $\mathbb{R}^n$. Given any initial n-vector $\mathbf{x}_0$, equation (6.41) in the general case becomes

$$\mathbf{x}_0 = c_1\mathbf{v}_1 + c_2\mathbf{v}_2 + \cdots + c_n\mathbf{v}_n, \tag{6.44}$$

where the coordinates $c_1, c_2, \ldots, c_n$ of $\mathbf{x}_0$ relative to $\mathcal{B}$ are unique.

Equation (6.39) is true in the general case and (6.41) generalizes into the statement

$$\mathbf{x}_k = c_1\lambda_1^k\mathbf{v}_1 + c_2\lambda_2^k\mathbf{v}_2 + \cdots + c_n\lambda_n^k\mathbf{v}_n. \tag{6.45}$$

The long-term evolution of the dynamical system is determined by (6.45), where the unique scalars $c_1, c_2, \ldots, c_n$ are given by (6.44).

The magnitude (in absolute value) of each eigenvalue λ relative to 1 has the following effect:

$$\begin{array}{llll} |\lambda| < 1 & \Rightarrow \lambda^k \to 0 & \text{as } k \to \infty & \text{(Decline)} \\ |\lambda| = 1 & \Rightarrow \lambda^k = \pm 1 & \text{as } k = 1, 2, 3, \ldots & \text{(Oscillation)} \\ |\lambda| > 1 & \Rightarrow \lambda^k \to \infty & \text{as } k \to \infty & \text{(Growth)} \end{array}$$

and of course equation (6.45) has to accommodate a mixture of the these effects.

Definition 6.7

Spectral radius

Let $\lambda_1, \lambda_2, \ldots, \lambda_n$ be the n eigenvalues of an $n \times n$ matrix $\mathbf{A}$. The number ρ (Greek letter rho) defined by

$$\rho(\mathbf{A}) = \max\{\, |\lambda_i|,\ i = 1, 2, \ldots, n \,\}$$

is called the *spectral radius* of $\mathbf{A}$.

The number $\rho(\mathbf{A})$ is defined for real or complex eigenvalues.

We now return to the analysis of the models described in Section 2.4.

■ ILLUSTRATION 6.4

Car Rentals Revisited

Review Illustration 2.8 in Section 2.4. The step matrix for this model is

$$\mathbf{A} = \begin{bmatrix} 0.6 & 0.1 & 0.2 \\ 0.1 & 0.7 & 0.1 \\ 0.1 & 0.1 & 0.6 \end{bmatrix}.$$

Using the command $[\mathbf{P},\ \mathbf{D}] = \mathbf{eig}(\mathbf{A})$, we have

$$\mathbf{P} = [\,\mathbf{v}_1\ \ \mathbf{v}_2\ \ \mathbf{v}_3\,] = \begin{bmatrix} -0.5992 & -0.8469 & 0.6427 \\ -0.6478 & 0.1405 & -0.7168 \\ -0.4704 & 0.5129 & 0.2704 \end{bmatrix},$$

$$\mathbf{D} = \begin{bmatrix} 0.8651 & 0 & 0 \\ 0 & 0.4623 & 0 \\ 0 & 0 & 0.5726 \end{bmatrix}$$

and $\mathbf{AP} = \mathbf{PD}$. The columns $\mathbf{v}_1, \mathbf{v}_2, \mathbf{v}_3$ of $\mathbf{P}$ are normalized eigenvectors associated, respectively, with eigenvalues $\lambda_1 \simeq 0.87$, $\lambda_2 \simeq 0.46$, $\lambda_3 \simeq 0.57$ appearing on the diagonal of $\mathbf{D}$. The dominant eigenvalue is λ_1. The eigenvalues are distinct in this case and so the eigenvectors $\mathbf{v}_1, \mathbf{v}_2, \mathbf{v}_3$ are linearly independent. Hence, $\mathcal{B} = \{\mathbf{v}_1, \mathbf{v}_2, \mathbf{v}_3\}$ is a basis for $\mathbb{R}^3$. Equation (6.44) in this case becomes

$$\mathbf{x}_0 = c_1\mathbf{v}_1 + c_2\mathbf{v}_2 + c_3\mathbf{v}_3 \tag{6.46}$$

for unique scalars c_1, c_2, c_3 determined by $\mathbf{x}_0$ and equation (6.45) becomes

$$\mathbf{x}_k = c_1(0.87)^k + c_2(0.57)^k\mathbf{v}_2 + c_3(0.46)^k\mathbf{v}_3. \tag{6.47}$$

The spectral radius is $\rho(\mathbf{A}) = 0.87$. The dominant eigenvalue is $\lambda_1 = 0.87 < 1$, which indicates that for any initial step vector $\mathbf{x}_0$, each component in $\mathbf{x}_k$ tends to zero as $k \to \infty$, showing that the entire rental fleet will be depleted in the long run for any initial distribution of cars. ■

■ ILLUSTRATION 6.5

Age-structured Population Models Revisited

Review Illustration 2.9 in Section 2.4. The step matrix for this model is

$$\mathbf{A} = \begin{bmatrix} 0 & 0 & 1 & 1.5 & 1 \\ 0.4 & 0 & 0 & 0 & 0 \\ 0 & 0.6 & 0 & 0 & 0 \\ 0 & 0 & 0.7 & 0 & 0 \\ 0 & 0 & 0 & 0.5 & 0 \end{bmatrix}.$$

Using MATLAB and approximating to two decimal places, we have

$$\mathbf{P} = [\,\mathbf{v}_1\ \ \mathbf{v}_2\ \ \mathbf{v}_3\ \ \mathbf{v}_4\ \ \mathbf{v}_5\,] = \begin{bmatrix} 0.84 & 0.66 - 0.29i & 0.66 + 0.29i & 0.03 - 0.29i & 0.03 + 0.29i \\ 0.39 & -0.19 - 0.37i & -0.19 + 0.37i & -0.14 + 0.22i & -0.14 - 0.22i \\ 0.27 & -0.31 + 0.17i & -0.31 - 0.17i & 0.31 - 0.18i & 0.31 + 0.18i \\ 0.22 & 0.19 + 0.30i & 0.19 - 0.30i & -0.56 + 0.03i & -0.56 - 0.03i \\ 0.13 & 0.20 - 0.14i & 0.20 + 0.14i & 0.58 + 0.26i & 0.58 - 0.26i \end{bmatrix}$$

$$\mathbf{D} = \begin{bmatrix} 0.87 & 0 & 0 & 0 & 0 \\ 0 & -0.06+0.70i & 0 & 0 & 0 \\ 0 & 0 & -0.06-0.70i & 0 & 0 \\ 0 & 0 & 0 & -0.40+0.20i & 0 \\ 0 & 0 & 0 & 0 & -0.40-0.20i \end{bmatrix}.$$

The only real eigenvalue is $\lambda_1 = 0.87$. The other four eigenvalues occur in two complex conjugate pairs: If $\lambda_2 = -0.06 + 0.70i$, then $\lambda_3 = \overline{\lambda_2} = -0.06 - 0.70i$ (change i to $-i$). Similarly, $\lambda_4 = -0.40 + 0.20i$ and $\lambda_5 = \overline{\lambda_4} = -0.40 - 0.20i$. The moduli of the complex eigenvalues are 0.70, 0.70, 0.44, 0.44 and so $\rho(\mathbf{A}) = 0.87$. In this case equation (6.45) becomes

$$\mathbf{x}_k = (0.87)^k \mathbf{v}_1 + \lambda_2^k \mathbf{v}_2 + \overline{\lambda_2}^k \mathbf{v}_3 + \lambda_4^k \mathbf{v}_4 + \overline{\lambda_4}^k \mathbf{v}_5$$

and the sequence $\mathbf{x}_k \to \mathbf{0}$ as $k \to \infty$. ■

Linear Recurrence Relations

We close this section with a different application of diagonalization. Consider an infinite *sequence* of terms $S = \{x_k\}$ given by

$$x_0, \ x_1, \ x_2, \ \ldots \ , x_k, \ \ldots .$$

We call x_0 the first term, x_1 the second term, and so on. The general $(k+1)$st term is denoted by x_k.

The terms of a sequence may be real or complex numbers, matrices, letters of the alphabet, other sequences, operations, and so on. Infinite sequences and series of numbers appear in applications to engineering, science, and computer science.

The terms in a sequence can sometimes be defined by a *recurrence relation* (formula). The *arithmetic progression* (or sequence) 1, 3, 5, ... is defined *recursively* by the rule

$$x_0 = 1, \quad x_{k+1} = x_k + 2, \quad k = 0, 1, 2, \ldots$$

and the *geometric progression* 4, −8, 16, ... is defined by the recurrence relation

$$x_0 = 4, \quad x_{k+1} = -2x_k, \quad k = 0, 1, 2, \ldots .$$

Second-Order Relations

Consider the sequence $S = \{x_k\}$ defined by

$$x_0 = a, \quad x_1 = b, \quad x_{k+2} = px_{k+1} + qx_k, \quad n = 0, 1, 2, \ldots , \qquad (6.48)$$

where a, b, p, and q are fixed numbers. We call (6.48) a *second-order* recurrence relation because the value x_{k+2} is determined by two previous values x_{k+1} and x_k. The values x_0 and x_1 are called *initial conditions* (they start off the sequence). Our goal is to find a closed form expression (or formula) for the kth term x_k that depends only on a, b, p, and q.

Applying Diagonalization

Define a sequence of 2-vectors (step vectors)

$$\mathbf{u}_k = \begin{bmatrix} x_k \\ x_{k+1} \end{bmatrix}, \quad k = 0, 1, 2 \dots . \tag{6.49}$$

Equation (6.48) can then be written in the form

$$\mathbf{u}_{k+1} = \begin{bmatrix} x_{k+1} \\ x_{k+2} \end{bmatrix} = \begin{bmatrix} 0 & 1 \\ q & p \end{bmatrix} \begin{bmatrix} x_k \\ x_{k+1} \end{bmatrix} = \mathbf{A}\mathbf{u}_k, \quad \text{where} \quad \mathbf{A} = \begin{bmatrix} 0 & 1 \\ q & p \end{bmatrix}.$$

Thus (6.49) can be viewed as a discrete dynamical system with step matrix $\mathbf{A}$ and as usual we have

$$\mathbf{u}_k = \mathbf{A}^k \mathbf{u}_0, \quad k = 0, 1, 2, \dots .$$

If $\mathbf{A}$ happens to be diagonalizable, then $\mathbf{A} = \mathbf{PDP}^{-1}$ for appropriate $\mathbf{P}$ and $\mathbf{D}$ and so

$$\mathbf{u}_k = \mathbf{PD}^k\mathbf{P}^{-1}\mathbf{u}_0, \quad k = 0, 1, 2 \dots ,$$

which gives a closed form expression for $\mathbf{u}_k$.

EXAMPLE 1

The Fibonacci Sequence

The sequence of positive integers $S = \{1, 1, 2, 3, 5, 8, \dots\}$ defined by the second-order recurrence relation

$$f_0 = f_1 = 1, \quad f_{k+2} = f_{k+1} + f_k, \quad k = 0, 1, 2, \dots$$

is called the *Fibonacci sequence*. Our goal is to apply the preceding theory to find a closed form expression for f_k. We have $p = q = 1$ in (6.48). Following (6.49), let $\mathbf{u}_k = \begin{bmatrix} f_k \\ f_{k+1} \end{bmatrix}$, for $k = 0, 1, 2, \dots$. The step matrix is $\mathbf{A} = \begin{bmatrix} 0 & 1 \\ 1 & 1 \end{bmatrix}$ and its characteristic polynomial is given by

$$|\mathbf{A} - \lambda\mathbf{I}_2| = \begin{vmatrix} -\lambda & 1 \\ 1 & 1-\lambda \end{vmatrix} = (-\lambda)(1-\lambda) - 1 = \lambda^2 - \lambda - 1,$$

so that $\lambda^2 - \lambda - 1 = 0$ is the characteristic equations of $\mathbf{A}$. Using the quadratic formula, $\mathbf{A}$ has two distinct eigenvalues

$$\lambda_1 = \frac{1+\sqrt{5}}{2}, \quad \lambda_2 = \frac{1-\sqrt{5}}{2},$$

which implies that $\mathbf{A}$ is diagonalizable. The columns of a diagonalizing matrix $\mathbf{P}$ for $\mathbf{A}$ are eigenvectors associated with λ_1 and λ_2, respectively. You may check that

$$\mathbf{P} = \begin{bmatrix} \lambda_1 - 1 & \lambda_2 - 1 \\ 1 & 1 \end{bmatrix}, \quad \mathbf{P}^{-1} = \frac{1}{\sqrt{5}} \begin{bmatrix} 1 & 1-\lambda_2 \\ -1 & \lambda_1 - 1 \end{bmatrix}, \quad \mathbf{D} = \begin{bmatrix} \lambda_1 & 0 \\ 0 & \lambda_2 \end{bmatrix}.$$

Thus, for each nonnegative integer k, we have $\mathbf{u}_k = \mathbf{PD}^k\mathbf{P}^{-1}\mathbf{u}_0$. Multiply out $\mathbf{PD}^k$ first, then $\mathbf{P}^{-1}\mathbf{u}_0$, using the relationships such as $2 - \lambda_2 = 1 + \lambda_1$, and then form the whole product, noting that $\lambda^3 - \lambda^2 = \lambda$. We find that the first component in $\mathbf{u}_k$ is

$$f_k = \frac{1}{\sqrt{5}}(\lambda_1^{k+1} - \lambda_2^{k+1}) = \frac{1}{\sqrt{5}}\left(\left(\frac{1+\sqrt{5}}{2}\right)^{k+1} - \left(\frac{1-\sqrt{5}}{2}\right)^{k+1}\right). \quad (6.50)$$

Note that f_k is an integer for each k, although this fact is not at all obvious from the formula shown on the right side of (6.50). ■

Historical Notes

Leonardo of Pisa (1170–1250), also known as **Fibonacci** (son of Bonaccio). In his early years, Leonardo travelled extensively throughout the Mediterranean, coming into contact with the Hindu-Arabic methods of calculation used at that time in North Africa. Promoting these methods in Italy was one rationale for his book *Liber Abaci* (1202), which included the following famous rabbit problem: Every pair of newborn rabbits begins breeding in their second month and produces new pair in every month thereafter. Beginning with an initial newborn pair in month 0, how many pairs f_k will there be after k months? We have $f_0 = f_1 = 1$, $f_2 = 1 + 1$, $f_3 = f_2 + f_1 = 3$, $f_4 = f_3 + f_2 = 5$, and so on.

Jacques Binet (1786–1856) French mathematician. The expression (6.50), known as Binet's formula, was given by Binet in the 1840s, although Euler is said to have published the same result some 80 years earlier. Binet made important contributions to mathematics and science.

EXERCISES 6.3

1. The system of difference equations

$$\begin{cases} x_k = x_{k-1} + 2y_{k-1} \\ y_k = 2x_{k-1} + y_{k-1} \end{cases} \quad k = 0, \pm 1, \pm 2, \cdots \pm 5$$

defines a discrete dynamical system where $x_0 = y_0 = 10$. Find x_4 and y_4.

2. The system of difference equations

$$\begin{cases} x_k = x_{k-1} - 2y_{k-1} \\ y_k = -2x_{k-1} + y_{k-1} \end{cases} \quad k = 0, \pm 1, \cdots$$

defines a discrete dynamical system with $x_0 = y_0 = 1$. Write down the step matrix $\mathbf{A}$. Compute $\mathbf{u}_k$ for $k = 0, \pm 1, \pm 2, \pm 3$ and determine a formula for $\mathbf{u}_k$ for all k.

3. In a controlled experiment biologists used the 2-tuple $\mathbf{u}_k = (x_k, y_k)$ to record, respectively, the number of female and male salmon in year k. For year zero $(x_0, y_0) = (800, 300)$. Data from the field suggest the step from year $(k-1)$ to year k is approximated by the system of difference equations

$$\begin{cases} x_{k+1} = 0.9x_k + 0.2y_k \\ y_{k+1} = 0.3x_k + 0.8y_k \end{cases}$$

Write down the step matrix for this discrete dynamical system. Determine $\mathbf{u}_2$ and $\mathbf{u}_4$ approximately. Explore the behavior of the system for large k. What happens when machine computation is used to compute $\mathbf{u}_{42}$?

4. A system of difference equations is given by

$$\begin{cases} x_k = x_{k-1} + 2z_{k-1} \\ y_k = y_{k-1} \\ z_k = 2x_{k-1} + z_{k-1} \end{cases} \quad k = 0, \pm 1, \cdots \pm 20$$

with $x_0 = z_0 = 1$, $y_0 = 2$. Find (x_3, y_3, z_3) and (x_{-3}, y_{-3}, z_{-3}).

5. Fill in the details of the solution to the Fibonacci recurrence relation.

6. Solve the recurrence relation

$$x_0 = x_1 = 1, \quad x_{n+2} = 2x_{n+1} + 3x_n, \quad n \geq 0.$$

7. Solve the recurrence relation

$$x_0 = 1, \quad x_1 = 2, \quad x_{n+2} = -x_{n+1} + 2x_n, \quad n \geq 0.$$

8. Let p_i denote the female population of white-naped cranes in year k. Assume that the birth rate b and death rate d are proportional to the population size at each step. The model is then defined by the single difference equation

$$p_{k+1} = (1 + b - d)p_k.$$

Use the theory of this section to find a closed form expression for p_k when $b = 0.1$ and $d = 0.2$.

USING MATLAB

9. In month k, the number of cars at locations A, B, C is denoted by x_k, y_k, z_k, respectively. Initially $x_0 = 500$, $y_0 = 450$, $z_0 = 600$. The probability that a random car is rented at the beginning of the month from one location and dropped off at the end of the month at one of the three locations is shown in the state diagram in Figure 6.4.

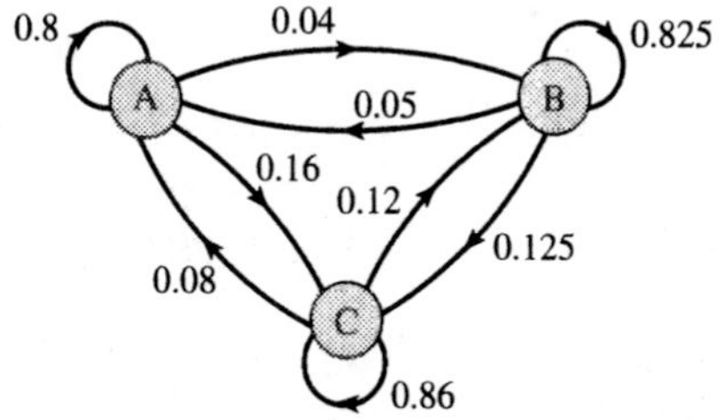

Figure 6.4 State diagram for car rental.

Investigate the long-term behavior of this discrete dynamical system. In what month is there likely to be less than 407 cars at A? Describe the likely distribution of cars in month 20.

10. The linear relationship in Example 1 can be found from the three initial vectors $\mathbf{u}_0$, $\mathbf{u}_1$, $\mathbf{u}_3$. Write an M-file that will solve two linear systems in order to find the entries in the 2×2 step matrix $\mathbf{A}$. Write an M-file to carry out the computation for a discrete dynamical system with a 3×3 step matrix. How many initial vectors are required in this case?

11. The system of difference equations that define the general predator-prey model is shown, where p_k and q_k denote the populations of predator and prey, respectively, at step k.

$$\begin{cases} p_{k+1} = ap_k + bq_k \\ q_{k+1} = -kp_k + cq_k \end{cases}$$

Write an M-file whose purpose is to explore the long-term behavior of the model, where a, b, c are three fixed input parameters and the parameter k represents the kill rate, which is allowed to vary. Test your program using the parameters from the harp seal–cod model in Illustration 6.3.

6.4 Markov Chains

Probability and statistics have become vital tools to the engineer, scientist, and people working in other applied fields. This section gives a brief introduction to probability in the context of the generic model, which was invented by Andrei Markov and which is used today to model a wide variety of practical situations. This theory is part of an important subfield of mathematics called *stochastics*.[5]

We begin with a few very basic concepts. The terms *experiment* and *outcome* are used in a very general sense. An experiment may consist of the simple act of tossing a coin or the demanding task of finding the mass of an electron. The outcome of an experiment could be, for example, the result of a measurement or the *state* of having obtained a head after tossing a coin.

Definition 6.8

Probability

The *probability* that an experiment will have an outcome O is measured by a real number $p = p(O)$, where $0 \le p \le 1$. The value $p = 0$ represents impossibility and $p = 1$ represents certainty.[6]

[5] Originating from the Greek word *stochos*, meaning "to guess."

[6] A 6/49 lottery is an experiment that chooses six numbers from the integers $1, 2, \ldots, 49$ and has 13,983,816 possible outcomes. The probability p of having the correct six numbers is $p = 0$ if you own no ticket and is $p = 1/13983816 = 7.1511 \times 10^{-8}$ (small, but not zero) if you own one ticket.

We refer to one (or many) of the possible outcomes of an experiment as an *event*.

Consider, for example, the experiment of drawing one ball from a bag containing 2 red balls and 3 black balls. The total possible outcomes are listed in a set $\mathcal{S} = \{R_1, R_2, B_1, B_2, B_3\}$, which is called the *state space* for the experiment. The event of drawing a red ball is denoted by $\mathcal{R} = \{R_1, R_2\}$, which is a subset of $\mathcal{S}$. The probability $p(\mathcal{R})$ of drawing a red ball is defined by the quotient

$$p(\mathcal{R}) = \frac{\text{Number of possible outcomes of the event } \mathcal{R}}{\text{Total number of outcomes of the experiment}} \qquad (6.51)$$

$$= \frac{2}{5} = 0.4$$

The outcome of the experiment will be either red or black and so the probability of obtaining a black ball is $p(\mathcal{B}) = 1 - p(\mathcal{R}) = 1 - 0.4 = 0.6$, which can be calculated directly by forming the quotient corresponding to (6.51) for the event $\mathcal{B} = \{B_1, B_2, B_3\}$.

Our first illustration introduces all the concepts and terminology required for the definition of a *Markov chain*, which follows after.

■ ILLUSTRATION 6.6

Simplified Weather Forecasting

In certain regions on earth, the weather patterns are stable enough to be able to estimate what the state of the weather will be tomorrow given the state of the weather today.

Assume that the weather on any given day can be in one (and only one) of two possible states: dull (D) or sunny (S). If the weather is dull today, there is a 40% chance that it will be dull tomorrow, and if the weather is sunny today, there is an 80% chance that it will be sunny tomorrow. Thus the climate is such that there is less than an even chance that a dull day will follow a dull day and a very good chance that a sunny day will follow a sunny day.

An experiment consists of checking the state of the weather on a particular day. The *transition* from the current state to the likely state tomorrow can be described very elegantly using the *state diagram* in Figure 6.5. State diagrams are weighted directed graphs (see Section 2.4).

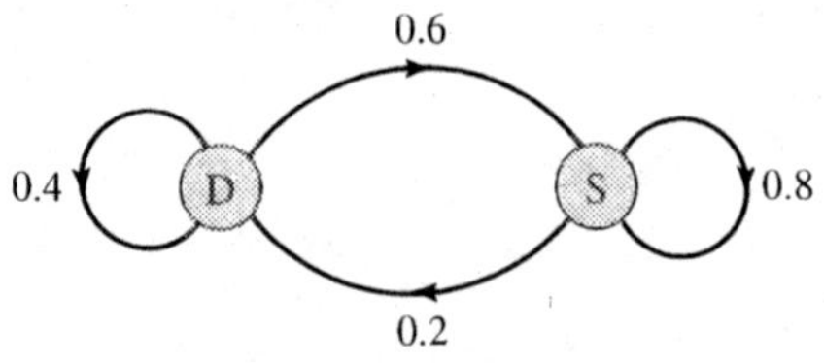

Figure 6.5 State diagram.

Alternatively, the information in the state diagram can be recorded in the following 2×2 *transition matrix* **T**.

$$\begin{array}{c} \\ D \\ S \end{array} \begin{array}{c} \begin{array}{cc} D & S \end{array} \\ \begin{bmatrix} 0.4 & 0.2 \\ 0.6 & 0.8 \end{bmatrix} \end{array} \quad \Rightarrow \quad \mathbf{T} = \begin{bmatrix} 0.4 & 0.2 \\ 0.6 & 0.8 \end{bmatrix} \tag{6.52}$$

Caution! The term *transition matrix* is also used in Section 7.3 for the matrix that defines change of basis in a vector space.

The entries in **T** are probabilities, and each column sum is 1, which indicates that on any day the weather must be either dull or sunny (but not both). Interpret the table on the left side of (6.52) column by column. First column: Given D today, the probability of D tomorrow is $p(\text{D}) = 0.4$ and the probability of a sunny day tomorrow is $p(\text{S}) = 0.6$. Interpret column 2 similarly.

We will now describe a *chain* of experiments or a *process*. Assume that the state of the weather is sunny today (on day zero). This information is recorded by the initial *state vector* $\mathbf{u}_0$, where

$$\mathbf{u}_0 = \begin{bmatrix} 0 \\ 1 \end{bmatrix} \begin{array}{l} \leftarrow p(D) \\ \leftarrow p(S) \end{array} .$$

The weather tomorrow (on day 1) is predicted by the entries in the next state vector $\mathbf{u}_1$, which is obtained by matrix-column vector multiplication, thus:

$$\mathbf{T}\mathbf{u}_0 = \begin{bmatrix} 0.4 & 0.2 \\ 0.6 & 0.8 \end{bmatrix} \begin{bmatrix} 0 \\ 1 \end{bmatrix} = \begin{bmatrix} 0.2 \\ 0.8 \end{bmatrix} = \mathbf{u}_1 .$$

The first component in $\mathbf{u}_1$ says that there is a 20% chance that the weather will be dull tomorrow and the second component says there is an 80% chance of a sunny day tomorrow. This information we already know.

The likely state of the weather on day 2 is described by the components in the state vector $\mathbf{u}_2$, obtained by multiplying $\mathbf{u}_1$ by **T**, thus:

$$\mathbf{T}\mathbf{u}_1 = \begin{bmatrix} 0.4 & 0.2 \\ 0.6 & 0.8 \end{bmatrix} \begin{bmatrix} 0.2 \\ 0.8 \end{bmatrix} = \begin{bmatrix} 0.24 \\ 0.76 \end{bmatrix} = \mathbf{u}_2,$$

showing, for example, that there is a 28% chance day 2 will be dull given that day 0 is sunny. The chain of experiments can be continued indefinitely, and we obtain a sequence of state vectors

$$\mathbf{u}_0 \text{ (initial)} \qquad \mathbf{u}_1,\ \mathbf{u}_2,\ \ldots$$

computed by repeated multiplication by **T**. This model is a discrete dynamical system (Sections 2.4 and 6.3) with step matrix **T** and we have

$$\mathbf{u}_{k+1} = \mathbf{T}\mathbf{u}_k \qquad \text{and} \qquad \mathbf{u}_k = \mathbf{T}^k \mathbf{u}_0, \quad k = 0, 1, 2, \ldots .$$

It is interesting that by day 6 the process has *converged* to a *steady state* in the following sense: Computing $\mathbf{T}^6$, we have

$$\mathbf{T}^6 \simeq \begin{bmatrix} 0.25 & 0.25 \\ 0.75 & 0.75 \end{bmatrix} = [\,\mathbf{w} \;\; \mathbf{w}\,]$$

and in fact $\mathbf{T}^k \simeq [\,\mathbf{w} \;\; \mathbf{w}\,]$, for all $k \geq 6$. Each column of $\mathbf{T}^k$ is approximately the same state vector $\mathbf{w}$. We have

$$\mathbf{u}_6 = \mathbf{T}^6\mathbf{u}_0 \simeq \begin{bmatrix} 0.25 & 0.25 \\ 0.75 & 0.75 \end{bmatrix}\begin{bmatrix} 0 \\ 1 \end{bmatrix} = \begin{bmatrix} 0.25 \\ 0.75 \end{bmatrix} = \mathbf{w}$$

and hence $\mathbf{T}^k\mathbf{w} \simeq \mathbf{w}$, for all $k \geq 6$. In fact, $\mathbf{Tw} = \mathbf{w}$ and we will comment on the computation, namely

$$\mathbf{Tw} = \begin{bmatrix} 0.4 & 0.2 \\ 0.6 & 0.8 \end{bmatrix}\begin{bmatrix} 0.25 \\ 0.75 \end{bmatrix} = \begin{bmatrix} 0.4(0.25) + 0.2(0.75) \\ 0.6(0.25) + 0.8(0.75) \end{bmatrix}$$

$$= \begin{bmatrix} 0.10 + 0.15 \\ 0.15 + 0.60 \end{bmatrix} = \begin{bmatrix} 0.25 \\ 0.75 \end{bmatrix}.$$

The first entry in $\mathbf{Tw}$ is the sum of two probabilities: $p = 0.4(0.25)$ and $q = 0.2(0.75)$. To explain p, if there is a 25% chance that the weather is dull today, and a 40% chance that the weather is dull tomorrow if it is dull today, then there is a 10% chance that the weather will be dull tomorrow—probabilities are multiplied in this case (Appendix B: Toolbox). The value q is interpreted similarly. On any day, the weather is either sunny or dull (exclusive states), and so the probability that the weather will be dull tomorrow, corresponding to the first entry in $\mathbf{Tw}$, will be the sum $p + q$. As an exercise, interpret the second entry in $\mathbf{Tw}$.

We may deduce that, on average, we are three times more likely to have a sunny day than a dull day. ■

Caution! Not every Markov chain converges to a steady state, as in Illustration 6.6.

Definition 6.9

Markov chain

Consider an experiment that has n possible outcomes, called *states*, denoted by

$$O_1,\; O_2,\; \ldots,\; O_n\,. \tag{6.53}$$

Repeating the experiment over and over defines a *chain of experiments* or *process* which is described by a sequence of *state vectors* (these are n-vectors)

$$\mathbf{u}_0,\ (\text{initial})\quad \mathbf{u}_1,\ \mathbf{u}_2,\ \ldots\,. \tag{6.54}$$

Initially the process is in one (and only one) of the n possible states (6.53), and this remains true after each experiment. On each experiment, the probability of passing from current state O_j to state O_i after the experiment is given by the transition probabilities t_{ij}, where $1 \leq i, j \leq n$. The next state depends *only* on the present state and the transition probabilities and is *independent* of past states.

Caution! We repeat, the next state in a Markov chain does not depend on past states.

The *transition matrix* $\mathbf{T} = [t_{ij}]$ for a Markov chain is constructed from the n^2 transition probabilities t_{ij}, where $0 \leq t_{ij} \leq 1$. For $n = 2$ we have the general form

$$\text{To}\left\{\begin{array}{c} O_1 \\ O_2 \end{array}\right. \overbrace{\begin{array}{cc} O_1 & O_2 \\ t_{11} & t_{12} \\ t_{21} & t_{22} \end{array}}^{\text{From}} \qquad \begin{array}{c} O_1 \\ \downarrow t_{11} \\ O_1 \end{array} \qquad \begin{array}{c} O_1 \\ \downarrow t_{21} \\ O_2 \end{array} \qquad \mathbf{T} = \begin{bmatrix} t_{11} & t_{12} \\ t_{21} & t_{22} \end{bmatrix}.$$

State O_1 must become either state O_1 or O_2 and so we must have $t_{11} + t_{21} = 1$ in column 1 of $\mathbf{T}$. Similarly, the sum of column 2 in $\mathbf{T}$ must be 1. In the $n \times n$ case, each column in a transition matrix has sum equal to 1.

The theory of stochastics uses some special vectors and matrices, which we will now define and investigate.

Definition 6.10

Stochastic vectors, stochastic matrices

A column vector is called *stochastic* if each of its entries is a nonnegative real number and the column sum is equal to 1. A square matrix is called (column) *stochastic* if each of its columns is a stochastic vector. An $n \times n$ matrix $\mathbf{P}$ is row stochastic if $\mathbf{P}^\mathsf{T}$ is column stochastic.

Examples of a stochastic vector $\mathbf{u}$ and 2×2 stochastic matrices $\mathbf{P}$ and $\mathbf{Q}$ are:

$$\mathbf{u} = \begin{bmatrix} 0.7 \\ 0 \\ 0.3 \end{bmatrix}, \qquad \mathbf{P} = \begin{bmatrix} 1 & 0 \\ 0 & 1 \end{bmatrix}, \qquad \mathbf{Q} = \begin{bmatrix} 0.4 & 0.25 \\ 0.6 & 0.75 \end{bmatrix}.$$

The definition of a stochastic vector $\mathbf{u}$ implies that each component r in $\mathbf{u}$ satisfies $0 \leq r \leq 1$ and so we refer to $\mathbf{u}$ as a *probability vector.*

Theorem 6.7 **Stochastic Matrix Products**

Let $\mathbf{P}$ *be a column stochastic* $n \times n$ *matrix and let* $\mathbf{u}$ *be a stochastic* n*-vector. Then the product* $\mathbf{Pu}$ *is a column stochastic* n*-vector.*

Proof We will illustrate the proof when $\mathbf{P}$ is 2×2. Let

$$\mathbf{P} = \begin{bmatrix} a & b \\ c & d \end{bmatrix} \quad \text{and} \quad \mathbf{u} = \begin{bmatrix} x \\ y \end{bmatrix},$$

where all entries are nonnegative and $a + c = 1$, $b + d = 1$, $x + y = 1$. Then

$$\mathbf{Pu} = \begin{bmatrix} a & b \\ c & d \end{bmatrix} \begin{bmatrix} x \\ y \end{bmatrix} = \begin{bmatrix} ax + by \\ cx + dy \end{bmatrix} = \mathbf{v}.$$

The components in $\mathbf{v}$ are nonnegative and

$$(ax + by) + (cx + dy) = (a + c)x + (b + d)y = (1)\,x + (1)\,y = 1.$$

Hence $\mathbf{v}$ is stochastic. The proof for the $n \times n$ case follows the same idea: Multiply out $\mathbf{Pu}$ and use the column sums in $\mathbf{P}$ first, then the column sum in $\mathbf{u}$. ■

It follows from Theorem 6.7 that if $\mathbf{P}$ and $\mathbf{Q}$ are $n \times n$ column stochastic matrices, then the product $\mathbf{PQ}$ is column stochastic. The next result is a consequence of this fact.

Theorem 6.8 **Powers of Stochastic Matrices**

If $\mathbf{P}$ *is a column stochastic matrix, then* $\mathbf{P}^k$ *is column stochastic for* $k = 1, 2, 3, \ldots$.

The Stationary Vector

Note that $\lambda = 1$ is an eigenvalue of any stochastic transition matrix $\mathbf{T}$ (Exercises 6.1, 33). If $\mathbf{w}$ is a stochastic eigenvector of $\mathbf{T}$ associated with $\lambda = 1$, then

$$\mathbf{Tw} = \mathbf{w} \tag{6.55}$$

and the state vector $\mathbf{w}$ is called *invariant*, meaning that the probability distribution before and after an experiment is exactly the same. If $\mathbf{u}_0 = \mathbf{w}$ is the initial state vector, then the Markov process looks like $\mathbf{w}, \mathbf{w}, \mathbf{w}, \ldots$, but if $\mathbf{u}_0$ is any other initial state vector, it may or may not be the case that $\mathbf{u}_k = \mathbf{T}^k\mathbf{u}_0$ approaches $\mathbf{w}$, as k increases, as was the case in Illustration 6.6.

Definition 6.11 *Stationary vector*

Let $\mathbf{T}$ be the transition matrix for a Markov chain. If the sequence of state vectors (6.54) approaches a vector $\mathbf{w}$ such that $\mathbf{Tw} = \mathbf{w}$, then $\mathbf{w}$ is called a *stationary* or *equilibrium* vector for the Markov chain and we say that the Markov chain *converges* to a *steady state* $\mathbf{w}$.

If a steady state $\mathbf{w}$ exists, then $\mathbf{w}$ is an eigenvector of $\mathbf{T}$ associated with the eigenvalue $\lambda = 1$, and if $\mathbf{w}$ does not exist, then the Markov chain is said to *diverge*. We now introduce an important class of transition matrices that define convergent Markov chains.

Definition 6.12

A column stochastic matrix $\mathbf{T}$ is called *regular* if there is a positive integer k for which $\mathbf{T}^k$ has all positive entries, and we write $\mathbf{T}^k > 0$ when this is the case. A Markov chain is called *regular* if its transition matrix is regular.

EXAMPLE 1

Regular and Nonregular Transition Matrices

Each matrix $\mathbf{T}$ shown in (6.56) is column stochastic. The first matrix is clearly regular and the second matrix is regular because $\mathbf{T}^2 > 0$. The last matrix is not regular (Exercises 6.4).

$$\begin{bmatrix} 0.4 & 0.3 \\ 0.6 & 0.7 \end{bmatrix}, \quad \begin{bmatrix} 0 & 0.3 \\ 1 & 0.7 \end{bmatrix}, \quad \begin{bmatrix} 0.4 & 0 \\ 0.6 & 1 \end{bmatrix} \tag{6.56}$$

■

Theorem 6.9 **Convergence of Regular Markov Chains**

Let $\mathbf{T}$ *be the transition matrix for a regular Markov chain. Then*

(a) The powers $\mathbf{T}^k$ approach a stochastic matrix $\mathbf{S} = [\mathbf{w}\ \mathbf{w} \cdots \mathbf{w}]$ that has all columns equal to the same stochastic vector $\mathbf{w}$.
(b) The vector $\mathbf{w}$ is unique and has positive components.
(c) For any initial state vector $\mathbf{u}_0$, the vectors $\mathbf{u}_k = \mathbf{T}^k\mathbf{u}_0$ approach $\mathbf{w}$ as k becomes large.

EXAMPLE 2

Convergence of a Regular Markov Chain

We will use the regular stochastic matrix $\mathbf{T}$ shown below to illustrate Theorem 6.9. Using MATLAB to compute powers of the transition matrix

$$\mathbf{T} = \begin{bmatrix} 0 & 0.3 \\ 1 & 0.7 \end{bmatrix}, \tag{6.57}$$

we have

$$\mathbf{T} = \begin{bmatrix} 0.0000 & 0.3000 \\ 1.0000 & 0.70000 \end{bmatrix}, \qquad \mathbf{T}^2 = \begin{bmatrix} 0.3000 & 0.21000 \\ 0.7000 & 0.79000 \end{bmatrix},$$

$$\mathbf{T}^4 = \begin{bmatrix} 0.2370 & 0.2289 \\ 0.7630 & 0.7711 \end{bmatrix}, \qquad \mathbf{T}^7 = \begin{bmatrix} 0.2306 & 0.2308 \\ 0.7694 & 0.7692 \end{bmatrix},$$

$$\mathbf{T}^8 = \begin{bmatrix} 0.2308 & 0.2308 \\ 0.7692 & 0.7692 \end{bmatrix} \simeq [\mathbf{w}\ \mathbf{w}] = \mathbf{S}.$$

The powers $\mathbf{T}^k$ become closer and closer to $\mathbf{S} = [\mathbf{w}\ \mathbf{w}]$, where (to four significant figures)

$$\mathbf{w} = \begin{bmatrix} 0.2308 \\ 0.7692 \end{bmatrix} \tag{6.58}$$

and for $k \geq 8$, $\mathbf{T}^k$ has all columns equal to $\mathbf{w}$. The power k for which $\mathbf{T}^k = \mathbf{S}$ depends on the Markov chain. If k is relatively small, we say that the convergence is *fast*, and if k is relatively large, we say that the convergence is *slow* (Exercises 6.4). ■

Finding the Steady State

Theorem 6.9 applies to any regular Markov chain. A unique stationary vector $\mathbf{w}$ exists and $\mathbf{Tw} = \mathbf{w}$. There are two methods to find $\mathbf{w}$. The first involves computing powers $\mathbf{T}^k$, and the second involves solving a homogeneous linear system (S) derived as follows:

$$\mathbf{Tw} = \mathbf{Iw} \quad \Rightarrow \quad \mathbf{Tw} = \mathbf{Iw} \quad \Rightarrow \quad (S) \quad (\mathbf{T} - \mathbf{I})\mathbf{w} = \mathbf{0} \tag{6.59}$$

In practice we solve (S) by reducing $\mathbf{M} = \mathbf{T} - \mathbf{I}$.

EXAMPLE 3

Computing the Stationary Vector

We will find the stationary vector for the regular Markov chain defined by $\mathbf{T}$ in (6.57). Let $\mathbf{w} = \begin{bmatrix} p \\ q \end{bmatrix}$ in (6.59), where $p + q = 1$. Reduce $\mathbf{M} = \mathbf{T} - \mathbf{I}_2$ as follows:

$$\mathbf{T} = \begin{bmatrix} 0 & 0.3 \\ 1 & 0.7 \end{bmatrix}, \qquad \mathbf{M} = \mathbf{T} - \mathbf{I}_2 = \begin{bmatrix} -1 & 0.3 \\ 1 & -0.3 \end{bmatrix} \sim \begin{bmatrix} 1 & -0.3 \\ 0 & 0 \end{bmatrix} = \mathbf{M}^*.$$

Let $q = t$ be a real parameter and then $p = 0.3t$. We have

$$p + q = 0.3t + t = 1 \ \Rightarrow \ t = \frac{10}{13} \ \Rightarrow \ \mathbf{w} = \frac{1}{13}\begin{bmatrix} 3 \\ 10 \end{bmatrix}$$

$$\simeq \begin{bmatrix} 0.2308 \\ 0.7692 \end{bmatrix} \text{ (to four figures),}$$

which agrees with the vector $\mathbf{w}$ shown in (6.58). ■

■ ILLUSTRATION 6.7

Workforce Mobility

Three competing computer companies labeled A, B, C, initially have work forces of 600, 500, 400 employees, respectively. The probability that a random worker will move from one company to another during a given year is shown in the state diagram in Figure 6.6.

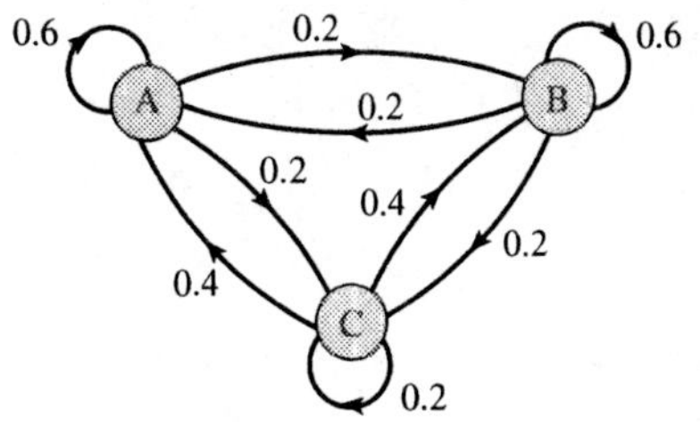

Figure 6.6 State diagram for workforce mobility.

Assuming that the total workforce remains constant, our goal is to describe the situation in future years.

The total workforce of all three companies is $N = 1500$. Let x_k, y_k, z_k denote the probabilities that a randomly chosen worker will work for A, B, C, respectively, in year k. Initially ($k = 0$) $x_0 = \frac{6}{15}$, $y_0 = \frac{5}{15}$, $z_0 = \frac{4}{15}$ are such that $x_0 + y_0 + z_0 = 1$. The workforce remains constant (every person works for some company) and so $x_k + y_k + z_k = 1$ for every $k = 1, 2, \ldots$. The initial state vector $\mathbf{u}_0$ and kth state vector $\mathbf{u}_k$ are

$$\mathbf{u}_0 = \begin{bmatrix} \frac{6}{15} \\ \frac{5}{15} \\ \frac{4}{15} \end{bmatrix}, \quad \mathbf{u}_k = \begin{bmatrix} x_k \\ y_k \\ z_k \end{bmatrix}.$$

Using Figure 6.6, the 3×3 transition matrix $\mathbf{T}$ for this process is

$$\mathbf{T} = \begin{bmatrix} \frac{3}{5} & \frac{1}{5} & \frac{2}{5} \\ \frac{1}{5} & \frac{3}{5} & \frac{2}{5} \\ \frac{1}{5} & \frac{1}{5} & \frac{1}{5} \end{bmatrix}$$

and we have

$$\mathbf{T}\mathbf{u}_0 = \begin{bmatrix} \frac{3}{5} & \frac{1}{5} & \frac{2}{5} \\ \frac{1}{5} & \frac{3}{5} & \frac{2}{5} \\ \frac{1}{5} & \frac{1}{5} & \frac{1}{5} \end{bmatrix} \begin{bmatrix} \frac{6}{15} \\ \frac{5}{15} \\ \frac{4}{15} \end{bmatrix} = \begin{bmatrix} \frac{3}{5}\frac{6}{15} + \frac{1}{5}\frac{5}{15} + \frac{2}{5}\frac{4}{15} \\ \frac{1}{5}\frac{6}{15} + \frac{3}{5}\frac{5}{15} + \frac{2}{5}\frac{4}{15} \\ \frac{1}{5}\frac{6}{15} + \frac{1}{5}\frac{5}{15} + \frac{1}{5}\frac{4}{15} \end{bmatrix} = \begin{bmatrix} x_1 \\ y_1 \\ z_1 \end{bmatrix} = \mathbf{u}_1.$$

The entry x_1 represents the probability that a worker randomly chosen from the total workforce of $N = 1500$ workers will be employed at A after one step. We have

$$x_1 = \frac{3}{5} \cdot \frac{6}{15} + \frac{1}{5} \cdot \frac{5}{15} + \frac{2}{5} \cdot \frac{4}{15} = \frac{31}{75} = 0.4133. \tag{6.60}$$

The first term in the sum (6.60) is explained as follows. The probability that a worker will be at A initially is $\frac{6}{15}$ and the probability of staying at A is $\frac{3}{5}$. Hence the probability of a worker being at A after one step is $\frac{3}{5} \cdot \frac{6}{15} = \frac{9}{25}$ (multiply the probabilities). The other terms in (6.60) are explained similarly.

The first few state vectors are

$$\mathbf{u}_1 = \begin{bmatrix} 0.4133 \\ 0.3867 \\ 0.2000 \end{bmatrix}, \quad \mathbf{u}_2 = \begin{bmatrix} 0.4053 \\ 0.3947 \\ 0.2000 \end{bmatrix}, \quad \mathbf{u}_3 = \begin{bmatrix} 0.4021 \\ 0.3979 \\ 0.2000 \end{bmatrix}, \ldots .$$

The Expected Number

If a coin is tossed $N = 10$ times, we expect on average $E = 0.5N = 5$ heads (and 5 tails). Thus $E = 5$ is the *expected number* of heads.

For the workforce, a sequence of *distribution vectors* $\mathbf{v}_0 = N\mathbf{u}_0$, $\mathbf{v}_1 = N\mathbf{u}_1, \ldots$ carries the expected number of workers at each company at each stage. The initial and first distribution vectors are

$$\mathbf{v}_0 = 1500 \begin{bmatrix} 0.4000 \\ 0.3333 \\ 0.2667 \end{bmatrix} = \begin{bmatrix} 600 \\ 500 \\ 300 \end{bmatrix}, \qquad \mathbf{v}_1 = 1500 \begin{bmatrix} 0.4133 \\ 0.3867 \\ 0.2000 \end{bmatrix} = \begin{bmatrix} 620 \\ 580 \\ 300 \end{bmatrix}.$$

Computing the Steady State

Referring to (6.59), we reduce $\mathbf{M} = \mathbf{T} - \mathbf{I}_3$ as follows:

$$\mathbf{M} = \mathbf{T} - \mathbf{I}_3 = \frac{1}{5} \begin{bmatrix} -2 & 1 & 2 \\ 1 & -2 & 2 \\ 1 & 1 & -4 \end{bmatrix} \sim \begin{bmatrix} 1 & 0 & -2 \\ 0 & 1 & -2 \\ 0 & 0 & 0 \end{bmatrix}.$$

The linear system $\mathbf{Mx} = \mathbf{0}$ has solution $\mathbf{x} = (2t, 2t, t)$, where t is a real parameter, and the condition $2t+2t+t = 1$ gives $t = \frac{1}{5}$. Hence $\mathbf{w} = [\,\frac{2}{5}\ \frac{2}{5}\ \frac{1}{5}\,]^{\mathsf{T}}$ is the stationary vector. The total workforce is $N = 1500$, which gives the expected steady-state distribution $\mathbf{v} = N\mathbf{w} = [\,600\ 600\ 300\,]^{\mathsf{T}}$. ■

Divergence

A Markov chain that does not approach a steady state is said to *diverge*. Some nonregular Markov chains diverge and some converge. Consider the Markov chain defined by the following transition matrix $\mathbf{T}$.

$$\mathbf{T} = \begin{bmatrix} 0.4 & 0 \\ 0.6 & 1 \end{bmatrix}, \quad \mathbf{w} = \begin{bmatrix} 0 \\ 1 \end{bmatrix} \quad \mathbf{u}_0 = \begin{bmatrix} t_0 \\ q_0 \end{bmatrix}$$

The transition matrix $\mathbf{T}$ is not regular (Exercises 6.4). However, $\mathbf{w}$ is the stationary vector. From any initial state $\mathbf{u}_0$ we find that $\mathbf{u}_k$ approaches $\mathbf{w}$.

Remarks

For technical reasons, people working in the field of stochastics usually define t_{ij} to be the probability of transition from state O_i to state O_j, and the transition matrix $\mathbf{T} = [t_{ij}]$ is then *row stochastic*. A Markov chain becomes a sequence of row stochastic vectors $\mathbf{u}_0, \mathbf{u}_1, \ldots$, and transition from state $\mathbf{u}_{k-1}$ to state $\mathbf{u}_k$ is given by the row vector-matrix product $\mathbf{u}_{k-1}\mathbf{T} = \mathbf{u}_k$. For example, for the following 2 × 2 transition matrix **T** and $\mathbf{u}_0 = [\,0.2 \;\; 0.8\,]$, we have

$$\mathbf{u}_0\mathbf{T} = [\,0.2 \;\; 0.8\,]\begin{bmatrix} 0.6 & 0.4 \\ 0.3 & 0.7 \end{bmatrix} = [\,0.36 \;\; 0.64\,] = \mathbf{u}_1.$$

Application of Markov Chains

In [1], Cleve Moler writes about the world's largest matrix computation. The Google™ search engine on the Web uses the PageRank™ algorithm developed by Google's founders, Larry Page and Sergey Brin. The algorithm assigns a probability p, its PageRank, to every page on the Web. The PageRank is used to determine an ordering for the most likely relevant pages from any query.

The PageRank is determined entirely by the link structure and is independent of actual queries or page content. It is recomputed on a regular basis. The computation of PageRank starts out with an $n \times n$ binary matrix $\mathbf{G} = [g_{ij}]$, where **G** is the incidence matrix of the graph whose vertices are the pages of the Web.

The entry in $g_{ij} = 1$ if there is a hyperlink between page i and page j. The size of n is approaching 3 billion and increasing rapidly on a daily basis. The matrix requires to be partitioned in order to fit into any computer.

A stochastic matrix **A** is constructed from **G** and **A** is the transition matrix of a Markov chain. The entries in **A** are all nonnegative, and by the Perron–Frobenius theorem there is an n-vector **x** such that $\mathbf{Ax} = \mathbf{x}$, which is unique up to a scaling factor. When scaled, the components of the state vector **x** are the PageRanks. For more references and links see [1].

REFERENCES

1. Moler, C.,"The World's Largest Matrix Computation," *MATLAB News and Notes*, The MathWorks, Natick, MA, 01760-2098, USA, pp. 12–13, October 2002.

EXERCISES 6.4

1. Rework Illustration 6.6 using the transition matrix

$$\mathbf{T} = \begin{bmatrix} 0.2 & 0.9 \\ 0.8 & 0.1 \end{bmatrix}$$

Find the stationary vector **w** by solving a linear system and verify that $\mathbf{T}^k \simeq [\mathbf{w}, \mathbf{w}]$ for some k.

2. A switch is in one of two states: $O_1 = On$ or $O_2 = Off$. If the switch is turned every hour, find the transition matrix **T** for the resulting Markov chain. Find an eigenvector **w** of **T** corresponding to the eigenvalue $\lambda = 1$ of **T**. Starting from an initial position of O_1, describe the chain of events. Is **w** a stationary vector for this Markov process?

3. A car rental agency has two locations labeled A (airport) and D (downtown). The state diagram in Figure 6.4.1 gives probability that a random car rented from some location today will be returned to the same or some other location tomorrow.

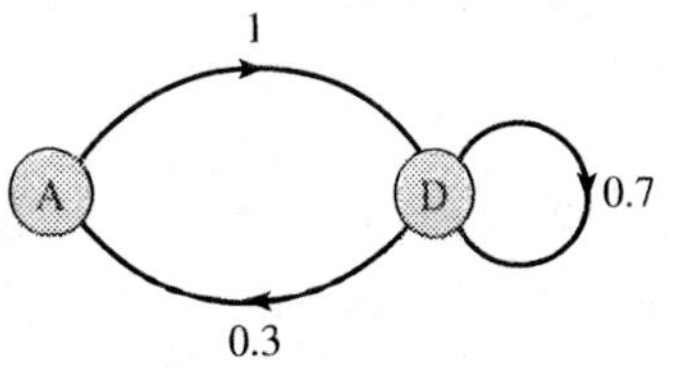

Interpret the state diagram. If there are 100 cars at A and 200 at D on the beginning of day 0, find the distribution of cars on day 3. Determine if a steady state exists, and if so, find it.

4. A survey conducted in 2003 recorded changes in land use over the previous five-year period. Land in a specific region was placed into one of three categories: forest (F), agricultural (A), urban (U). The result of the survey is shown in the following table.

	F	A	U
F	90%	5%	0%
A	10%	92%	0%
U	0%	3%	100%

The first column of the table shows that 90% of forest remained forest, while 10% of forest became agricultural land and none became urban.

(a) In your own words, explain the columns labeled (A) and (U) in the table.

(b) Assuming that the initial distribution of land is (F) 30%, (A) 50%, (U) 20% and that the trend found in the survey remains constant, find the distribution of land use after 5 years, 10 years, 15 years.

(c) Starting from any initial distribution, show that all land will eventually become urban.

(d) How long after 2003 will more than 50% of the land be urban?

5. Assume that a person lives either downtown or in the suburbs. At the beginning of each year, the probability of a person moving from downtown to suburbs is 0.2 and from suburbs to downtown is 0.3. Use this information to define a Markov chain $\mathbf{u}_0, \mathbf{u}_1, \ldots$, find the transition matrix $\mathbf{T}$, and draw the state diagram. If initially a randomly chosen person lives downtown, find the state vectors $\mathbf{u}_1$ $\mathbf{u}_2$, $\mathbf{u}_3$ and interpret the $(1, 1)$-entry in $\mathbf{u}_3$. Why would you expect a steady state to exist? Find the steady state.

6. If $\mathbf{P}$ and $\mathbf{Q}$ are $n \times n$ column stochastic matrices, show that $\mathbf{PQ}$ is column stochastic.

7. Just for fun! Let's do the time warp[7] again: "It's just a step to the left and a jump to the right!" Mark off positions O_1, O_2, O_3, O_4 at points $1, 2, 3, 4$ on the real axis. You at O_3. Begin. On every nanosecond $(= 10^{-9}$ seconds) toss a coin and step one position to the left if head or jump one position to the right if tail. If you reach O_1 or O_4, you disappear. Find the state vectors $\mathbf{u}_0$, $\mathbf{u}_1$, $\mathbf{u}_3$. What is the approximate probability of reaching O_2 in exactly 10 steps? Answer the same questions for other positions on the line.

8. Consider the airline *WestAir* (WA) and their competition (C). If the last flight by a customer was with (WA) there is a 60% chance that their next flight will be with (WA) and a 40% chance that they will switch to (C). Similarly, if the last flight was with (C) there is a 80% chance that their next flight will be with (C) and a 20% chance that they will switch to (WA). Assume a customer makes a flight twice a year and that currently (WA) has a 20% share of the market. Find the market share for (WA) after two years and the market share in the long term.

9. The 2×2 transition matrices for Markov chains have the form

$$\mathbf{T} = \begin{bmatrix} 1-a & b \\ a & 1-b \end{bmatrix}$$

for real scalars a and b, $0 \le a, b \le 1$.

(a) Show that $\mathbf{T}$ has eigenvalues $\lambda_1 = 1$ and $\lambda_2 = 1 - a - b$ with associated eigenvectors $\mathbf{v}_1 = \begin{bmatrix} b \\ a \end{bmatrix}$ and $\mathbf{v}_2 = \begin{bmatrix} 1 \\ -1 \end{bmatrix}$, respectively.

(b) For what values of a and b is $\mathbf{T}$ regular?

(c) Write any initial state vector $\mathbf{u}_0$ in the form $c_1\mathbf{v}_1 + c_2\mathbf{v}_2$ and show that $\mathbf{u}_k = \mathbf{T}^k\mathbf{u}_0$ approaches $c_1\mathbf{v}_1$ as k becomes large.

(d) Find the stationary vectors in the cases (i) a and b nonzero, (ii) $a = 0$, (iii) $b = 0$, (iv) $a = b = 0$.

10. Suppose that $\mathbf{S} = [\mathbf{w}\ \mathbf{w}\ \cdots\ \mathbf{w}]$ is an $n \times n$ column stochastic matrix and that $\mathbf{u}$ is a stochastic n-vector. Show that $\mathbf{Su} = \mathbf{w}$.

11. Show that if the transition matrix $\mathbf{T}$ for a Markov process is regular, then any initial vector $\mathbf{u}$ converges to the stationary vector $\mathbf{w}$. *Hint*: Use the previous exercise.

12. Show that the last matrix $\mathbf{T}$ in (6.56) from Example 1 is not regular. More generally, prove that any $n \times n$ column stochastic matrix $\mathbf{T}$ cannot be regular if $\mathbf{T}$ has a 1 on its principal diagonal.

13. Three companies labeled A, B, C introduce new brands of computers simultaneously onto the market. Initially the market shares are 0.4, 0.2, 0.4 for A, B, C, respectively.

[7] The reference is to a song from the cult movie, *The Rocky Horror Picture Show*.

At the start of year 2, A has retained 85% of its customers, lost 5% to B, and lost 10% to C. Company B has retained 75% of its customers, lost 15% to A, and lost 10% to C. Company C has retained 90% of its customers, lost 5% to A, and lost 5% to B. Assume that buying habits do not change and that the market does not expand or contract. What share of the market will be held by each company at the end of two years? What will the final equilibrium market shares be?

Exercises 14–15. Find the stationary vector for each stochastic matrix.

14. $\begin{bmatrix} 0.33 & 0.21 \\ 0.67 & 0.79 \end{bmatrix}$

15. $\begin{bmatrix} 0.1 & 0.6 & 0.1 \\ 0.8 & 0 & 0.2 \\ 0.1 & 0.4 & 0.7 \end{bmatrix}$

16. *Genetics.* Suppose there are two types of gene labeled G and g. A parent carries two genes and can have *genotype* (G,G) or (G,g) or (g,g). An offspring inherits one gene from each of its parents' pairs of genes to form its own genotype. We assume that if one parent is of genotype (G,g) and the other of genotype (g,g), the offspring will be of genotype (G,g) or (g,g) with equal probability. In case both parents are of the same genotype (G,g), then the probability that the offspring will be of type (G,G) or (g,g) is $\frac{1}{4}$ in each case and of type (G,g) it will be $\frac{1}{2}$.

In a certain experiment one parent is always of type (G,G) while the other parent has a distribution state $\mathbf{u} = [\,x \;\; y \;\; z\,]^{\mathsf{T}}$ for the three types (G,G), (G,g), and (g,g), respectively. The transition matrix from one generation to the next is then given by the matrix

	(G,G)-(G,G)	(G,G)-(G,g)	(G,G)-(g,g)
(G,G)	1	$\frac{1}{2}$	0
(G,g)	0	$\frac{1}{2}$	1
(g,g)	0	0	0

The rows denote genotypes of offspring, and the columns denote genotypes of parents.

Use the results on matrix powers to find the kth distribution state $\mathbf{u}_k = [\,x_k \;\; y_k \;\; z_k\,]^{\mathsf{T}}$, where $x_k,\ y_k,\ z_k$ are the fractions of individuals of types (G,G), (G,g), (g,g), respectively, and $\mathbf{u}_0$ is the initial distribution. Show that $\mathbf{u}_k$ approaches the stationary vector $\mathbf{w}$ as k tends to infinity.

17. Refer to the previous exercise and suppose both parents are of the same genotype. Confirm that the transition matrix $\mathbf{P}$ for this Markov chain is

$$\mathbf{P} = \begin{bmatrix} 1 & \frac{1}{4} & 0 \\ 0 & \frac{1}{2} & 0 \\ 0 & \frac{1}{4} & 1 \end{bmatrix}.$$

Is $\mathbf{P}$ regular? Use mathematical induction to show that $\mathbf{P}^k$ is given by the equation

$$\begin{bmatrix} 1 & \frac{1}{4} & 0 \\ 0 & \frac{1}{2} & 0 \\ 0 & \frac{1}{4} & 1 \end{bmatrix}^k = \begin{bmatrix} 1 & \frac{1}{2} - (\frac{1}{2})^{k+1} & 0 \\ 0 & (\frac{1}{2})^k & 0 \\ 0 & \frac{1}{2} - (\frac{1}{2})^{k+1} & 1 \end{bmatrix}.$$

If the initial probability vector is $\mathbf{u}_0 = [\,x_0 \;\; y_0 \;\; z_0\,]^{\mathsf{T}}$, find the kth distribution state and the vector $\mathbf{w}$ such that $\mathbf{u}_k \to \mathbf{w}$ as k tends to infinity.

18. Acetaminophen is an analgesic and fever-reducing medicine. A leading pharmaceutical company X had to withdraw its supplies due to product tampering at the stores. When it reissued its supplies of acetaminophen some time later, its market share with respect to its competitor, company Y was zero. However, company X had 65% loyalty rating—of the people who bought this month, 65% would buy again next month, as opposed to buying from company Y. Company Y had a 35% loyalty rating. Investigate what happened to sales in the ensuing months.

19. A *stochastic process* is generally any process which involves probability and a Markov chain is a particular stochastic process. Not every stochastic process is a Markov chain. Consider a bag containing 5 red and 5 black balls and the experiment of drawing one ball from the bag. Perform the experiment repeatedly, each time without replacing the drawn ball. Work out the probabilities for a sequence of three experiments. Explain why this is not a Markov chain. How can the process be modified so that it is a Markov chain?

USING MATLAB

20. *Time* magazine (TM) is analyzing the trend in annual subscription numbers. Market research suggests that a group of subscribers switch from year to year between TM and its major competitor (MC) according to the transition matrix shown below.

		Last Year	
		TM	MC
This Year	TM	0.85	0.10
	MC	0.15	0.90

Suppose initially that TM has 30% market share.

(a) Interpret the transition matrix.

(b) What are the market shares for TM and MC after two years?

(c) Find the steady state vector **w** by hand calculation.

(d) Compute the powers $\mathbf{T}^k$ of the transition matrix **T** and find the smallest integer k for which $\mathbf{T}^k = [\,\mathbf{w}\ \ \mathbf{w}\,]$ to four decimal places. Would you consider the convergence to be fast or slow?

(e) Investigate how the speed of convergence to the steady state varies with the initial state vector **x**.

21. At the beginning of each week a market research company uses the same group of people to test preference for Coke or Pepsi. At every tasting Coke retains $\frac{1}{3}$ of its share of the test group while Pepsi retains $\frac{1}{2}$ of its share. Initially, $\frac{4}{10}$ of the test group prefer Coke. Define the transition matrix **T** of a Markov chain that models this situation. Draw the state diagram. Determine how the test group is shared between Coke and Pepsi after 4 weeks. Does a steady state **w** exist? Compute the exact steady state by hand and confirm by taking powers of **T**. Would you describe the convergence to the steady state to be slow or fast?

22. Investigate the relative speed of convergence of a regular Markov chain. To do this, write an M-file that will generate random regular transition matrices of order $n = 2, 3, \ldots$. Test the speed of convergence and record the results. Draw conclusions.

23. Write an M-file which computes the state vectors for a Markov chain that results from choosing a ball (with replacement) from a bag containing 3 red, 4 black, 5 white and 6 green balls. Assume the initial state is red.

6.5 Systems of Linear Differential Equations

This section requires a knowledge of calculus.

Exponential Growth

Suppose that $y = y(t)$ denotes the number of bacteria present in a culture[8] at time t hours. It is known that, under ideal conditions, the instantaneous rate of increase of the function $y(t)$ with respect to t is proportional to the number y of bacteria present at time t. The growth in the number of bacteria is modeled by the *first-order differential equation*

$$y'(t) = k\,y(t), \tag{6.61}$$

where k is a constant of proportionality and the *derivative* $y'(t) = dy/dt$ is the instantaneous rate of change of y with respect to t.

The solution to (6.61) is obtained in the following standard way. Assuming $y(t) \neq 0$, we have from (6.61)

$$\frac{y'(t)}{y(t)} = k \quad \Rightarrow \quad \int \frac{y'(t)}{y(t)}\,dt = \int k\,dt = kt + c,$$

where C is a constant of integration. Hence

$$\ln|y(t)| = kt + c, \tag{6.62}$$

[8] A *bacterium* (plural, *bacteria*) is a single-celled microscopic organism that is found almost everywhere. These organisms are kept alive in the laboratory in a liquid environment called a culture.

where ln is the natural logarithm (base $e \simeq 2.71828$), and c is the constant of integration. Taking exponentials on both sides of (6.62), we obtain

$$|y(t)| = e^{\ln y(t)} = e^{kt+c} = e^c e^{kt}.$$

But $y(t)$ is a nonnegative number of bacteria, and so

$$y(t) = \alpha e^{kt}, \tag{6.63}$$

where $\alpha = e^c$ is a positive constant. Equation (6.63) is the *general solution* to (6.61). Putting $t = 0$ in (6.63), we have $y(0) = \alpha$, giving

$$y(t) = y(0)e^{kt}, \tag{6.64}$$

which is a *particular solution* to (6.61), where $y(0)$ is the number of bacteria present in the culture when $t = 0$. For this model, $k > 0$ and the function e^{kt} increases with time t. We call this *exponential growth.*

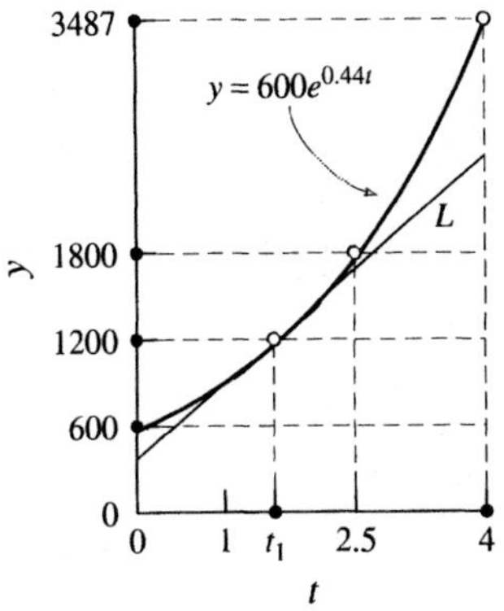

Figure 6.7 Exponential growth.

Refer to Figure 6.7. Suppose it is observed that y increases from 600 to 1800 organisms in 2.5 hours. At time $t = 0$ we have $y(0) = 600$ and then $y(2.5) = 1800$. Solving (6.64) for k by taking logarithms, we have

$$k = \frac{\ln(1800) - \ln(600)}{2.5} \simeq 0.44$$

and k is positive as expected. The number of bacteria present after 4 hours is $y = 600e^{1.76} \simeq 3487$. The value of dy/dt for any t is the slope of the tangent line to the graph of $y(t)$. For example, $y = 1200$ when $t_1 \simeq 1.58$ hours and so $dy/dt \simeq (0.44)(1.58) = 0.70$ is the slope of the line L when $t = t_1$. The angle of inclination of L is $\theta = \arctan(0.7) \simeq 0.6107$ radians, or about 35°.

Exponential Decay

Suppose that $y(t)$ is a function of t that satisfies equation (6.61). The case when $k > 0$ has been illustrated immediately above. If $k = 0$, then the solution to (6.61) is the constant function $y = y(0)$. If $k < 0$, then the instantaneous rate of *decrease* of $y(t)$ with respect to t is proportional to $y(t)$ for any t and the function e^{kt} is decreasing. In this case equation (6.61) models the *exponential decay* of the quantity $y(t)$. Examples of this type of phenomenon include the radioactive decay[9] of carbon-14 and the loss of temperature over time of a hot body according to Newton's law of cooling.[10]

Systems of Linear Differential Equations

We now turn to the case when $y_1(t), y_2(t), \ldots, y_n(t)$ are n functions of t and the instantaneous rate of change of each function with respect to t is a linear

[9] A *half-life* is the amount of time it takes one half of a radioactive sample to decompose. The half-life of carbon-14 is about 5570 years.

[10] Newton's law states that the rate of decrease in temperature of a body is proportional to the difference in temperature between the body and the temperature of the surrounding medium.

combination of the other functions. The general pattern is

$$\begin{cases} \dfrac{dy_1}{dt} = a_{11}y_1 + \cdots + a_{1n}y_n \\ \vdots \qquad \vdots \qquad \qquad \vdots \\ \dfrac{dy_n}{dt} = a_{n1}y_1 + \cdots + a_{nn}y_n \end{cases} \tag{6.65}$$

To obtain the matrix form of (6.65), define

$$\frac{d\mathbf{y}}{dt} = \begin{bmatrix} dy_1/dt \\ \vdots \\ dy_n/dt \end{bmatrix}, \quad \mathbf{y} = \begin{bmatrix} y_1 \\ \vdots \\ y_n \end{bmatrix}, \quad \mathbf{A} = [a_{ij}] = \begin{bmatrix} a_{11} & \cdots & a_{1n} \\ \vdots & \ddots & \vdots \\ a_{n1} & \cdots & a_{nn} \end{bmatrix}$$

and then (6.65) has the compact form

$$\frac{d\mathbf{y}}{dt} = \mathbf{A}\mathbf{y}. \tag{6.66}$$

Method of Solution

Make the following initial observation. Equation (6.66) is a generalized form of the growth/decay equation (6.61) that has general solution (6.64). Let $\mathbf{y} = \mathbf{v}e^{kt}$, where $\mathbf{v}$ is an n-vector that is independent of t. Substituting $\mathbf{y}$ into (6.66), we have

$$\frac{d\mathbf{y}}{dt} = k\mathbf{v}e^{kt} \quad \text{and} \quad \mathbf{A}\mathbf{y} = \mathbf{A}\mathbf{v}e^{kt} \quad \Rightarrow \quad (k\mathbf{v})e^{kt} = (\mathbf{A}\mathbf{v})e^{kt}$$

and canceling e^{kt}, which is nonzero for all t, we have

$$\mathbf{A}\mathbf{v} = k\mathbf{v}, \tag{6.67}$$

which shows that $\mathbf{y} = \mathbf{v}e^{kt}$ is a particular solution to (6.66) and that k is an eigenvalue of $\mathbf{A}$ with corresponding eigenvector $\mathbf{v}$. Thus, solving (6.66) for $\mathbf{y}$ involves finding the eigenvalues and eigenvectors of $\mathbf{A}$. We will need to consider the cases when $\mathbf{A}$ is diagonalizable and when $\mathbf{A}$ is not diagonalizable.

The method of solution of (6.66), illustrated next when $n = 2$, extends by analogy to any n. Setting $n = 2$ in (6.65), we have

$$\begin{cases} \dfrac{dy_1}{dt} = a_{11}y_1 + a_{12}y_2 \\ \dfrac{dy_2}{dt} = a_{21}y_1 + a_{22}y_2 \end{cases} \qquad \mathbf{A} = \begin{bmatrix} a_{11} & a_{21} \\ a_{21} & a_{22} \end{bmatrix}. \tag{6.68}$$

Case I

When $\mathbf{A}$ is diagonalizable there exist eigenvalues λ_1 and λ_2 and associated linearly independent eigenvectors $\mathbf{v}_1$ and $\mathbf{v}_2$ that are used to construct a diagonalizing matrix $\mathbf{P} = [\,\mathbf{v}_1 \;\; \mathbf{v}_2\,]$ such that

$$\mathbf{P}^{-1}\mathbf{A}\mathbf{P} = \mathbf{D} = \begin{bmatrix} \lambda_1 & 0 \\ 0 & \lambda_2 \end{bmatrix}. \tag{6.69}$$

Introduce new functions $z_1(t)$ and $z_2(t)$ such that

$$\mathbf{z} = \begin{bmatrix} z_1(t) \\ z_2(t) \end{bmatrix} = \mathbf{P}^{-1}\begin{bmatrix} y_1 \\ y_2 \end{bmatrix} = \mathbf{P}^{-1}\mathbf{y} \quad \text{or, equivalently,} \quad \mathbf{y} = \mathbf{Pz}.$$

The entries in $\mathbf{P}$ are independent of t and so differentiating $\mathbf{y} = \mathbf{Pz}$, and using (6.66), we have

$$\mathbf{P}\frac{d\mathbf{z}}{dt} = \frac{d\mathbf{y}}{dt} = \mathbf{Ay} = \mathbf{APz} \quad \Rightarrow \quad \frac{d\mathbf{z}}{dt} = \mathbf{P}^{-1}\mathbf{APz}. \tag{6.70}$$

Using (6.69) with (6.70), we obtain a new system $d\mathbf{z}/dt = \mathbf{Dz}$, where

$$\frac{d\mathbf{z}}{dt} = \begin{bmatrix} dz_1/dt \\ dz_2/dt \end{bmatrix} = \begin{bmatrix} \lambda_1 & 0 \\ 0 & \lambda_2 \end{bmatrix}\begin{bmatrix} z_1 \\ z_2 \end{bmatrix} = \mathbf{Dz} \quad \Rightarrow \quad \begin{cases} \dfrac{dz_1}{dt} = \lambda_1 z_1 \\ \dfrac{dz_2}{dt} = \lambda_2 z_2. \end{cases} \tag{6.71}$$

The process of diagonalization *decouples* the system in (6.68) and we obtain two standard growth-decay differential equations in (6.71) that have general solutions

$$z_1 = z_1(0)e^{\lambda_1 t}, \quad z_2 = z_2(0)e^{\lambda_2 t},$$

where $z_1(0)$ and $z_2(0)$ are scalars determined by the initial conditions. Returning to the transformation $\mathbf{y} = \mathbf{Pz} = [\,\mathbf{v}_1 \;\; \mathbf{v}_2\,]\,\mathbf{z}$, we can now write

$$\mathbf{y} = [\,\mathbf{v}_1 \;\; \mathbf{v}_2\,]\begin{bmatrix} z_1(0)e^{\lambda_1 t} \\ z_2(0)e^{\lambda_2 t} \end{bmatrix} = z_1(0)\mathbf{v}_1 e^{\lambda_1 t} + z_2(0)\mathbf{v}_2 e^{\lambda_2 t} \tag{6.72}$$

and (6.72) is the *general solution* to (6.68). If initial conditions $y_1(0)$ and $y_2(0)$ are known, then $\mathbf{y}(0) = \mathbf{Pz}(0)$ and so

$$\begin{bmatrix} y_1(0) \\ y_2(0) \end{bmatrix} = \mathbf{P}\begin{bmatrix} z_1(0) \\ z_2(0) \end{bmatrix} \quad \Rightarrow \quad \begin{bmatrix} z_1(0) \\ z_2(0) \end{bmatrix} = \mathbf{P}^{-1}\begin{bmatrix} y_1(0) \\ y_2(0) \end{bmatrix}$$

determines the scalars $z_1(0)$, $z_2(0)$, which, when substituted into (6.72), give a *particular solution* to (6.68) with initial conditions $\mathbf{y}(0) = \begin{bmatrix} y_1(0) \\ y_2(0) \end{bmatrix}$.

EXAMPLE 1

Illustrating Case I

We will find a particular solution of the system

$$\begin{cases} \dfrac{dy_1}{dt} = 2y_1 + y_2 \\ \dfrac{dy_2}{dt} = y_1 + 2y_2 \end{cases}$$

subject to the initial conditions $y_1(0) = 2$ and $y_2(0) = 0$. Let

$$\mathbf{y} = \begin{bmatrix} y_1 \\ y_2 \end{bmatrix} \qquad \mathbf{A} = \begin{bmatrix} 2 & 1 \\ 1 & 2 \end{bmatrix}.$$

The characteristic polynomial of $\mathbf{A}$ is

$$\det(\lambda\mathbf{I} - \mathbf{A}) = \begin{vmatrix} \lambda - 2 & -1 \\ -1 & \lambda - 2 \end{vmatrix} = \lambda^2 - 4\lambda + 3 = (\lambda - 1)(\lambda - 3),$$

giving eigenvalues $\lambda_1 = 1$ and $\lambda_2 = 3$ and associated eigenvectors $\mathbf{v}_1$, $\mathbf{v}_2$ that form a diagonalizing matrix $\mathbf{P}$, where

$$\mathbf{v}_1 = \begin{bmatrix} 1 \\ -1 \end{bmatrix}, \quad \mathbf{v}_2 = \begin{bmatrix} 1 \\ 1 \end{bmatrix}, \quad \mathbf{P} = [\,\mathbf{v}_1 \ \mathbf{v}_2\,] = \begin{bmatrix} 1 & 1 \\ -1 & 1 \end{bmatrix}.$$

Let $\mathbf{y} = \mathbf{Pz}$ and then $\dfrac{d\mathbf{z}}{dt} = \begin{bmatrix} 1 & 0 \\ 0 & 3 \end{bmatrix}\mathbf{z}$ so that $\mathbf{z} = \begin{bmatrix} z_1(0)e^t \\ z_2(0)e^{3t} \end{bmatrix}$. Hence

$$\mathbf{y} = \begin{bmatrix} y_1 \\ y_2 \end{bmatrix} = \begin{bmatrix} 1 & 1 \\ -1 & 1 \end{bmatrix}\begin{bmatrix} z_1(0)e^t \\ z_2(0)e^{3t} \end{bmatrix} = \begin{bmatrix} z_1(0)e^t + z_2(0)e^{3t} \\ -z_1(0)e^t + z_2(0)e^{3t} \end{bmatrix},$$

giving

$$\begin{cases} y_1 = z_1(0)e^t + z_2(0)e^{3t} \\ y_2 = -z_1(0)e^t + z_2(0)e^{3t}, \end{cases}$$

which is the general solution to the problem. We find that $\mathbf{P}^{-1} = \dfrac{1}{2}\begin{bmatrix} 1 & -1 \\ 1 & 1 \end{bmatrix}$ and so

$$\begin{bmatrix} z_1(0) \\ z_2(0) \end{bmatrix} = \frac{1}{2}\begin{bmatrix} 1 & -1 \\ 1 & 1 \end{bmatrix}\begin{bmatrix} y_1(0) \\ y_2(0) \end{bmatrix} = \frac{1}{2}\begin{bmatrix} 1 & -1 \\ 1 & 1 \end{bmatrix}\begin{bmatrix} 2 \\ 0 \end{bmatrix} = \begin{bmatrix} 1 \\ 1 \end{bmatrix}$$

which gives the particular solution

$$\begin{cases} y_1 = e^t + e^{3t} \\ y_2 = -e^t + e^{3t}. \end{cases}$$

■

When $\mathbf{A}$ is $n \times n$ and diagonalizable, $\mathbf{A}$ has n eigenvalues $\lambda_1, \lambda_2, \ldots, \lambda_n$ (including possibly repeated values) and associated linearly independent eigenvectors $\mathbf{v}_1, \mathbf{v}_2, \ldots, \mathbf{v}_n$. Then the general solution to (6.66) is

$$\mathbf{y} = z_1(0)\mathbf{v}_1e^{\lambda_1 t} + z_2(0)\mathbf{v}_2e^{\lambda_2 t} + \cdots + \alpha_n\mathbf{v}_ne^{\lambda_n t},$$

where the scalars $\alpha_1, \alpha_2, \ldots, \alpha_n$ can be determined from n initial conditions

$$y_1(0) = c_1, \ y_2(0) = c_n, \ \ldots, \ y_n(0) = c_n.$$

Case II

Consider the system (6.68) in the case when $\mathbf{A}$ is not diagonalizable. In this case some eigenvalue λ is repeated and its geometric multiplicity g_λ is less than its algebraic multiplicity a_λ. We will illustrate just one particular case. Suppose λ is an eigenvalue of $\mathbf{A}$ with $g_\lambda = 1$ and $a_\lambda = 2$. Let $\mathbf{v}$ be an eigenvector of $\mathbf{A}$ associated with λ. We know already that $\mathbf{y} = \mathbf{v}e^{\lambda t}$ is a particular solution to (6.66). To obtain a second particular solution associated with the same λ, consider

$$\mathbf{y} = \mathbf{v}te^{\lambda t} + \mathbf{u}e^{\lambda t} \tag{6.73}$$

for some vector $\mathbf{u}$ soon to be determined. Then, using the product rule for differentiation, we have

$$\left(\frac{d\mathbf{y}}{dt} =\right) \quad \mathbf{v}e^{\lambda t} + \lambda\mathbf{v}te^{\lambda t} + \lambda\mathbf{u}e^{\lambda t} = \mathbf{A}\mathbf{v}te^{\lambda t} + \mathbf{A}\mathbf{u}e^{\lambda t} \quad (= \mathbf{A}\mathbf{y}), \tag{6.74}$$

giving

$$(\mathbf{A}\mathbf{v} - \lambda\mathbf{v})te^{\lambda t} + (\mathbf{A}\mathbf{u} - \lambda\mathbf{u} - \mathbf{v})e^{\lambda t} = \mathbf{0},$$

which must hold for all t and so the bracketed terms are both zero; that is,

$$\mathbf{A}\mathbf{v} = \lambda\mathbf{v} \tag{6.75}$$

$$(\mathbf{A} - \lambda\mathbf{I}_2)\mathbf{u} = \mathbf{v}. \tag{6.76}$$

Equation (6.75) confirms that λ is an eigenvalue of $\mathbf{A}$ with associated eigenvector $\mathbf{v}$. The solution to the nonhomogeneous linear system (6.76) determines the vector $\mathbf{u}$ that appears in (6.73). The total contribution to the general solution from the eigenvalue λ is

$$\alpha\mathbf{v}e^{\lambda t} + \beta(\mathbf{v}te^{\lambda t} + \mathbf{u}e^{\lambda t}) = (\alpha\mathbf{v} + \beta\mathbf{u})e^{\lambda t} + \beta\mathbf{v}te^{\lambda t}$$

for constants α and β.

EXAMPLE 2

Illustrating Case II

We will find the general solution of the system

$$\begin{cases} \dfrac{dy_1}{dt} = 3y_1 + y_2 \\[2ex] \dfrac{dy_2}{dt} = -y_1 + y_2. \end{cases} \tag{6.77}$$

The matrix $\mathbf{A} = \begin{bmatrix} 3 & 1 \\ -1 & 1 \end{bmatrix}$ has an eigenvalue $\lambda = 2$ with $g_2 = 1$ and $a_2 = 2$ and an associated eigenvector $\mathbf{v} = \begin{bmatrix} 1 \\ -1 \end{bmatrix}$. A particular solution is $\mathbf{y}_1 = \mathbf{v}e^{2t}$.

For a second solution, let $\mathbf{y}_2 = \mathbf{v}te^{2t} + \mathbf{u}e^{2t}$, where $\mathbf{u} = [u_1, u_2]^{\mathsf{T}}$ is the solution to the linear system

$$\begin{cases} u_1 + u_2 = 1 \\ -u_1 - u_2 = -1. \end{cases}$$

Solving, we have $u_1 = 1 - s$, $u_2 = s$, where s is a parameter. Choose, for example, $s = 0$, and then the general solution to (6.77) is

$$\mathbf{y} = \alpha\mathbf{v}e^{2t} + \beta(\mathbf{v}te^{2t} + \mathbf{u}e^{2t})$$

or

$$\begin{bmatrix} y_1 \\ y_2 \end{bmatrix} = \left(\alpha \begin{bmatrix} 1 \\ -1 \end{bmatrix} + \beta \begin{bmatrix} 1 \\ 0 \end{bmatrix}\right) e^{2t} + \beta \begin{bmatrix} 1 \\ -1 \end{bmatrix} te^{2t}.$$ ■

EXERCISES 6.5

Exercises 1–4. Solve each system of linear differential equations, where $y' = dy/dt$. Find a particular solution given initial conditions $y_1(0) = 2$, $y_2(0) = 0$.

1. $\begin{cases} y_1' = 0.5y_1 + 0.5y_2 \\ y_2' = 0.5y_1 + 0.5y_2 \end{cases}$

2. $\begin{cases} y_1' = y_1 + 3y_2 \\ y_2' = 3y_1 + y_2 \end{cases}$

3. $\begin{cases} y_1' = -y_2 \\ y_2' = -y_1 \end{cases}$

4. $\begin{cases} y_1' = 0.5y_1 \\ y_2' = y_1 - 0.5y_2 \end{cases}$

Exercises 5–6. Find the general solution of each system.

5. $\begin{bmatrix} y_1' \\ y_2' \end{bmatrix} = \begin{bmatrix} 3 & 9 \\ -1 & -3 \end{bmatrix} \begin{bmatrix} y_1 \\ y_2 \end{bmatrix}$

6. $\begin{bmatrix} y_1' \\ y_2' \end{bmatrix} = \begin{bmatrix} 2 & 1 \\ 0 & 2 \end{bmatrix} \begin{bmatrix} y_1 \\ y_2 \end{bmatrix}$

Exercises 7–8. Find the general solution to each system and a particular solution given initial conditions $y_1(0) = 3$, $y_2(0) = -1$, $y_3(0) = 1$.

7. $\begin{cases} y_1' = y_2 + y_3 \\ y_2' = y_1 + y_3 \\ y_3' = y_1 + y_2 \end{cases}$

8. $\begin{cases} y_1' = y_1 + y_2 - y_3 \\ y_2' = y_1 - y_2 + y_3 \\ y_3' = -y_1 + y_2 + y_3 \end{cases}$

9. Let $\mathbf{y}' = \mathbf{A}\mathbf{y}$ denote the given system (S). Verify that the eigenvalues of $\mathbf{A}$ are $\lambda = -1, 1, 1$. Find the general solution.

$$(S) \begin{cases} y_1' = y_2 \\ y_2' = y_1 \\ y_3' = y_3 \end{cases}$$

10. Let $\mathbf{y}' = \mathbf{A}\mathbf{y}$ denote the given system (S). Verify that the eigenvalues of $\mathbf{A}$ are $\lambda = 3$ (with multiplicity 3) and that the eigenspace E_3 has dimension 1, spanned by an eigenvector $\mathbf{v}_1 = (9, 3, 1)$. Find vectors $\mathbf{v}_2$ and $\mathbf{v}_3$ for which

$$\mathbf{y} = \left(\mathbf{v}_1 \frac{t^2}{2} + \mathbf{v}_2 t + \mathbf{v}_3\right) e^{\lambda t}$$

is a particular solution to (S).

$$(S) \begin{cases} y_1' = 9y_1 - 27y_2 + 27y_3 \\ y_2' = y_1 \\ y_3' = y_2 \end{cases}$$

Hint: Show that the problem reduces to solving the linear systems $(\mathbf{A} - \lambda\mathbf{I})\mathbf{v}_2 = \mathbf{v}_1$ and $(\mathbf{A} - \lambda\mathbf{I})\mathbf{v}_3 = \mathbf{v}_2$. Verify that the general solution of (S) is

$$\mathbf{y} = \alpha \begin{bmatrix} 9 \\ 3 \\ 1 \end{bmatrix} e^{3t} + \beta \left(\begin{bmatrix} 6 \\ 1 \\ 0 \end{bmatrix} t + \begin{bmatrix} 1 \\ 0 \\ 0 \end{bmatrix}\right) e^{3t}$$

$$+\gamma \left(\begin{bmatrix} 9 \\ 3 \\ 1 \end{bmatrix} \frac{t^2}{2} + \begin{bmatrix} 6 \\ 1 \\ 0 \end{bmatrix} t + \begin{bmatrix} 1 \\ 0 \\ 0 \end{bmatrix}\right) e^{3t},$$

where α, β, γ are scalars.

USING MATLAB

11. Consider the system of linear differential equations

$$(S) \begin{cases} y_1' = 5y_1 + 8y_2 + 16y_3 \\ y_2' = 4y_1 + y_2 + 8y_3 \\ y_3' = -4y_1 - 4y_2 - 11y_3 \end{cases}$$

Write an M-file that will compute the solution

$$\mathbf{y} = \begin{bmatrix} y_1 \\ y_2 \\ y_3 \end{bmatrix} = \begin{bmatrix} a_1 \\ a_2 \\ a_3 \end{bmatrix} e^{\lambda_1 t} + \begin{bmatrix} b_1 \\ b_2 \\ b_3 \end{bmatrix} e^{\lambda_2 t}.$$

Plot the solutions in the same figure window and label the graphs. Consult **help gtext**, and related commands.

12. Solve the system (S) by hand and find a particular solution given the initial conditions $y_1(0) = 500 \times 10^3$,

$y_2(0) = 300 \times 10^3$.

$$(S)\quad \begin{cases} y_1' = 0.6y_1 + 0.5y_2 \\ y_2' = -0.24y_1 + 1.4y_2 \end{cases}$$

Plot the solutions in the same figure window. Compare the continuous model with the discrete dynamical system (6.1). Compare the population in year 4 for continuous and discrete models.

13. Write an M-file to plot the growth of bacteria as in Figure 6.7.

14. For Exercises 1, 3, 5, 7, 9, write an M-file to plot the solutions in the same figure window.

CHAPTER 6 REVIEW

State whether each statement is true or false as stated. Provide a clear reason for your answer. If a false statement can be modified to make it true, then do so.

1. $\lambda = -1$ is an eigenvalue of $\mathbf{A} = \begin{bmatrix} 0 & 2 \\ 3 & 4 \end{bmatrix}$.

2. The vector $\mathbf{x} = \begin{bmatrix} 2 \\ 1 \end{bmatrix}$ is an eigenvector of $\mathbf{A} = \begin{bmatrix} 3 & -2 \\ 1 & 0 \end{bmatrix}$.

3. The sum of the eigenvalues of a square matrix $\mathbf{A}$ is $-$ trace $\mathbf{A}$.

4. The characteristic polynomial of $\begin{bmatrix} 1 & -2 \\ 1 & 1 \end{bmatrix}$ is $\lambda^2 - 2\lambda - 1$.

5. If $\lambda = 0$ is an eigenvalue of the square matrix $\mathbf{A}$, then $\mathbf{A}$ is invertible.

6. The nonzero vector $\mathbf{x}$ in $\mathbb{R}^n$ is an eigenvector of the matrix $\mathbf{x}\mathbf{x}^{\mathsf{T}}$.

7. The algebraic multiplicity of an eigenvalue is always less than or equal to its geometric multiplicity.

8. The matrix $\mathbf{P} = \begin{bmatrix} 1 & 1 \\ 2 & 1 \end{bmatrix}$ is a diagonalizing matrix for $\mathbf{A} = \begin{bmatrix} 4 & -1 \\ 2 & 1 \end{bmatrix}$.

9. Every invertible square matrix is diagonalizable.

10. Every diagonalizable matrix is invertible.

11. Applying an ERO to the matrix $\mathbf{A}$ does not change the eigenvalues of $\mathbf{A}$.

12. If $\mathbf{A}$ is 3×3 and has eigenvectors $[1\ 2\ -1]^{\mathsf{T}}$, $[1\ -3\ 1]^{\mathsf{T}}$, $[0\ -1\ -2]^{\mathsf{T}}$, then $\mathbf{A}$ is diagonalizable.

13. If $\mathbf{P}^{-1}\mathbf{AP} = \mathbf{D}$ for some diagonal matrix $\mathbf{D}$, then $\mathbf{D}$ is unique.

14. Diagonalization fails when $\mathbf{A}$ has only one eigenvector.

15. The matrix $\mathbf{A} = \begin{bmatrix} 0 & 2 \\ 2 & 0 \end{bmatrix}$ is diagonalizable.

16. There exist square matrices that have no eigenvalues.

17. Two square matrices of the same order can have the same characteristic polynomial.

18. The matrix $\begin{bmatrix} 1 & 1 \\ 0 & 1 \end{bmatrix}$ is defective.

19. A square matrix can never be both column stochastic and row stochastic.

20. The transition matrix for a regular Markov chain has eigenvalues less than or equal to 1.

Giuseppe Peano (1858–1932)

Peano was born near the town of Cuneo, Piemonte, in Northern Italy and studied mathematics at the University of Torino (Turin), where he was appointed professor in 1880.

Peano founded the field of *symbolic logic*, and the notation he developed in this field is standard today. His ultimate goal was to derive all mathematics from fundamental principles (axioms), an idea that would continue through the work of Bertrand Russell (1872–1970), Alfred North Whitehead (1861–1947), and others. In 1890 Peano published a set of axioms, now known as *Peano's axioms*, that define the natural numbers $1, 2, 3, \ldots$ in terms of sets and from which the real number system can be developed. He founded *Rivista di matematica* in 1891, a mathematical journal devoted to logic and related topics.

Peano gave new definitions for the length of an arc and the area of a curved surface. In 1890 he defined a continuous curve that passed through all the points in the unit square—something quite revolutionary and unbelievable at that time. This work is now part of a field of mathematics known as *topology*.

The concept of an abstract vector space became widely known during the 1920s through the publications of Hermann Weyl (1885–1955) and others. However, it is believed that it was Peano who gave the first axiomatic definition of a vector space in his book *Calcolo Geometrico*, published in Torino, Italy in 1888. Peano considered a set of abstract *objects* satisfying four axioms, introduced the idea of *independence* of objects, and defined the *dimension* of the space to be the maximum number of independent objects. Peano credited the work of Hermann Grassmann (1809–1877) during the 1840s and William Rowan Hamilton (1805–1865) in shaping his ideas.

Peano was interested in linguistics and in creating new languages. In 1903 he invented an artificial language, which he called he called *latino sine flexione* (later renamed *Interlingua*), and wrote a vocabulary compiled from words in English, French, German, and Latin.

Chapter 7

VECTOR SPACES

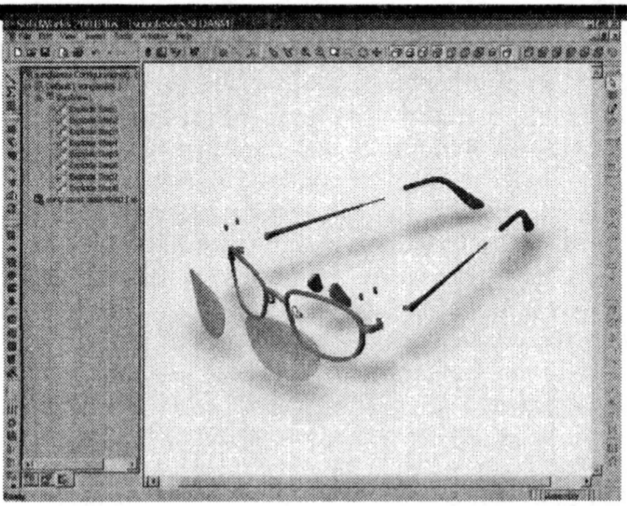

Introduction

The theory of vector spaces is one of many *axiomatic theories* of abstract algebra. To define a vector space we begin with a set of *objects*,[1] denoted by $\mathcal{V}$, the nature of which is purposely unspecified, as in Figure 7.1.

We assume that two algebraic operations, called *addition* and *scalar multiplication*, can be defined on $\mathcal{V}$ and that these operations satisfy the 10 *axioms* listed in Definition 7.1. Any set of objects that satisfies the definition can be called a *vector space*.

Particular examples of vector spaces are obtained by specifying what type of objects are in $\mathcal{V}$ and how the operations of addition and scalar multiplication are defined. We then check that the axioms in Definition 7.1 are satisfied. Some standard examples of vector spaces are given in Section 7.1. These include the sets $\mathbb{R}^{m\times n}$ of $m \times n$ matrices (Chapter 2) and the spaces $\mathbb{R}^n$ and their subspaces (Chapter 3). Other examples of vector spaces are defined using functions and polynomials as objects.

Algebraic properties of an abstract vector space are developed in a *systematic* way. By this we mean that the truth of each new property must be proved in a rigorous way using only the axioms or other properties that have already been established. Nothing can be assumed and each step in a proof must be justified. For example, consider the following simple property, which is labeled (**P4**) in Section 7.1.

$0\mathbf{v} = \mathbf{0}$ Scaling any object $\mathbf{v}$ in a vector space $\mathcal{V}$ by the scalar $s = 0$ results in the zero object $\mathbf{0}$ in $\mathcal{V}$. (7.1)

[1] It is traditional to refer to a member of $\mathcal{V}$ as a *vector*. However, by using the term *object* we hope to avoid identification with an $n \times 1$ matrix (column vector), which is only one of the many types of objects to be considered.

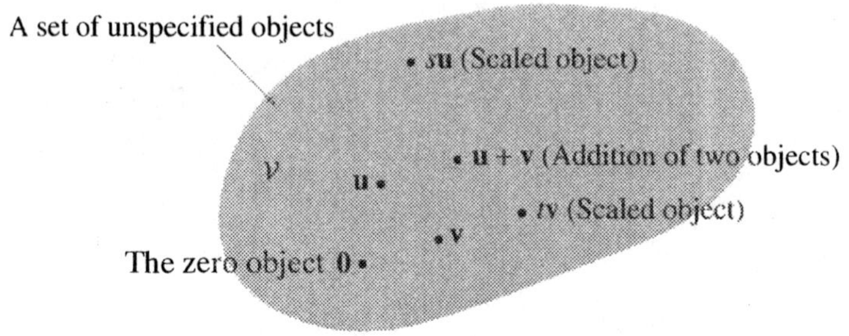

Figure 7.1 Schematic picture of a vector space.

This fact, which is trivial to prove in the vector space $\mathbb{R}^{m\times n}$ (multiplying all entries in a matrix by the scalar 0 gives the zero matrix $\mathbf{O}$), requires formal proof in the abstract setting because the object $\mathbf{v}$ is unspecified and the statement (7.1) is not part of Definition 7.1. Property (7.1) is proved in Illustration 7.1.

The theory of vector spaces is a *unifying* theory. Once a property is proved in the abstract setting, it no longer needs to be reproved in particular cases. In this way, vector space theory develops (unifies) the same algebraic concepts and properties that are found in a diverse collection of particular examples.

In many respects, the 19th century may be considered a golden age of mathematics. More advancement was made in this period in all fields, both qualitatively and quantitatively, than in any other century before it. The work of Cayley, Grassmann, Hamilton, Hilbert, Peano, and many others in the 19th century opened up the development of abstract algebra in the 20th century. In the last hundred years mathematicians have invented and developed many abstract algebraic structures (vector spaces, groups, rings, fields, modules, and so on), each one built up from a set of defining axioms. The development continues today, and there are important applications. Group theory, for example, has become a necessary prerequisite for understanding fundamental applications in physics and chemistry.

You are already familiar with many vector space concepts, such as *subspace*, *linear combination*, *span*, *linear independence*, *basis*, *dimension*, and so on, that have been discussed in Chapter 3 in the context of the particular vector space $\mathbb{R}^n$ and its subspaces. Our goal in this chapter is to extend these concepts into the general setting, making comparisons with the work in Chapter 3 and drawing on a wider collection of examples for illustration. Some proofs of theorems in Chapter 3 are valid in general by simply regarding a column vector $\mathbf{v}$ as a general object. Those proofs will not be repeated here.

7.1 Vector Spaces and Subspaces

Definition 7.1

Vector space

Let $\mathcal{V}$ be a set containing one or more *objects* $\mathbf{u}, \mathbf{v}, \mathbf{w}, \ldots$, and assume two operations, called *addition* and *scalar multiplication*, are defined on the objects in $\mathcal{V}$. The addition of two objects $\mathbf{u}$ and $\mathbf{v}$ is denoted by $\mathbf{u}+\mathbf{v}$ and the multiplication of an object $\mathbf{v}$ by a scalar s is denoted by $s\mathbf{v}$. Then $\mathcal{V}$ is called a *vector space* if the following 10 axioms are satisfied for all objects $\mathbf{u}$, $\mathbf{v}$, $\mathbf{w}$ in $\mathcal{V}$ and all scalars s, t.

Axioms of Closure

(**C1**) $\mathbf{u}+\mathbf{v}$ is an object in $\mathcal{V}$.
(**C2**) $s\mathbf{v}$ is an object in $\mathcal{V}$.

Axioms for Addition

(**A1**) $\mathbf{u}+\mathbf{v}=\mathbf{v}+\mathbf{u}$ (commutative law of addition)
(**A2**) $(\mathbf{u}+\mathbf{v})+\mathbf{w}=\mathbf{u}+(\mathbf{v}+\mathbf{w})$ (associative law of addition)
(**A3**) The set $\mathcal{V}$ contains a zero object, denoted by $\mathbf{0}$, with the property that $\mathbf{v}+\mathbf{0}=\mathbf{v}$, for every $\mathbf{v}$ in $\mathcal{V}$. We call $\mathbf{0}$ the *additive identity* for $\mathcal{V}$.
(**A4**) To each object $\mathbf{v}$ in $\mathcal{V}$ there corresponds an object $-\mathbf{v}$ in $\mathcal{V}$ with the property that $\mathbf{v}+(-\mathbf{v})=\mathbf{0}$. We call $-\mathbf{v}$ the *additive inverse* of $\mathbf{v}$.

Axioms for Scalar Multiplication

(**A5**) $(s+t)\mathbf{v}=s\mathbf{v}+t\mathbf{v}$
(**A6**) $s(\mathbf{u}+\mathbf{v})=s\mathbf{u}+s\mathbf{v}$
(**A7**) $(st)\mathbf{v}=s(t\mathbf{v})$
(**A8**) $1\mathbf{v}=\mathbf{v}$

Axioms (**C1**) and (**C2**) say that $\mathcal{V}$ is *closed with respect to addition and scalar multiplication*—adding two objects in $\mathcal{V}$ results in an object in $\mathcal{V}$, multiplying an object in $\mathcal{V}$ by any scalar results in an object in $\mathcal{V}$.

Addition is a *binary* operation, acting on two objects in $\mathcal{V}$ at a time. Axiom (**A1**) says that the order in which the addition of $\mathbf{u}$ and $\mathbf{v}$ takes place is immaterial and the final result is the same. Axiom (**A2**) is concerned with forming the sum $\mathbf{u}+\mathbf{v}+\mathbf{w}$ of three objects. It says that we are at liberty to form $(\mathbf{u}+\mathbf{v})$ first and add $\mathbf{w}$, or form $(\mathbf{v}+\mathbf{w})$ first and add $\mathbf{u}$. The final result is the same.

We must make the following important distinction. If the scalars in Definition 7.1 are real numbers, then $\mathcal{V}$ is called a *real vector space*. If the scalars are complex numbers, then $\mathcal{V}$ is called a *complex vector space*. A brief comparison of the real and complex cases is given in Section 8.5. This chapter deals with real vector spaces.

Some Standard Examples

The unifying power of vector space theory can only be appreciated by reviewing a rich assortment of examples of vector spaces. Some of the standard examples are given next and other examples are introduced later in the section. In each case, a nonempty set of objects is specified together with the operations of addition and scalar multiplication that are to be used for that set. If the ten axioms in Definition 7.1 are satisfied, then $\mathcal{V}$ together with its operations can be called a vector space.

EXAMPLE 1

The Vector Spaces $\mathbb{R}^{m \times n}$

For each pair of positive integers (m, n), the set $\mathcal{V} = \mathbb{R}^{m \times n}$ of real $m \times n$ matrices (entries are real numbers) is a vector space. The objects in $\mathcal{V}$ are matrices and the operations on $\mathcal{V}$ are matrix operations—if $\mathbf{A}$ and $\mathbf{B}$ are in $\mathcal{V}$, add corresponding entries to get $\mathbf{A}+\mathbf{B}$ and multiply all the entries in $\mathbf{A}$ by the scalar s to get $s\mathbf{A}$. The closure axioms are satisfied because adding two $m \times n$ matrices results in an $m \times n$ matrix and scaling an $m \times n$ matrix results in an $m \times n$ matrix. The zero object in $\mathcal{V}$ is the $m \times n$ matrix $\mathbf{O}$ with all entries equal to zero. The additive inverse of $\mathbf{A}$ is $-\mathbf{A}$, which is obtained by negating all the entries in $\mathbf{A}$. Verifying the axioms (**A1**)–(**A8**) are satisfied is straightforward (see Section 2.1). ■

Example 1 includes the following two special cases:

(a) For each n, the set $\mathbb{R}^n$ of $n \times 1$ matrices (column vectors) is identical with $\mathbb{R}^{n \times 1}$ and so $\mathbb{R}^n$ is a vector space,

(b) For each n, the set of $1 \times n$ matrices (row vectors) is identical with $\mathbb{R}^{1 \times n}$ and is called the *vector space of row vectors* with n components.

Although the vector spaces $\mathbb{R}^{n \times 1}$ and $\mathbb{R}^{1 \times n}$ are formally different (their objects are not the same), they are nevertheless "structurally" the same—for this reason the spaces are said to be *isomorphic* (see Section 7.3 and Exercises 7.1).

EXAMPLE 2

Real-valued Functions of a Real Variable

The set of all real-valued functions defined on the real line $\mathbb{R}$ is denoted by $\mathbf{F}(-\infty, \infty)$. The objects in $\mathbf{F}(-\infty, \infty)$ are functions. Addition and scalar multiplication of functions are *pointwise* operations. For example, consider the objects

$$\mathbf{f}_1 = t, \qquad \mathbf{f}_2 = \cos t, \qquad \mathbf{f}_3 = e^{-t}.$$

The sum $\mathbf{g} = \mathbf{f}_1 + \mathbf{f}_2$ and scalar multiplication $\mathbf{h} = -2\mathbf{f}_3$, where $s = -2$, are defined by performing the arithmetic operations

$$\mathbf{g}(t) = \mathbf{f}_1(t) + \mathbf{f}_2(t) = t + \cos t \quad \text{and} \quad \mathbf{h}(t) = -2\,\mathbf{f}_3(t) = -2e^{-t}, \tag{7.2}$$

for *each* real value t; that is, at every point on the t-axis. The closure axioms are satisfied because $\mathbf{f}+\mathbf{g}$ and $s\mathbf{f}$ are real-valued functions for all $\mathbf{f}$, $\mathbf{g}$ in $\mathbf{F}$ and all scalars s. Each object $\mathbf{f}$ in $\mathbf{F}$ may be visualized by its graph $y = \mathbf{f}(t)$. The zero object $\mathbf{0}$ in $\mathbf{F}(-\infty, \infty)$ is the *zero function* defined by $\mathbf{0}(t) = 0$, for all t, and $\mathbf{0}$ is visualized as the t-axis. The additive inverse of $\mathbf{f}$ is the object $-\mathbf{f}$ that is visualized as the reflection of the graph of $y = \mathbf{f}(t)$ in the t-axis. It is routine to check that the 10 axioms in Definition 7.1 are satisfied and consequently $\mathbf{F}(-\infty, \infty)$ is a vector space.

Let a and b be real numbers with $a < b$. The set of all real numbers t such that $a \le t \le b$ is called a *closed interval* (the endpoints are included) on the real line $\mathbb{R}$ and is denoted by $[a, b]$. The set of real numbers t such that $a < t < b$ is called an *open interval* and denoted by (a, b).

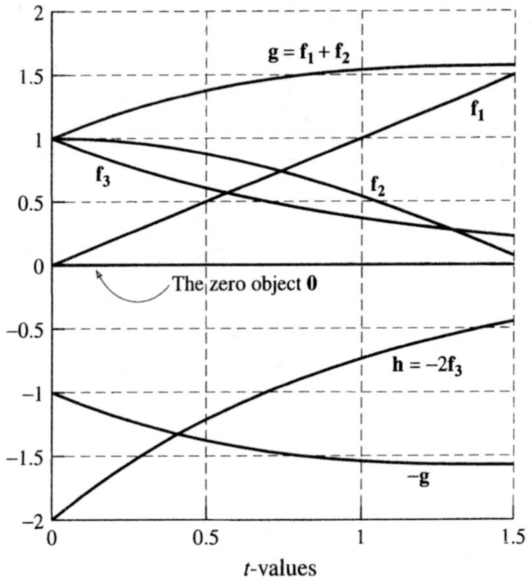

Figure 7.2 Some functions in the vector space **F**[0, 1.5].

The set **F**$[a, b]$ of all real-valued functions defined on the interval $[a, b]$ is a vector space using the same operations as in **F**$(-\infty, \infty)$. Note that Definition 7.1 is automatically satisfied (explain). Similarly, **F**(a, b) is a vector space.

If the domain of each function in (7.2) is restricted to the interval [0, 1.5], then the functions become objects in **F**[0, 1.5]. Figure 7.2 shows a plot of these functions. The object **g** and its additive inverse $-\mathbf{g}$ are shown and we see that $\mathbf{g} + (-\mathbf{g}) = \mathbf{0}$ (the zero object), thus illustrating axiom (**A4**). ■

Specifying a set of objects $\mathcal{V}$ and corresponding operations on objects in $\mathcal{V}$ does not guarantee, ipso facto, that the structure will be a vector space. Quite often some of the 10 axioms are satisfied while others fail, as we show next.

EXAMPLE 3

Failure of Some Axioms

Consider the set $\mathcal{V}$ of ordered pairs (x, y) of real numbers with the operations of addition, denoted by $+$, and multiplication by any scalar s defined as follows:

$$(x_1, y_1) + (x_2, y_2) = (x_1 + x_2, y_1 + y_2), \quad s(x, y) = (2sx, 2sy).$$

The closure axioms are clearly satisfied. The definition of addition is the usual (componentwise) operation on ordered pairs and axioms (**A1**) through (**A4**) are satisfied—$(0, 0)$ is the zero object in $\mathcal{V}$ and $(-x, -y)$ is the additive inverse of (x, y). Axioms (**A5**) and (**A6**) are satisfied while (**A7**) fails (Exercises 7.1). Clearly, (**A8**) fails because $1(x, y) = (2x, 2y) \neq (x, y)$. Thus $\mathcal{V}$ is not a vector space. ■

Elementary Algebraic Properties

The abstract nature of a vector space dictates that we cannot use our knowledge about some particular vector space to assert that an algebraic property is true in general. Remember that the objects are unspecified! Proof is required in every case.

The elementary properties listed next are true for any vector space $\mathcal{V}$ and are proved either directly from Definition 7.1 or using properties already established. Let $\mathbf{u}$, $\mathbf{v}$, $\mathbf{w}$ be objects in $\mathcal{V}$ and let s be any scalar.

(**P1**) *Cancelation Law* If $\mathbf{u}+\mathbf{w}=\mathbf{v}+\mathbf{w}$ or $\mathbf{w}+\mathbf{u}=\mathbf{w}+\mathbf{v}$, then $\mathbf{u}=\mathbf{v}$.
(**P2**) If $\mathbf{u}+\mathbf{v}=\mathbf{u}$, then $\mathbf{v}=\mathbf{0}$.
(**P3**) If $\mathbf{u}+\mathbf{v}=\mathbf{0}$, then $\mathbf{v}=-\mathbf{u}$.
(**P4**) $0\mathbf{v}=\mathbf{0}$
(**P5**) $s\mathbf{0}=\mathbf{0}$
(**P6**) $(-1)\mathbf{v}=-\mathbf{v}$
(**P7**) $s\mathbf{v}=\mathbf{0}$ implies either $s=0$ or $\mathbf{v}=\mathbf{0}$.
(**P8**) The zero object is unique.
(**P9**) The additive inverse $-\mathbf{v}$ of $\mathbf{v}$ is unique.

We illustrate next the typical method of proof required for these and similar properties.

▪ ILLUSTRATION 7.1

Proving (**P2**) *and* (**P4**)

We prove (**P2**), justifying each step. Given $\mathbf{u}+\mathbf{v}=\mathbf{u}$, we have

$$-\mathbf{u}+(\mathbf{u}+\mathbf{v})=-\mathbf{u}+\mathbf{u} \quad \text{adding } -\mathbf{u} \text{ to both sides}$$

$$(-\mathbf{u}+\mathbf{u})+\mathbf{v}=-\mathbf{u}+\mathbf{u} \quad \text{by } (\mathbf{A2})$$

$$\mathbf{0}+\mathbf{v}=\mathbf{0} \quad \text{by } (\mathbf{A4})$$

$$\mathbf{v}=\mathbf{0} \quad \text{by } (\mathbf{A3}).$$

QED (quod erat demonstrandum, meaning *that which was to be proved*).

We will prove (**P4**) from the axioms using (**P2**). Expand $(1+0)\mathbf{v}$ in two different ways.

$$(1+0)\mathbf{v}=1\mathbf{v}+0\mathbf{v} \quad \text{by } (\mathbf{A5})$$

$$(1+0)\mathbf{v}=1\mathbf{v} \quad \text{because } 1+0=1$$

Hence $1\mathbf{v}+0\mathbf{v}=1\mathbf{v}$ and by (**P2**), we have $0\mathbf{v}=\mathbf{0}$. QED. ▪

Subspaces

Some subsets of a vector space $\mathcal{V}$ are vector spaces in their own right using the operations of addition and scalar multiplication they inherited from $\mathcal{V}$. For example, we have seen already in Chapter 3 that the set of all scalar multiples of a column vector $\mathbf{v}$ in $\mathbb{R}^n$ is a one-dimensional subspace of $\mathbb{R}^n$ and later in this section it will be seen that the set of all $n\times n$ upper triangular matrices is a subspace of the vector space $\mathbb{R}^{n\times n}$ of all $n\times n$ matrices.

Definition 7.2

Subspace

Let $\mathcal{V}$ be a vector space and let $\mathcal{U}$ be a set containing one or more objects from $\mathcal{V}$. Then $\mathcal{U}$ is a subspace of $\mathcal{V}$ if the following two conditions are satisfied:

(S1) $\mathbf{u} + \mathbf{v}$ belongs to $\mathcal{U}$ for all objects $\mathbf{u}$ and $\mathbf{v}$ in $\mathcal{U}$. We say that $\mathcal{U}$ is *closed with respect to addition*,

(S2) $s\mathbf{v}$ belongs to $\mathcal{U}$ for all objects $\mathbf{v}$ in $\mathcal{U}$ and all scalars s. We say that $\mathcal{U}$ is *closed with respect to scalar multiplication.*

Figure 7.3 shows a schematic picture of a subspace $\mathcal{U}$ in an abstract vector space $\mathcal{V}$.

It remains to be shown that a subspace $\mathcal{U}$ satisfies the 10 axioms in Definition 7.1 and so $\mathcal{U}$ is a vector space in its own right. Only axioms (**A3**) and (**A4**) are of concern—the rest follow easily because these axioms are valid in $\mathcal{V}$ and so, given the closure conditions on $\mathcal{U}$, are valid in $\mathcal{U}$.

Theorem 7.1

Verifying Axioms (A3) and (A4) for a Subspace $\mathcal{U}$

Let $\mathcal{U}$ be a subspace of the vector space $\mathcal{V}$.

(a) The zero object $\mathbf{0}$ in $\mathcal{V}$ belongs to $\mathcal{U}$. Hence (**A3**) is satisfied.

(b) If $\mathbf{v}$ is an object in $\mathcal{U}$, then $-\mathbf{v}$ is also in $\mathcal{U}$. Hence (**A4**) is satisfied.

Proof (a) By (**S1**), the scalar multiple of an object $\mathbf{v}$ in $\mathcal{U}$ is again $\mathcal{U}$. Hence, multiplying by the scalar $s = 0$, we have $0\mathbf{v} = \mathbf{0}$, by property (**P4**), and so $\mathbf{0}$ is in $\mathcal{U}$. But $\mathbf{v} + \mathbf{0} = \mathbf{v}$ for every object in $\mathcal{V}$ and so the same property holds for all objects in $\mathcal{U}$. Using (**P6**), we have $(-1)\mathbf{v} = -\mathbf{v}$ is in $\mathcal{U}$ and so (b) follows.

Many interesting vector spaces appear as subspaces of the standard vector spaces mentioned so far. In particular, the vector space $\mathbf{F}(-\infty, \infty)$ contains many sets of specialized functions—continuous, differentiable, polynomial, and so on, that are used in defining subspaces of $\mathbf{F}(-\infty, \infty)$.

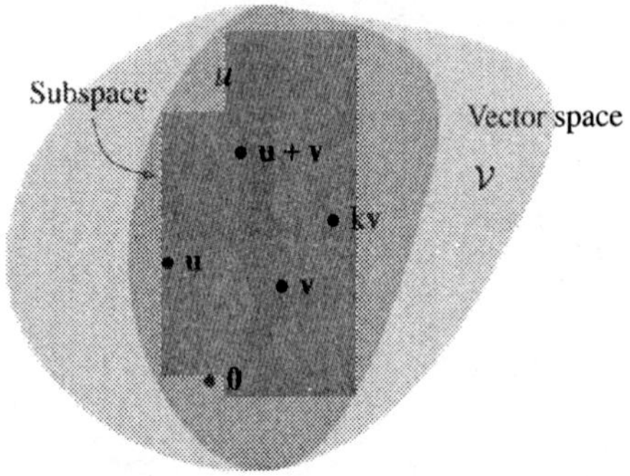

Figure 7.3 Subspace of a vector space.

EXAMPLE 4

Polynomials

Let n be a positive integer. The symbol $\mathbf{P}_n$ denotes the set of all polynomial functions

$$\mathbf{p} = a_0 + a_1 t + \cdots + a_n t^n, \tag{7.3}$$

defined for all real values of t. The coefficients $a_0, \ldots, a_n$ are real numbers and the positive integer n, called the *degree* of $\mathbf{p}$, is the highest power of t occurring in (7.3). Note that some (or even all) of the coefficients in (7.3) can be zero and so $\mathbf{P}_n$ actually contains all polynomials of degree n or less. For example, $\mathbf{P}_2$ contains all polynomials of the form $\mathbf{p} = a_0 + a_1 t + a_2 t^2$ and so, for example, the polynomials

$$\mathbf{p}_1 = 6, \quad \mathbf{p}_2 = 2 + 3t, \quad \mathbf{p}_3 = (1-t)^2, \quad \mathbf{p}_4 = -t + 3t^2,$$

all belong to $\mathbf{P}_2$. In this case, we have

$$\begin{aligned}\mathbf{p}_1 + \mathbf{p}_3 &= (6 + 0t + 0t^2) + (1 - 2t + t^2)\\ &= (6+1) + (0-2)t + (0+1)t^2 = 7 - 2t + t^2\\ \text{and} \quad -2\mathbf{p}_4 &= -2(0 - t + 3t^2) = 2t - 6t^2.\end{aligned}$$

Consider objects $\mathbf{p}$ and $\mathbf{q}$ in $\mathbf{P}_n$, where

$$\mathbf{p} = a_0 + a_1 t + \cdots + a_n t^n \quad \text{and} \quad \mathbf{q} = b_0 + b_1 t + \cdots + b_n t^n.$$

The operations of addition and multiplication by a scalar s in $\mathbf{P}_n$ are defined by

$$\begin{aligned}\mathbf{p} + \mathbf{q} &= (a_0 + b_0) + (a_1 + b_1)t + \cdots + (a_n + b_n)t^n\\ s\mathbf{p} &= sa_0 + sa_1 t + \cdots + sa_n t^n,\end{aligned} \tag{7.4}$$

which shows that the axioms of closure in $\mathbf{P}_n$ are satisfied because addition of two polynomials of degree at most n results in a polynomial of degree at most n, and similarly for scalar multiplication. The zero object in $\mathbf{P}_n$ is the polynomial $\mathbf{0} = 0 + 0t + \cdots + 0t^n$, for all values of t and the additive inverse of $\mathbf{p} = a_0 + a_1 t + \cdots + a_n t^n$ is $-\mathbf{p} = -a_0 - a_1 t - \cdots - a_n t^n$.

If we think of $\mathbf{p} = \mathbf{p}(t)$ as a real-valued function defined for all real values t, then the objects in $\mathbf{P}_n$ can be visualized geometrically as the graphs of functions, as in Figure 7.2. Addition and scalar multiplication are pointwise operations on functions and the zero object in $\mathbf{P}_n$ is the zero function, visualized as the t-axis. Thus $\mathbf{P}_n$ is a subset of $\mathbf{F}(-\infty, \infty)$ and, furthermore, (7.4) shows that conditions (**S1**) and (**S2**) are satisfied, indicating that $\mathbf{P}_n$ is a subspace of $\mathbf{F}(-\infty, \infty)$. ■

EXAMPLE 5

Subspaces of Upper Triangular Matrices

Consider the set $\mathcal{U}$ of upper triangular 2×2 matrices. Then $\mathcal{U}$ is a subset of $\mathbb{R}^{2\times 2}$ consisting of all matrices of the form

$$\mathbf{A} = \begin{bmatrix} a & b \\ 0 & c \end{bmatrix}, \quad \text{where } a, b, c \text{ are real numbers.} \tag{7.5}$$

Checking the conditions in Definition 7.2, we have

(S1) If $\mathbf{A}_1$, $\mathbf{A}_2$ are in $\mathcal{U}$, then

$$\mathbf{A}_1 + \mathbf{A}_2 = \begin{bmatrix} a_1 & b_1 \\ 0 & c_1 \end{bmatrix} + \begin{bmatrix} a_2 & b_2 \\ 0 & c_2 \end{bmatrix} = \begin{bmatrix} a_1 + a_2 & b_1 + b_2 \\ 0 & c_1 + c_2 \end{bmatrix},$$

and so $\mathbf{A}_1 + \mathbf{A}_2$ is in $\mathcal{U}$.

(S2) Let $\mathbf{A}$ be in $\mathcal{U}$ and let s be any scalar. Then

$$s\mathbf{A} = s\begin{bmatrix} a & b \\ 0 & c \end{bmatrix} = \begin{bmatrix} sa & sb \\ 0 & sc \end{bmatrix},$$

and so $s\mathbf{A}$ is in $\mathcal{U}$.

Hence, $\mathcal{U}$ is a subspace of $\mathbb{R}^{2\times 2}$ and, more generally, the set of all $n \times n$ upper triangular matrices is a subspace of $\mathbb{R}^{n\times n}$. ■

Every vector space $\mathcal{V}$ contains two special subspaces that correspond to extreme possibilities. The subset $\{\mathbf{0}\}$ consisting of only the zero object from $\mathcal{V}$ is a subspace of $\mathcal{V}$. We call this the *zero*, or *trivial*, subspace. Also, $\mathcal{V}$ is a subspace of itself and we refer to $\mathcal{V}$ as the *whole space*. Any subspace that is neither the zero subspace nor the whole space is called a *proper subspace* of $\mathcal{V}$. For example, the subspace $\mathcal{U}$ in Example 5 is a proper subspace of $\mathbb{R}^{2\times 2}$.

Historical Notes

Hermann Grassmann (1809–1877) Polish mathematician. We believe that Grassmann's most important work, *Die lineale Ausdehnungslehre, ein neuer Zweig der Mathematik* (1844), is the origin of many of the concepts in linear algebra as we know it today. His papers were notoriously difficult to understand and he consequently became disappointed with the interest his work generated. He wrote on a wide variety of scientific subjects and in later years produced a Sanskrit dictionary that is still used today.

Lord Bertrand Arthur William Russell (1872–1970) English mathematician, logician, philosopher, and writer. Russell was a Fellow and lecturer at Cambridge University (1910–1916) and was dismissed and later jailed (1918) for his pacifist statements during World War I. He is famous for his outstanding and fundamental work in the foundations of mathematics (logic). In 1902 he stated *Russell's Paradox* in axiomatic set theory and in 1919 gave the following popular version of the paradox:

> There is a village in which the barber shaves every man who does not shave himself. Does the barber shave himself?

Russell was awarded the Nobel Prize for Literature in 1950 "in recognition of his varied and significant writings in which he champions humanitarian ideals and freedom of thought."

Alfred North Whitehead (1861–1947) English mathematician. Whitehead coauthored the work *Principia Mathematica* (3 Vols., 1910–1913) with Bertrand Russell and took axiomatic systems to the extreme. This great work, based on Peano's axioms, attempted (unsuccessfully) to show that all of pure mathematics can be derived from propositional logic alone. Whitehead is quoted as saying that "mathematics as a science, commenced when first someone, probably a Greek, proved propositions about 'any' things or about 'some' things, without specifications of definite particular things."

EXERCISES 7.1

1. Refer to Example 3. Prove that axioms (**A5**) and (**A6**) are satisfied and that (**A7**) fails.
2. Let $\mathcal{U}$ be a subset of a vector space $\mathcal{V}$ containing one or more objects. Show that $\mathcal{U}$ is a subspace of $\mathcal{V}$ if and only if $\alpha\mathbf{u} + \beta\mathbf{v}$ is in $\mathcal{U}$ for all objects $\mathbf{u}$ and $\mathbf{v}$ in $\mathcal{U}$ and all scalars α and β.

Exercises 3–9. *Prove each property using the given axioms in order, when specified.*

3. (**P1**): Use (**A2**), (**A4**), (**A3**).
4. (**P3**): Use (**P1**).
5. (**P5**): Expand $s(\mathbf{0} + \mathbf{0})$ in two ways and use (**P2**).
6. (**P6**): Expand $[1 + (-1)]\mathbf{v}$ in two ways and use (**P3**).
7. (**P7**)
8. (**P8**)
9. (**P9**)

Exercises 10–25. *In each case a subset $\mathcal{U}$ of $\mathbb{R}^{2\times 2}$ is defined, where a, b, c, d are any real scalars. Determine if $\mathcal{U}$ is a subspace of $\mathbb{R}^{2\times 2}$. When possible, generalize to the vector space $\mathbb{R}^{n\times n}$ of $n \times n$ matrices.*

10. The scalar matrices $\begin{bmatrix} a & 0 \\ 0 & a \end{bmatrix}$
11. The diagonal matrices $\begin{bmatrix} a & 0 \\ 0 & b \end{bmatrix}$
12. The upper triangular matrices $\begin{bmatrix} a & a \\ 0 & a \end{bmatrix}$
13. The lower triangular matrices $\begin{bmatrix} a & 0 \\ b & c \end{bmatrix}$
14. All matrices of the form $\begin{bmatrix} a & b \\ c & 0 \end{bmatrix}$
15. All matrices of the form $\begin{bmatrix} 0 & b \\ c & a \end{bmatrix}$
16. All invertible matrices
17. All singular matrices
18. All matrices $\mathbf{A}$ such that $\mathbf{A}^k = \mathbf{O}$, for some positive integer k. These are the 2×2 nilpotent matrices.
19. All matrices $\mathbf{A}$ such that $\mathbf{A}^2 = \mathbf{A}$. These are the 2×2 idempotent matrices.
20. All matrices $\begin{bmatrix} a & b \\ c & d \end{bmatrix}$ such that $a + d = 0$
21. All matrices of the form $\begin{bmatrix} a & b \\ b & a \end{bmatrix}$
22. All matrices of the form $\begin{bmatrix} a & b \\ 0 & 0 \end{bmatrix}$
23. The symmetric matrices $\begin{bmatrix} a & b \\ b & c \end{bmatrix}$
24. The skew-symmetric matrices $\begin{bmatrix} 0 & b \\ -b & 0 \end{bmatrix}$
25. All matrices of the form $\begin{bmatrix} a & b \\ -b & a \end{bmatrix}$

Exercises 26–33. *Determine if the given subset $\mathcal{U}$ is a subspace of the vector space indicated (a, b, c are arbitrary real numbers).*

26. $\mathbf{P}_2$: $\mathcal{U} = \{at + bt^2\}$
27. $\mathbf{P}_2$: $\mathcal{U} = \{a + bt + ct^2\}$, where $a + 2b + 3c = 0$
28. $\mathbf{P}_4$: $\mathcal{U} = \{a + bt\}$
29. $\mathbf{P}_n$: $\mathcal{U} = \{\mathbf{p} \mid \mathbf{p}(0) = 0\}$
30. The subset of $\mathbf{F}(-\infty, \infty)$ consisting of all functions $\mathbf{f}$ such that $\mathbf{f}(0) = 0$
31. The subset of $\mathbf{F}(-\infty, \infty)$ consisting of all constant functions.
32. The subset of $\mathbf{F}(-\infty, \infty)$ consisting of all functions $\mathbf{f}$ such that $\mathbf{f}(0) = 1$.
33. The set of all differentiable functions in $\mathbf{F}[0, 1]$.

Exercises 34–37. *Let $\mathcal{V}$ be the set of ordered pairs of real numbers (x, y). Addition on $\mathcal{V}$ is defined by $(x_1, y_1) + (x_2, y_2) = (x_1 + x_2, y_1 + y_2)$. Scalar multiplication is defined by the condition given in each exercise. Explain why the set $\mathcal{V}$ with* *the given operations is not vector space. List the axioms that fail.*

34. $s(x, y) = (0, 0)$

35. $s(x, y) = (sx, 0)$

36. $s(x, y) = \left(\frac{x}{s}, \frac{y}{s}\right)$, $s \neq 0$, $0(x, y) = (0, 0)$

37. $s(x, y) = (x, y)$

38. $\mathbf{P}_n$ is the vector space of all polynomials of degree n or less. A polynomial $\mathbf{p}$ is called *even* if $\mathbf{p}(-t) = \mathbf{p}(t)$ for all real t, and it is called *odd* if $\mathbf{p}(-t) = -\mathbf{p}(t)$ for all real t. Let $\mathcal{E}_n$ and $\mathcal{O}_n$ be the subsets of even and odd polynomials in $\mathbf{P}_n$, respectively. Show that $\mathcal{E}_n$ and $\mathcal{O}_n$ are subspaces of $\mathbf{P}_n$.

39. Let $\mathcal{U}$ and $\mathcal{W}$ be subspaces of a vector space $\mathcal{V}$. Define subsets $\mathcal{U} + \mathcal{W}$ and $\mathcal{U} \cap \mathcal{W}$ of $\mathcal{V}$ as follows:

$$\mathcal{U} + \mathcal{W} = \{\mathbf{v} | \mathbf{v} = \mathbf{u} + \mathbf{w},\ \mathbf{u} \text{ in } \mathcal{U} \text{ and } \mathbf{w} \text{ in } \mathcal{W}\}$$

$$\mathcal{U} \cap \mathcal{W} = \{\mathbf{v} \mid \mathbf{v} \text{ in } \mathcal{U} \text{ and in } \mathcal{W}\}.$$

Show that $\mathcal{U} + \mathcal{W}$ and $\mathcal{U} \cap \mathcal{W}$ are both subspaces of $\mathcal{V}$.

40. Refer to the previous exercise. Let $\mathcal{U}$ and $\mathcal{W}$ be subsets of $\mathbb{R}^3$ defined as follows:

$$\mathbf{v} = \begin{bmatrix} x_1 \\ x_2 \\ x_3 \end{bmatrix}, \quad \begin{aligned} \mathcal{U} &= \{\mathbf{v} \mid \ x_1 + 2x_2 + x_3 = 0\} \\ \mathcal{W} &= \{\mathbf{v} \mid -2x_1 - 2x_2 + 3x_3 = 0\}. \end{aligned}$$

Show that $\mathcal{U}$ and $\mathcal{W}$ are subspaces of $\mathbb{R}^3$ and describe the subspaces $\mathcal{U} + \mathcal{W}$ and $\mathcal{U} \cap \mathcal{W}$.

41. If possible, find two subspaces $\mathcal{U}$ and $\mathcal{W}$ of $\mathbb{R}^{2\times 2}$ that have only the zero object in common and such that $\mathcal{U} + \mathcal{W} = \mathbb{R}^{2\times 2}$.

42. Let $\mathcal{U}$ and $\mathcal{W}$ be two subspaces of a vector space $\mathcal{V}$. Give an example to show that the union $\mathcal{U} \cup \mathcal{W}$ consisting of all objects in $\mathcal{U}$ and in $\mathcal{W}$ may not be a subspace of $\mathcal{V}$.

43. Consider the subset $\mathcal{U}$ of $\mathbb{R}^{n\times n}$ consisting of all matrices $\mathbf{A}$ such that trace $\mathbf{A} = 0$ (see Exercises 6.1 for properties of trace). Show that $\mathcal{U}$ is a proper subspace of $\mathbb{R}^{2\times 2}$.

44. Show that the set of all 2×2 matrices $\mathbf{A}$ such that $\det(\mathbf{A}) \geq 0$ is not a subspace of $\mathbb{R}^{2\times 2}$. What happens if the condition on $\mathbf{A}$ is modified to read $\det(\mathbf{A}) = 0$ or $\det(\mathbf{A}) \leq 0$?

45. Let $\mathcal{V}$ be the set of vectors in $\mathbb{R}^2$ with the following nonstandard operations of addition (denoted by $\oplus$) and scalar multiplication (denoted by $*$) are defined on $\mathcal{V}$ via the equations

$$\begin{bmatrix} x_1 \\ y_1 \end{bmatrix} \oplus \begin{bmatrix} x_2 \\ y_2 \end{bmatrix} = \begin{bmatrix} x_1 + x_2 \\ y_1 + y_2 \end{bmatrix} + \begin{bmatrix} 1 \\ 0 \end{bmatrix}$$

$$s * \begin{bmatrix} x \\ y \end{bmatrix} = \begin{bmatrix} sx \\ sy \end{bmatrix} + \begin{bmatrix} s-1 \\ 0 \end{bmatrix}.$$

The operations are the standard ones followed by a translation parallel to the x_1-axis. Show that $\mathcal{V}$ is a vector space.

46. Explain the popular version of Russell's paradox that is stated at the end of this section.

Exercises 47–50. *Label each statement as either true (T) or false (F). If a false statement can be made true by modifying the wording, then do so.*

47. If $\mathbf{v}$ is a nonzero object in a vector space $\mathcal{V}$ and $\alpha\mathbf{v} = \beta\mathbf{v}$ for some scalars α and β, then $\alpha = \beta$.

48. The set $\mathcal{S}$ of matrices of the form $\begin{bmatrix} a & 2 \\ b & c \end{bmatrix}$ is not a subspace of $\mathbb{R}^{2\times 2}$.

49. $\mathbf{F}[1, 2]$ is a subspace of the vector space $\mathbf{F}$.

50. The set of functions $\mathbf{f}$ in $\mathbf{F}$ such that $\mathbf{f}(x) = -\mathbf{f}(x)$ for every real value x is a subspace of $\mathbf{F}$.

USING MATLAB

51. Write an M-file to plot the polynomials

$$\mathbf{p}_1 = 6, \quad \mathbf{p}_2 = 2 + 3t, \quad \mathbf{p}_3 = (1-t)^2, \quad \mathbf{p}_4 = -t + 3t^2,$$

in the same figure window and label the graphs. Add the plots of $\mathbf{p}_5 = -2\mathbf{p}_2$, $\mathbf{p}_6 = 3\mathbf{p}_2 + 4\mathbf{p}_3$, and $\mathbf{p}_7 = \mathbf{p}_1 + \mathbf{p}_2 + \mathbf{p}_3 + \mathbf{p}_4$.

52. Reproduce Figure 7.2.

7.2 Linear Independence, Basis, Dimension

The concepts of span, linear independence, basis, and dimension were studied in Sections 3.1 and 3.2 in the context $\mathbb{R}^n$ and its subspaces. Some theorems and proofs in Chapter 3 apply to any vector space if the column vector $\mathbf{v}$ is regarded as an unspecified object. We will not repeat the details of Chapter 3 here, but rephrase the concepts from Sections 3.1 and 3.2 in the context of an abstract vector space, drawing on examples of different vector spaces for illustration. Along the way we will fill in some proofs of theorems left outstanding from Section 3.2.

Linear Combinations

Consider the unspecified objects $\mathbf{v}_1$ and $\mathbf{v}_2$ in a vector space $\mathcal{V}$ and scalars x_1 and x_2. By (**C2**), $x_1\mathbf{v}_1$ and $x_2\mathbf{v}_2$ are objects in $\mathcal{V}$, and by (**C1**), $x_1\mathbf{v}_1 + x_2\mathbf{v}_2$ is an object in $\mathcal{V}$. Given a third object $\mathbf{v}_3$ in $\mathcal{V}$ and scalar x_3, the same argument shows that

$$(x_1\mathbf{v}_1 + x_2\mathbf{v}_2) + x_3\mathbf{v}_3 = x_1\mathbf{v}_1 + x_2\mathbf{v}_2 + x_3\mathbf{v}_3$$

is an object in $\mathcal{V}$, and so on—we can add a scalar multiple of a new object each time until a (finite) sum with k summands is achieved.

Definition 7.3

Linear combination, span

Let $\mathcal{S} = \{\mathbf{v}_1, \mathbf{v}_2, \ldots, \mathbf{v}_k\}$ be a set of objects in a vector space $\mathcal{V}$. The object

$$x_1\mathbf{v}_1 + x_2\mathbf{v}_2 + \cdots + x_k\mathbf{v}_k, \tag{7.6}$$

is called a *linear combination* of $\mathbf{v}_1, \mathbf{v}_2, \ldots, \mathbf{v}_k$ using scalars $x_1, x_2, \ldots, x_k$. The set of *all* linear combinations (7.6), as the scalars $x_1, x_2, \ldots, x_k$ range over all possible real values, is called the *span* of $\{\mathbf{v}_1, \mathbf{v}_2, \ldots, \mathbf{v}_k\}$ and is denoted by either

$$\text{span}\{\mathbf{v}_1, \mathbf{v}_2, \ldots, \mathbf{v}_k\} \quad \text{or} \quad \text{span}\,\mathcal{S}.$$

Setting $x_1 = 1$ and $x_2 = 0, \cdots, x_k = 0$, shows that $\mathbf{v}_1$ is in span $\mathcal{S}$, and applying the same argument to the other objects $\mathbf{v}_2, \cdots, \mathbf{v}_k$ shows that every object in $\mathcal{S}$ is in span $\mathcal{S}$; that is, $\mathcal{S}$ is a *subset* of span $\mathcal{S}$.

EXAMPLE 1

Polynomials in $\mathbf{P}_2$

Let $\mathbf{p}_1 = 4$, $\mathbf{p}_2 = 1 + 3t$, $\mathbf{p}_3 = 1 - t$ and let $x_1 = 2$, $x_2 = 3$, $x_3 = 5$. Then

$$x_1\mathbf{p}_1 + x_2\mathbf{p}_2 + x_3\mathbf{p}_3 = 2(4) + 3(1 + 3t) + 5(1 - t) = 16 + 4t,$$

which shows that $\mathbf{p} = 16 + 4t$ is a linear combination of $\{\mathbf{p}_1, \mathbf{p}_2, \mathbf{p}_3\}$ and consequently that $\mathbf{p}$ is in span$\{\mathbf{p}_1, \mathbf{p}_2, \mathbf{p}_3\}$. ■

Consider a set of objects $\mathcal{S} = \{\mathbf{v}_1, \mathbf{v}_2, \ldots, \mathbf{v}_k\}$ in $\mathcal{V}$ and a given object $\mathbf{v}$ in $\mathcal{V}$. If

$$x_1\mathbf{v}_1 + x_2\mathbf{v}_2 + \cdots + x_k\mathbf{v}_k = \mathbf{v}, \tag{7.7}$$

for some scalars $x_1, x_2, \ldots, x_k$, then $\mathbf{v}$ is a linear combination of $\mathbf{v}_1, \mathbf{v}_2, \ldots, \mathbf{v}_k$; that is, $\mathbf{v}$ is in span $\mathcal{S}$. If there are no scalars $x_1, x_2, \ldots, x_k$ that satisfy (7.7), then $\mathbf{v}$ is not a linear combination of $\mathbf{v}_1, \mathbf{v}_2, \ldots, \mathbf{v}_k$; that is, $\mathbf{v}$ is not in span $\mathcal{S}$.

EXAMPLE 2

The Vector Space $\mathbb{R}^{2\times 2}$

Consider the objects shown next (matrices) in the vector space $\mathbb{R}^{2\times 2}$.

$$\mathbf{A}_1 = \begin{bmatrix} -1 & 0 \\ 2 & 1 \end{bmatrix}, \quad \mathbf{A}_2 = \begin{bmatrix} 1 & 1 \\ 1 & 0 \end{bmatrix}, \quad \mathbf{B} = \begin{bmatrix} 1 & 0 \\ 0 & 1 \end{bmatrix}$$

We ask if $\mathbf{B}$ is a linear combination of $\mathbf{A}_1$ and $\mathbf{A}_2$. The equation corresponding to (7.7) is

$$x_1\mathbf{A}_1 + x_2\mathbf{A}_2 = \mathbf{B} \quad \Rightarrow \quad x_1\begin{bmatrix} -1 & 0 \\ 2 & 1 \end{bmatrix} + x_2\begin{bmatrix} 1 & 1 \\ 1 & 0 \end{bmatrix} = \begin{bmatrix} 1 & 0 \\ 0 & 1 \end{bmatrix}. \tag{7.8}$$

Equating corresponding entries in (7.8) gives the 4×2 linear system

$$(S) \quad \begin{cases} -x_1 + x_2 = 1 \\ \phantom{-x_1 +{}} x_2 = 0 \\ 2x_1 + x_2 = 0 \\ x_1 \phantom{{}+x_2} = 1 \end{cases},$$

which is clearly inconsistent. Hence, $\mathbf{B}$ is not a linear combination of $\mathbf{A}_1$ and $\mathbf{A}_2$; that is, $\mathbf{B}$ is not in span$\{\mathbf{A}_1, \mathbf{A}_2\}$. ■

Part (a) of the next result was stated without proof in Section 3.1.

Theorem 7.2 **The Span of a Set of Objects is a Subspace**

Let $\mathcal{S} = \{\mathbf{v}_1, \mathbf{v}_2, \ldots, \mathbf{v}_k\}$ be a set of objects in a vector space $\mathcal{V}$. Then

(a) span $\mathcal{S}$ is a subspace of $\mathcal{V}$.
(b) span $\mathcal{S}$ is the smallest subspace of $\mathcal{V}$ that contains $\mathcal{S}$.

Proof For (a), note that $0\mathbf{v}_1 + 0\mathbf{v}_2 + \cdots + 0\mathbf{v}_k = \mathbf{0}$ is in span $\mathcal{S}$ and

$$(x_1\mathbf{v}_1 + \cdots + x_k\mathbf{v}_k) + (y_1\mathbf{v}_1 + \cdots + y_k\mathbf{v}_k) = (x_1 + y_1)\mathbf{v}_1 + \cdots + (x_k + y_k)\mathbf{v}_k$$
$$s(x_1\mathbf{v}_1 + \cdots + x_k\mathbf{v}_k) = sx_1\mathbf{v}_1 + \cdots + sx_k\mathbf{v}_k,$$

showing that span $\mathcal{S}$ is closed under addition and scalar multiplication and is therefore a subspace of $\mathcal{V}$. For (b), suppose $\mathcal{U}$ is a subspace of $\mathcal{V}$ that contains $\mathcal{S}$. Then $\mathcal{U}$ is closed under addition and scalar multiplication and so contains all linear combinations of objects in $\mathcal{S}$; that is, every vector in span $\mathcal{S}$ is in $\mathcal{U}$. Thus span $\mathcal{S}$ is a subset of any subspace $\mathcal{U}$ that contains $\mathcal{S}$.

—

EXAMPLE 3

A Proper Subspace of $\mathbb{R}^{2\times 2}$

Refer to Example 2. Let $\mathcal{S} = \{\mathbf{A}_1, \mathbf{A}_2\}$ and consider span $\mathcal{S}$; that is, the set of all linear combinations $x_1\mathbf{A}_1 + x_1\mathbf{A}_2$, for all real values x_1, x_2. We have

$$x_1\mathbf{A}_1 + x_1\mathbf{A}_2 = x_1\begin{bmatrix} -1 & 0 \\ 2 & 1 \end{bmatrix} + x_2\begin{bmatrix} 1 & 1 \\ 1 & 0 \end{bmatrix} = \begin{bmatrix} -x_1 + x_2 & x_2 \\ 2x_1 + x_2 & x_1 \end{bmatrix}.$$

Then span $\mathcal{S}$ consists of all 2×2 matrices of the form

$$\begin{bmatrix} -x_1 + x_2 & x_2 \\ 2x_1 + x_2 & x_1 \end{bmatrix}.$$

By Theorem 7.2, span $\mathcal{S}$ is the smallest subspace $\mathbb{R}^{2\times 2}$ containing the objects $\mathbf{A}_1$ and $\mathbf{A}_2$. Assigning real values $x_1 = 3.5$ and $x_2 = -1.4$, say, we conclude that the object

$$\mathbf{A} = \begin{bmatrix} -4.9 & -1.4 \\ 5.6 & 3.5 \end{bmatrix}$$

is in span $\mathcal{S}$. Clearly, span $\mathcal{S}$ is not the zero subspace and Example 2 shows that the object $\mathbf{B}$ is not in span $\mathcal{S}$. Hence span $\mathcal{S}$ is a proper subspace of $\mathbb{R}^{2\times 2}$. ▪

Definition 7.4

Spanning set

Let $\mathcal{U}$ be a subspace of a vector space $\mathcal{V}$ and let $\mathcal{S}$ be a subset of $\mathcal{U}$. If span $\mathcal{S} = \mathcal{U}$, then we say that $\mathcal{S}$ is a *spanning set* for $\mathcal{U}$.

Finite and Infinite Spanning Sets

Refer to Definition 7.4. If we consider a *finite*[2] subset $\mathcal{S} = \{\mathbf{v}_1, \mathbf{v}_2, \ldots, \mathbf{v}_k\}$ in $\mathcal{U}$, then span $\mathcal{S} = \mathcal{U}$ means that every object $\mathbf{v}$ in $\mathcal{U}$ can be expressed as the linear combination (7.7) for some set of scalars $x_1, x_2, \ldots, x_k$. We must also allow for $\mathcal{S}$ to be an *infinite* set, and in such cases span $\mathcal{S} = \mathcal{U}$ means that every object $\mathbf{v}$ in $\mathcal{U}$ can be expressed as a linear combination of some finite collection of objects in $\mathcal{S}$— keep in mind that the collection may vary depending on $\mathbf{v}$. These two cases are illustrated next.

EXAMPLE 4

Spanning the Vector Space of Polynomials $\mathbf{P}_2$

A typical object $\mathbf{p}$ in $\mathbf{P}_2$ has the form

$$\mathbf{p} = a_0 + a_1 t + a_2 t^2, \quad \text{where } a_0, a_1, a_2 \text{ are real numbers.} \tag{7.9}$$

Consider $\mathcal{S} = \{\mathbf{p}_1, \mathbf{p}_2, \mathbf{p}_3\}$, where

$$\mathbf{p}_1(t) = 1, \quad \mathbf{p}_2(t) = t, \quad \mathbf{p}_3(t) = t^2,$$

for all real t. Then every linear combination $a_0\mathbf{p}_0 + a_1\mathbf{p}_1 + a_2\mathbf{p}_2$ is in $\mathbf{P}_2$, showing that span $\mathcal{S}$ is a subset of $\mathbf{P}_2$. Also, every object in $\mathbf{P}_2$ has the form (7.9) and so belongs to span $\mathcal{S}$, showing that $\mathbf{P}_2$ is a subset of span $\mathcal{S}$. Thus span $\mathcal{S} = \mathbf{P}_2$. Note that the vector space $\mathbf{P}_2$, which contains infinitely many objects, can be "built up" by taking linear combinations of objects from the finite set $\{1, t, t^2\}$. ▪

[2] For example, the set $\{1, 2, 3\}$ is *finite* because it contains three objects. By contrast, the set of all positive integers $\{1, 2, 3, \ldots\}$ is *infinite*.

EXAMPLE 5

The Vector Space of All Polynomials

Consider the set **P** of all objects (polynomials)

$$\mathbf{p}(t) = a_0 + a_1 t + \cdots + a_n t^k,$$

where k is a nonnegative integer that is not constant. Then **P** is a vector space (see Exercises 7.2). Note the difference between **P** and $\mathbf{P}_n$, the vector space of polynomials of degree n or less (n is constant). The set **P** contains polynomials of the type

$$4\ (=4t^0), \quad \mathbf{p}_1 = 1 + 3t - t^5, \quad \mathbf{p}_2 = 2 - t^9, \quad \mathbf{p}_3 = 2t^{2000}, \ldots, \tag{7.10}$$

and so **P** contains polynomials of arbitrary large degree. The set $\mathcal{S} = \{1, t, t^2, t^3, \ldots\}$ is an infinite spanning set for **P** because any object in **P** is a linear combination of a finite (but variable) set of objects from $\mathcal{S}$. **P** is not spanned by any finite set of objects from **P**. ■

EXAMPLE 6

A Spanning Set for a Subspace of $\mathbb{R}^{2\times2}$

Let $\mathcal{U}$ be the set of all upper triangular matrices in the vector space $\mathbb{R}^{2\times2}$. The objects (matrices) in $\mathcal{U}$ have the form

$$\mathbf{A} = \begin{bmatrix} a & b \\ 0 & c \end{bmatrix}, \quad \text{where } a, b, c \text{ are real numbers.} \tag{7.11}$$

Example 5 in Section 7.1 shows that $\mathcal{U}$ is a subspace of $\mathbb{R}^{2\times2}$. Consider the set $\{\mathbf{E}_1, \mathbf{E}_2, \mathbf{E}_3\}$ defined by

$$\mathbf{E}_1 = \begin{bmatrix} 1 & 0 \\ 0 & 0 \end{bmatrix}, \quad \mathbf{E}_2 = \begin{bmatrix} 0 & 1 \\ 0 & 0 \end{bmatrix}, \quad \mathbf{E}_3 = \begin{bmatrix} 0 & 0 \\ 0 & 1 \end{bmatrix}.$$

Suppose x_1, x_2, x_3 are any scalars. Then the equation

$$x_1\mathbf{E}_1 + x_2\mathbf{E}_2 + x_3\mathbf{E}_3 = x_1\begin{bmatrix} 1 & 0 \\ 0 & 0 \end{bmatrix} + x_2\begin{bmatrix} 0 & 1 \\ 0 & 0 \end{bmatrix} x_3\begin{bmatrix} 0 & 0 \\ 0 & 1 \end{bmatrix} = \begin{bmatrix} x_1 & x_2 \\ 0 & x_3 \end{bmatrix}$$

shows that any linear combination of $\{\mathbf{E}_1, \mathbf{E}_2, \mathbf{E}_3\}$ results in a matrix in $\mathcal{U}$, and an arbitrary matrix in $\mathcal{U}$ can be written as such a linear combination. Thus $\mathcal{U} = \text{span}\{\mathbf{E}_1, \mathbf{E}_2, \mathbf{E}_3\}$. ■

Linear Independence

We will restate the following fundamental definition from Section 3.2.

Definition 7.5

Linear independence, linear dependence

A set of objects $\mathcal{S} = \{\mathbf{v}_1, \mathbf{v}_2, \ldots, \mathbf{v}_k\}$ in a vector space $\mathcal{V}$ is *linearly independent* if the equation

$$x_1\mathbf{v}_1 + x_2\mathbf{v}_2 + \cdots + x_k\mathbf{v}_k = \mathbf{0} \tag{7.12}$$

has *only* the zero (or trivial) solution $(x_1, x_2, \ldots, x_k) = (0,0, \ldots ,0)$. The set $\mathcal{S}$ is called *linearly dependent* if (7.12) has a nonzero solution

$(x_1, x_2, \ldots, x_k)$; that is, a solution in which at least one of the scalars $x_1, x_2, \ldots, x_k$ is nonzero.

EXAMPLE 7

Linear Dependence in the Vector Space $\mathbb{R}^{2\times 2}$ of Matrices

Let $\mathcal{S} = \{\mathbf{A}_1, \mathbf{A}_2, \mathbf{A}_3\}$ be a set of objects (matrices) in $\mathbb{R}^{2\times 2}$, where

$$\mathbf{A}_1 = \begin{bmatrix} 1 & 0 \\ 3 & 2 \end{bmatrix}, \quad \mathbf{A}_2 = \begin{bmatrix} -1 & 2 \\ 3 & 2 \end{bmatrix}, \quad \mathbf{A}_3 = \begin{bmatrix} 5 & -6 \\ -3 & -2 \end{bmatrix}.$$

Equation (7.12), which is the standard test for linear independence, in this case takes the form $x_1\mathbf{A}_1 + x_2\mathbf{A}_2 + x_3\mathbf{A}_3 = \mathbf{O}$; that is,

$$x_1\begin{bmatrix} 1 & 0 \\ 3 & 2 \end{bmatrix} + x_2\begin{bmatrix} -1 & 2 \\ 3 & 2 \end{bmatrix} + x_3\begin{bmatrix} 5 & -6 \\ -3 & -2 \end{bmatrix} = \begin{bmatrix} 0 & 0 \\ 0 & 0 \end{bmatrix} = \mathbf{O}, \qquad (7.13)$$

$$\text{implies} \quad \begin{bmatrix} x_1 - x_2 + 5x_3 & x_2 - 6x_3 \\ 3x_1 + 3x_2 - 3x_3 & 2x_1 + 2x_2 - 2x_3 \end{bmatrix} = \begin{bmatrix} 0 & 0 \\ 0 & 0 \end{bmatrix}, \qquad (7.14)$$

where $\mathbf{O}$ is the zero object in $\mathbb{R}^{2\times 2}$. Equating corresponding entries in (7.14) gives the 4×3 homogeneous linear system

$$(S)\quad \begin{cases} x_1 - x_2 + 5x_3 = 0 & (1,1)\text{-entries} \\ x_2 - 6x_3 = 0 & (1,2)\text{-entries} \\ 3x_1 + 3x_2 - 3x_3 = 0 & (2,1)\text{-entries} \\ 2x_1 + 2x_2 - 2x_3 = 0 & (2,2)\text{-entries} \end{cases}$$

that has the general solution $(x_1, x_2, x_3) = (-2t, 3t, t)$, where t is a real parameter. Choosing $t = -1$ (a convenient nonzero value), we obtain $2\mathbf{A}_1 - 3\mathbf{A}_2 - \mathbf{A}_3 = \mathbf{O}$, showing that $\mathcal{S}$ is linearly dependent because (7.13) has a nonzero solution—*all* coefficients happen to be nonzero in this case. ■

EXAMPLE 8

Linear Independence in the Function Space $\mathbf{F}(-\infty, \infty)$

Consider the set of objects (functions) $\mathcal{S} = \{t, t^2, e^{3t}\}$ in the vector space $\mathbf{F}(-\infty, \infty)$. Equation (7.12) in this case becomes

$$x_1 t + x_2 t^2 + x_3 e^{3t} = \mathbf{0}, \qquad (7.15)$$

where $\mathbf{0}$ is the zero function, and (7.15) must be satisfied for all t. Substituting three convenient values of t into (7.15) results in a 3×3 linear system (S). For example,

$$(S)\quad \begin{cases} x_3 = 0 & (t = 0) \\ x_1 + x_2 + e^3 x_3 = 0 & (t = 1), \\ 2x_2 + 4x_2 + e^6 x_3 = 0 & (t = 2) \end{cases}$$

which has only the solution is $x_1 = x_2 = x_3 = 0$, showing that $\mathcal{S}$ is linearly independent. ■

The next result is stated and proved in Section 3.2.

Theorem 7.3 **Characterization of Linear Dependence**

Let $\mathcal{V}$ be a vector space and let $\mathcal{S} = \{\mathbf{v}_1, \mathbf{v}_2, \ldots, \mathbf{v}_k\}$ be a subset of $\mathcal{V}$. Then $\mathcal{S}$ is linearly dependent if and only if at least one object in $\mathcal{S}$ is a linear combination of other objects in $\mathcal{S}$.

Some Useful Facts

The following facts, valid in any vector space $\mathcal{V}$, have been already discussed in Section 3.2 in the context of $\mathbb{R}^n$. However, some of the proofs require a little more thought when written in the general case.

(L1) Any set of objects in $\mathcal{V}$ that contains the zero object is linearly dependent.
(L2) Any nonzero object $\mathbf{v}$ in $\mathcal{V}$ is linearly independent.
(L3) A set consisting of two objects $\mathbf{u}$ and $\mathbf{v}$ in $\mathcal{V}$ is linearly dependent if and only if one object is a scalar multiple of the other.
(L4) Any subset of a linearly independent set is linearly independent.
(L5) Adding an object to a linearly dependent set results in a linearly dependent set.

Proof

(L1) Consider the set of objects $\mathbf{v}_1, \mathbf{v}_2, \ldots, \mathbf{v}_k, \mathbf{0}$ in $\mathcal{V}$. By property (**P5**) in Section 7.1, $s\mathbf{0} = \mathbf{0}$ for all scalars s. We have $0\mathbf{v}_1 + 0\mathbf{v}_2 + \cdots + 0\mathbf{v}_k + s\mathbf{0} = \mathbf{0}$, for any nonzero s, showing that (7.12) has a nonzero solution.
(L2) By (**P7**) in Section 7.1, $s\mathbf{v} = \mathbf{0}$ implies $s = 0$ because $\mathbf{v}$ is nonzero.
(L3) This is Theorem 7.3 with $k = 2$.

The proofs of (**L4**) and (**L5**) from Exercises 3.2 are valid in general.

EXAMPLE 9 **A Review of Linear Independence and Dependence in $\mathbf{P}_1$**

Any object in $\mathbf{P}_1$ has the form $\mathbf{p} = a_0 + a_1 t$ (a polynomial of degree 1). Consider

$$\mathbf{0} = 0 + 0t, \qquad \mathbf{p}_1 = 1 - 2t, \qquad \mathbf{p}_2 = -3 + 6t, \qquad \mathbf{p}_3 = 4 + 2t.$$

Refer to the basic facts stated immediately preceding. Each of the sets $\{\mathbf{p}_1\}$, $\{\mathbf{p}_2\}$, $\{\mathbf{p}_3\}$ is linearly independent. The set $\{\mathbf{p}_1, \mathbf{p}_3\}$ is linearly independent because neither object is a scalar multiple of the other and $\{\mathbf{p}_2, \mathbf{p}_3\}$ is linearly independent for the same reason. The set $\{\mathbf{0}, \mathbf{p}_1, \mathbf{p}_3\}$ is linearly dependent (contains the zero object) and $\{\mathbf{p}_1, \mathbf{p}_2\}$ is linearly dependent because $\mathbf{p}_2 = -3\mathbf{p}_1$. Hence $\{\mathbf{p}_1, \mathbf{p}_2, \mathbf{p}\}$ is linearly dependent for any object $\mathbf{p}$ in $\mathbf{P}_1$. We will see in due course that three or more objects in $\mathbf{P}_1$ are necessarily linearly dependent because $\dim \mathbf{P}_1 = 2$. ■

Basis

Definition 7.6 *Basis for a vector space*

A *basis* for a vector space $\mathcal{V}$ is a set of objects $\mathcal{B}$ in $\mathcal{V}$ such that $\mathcal{B}$ is linearly independent and spans $\mathcal{V}$.

Standard Bases

A basis $\mathcal{B}$ for a vector space is not unique. Of the many choices for $\mathcal{B}$, some are more natural or convenient. For example, the standard basis for $\mathbb{R}^n$ consisting of the columns $\mathbf{e}_1, \mathbf{e}_2, \ldots, \mathbf{e}_n$ of the identity matrix $\mathbf{I}_n$ is orthonormal and has computational advantages (Sections 3.2 and 4.3). The next two examples describe standard bases for other vector spaces.

EXAMPLE 10

Standard Bases for $\mathbb{R}^{m\times n}$

The standard basis for $\mathbb{R}^{m\times n}$ consists of all possible $m \times n$ matrices that have a single entry equal to 1 and all other entries equal to zero. There are mn such matrices. For example, the standard basis for $\mathbb{R}^{2\times 2}$ consists of the following $mn = 4$ binary matrices (entries 0 or 1).

$$\mathbf{E}_1 = \begin{bmatrix} 1 & 0 \\ 0 & 0 \end{bmatrix}, \quad \mathbf{E}_2 = \begin{bmatrix} 0 & 1 \\ 0 & 0 \end{bmatrix}, \quad \mathbf{E}_3 = \begin{bmatrix} 0 & 0 \\ 1 & 0 \end{bmatrix}, \quad \mathbf{E}_4 = \begin{bmatrix} 0 & 0 \\ 0 & 1 \end{bmatrix}$$

It is a routine exercise to check that the set $\{\mathbf{E}_1, \mathbf{E}_2, \mathbf{E}_3, \mathbf{E}_4\}$ is linearly independent and spans $\mathbb{R}^{2\times 2}$. ■

EXAMPLE 11

Standard Basis for $\mathbf{P}_n$

The standard basis for $\mathbf{P}_n$ consists of the $n + 1$ polynomials defined by

$$\mathbf{p}_0 = 1\ (= t^0), \quad \mathbf{p}_1 = t, \quad \mathbf{p}_2 = t^2, \quad \ldots, \quad \mathbf{p}_n = t^n. \tag{7.16}$$

It is routine to check that the set of $n + 1$ polynomials in (7.16) is linearly independent and spans $\mathbf{P}_n$. ■

We now embark on a sequence of results that are steps toward a deeper understanding of the meaning and construction of a basis. The first of these relates to Theorem 3.6 in Section 3.2.

Theorem 7.4 **Linear Dependence and Span**

Let $\mathcal{V}$ be a vector space with a basis $\mathcal{B} = \{\mathbf{v}_1, \mathbf{v}_2, \ldots, \mathbf{v}_k\}$.

(a) A subset of $\mathcal{V}$ containing more that k objects is linearly dependent.

(b) A subset of $\mathcal{V}$ containing fewer than k objects does not span $\mathcal{V}$.

Proof (a) Consider a set of objects $\mathcal{S} = \{\mathbf{u}_1, \mathbf{u}_2, \ldots, \mathbf{u}_n\}$ in $\mathcal{V}$, where $k < n$. Each object in $\mathcal{S}$ is a linear combination of basis objects, and so we obtain the following system of equations, where a_{ij} are the scalars used to form the linear combinations.

$$\begin{cases} \mathbf{u}_1 = a_{11}\mathbf{v}_1 + a_{12}\mathbf{v}_2 + \cdots + a_{k1}\mathbf{v}_k \\ \mathbf{u}_2 = a_{12}\mathbf{v}_1 + a_{22}\mathbf{v}_2 + \cdots + a_{k2}\mathbf{v}_k \\ \quad\vdots \qquad \vdots \qquad \vdots \qquad \vdots \\ \mathbf{u}_n = a_{1n}\mathbf{v}_1 + a_{2n}\mathbf{v}_2 + \cdots + a_{kn}\mathbf{v}_k \end{cases} \tag{7.17}$$

We will test $\mathcal{S}$ for linear independence in the usual way by solving the equation $x_1\mathbf{u}_1+x_2\mathbf{u}_2+\cdots+x_n\mathbf{u}_n=\mathbf{0}$. To form this equation, multiply the first equation in (7.17) by x_1, the second by x_2, and so on, and gather terms $\mathbf{v}_1$, $\mathbf{v}_2$, and so on, to obtain

$$\begin{aligned}(x_1a_{11}+x_2a_{12}+\cdots+x_na_{1n})\mathbf{v}_1 \\ +(x_1a_{12}+x_2a_{22}+\cdots+x_na_{2n})\mathbf{v}_2+\cdots \\ +(x_1a_{k1}+x_2a_{k2}+\cdots+x_na_{kn})\mathbf{v}_k=\mathbf{0}.\end{aligned} \tag{7.18}$$

However, $\mathcal{B}$ is linearly independent and so all the coefficients in (7.18) are zero. We obtain a $k\times n$ homogeneous linear system

$$\begin{cases} x_1a_{11}+x_2a_{12}+\cdots+x_na_{1n}=0 \\ x_1a_{12}+x_2a_{22}+\cdots+x_na_{2n}=0 \\ \vdots \qquad \vdots \qquad \vdots \qquad \vdots \\ x_1a_{k1}+x_2a_{k2}+\cdots+x_na_{kn}=0 \end{cases}, \tag{7.19}$$

which has a nonzero solution $x_1, x_2, \ldots, x_n$ because $k<n$. Thus $\mathcal{S}$ is linearly dependent.

(b) Consider a set of objects $\mathcal{S}=\{\mathbf{u}_1,\mathbf{u}_2,\ldots,\mathbf{u}_n\}$ in $\mathcal{V}$, where $n<k$ and suppose that $\mathcal{S}$ spans $\mathcal{V}$. Then every object in $\mathcal{B}$ is a linear combination of objects from $\mathcal{S}$, and we obtain a system of equations

$$\begin{cases} \mathbf{v}_1=a_{11}\mathbf{u}_1+a_{12}\mathbf{u}_2+\cdots+;\ a_{n1}\mathbf{u}_n \\ \mathbf{v}_2=a_{12}\mathbf{u}_1+a_{22}\mathbf{u}_2+\cdots+;\ a_{n2}\mathbf{u}_n \\ \vdots \quad \vdots \quad \vdots \quad \vdots \\ \mathbf{v}_k=a_{1k}\mathbf{u}_1+a_{2k}\mathbf{u}_2+\cdots+;\ a_{nk}\mathbf{u}_n \end{cases}, \tag{7.20}$$

where the a_{ij} are scalars used to form these linear combinations. Incidentally, (7.20) can be obtained from (7.17) by interchanging $\mathbf{u}$ and $\mathbf{v}$ and n and k. We will show that the equation $x_1\mathbf{v}_1+x_2\mathbf{v}_2+\cdots+x_k\mathbf{v}_k=\mathbf{0}$ has a nonzero solution. Proceeding exactly as in (a), multiplying the first equation in (7.20) by x_1, the second by x_2 and so on, we arrive at an $n\times k$ homogeneous linear system given by (7.19) with n and k interchanged, and this system has a nonzero solution $x_1, x_2, \ldots, x_k$ because $n<k$. However, $\mathcal{B}$ is linearly independent—this contradiction means that our opening assumption that $\mathcal{S}$ spans $\mathcal{V}$ must have been false.

Notice that, although Theorem 7.4 applies to any vector space, its proof ultimately depends on the solution of homogeneous linear systems of real scalars (or complex scalars in the case of a complex space vector).

Theorem 7.5 **Number of Objects in a Basis**

Let $\mathcal{V}$ be a vector space and let $\mathcal{B}$ be a basis for $\mathcal{V}$ containing k objects. Then the number of objects in any other basis is k.

Proof Let $\mathcal{B}'$ be another basis for $\mathcal{V}$ containing k' objects. We apply the contrapositive of Theorem 7.4(a). Because the set $\mathcal{B}'$ is linearly independent, we have $k' \leq k$, and, reversing the roles of $\mathcal{B}$ and $\mathcal{B}'$, we have $k \leq k'$. Thus $k = k'$.

Definition 7.7

Dimension

A vector space is called *finite-dimensional* if it has a basis consisting of finitely many objects. The number of objects in any basis for $\mathcal{V}$, denoted by dim $\mathcal{V}$, is called the *dimension* of $\mathcal{V}$. The dimension of the zero subspace is defined to be zero. A vector space is called *infinite-dimensional* if it does not have a finite basis.

The standard bases for standard examples of vector spaces provide immediately the dimension of these spaces. For example, dim $\mathbb{R}^n = n$ (from Section 3.2), dim $\mathbb{R}^{m\times n} = mn$ (from Example 10), and dim $\mathbf{P}_n = n + 1$ (from Example 11)—these spaces are finite-dimensional. The space $\mathbf{P}$ of all polynomials is infinite-dimensional because it has no finite basis.

Constructing Bases

The next two theorems are fundamental in constructing a basis from a given set of objects.

Theorem 7.6 **Constructing a Basis from a Linearly Independent Set**

Let $\mathcal{S} = \{\mathbf{v}_1, \mathbf{v}_2, \ldots, \mathbf{v}_p\}$ be a linearly independent set of objects in a vector space $\mathcal{V}$ with dim $\mathcal{V} = k$ and let with $\mathbf{v}$ be an object in $\mathcal{V}$ that is not in span $\mathcal{S}$. Then the set $\mathcal{S}' = \{\mathbf{v}_1, \mathbf{v}_2, \ldots, \mathbf{v}_p, \mathbf{v}\}$ is linearly independent. The process may be repeated (if necessary) until a basis for $\mathcal{V}$ is found.

Proof We prove the contrapositive (Appendix B: Toolbox) of the implication in the theorem which is: If $\mathcal{S}'$ is linearly dependent, then $\mathbf{v}$ is in span $\mathcal{S}$. If $\mathcal{S}'$ is linearly dependent, then the equation

$$x_1\mathbf{v}_1 + x_2\mathbf{v}_2 + \cdots + x_p\mathbf{v}_p + x\mathbf{v} = \mathbf{0} \tag{7.21}$$

has a nonzero solution $x_1, x_2, \ldots, x_p, x$. If $x = 0$, then some other coefficient in (7.21) must be nonzero, giving a nonzero solution to

$$x_1\mathbf{v}_1 + x_2\mathbf{v}_2 + \cdots + x_p\mathbf{v}_p = \mathbf{0},$$

which contradicts the fact that $\mathcal{S}$ is linearly independent. Hence $x \neq 0$, allowing us to use (7.21) to express $\mathbf{v}$ in the form

$$\mathbf{v} = -\left(\frac{x_1}{x}\right)\mathbf{v}_1 - \left(\frac{x_2}{x}\right)\mathbf{v}_2 - \cdots - \left(\frac{x_p}{x}\right)\mathbf{v}_p,$$

which shows that $\mathbf{v}$ belongs to span $\mathcal{S}$. The process of adjoining to $\mathcal{S}$ one object at a time must eventually yield a linearly independent spanning set (basis) for $\mathcal{V}$ containing k objects because, by Theorem 7.4(a), a set containing more that k objects from $\mathcal{V}$ is linearly dependent.

Theorem 7.7 **Constructing a Basis from a Linearly Dependent Spanning Set**

Let $\mathcal{V}$ be a vector space with dim $\mathcal{V} = k$ and let $\mathcal{S} = \{\mathbf{v}_1, \mathbf{v}_2, \ldots, \mathbf{v}_p\}$ be a linearly dependent set of nonzero objects such that span $\mathcal{S} = \mathcal{V}$. *It is possible to delete an object* $\mathbf{v}$ *from* $\mathcal{S}$ *to obtain a smaller set* $\mathcal{S}'$ *such that* span $\mathcal{S}' = \mathcal{V}$. *The process may be repeated (if necessary) until a basis for* $\mathcal{V}$ *is found.*

Proof The following illustration with $p = 4$ will show the idea of the general proof. Suppose $\mathcal{S} = \{\mathbf{v}_1, \mathbf{v}_2, \mathbf{v}_3, \mathbf{v}_4\}$ is a linearly dependent set of nonzero objects that spans $\mathcal{V}$. By Theorem 7.4(b), we have, $k \leq p$, and by Theorem 7.3, some object, say $\mathbf{v}_3$, is a linear combination of other objects in $\mathcal{S}$. Suppose that $\mathbf{v}_3 = y_1\mathbf{v}_1 + y_2\mathbf{v}_2 + 0\mathbf{v}_4, \ldots,$ for some scalars y_1 and y_2. Delete $\mathbf{v} = \mathbf{v}_3$ from $\mathcal{S}$ to obtain the smaller set $\mathcal{S}' = \{\mathbf{v}_1, \mathbf{v}_2, \mathbf{v}_4\}$. Because $\mathcal{S}$ spans $\mathcal{V}$, every object $\mathbf{u}$ in $\mathcal{V}$ can be expressed in the form $\mathbf{u} = x_1\mathbf{v}_1 + x_2\mathbf{v}_2 + x_3\mathbf{v}_3 + x_4\mathbf{v}_4$ for some scalars x_1, x_2, x_3, x_4. Substituting for $\mathbf{v}_3$, we have

$$\begin{aligned}\mathbf{u} &= x_1\mathbf{v}_1 + x_2\mathbf{v}_2 + x_3\mathbf{v}_3 + x_4\mathbf{v}_4 \\ &= x_1\mathbf{v}_1 + x_2\mathbf{v}_2 + x_3(y_1\mathbf{v}_1 + y_2\mathbf{v}_2) + x_4\mathbf{v}_4 \\ &= (x_1 + x_3y_1)\mathbf{v}_1 + (x_2 + x_3y_2)\mathbf{v}_2 + x_4\mathbf{v}_4,\end{aligned}$$

showing that $\mathbf{u}$ is a linear combination of $\{\mathbf{v}_1, \mathbf{v}_2, \mathbf{v}_4\}$ alone and so span $\mathcal{S}' = \mathcal{V}$. If $\mathcal{S}'$ is linearly independent, then $\mathcal{S}'$ is a basis for $\mathcal{V}$; otherwise $\mathcal{S}'$ is linearly dependent and the same process is applied to $\mathcal{S}'$. Each set consisting of a single object in $\mathcal{S}$ is linearly independent, and so at some point we will arrive at a linearly independent spanning set for $\mathcal{V}$ that contains k objects. In the extreme situation, the basis consists of a single nonzero object.

Remarks

Observe that the basis formed using Theorem 7.6 is build up from a given linearly independent set $\mathcal{S}$ which we are at liberty to choose—the constructed basis will certainly contain $\mathcal{S}$. Theorem 7.6 does not offer the same advantage—we do not know, a priori, exactly which objects in $\mathcal{S}$ will form the basis. Here is a useful technique that addresses this concern.

Let $\mathcal{V}$ be an k-dimensional vector space and let $\mathcal{B} = \{\mathbf{b}_1, \mathbf{b}_2, \ldots, \mathbf{b}_k\}$ be a basis for $\mathcal{V}$. Select a linearly independent set of objects $\{\mathbf{v}_1, \mathbf{v}_2, \ldots, \mathbf{v}_p\}$ that does not span $\mathcal{V}$ and form the *ordered* set

$$\mathcal{S} = \{\mathbf{v}_1, \mathbf{v}_2, \ldots, \mathbf{v}_p, \mathbf{b}_1, \mathbf{b}_2, \ldots, \mathbf{b}_k\}. \tag{7.22}$$

Then $\mathcal{S}$ spans $\mathcal{V}$ because $\mathcal{B}$ already does. By Theorem 7.4(a), $\mathcal{S}$ is linearly dependent because it contains more than k objects. By Theorem 7.7, we can delete some object $\mathbf{u}$ from $\mathcal{S}$ that is a linear combination of its predecessors (Exercises 7.2, 40) and the smaller set still spans $\mathcal{V}$. But $\mathbf{u}$ cannot come from the list $\mathbf{v}_1, \mathbf{v}_2, \ldots, \mathbf{v}_k$ because this set is linearly independent and so $\mathbf{u} = \mathbf{b}_i$, for some i. Continue this process step-by-step until a basis is reached consisting of $\mathbf{v}_1, \mathbf{v}_2, \ldots, \mathbf{v}_p$ together with $k - p$ objects from $\mathcal{B}$.

Checking that a set of objects is a basis for a finite-dimensional vector space is made easier if we already know the dimension of the space. The next result is therefore very useful in practice.

Theorem 7.8 **The Basis Theorem**

Let $\mathcal{V}$ be a vector space of dimension k, where $k > 0$, and let $\mathcal{S}$ be a subset of $\mathcal{V}$ containing k objects.

(a) If $\mathcal{S}$ spans $\mathcal{V}$, then $\mathcal{S}$ is linearly independent, and so $\mathcal{S}$ is a basis.
(b) If $\mathcal{S}$ is linearly independent, then $\mathcal{S}$ spans $\mathcal{V}$, and so $\mathcal{S}$ is a basis.

Proof (a) If $\mathcal{S}$ spans $\mathcal{V}$ and is linearly dependent, then Theorem 7.7 shows that $\mathcal{S}$ contains a basis with fewer than k objects, which is a contradiction. (b) If $\mathcal{S}$ is linearly independent and does not span $\mathcal{V}$, then Theorem 7.6 says that $\mathcal{S}$ can be extended to a basis containing more than k objects, which is a contradiction.

EXAMPLE 12 **Recognizing a Basis for $\mathbf{P}_3$**

We have dim $\mathbf{P}_3 = 4$. If we verify that the set of objects $\mathcal{B}$ in $\mathbf{P}_3$ defined by

$$\mathcal{B} = \{1, t, t^2, 1 + t + t^2 + t^3\}$$

is linearly independent, then by Theorem 7.8(b), $\mathcal{B}$ is a basis for $\mathbf{P}_3$. Solving $x_1(1) + x_2(t) + x_3(t^2) + x_4(1 + t + t^2 + t^3) = 0 + 0t + 0t^2 + 0t^3$, we have $x_1 + x_4 = x_2 + x_4 = x_3 + x_4 = 0$, but $x_4 = 0$ and so $x_1 = x_2 = x_3 = 0$; that is, $\mathcal{B}$ is linearly independent. ■

One last word in this section regarding subspaces. Suppose $\mathcal{V}$ is an k-dimensional vector space and let $\mathcal{U}$ be a subspace of $\mathcal{V}$. Any basis for $\mathcal{V}$ spans $\mathcal{V}$ and also spans $\mathcal{U}$ and so, by Theorem 7.5, $\mathcal{U}$ has a basis containing $m \le k$ objects. Thus $\mathcal{U}$ is finite-dimensional and dim $\mathcal{U} \le$ dim $\mathcal{V}$.

EXERCISES 7.2

Exercises 1–6. *Determine if the given object is a linear combination of objects in the given set $\mathcal{S}$.*

1. $\mathbf{M} = \begin{bmatrix} 1 & 2 \\ 3 & 4 \end{bmatrix}$, $\mathcal{S}: \begin{bmatrix} 1 & 2 \\ 0 & 3 \end{bmatrix}, \begin{bmatrix} 0 & 1 \\ 2 & 0 \end{bmatrix}$

2. $\mathbf{M} = \begin{bmatrix} 4 & 7 \\ 1 & 10 \end{bmatrix}$, $\mathcal{S}: \begin{bmatrix} 6 & 1 \\ 0 & 0 \end{bmatrix}, \begin{bmatrix} -1 & 2 \\ -1 & 3 \end{bmatrix}, \begin{bmatrix} 2 & 2 \\ 2 & 2 \end{bmatrix}$

3. $\mathbf{f} = 1$, $\mathcal{S}: \sin^2 t, \ \cos^2 t$

4. $\mathbf{f} = \sin 2t$, $\mathcal{S}: \sin t, \ \cos t$

5. $\mathbf{p} = 5 + 4t + t^2$, $\mathcal{S}: -1 + t + t^2, 1 + 2t + t^2$

6. $\mathbf{p} = 5 + 4t + t^2$, $\mathcal{S}: -1 + t + t^2, 1 + 2t + t^2$

Exercises 7–8. *Determine whether the given subset $\mathcal{S}$ spans the vector space indicated.*

7. $\mathcal{S} = \{\mathbf{M}_1, \mathbf{M}_2, \mathbf{M}_3, \mathbf{M}_4\} \subset \mathbb{R}^{2\times2}$, where

$$\mathbf{M}_1 = \begin{bmatrix} 1 & 1 \\ 1 & 1 \end{bmatrix}, \quad \mathbf{M}_2 = \begin{bmatrix} 1 & 0 \\ 0 & 1 \end{bmatrix},$$

$$\mathbf{M}_3 = \begin{bmatrix} 1 & -1 \\ 1 & 0 \end{bmatrix}, \quad \mathbf{M}_4 = \begin{bmatrix} 0 & 1 \\ -1 & 0 \end{bmatrix}$$

8. $\mathcal{S} = \{1 - t + t^2, 1 + 2t\} \subset \mathbf{P}_2$

Exercises 9–12. *Determine whether the given set of objects spans the vector space indicated.*

9. $\mathbf{p}_1 = 2$, $\mathbf{p}_2 = 1 + t + 2t^2$, $\mathbf{P}_2$

10. $\mathbf{p}_1 = 3$, $\mathbf{p}_2 = 1 + t$, $\mathbf{p}_3 = 1 - 2t^2$, $\mathbf{P}_2$

11. $\begin{bmatrix} 1 & 1 \\ 0 & 0 \end{bmatrix}, \begin{bmatrix} 0 & 0 \\ 1 & 1 \end{bmatrix}, \begin{bmatrix} 0 & 1 \\ 1 & 0 \end{bmatrix}, \begin{bmatrix} 1 & 0 \\ 0 & 0 \end{bmatrix}$, $\mathbb{R}^{2\times2}$

12. $\begin{bmatrix} 1 & 1 \\ 1 & 0 \end{bmatrix}, \begin{bmatrix} 0 & 1 \\ 1 & 1 \end{bmatrix}, \begin{bmatrix} 0 & 0 \\ 1 & 0 \end{bmatrix}, \begin{bmatrix} 1 & 0 \\ 0 & 1 \end{bmatrix}, \mathbb{R}^{2\times 2}$

Exercises 13–14. *Determine if the given function $p(t)$ is in the vector space spanned by the given set of functions $\mathcal{S}$.*

13. $\mathbf{p}(t) = 3 + t$, $\mathcal{S} = \{1 + t, 1 - t + t^2\}$

14. $\mathbf{p}(t) = 1 + 5t + t^2$, $\mathcal{S} = \{2 + t^2, -1 + 4t + t^2\}$

Exercises 15–16. *Show that the vector space $\mathbf{P}_2$ is spanned by the given set of vectors.*

15. $\mathbf{p}_1 = 2$, $\mathbf{p}_2 = 1 + t$, $\mathbf{p}_3 = t^2$

16. $\mathbf{p}_1 = 1$, $\mathbf{p}_2 = t$, $\mathbf{p}_3 = (1 + t)^2$

17. Find all possible ways of representing the object $\mathbf{p} = 4 - 2t + 4t^2$ as a linear combination of the vectors $\mathbf{p}_1 = 1 + t + t^2$, $\mathbf{p}_2 = 1 - t + t^2$, $\mathbf{p}_3 = 2 + 2t^2$. Is the representation of $\mathbf{p}$ unique? Explain.

Exercises 18–29. *In each case, determine if the set of objects $\mathcal{S}$ is linearly independent. When $\mathcal{S}$ is linearly dependent, express some object in $\mathcal{S}$ as a linear combination of other objects in $\mathcal{S}$.*

18. $\mathbb{R}^{2\times 2}$: $\mathbf{M}_1 = \begin{bmatrix} 1 & 0 \\ 0 & 1 \end{bmatrix}$, $\mathbf{M}_2 = \begin{bmatrix} 1 & 1 \\ 0 & 1 \end{bmatrix}$, $\mathbf{M}_3 = \begin{bmatrix} 0 & -1 \\ -1 & 0 \end{bmatrix}$

19. $\mathbb{R}^{2\times 2}$: $\mathbf{M}_1 = \begin{bmatrix} 1 & 1 \\ 1 & 0 \end{bmatrix}$, $\mathbf{M}_2 = \begin{bmatrix} 1 & 0 \\ 0 & 1 \end{bmatrix}$, $\mathbf{M}_3 = \begin{bmatrix} 0 & 1 \\ 1 & 1 \end{bmatrix}$, $\mathbf{M}_4 = \begin{bmatrix} 1 & 1 \\ 1 & 1 \end{bmatrix}$

20. $\mathbb{R}^{2\times 2}$: $\mathbf{M}_1 = \begin{bmatrix} 0 & 0 \\ 0 & 0 \end{bmatrix}$, $\mathbf{M}_2 = \begin{bmatrix} 1 & 2 \\ 0 & 1 \end{bmatrix}$, $\mathbf{M}_3 = \begin{bmatrix} 1 & 0 \\ 1 & 0 \end{bmatrix}$, $\mathbf{M}_4 = \begin{bmatrix} 1 & -1 \\ -1 & 1 \end{bmatrix}$

21. $\mathbb{R}^{2\times 2}$: $\mathbf{M}_1 = \begin{bmatrix} 1 & 0 \\ 0 & 2 \end{bmatrix}$, $\mathbf{M}_2 = \begin{bmatrix} 1 & 1 \\ -1 & 1 \end{bmatrix}$, $\mathbf{M}_3 = \begin{bmatrix} 5 & 3 \\ -3 & 5 \end{bmatrix}$, $\mathbf{M}_4 = \begin{bmatrix} 2 & 1 \\ -1 & 3 \end{bmatrix}$

22. $\mathbf{F}(-\infty, \infty)$: $\mathbf{f}_1 = \sin t$, $\mathbf{f}_2 = \cos t$

23. $\mathbf{F}(-\infty, \infty)$: $\mathbf{f}_1 = \sin^2 t$, $\mathbf{f}_2 = \cos^2 t$, $\mathbf{f}_3 = \cos 2t$

24. $\mathbf{F}(-\infty, \infty)$: $\mathbf{f}_1 = e^t$, $\mathbf{f}_2 = e^{2t}$

25. $\mathbf{F}(-\infty, \infty)$: $\mathbf{f}_1 = \ln t$, $\mathbf{f}_2 = \ln 3t$, $\mathbf{f}_3 = \ln 3$

26. $\mathbf{P}_2$: $\mathbf{p}_1 = 1 - t$, $\mathbf{p}_2 = 1 + t + t^2$, $\mathbf{p}_3 = 3 - t + t^2$

27. $\mathbf{P}_2$: $\mathbf{p}_1 = 1 + 2t^2$, $\mathbf{p}_2 = 1 - 2t^2$, $\mathbf{p}_3 = 1 + t + t^2$

28. $\mathbf{P}_2$: $\mathbf{p}_1 = 1$, $\mathbf{p}_2 = 1 + t$, $\mathbf{p}_3 = 1 + t + t^2$

29. $\mathbf{P}_3$: $\mathbf{p}_1 = 3 - 2t$, $\mathbf{p}_2 = 1 + 2t + 3t^3$, $\mathbf{p}_3 = t - 2t^3$

Exercises 30–33. *Extend the given set of objects $\mathcal{S}$ to a basis by adjoining objects from the standard basis for the given vector space.*

30. $\mathbb{R}^{2\times 2}$: $\mathbf{M}_1 = \begin{bmatrix} 1 & 1 \\ 1 & 0 \end{bmatrix}$, $\mathbf{M}_2 = \begin{bmatrix} 1 & 0 \\ 0 & 1 \end{bmatrix}$, $\mathbf{M}_3 = \begin{bmatrix} 0 & 1 \\ 1 & 1 \end{bmatrix}$

31. $\mathbb{R}^{2\times 2}$: $\mathbf{M}_1 = \begin{bmatrix} 1 & 1 \\ 0 & 0 \end{bmatrix}$, $\mathbf{M}_2 = \begin{bmatrix} 0 & 0 \\ 1 & 1 \end{bmatrix}$, $\mathbf{M}_3 = \begin{bmatrix} -1 & 0 \\ 0 & -1 \end{bmatrix}$

32. $\mathbf{P}_2$: $1 - t, 1 + t$

33. $\mathbf{P}_3$: $1 - t^3, t - t^2$

Exercises 34–35. *Find a basis for the given subspace $\mathcal{U}$ and determine $\dim \mathcal{U}$.*

34. $\mathbf{P}_3$: $\mathcal{U}$ consists of all polynomials $\mathbf{p}$ of degree 3 or less with $\mathbf{p}(1) = 0$.

35. $\mathcal{U} = \operatorname{span} \mathcal{S}$, where $\mathcal{S} = \{1 - t, t - t^2, t^2 + t^3, t^3\}$.

Exercises 36–37. *Describe the subspace $\mathcal{U}$ of $\mathbb{R}^{2\times 2}$ spanned by the set of objects $\mathcal{S}$. Find a basis for $\mathcal{U}$ and determine $\dim \mathcal{U}$.*

36. $\mathcal{S}$: $\mathbf{M}_1 = \begin{bmatrix} 1 & 0 \\ 0 & 1 \end{bmatrix}$, $\mathbf{M}_2 = \begin{bmatrix} 0 & 0 \\ 1 & 0 \end{bmatrix}$, $\mathbf{M}_3 = \begin{bmatrix} 0 & 1 \\ 0 & 0 \end{bmatrix}$

37. $\mathcal{S}$: $\mathbf{M}_1 = \begin{bmatrix} 1 & 0 \\ 0 & 1 \end{bmatrix}$, $\mathbf{M}_2 = \begin{bmatrix} 1 & 0 \\ 1 & 0 \end{bmatrix}$, $\mathbf{M}_3 = \begin{bmatrix} 0 & 1 \\ 1 & 1 \end{bmatrix}$

38. Consider the set of those diagonal matrices $\mathbf{D}$ in $\mathbb{R}^{2\times 2}$ whose main diagonal entries sum to zero. Show that $\mathbf{D}$ is a proper subspace of $\mathbb{R}^{2\times 2}$ and find a basis for $\mathbf{D}$.

39. Show that the set $\mathcal{U}$ of all polynomials $\mathbf{p} = a + bt + ct^2$ such that $a + b + c = 0$ is a subspace of $\mathbf{P}_2$. Find a basis for $\mathcal{U}$.

40. Prove that an ordered list of objects $\mathcal{S} = \{\mathbf{v}_1, \mathbf{v}_2, \ldots, \mathbf{v}_k\}$ is linearly dependent if and only if some object in the list $\mathcal{S}$ is a linear combination of objects that precede it in the list. *Hint*: Suppose $\mathcal{S}$ is linearly dependent and let j be the largest subscript in equation (7.12) for which $x_j \neq 0$.

41. Show that $\mathbf{F}(-\infty, \infty)$ is a vector space. Let $\mathbf{f} = f(x)$, $\mathbf{g} = g(x)$, $\mathbf{h} = h(x)$ be objects in $\mathbf{F}(-\infty, \infty)$ that are twice differentiable. The determinant function $W(x)$ defined by

$$W(x) = \begin{vmatrix} f(x) & g(x) & h(x) \\ f'(x) & g'(x) & h'(x) \\ f''(x) & g''(x) & h''(x) \end{vmatrix}$$

is called the *Wronskian* of $\mathbf{f}, \mathbf{g}, \mathbf{h}$. Show that if $W(x) \neq \mathbf{0}$, where $\mathbf{0}$ is the zero function, then the set $\{\mathbf{f}, \mathbf{g}, \mathbf{h}\}$ is linearly independent. This concept is applied in the study of differential equations.

Exercises 42–43. *Determine if the given set of polynomials is a basis for the given vector space of polynomials.*

42. $\mathbf{P}_3$: $1,\ 1+t,\ 1+t+t^2,\ 1+t+t^2+t^3$

43. $\mathbf{P}_4$: $1,\ 1-t,\ t-t^2,\ t^2-t^3,\ t^3-t^4$

USING MATLAB

44. Find a basis for the subspace $\mathcal{U}$ of $\mathbb{R}^{2\times 2}$ spanned by the vectors

$$\begin{bmatrix} 1 & -1 \\ 1 & 5 \end{bmatrix} \begin{bmatrix} 1 & -4 \\ 5 & -2 \end{bmatrix} \begin{bmatrix} 4 & -8 \\ -10 & 14 \end{bmatrix} \begin{bmatrix} 1 & -5 \\ -7 & 1 \end{bmatrix}.$$

State the dimension of $\mathcal{U}$.

45. Show that the set $\mathcal{B} = \{1, 1-t, (1-t)^2, (1-t)^3)\}$ is a basis for $\mathbf{P}_3$. Find the coordinates of the polynomials below relative to $\mathcal{B}$.

(a) $2 + t + 2t^2 - 3t^3$

(b) $2 - t^2 + t^3$

Exercises 46–49. *Use the Symbolic Math Toolbox and the Wronskian to show that the given set $\mathcal{S}$ is linearly independent.*

46. $\mathcal{S} = \{1, 2x, 3x^2\}$

47. $\mathcal{S} = \{\cos x, \sin x, x\cos x\}$

48. $\mathcal{S} = \{2, 3x, 4e^x\}$

49. $\mathcal{S} = \{xe^x, x^2e^{2x}, x^3e^{3x}\}$

Exercises 50–51. *Determine if the matrix $\mathbf{M}$ is a linear combination of the given matrices.*

50. $\mathbf{M} = \begin{bmatrix} 10 & 14 \\ 10 & -20 \end{bmatrix}$: $\begin{bmatrix} 3 & 6 \\ 9 & -12 \end{bmatrix}, \begin{bmatrix} 0 & 9 \\ 3 & 6 \end{bmatrix}, \begin{bmatrix} 4 & 8 \\ 0 & 0 \end{bmatrix}$

51. $\mathbf{M} = \begin{bmatrix} 16 & 4 \\ 28 & 40 \end{bmatrix}$: $\begin{bmatrix} 3 & 3 \\ 3 & 3 \end{bmatrix}, \begin{bmatrix} 9 & 1 \\ 0 & 0 \end{bmatrix}, \begin{bmatrix} 4 & 4 \\ -8 & -12 \end{bmatrix}$

7.3 Coordinates, Linear Transformations

We begin by noting that Theorem 3.8 in Section 3.2 and its proof are valid in the context of an abstract vector space. This theorem, restated next, is fundamental to defining a *coordinate system* in a vector space $\mathcal{V}$ relative to a given basis for $\mathcal{V}$.

Theorem 7.9 **Unique Representation Relative to a Basis**

Let $\mathcal{V}$ be an n–dimensional vector space and let $\mathcal{B} = \{\mathbf{v}_1, \mathbf{v}_2, \ldots, \mathbf{v}_n\}$ be an ordered[3] basis for $\mathcal{V}$. Then every object $\mathbf{v}$ in $\mathcal{V}$ can be written in only one way as a linear combination of basis objects, namely,

$$\mathbf{v} = x_1\mathbf{v}_1 + x_2\mathbf{v}_2 + \cdots + x_n\mathbf{v}_n, \tag{7.23}$$

where the scalars $x_1, x_2, \ldots, x_n$ are unique.

[3] The order in which the objects appear in the list matters: $\{\mathbf{v}_1, \mathbf{v}_2\}$ is different from $\{\mathbf{v}_2, \mathbf{v}_1\}$.

Coordinates

Refer to Theorem 7.9. With every object $\mathbf{v}$ in $\mathcal{V}$ is associated a unique column vector in $\mathbb{R}^n$, denoted by $[\mathbf{v}]_{\mathcal{B}}$, with components $x_1, x_2, \ldots, x_n$. We have

$$[\mathbf{v}]_{\mathcal{B}} = \begin{bmatrix} x_1 \\ x_2 \\ \vdots \\ x_n \end{bmatrix}. \tag{7.24}$$

Definition 7.8

Coordinates

The column vector $[\mathbf{v}]_{\mathcal{B}}$ shown in (7.24) is called the *coordinate vector* of $\mathbf{v}$ relative to the ordered basis $\mathcal{B}$, and $x_1, x_2, \ldots, x_n$ are called the *coordinates* of $\mathbf{v}$ relative to $\mathcal{B}$.

Caution! Changing the order of the basis objects will of course change the order in which the coordinates appear in the coordinate vector. Hence order is important.

EXAMPLE 1

A Coordinate System for a Subspace of $\mathbb{R}^{2\times 2}$

The set $\mathcal{D}$ of all diagonal matrices is a subspace of $\mathbb{R}^{2\times 2}$ and the standard basis $\mathcal{B}$ for $\mathcal{D}$, shown next, confirms that dim $\mathcal{D} = 2$.

$$\mathcal{B} = \{\mathbf{D}_1, \mathbf{D}_2\}, \quad \text{where} \quad \mathbf{D}_1 = \begin{bmatrix} 1 & 0 \\ 0 & 0 \end{bmatrix}, \quad \mathbf{D}_2 = \begin{bmatrix} 0 & 0 \\ 0 & 1 \end{bmatrix}$$

Figure 7.4 shows a schematic of $\mathcal{D}$, where $\mathbf{O}$ is the zero object in $\mathbb{R}^{2\times 2}$.

Every object $\mathbf{D}$ in $\mathcal{D}$ is a unique linear combination of basis objects and its coordinate vector $[\mathbf{D}]_{\mathcal{B}}$ relative to $\mathcal{B}$ is shown next.

$$\mathbf{D} = \begin{bmatrix} x_1 & 0 \\ 0 & x_2 \end{bmatrix} = \begin{bmatrix} x_1 & 0 \\ 0 & 0 \end{bmatrix} + \begin{bmatrix} 0 & 0 \\ 0 & x_2 \end{bmatrix} = x_1\mathbf{D}_1 + x_2\mathbf{D}_2 \quad \Rightarrow \quad [\mathbf{D}]_{\mathcal{B}} = \begin{bmatrix} x_1 \\ x_2 \end{bmatrix}$$

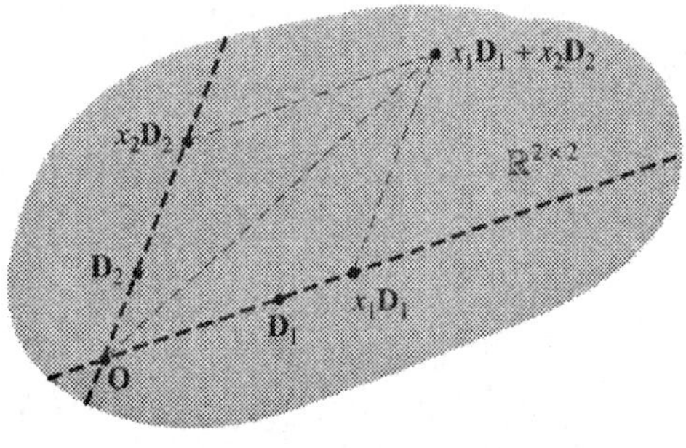

Figure 7.4 Schematic of a coordinate system in $\mathcal{D}$.

We will see in due course that $\mathcal{D}$ has the same structure as (is *isomorphic* to) the plane $\mathbb{R}^2$. ■

Let $\mathcal{V}$ be an n-dimensional vector space and let $\mathcal{B} = \{\mathbf{v}_1, \mathbf{v}_2, \ldots, \mathbf{v}_n\}$ be an ordered basis for $\mathcal{V}$. The uniqueness property in Theorem 7.9 allows us to define a transformation (function) $G_{\mathcal{B}} : \mathcal{V} \to \mathbb{R}^n$ from $\mathcal{V}$ to $\mathbb{R}^n$ given by the rule

$$G_{\mathcal{B}}(\mathbf{v}) = [\mathbf{v}]_{\mathcal{B}}, \quad \text{for all } \mathbf{v} \text{ in } \mathcal{V}.$$

Then $G_{\mathcal{B}}$ transforms (or maps) each object $\mathbf{v}$ in $\mathcal{V}$ to its coordinate vector in $\mathbb{R}^n$, or, equivalently, $[\mathbf{v}]_{\mathcal{B}}$ is the *image* of $\mathbf{v}$ under the mapping $G_{\mathcal{B}}$. Note that $G_{\mathcal{B}}$ depends on the basis $\mathcal{B}$.

EXAMPLE 2

Visualizing the Coordinate Mapping in $\mathbf{P}_2$

Consider the polynomial $\mathbf{p} = 2 + 4t - 2t^2$ in the vector space $\mathbf{P}_2$ of all polynomials of degree 2 or less and the ordered set of objects $\mathcal{B}$, where

$$\mathcal{B} : \mathbf{p}_1 = 2, \quad \mathbf{p}_2 = -1 + 2t, \quad \mathbf{p}_3 = 1 + t^2.$$

Refer to Figure 7.5. The coordinate mapping $G_{\mathcal{S}} : \mathbf{P}_2 \to \mathbb{R}^3$ relative to the standard basis $\mathcal{S} = \{1, t, t^2\}$ for $\mathbf{P}_2$ is given by $G_{\mathcal{S}}(\mathbf{p}) = [\mathbf{p}]_{\mathcal{S}}$ and we have

$$[\mathbf{p}_1]_{\mathcal{S}} = \begin{bmatrix} 2 \\ 0 \\ 0 \end{bmatrix}, \quad [\mathbf{p}_2]_{\mathcal{S}} = \begin{bmatrix} -1 \\ 2 \\ 0 \end{bmatrix}, \quad [\mathbf{p}_3]_{\mathcal{S}} = \begin{bmatrix} 1 \\ 0 \\ 1 \end{bmatrix}, \quad [\mathbf{p}]_{\mathcal{S}} = \begin{bmatrix} 2 \\ 4 \\ -2 \end{bmatrix}.$$

The set $\mathcal{B}$ is a basis for $\mathbf{P}_2$ (verify). If $\mathbf{p}(t) = a + bt + ct^2$ is any polynomial in $\mathbf{P}_2$, then $\mathbf{p}$ can be expressed in the form $\mathbf{p} = x_1\mathbf{p}_1 + x_2\mathbf{p}_2 + x_3\mathbf{p}_3$; that is,

$$a + bt + ct^2 = x_1(2) + x_2(-1 + 2t) + x_3(1 + t^2), \tag{7.25}$$

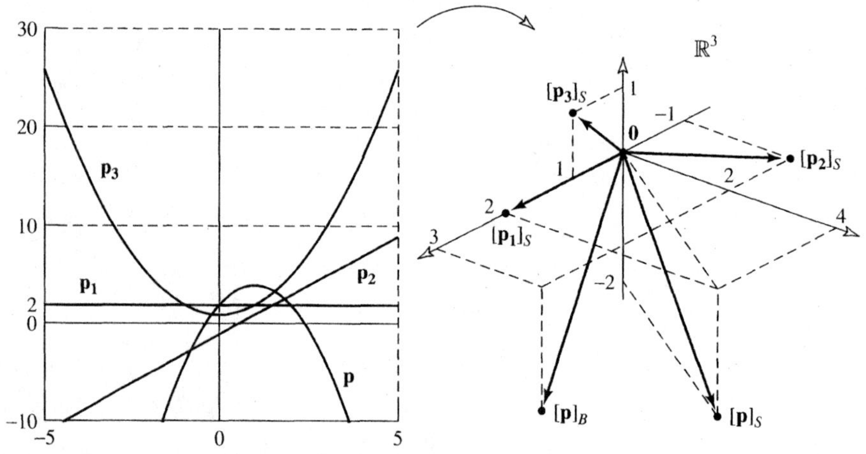

Figure 7.5 Visualizing the coordinate mapping in $\mathbf{P}_2$.

where the coordinates x_1, x_2, x_3 of $\mathbf{p}$ relative to $\mathcal{B}$ are unique. Solving (7.25), the coordinate mapping $G_{\mathcal{B}} : \mathbf{P}_2 \to \mathbb{R}^3$ relative to $\mathcal{B}$ is defined by

$$G_{\mathcal{B}}(\mathbf{p}) = [\,\mathbf{p}\,]_{\mathcal{B}} = \begin{bmatrix} x_1 \\ x_2 \\ x_3 \end{bmatrix} = \begin{bmatrix} (2a+b-2c)/4 \\ b/2 \\ c \end{bmatrix},$$

for all $\mathbf{p}$ in $\mathbf{P}_2$. For example, the coordinates of the polynomial $\mathbf{p} = 2+4t-2t^2$ relative to $\mathcal{B}$ are $[\,\mathbf{p}\,]_{\mathcal{B}} = [3, 2, -2]^{\mathsf{T}}$. ■

In order to say more about the coordinate mapping, and its properties we need to recall the main concepts connected with linear transformations from Section 3.4 and rephrase them in the general setting.

Definition 7.9

Linear transformation

Let $\mathcal{V}$ and $\mathcal{W}$ be vector spaces. A transformation $T : \mathcal{V} \to \mathcal{W}$ with domain $\mathcal{V}$ and codomain $\mathcal{W}$ is called *linear* if two conditions are satisfied:

$$T(\mathbf{u}+\mathbf{v}) = T(\mathbf{u}) + T(\mathbf{v}), \tag{7.26}$$

$$T(c\mathbf{v}) = cT(\mathbf{v}), \tag{7.27}$$

for all objects $\mathbf{u}$, $\mathbf{v}$ in $\mathcal{V}$ and scalars c.

Caution! The symbol $+$ is used (ambiguously) for the operation of addition, both in $\mathcal{V}$ and in $\mathcal{W}$. In (7.26), $\mathbf{u}+\mathbf{v}$ is the sum of objects $\mathbf{u}$ and $\mathbf{v}$ in $\mathcal{V}$, while $T(\mathbf{u}) + T(\mathbf{v})$ is the sum of objects $T(\mathbf{u})$ and $T(\mathbf{v})$ in $\mathcal{W}$. A similar remark applies to scalar multiplication given by (7.27).

Putting $c = 0$ in (7.27) and using (**P4**) from Section 7.1, we have $T(\mathbf{0}) = 0T(\mathbf{v}) = \mathbf{0}$, showing that any linear transformation T maps the zero object in $\mathcal{V}$ to the zero object in $\mathcal{W}$.

Theorem 3.6 in Section 3.4 and its proof translate into the general setting, namely, a transformation $T : \mathcal{V} \to \mathcal{W}$ is linear if and only if the equation

$$T(x_1\mathbf{v}_1 + x_2\mathbf{v}_2) = x_1T(\mathbf{v}_1) + x_2T(\mathbf{v}_2) \tag{7.28}$$

is true for all objects $\mathbf{v}_1$, $\mathbf{v}_2$ in $\mathcal{V}$ and all scalars x_1 and x_2. Using mathematical induction, equation (7.28) extends to k summands, namely

$$T(x_1\mathbf{v}_1 + x_2\mathbf{v}_2 + \cdots + x_k\mathbf{v}_k) = x_1T(\mathbf{v}_1) + x_2T(\mathbf{v}_2) + \cdots + x_kT(\mathbf{v}_k). \tag{7.29}$$

Suppose $T : \mathcal{V} \to \mathcal{W}$ is a linear transformation on $\mathcal{V}$. Then

(a) T is *one-to-one* if $T(\mathbf{v}_1) = T(\mathbf{v}_2)$ implies $\mathbf{v}_1 = \mathbf{v}_2$ for all $\mathbf{v}_1$, $\mathbf{v}_2$ in $\mathcal{V}$.
(b) T is *onto* if for each object $\mathbf{w}$ in $\mathcal{W}$ there is an object $\mathbf{v}$ in $\mathcal{V}$ such that $T(\mathbf{v}) = \mathbf{w}$.
(c) T is an *isomorphism* if T is both one-to-one and onto.

The vector spaces $\mathcal{V}$ and $\mathcal{W}$ are called *isomorphic* (*iso* meaning "same" and *morphic* meaning "structure") if there exists an isomorphism T from $\mathcal{V}$ onto

$\mathcal{W}$. For example, generalizing Example 1, the vector space $\mathcal{D}$ of all diagonal matrices in $\mathbb{R}^{n\times n}$ is isomorphic to $\mathbb{R}^n$ (Exercises 7.3). More generally, the next key result shows that any n-dimensional vector space is isomorphic with $\mathbb{R}^n$.

Theorem 7.10 **Coordinate Mapping Theorem**

Let $\mathcal{V}$ be an n-dimensional vector space and let $\mathcal{B} = \{\mathbf{v}_1, \mathbf{v}_2, \ldots, \mathbf{v}_n\}$ be an ordered basis for $\mathcal{V}$. Then the coordinate mapping $G_{\mathcal{B}} : \mathcal{V} \to \mathbb{R}^n$ defined by $G_{\mathcal{B}}(\mathbf{v}) = [\mathbf{v}]_{\mathcal{B}}$ is a linear transformation that is one-to-one and onto. Hence $G_{\mathcal{B}}$ is an isomorphism.

Proof

(a) Let $x_1, x_2, \ldots, x_n$ and $y_1, y_2, \ldots, y_n$ denote, respectively, the (unique) coordinates of objects $\mathbf{v}_1$ of $\mathbf{v}_2$ in $\mathcal{V}$ relative to $\mathcal{B}$. Then

$$[\mathbf{v}_1 + \mathbf{v}_2]_{\mathcal{B}} = \begin{bmatrix} x_1 + y_1 \\ x_2 + y_2 \\ \vdots \\ x_n + y_n \end{bmatrix} = \begin{bmatrix} x_1 \\ x_2 \\ \vdots \\ x_n \end{bmatrix} + \begin{bmatrix} y_1 \\ y_2 \\ \vdots \\ y_n \end{bmatrix} = [\mathbf{v}_1]_{\mathcal{B}} + [\mathbf{v}_2]_{\mathcal{B}},$$

showing that $G_{\mathcal{B}}(\mathbf{v}_1 + \mathbf{v}_2) = G_{\mathcal{B}}(\mathbf{v}_1) + G_{\mathcal{B}}(\mathbf{v}_2)$. Also, if c is any scalar, then

$$[c\mathbf{v}_1]_{\mathcal{B}} = \begin{bmatrix} cx_1 \\ cx_2 \\ \vdots \\ cx_n \end{bmatrix} = c\begin{bmatrix} x_1 \\ x_2 \\ \vdots \\ x_n \end{bmatrix} = c[\mathbf{v}_1]_{\mathcal{B}},$$

showing that $G_{\mathcal{B}}(c\mathbf{v}_1) = cG_{\mathcal{B}}(\mathbf{v}_1)$. Hence, Definition 7.9, $G_{\mathcal{B}}$ is linear.

(b) The property $G_{\mathcal{B}}(\mathbf{v}_1) = G_{\mathcal{B}}(\mathbf{v}_2)$ implies that $\mathbf{v}_1 = \mathbf{v}_2$ follows from the uniqueness of representation in Theorem 7.9. Hence $G_{\mathcal{B}}$ is one-to-one.

(c) For every column vector $\mathbf{x}$ in $\mathbb{R}^n$ there is an object $\mathbf{v}$ defined by (7.23), where the scalars $x_1, x_2, \ldots, x_n$ are the components from $\mathbf{x}$, that has property $G_{\mathcal{B}}(\mathbf{v}) = [\mathbf{v}]_{\mathcal{B}} = \mathbf{x}$. Hence $G_{\mathcal{B}}$ is onto.

The mapping $G_{\mathcal{B}}$ is therefore an isomorphism.

Equation (7.29) implies that, for any objects $\mathbf{v}_1, \mathbf{v}_2, \ldots, \mathbf{v}_k$ in $\mathcal{V}$, we have

$$[x_1\mathbf{v}_1 + x_2\mathbf{v}_2 + \cdots + x_k\mathbf{v}_k]_{\mathcal{B}} = x_1[\mathbf{v}_1]_{\mathcal{B}} + x_2[\mathbf{v}_2]_{\mathcal{B}} + \cdots + x_k[\mathbf{v}_k]_{\mathcal{B}}. \quad (7.30)$$

It follows from Theorem 7.10 that any two n-dimensional vector spaces are isomorphic (Exercises 7.3).

Change of Basis

Let $\mathcal{V}$ be an n-dimensional vector space and suppose that $\mathcal{C} = \{\mathbf{v}_1, \mathbf{v}_2, \ldots, \mathbf{v}_n\}$ is an ordered basis for $\mathcal{V}$, which we will regard as the *current* basis. Let $\mathcal{B} = \{\mathbf{u}_1, \mathbf{u}_2, \ldots, \mathbf{u}_n\}$ be an ordered basis for $\mathcal{V}$, which we will regard as the

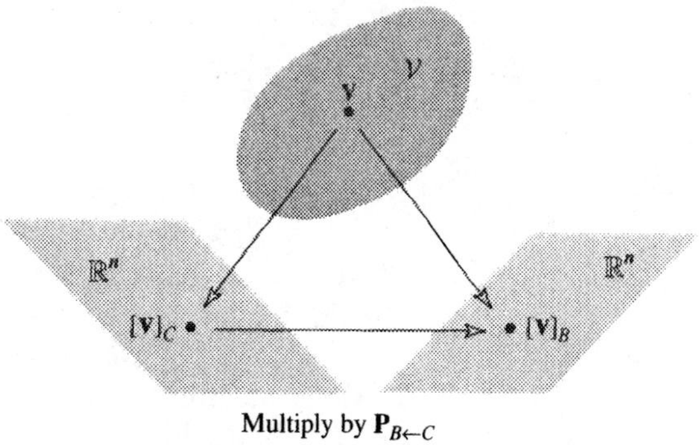

Figure 7.6 Change of basis in a vector space $\mathcal{V}$.

new basis. Assuming that the coordinates of an object $\mathbf{v}$ in $\mathcal{V}$ relative to the current basis $\mathcal{C}$ are known, our goal is to compute the coordinates of $\mathbf{v}$ relative to $\mathcal{B}$. Proceed as follows and refer to Figure 7.6.

Using the notation in (7.23), we have

$$\mathbf{v} = x_1\mathbf{v}_1 + x_2\mathbf{v}_2 + \cdots + x_n\mathbf{v}_n \qquad \Rightarrow \qquad [\mathbf{v}]_{\mathcal{C}} = \begin{bmatrix} x_1 \\ x_2 \\ \vdots \\ x_n \end{bmatrix},$$

and using the linearity of the coordinate mapping (7.30), we have

$$\begin{aligned} [\mathbf{v}]_{\mathcal{B}} &= [x_1\mathbf{v}_1 + x_2\mathbf{v}_2 + \cdots + x_n\mathbf{v}_n]_{\mathcal{B}} \\ &= x_1[\mathbf{v}_1]_{\mathcal{B}} + x_2[\mathbf{v}_2]_{\mathcal{B}} + \cdots + x_n[\mathbf{v}_n]_{\mathcal{B}} \\ &= [\,[\mathbf{v}_1]_{\mathcal{B}} \; [\mathbf{v}_2]_{\mathcal{B}} \; \cdots \; [\mathbf{v}_n]_{\mathcal{B}}\,][\mathbf{v}]_{\mathcal{C}}. \end{aligned} \tag{7.31}$$

From (7.31), we have

$$\begin{array}{ccc} \text{New basis } \mathcal{B} & \overset{\text{Transition}}{\longleftarrow} & \text{Current basis } \mathcal{C} \\ [\mathbf{v}]_{\mathcal{B}} = \mathbf{P}_{\mathcal{B} \leftarrow \mathcal{C}}[\mathbf{v}]_{\mathcal{C}} & & [\mathbf{v}]_{\mathcal{C}} \end{array} \tag{7.32}$$

where

$$\mathbf{P}_{\mathcal{B} \leftarrow \mathcal{C}} = [\,[\mathbf{v}_1]_{\mathcal{B}} \quad [\mathbf{v}_2]_{\mathcal{B}} \quad \cdots \quad [\mathbf{v}_n]_{\mathcal{B}}\,].$$

The matrix $\mathbf{P}_{\mathcal{B} \leftarrow \mathcal{C}}$ in (7.32) is called the *transition matrix* for change of basis from $\mathcal{C}$ to $\mathcal{B}$.

The matrix $\mathbf{P}_{\mathcal{B} \leftarrow \mathcal{C}}$ is invertible because its columns are the images of the linearly independent set $\mathcal{C}$ under the isomorphism $G_{\mathcal{B}}$ and are therefore linearly independent (Exercises 7.3). Thus the transition matrix from $\mathcal{B}$ back to $\mathcal{C}$ is given by

$$[\mathbf{v}]_{\mathcal{C}} = \mathbf{P}^{-1}_{\mathcal{B} \leftarrow \mathcal{C}}[\mathbf{v}]_{\mathcal{B}}. \tag{7.33}$$

Matrix Representation of a Linear Transformation

Refer to Theorem 3.18 in Section 3.4. It was shown there that for each linear transformation $T : \mathbb{R}^n \to \mathbb{R}^m$ there is a unique $m \times n$ matrix $\mathbf{A} = [T(\mathbf{e}_1)\ T(\mathbf{e}_2)\ \ldots\ T(\mathbf{e}_n)]$ such that $T(\mathbf{x}) = \mathbf{Ax}$, for all column vectors $\mathbf{x}$ in $\mathbb{R}^n$. Note that $\{\mathbf{e}_1, \mathbf{e}_2, \ldots, \mathbf{e}_n\}$ is the standard basis for $\mathbb{R}^n$ and for this reason we sometimes refer to $\mathbf{A}$ as the *standard matrix* for T. Although $\mathbf{A}$ is unique relative to the standard basis, the same linear transformation can be defined by $T(\mathbf{x}) = \mathbf{Bx}$, where the $m \times n$ matrix $\mathbf{B}$ is defined relative to some other basis $\mathcal{B}$ for $\mathbb{R}^n$. We will shortly describe the connection between the matrices $\mathbf{A}$ and $\mathbf{B}$. However, we first move the discussion into the general context by showing that any linear transformation $T : \mathcal{V} \to \mathcal{W}$ between finite-dimensional vector spaces $\mathcal{V}$ and $\mathcal{W}$ is represented by a matrix.

Let $\mathcal{V}$ and $\mathcal{W}$ be finite-dimensional vector spaces and let $\mathcal{C} = \{\mathbf{v}_1, \mathbf{v}_2, \ldots, \mathbf{v}_n\}$ and $\mathcal{B} = \{\mathbf{w}_1, \mathbf{w}_2, \ldots, \mathbf{w}_m\}$ be ordered bases for $\mathcal{V}$ and $\mathcal{W}$, respectively.

Refer to Figure 7.7. Consider any object $\mathbf{v}$ in $\mathcal{V}$ and let $\mathbf{w} = T(\mathbf{v})$. The goal is to compute the coordinates of $\mathbf{w}$ relative to $\mathcal{B}$ using the coordinates of $\mathbf{v}$ relative to $\mathcal{C}$ and the coordinates of the images $T(\mathbf{v}_j)$ $(1 \leq j \leq n)$ relative to $\mathcal{B}$.

Let $\mathbf{x} = [x_1\ x_2\ \cdots\ x_n]^{\mathsf{T}}$ be the coordinate vector for $\mathbf{v}$ relative to $\mathcal{C}$; that is, $\mathbf{x} = [\mathbf{v}]_{\mathcal{C}}$. Using (7.29), we have

$$\mathbf{v} = x_1\mathbf{v}_1 + x_2\mathbf{v}_2 + \cdots + x_n\mathbf{v}_n,$$

$$T(\mathbf{v}) = x_1T(\mathbf{v}_1) + x_2T(\mathbf{v}_2) + \cdots + x_nT(\mathbf{v}_n). \tag{7.34}$$

For $1 \leq j \leq n$, let $a_{1j}, a_{2j}, \ldots, a_{mj}$ be the coordinates of $T(\mathbf{v}_j)$ relative to $\mathcal{B}$. Then $[T(\mathbf{v}_j)]_{\mathcal{B}} = [a_{1j}\ a_{2j}\ \cdots\ a_{mj}]^{\mathsf{T}}$ and

$$T(\mathbf{v}_j) = a_{1j}\mathbf{w}_1 + a_{2j}\mathbf{w}_2 + \cdots + a_{mj}\mathbf{w}_m. \tag{7.35}$$

For each j, substituting the expression for $T(\mathbf{v}_j)$ from (7.35) into (7.34), we have

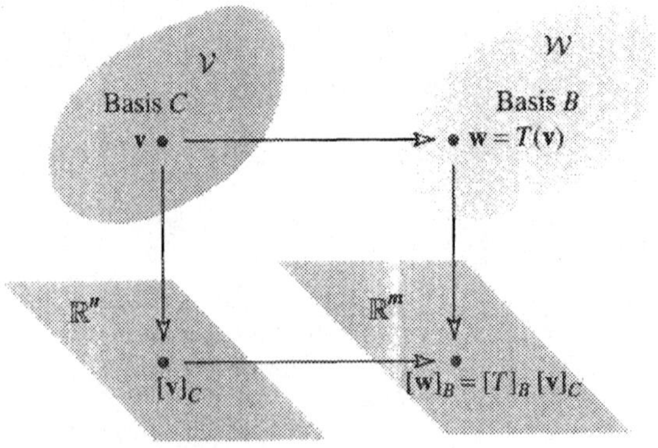

Figure 7.7 The matrix of a linear transformation.

$$\begin{aligned} T(\mathbf{v}) = x_1(&a_{11}\mathbf{w}_1 + a_{21}\mathbf{w}_2 + \cdots + a_{i1}\mathbf{w}_i + \cdots + a_{m1}\mathbf{w}_m) \\ &+ x_2(a_{12}\mathbf{w}_1 + a_{22}\mathbf{w}_2 + \cdots + a_{i2}\mathbf{w}_i + \cdots + a_{m2}\mathbf{w}_m) \\ &\vdots \\ &+ x_n(a_{1n}\mathbf{w}_1 + a_{2n}\mathbf{w}_2 + \cdots + a_{in}\mathbf{w}_i + \cdots + a_{mn}\mathbf{w}_m). \end{aligned} \tag{7.36}$$

Define an $m \times n$ matrix $[T]_{\mathcal{B}}$ as follows:

$$\begin{aligned} [T]_{\mathcal{B}} &= \begin{bmatrix} a_{11} & a_{12} & \cdots & a_{1n} \\ a_{21} & a_{22} & \cdots & a_{2n} \\ \vdots & \vdots & & \vdots \\ a_{m1} & a_{m2} & \cdots & a_{mn} \end{bmatrix} \\ &= [\,[T(\mathbf{v}_1)]_{\mathcal{B}}\ [T(\mathbf{v}_2)]_{\mathcal{B}}\ \cdots\ [T(\mathbf{v}_n)]_{\mathcal{B}}\,]. \end{aligned} \tag{7.37}$$

Collecting terms in (7.36) and referring to (7.37), we see that the coefficient of $\mathbf{w}_i$ is the product $[\,a_{i1}\ a_{i2}\ \cdots\ a_{in}\,]\mathbf{x}$, of row i in $[T]_{\mathcal{B}}$ with the coordinate vector of $\mathbf{v}$ relative to $\mathcal{C}$ and so

$$[T(\mathbf{v})]_{\mathcal{B}} = [T]_{\mathcal{B}}[\mathbf{v}]_{\mathcal{C}}. \tag{7.38}$$

As in the case of the standard matrix, it can be shown (Exercises 7.3) that $[T]_{\mathcal{B}}$ is unique; that is, it is the only $m \times n$ matrix satisfying (7.38) for all $\mathbf{v}$ in $\mathcal{V}$.

The notation $[T]_{\mathcal{B}}$ reminds us that the columns in (7.37) are formed relative to the basis $\mathcal{B}$. A linear transformation $T : \mathcal{V} \to \mathcal{W}$ is well-defined if given the coordinate vector $[\mathbf{v}]_{\mathcal{C}}$ of each object $\mathbf{v}$ in $\mathcal{V}$, we can compute the corresponding coordinate vector $[T(\mathbf{v})]_{\mathcal{B}}$ of its image $T(\mathbf{v})$ in $\mathcal{W}$. Thus the real $m \times n$ matrix $[T]_{\mathcal{B}}$ *represents* T because the relationship between these sets of coordinates is given by (7.38).

Notice that the standard matrix representing the linear transformation $T : \mathbb{R}^n \to \mathbb{R}^m$ relative to the standard basis has exactly the form (7.37).

EXAMPLE 3

A Linear Transformation on $\mathbf{P}_2$

Define a linear transformation $T : \mathbf{P}_1 \to \mathbf{P}_2$ by the formula $T(\mathbf{p}) = \mathbf{p}(4 + 2t + 3t^2)$, for all objects $\mathbf{p}$ in $\mathbf{P}_1$. Let $\mathcal{C} = \{1, t\}$ and $\mathcal{B} = \{1, t, t^2\}$ be the standard bases for $\mathbf{P}_1$ and $\mathbf{P}_2$, respectively. The images of basis objects in $\mathcal{C}$ are: $T(1) = 1$ and $T(t) = 4 + 2t + 3t^2$ and we use these to construct the columns of $[T]_{\mathcal{B}}$. We have

$$[T]_{\mathcal{B}} = [\,[T(1)]_{\mathcal{B}}\ [T(t)]_{\mathcal{B}}\,] = \begin{bmatrix} 1 & 4 \\ 0 & 2 \\ 0 & 3 \end{bmatrix}.$$

Let $\mathbf{p} = a + bt$. Then by (7.38), we have

$$[T(\mathbf{v})]_{\mathcal{B}} = [T]_{\mathcal{B}}[\mathbf{p}]_{\mathcal{C}} = \begin{bmatrix} 1 & 4 \\ 0 & 2 \\ 0 & 3 \end{bmatrix}\begin{bmatrix} a \\ b \end{bmatrix} = \begin{bmatrix} a + 4b \\ 2b \\ 3b \end{bmatrix}$$

and by direct computation, we have $T(\mathbf{p}) = a + b(4 + 2t + 3t^2) = (a + 4b) + 2bt + 3bt^2$. ■

Similar Matrices

Consider an n-dimensional vector space $\mathcal{V}$ and a linear operator (transformation) $T : \mathcal{V} \to \mathcal{V}$ that maps $\mathcal{V}$ to itself. Let $\mathcal{C} = \{\mathbf{v}_1, \mathbf{v}_2, \ldots, \mathbf{v}_n\}$ be an ordered basis for $\mathcal{V}$. Referring to the preceding theory, we have $\mathcal{W} = \mathcal{V}$ and $\mathcal{B} = \mathcal{C}$. The operator T is defined by an $n \times n$ matrix $[T]_\mathcal{C}$ and equation (7.38) becomes

$$[T(\mathbf{v})]_\mathcal{C} = [T]_\mathcal{C}[\mathbf{v}]_\mathcal{C}, \tag{7.39}$$

where the columns of $[T]_\mathcal{C}$ are the coordinate vectors of $T(\mathbf{v}_j)$ relative to $\mathcal{C}$. Suppose that $\mathcal{B} = \{\mathbf{u}_1, \mathbf{u}_2, \ldots, \mathbf{u}_n\}$ is a second ordered basis for $\mathcal{V}$. Then T is defined similarly by a matrix $[T]_\mathcal{B}$ and the equation corresponding to (7.39) is

$$[T(\mathbf{v})]_\mathcal{B} = [T]_\mathcal{B}[\mathbf{v}]_\mathcal{B}, \tag{7.40}$$

where the columns of $[T]_\mathcal{B}$ are the coordinate vectors of $T(\mathbf{u}_j)$ relative to $\mathcal{B}$.

Our goal is to describe the relationship between $[T]_\mathcal{C}$ and $[T]_\mathcal{B}$.

Theorem 7.11 **Similarity**

Using the preceding notation, we have

$$[T]_\mathcal{B} = \mathbf{P}[T]_\mathcal{C}\mathbf{P}^{-1}, \tag{7.41}$$

where $\mathbf{P} = \mathbf{P}_{\mathcal{B}\leftarrow\mathcal{C}}$ *is the transition matrix from* $\mathcal{C}$ *to* $\mathcal{B}$.

Proof Because $\mathcal{V}$ is isomorphic to $\mathbb{R}^n$, we can regard $[\mathbf{v}]_\mathcal{C}$ as an arbitrary vector in $\mathbb{R}^n$. Substituting $T(\mathbf{v})$ for $\mathbf{v}$ in (7.32), we have

$$[\mathbf{v}]_\mathcal{B} = \mathbf{P}[\mathbf{v}]_\mathcal{C} \quad \Rightarrow \quad \begin{aligned} [T(\mathbf{v})]_\mathcal{B} &= \mathbf{P}[T(\mathbf{v})]_\mathcal{C} \\ &= \mathbf{P}[T]_\mathcal{C}[\mathbf{v}]_\mathcal{C} \quad \text{[by (7.39)]} \end{aligned}$$

and

$$\begin{aligned} [T(\mathbf{v})]_\mathcal{B} &= [T]_\mathcal{B}[\mathbf{v}]_\mathcal{B} \quad \text{[by (7.40)]} \\ &= [T]_\mathcal{B}\mathbf{P}[\mathbf{v}]_\mathcal{C} \quad \text{[by (7.32)]} \end{aligned}$$

Hence, from the right sides of the preceding equations,

$$\mathbf{P}[T]_\mathcal{C}[\mathbf{v}]_\mathcal{C} = [T]_\mathcal{B}\mathbf{P}[\mathbf{v}]_\mathcal{C},$$

and we conclude that $\mathbf{P}[T]_\mathcal{C} = [T]_\mathcal{B}\mathbf{P}$ because $[\mathbf{v}]_\mathcal{C}$ is arbitrary (Exercises 7.3). The result follows since $\mathbf{P}$ is invertible.

Remarks

As with the theory of change of basis explained earlier in the section, suppose that $\mathcal{C}$ is regarded as the current basis and that the matrix $[T]_{\mathcal{C}}$ representing T is known. If $\mathcal{B}$ is regarded as the new basis, then the matrix $[T]_{\mathcal{B}}$ which represents T relative to $\mathcal{B}$ is given by (7.41).

Definition 7.10

Similar matrices

Let $\mathbf{B}$ and $\mathbf{C}$ be two $n \times n$ matrices. Then $\mathbf{B}$ is *similar* to $\mathbf{C}$, denoted by $\mathbf{B} \sim \mathbf{C}$, if there exists an invertible $n \times n$ matrix $\mathbf{P}$ such that $\mathbf{B} = \mathbf{PCP}^{-1}$.

Connecting Definition 7.10 with (7.41), we see that a pair of similar $n \times n$ matrices define the same linear operator on an n-dimensional vector space $\mathcal{V}$ relative to suitably chosen bases $\mathcal{C}$ & $\mathcal{B}$.

Some properties of similar matrices are given in Exercises 7.3. However, we will mention here that $\mathbf{B}$ is similar to $\mathbf{C}$ implies that $\mathbf{C}$ is similar to $\mathbf{B}$, and so similarity is a symmetric relationship.

EXAMPLE 4

A Linear Operator on $\mathbb{R}^2$

Consider a linear operator T on $\mathbb{R}^2$ defined by $T(x, y) = (x - y, x + y)$, and let $\mathcal{C}$ and $\mathcal{B}$ be bases for $\mathbb{R}^2$ as shown.

$$\mathcal{C}: \ \mathbf{v}_1 = \begin{bmatrix} 1 \\ 2 \end{bmatrix}, \mathbf{v}_2 = \begin{bmatrix} -1 \\ 3 \end{bmatrix}, \quad \mathcal{B}: \ \mathbf{u}_1 = \begin{bmatrix} 1 \\ 1 \end{bmatrix}, \mathbf{v}_1 = \begin{bmatrix} -1 \\ 0 \end{bmatrix}$$

We have

$$T(\mathbf{v}_1) = \begin{bmatrix} -1 \\ 3 \end{bmatrix}, \quad T(\mathbf{v}_2) = \begin{bmatrix} -4 \\ 2 \end{bmatrix}, \quad T(\mathbf{u}_1) = \begin{bmatrix} 0 \\ 2 \end{bmatrix}, \quad T(\mathbf{u}_1) = \begin{bmatrix} -1 \\ -1 \end{bmatrix}.$$

The coordinate vectors $[T(\mathbf{v}_1)]_{\mathcal{C}}$, $[T(\mathbf{v}_2)]_{\mathcal{C}}$ are found from the reduced form of the matrix $[\,\mathbf{v}_1 \ \mathbf{v}_2 \mid T(\mathbf{v}_1) \ T(\mathbf{v}_2)\,]$ and a similar computation (replacing $\mathbf{v}$ by $\mathbf{u}$) gives $[T(\mathbf{u}_1)]_{\mathcal{B}}$, $[T(\mathbf{u}_2)]_{\mathcal{B}}$. Hence

$$[T]_{\mathcal{C}} = \begin{bmatrix} 0 & -2 \\ 1 & 2 \end{bmatrix}, \quad [T]_{\mathcal{B}} = \begin{bmatrix} 2 & -1 \\ 2 & 0 \end{bmatrix}.$$

The transition matrix $\mathbf{P} = \mathbf{P}_{\mathcal{B}\leftarrow\mathcal{C}}$ is obtained from the reduced form of $[\,\mathbf{u}_1 \ \mathbf{u}_1 \mid \mathbf{v}_1 \ \mathbf{v}_2\,]$ and from this calculation we have

$$\mathbf{P} = \begin{bmatrix} 2 & 3 \\ 1 & 4 \end{bmatrix} \quad \text{and} \quad \mathbf{P}^{-1} = \begin{bmatrix} 0.8 & -0.6 \\ -0.2 & 0.4 \end{bmatrix}$$

and so

$$\mathbf{P}[T]_{\mathcal{C}}\mathbf{P}^{-1} = \begin{bmatrix} 2 & 3 \\ 1 & 4 \end{bmatrix}\begin{bmatrix} 0 & -2 \\ 1 & 2 \end{bmatrix}\begin{bmatrix} 0.8 & -0.6 \\ -0.2 & 0.4 \end{bmatrix} = \begin{bmatrix} 2 & -1 \\ 2 & 0 \end{bmatrix} = [T]_{\mathcal{B}}.$$

■

Diagonalization

We now revisit Section 6.2 to explain the connection between eigenvalue problems and similarity. We say that an $n \times n$ matrix $\mathbf{A}$ is diagonalizable if there exists an invertible $n \times n$ matrix $\mathbf{P}$ such that $\mathbf{P}^{-1}\mathbf{A}\mathbf{P} = \mathbf{D}$, where $\mathbf{D}$ is diagonal. According to Theorem 6.3, $\mathbf{A}$ is diagonalizable if and only if $\mathbf{A}$ has n linearly independent eigenvectors—these vectors form the columns of the diagonalizing matrix $\mathbf{P}$ and the corresponding eigenvalues lie on the diagonal of $\mathbf{D}$. This means that $\mathbf{A}$ is diagonalizable if and only if $\mathbf{A}$ is similar to the diagonal matrix $\mathbf{D}$. Moreover, $\mathbf{A}$ and $\mathbf{D}$ define the same linear operator T on $\mathbb{R}^n$ and so choosing to use the basis for $\mathbb{R}^n$ consisting of the n linearly independent eigenvectors of $\mathbf{A}$ makes the representation of T the simplest possible.

EXAMPLE 5

Diagonalization and Similarity

The details of the following illustration are left for Exercises 7.3. The matrix

$$\mathbf{A} = \begin{bmatrix} 1 & 1 \\ -2 & 4 \end{bmatrix}$$

has eigenvalues $\lambda_1 = 2$ and $\lambda_2 = 3$ and is diagonalizable, with diagonalizing matrix $\mathbf{P}$, where

$$\mathbf{P} = \begin{bmatrix} 1 & 1 \\ 1 & 2 \end{bmatrix} \quad \text{and} \quad \mathbf{v}_1 = \begin{bmatrix} 1 \\ 1 \end{bmatrix}, \quad \mathbf{v}_2 = \begin{bmatrix} 1 \\ 2 \end{bmatrix}$$

are linearly independent eigenvectors of $\mathbf{A}$ associated with λ_1 and λ_2, respectively. We have

$$\mathbf{P}^{-1} = \begin{bmatrix} 2 & -1 \\ -1 & 1 \end{bmatrix}, \quad \mathbf{D} = \begin{bmatrix} 2 & 0 \\ 0 & 3 \end{bmatrix} \quad \text{and} \quad \mathbf{P}^{-1}\mathbf{A}\mathbf{P} = \mathbf{D}.$$

The goal is to understand the connection between the linear operators $T_{\mathbf{A}}$ and $T_{\mathbf{D}}$ on $\mathbb{R}^2$ represented by the matrices $\mathbf{A}$ and $\mathbf{D}$, respectively. We have

$$\mathbf{x} = \begin{bmatrix} x_1 \\ x_2 \end{bmatrix} \quad \Rightarrow \quad T_{\mathbf{A}}(\mathbf{x}) = \mathbf{A}\mathbf{x} = \begin{bmatrix} x_1 + x_2 \\ -2x_1 + 4x_2 \end{bmatrix} \quad \text{and}$$

$$T_{\mathbf{D}}(\mathbf{x}) = \mathbf{D}\mathbf{x} = \begin{bmatrix} 2x_1 \\ 3x_2 \end{bmatrix}$$

for all $\mathbf{x}$ in $\mathbb{R}^2$. Consider a fixed vector $\mathbf{u}$ in $\mathbb{R}^2$ and its image $T_{\mathbf{A}}(\mathbf{u})$. For example,

$$\mathbf{u} = \begin{bmatrix} 2 \\ 0 \end{bmatrix} \quad \text{and} \quad T_{\mathbf{A}}(\mathbf{u}) = \mathbf{A}\mathbf{u} = \begin{bmatrix} 2 \\ -4 \end{bmatrix}.$$

Calculating coordinates of $\mathbf{u}$ and $T_{\mathbf{A}}(\mathbf{u})$ relative to the basis $\mathcal{B} = \{\mathbf{v}_1, \mathbf{v}_2\}$ for $\mathbb{R}^2$, we have

$$[\,\mathbf{u}\,]_{\mathcal{B}} = \begin{bmatrix} 4 \\ -2 \end{bmatrix}, \quad [\,T_{\mathbf{A}}(\mathbf{u})\,]_{\mathcal{B}} = \begin{bmatrix} 8 \\ -6 \end{bmatrix} \Rightarrow T_{\mathbf{D}}([\,\mathbf{u}\,]_{\mathcal{B}}) = \mathbf{D}[\,T_{\mathbf{A}}(\mathbf{u})\,]_{\mathcal{B}}.$$

Although the example is small-scale, the advantage of working with a basis of eigenvectors is clear—relative to $\mathcal{B}$, the image of a vector $\mathbf{x}$ under $T_{\mathbf{D}}$ is obtained by scaling each component by each eigenvalue. ■

Topics for Further Study

We will round out this section by mentioning briefly some further connections between Section 3.4 and the theory of linear transformations on general vector spaces. The details are left for Exercises 7.3.

Consider a linear transformation $T : \mathcal{V} \to \mathcal{W}$ on a vector space $\mathcal{V}$ to a vector space $\mathcal{W}$.

Kernel and Range

The subset of $\mathcal{V}$, denoted by $\ker T$, defined by

$$\ker T = \{\mathbf{v} \text{ in } \mathcal{V} | T(\mathbf{v}) = 0\}$$

is called the *kernel* of T. The subset of $\mathcal{W}$, denoted by $\operatorname{ran} T$, defined by

$$\operatorname{ran} T = \{\mathbf{w} \text{ in } \mathcal{W} | T(\mathbf{v}) = \mathbf{w} \text{ for some } \mathbf{v} \text{ in } \mathcal{V}\}$$

is called the *range* of T. Then $\ker T$ is the set of all objects in $\mathcal{V}$ that are mapped by T onto the zero object in $\mathcal{W}$ and $\operatorname{ran} T$ is the set of all images under T of objects in $\mathcal{V}$.

Then $\ker T$ is a subspace of $\mathcal{V}$ and $\operatorname{ran} T$ is a subspace of $\mathcal{W}$. The dimension of $\ker T$ is called the *nullity* of T and the dimension of $\operatorname{ran} T$ is called the *rank* of T. The connections between the nullity and rank of a linear transformation T and the nullity and rank of a matrix that represents T will be given shortly.

If T is the *zero transformation* defined by $T(\mathbf{v}) = \mathbf{0}$ for all $\mathbf{v}$ in $\mathcal{V}$, then $\ker T = \mathcal{V}$ and $\operatorname{ran} T = \{\mathbf{0}\}$. If T is onto, then $\operatorname{ran} T = \mathcal{W}$.

Theorem 7.12 **Nullity and Rank**

Let $T : \mathcal{V} \to \mathcal{W}$ be a linear transformation on an n-dimensional vector space $\mathcal{V}$ to a vector space $\mathcal{W}$. Then nullity $T + \operatorname{rank} T = n$.

We can recognize one-to-one linear transformations using the kernel.

Theorem 7.13 **Characterizing One-to-One**

Let $T : \mathcal{V} \to \mathcal{W}$ be a linear transformation on a vector space $\mathcal{V}$ to a vector space $\mathcal{W}$. Then T is one-to-one if and only if $\ker T = \{\mathbf{0}\}$*; that is, the nullity of T is zero.*

Consider a linear transformation $T : \mathcal{V} \to \mathcal{W}$ from an n-dimensional vector space $\mathcal{V}$ with basis $\mathcal{C}$ to an m-dimensional vector space $\mathcal{W}$ with basis $\mathcal{B}$ and let $[T]_{\mathcal{B}}$ be the $m \times n$ matrix that represents T relative to $\mathcal{B}$. Recall that $G_{\mathcal{C}}$ and $G_{\mathcal{B}}$ denote the coordinate mappings from $\mathcal{V}$ onto $\mathbb{R}^n$ and from $\mathcal{W}$ onto $\mathbb{R}^m$, respectively. Using the fact that $G_{\mathcal{C}}$ and $G_{\mathcal{B}}$ are isomorphisms, we have

$$\text{An object } \mathbf{v} \text{ is in } \ker T \text{ if and only if } [\mathbf{v}]_{\mathcal{B}} \text{ is in null}[T]_{\mathcal{B}} \tag{7.42}$$

$$\text{An object } \mathbf{w} \text{ is in } \operatorname{ran} T \text{ if and only if } [\mathbf{w}]_{\mathcal{C}} \text{ is in } \operatorname{col}[T]_{\mathcal{B}} \tag{7.43}$$

Using (7.42) and (7.43), it is now possible to describe $\ker T$ *and* $\operatorname{ran} T$ *by defining bases for these two subspaces. Using the above notation, let* $r = \operatorname{rank}[T]_{\mathcal{B}}$. *Then $\nu = n - r$ is the nullity of $[T]_{\mathcal{B}}$.*

Theorem 7.14 **Kernel and Range**

(a) If the null space of $[T]_{\mathcal{B}}$ has a basis $\mathbf{x}_1, \mathbf{x}_2, \ldots, \mathbf{x}_\nu$, then $\ker T$ is of dimension ν and has a basis $\{G_{\mathcal{C}}^{-1}(\mathbf{x}_1), G_{\mathcal{C}}^{-1}(\mathbf{x}_2), \ldots, G_{\mathcal{C}}^{-1}(\mathbf{x}_\nu)\}$.
(b) If the column space of $[T]_{\mathcal{B}}$ has a basis $\mathbf{y}_1, \mathbf{y}_2, \ldots, \mathbf{y}_r$, then $\operatorname{col} T$ is of dimension r and has a basis $\{G_{\mathcal{B}}^{-1}(\mathbf{y}_1), G_{\mathcal{B}}^{-1}(\mathbf{y}_2), \ldots, G_{\mathcal{B}}^{-1}(\mathbf{y}_r)\}$.

EXERCISES 7.3

Exercises 1–2. *Check that $\mathcal{B}$ is a basis for the given vector space and find the coordinates of the given object relative to $\mathcal{B}$.*

1. $\mathbf{P}_1$, $\mathcal{B} = \{1 - t, 1 + t\}$, $\mathbf{p} = 3 - 6t$
2. $\mathbf{P}_2$, $\mathcal{B} = \{t, 1 + t, -t + t^2\}$, $\mathbf{p} = 3 + 16t - 6t^2$

Exercises 3–5. *Let **P** be an $n \times n$ matrix. Show that the given linear operator T on $\mathbb{R}^{n\times n}$ defined by mapping **A** to $T(\mathbf{A})$, for all **A** in $\mathbb{R}^{n\times n}$ is linear.*

3. $T(\mathbf{A}) = \mathbf{AP} + \mathbf{PA}$
4. $T(\mathbf{A}) = \mathbf{AP} - \mathbf{PA}$
5. $T(\mathbf{A}) = \mathbf{P}^{-1}\mathbf{AP}$, where **P** is invertible.
6. Let **B** be an $m \times n$ matrix. Show that the transformation $T_{\mathcal{B}} : \mathbb{R}^{n\times p} \to \mathbb{R}^{m\times p}$ given by $T_{\mathcal{B}}(\mathbf{A}) = \mathbf{BA}$, for all objects **A** in $\mathbb{R}^{n\times p}$ is linear. Take $n = m = 3$ and $p = 2$. Find the matrix $[T]_S$ that represents T relative to the standard bases.
7. Let $\mathcal{V}$ and $\mathcal{W}$ be n-dimensional vector spaces. (a) Show that $\mathbb{R}^n$ is isomorphic to $\mathcal{V}$. (b) Show that $\mathcal{V}$ and $\mathcal{W}$ are isomorphic.
8. Show that the vector space of all diagonal matrices in $\mathbb{R}^{n\times n}$ is isomorphic to $\mathbb{R}^n$.
9. *Algebra of Linear Transformations.* Let S and T be linear transformations from the vector space $\mathcal{V}$ to the vector space $\mathcal{W}$ and let α be any scalar. Show that the transformations $S + T$ and αT on $\mathcal{V}$ defined by
$$(S + T)(\mathbf{v}) = S(\mathbf{v}) + T(\mathbf{v}),$$
$$(\alpha T)(\mathbf{v}) = \alpha T(\mathbf{v}),$$
for all **v** in $\mathcal{V}$, are linear.
10. Let $\mathcal{U}$, $\mathcal{V}$, $\mathcal{W}$ be vector spaces and let $S : \mathcal{U} \to \mathcal{V}$ and let $T : \mathcal{V} \to \mathcal{W}$ be linear transformations on $\mathcal{U}$ and $\mathcal{V}$, respectively. Show that the *composition* $T \circ S$ (note the order) mapping from $\mathcal{U}$ to $\mathcal{W}$ is linear, where
$$(T \circ S)(\mathbf{v}) = T(S(\mathbf{v}))$$
for all **v** in $\mathcal{V}$.

Exercises 11–15. *Let $T : \mathcal{V} \to \mathcal{W}$ be a linear transformation between vector spaces $\mathcal{V}$ and $\mathcal{W}$ and let $\mathcal{S} = \{\mathbf{v}_1, \mathbf{v}_2, \ldots, \mathbf{v}_k\}$ be subset of $\mathcal{V}$. Prove each statement.*

11. If $\mathcal{S}$ is a linearly dependent, then $\{T(\mathbf{v}_1), T(\mathbf{v}_2), \ldots, T(\mathbf{v}_k)\}$ is linearly dependent. State the contrapositive of this implication.

12. If T is onto and $\mathcal{S}$ spans $\mathcal{V}$, then $\{T(\mathbf{v}_1), T(\mathbf{v}_2), \ldots, T(\mathbf{v}_k)\}$ spans $\mathcal{W}$.

13. If T is one-to-one, then $\dim \mathcal{V} \leq \dim \mathcal{W}$.

14. If T is onto, then $\dim \mathcal{V} \geq \dim \mathcal{W}$.

15. If T is one-to-one and onto and $\mathcal{S}$ is a basis for $\mathcal{V}$, then $\{T(\mathbf{v}_1), T(\mathbf{v}_2), \ldots, T(\mathbf{v}_k)\}$ is a basis for $\mathcal{W}$.

16. Use previous exercises to show that the transition matrix $\mathbf{P}_{\mathcal{B}\leftarrow\mathcal{C}}$ from basis $\mathcal{C}$ to basis $\mathcal{B}$ is invertible.

17. Give the details of why, in Theorem 7.11, the equation $\mathbf{P}[T]_{\mathcal{C}}[\mathbf{v}]_{\mathcal{C}} = [T]_{\mathcal{B}}\mathbf{P}[\mathbf{v}]_{\mathcal{C}}$ implies that $\mathbf{P}[T]_{\mathcal{C}} = [T]_{\mathcal{B}}\mathbf{P}$. *Hint*: Consult Theorem B.3 in Appendix B: Toolbox.

18. Show that the matrix $[\,T\,]_{\mathcal{B}}$ in (7.38) is unique.

Exercises 19–20. Find the matrix representation of each linear operator on $\mathbf{P}_3$ *relative to the standard basis* $\{1, t, t^2, t^3\}$ *for* $\mathbf{P}_3$.

19. $T(\mathbf{p}) = \mathbf{p} - \mathbf{p}(0)$

20. $T(\mathbf{p}) = \mathbf{p}(0) + \mathbf{p}(1)t + \mathbf{p}(2)t^2$

21. Refer to Definition 7.10 and Appendix B: Toolbox for a discussion of equivalence relations. Prove the following properties for $n \times n$ matrices **B**, **C**, **D**.

(a) $\mathbf{A} \sim \mathbf{A}$

(b) $\mathbf{A} \sim \mathbf{B}$ implies $\mathbf{B} \sim \mathbf{A}$

(c) $\mathbf{A} \sim \mathbf{B}$ and $\mathbf{B} \sim \mathbf{C}$ implies $\mathbf{A} \sim \mathbf{C}$

Thus, $\sim$ is an equivalence relation on the set $\mathbb{R}^{n\times n}$ of all $n \times n$ matrices. $\mathbb{R}^{n\times n}$ is *partitioned* into disjoint equivalence classes—any two matrices belonging to the same class are similar and matrices from different equivalence classes are not similar.

Exercises 22–24. *Let* **B** *and* **C** *be* $n\times n$ *similar matrices. Prove each property.*

22. $\det(\mathbf{B}) = \det(\mathbf{C})$ and $\operatorname{trace}\mathbf{B} = \operatorname{trace}\mathbf{C}$.

23. **B** and **C** have the same set of eigenvalues.

24. $\operatorname{rank}\mathbf{B} = \operatorname{rank}\mathbf{C}$ and $\nu(\mathbf{B}) = \nu(\mathbf{C})$, where ν denotes the nullity of a matrix.

Exercises 25–26. Show that the matrices are not similar.

25. $\begin{bmatrix} 1 & 2 \\ 3 & 4 \end{bmatrix}, \quad \begin{bmatrix} 2 & 2 \\ 2 & 2 \end{bmatrix}$

26. $\begin{bmatrix} 1 & 2 \\ 3 & 4 \end{bmatrix}, \quad \begin{bmatrix} 2 & 1 \\ 1 & 3 \end{bmatrix}$

27. Show that the trace, determinant, rank and eigenvalues of the given matrices are identical. Show that the matrices are similar by finding an invertible matrix **P** such that $\mathbf{P}^{-1}\mathbf{A}\mathbf{P} = \mathbf{B}$.

$$\mathbf{A} = \begin{bmatrix} 1 & 2 \\ 2 & 4 \end{bmatrix}, \quad \mathbf{B} = \begin{bmatrix} 1 & 1 \\ 4 & 4 \end{bmatrix}$$

28. Let **A** and **B** be $n \times n$ similar matrices. Show that $\mathbf{A} + \alpha\mathbf{I}_n$ is similar to $\mathbf{B} + \alpha\mathbf{I}_n$, for any scalar α.

29. Let $T : \mathcal{V} \to \mathcal{W}$ be a linear transformation on a vector space $\mathcal{V}$ to a vector space $\mathcal{W}$. Show that $\ker T$ is a subspace of $\mathcal{V}$ and that $\operatorname{ran} T$ is a subspace of $\mathcal{W}$.

Exercises 30–32. For each linear operator T defined on $\mathbf{P}_2$, *find a basis for (a)* $\ker T$, *(b)* $\operatorname{ran} T$.

30. $T(a + bt + ct^2) = (b + c) + (a + b + c)t + (c - 2b)t^2$

31. $T(a + bt + ct^2) = b + (a + b)t + (a + b + c)t^2$

32. $T(\mathbf{p}) = (1+t)\mathbf{p}' - 3\mathbf{p}$, where $\mathbf{p}'$ is the first derivative of **p**.

33. Consider a transformation $T : \mathbf{P}_n \to \mathbf{P}_n$ defined by $T(\mathbf{p}) = \mathbf{p} + t\mathbf{p}'$, where $\mathbf{p}'$ is the first derivative of **p**. Show that T is linear. Show also that T is one-to-one and onto.

34. Let p and q be fixed scalars. Consider the linear transformation T defined by

$$T(a + bt + ct^2) = \begin{bmatrix} pa & c \\ c & qb \end{bmatrix},$$

from the vector space $\mathbf{P}_2$ to the vector space $\mathcal{S}_{2\times 2}$ of all 2×2 symmetric matrices. Using the standard bases for $\mathbf{P}_2$ and $\mathcal{S}_{2\times 2}$, find the matrix $[T]_{\mathcal{S}}$ that represents T. Find $\ker T$ and $\operatorname{ran} T$ and give bases for these subspaces.

Exercises 35–38. Label each statement as either true (T) of false (F). If a false statement can be made true by modifying the working, then do so.

35. If $\mathbf{B}$ is a fixed $n \times n$ matrix, then the linear operator on $\mathbb{R}^n$ defined by $T(\mathbf{A}) = \mathbf{AB} - \mathbf{BA}$ is one-to-one.

36. If T is a linear operator on the vector space $\mathcal{V}$ such that $\ker T = \operatorname{ran} T$ then $\dim \mathcal{V}$ is even.

37. Let $T : \mathcal{V} \to \mathcal{W}$ be a linear transformation between vector spaces $\mathcal{V}$ and $\mathcal{W}$. If T is one-to-one and $\dim \mathcal{V} = \dim \mathcal{W}$, then T is onto.

38. The vector space $\mathbf{P}_2$ is isomorphic with the vector space of real 2×2 symmetric matrices.

USING MATLAB

39. Check that the set $\mathcal{B}$ is a basis for $\mathbb{R}^{2\times 2}$ and find the coordinates of the object $\mathbf{A}$ relative to $\mathcal{B}$.

$$\mathcal{B}: \begin{bmatrix} 1 & 1 \\ 1 & 1 \end{bmatrix}, \begin{bmatrix} 0 & -2 \\ 0 & 0 \end{bmatrix}, \begin{bmatrix} 0 & 0 \\ 1 & 1 \end{bmatrix}, \begin{bmatrix} 0 & 1 \\ 1 & 1 \end{bmatrix},$$

$$\mathbf{A} = \begin{bmatrix} -2 & 9 \\ 4 & 4 \end{bmatrix}$$

40. Consider the basis $\mathcal{B}$ for $\mathbb{R}^{2\times 2}$ in the previous exercise and the linear operator T on $\mathbb{R}^{2\times 2}$ defined by $T(\mathbf{A}) = \mathbf{A}^{\mathsf{T}}$ for all objects $\mathbf{A}$ in $\mathbb{R}^{2\times 2}$. Find the matrix representation of T relative to $\mathcal{B}$ and compare with the matrix representation of T relative to the standard basis for $\mathbb{R}^{2\times 2}$.

41. *Project.* Determine if the matrices

$$\mathbf{A} = \begin{bmatrix} 1 & -1 & 0 \\ 0 & 2 & 1 \\ 1 & 1 & 0 \end{bmatrix}, \quad \mathbf{B} = \begin{bmatrix} 3 & 3 & -3.5 \\ -1 & -1 & 1.5 \\ 1 & 3 & 1 \end{bmatrix}$$

are similar.

42. *Project.* Suppose $T : \mathbb{R}^n \to \mathbb{R}^n$ is a linear operator on $\mathbb{R}^n$ such that $\operatorname{rank} T = r$. Let $\mathbf{v}_1, \mathbf{v}_2 \dots, \mathbf{v}_r$ be a basis for $\operatorname{ran} T$ and let $T(\mathbf{e}_j) = a_{j1}\mathbf{v}_1 + a_{j2}\mathbf{v}_2 + \cdots + a_{jr}\mathbf{v}_r$, where $\mathbf{e}_j$ is the $j$the column of $\mathbf{I}_n$. Show that T can be represented by the product

$$T(\mathbf{v}) = [\, \mathbf{v}_1 \ \mathbf{v}_2 \ \cdots \ \mathbf{v}_r \,] \begin{bmatrix} a_{11} & \cdots & a_{n1} \\ \dots & \ddots & \dots \\ a_{1r} & \cdots & a_{nr} \end{bmatrix} \mathbf{v},$$

for all $\mathbf{v}$ in $\mathbb{R}^n$. Work out the details of this decomposition when T is represented by magic(4).

CHAPTER 7 REVIEW

State whether each statement is true or false as stated. Provide a clear reason for your answer. If a false statement can be modified to make it true, then do so.

1. The zero object $\mathbf{0}$ in a vector space is unique.
2. $\mathbf{F}[a, b]$ is a subspace of $\mathbf{F}(-\infty, \infty)$.
3. The set of all functions $\mathbf{f}$ in $\mathbf{F}(-\infty, \infty)$ such that $\mathbf{f}(-1) = \mathbf{f}(1)$ is a subspace.
4. The set $\{1 - t, 2t - 2t^2, t^2, 1 + t^2\}$ in $\mathbf{P}_2$ is linearly dependent.
5. The set $\{\cos^2\theta, \sin^2\theta, 1\}$ in $\mathbf{F}(-\infty, \infty)$ is linearly independent.
6. If the set $\mathcal{S} = \{\mathbf{f}_1, \mathbf{f}_2\}$ is linearly independent in $\mathbf{F}(-\infty, \infty)$ then $\mathcal{S}$ is linearly independent in $\mathbf{F}[0, 1]$.
7. The set $\{t, 1 + t\}$ is a basis for $\mathbf{P}_1$.
8. If the set $\{\mathbf{v}_1, \mathbf{v}_2, \mathbf{v}_3\}$ is a basis for a vector space $\mathcal{V}$, then $\{\mathbf{v}_1, \mathbf{v}_1 + \mathbf{v}_2, \mathbf{v}_1 + \mathbf{v}_2 + \mathbf{v}_3\}$ is also a basis for $\mathcal{V}$.
9. If some object in a vector space $\mathcal{V}$ is a linear combination of objects $\mathbf{v}_1, \mathbf{v}_2, \dots, \mathbf{v}_k$, then some subset of $\mathbf{v}_1, \mathbf{v}_2, \dots, \mathbf{v}_k$ is a basis for $\mathcal{V}$.
10. If the set of objects $\{\mathbf{v}_1, \mathbf{v}_2, \mathbf{v}_3\}$ in a vector space $\mathcal{V}$ is linearly independent, then $\{\mathbf{v}_2, \mathbf{v}_3\}$ does not span $\mathcal{V}$.
11. A set consisting of one nonzero object in a vector space $\mathcal{V}$ can be extended to a basis for $\mathcal{V}$.
12. If every object in a vector space $\mathcal{V}$ can be written uniquely as a linear combination of objects from a subset $\mathcal{B}$ of $\mathcal{V}$, then $\mathcal{B}$ is a basis for $\mathcal{V}$.
13. The vector space of $n \times n$ upper triangular matrices has dimension $n^2/2$.
14. The vector space of $n \times n$ skew-symmetric matrices has dimension $n(n + 1)/2$.
15. The vector spaces $\mathbb{R}^{mn}$ and $\mathbb{R}^{m\times n}$ are isomorphic.
16. The subspace of $\mathbf{F}(-\infty, \infty)$ spanned by the set $\{\cos t, \sin t\}$ is isomorphic with $\mathbb{R}^2$.

17. The coordinate vector of the object $\begin{bmatrix} 1 & 2 \\ 3 & 4 \end{bmatrix}$ in $\mathbb{R}^{2\times 2}$ relative to any basis is a 4-vector.

18. If ordered bases for $\mathbf{P}_2$ are defined by $\mathcal{C} = \{1, t\}$ and $\mathcal{B} = \{1 - t, 1 + t\}$, then $\mathbf{P}_{\mathcal{B}\leftarrow\mathcal{C}} = 0.5 \begin{bmatrix} 1 & -1 \\ 1 & 1 \end{bmatrix}$.

19. The matrices $\begin{bmatrix} 3 & 4 \\ 1 & 2 \end{bmatrix}$ and $\begin{bmatrix} 1 & 2 \\ 3 & 4 \end{bmatrix}$ are similar.

20. The matrices $\begin{bmatrix} 2 & 2 \\ 6 & -1 \end{bmatrix}$ and $\begin{bmatrix} 1 & 2 \\ 3 & 2 \end{bmatrix}$ are similar.

Leonhard Euler (pronounced "oiler") was born in Basel, Switzerland, the son of a Calvinist pastor. At university he studied theology and philosophy in preparation for the ministry, but the course of his life changed under the influence of Johann Bernoulli (1667–1748), father of the famous Bernoulli family of mathematicians and physicists.

Euler was appointed to the Saint Petersburg Academy of Science, Russia, in 1727, where his international reputation as a mathematician and scientist grew. During this period he married, had 13 children and lost sight in one eye. He joined the Berlin Academy of Science in 1741, where his impressive output of significant research continued over the next two decades. In 1755 he was elected as a foreign member of the Paris Academy of Science, one of many honors, prizes, and awards that he would receive during his lifetime. On his return to Saint Petersburg in 1766, Euler became totally blind, but, aided by a prodigious memory, he continued to dictate his discoveries in mathematics, physics, and astronomy.

Euler is considered to be the leading mathematician of the 18th century and the most prolific writer of mathematics of all time. About 866 books and articles now fill 80 large volumes. His name is connected with most branches of mathematics, and in physics, particularly with the theory of optics, electricity and magnetism, and planetary motion. His solution to particular problems often created whole new fields of mathematics, as was the case with graph theory and combinatorial topology. The Laplace equation, Fourier coefficients, Lagrangian multipliers, and the gamma function can all be traced back to the writings of Leonhard Euler. At his death, Euler's office was filled with unpublished articles that would be published by the Saint Petersburg Academy over the next 50 years.

Much of the notation and terminology that Euler introduced has become standard today. These include the notation $f(x)$ for a function, concepts in analytic trigonometry, quadratic surfaces, the theory of investment and annuities, linear differential equations, and the theory of complex numbers. His books and papers were written with a clarity, detail, and completeness that serve as a model of style for the textbooks of today.

Chapter 8

COMPLEX NUMBERS

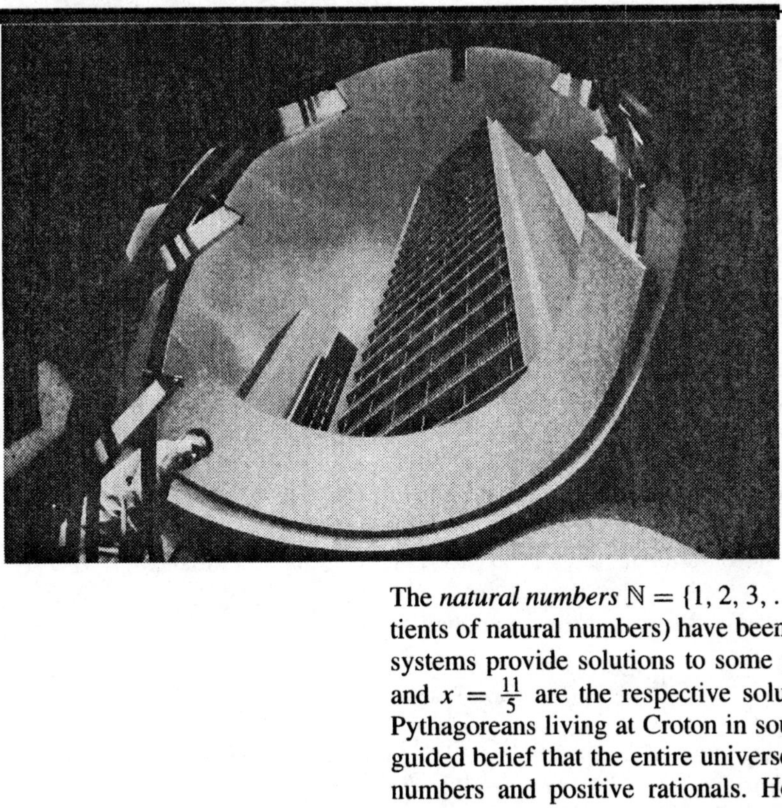

Introduction

The development of number systems has taken place over thousands of years and is tied closely to the search for solutions to *polynomial equations*, such as

$$x - 4 = 0, \qquad 5x - 11 = 0, \cdots \qquad \text{(linear)} \qquad (8.1)$$

$$x^2 - 2 = 0, \quad x^2 - 2x + 2 = 0, \cdots \qquad \text{(quadratic)} \qquad (8.2)$$

$$x^3 - 5 = 0, \quad x^3 - 15x - 4 = 0, \cdots \qquad \text{(cubic)} \qquad (8.3)$$

The *natural numbers* $\mathbb{N} = \{1, 2, 3, \ldots\}$ and the *positive rational numbers* (quotients of natural numbers) have been known for over 2000 years. These number systems provide solutions to some polynomial equations. For example, $x = 4$ and $x = \frac{11}{5}$ are the respective solutions to the linear equations in (8.1). The Pythagoreans living at Croton in southeastern Italy about 550 B.C. had the misguided belief that the entire universe was based on the properties of the natural numbers and positive rationals. However, their world changed dramatically with their discovery of a length x that satisfied the equation $x^2 - 2 = 0$ and that was not a positive rational.[1] From that time forward it became necessary to accept the existence of numbers which are *incommensurable*, or *irrational*. *Negative numbers* were not known at the time of the Pythagoreans and would not be fully incorporated into number systems until the 16th century.

[1] See Historical Notes at the end of this Introduction.

Some important sets of real numbers are defined next.

Integers	$\mathbb{Z} = \{ \ldots, -2, -1, 0, 1, 2, \ldots \}$
Rationals	$\mathbb{Q} = \{ r = \dfrac{m}{n} : m, n \text{ integers}, n \neq 0 \}$
Irrationals	$\mathbb{P}$

Clearly, $\mathbb{N} \subset \mathbb{Z}$ and each integer m can be written as a rational $m = m/1$ so that $\mathbb{Z} \subset \mathbb{Q}$. A *real number* is either rational or irrational and so the set of real numbers $\mathbb{R}$ is the union of the sets $\mathbb{Q}$ and $\mathbb{P}$. Using the decimal system, each rational number can be characterized by its decimal expansion, which either terminates or repeats in finite blocks of digits. Irrational numbers are therefore those real numbers that have nonterminating *and* nonrepeating decimal expansions. For example, the decimal expansions

$$\frac{33}{100} = 0.33000\ldots, \quad \frac{1}{3} = 0.\dot{3} = 0.3333\ldots, \quad e = 2.71828182845905\ldots$$

show two rational numbers and an irrational number e (the exponential number). Each irrational number can be viewed as the *limit* of an infinite sequence of rational numbers. For example, the irrational number $\sqrt{2} = 1.4142135623731\ldots$ is the limit of the sequence of rational numbers $1.4,\ 1.41, 1.414,\ 1.4142,\ \ldots$.

Babylonian mathematicians circa 400 B.C. were able to solve quadratic equations by completing the square. Applying their quadratic formula to the second equation in (8.2) gives solutions $x = 1 \pm \sqrt{-1}$, but the significance of the symbol $\sqrt{-1}$ would not be fully explained for over 1000 years after their time.

In 16th-century Italy, the problem of finding algebraic (closed form) solutions to cubic equations was solved by the Italian mathematicians Niccolo Fontana Tartaglia (circa 1499–1557) and Girolamo Cardano (1501–1576). However, certain cubics, such as the second equation in (8.3), were exceptional. The Cardano–Tartaglia formula applied to this equation gives

$$x = \sqrt[3]{2 + \sqrt{-121}} + \sqrt[3]{2 - \sqrt{-121}}, \tag{8.4}$$

but the meaning of the symbol $\sqrt{-121}$ could not be explained satisfactorily at the time[2] or why (8.4) includes the obvious root $x = 4$. The formula (8.4) actually gives the three roots of $x^3 - 15x - 4$, which turn out to be all real in this case, as seen in Figure 8.1.

The theory of roots of negative numbers was finally resolved by the development of the *complex number system* $\mathbb{C}$, which includes the real number system $\mathbb{R}$ and which in a certain sense is complete.[3] Gauss coined the term *complex number* and is credited with developing their general use.

[2] Rafael Bombelli (circa 1526–1573) gave an explanation that in some sense recognized (in modern language) that the sum of complex conjugate pairs is a real number.

[3] *The fundamental theorem of algebra* proved by Gauss (in four different ways!) states that a polynomial with complex coefficients of degree n has n real or complex (possibly repeated) roots. Hence, complex numbers suffice in this regard.

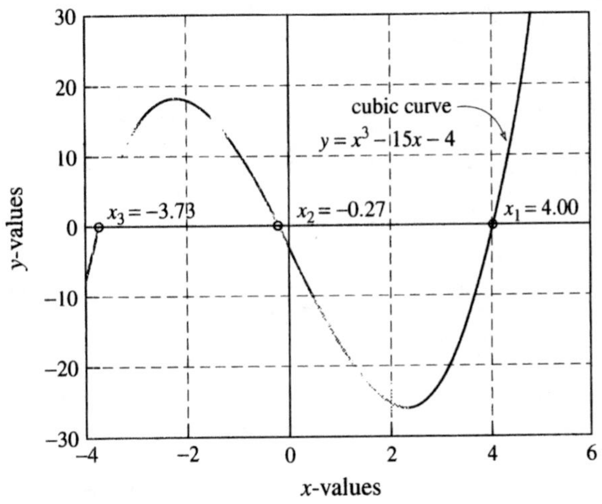

Figure 8.1 Roots of the cubic $x^3 - 15x - 4$.

Complex numbers appear in applications throughout engineering and science. This chapter presents the elementary properties of complex numbers and outlines connections with linear algebra.

Historical Notes

The Pythagoreans were a sect of Greek mathematicians. They were vegetarians, were politically conservative, and believed that their philosophical and mathematical studies were the moral basis for their lives. They are known mainly for developing properties of the positive integers (number theory), plane and solid geometry and the mathematics of music.

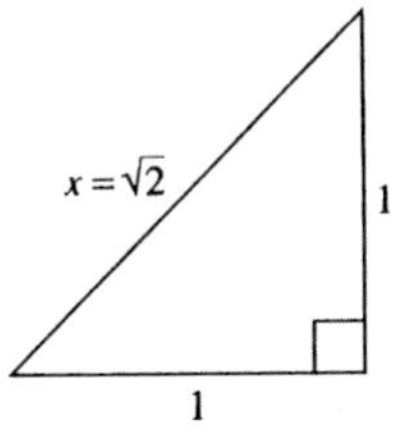

Existence of irrational numbers. Let x be the length of the hypotenuse in the right–angled isosceles triangle opposite. By the Pythagorean theorem, $x^2 = 1^2 + 1^2 = 2$. Suppose $x = \frac{m}{n}$, where m and n are natural numbers with no common divisor (we cancel any common factors). Then $\frac{m^2}{n^2} = 2 \Rightarrow m^2 = 2n^2$. Hence, m^2 is even, so m is even. Write $m = 2k$, where k is a natural number. Then $m^2 = 4k^2 = 2n^2 \Rightarrow n^2 = 2k^2$. Hence, n^2 is even, so n is even. But then m and n have a common factor 2, which contradicts the assumption. The number $x = \sqrt{2}$ is not rational.

8.1 Algebraic Theory

A number z is called a *complex number* if it has the form

$$z = a + bi, \tag{8.5}$$

where a and b are real numbers and the symbol i satisfies the equation $i^2 = -1$. The complex number z in (8.5) is said to be expressed in *standard form*.

We call a the *real part* of z and b the *imaginary part* of z and write

$$a = \text{Re}\, z, \qquad b = \text{Im}\, z.$$

For example, when $z = 2 - 3i$, we have $2 = \text{Re}\, z$ and $-3 = \text{Im}\, z$. The set of all complex numbers is denoted by $\mathbb{C}$.

A complex number z is called *purely real* if $b = 0$ in (8.5), and we write $z = a$ in place of $z = a + 0i$. A complex number z is called *purely imaginary* if $a = 0$ in (8.5), and we write $z = bi$ in place of $z = 0 + bi$. Setting $a = b = 0$ in (8.5) gives the *zero complex number* $z = 0 + 0i$, which is usually written in the abbreviated form $z = 0$.

It may be useful to think of a complex number $a + bi$ as a polynomial of first degree in the symbol i. From this point of view, complex numbers are manipulated as if they are polynomials, replacing i^2 by -1 when required.

Equality, Addition, Subtraction, Multiplication

Let $a + bi$ and $c + di$ be two complex numbers. Then

$$a + bi = c + di \quad \text{if and only if} \quad a = c \quad \text{and} \quad b = d. \tag{8.6}$$

In other words, two complex numbers are equal exactly when their real and imaginary parts are equal.

The sum and difference of $a + bi$ and $c + di$ are defined by

$$(a + bi) + (c + di) = (a + c) + (b + d)i \tag{8.7}$$

$$(a + bi) - (c + di) = (a - c) + (b - d)i. \tag{8.8}$$

The product of $a + bi$ and $c + di$ is computed as follows:

$$\begin{aligned}(a + bi)(c + di) &= ac + adi + bci + bdi^2 \quad (i^2 = -1) \\ &= (ac - bd) + (ad + bc)i. \end{aligned} \tag{8.9}$$

Taking the sum, difference, and product of two complex numbers returns in each case a number of the form (8.5) and so $\mathbb{C}$ is said to be *closed* with respect to these operations.

EXAMPLE 1

Two Calculations

(a) Express $(7 + 6i)(2 - 3i)$ in standard form. Formula (8.9) can be used, but as an alternative, simply multiply out, collect real and imaginary parts, and substitute -1 for i^2 when required. We have

$$\begin{aligned}(7 + 6i)(2 - 3i) &= 14 - 18i^2 + 12i - 21i \\ &= (14 + 18) + (12 - 21)i = 32 - 9i.\end{aligned}$$

(b) Find real numbers x, y such that

$$(1 + 2i)x + (3 - 5i)y = 1 - 3i.$$

Collecting real and imaginary parts, we have $(x + 3y) + (2x - 5y)i = 1 - 3i$. Equating real and imaginary parts on both sides of this equation gives a

2×2 linear system

$$(S) \quad \begin{cases} x + 3y = 1 \\ 2x - 5y = -3 \end{cases}.$$

Using elimination or Cramer's rule, we have $x = -\dfrac{4}{11}$, $y = \dfrac{5}{11}$. ■

The Complex Conjugate

The *complex conjugate* of the number $z = a + bi$ is the number $\overline{z} = a - bi$ that is obtained by replacing i by $-i$ in (8.5). The symbol $\overline{z}$ is pronounced "zee bar." The real and imaginary parts of z may be expressed in terms of z and $\overline{z}$ as follows:

$$z + \overline{z} = (a + bi) + (a - bi) = 2a \;\Rightarrow\; a = \operatorname{Re} z = \frac{1}{2}(z + \overline{z}) \qquad (8.10)$$

$$z - \overline{z} = (a + bi) - (a - bi) = 2bi \;\Rightarrow\; b = \operatorname{Im} z = \frac{1}{2i}(z - \overline{z}). \qquad (8.11)$$

Note that z is (purely) real if and only if $\operatorname{Im} z = 0$ and, from (8.11), if and only if $z = \overline{z}$.

The Modulus

The *modulus* (or *absolute value*) of $z = a + bi$ is the *nonnegative* real number $|z|$ defined by

$$|z| = \sqrt{a^2 + b^2}. \qquad (8.12)$$

For example, for $z = 2 - 3i$ we have $|z| = \sqrt{2^2 + (-3)^2} = \sqrt{13} \simeq 3.6056$.

Note that $|z| = 0$ if and only if $z = 0$.

Multiplying out, $z\overline{z} = (a + bi)(a - bi) = a^2 + b^2$. Hence,

$$|z|^2 = z\overline{z} \qquad \text{or} \qquad |z| = \sqrt{z\overline{z}}. \qquad (8.13)$$

Quotients

The real number $x = s/t$, where $t \neq 0$, is the unique solution to the equation $tx = s$ and x is the called the *quotient* or *division* of s and t. Now suppose $z_1 = a + bi$ and $z_2 = c + di$ are two complex numbers such that $z_2 \neq 0$, meaning that the real and imaginary parts of z_2 are not *both* zero. The complex quotient $z = z_1/z_2$ is the unique solution to the equation $z_2 z = z_1$. We can find the standard form for z_1/z_2 by multiplying numerator and denominator of the quotient by the conjugate of z_2. We have

$$\frac{z_1}{z_2} = \frac{a + bi}{c + di} = \frac{(a + bi)(c - di)}{(c + di)(c - di)} = \left(\frac{ac + bd}{c^2 + d^2}\right) + \left(\frac{bc - ad}{c^2 + d^2}\right) i$$

and so

$$z = \frac{z_1}{z_2} \quad \Rightarrow \quad \frac{a + bi}{c + di} = \left(\frac{ac + bd}{c^2 + d^2}\right) + \left(\frac{bc - ad}{c^2 + d^2}\right) i. \qquad (8.14)$$

Note that the quotient of two complex numbers is a complex number; that is, $\mathbb{C}$ is closed with respect to taking quotients.

Formula (8.14) gives the quotient $z = z_1/z_2$ in standard form. However, when forming a quotient it is often easier to multiply numerator and denominator of the quotient by the complex conjugate of the denominator, in the following way,

$$z = \frac{z_1\,\overline{z}_2}{z_2\,\overline{z}_2} = \frac{z_1\overline{z}_2}{|z_2|^2},$$

where the denominator $|z_2|^2$ is purely real. This process, called *rationalizing the denominator*, gives (8.14).

EXAMPLE 2

Basic Calculations

(a) To find the standard form of a quotient we rationalize the denominator as follows:

$$\frac{(2-3i)}{1+2i} = \frac{(2-3i)(1-2i)}{(1+2i)(1-2i)} = \frac{(-4-7i)}{5} = -0.8 - 1.4i.$$

(b) Consider the products $i^2 = -1$, $i^3 = i^2 i = -i$, $i^4 = (i^3)i = (-i)i = 1$, $i^5 = (i^4)i = i$, and also the quotients $i^{-1} = 1/i = -i$, $i^{-2} = 1/i^2 = -1$, $i^3 = 1/i^3 = i$, and so on. The general pattern is this: If n is an integer, we have

$$i^n = \begin{cases} 1 & \text{if } n = 4k+0, \\ i & \text{if } n = 4k+1, \\ -1 & \text{if } n = 4k+2, \\ -i & \text{if } n = 4k+3, \end{cases} \tag{8.15}$$

where k is an integer. Note that dividing any integer by 4 always leaves a remainder of 0, 1, 2, or 3. For example, $i^{15} = -i$ because $n = 15 = 4 \cdot 3 + 3$ and $i^{-6} = -1$ because $n = -6 = 4 \cdot (-2) + 2$. Note that $n = 0 = 4 \cdot 0$ and so (8.15) gives $i^0 = 1$.

(c) We will simplify an expression and put it into standard form. Using the binomial expansion[4] for $(2+i)^3$, we have

$$\begin{aligned} &i^{15}(1+i)(2-i)(5+2i) - (2+i)^3 \\ &= -i(1+i)(12-i) - (8 + 12i + 6i^2 + i^3) \\ &= -i(13+11i) - (2+11i) \\ &= 11 - 2 - 13i - 11i = 9 - 24i. \end{aligned}$$

▪

Square Roots

A complex number z such that $z^2 = w$ is called a *square root* of w. If z is a square root of w, then $(-z)^2 = w$ also and hence a nonzero complex number w has two square roots, z and $-z$. To compute z, suppose $w = a + bi$ and $z = x + yi$. Then $z^2 = w$ becomes

$$(x+yi)^2 = x^2 - y^2 + 2xyi = a + bi. \tag{8.16}$$

4 For numbers p and q, we have $(p+q)^3 = p^3 + 3p^2q + 3pq^2 + q^3$.

Equating real and imaginary parts in (8.16) gives

$$x^2 - y^2 = a \quad \text{and} \quad 2xy = b. \tag{8.17}$$

Expanding $(x^2 + y^2)^2$, introducing a term $-2x^2y^2 + 2x^2y^2$ $(= 0)$, and using (8.17), we obtain

$$\begin{aligned}(x^2 + y^2)^2 &= x^4 + y^4 + 2x^2y^2 = x^4 + y^4 + 2x^2y^2 - 2x^2y^2 + 2x^2y^2 \\ &= (x^2 - y^2)^2 + 4x^2y^2 = a^2 + b^2\end{aligned}$$

giving

$$x^2 + y^2 = \sqrt{a^2 + b^2}, \tag{8.18}$$

taking only the positive root because $x^2 + y^2 \geq 0$. Solving (8.18) with the first equation in (8.17) gives expressions for x^2 and y^2 as follows:

$$x^2 = \frac{1}{2}(a + \sqrt{a^2 + b^2}) \quad \text{and} \quad y^2 = \frac{1}{2}(-a + \sqrt{a^2 + b^2}). \tag{8.19}$$

The values of x^2 and y^2 in (8.19) are actually nonnegative (explain!) and so x has two values (the positive and negative root) and so does y. But the second equation in (8.17) tells us that if $b > 0$, then x and y must have the same sign (both $+$ or both $-$), and if $b < 0$, then x and y must have opposite signs. These conditions ensure that there are only two values for w.

EXAMPLE 3

Finding a Square Root

We let $z = 5 - 12i$ and compute $w = \sqrt{5 - 12i}$. Using the notation immediately preceding, $a = 5$ and $b = -12$. With these values, the equations in (8.16) become

$$x^2 = \frac{1}{2}(5 + \sqrt{5^2 + (-12)^2}) = \frac{1}{2}(5 + \sqrt{169}) = 9 \quad \text{and}$$

$$y^2 = \frac{1}{2}(-5 + \sqrt{169}) = 4,$$

giving the values $x = \pm 3$ and $y = \pm 2$. However, $b = -12 < 0$ in this case, showing that x and y have opposite signs. The two roots are $w_1 = 3 - 2i$ and $w_2 = -3 + 2i$. Observe that $w_1 = -w_2$ here, which is the general rule when computing *any* square root. ■

Caution! The method of finding square roots just described is only of theoretical interest. In fact, complex nth-roots are found much more efficiently using the *polar form*, which appears in Section 8.3.

The Quadratic Formula

The quadratic equation in z

$$\alpha z^2 + \beta z + \gamma = 0, \tag{8.20}$$

where α ($\neq 0$), β, γ are complex numbers, is solved by *completing the square* and there are two solutions (or roots) given by the *quadratic formula*

$$z = \frac{-\beta \pm \sqrt{\beta^2 - 4\alpha\gamma}}{2\alpha}, \tag{8.21}$$

where $\pm\sqrt{\beta^2 - 4\alpha\gamma}$ are the two roots of the number $\beta^2 - 4\alpha\gamma$.

EXAMPLE 4

Solving a Quadratic Equation

We wish to solve the complex quadratic equation

$$z^2 + 2z - i = 0. \tag{8.22}$$

Set $\alpha = 1$, $\beta = 2$ and $\gamma = -i$ in (8.20). The roots given by (8.21) are

$$z = \frac{-2 \pm \sqrt{4 + 4i}}{2} = -1 \pm \sqrt{1 + i}. \tag{8.23}$$

We need to compute the complex root $\sqrt{1+i}$ in (8.23). Solving $(x + yi)^2 = 1 + i$, as in (8.16) with $a = b = 1$, the solution given by (8.19) is

$$x^2 = \frac{1}{2}(1 + \sqrt{2}) \sim 1.2071, \quad \text{and} \quad y^2 = \frac{1}{2}(-1 + \sqrt{2}) \sim 0.2071.$$

Hence, $x \simeq \pm 1.0987$ and $y \simeq \pm 0.4551$. But x and y must have the same sign because $b = 1 > 0$ and so the two roots of (8.22) are (to four decimal places)

$$z_1 \simeq 0.0987 + 0.4551i \quad \text{and} \quad z_2 \simeq -2.0987 - 0.4551i.$$ ■

Properties of the Modulus

Let z and w be given complex numbers.

(P1) $|zw| = |z||w|$, and in particular, $|z^2| = |z|^2$.

Proof $|zw|^2 = (zw)\overline{(zw)} = zw(\overline{z})(\overline{w}) = |z|^2|w|^2$. Take positive square roots.

(P2) $|z^n| = |z|^n, \quad n = 0, 1, 2, \ldots$

Proof Use mathematical induction and **(P1)**.

(P3) For $z \neq 0$, we have $\left|\frac{1}{z}\right| = \frac{1}{|z|}$.

Proof Using **(P1)** with $w = 1/z$, we have $1 = |z||w|$ implies $|w| = 1/|z|$.

(P4) For $w \neq 0$, we have $\left|\frac{z}{w}\right| = \frac{|z|}{|w|}$.

Proof Using (**P1**) and (**P3**), we have

$$\left|\frac{z}{w}\right| = \left|z\frac{1}{w}\right| = |z|\left|\frac{1}{w}\right| = |z|\frac{1}{|w|} = \frac{|z|}{|w|}.$$

(**P5**) Let $z = x + yi$. Then $x = \operatorname{Re} z \le |z|$ and $y = \operatorname{Im} z \le |z|$.

Proof $\operatorname{Re} z = x \le |x| = \sqrt{x^2} \le \sqrt{x^2 + y^2} = |z|$. Use a similar argument for y.

EXAMPLE 5

Modulus of a Quotient

Using properties (**P1**)–(**P4**), we have

$$\left|\frac{(1+2i)^3}{(1-2i)^2}\right| = \frac{|(1+2i)^3|}{|(1-2i)^2|} = \frac{|1+2i|^3}{|1-2i|^2} = \frac{(\sqrt{5})^3}{(\sqrt{5})^2} = \sqrt{5}.$$ ■

■ ILLUSTRATION 8.1

Identities

The equation

$$|z + w|^2 + |z - w|^2 = 2|z|^2 + 2|w|^2 \tag{8.24}$$

is true for any complex numbers z and w. We call an equation such as (8.24) an *identity*. To prove that it is valid, argue as follows:

$$\begin{aligned}|z + w|^2 + |z - w|^2 &= (z + w)\overline{(z + w)} + (z - w)\overline{(z - w)} \\ &= (z + w)(\bar{z} + \bar{w}) + (z - w)(\bar{z} - \bar{w}) \\ &= z\bar{z} + z\bar{w} + w\bar{z} + w\bar{w} + z\bar{z} - z\bar{w} - w\bar{z} + w\bar{w} \\ &= 2(z\bar{z} + w\bar{w}) = 2|z|^2 + 2|w|^2.\end{aligned}$$ ■

Caution! Note that $|z + w|$ and $|z| + |w|$ are not equal for all z and w. For example,

$$z = 1 + i, \quad w = -i \quad \Rightarrow \quad |z + w| = 1 < \sqrt{2} + 1 = |z| + |w|.$$

Theorem 8.1 **Triangle Inequality**

For any complex numbers z and w, we have

$$|z + w| \le |z| + |w| \tag{8.25}$$

Proof We use (**P5**) as follows:

$$|z+w|^2 = (z+w)\overline{(z+w)} = z\bar{z} + z\overline{w} + \bar{z}w + w\overline{w} = |z|^2 + z\overline{w} + \overline{(z\overline{w})} + |w|^2$$
$$= |z|^2 + 2Re(z\overline{w}) + |w|^2$$
$$\le |z|^2 + 2|z\overline{w}| + |w|^2 = |z|^2 + 2|z||w| + |w|^2 = (|z| + |w|)^2.$$

Take positive square roots to complete the proof.

Mathematical induction extends inequality (8.25) to three or more complex numbers. For any positive integer k, we have

$$|z_1 + z_2 + \cdots + z_k| \le |z_1| + |z_2| + \cdots + |z_k|.$$

EXERCISES 8.1

Exercises 1–12. Express each complex number in standard form.

1. $7 + 5i + 2(3 - 4i)$
2. $7 + 4i + i(6 - 5i)$
3. $(7 + 5i)(3 - 4i)$
4. $(3 - 4i)(5 - 6i)$
5. $(\sqrt{2} + 3i)(\sqrt{2} - 4i)$
6. $(1 + i)^4$
7. $(2 + 3i)^2$
8. $(\sqrt{3} + i\sqrt{2})(\sqrt{3} - i\sqrt{2})$
9. $(4 + i)^3$
10. $(2 + i)^3(2 - i)$
11. $(3 + 4i)^2(4 + 3i) - (2 + 3i)^2$
12. $5i(1 + i)$

Exercises 13–30. Rationalize denominators and express in standard form.

13. $(1 + \frac{1}{i})^2 + (1 - \frac{1}{i})^2$
14. $\dfrac{7 - 9i}{1 + i}$
15. $\dfrac{1 + i}{1 - i}$
16. $\dfrac{5}{i}$
17. $\dfrac{5 + 3i}{8 - 2i}$
18. $\dfrac{2 - 3i}{1 + 2i}$
19. $\dfrac{(1 + i)^2}{1 - i}$
20. $\dfrac{1}{(4 + 2i)(2 - 3i)}$
21. $\dfrac{1 + i}{1 - i} + \dfrac{1 - i}{1 + i}$
22. $\dfrac{5 + 2i}{5 - 2i} + \dfrac{5 - 2i}{5 + 2i}$
23. $\dfrac{(1 + 2i)(4 - 5i)}{2 + 3i}$
24. $\dfrac{(4 + i)(5 - 2i)}{(4 - i)(5 + 2i)}$
25. $\dfrac{1}{(1 + i)^2} + \dfrac{1}{(1 - i)^2}$
26. $\dfrac{1}{(1 + i)^4} + \dfrac{1}{(1 - i)^4}$
27. $\dfrac{(\sqrt{3} + i)(1 + i\sqrt{3})}{(1 + i)^2}$
28. $\left(-\dfrac{1}{2} + i\dfrac{\sqrt{3}}{2}\right)^3$
29. $\dfrac{1}{i^3(3 - 4i)}$
30. $\left(\dfrac{1 + i}{1 - i}\right)^3$

Exercises 31–36. Solve for real numbers a and b.

31. $(2 - i)a + (1 + 3i)b + 2 = 0$
32. $a(5 + 4i) + b(4 + 3i) + 1 = 0$
33. $(4 + 3i)a + (3 - 4i)b = 7 - i$
34. $2ia + (5 + 3i)b = 11 - 5i$
35. $(5 + 3i)a + (2 + 5i)b = 10 + 11i$
36. $\dfrac{5 + ai}{8 - 2i} = \dfrac{b + 4i}{5 + 3i}$

Exercises 37–38. Solve each linear system for the real unknowns x, y, z, w.

37. $(S)\quad \begin{cases} (1 + i)x + (1 + 2i)y + (1 + 3i)z + (1 + 4i)w = 1 + 5i \\ (3 - i)x + (4 - 2i)y + (1 + i)z + 4iw = 2 - i \end{cases}$

38. $(S)\quad \begin{cases} (1 + 2i)x + (2 + i)y + (3 + 2i)z + (4 + 3i)w = 5 + i \\ (3 + 4i)x + (2 + 3i)y + (1 + 2i)z + (2 + i)w = 1 - 5i \end{cases}$

Exercises 39–44. Find the modulus of each complex number.

39. $4 + 3i$
40. $1 + i$
41. $-i$
42. -3
43. $2 - i$
44. $\dfrac{(1 + i)}{(1 - i)}$

Exercises 45–50. Find all the complex numbers $z = a + ib$ that satisfy the given equation.

45. $z^2 = 2 + 2i$
46. $z^2 = i$

47. $z^2 = 3 + 4i$ **48.** $z = \bar{z}^2$

49. $iz + |z| = 2 + i$ **50.** $z^2 - iz = 0$

Exercises 51–56. Solve each equation for the complex number z.

51. $2z^2 + (1 - 7i)z = 12$ **52.** $z^2 - 6z + 25 = 0$

53. $z^2 + 2 - 2i\sqrt{3} = 0$

54. $z^2 - (2 + i)z + (-1 + 7i) = 0$

55. $z^2 - (1 - i)z - 2i = 0$

56. $z^4 - 2z^2 + 4 = 0$

57. Let $z = a + ib$ and $w = c + id$. Prove that

$$\overline{z \pm w} = \bar{z} \pm \bar{w}, \qquad \overline{zw} = \bar{z}\,\bar{w}, \qquad \overline{\left(\frac{z}{w}\right)} = \frac{\bar{z}}{\bar{w}}.$$

58. Write the complex number $z = \dfrac{1 - it}{t - i}$ in standard form $z = x + iy$ and prove that $x^2 + y^2 = 1$. Interpret geometrically.

59. Expand $\frac{1}{2}(1 - i)((1 - ix)^5 + i(1 + ix)^5)$ in powers of the real variable x.

Exercises 60–68. Compute each modulus.

60. $|5i(1 + i)|$ **61.** $|(2 + 3i)(3 - 4i)|$

62. $\left|\dfrac{1}{i}\right|$ **63.** $|(1 + i)^2(\sqrt{3} - i)^5|$

64. $|(2 - i)^6|$ **65.** $\left|\left(\dfrac{6 - 3i}{2 + i\sqrt{6}}\right)^6\right|$

66. $\left|\dfrac{(2 - 3i)^2}{(1 + 2i)^3}\right|$ **67.** $\left|\dfrac{(2 - i)(2 + i)}{(1 - i\sqrt{3})(1 + i\sqrt{3})}\right|$

68. $\left|\dfrac{(1 + 2i)^{12}}{(1 - 2i)^9}\right|$

69. Solve the equation $z^2 = -8 + 6i$ for z.

70. Let $z = x + iy$ and compute $r = \left|\dfrac{z - i}{z + i}\right|$. Show that $r < 1$ when $0 < y$.

71. Let $z = x + iy$. Simplify $\left(\dfrac{z}{\bar{z}}\right)^2 - \left(\dfrac{\bar{z}}{z}\right)^2$.

72. Let $z = \cos\theta + i\sin\theta$. Show that $|2z + z^2|^2 = 5 + 4\cos\theta$.

73. Let z be a nonzero complex number whose imaginary part is nonzero. Show that $\dfrac{i(z - |z|)}{z + |z|}$ is purely real. *Hint*: $z - \bar{z} = 2i\,\mathrm{Im}\,z$.

74. Consider distinct complex numbers z_1 and z_2 such that $|z_1| = |z_2|$. Prove that $\dfrac{z_1 + z_2}{z_1 - z_2}$ is purely imaginary.

75. Prove that $|z_1 + z_2| = |z_1| + |z_2|$ if and only if $z_1 = kz_2$, where $k \geq 0$ is a real number.

76. Let z_1, z_2, z_3 be any complex numbers. Substitute $z_1 = z_3 - z_2$ into the triangle inequality and show that $||z_3| - |z_2|| \leq |z_3 - z_2|$.

77. Use standard inequalities to prove that if z_1 and z_2 are complex numbers such that $z_1 + z_2 = 1$, then $|z_1| + |z_2| \geq 1$, with equality holding if and only if z_1 and z_2 are both real and nonnegative.

78. Suppose z_1, z_2, z_3 are complex numbers such that $|z_2| \neq |z_3|$. Prove that

$$\left|\frac{z_1}{z_2 + z_3}\right| \leq \frac{|z_1|}{||z_2| - |z_3||}.$$

USING MATLAB

Consult online help for the commands **conj**, **abs**, **roots**, *and related commands.*

79. Use the command **roots** to check hand calculations in Exercises 45–56. *Hint*: The polynomial $z^2 - 2 - 2i$ is expressed as the row vector $[1, 0, -2 - 2i]$.

80. Use MATLAB to check hand calculations in this exercise set.

8.2 Geometric Theory

The Complex Plane

We visualize complex numbers geometrically as points in the *complex plane*. Refer to Figure 8.2. Imagine the plane with a rectangular coordinate system as the frame of reference and axes labeled Re and Im. This is called the

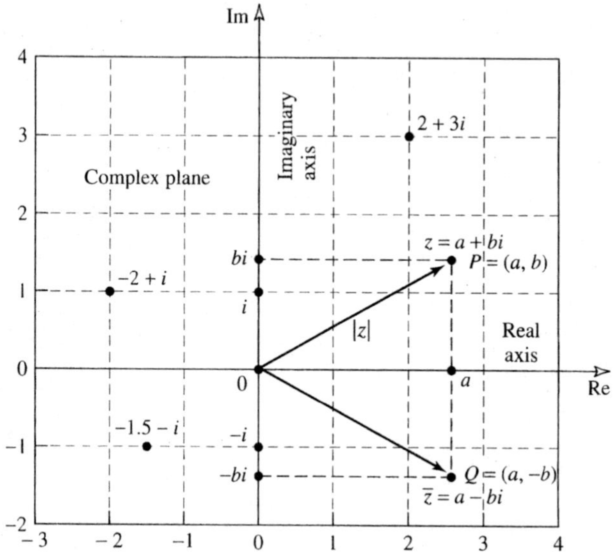

Figure 8.2 The complex plane.

complex plane. Each complex number $z = a + bi$ corresponds to a unique point $P = (a, b)$ in the plane, and conversely, each point $P = (a, b)$ in the plane corresponds to a unique complex number $z = a + bi$.

The purely real numbers $z = a + 0i$ are points on the *real axis* Re, and the purely imaginary numbers $z = 0 + bi$ are points on the *imaginary axis* Im. The complex number $z = 0$ corresponds to the origin $(0, 0)$.

Each nonzero complex number $z = a + bi$ is represented by the point P in the complex plane that is the terminal point of a directed line segment (arrow) $\overrightarrow{OP}$ of length $|z| = \sqrt{a^2 + b^2}$. Reflecting P in the real axis gives a point Q that represents $\bar{z} = a - bi$, the complex conjugate of z. Note that $|\overrightarrow{OP}| = |\overrightarrow{OQ}|$; that is, $|z| = |\bar{z}|$.

Geometric Addition, Subtraction, Negation

Refer to Figure 8.3. Algebraic operations with complex numbers are seen as vector operations in the complex plane. The complex numbers $z_1 = a_1 + b_1 i$ and $z_2 = a_2 + b_2 i$ are represented by the vectors (arrows) $\overrightarrow{OP}$ and $\overrightarrow{OQ}$, respectively. Using the parallelogram law, we have

$$\overrightarrow{OR} = \overrightarrow{OP} + \overrightarrow{OQ} = \overrightarrow{OP} + \overrightarrow{PR} = \overrightarrow{OQ} + \overrightarrow{QR}$$

and so the vector $\overrightarrow{OR}$ defined by the diagonal in parallelogram $OPRQ$ represents the sum $z_1 + z_2 = (a_1 + a_2) + (b_1 + b_2)i$. The vector $\overrightarrow{OT}$ represents $-z_2 = -a_2 - b_2 i$, which is the negative of z_2 and so

$$\overrightarrow{OS} = \overrightarrow{OT} + \overrightarrow{TS} \quad (= \overrightarrow{OP} - \overrightarrow{OQ}),$$

showing that $\overrightarrow{OS}$ represents the difference $z_1 - z_2 = (a_1 - a_2) + (b_1 - b_2)i$.

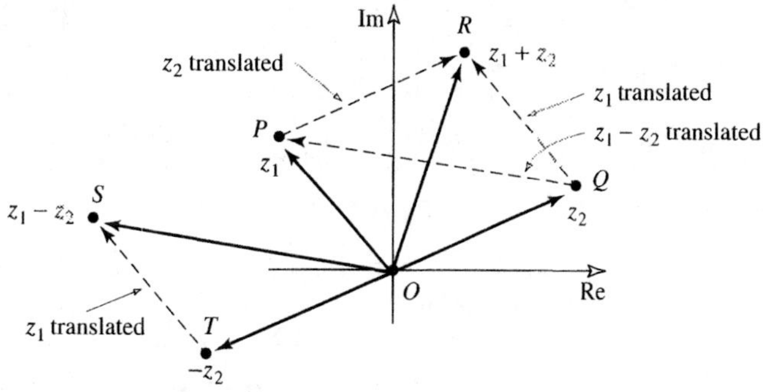

Figure 8.3 Complex numbers and vector operations.

Interplay between Algebra and Geometry

Viewing an algebraic expression (equation, identity, inequality, and so on) as a geometrical relationship in the complex plane often provides a different perspective on the meaning of the algebraic expression. Conversely, interpreting geometric relationships algebraically can express geometric relationships in an economical way. This interplay between the algebra and geometry of complex numbers is considered by mathematicians to be one of the most elegant theories in the whole of mathematics. Here are some examples.

EXAMPLE 1

Circles, Disks, Unbounded Regions

Refer to Figure 8.4. The complex numbers z for which $\operatorname{Re} z \geq 2$ lie in a region H called a *half-plane* that consists of all points on and to the right of the line $a = 2$. The region H is *unbounded* because there are complex numbers in H of arbitrarily large modulus.

The complex numbers z for which $|z| = r$, where $r \geq 0$, all lie on a circle C, center O, radius r. When $r = 1$ we call C the *unit circle*.

Let w be a fixed complex number. The points z in the complex plane for which $|z - w| = r$ lie on a circle C, center w, radius r. The points z for which $|z - w| \leq r$ lie inside and on C and define a region D in the

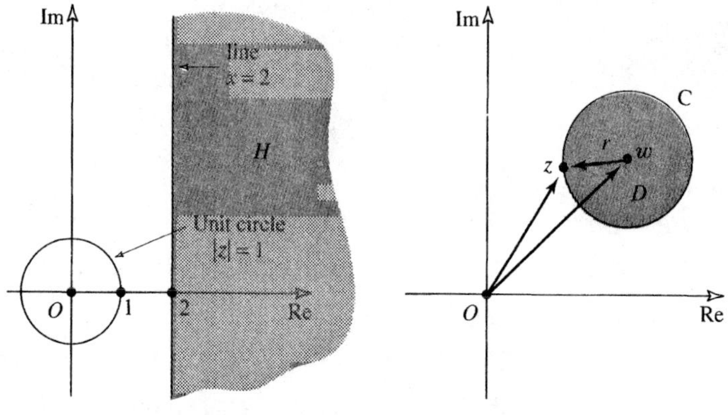

Figure 8.4 Half-plane H and disk D.

complex plane that we call a *disk*. The disk is called *closed* (*open*) if its bounding circle is (is not) included. The points z for which $|z - w| > r$ lie outside C and define a region E containing all points in the complex plane outside D. ■

Complex Functions

A *complex function* is a rule, denoted by f, that assigns to every complex number z in a subset $\mathcal{D}$ of the complex plane a corresponding complex number w. We write $w = f(z)$ and we call w the *image* of z under f. The set $\mathcal{D}$ is called the *domain* of f and the set of all numbers w such that $w = f(z)$ for some z in $\mathcal{D}$ is called the *range* of f. We write $f : \mathcal{D} \to \mathcal{W}$ and we refer to f as a *mapping* or a *transformation* on $\mathcal{D}$. For example, $w = f(z) = \bar{z}$ defines a transformation whose domain $\mathcal{D}$ is the whole complex plane. Each point z is mapped to its complex conjugate $w = \bar{z}$. Geometrically, the image w of z is located by reflecting z in the real axis. Transformations of the complex plane play an important role in complex analysis.

EXAMPLE 2

A Complex Function

The function

$$w = f(z) = \frac{1}{z - 1} \tag{8.26}$$

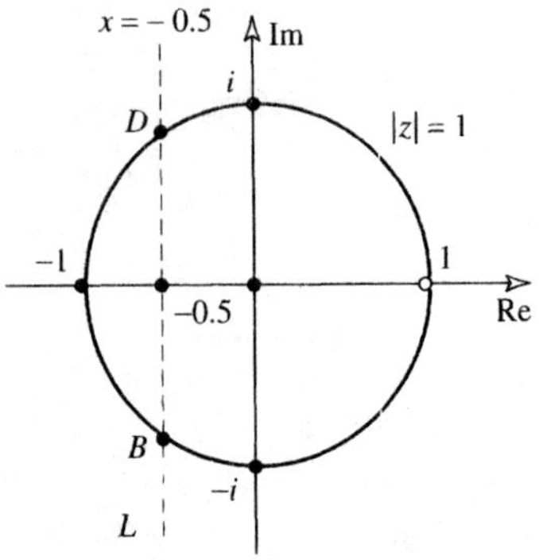

Figure 8.5 Example 2.

defines a transformation whose domain $\mathcal{D}$ is the set of all complex numbers $z \neq 1$. Refer to Figure 8.5. We visualize $\mathcal{D}$ as the complex plane with one point $z = 1$ deleted. Our goal is to describe the set of points w that are images of those points z in $\mathcal{D}$ that lie on the unit circle. For such a point z, we have

$$w = \frac{1}{z - 1} \quad \Rightarrow \quad z - 1 = \frac{1}{w} \quad \Rightarrow \quad z = \frac{1}{w} + 1 = \frac{w + 1}{w}.$$

Hence $\left|\dfrac{w + 1}{w}\right| = |z| = 1$ and so $|w + 1| = |w|$. Recall that $|w - (-1)|$ is the distance between w and -1 and $|w|$ is the distance between w and 0. Hence the condition $|w + 1| = |w|$ locates w on the vertical line L with equation $x = -0.5$. Any number $z \neq 1$ on the unit circle is mapped by f to the point w on L determined by (8.26). ■

EXAMPLE 3

A Transformation on the Unit Circle

The domain $\mathcal{D}$ of the function

$$w = f(z) = \frac{2z + 1}{z + 2} \tag{8.27}$$

is the set of all complex numbers $z \neq -2$. We visualize $\mathcal{D}$ as the complex plane with the point $z = -2$ deleted. The function $f : \mathcal{D} \to \mathcal{W}$ can be visualized by forming two copies of the complex plane, called the z-plane and the w-plane, and by plotting the image w in the w-plane of each point z in $\mathcal{D}$.

Our goal is to show that the image w of each point z on the unit circle in the z-plane lies on the unit circle in the w-plane. Consider z such that $|z| = 1$ and let $w = f(z)$. To show $|w| = 1$, consider $|w|^2$, as follows:

$$|w|^2 = \left|\frac{2z+1}{z+2}\right|^2 = \frac{|2z+1|^2}{|z+2|^2} = \frac{(2z+1)(2\bar{z}+1)}{(z+2)(\bar{z}+2)} = \frac{4|z|^2+2(z+\bar{z})+1}{|z|^2+2(z+\bar{z})+4}$$

$$= \frac{2(z+\bar{z})+5}{2(z+\bar{z})+5} = 1,$$

and taking the positive square root gives $|w| = 1$. Hence, the transformation (8.27) maps the unit circle onto itself. In particular, $z = i$ is mapped to $w = \frac{4}{5} + \frac{3}{5}i$ and $z = 1$ is mapped to $w = 1$. We say that the point $z = 1$ is *invariant* under the mapping f. ■

EXAMPLE 4

A Transformation on the Complex Plane

The domain of the function $f(z) = 4z^2 - 3(1+i)z$ is the whole complex plane. Let $z = x + yi$. Then

$$\begin{aligned} f(z) &= 4(x+yi)^2 - 3(1+i)(x+yi) \\ &= (4x^2 - 4y^2 - 3x + 3y) + (8xy - 3y - 3x)i \\ &= p + qi. \end{aligned}$$

Hence, $w = f(z)$, where $w = p + qi$ and

$$4x^2 - 4y^2 - 3x + 3y = p \quad \text{and} \quad 8xy - 3y - 3x = q.$$

The transformation f can be visualized by keeping p and q fixed. If $p = q = 1$, for example, each point (x, y) in the z-plane lying at the intersection of curves defined by

$$4x^2 - 4y^2 - 3x + 3y = 1 \quad \text{and} \quad 8xy - 3y - 3x = 1.$$

is mapped onto the point $(1, 1)$ in the w-plane. ■

Historical Notes

Jean Robert Argand (1768–1822) Amateur Swiss mathematician. The first geometrical interpretation of complex numbers was given by Caspar Wessel (1745–1818) about 1797, but credit is usually given to Argand who published his work on the complex plane privately in 1806. The complex plane is known as the *Argand diagram* or the *Gaussian plane*.

EXERCISES 8.2

1. Locate each complex number in the complex plane.

(i) 1 (ii) -1 (iii) $3i$

(iv) $-i$ (v) $1-i$ (vi) $2+2i$

(vii) $\frac{5}{i}$ (viii) $\frac{1+i\sqrt{3}}{1-i\sqrt{3}}$ (ix) $\frac{1}{1+i}$

2. Let P represent the complex number $z = 1 - i$. Locate each of the following complex numbers in the complex plane.

(i) $2z$ (ii) $-3z$ (iii) $z+2$

(iv) $z-6$ (v) $3z+7$ (vi) $6-2z$

(vii) $z-2-i$ (viii) $z-4-3i$ (ix) z^2

3. For each given complex number z, locate the points z and $1/z$ in the complex plane.

(i) 2 (ii) i (iii) $-i$ (iv) $1+i$ (v) $3-4i$

4. Indicate the region in the complex plane corresponding to all points z that satisfy the given condition.

(i) $|z| = 11$ (ii) $|z+3| = 2$

(iii) $|z-3+4i| = 5$ (iv) $|z-7-12i| = 13$

(v) $|z+1+i| = 4$ (vi) $|2z+1| = 3$

(vii) $|i-2z| = 5$ (viii) $1 < |z+2-3i| < 2$

(ix) $\operatorname{Re} z > 4$

5. Suppose $|z| = 1$. In each case describe the set of points in the complex plane that correspond to the given expression.

(i) $3z$ (ii) $z+2$

(iii) $3z+7$ (iv) $\frac{1}{z}$

6. Find the maximum and minimum values of the given expression in z.

(i) $|z+3|$ such that $|z| \le 1$

(ii) $|z-4|$ such that $|z+3i| \le 1$

7. Let

$$w = f(z) = \frac{3z-1}{z-3}.$$

Show that the transformation $w = f(z)$ maps the unit circle onto itself. Draw two copies of the complex plane, labeling these the "z-plane" and "w-plane." Plot z and its image w when $z = 1, -1, i, -i$.

8. Show that the three numbers 1 and $(-1 \pm i\sqrt{3})/2$ form the vertices of an equilateral triangle inscribed in the circle $|z| = 1$. Draw a diagram. *Hint*: Use the vector properties of complex numbers.

9. Show that the complex number

$$z = \frac{1}{\cos\theta - 1 + i\sin\theta}$$

is defined whenever $\theta \neq 2\pi k$, where k is any integer. Compute the standard form $z = x + iy$ and prove that $x = -\frac{1}{2}$. Give a geometric description of all points z in the complex plane for which z is defined; that is, find the *locus* of all such points z.

10. Express the complex number

$$z = \frac{3}{2 + \cos\theta + i\sin\theta}$$

in standard form $z = x + iy$. Prove that $x^2 + y^2 = 4x - 3$ and show that all the numbers z (allowing θ to vary) lie on the circle in the complex plane, center $(2, 0)$ and radius $r = 1$.

11. If $w = z + \frac{1}{z}$ is real, show that either z is real or $|z| = 1$. *Hint*: consider $w - \overline{w}$.

12. Consider Example 2 and Figure 8.5.

(a) Find the points z such that $f(z) = z$. These points are *invariant* under the mapping f. Find points A and C such that A is mapped onto B and C onto D.

(b) Find the image under f of the points -1, i, $-i$, $\frac{1}{\sqrt{2}} + i\frac{1}{\sqrt{2}}$, $-\frac{1}{\sqrt{2}} + i\frac{1}{\sqrt{2}}$.

USING MATLAB

Consult online help for the command **plot**, *and related commands.*

13. Write an M-file to plot the numbers in Exercise 1 in the complex plane.

14. Plot the equilateral triangle and unit circle in Exercise 8.

15. *Project*. Study the transformation $w = f(z) = \frac{1}{z-2}$. Choose a few points in the z-plane and plot their images in the w-plane. Plot the image in the w-plane of various regions of the z-plane such as the disk $|z| \le 1$ and the line $2\operatorname{Re} z + \operatorname{Im} z = -1$.

8.3 Polar Form

Refer to Figure 8.6. Consider a nonzero complex number $z = a+bi$ represented in the complex plane by the point P. We have $|z| = r > 0$. The vector $\overrightarrow{OP}$ also represents z and $r = |\overrightarrow{OP}|$. Let θ denote the angle formed by turning the positive real axis counterclockwise to coincide with the vector $\overrightarrow{OP}$. By convention, the angle θ so formed is considered positive.

The length of the line segment AP is b, and so in ΔOAP we have $a = r\cos\theta$ and $b = r\sin\theta$. Substituting into $z = a + ib$ (noting that $ib = bi$) gives

$$\begin{aligned} z &= r\cos\theta + ir\sin\theta \\ &= r(\cos\theta + i\sin\theta). \end{aligned} \tag{8.28}$$

We call (8.28) the *polar form* or the *modulus-argument form* of z.

For all θ, we have $\cos(-\theta) = \cos\theta$ (cosine is an even function of θ) and $\sin(-\theta) = -\sin\theta$ (sine is an odd function of θ), and so the polar form of the complex conjugate $\bar{z} = a - bi$ is

$$\bar{z} = r(\cos\theta - i\sin\theta) = r\big(\cos(-\theta) + i\sin(-\theta)\big).$$

The angle θ shown in Figure 8.6 is not the only angle for which (8.28) is true. Adding $2\pi, 4\pi, \ldots$ (positive revolutions) or $-2\pi, -4\pi, \ldots$ (negative revolutions) to θ gives the same value of z because

$$\cos(\theta \pm 2k\pi) = \cos\theta, \qquad \sin(\theta \pm 2k\pi) = \sin\theta$$

due to the periodicity of the trigonometric functions. Any angle chosen from the set $\theta \pm 2k\pi$, $k = 0, \pm 1, \pm 2, \ldots$ is called an *argument* or an *amplitude* of z and denoted $\arg z$. We say that $\arg z$ is only determined up to an integer multiple of 2π radians.

The Principal Argument

For any nonzero complex number z there is a unique angle θ such that

$$z = r(\cos\theta + i\sin\theta), \quad \text{where } -\pi < \theta \leq \pi.$$

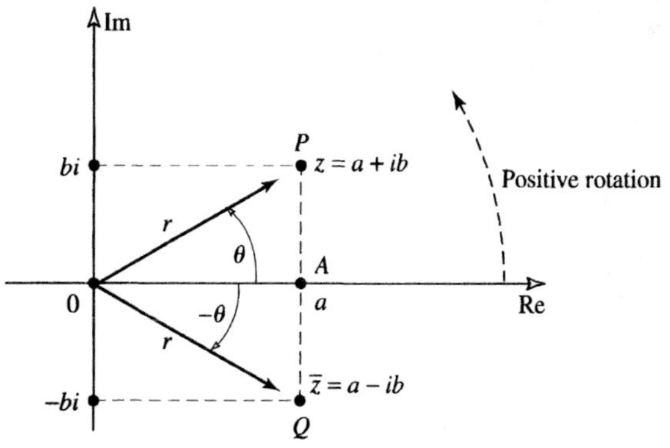

Figure 8.6 Polar form of a complex number.

We call θ *the principal argument* of z and denote it by Arg z (as opposed to arg z). For example,

$$\text{Arg } 2 = 0, \quad \text{Arg } (-2) = \pi, \quad \text{Arg } i = \frac{\pi}{2}, \quad \text{Arg } (-3i) = -\frac{\pi}{2}.$$

Suppose $z = a + bi$ with $a \neq 0$. Dividing $b = r\sin\theta$ by $a = r\cos\theta$, we have

$$\tan\theta = \frac{b}{a} \quad \Rightarrow \quad \theta = \arctan\left(\frac{b}{a}\right). \tag{8.29}$$

Equation (8.29) gives infinitely many values for θ. Which one is Arg z? Let $\alpha = \text{Arctan}(b/a)$ denote the principal value of arctan. By definition, α lies in the range $-\pi/2 < \alpha < \pi/2$. Then the value of Arg z is given by

$$\text{Arg } z = \theta = \begin{cases} \alpha, & \text{if } a > 0 \\ \alpha + \pi & \text{if } a < 0, \quad b \geq 0 \\ \alpha - \pi & \text{if } a < 0, \quad b < 0 \end{cases}. \tag{8.30}$$

When $a = 0$, $b \neq 0$ (that is, z is purely imaginary) we have

$$\text{Arg } z = \begin{cases} \pi/2, & \text{if } b > 0 \\ -\pi/2, & \text{if } b < 0 \end{cases}$$

The relationship between arg z and Arg z is given by

$$\arg z = \text{Arg } z + 2k\pi, \quad \text{where, } k = 0, \pm 1, \pm 2, \ldots .$$

EXAMPLE 1

Converting to Polar Form

Refer to Figure 8.7.

(a) Let $z = \sqrt{3} + i$. Then $a = \sqrt{3}$, $b = 1$, $r = |z| = \sqrt{3+1} = 2$, so we have $\cos\theta = \frac{\sqrt{3}}{2}$, $\sin\theta = \frac{1}{2}$, and using (8.30), we have

$$\theta = \text{Arg } z = \text{Arctan}\left(\frac{1}{\sqrt{3}}\right) = \frac{\pi}{6},$$

so that $z = 2(\cos\pi/6 + i\sin\pi/6)$.

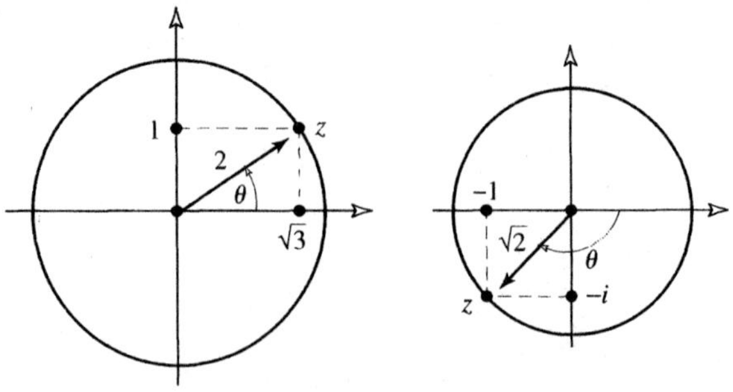

Figure 8.7 Example 1.

(b) Let $z = -1 - i$. Then $a = b = -1$, $r = |z| = \sqrt{1+1} = \sqrt{2}$, and we have

$$\cos\theta = -\frac{1}{\sqrt{2}} = \sin\theta, \quad \alpha = \text{Arctan}(1) = \frac{\pi}{4},$$

$$\theta = \text{Arg } z = \alpha - \pi = -\frac{3\pi}{4}.$$

Hence, $z = \sqrt{2}\,(\cos(-3\pi/4) + i\sin(-3\pi/4)) = \sqrt{2}\,(\cos 3\pi/4 - i\sin 3\pi/4)$. ■

EXAMPLE 2

The Unit Circle

Every complex number z on the unit circle $|z| = 1$ has the form

$$z = \cos\theta + i\sin\theta,$$

where θ is any argument of z. For any angle α, the number $w = -\sin\alpha + i\cos\alpha$ also lies on the unit circle. Because $-\sin\alpha = \sin(-\alpha) = \cos(\alpha + \pi/2)$ and $\cos\alpha = \sin(\alpha + \pi/2)$ we can write w in the form $w = \cos(\alpha + \pi/2) + i\sin(\alpha + \pi/2)$, where $\alpha + \pi/2$ is an argument of w. ■

Products and Quotients

The polar form of products and quotients of complex numbers is particularly useful in applications.

Theorem 8.2

Product in Polar Form

Let $z_1 = r_1(\cos\theta_1 + i\sin\theta_1)$ *and* $z_2 = r_2(\cos\theta_2 + i\sin\theta_2)$. *Then*

$$z_1 z_2 = r_1 r_2\,(\cos(\theta_1 + \theta_2) + i\,\sin(\theta_1 + \theta_2))\,.$$

Proof Using the addition laws for $\cos(\theta_1 + \theta_2)$ and $\sin(\theta_1 + \theta_2)$ we have

$$\begin{aligned} z_1 z_2 &= r_1 r_2\,((\cos\theta_1\cos\theta_2 - \sin\theta_1\sin\theta_2) + i(\sin\theta_1\cos\theta_2 + \cos\theta_1\sin\theta_2)) \\ &= r_1 r_2\,(\cos(\theta_1 + \theta_2) + i\sin(\theta_1 + \theta_2))\,. \end{aligned}$$

If z_1, z_2 are nonzero, Theorem 8.2 shows that

$$\arg(z_1 z_2) = \arg z_1 + \arg z_2. \tag{8.31}$$

with the understanding that $\arg(z_1 z_2)$ is obtained by adding all possible values from the sets $\arg z_1$ and $\arg z_2$.

Caution! The equation Arg $(z_1 z_2) =$ Arg $z_1 +$ Arg z_2 is false for some pairs z_1, z_2. For example, let $z_1 = -1 + i$ and $z_2 = i$. Then $z_1 z_2 = -1 - i$ and Arg $z_1 z_2 = -3\pi/4$, while Arg $z_1 +$ Arg $z_2 = 3\pi/4 + \pi/2 = 5\pi/4$.

Figure 8.8 shows how to visualize a product geometrically. Multiplying z_1 by z_2 rotates the vector z_2 through the angle θ_1 and scales the result by r_1 to give a vector $z_1 z_2$ of length $r_1 r_2$. Moduli are multiplied. Rotating z_1 through an angle θ_2 and scaling the result by r_2 gives the same result because $z_1 z_2 = z_2 z_1$.

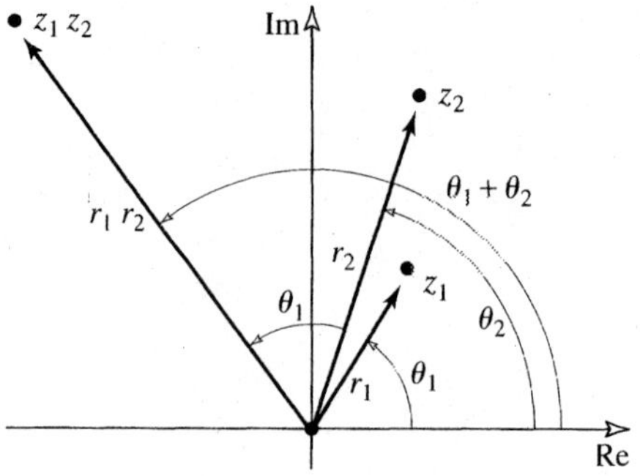

Figure 8.8 Multiplication of complex numbers.

Our next goal is to find the polar form of a quotient z_1/z_2 of two complex numbers, where $z_2 \neq 0$ of course. As a first step, let $z = r(\cos\theta + i\sin\theta)$. Rationalizing the denominator and using $\cos^2\theta + \sin^2\theta = 1$, we have

$$\begin{aligned}\frac{1}{\cos\theta + i\sin\theta} &= \left(\frac{1}{\cos\theta + i\sin\theta}\right)\left(\frac{\cos\theta - i\sin\theta}{\cos\theta - i\sin\theta}\right)\\ &= \cos\theta - i\sin\theta = \cos(-\theta) + i\sin(-\theta).\end{aligned}$$

Hence

$$\frac{1}{z} = \frac{1}{r}\,\frac{1}{(\cos\theta + i\sin\theta)} = \frac{1}{r}\big(\cos(-\theta) + i\sin(-\theta)\big). \tag{8.32}$$

Let $z_1 = r_1(\cos\theta_1 + i\sin\theta_1)$ and $z_2 = r_2(\cos\theta_2 + i\sin\theta_2)$. Then using (8.32), we have

$$\begin{aligned}\frac{z_1}{z_2} = z_1\frac{1}{z_2} &= \frac{r_1}{r_2}(\cos\theta_1 + i\sin\theta_1)\,\frac{1}{(\cos\theta_2 + i\sin\theta_2)}\\ &= \frac{r_1}{r_2}(\cos\theta_1 + i\sin\theta_1)(\cos(-\theta_2) + i\sin(-\theta_2)).\end{aligned}$$

Using Theorem 8.2, we obtain the *polar form for division*

$$\frac{z_1}{z_2} = \frac{r_1}{r_2}\big(\cos(\theta_1 - \theta_2) + i\sin(\theta_1 - \theta_2)\big). \tag{8.33}$$

Equation (8.33) gives

$$\arg\left(\frac{z_1}{z_2}\right) = \arg z_1 - \arg z_2, \quad \text{where } z_2 \neq 0,$$

which is interpreted in the same way as (8.31).

Theorem 8.3 **Powers of a Complex Number**

Let $z = r(\cos\theta + i\sin\theta)$. Then

$$z^n = r^n(\cos n\theta + i\sin n\theta), \quad n = 0, 1, 2, \ldots. \tag{8.34}$$

Proof The proof is by mathematical induction on n. The result is true for the initial value $n = 0$. We assume (this is the *inductive hypothesis*) that equation (8.34) is true for n; that is,

$$z^n = r^n(\cos n\theta + i \sin n\theta)$$

and prove that (8.34) is true for $n + 1$. Using Theorem 8.2 and the inductive hypothesis, we have

$$\begin{aligned} z^{n+1} &= (z^n)z = (r^n)r\,(\cos(n\theta + \theta) + i \sin(n\theta + \theta)) \\ &= r^{n+1}\,(\cos(n+1)\theta + i \sin(n+1)\theta)\,, \end{aligned}$$

as required. Hence (8.34) is true for all nonnegative integers n.

For a positive integer n, we have

$$z^{-n} = \frac{1}{z^n} = \frac{\cos 0 + i \sin 0}{r^n(\cos n\theta + i \sin n\theta)} = r^{-n}\big(\cos(-n\theta) + i \sin(-n\theta)\big). \quad (8.35)$$

We have now shown that equation (8.34) holds for *all* integers n. In particular, when z lies on the unit circle ($|z| = 1$), equation (8.34) gives the following result.

Theorem 8.4 **De Moivre**

$$(\cos\theta + i \sin\theta)^n = \cos n\theta + i \sin n\theta, \quad \textit{where } n \textit{ is an integer}$$

EXAMPLE 3 **Simplifying Polar Forms**

Using de Moivre's theorem, we have

$$\begin{aligned} \frac{(\cos 2\theta - i \sin 2\theta)^3}{(\cos 3\theta + i \sin 3\theta)^4} &= \frac{(\cos(-2\theta) + i \sin(-2\theta))^3}{(\cos 3\theta + i \sin 3\theta)^4} = \frac{\cos(-6\theta) + i \sin(-6\theta)}{\cos 12\theta + i \sin 12\theta} \\ &= \cos(-18\theta) + i \sin(-18\theta) = \cos 18\theta - i \sin 18\theta. \end{aligned}$$

■

EXAMPLE 4 **Simplifying Quotients and Powers**

$$\begin{aligned} \frac{(i - \sqrt{3})^5}{(1 - i)^2} &= \frac{2^5(\cos(5\pi/6) + i \sin(5\pi/6))^5}{(\sqrt{2})^2(\cos(-\pi/4) + i \sin(-\pi/4))^2} \\ &= 16\left(\cos\left(\frac{25\pi}{6} + \frac{\pi}{2}\right) + i \sin\left(\frac{25\pi}{6} + \frac{\pi}{2}\right)\right) \\ &= 16\left(\cos\frac{28\pi}{6} + i \sin\frac{28\pi}{6}\right) = 16\left(\cos\frac{2\pi}{3} + i \sin\frac{2\pi}{3}\right) \\ &= 16\left(-\frac{1}{2} + i\frac{\sqrt{3}}{2}\right) = 8(-1 + i\sqrt{3}) \end{aligned}$$

■

Shorthand Notation

The notation $\cos\theta + i\sin\theta = \operatorname{cis}\theta$ (read "$\cos\theta$ plus $i\sin\theta$") is standard shorthand used in technical calculations. As with trigonometric functions, the meaning of $\operatorname{cis}^2\theta$, for example, is $(\operatorname{cis}\theta)^2$. Here are some of the previous results written in short form.

$$\operatorname{cis}\theta_1 \operatorname{cis}\theta_2 = \operatorname{cis}(\theta_1+\theta_2), \quad \frac{\operatorname{cis}\theta_1}{\operatorname{cis}\theta_2} = \operatorname{cis}(\theta_1-\theta_2), \quad \overline{\operatorname{cis}\theta} = \operatorname{cis}(-\theta)$$

$$\frac{1}{\operatorname{cis}\theta} = \operatorname{cis}(-\theta), \qquad \operatorname{cis}^n\theta = \operatorname{cis}(n\theta) \quad \text{(de Moivre's theorem)}$$

EXAMPLE 5 **Shorthand Notation**

(a) $$\frac{\operatorname{cis}^4(3\theta)(\operatorname{cis}2\theta)^{-3}}{\operatorname{cis}^5\theta\operatorname{cis}(4\theta)} = \frac{\operatorname{cis}(12\theta)\operatorname{cis}(-6\theta)}{\operatorname{cis}(5\theta)\operatorname{cis}(4\theta)} = \frac{\operatorname{cis}(6\theta)}{\operatorname{cis}(9\theta)} = \operatorname{cis}(-3\theta).$$

(b)

$$\frac{(\sqrt{3}-i)^{14}}{(1-i)^3} = \frac{2^{14}\operatorname{cis}^{14}(-\pi/6)}{(\sqrt{2})^3\operatorname{cis}^3(-\pi/4)} = 2^{12}\sqrt{2}\,\frac{\operatorname{cis}(-7\pi/3)}{\operatorname{cis}(-3\pi/4)}$$

$$= 2^{12}\sqrt{2}\operatorname{cis}\left(-\frac{19\pi}{12}\right) = 2^{12}\sqrt{2}\operatorname{cis}\left(\frac{5\pi}{12}\right).$$ ■

Euler's Formula

Euler proved the fundamental identity

$$e^{i\theta} = \cos\theta + i\sin\theta, \tag{8.36}$$

where e is the exponential number. Hence any complex number z can be written in the compact form

$$z = re^{i\theta}. \tag{8.37}$$

Referring to Theorem 8.3 and the remark after the proof, we have

$$z = re^{i\theta} \quad \Rightarrow \quad z^n = r^n e^{in\theta}, \quad \text{for any integer } n.$$

Putting $z = 1$ in (8.37) gives the famous equation

$$e^{i\pi} + 1 = 0,$$

connecting five fundamental numbers of mathematics!

Infinite Series Expansions

Note that equation (8.36) follows from the infinite series expansions for e^z, $\cos\theta$, and $\sin\theta$.

$$e^z = \sum_{n=0}^{\infty} \frac{z^n}{n!} = 1 + z + \frac{z^2}{2!} + \frac{z^3}{3!} + \frac{z^4}{4!} + \cdots \quad \text{for real or complex } z.$$

Substituting $z = i\theta$ into e^z, we obtain (recall $0! = 1$)

$$e^{i\theta} = \sum_{n=0}^{\infty} \frac{(i\theta)^n}{n!} = 1 + i\theta - \frac{\theta^2}{2!} - \frac{i\theta^3}{3!} + \frac{\theta^4}{4!} + \frac{i\theta^5}{5!} - \frac{\theta^6}{6!} + \cdots$$

$$= \left(1 - \frac{\theta^2}{2!} + \frac{\theta^4}{4!} - \frac{\theta^6}{6!} + \cdots\right) + i\left(\theta - \frac{\theta^3}{3!} + \frac{\theta^5}{5!} - \cdots\right)$$

$$= \cos\theta + i\sin\theta,$$

using the series expansion for $\cos\theta$ and $\sin\theta$, namely

$$\cos\theta = \sum_{n=0}^{\infty} (-1)^n \frac{\theta^{2n}}{(2n)!} \quad \text{and} \quad \sin\theta = \sum_{n=0}^{\infty} (-1)^n \frac{\theta^{2n+1}}{(2n+1)!}.$$

Historical Notes

Abraham de Moivre (1667–1754). Born in France and educated in Belgium, de Moivre settled in England after fleeing from the persecution of the Huguenots (French Protestants) in France. His interest in mathematics started when, by accident, he saw a copy of *Principia Mathematica* by Sir Newton. De Moivre made important contributions to analytic geometry and the theory of probability through his publication *The Doctrine of Chance* (1718). However, his name is mostly remembered for the theorem that is stated in this section. Like Cardano, de Moivre is famous for predicting the actual day of his own death!

EXERCISES 8.3

1. Refer to Exercises 8.2, Problem 1(i)–(ix). Write the numbers in modulus-argument form.
2. Let α be any angle. Express each of the numbers in modulus-argument form.

   (i) $\cos\alpha - i\sin\alpha$ **(ii)** $\sin\alpha + i\cos\alpha$

   (iii) $\sin\alpha - i\cos\alpha$
3. If α is the principal argument of z, show that the principal argument of $\bar{z}$ is $-\alpha$.
4. Let $z = \cos\theta + i\sin\theta$. Express the following complex numbers in terms of θ.

   (i) $z + \frac{1}{z}$ **(ii)** $z - \frac{1}{z}$ **(iii)** $\frac{1}{\bar{z}}$

Exercises 5–22. Write in modulus-argument form.

5. $\left(\frac{1}{\sqrt{2}}(-1+i)\right)^4$ **6.** $(i - \sqrt{3})(1+i)^7$

7. $(1 + i\sqrt{3})^6(1-i)^4$ **8.** $i^3(1-i)^{17}$

9. $(1+i)^2(\sqrt{3} - i)^5$ **10.** $(1+i)^5 + (1-i)^5$

11. $\left(1 + \frac{i}{\sqrt{3}}\right)^8 + \left(1 - \frac{i}{\sqrt{3}}\right)^8$

12. $(\sqrt{3} - 1)^{382}$ **13.** $\left(\frac{i - \sqrt{3}}{1 - i}\right)^8$

14. $\left(\frac{\sqrt{3} + i}{1 - i}\right)^{12}$ **15.** $\frac{(\sqrt{3} + i)^9}{(1 + i\sqrt{3})^6}$

16. $\left(\frac{i}{1 - i\sqrt{3}}\right)^6$ **17.** $\left(\frac{-1 + i}{i + \sqrt{3}}\right)^9$

18. $\frac{(2+2i)^8}{(1 - i\sqrt{3})^6}$ **19.** $\left(\frac{1 + i\sqrt{3}}{\sqrt{3} - i}\right)^{18}$

20. $\left(\frac{i - \sqrt{3}}{1 - i}\right)^5$ **21.** $\frac{(1-i)^8}{(i - \sqrt{3})^9}$

22. $\left(\frac{1 - \sqrt{3}i}{1 + i}\right)^7$

Exercises 23–32. Write in polar form.

23. $\dfrac{1}{\cos 3\theta + i \sin 3\theta}$

24. $\dfrac{\cos 2\alpha + i \sin 2\alpha}{\cos \alpha + i \sin \alpha}$

25. $\dfrac{\cos \beta - i \sin \beta}{\cos \beta + i \sin \beta}$

26. $\dfrac{\cos 2\theta + i \sin 2\theta}{\cos 3\theta - i \sin 3\theta}$

27. $(\cos \theta - i \sin \theta)^3$

28. $\dfrac{(\cos \theta - i \sin \theta)^2}{(\cos \theta + i \sin \theta)^3}$

29. $\dfrac{(\operatorname{cis} 5\theta)^3 (\operatorname{cis} \theta)^{-3}}{(\operatorname{cis} 2\theta)^5 (\operatorname{cis} 3\theta)^2}$

30. $(\sin \frac{\pi}{5} + i \cos \frac{\pi}{5})^3$

31. $(\sin \theta - i \cos \theta)^3$

32. $\left(\cos(\beta + \frac{\pi}{2}) + i \sin(\beta + \frac{\pi}{2})\right)^4$

33. The function $w = 1/z$, defined for $z \neq 0$, is a fundamental transformation in the theory of complex numbers. Note that points outside the unit circle are mapped to points inside the unit circle, and vice versa. Points on the unit circle are mapped to points on the unit circle. Study the action of this transformation by locating images of points $z = 1, -1, i, -i, (-1 - i)/2, 1 + \sqrt{3}i$ in the complex plane.

USING MATLAB

Consult online help for the command **polar**, and related commands.

34. Write an M-file to plot a given complex number z (input) in standard and polar form. Test your program and use it to check hand calculations in Exercise 1 and Exercises 5–22 in this exercise set.

8.4 Extraction of Roots, Polynomials

We will consider the solution of the fundamental equation

$$z^n = a_0, \quad \text{where } n = 1, 2, 3, \ldots, \tag{8.38}$$

where a_0 is a given complex number and z is a complex variable. The method of solution uses the polar forms already introduced.

Let $a_0 = r_0(\operatorname{cis} \alpha)$ be a fixed nonzero complex number, where $r_0 = |a_0|$ and $\alpha = \operatorname{Arg} a_0$, and let $z = r(\operatorname{cis} \theta)$. Using de Moivre's formula, equation (8.38) becomes

$$r^n(\operatorname{cis} n\theta) = r_0(\operatorname{cis} \alpha). \tag{8.39}$$

Equating the modulus and arguments on both sides of (8.39), we have

$$r^n = r_0 \quad \text{and} \quad n\theta = \alpha + 2k\pi, \quad \text{where } k = 0, \pm 1, \pm 2, \ldots$$

because the difference $n\theta - \alpha$ can only be determined up to a multiple of 2π. Hence

$$r = \sqrt[n]{r_0}, \quad \text{and} \quad \theta = \frac{\alpha}{n} + \frac{2k\pi}{n}. \tag{8.40}$$

Notice that as k runs through the integers, (8.40) defines only n distinct complex numbers in the complex plane whose arguments are indexed by the integers 0 to $n - 1$. We have

$$\theta_0 = \frac{\alpha}{n}, \quad \theta_1 = \frac{\alpha}{n} + \frac{2\pi}{n}, \quad \ldots, \quad \theta_{n-1} = \frac{\alpha}{n} + \frac{2(n-1)\pi}{n}. \tag{8.41}$$

Observe that the value $k = n$ in (8.40) gives $\theta_0 + 2\pi$; that is, θ_0 plus a revolution, and $k = -1$ gives $\theta_{n-1} - 2\pi$; that is, θ_{n-1} minus a revolution, and so on. The n values for θ in (8.41) define the n-th *roots* of a_0 in (8.38), which are

$$z_k = \sqrt[n]{r_0}\left(\operatorname{cis} \frac{(\alpha + 2k\pi)}{n}\right), \quad k = 0, 1, 2, \ldots, n - 1. \tag{8.42}$$

The nth roots of a_0 all lie on the circle C in the complex plane, with center at the origin and radius $\sqrt[n]{r_0}$. If the basic root $z_0 = \sqrt[n]{r_0}\,\text{cis}(\alpha/n)$ is located on C, the other roots are located on C by adding multiples of $2\pi/n$ to α/n, which is the argument of z_0. The n-th roots are equally spaced around the circle C.

When $a_0 = 0$, equation (8.38) has only one (multiple) solution $z = 0$.

EXAMPLE 1

Fifth Roots

Find all the values of $(1-i)^{1/5}$. We have $1 - i = \sqrt{2}\,\text{cis}(-\pi/4)$. Hence,

$$z_k = (1-i)^{1/5} = 2^{1/10}\,\text{cis}\left(-\frac{\pi}{4} + \frac{2k\pi}{5}\right) = 2^{1/10}\,\text{cis}\left(\frac{(8k-1)\pi}{20}\right),$$

where $k = 0, 1, 2, 3, 4$. We therefore have five values:

$$z_0 = \sqrt[10]{2}\,\text{cis}\left(-\frac{\pi}{20}\right), \quad z_1 = \sqrt[10]{2}\,\text{cis}\left(\frac{7\pi}{20}\right), \qquad z_2 = \sqrt[10]{2}\,\text{cis}\left(\frac{3\pi}{4}\right),$$

$$z_3 = \sqrt[10]{2}\,\text{cis}\left(\frac{23\pi}{20}\right), \quad z_4 = \sqrt[10]{2}\,\text{cis}\left(\frac{31\pi}{20}\right).$$

Writing z_3 and z_4 with principal arguments, we get

$$z_3 = \sqrt[10]{2}\,\text{cis}\left(-\frac{17\pi}{20}\right), \qquad z_4 = \sqrt[10]{2}\,\text{cis}\left(-\frac{9\pi}{20}\right).$$

The numbers z_0, z_1, z_2, z_3, z_4 are equally spaced around a circle, center the origin and radius $\sqrt[10]{2}$. The angle between vectors representing consecutive numbers is $2\pi/5$. ■

An important special case arises when $a_0 = 1 = \text{cis}\,0$. Here $\alpha = \text{Arg}\,a_0 = 0$ and the n roots are called the n-th *roots of unity*, given as follows:

$$\sqrt[n]{1} = \text{cis}\,\frac{2\pi k}{n}, \quad k = 0, 1, 2, \ldots, n-1. \tag{8.43}$$

The root with $k = 1$ in (8.43), traditionally denoted by ω (Greek letter *omega*), is

$$\omega = \text{cis}\,\frac{2\pi}{n}.$$

Using de Moivre's theorem, the roots in equation (8.43) are

$$\omega^0\ (=1), \quad \omega, \quad \omega^2, \quad \ldots, \quad \omega^{n-1}.$$

Because ω satisfies the equation $\omega^n = 1$, the polynomial $\omega^n - 1$ has a factor $\omega - 1$ and dividing $\omega - 1$ into $\omega^n - 1$ gives a factorization

$$(\omega - 1)(\omega^{n-1} + \omega^{n-2} + \cdots \omega + 1) = 0.$$

But $\omega \neq 1$ and so

$$\omega^{n-1} + \omega^{n-2} + \cdots \omega + 1 = 0.$$

When plotted in the complex plane, the nth roots of unity form the vertices of a regular polygon of n sides, inscribed in the unit circle $|z| = 1$, with one vertex at $z = 1$. Figure 8.9 shows the polygons when $n = 3$ and $n = 6$.

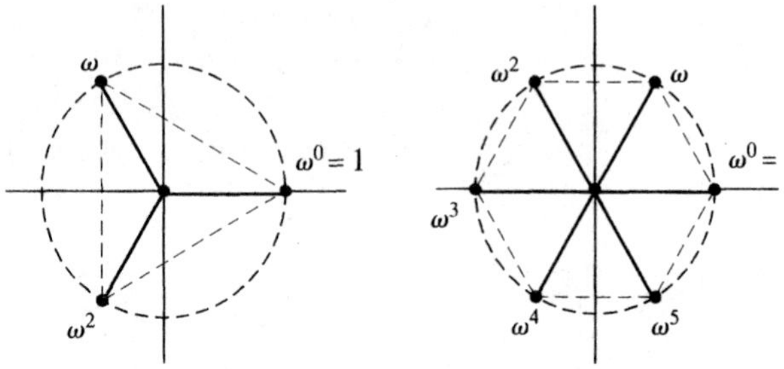

Figure 8.9 The nth roots of unity for $n = 3$ and $n = 6$.

Rational Powers

Let m and n be positive integers with no common factor. For the complex number a_0, we define $a_0^{m/n}$ $(= \sqrt[n]{a_0^m})$ to be any number z that satisfies the equation $z^n = a_0^m$. But, as seen above, there are n distinct roots to this equation, giving n possible values for $a_0^{m/n}$. Using (8.42), we have

$$a_0^{m/n} = \sqrt[n]{r_0^m}\left(\operatorname{cis}\frac{m}{n}(\alpha + 2\pi k)\right),$$

where $k = 0, 1, 2, \ldots, n-1$.

EXAMPLE 2

Solving a Polynomial Equation

The polynomial equation $z^9 + z^5 - z^4 - 1 = 0$ of degree 9 has a factorization

$$(z^5 - 1)(z^4 + 1) = 0$$

and therefore $z^5 = 1$ or $z^4 = -1$. Hence, $z^5 = \operatorname{cis} 0$ or $z^4 = \operatorname{cis} \pi$ and the roots are

$$z_p = \operatorname{cis}\left(\frac{2p\pi}{5}\right), \quad p = 0, 1, 2, 3, 4 \quad \text{and}$$

$$z_q = \operatorname{cis}\left(\frac{\pi}{4} + \frac{2q\pi}{4}\right) = \operatorname{cis}\frac{(2q+1)\pi}{4}, \quad q = 0,\ 1,\ 2,\ 3.$$

The roots $\operatorname{cis}\dfrac{(2q+1)\pi}{4}$ in standard form are

$$\operatorname{cis}\frac{\pi}{4} = \frac{1}{\sqrt{2}} + i\frac{1}{\sqrt{2}} = \frac{1}{\sqrt{2}}(1+i), \quad \operatorname{cis}\frac{3\pi}{4} = \frac{1}{\sqrt{2}}(-1+i)$$

and so on. Altogether, there are nine roots of the polynomial equation. ■

Polynomials

A *polynomial* is an expression of the form

$$p(z) = a_n z^n + a_{n-1} z^{n-1} + \cdots + a_1 z + a_0, \tag{8.44}$$

where the *coefficients* $a_n, a_{n-1}, \ldots, a_1, a_0$ are complex numbers and z is complex variable. When $a_n \neq 0$ the polynomial has *degree* n. For example,

$p(z) = (1+i)z^3 - z + i$ is a (cubic) polynomial of degree $n = 3$ with coefficients $a_3 = 1 + i$, $a_2 = 0$, $a_1 = -1$, $a_0 = i$.

The equation $p(z) = 0$ is called a *polynomial equation.* A complex number z_0 such that $p(z_0) = 0$ is called a *root* of $p(z)$ or *solution* of the polynomial equation $p(z) = 0$. At the beginning of this section we solved polynomial equations of the type $z^n = a_0$, or $p(z) = z^n - a_0 = 0$.

If $p(z)$ is a quadratic polynomial (degree 2), then $p(z) = a_2 z^2 + a_1 z + a_0$, and dividing by $a_2 \neq 0$ we have $p(z) = z^2 + bz + c$, where $b = a_1/a_2$ and $c = a_0/a_2$ are complex. By completing the square, $p(z)$ can be factored into two linear factors, as follows:

$$p(z) = \left(z + \frac{b + \sqrt{b^2 - 4c}}{2}\right)\left(z + \frac{b - \sqrt{b^2 - 4c}}{2}\right),$$

and the roots of the polynomial equation $p(z) = 0$ are

$$z_1 = \frac{-b + \sqrt{b^2 - 4c}}{2}, \qquad z_2 = \frac{-b - \sqrt{b^2 - 4c}}{2}. \tag{8.45}$$

The numbers $+\sqrt{b^2 - 4c}$ and $-\sqrt{b^2 - 4c}$ in (8.45) are the two (complex) roots of $b^2 - 4c$.

Theorem 8.5 Fundamental Theorem of Algebra

Every polynomial $p(z)$ of degree $n \geq 1$ has exactly n complex roots (counting repeated roots).

If z_0 is a root of $p(z)$, then $z - z_0$ is a factor of $p(z)$. Dividing $p(z)$ by $z - z_0$ gives $q(z)$, a polynomial of degree $n - 1$ such that $p(z) = (z - z_0)q(z)$. Repeating this process, $p(z)$ can be expressed as a product of linear factors.

When $p(z)$ has real coefficients, knowing one root gives immediately a second root (its complex conjugate) without extra calculation. We will prove this fact.

Theorem 8.6 Polynomials with Real Coefficients

Let $p(z) = a_n z^n + a_{n-1} z^{n-1} + \cdots + a_1 z + a_0$ be a polynomial with real coefficients. If z_0 is a root of $p(z)$, then so is its complex conjugate $\overline{z_0}$.

Proof Properties of the complex conjugate are used here. For each k we have $\overline{a_k} = a_k$ because the coefficients are real, and $\overline{z^k} = \overline{z}^k$. Thus,

$$\begin{aligned}\overline{p(z)} &= \overline{a_n z^n + a_{n-1} z^{n-1} + \cdots + a_1 z + a_0} \\ &= \overline{a_n z^n} + \overline{a_{n-1} z^{n-1}} + \cdots + \overline{a_1 z} + \overline{a_0} \\ &= a_n \overline{z}^n + a_{n-1} \overline{z}^{n-1} + \cdots + a_1 \overline{z} + a_0 \\ &= p(\overline{z}).\end{aligned}$$

Put $z = z_0$, then $p(z_0) = 0$ and so $p(\overline{z_0}) = \overline{p(z_0)} = \overline{0} = 0$.

Suppose the coefficients of $p(z)$ are real. We say that the roots of $p(z)$ appear in *complex conjugate pairs*. If z_0 is a (complex) root of $p(z)$, then $z - z_0$ and $z - \overline{z_0}$ are both factors of $p(z)$. Multiplying these factors, we obtain

$$(z - z_0)(z - \overline{z_0}) = z^2 - (z_0 + \overline{z_0})z + z_0\overline{z_0}. \tag{8.46}$$

But $z_0 + \overline{z_0} = 2\,\mathrm{Re}(z_0)$ and $z_0\overline{z_0} = |z_0|^2$ are both real quantities. Hence, the quadratic in (8.46) has real coefficients and is a factor of $p(z)$. To summarize, we will make a formal statement.

Theorem 8.7 **Factoring Polynomials**

A polynomial of degree $n \geq 1$ with real coefficients can be written as a product of linear and quadratic factors with real coefficients.

EXAMPLE 3 **Expressing a Polynomial as a Product of Factors**

The roots of the equation $z^5 - 1 = 0$ are $\operatorname{cis}\dfrac{2k\pi}{5}$, where $k = 0, 1, 2, 3, 4$.

These roots can be written in conjugate pairs as follows:

$$\operatorname{cis}\frac{2\pi}{5} \quad\text{and}\quad \operatorname{cis}-\left(\frac{2\pi}{5}\right)\left(=\operatorname{cis}\frac{8\pi}{5}\right), \quad \operatorname{cis}\frac{4\pi}{5} \quad\text{and}\quad \operatorname{cis}-\left(\frac{4\pi}{5}\right)\left(=\operatorname{cis}\frac{6\pi}{5}\right)$$

and one real root $\operatorname{cis} 0 = 1$.

Hence we have

$$z^5 - 1 = (z-1)\left(z - \operatorname{cis}\frac{2\pi}{5}\right)\left(z - \operatorname{cis}\left(-\frac{2\pi}{5}\right)\right)\left(z - \operatorname{cis}\frac{4\pi}{5}\right)\left(z - \operatorname{cis}\left(-\frac{4\pi}{5}\right)\right)$$

$$= (z-1)\left(z^2 - 2z\cos\frac{2\pi}{5} + 1\right)\left(z^2 - 2z\cos\frac{4\pi}{5} + 1\right).$$

■

EXAMPLE 4 **Finding Roots of a Polynomial**

Consider the polynomial

$$p(z) = z^4 - 6z^3 + 15z^2 - 18z + 10.$$

It so happens that the number $z = 2+i$ is a root of $p(z) = 0$. Then $\bar{z} = 2-i$ is also a root because the polynomial has real coefficients. Hence $(z-2-i)(z-2+i) = z^2 - 4z + 5$ is a factor and by long division,

$$z^4 - 6z^3 + 15z^2 - 18z + 10 = (z^2 - 4z + 5)(z^2 - 2z + 2).$$

The roots of $z^2 - 2z + 2$ are $1+i$ and $1-i$. Hence $p(z)$ has four roots: $2+i$, $2-i$, $1+i$, $1-i$. ■

EXERCISES 8.4

Exercises 1–6. Compute square roots of each number.

1. $\operatorname{cis} 2\theta$ **2.** $\cos 3\theta - i\sin 3\theta$

3. i **4.** $-i$

5. $1 - i$ **6.** 1

Exercises 7–12. Compute cube roots of each number.

7. $\operatorname{cis} 3\theta$ **8.** 1 **9.** i

10. $-8i$ **11.** $1+i$ **12.** -1

Exercises 13–20. Find the required roots in polar form. Plot the roots in the complex plane.

13. Cube roots of $-2+2i$ **14.** Fourth roots of -4

15. Sixth roots of $-i$ **16.** Fourth roots of $-1+i\sqrt{3}$

17. Fourth roots of $-1+i$ **18.** Sixth roots of $1-i$

19. Cube roots of $\operatorname{cis}\frac{5\pi}{4}$ **20.** Fifth roots of 32

Exercises 21–30. Solve each equation for z.

21. $z^2 - (3 - 2i)z + (5 - 5i) = 0$

22. $z^4 - 30z^2 + 289 = 0$

23. $1 + iz^5 = i$

24. $z^7 - z^6 + iz - i = 0$

25. $z^4 + 8 + 8\sqrt{3}i = 0$

26. $z^9 - z = 0$

27. $z = iz^7$

28. $z^3 + 8 = 0$

29. $z^6 + i = 0$

30. $z^9 - iz^8 - z + i = 0$

Exercises 31–34. Express each polynomial as a product of linear or quadratic factors with real coefficients.

31. $z^3 + 8$

32. $z^4 + 16$

33. $z^6 - 64$

34. $z^5 + 1$

35. If $z + i$ is a root of $z^4 - 6z^3 - 15z^2 - 6z - 16$, find all the roots.

36. If $2 + 3i$ is a root of $z^4 - 6z^3 + 23z^2 - 34z + 26$, find all the roots.

37. If $z - \sqrt{2} - i\sqrt{3}$ is a factor of $z^4 - 2\sqrt{2}z^3 + 6z^2 - 2\sqrt{2}z + 5$, find all the factors.

38. Solve the equation $z^6 + 2z^3 + 4 = 0$. *Hint*: Substitute $Z = z^3$.

39. Let p and q be real numbers. Prove that, if $z = x + iy$ is a complex root of $z^3 + pz + q$, then x is a real root of $8z^3 + 2pz - q$. *Hint*: The root z is complex, meaning that $y \neq 0$.

40. The cubic equation $z^3 - 15z - 4 = 0$ appears in the introduction to this chapter and its roots are plotted in Figure 8.1. Factor the cubic and find its roots. Prove that the roots just found are exactly the roots given by the Cardaro–Tartaglia formula shown in (8.4), namely

$$z = \sqrt[3]{2 + 11i} + \sqrt[3]{2 - 11i}$$

USING MATLAB

41. Use the Symbolic Math Toolbox to solve Problems 21–30.

42. *Plotting Roots*. Let ω be a cube root of unity. Suppose $\omega \neq 1$. Prove that $1 + \omega + \omega^2 = 0$ and find $(1 + 2\omega + 3\omega^2)(1 + 3\omega + 2\omega^2)$. Verify your computation by plotting the complex numbers $1, 2\omega, 3\omega^2, (1+2\omega+3\omega^2), (1 + 3\omega + 2\omega^2)$.

8.5 Linear Algebra: The Complex Case

In this section we return to linear algebra and review the main concepts in the case when scalars are complex numbers—we call this the *complex case*. The preceding chapters concentrated almost exclusively on the *real case* when scalars, coefficients, matrix entries, vector components, and so on are real numbers. The use of complex numbers as scalars was mentioned briefly in connection with eigenvalues.

The complex number system $\mathbb{C}$ is an extension of the real number system $\mathbb{R}$. Many concepts and definitions in linear algebra extend immediately from the real to the complex case without change, but there are striking exceptions and some adjustments are required. Certain concepts defined for real scalars need to be redefined in order to make sense in the complex case. The new definition, however, always agrees with the old one when all scalars are purely real.

The set of $m \times n$ matrices with complex entries is denoted by $\mathbb{C}^{m\times n}$. The space of n-vectors with complex components is denoted by $\mathbb{C}^n$. Note that $\mathbb{C}^n$ is identical to $\mathbb{C}^{n\times 1}$.

Matrices

An $m \times n$ matrix $\mathbf{A}$ is called *complex* if its entries are complex numbers. Of course, any real matrix (with purely real entries) is also complex. For example, the matrix $\mathbf{A}$, shown next, belongs to $\mathbb{C}^{2\times 3}$, and $\mathbf{v}$ is a vector in $\mathbb{C}^2$.

$$\mathbf{A} = \begin{bmatrix} i & 2 & 1+i \\ 1-i & 3i & 0 \end{bmatrix}, \qquad \mathbf{v} = \begin{bmatrix} 1+2i \\ 3 \end{bmatrix}$$

The basic operations on complex matrices—addition, subtraction, multiplication by a complex scalar, row-by-column multiplication, and transposition, are all performed as in the real case.

Linear Systems, Echelon Forms, Rank

Complex linear systems have complex unknowns, coefficients, and constant terms. The three elementary row operations (EROs) are used to reduce a complex matrix **A** to an echelon form **U** or to the reduced form $\mathbf{A}^*$. As in the real case, rank **A** is the number of nonzero rows in any echelon form **U** for **A** and complex linear systems can be solved by reduction of the augmented matrix $[\mathbf{A}|\,\mathbf{b}]$.

EXAMPLE 1

Solving a Complex Linear System

Our goal is to solve the linear system

$$(S)\qquad \begin{cases} iz_1 + (1+i)z_2 &= 2+3i \\ (1-i)z_1 + \qquad iz_2 &= 3+2i \end{cases}.$$

Reducing the augmented matrix $[\mathbf{A}|\,\mathbf{b}]$ of (S) to its reduced form **R**, we have

$$\left[\begin{array}{cc|c} i & 1+i & 2+3i \\ 1-i & i & 3+2i \end{array}\right] \underset{\frac{1}{i}\mathbf{r}_1 \to \mathbf{r}_1}{\sim} \left[\begin{array}{cc|c} 1 & 1-i & 3-2i \\ 1-i & i & 3+2i \end{array}\right]$$

$$\underset{\mathbf{r}_2-(1-i)\mathbf{r}_1 \to \mathbf{r}_2}{\sim} \left[\begin{array}{cc|c} 1 & 1-i & 3-2i \\ 0 & 3i & 2+7i \end{array}\right]$$

$$\underset{\frac{1}{3i}\mathbf{r}_2 \to \mathbf{r}_2}{\sim} \left[\begin{array}{cc|c} 1 & 1-i & 3-2i \\ 0 & 1 & \frac{7}{3}-\frac{2}{3}i \end{array}\right]$$

$$\underset{\mathbf{r}_1-(1-i)\mathbf{r}_2 \to \mathbf{r}_1}{\sim} \left[\begin{array}{cc|c} 1 & 0 & \frac{4}{3}+i \\ 0 & 1 & \frac{7}{3}-\frac{2}{3}i \end{array}\right].$$

We expect the unique solution $\begin{bmatrix} z_1 \\ z_2 \end{bmatrix} = \begin{bmatrix} \frac{4}{3}+i \\ \frac{7}{3}-\frac{2}{3}i \end{bmatrix}$ here because rank $\mathbf{A} = 2$, which is number of unknowns in (S). ■

Determinants

The definition of the determinant, denoted by either det(**A**) or $|\mathbf{A}|$, of an $n \times n$ matrix **A**, as given in Section 5.1, carries over to the complex case without change.

EXAMPLE 2

Evaluation of Complex Determinants

The rule $\begin{vmatrix} a & b \\ c & d \end{vmatrix} = ad - bc$ applies in the 2×2 complex case. For example,

$$\begin{vmatrix} 1-i & 2 \\ 3i & 2-i \end{vmatrix} = (1-i)(2-i) - 6i = 1 - 9i.$$

Determinants of higher order are evaluated using the standard algorithm, which reduces **A** to a triangular form **U**. Then det(**A**) is the product of diagonal entries in **U**. For example,

$$\begin{vmatrix} i & -2i & 0 \\ 1 & -i & 2 \\ -i & 0 & 3i \end{vmatrix} \quad \underset{\substack{\mathbf{r}_1 \leftrightarrow \mathbf{r}_2 \\ \mathbf{r}_2 - i\mathbf{r}_1 \to \mathbf{r}_2 \\ \mathbf{r}_3 + i\mathbf{r}_1 \to \mathbf{r}_3}}{\sim} \quad \begin{vmatrix} 1 & -i & 2 \\ 0 & -(1+2i) & -2i \\ 0 & 1 & 5i \end{vmatrix}$$

$$\underset{\substack{\mathbf{r}_2 \leftrightarrow \mathbf{r}_3 \\ \mathbf{r}_3 + (1+2i)\mathbf{r}_2 \to \mathbf{r}_3}}{\sim} \quad \begin{vmatrix} 1 & -i & 2 \\ 0 & 1 & 5i \\ 0 & 0 & -10+3i \end{vmatrix} = -10 + 3i.$$

■

Vectors, Linear Independence

A linear combination of complex vectors $\mathbf{v}_1, \mathbf{v}_2, \ldots, \mathbf{v}_k$ in $\mathbb{C}^n$ is formed using complex scalars $z_1, z_2, \ldots, z_k$ as follows:

$$z_1\mathbf{v}_1 + z_2\mathbf{v}_2 + \cdots + z_k\mathbf{v}_k.$$

The vectors $\mathbf{v}_1, \mathbf{v}_2, \ldots, \mathbf{v}_k$ are linearly independent if the equation

$$z_1\mathbf{v}_1 + z_2\mathbf{v}_2 + \cdots + z_k\mathbf{v}_k = \mathbf{0} \quad (= \text{the zero vector in } \mathbb{C}^n) \tag{8.47}$$

has only the trivial solution $z_j = 0 + 0i$, $1 \le j \le k$. The vectors are linearly dependent if there is a solution in which at least one z_j is nonzero.

One way to recognize the linear independence or dependence of $\mathbf{v}_1, \mathbf{v}_2, \ldots, \mathbf{v}_k$ is to form the $n \times k$ matrix $\mathbf{A} = [\,\mathbf{v}_1 \ \mathbf{v}_2 \ \cdots \ \mathbf{v}_k\,]$ and compute either rank **A** or solve $\mathbf{Az} = \mathbf{0}$ by reducing **A** to an echelon form. The coefficient matrix $\mathbf{A} = [\,\mathbf{v}_1 \ \mathbf{v}_2\,]$ in Example 1 has linearly independent columns because rank $\mathbf{A} = 2$, or alternatively, $\mathbf{v}_1$ and $\mathbf{v}_2$ are nonzero and neither vector is a scalar multiple of the other. As in the real case, $n + 1$ or more vectors in $\mathbb{C}^n$ are automatically linearly dependent and all the other theorems and properties of linearly independent (dependent) sets hold.

Dot Product, Norm

The interconnected concepts of *dot product* and *norm* for real (column) vectors appear in Section 4.1. We require that the norm $\|\,\mathbf{u}\,\|$ of a real n-vector **u** be real and nonnegative—the norm measures the physical length of a vector in $\mathbb{R}^2$ or $\mathbb{R}^3$. In the real case, we have $\|\,\mathbf{u}\,\|^2 = \mathbf{u}\cdot\mathbf{u} = \mathbf{u}^\mathsf{T}\mathbf{u}$, but this property cannot be extended to the complex case, as the following example shows:

$$\mathbf{u} = \begin{bmatrix} i \\ 1-i \end{bmatrix} \quad \Rightarrow \quad \mathbf{u}^\mathsf{T}\mathbf{u} = \begin{bmatrix} i & 1-i \end{bmatrix}\begin{bmatrix} i \\ 1-i \end{bmatrix}$$

$$= (-1) + (-2i) = -(1+2i),$$

and $-(1+2i)$ is not real. The concept of a *positive number* or *negative number* has no meaning for complex numbers. The definition of norm needs to be adjusted in the complex case, but we first introduce a more general concept.

Definition 8.1

Hermitian conjugate, adjoint

Let $\mathbf{A} = [a_{ij}]$ be an $m \times n$ complex matrix and let $\overline{\mathbf{A}} = [\overline{a}_{ij}]$ be formed by taking the complex conjugate of each entry in $\mathbf{A}$. Then the *conjugate transpose* operation, denoted by H, when applied to $\mathbf{A}$ returns an $n \times m$ matrix $\mathbf{A}^{\mathsf{H}}$ defined by

$$\mathbf{A}^{\mathsf{H}} = \overline{\mathbf{A}}^{\mathsf{T}}. \tag{8.48}$$

The matrix $\mathbf{A}^{\mathsf{H}}$ is called the *Hermitian conjugate* or *adjoint*[5] of $\mathbf{A}$.

For example,

$$\mathbf{A} = \begin{bmatrix} 0.5i & 6+2i & 2-3i \\ 1-i & 2.1 & 3+2i \end{bmatrix} \quad \Rightarrow \quad \mathbf{A}^{\mathsf{H}} = \begin{bmatrix} -0.5i & 1+i \\ 6-2i & 2.1 \\ 2+3i & 3-2i \end{bmatrix}.$$

Properties of the conjugate transpose operator H parallel those of the transpose operator T in the real case (Section 2.1).

Definition 8.2

Dot product

Let $\mathbf{u}$ and $\mathbf{v}$ be complex vectors in $\mathbb{C}^n$. The *dot product* of $\mathbf{u}$ and $\mathbf{v}$ (in that order) is defined by

$$\mathbf{u} \cdot \mathbf{v} = \mathbf{u}^{\mathsf{H}}\mathbf{v}. \tag{8.49}$$

The *norm* of $\mathbf{u}$ is defined by

$$\|\mathbf{u}\| = \sqrt{\mathbf{u} \cdot \mathbf{u}} = \sqrt{\mathbf{u}^{\mathsf{H}}\mathbf{u}}. \tag{8.50}$$

EXAMPLE 3

Dot Product, Norm

Consider the complex 2-vectors

$$\mathbf{u} = \begin{bmatrix} 1-i \\ 2i \end{bmatrix}, \quad \mathbf{v} = \begin{bmatrix} 3 \\ 1+i \end{bmatrix}.$$

[5] See the remarks on page 397.

Compute $\mathbf{u}\cdot\mathbf{v}$ and $\mathbf{v}\cdot\mathbf{u}$ as follows:

$$\mathbf{u}\cdot\mathbf{v} = \mathbf{u}^{\mathsf{H}}\mathbf{v} = \overline{\mathbf{u}}^{\mathsf{T}}\mathbf{v} = [\,1+i,\ -2i\,]\begin{bmatrix} 3 \\ 1+i \end{bmatrix} = 3+3i-2i+2 = 5+i,$$

$$\mathbf{v}\cdot\mathbf{u} = \mathbf{v}^{\mathsf{H}}\mathbf{u} = \overline{\mathbf{v}}^{\mathsf{T}}\mathbf{u} = [\,3,\ 1-i\,]\begin{bmatrix} 1-i \\ 2i \end{bmatrix} = 3-3i+2i+2 = 5-i$$

and

$$\|\mathbf{u}\|^2 = \mathbf{u}^{\mathsf{H}}\mathbf{u} = [\,1+i\ \ -2i\,]\begin{bmatrix} 1-i \\ 2i \end{bmatrix} = 6 \quad\Rightarrow\quad \|\mathbf{u}\| = \sqrt{6} \sim 2.4495,$$

$$\|\mathbf{v}\|^2 = \mathbf{v}^{\mathsf{H}}\mathbf{v} = [\,3\ \ 1-i\,]\begin{bmatrix} 3 \\ 1+i \end{bmatrix} = 11 \quad\Rightarrow\quad \|\mathbf{v}\| = \sqrt{11} \sim 3.3166.$$

For any complex vectors $\mathbf{u}$ and $\mathbf{v}$ in $\mathbb{C}^n$ we have $\mathbf{u}\cdot\mathbf{v} = \overline{\mathbf{v}\cdot\mathbf{u}}$, as seen in this example. Also, $\|\mathbf{u}\|$ is real and nonnegative (Exercises 8.2). ■

Hermitian Matrices

A real square matrix $\mathbf{A}$ is symmetric if $\mathbf{A}^{\mathsf{T}} = \mathbf{A}$. In the complex case, the transpose operation is replaced by the conjugate transpose operation and the concept of symmetric matrix extends to complex matrices in the following way.

Definition 8.3

Hermitian

An $n \times n$ matrix $\mathbf{A} = [a_{ij}]$ is called *Hermitian* if $\mathbf{A}^{\mathsf{H}} = \mathbf{A}$.

Comparing entries in $\mathbf{A}^{\mathsf{H}}$ and $\mathbf{A}$ when $i \neq j$, we have $\overline{a}_{ij}^{\mathsf{T}} = \overline{a}_{ji} = a_{ij}$, which says that two entries that are symmetric about the main diagonal form a complex conjugate pair. On the main diagonal, $\overline{a}_{ii} = a_{ii}$ and so all diagonal entries are necessarily real. It should now be clear how to recognize or construct a Hermitian matrix. For example,

$$\mathbf{A} = \begin{bmatrix} 4 & -i \\ i & 6 \end{bmatrix}, \qquad \begin{bmatrix} 1.2 & 1+i & 2-i\sqrt{3} \\ 1-i & 0 & -3i \\ 2+i\sqrt{3} & 3i & 4.5 \end{bmatrix}$$

are both Hermitian.

Eigenvalues

Recall the main definitions from Chapter 6. Suppose $\mathbf{A}$ is an $n \times n$ complex matrix. A real or complex scalar λ such that $\mathbf{Az} = \lambda\mathbf{z}$ for some nonzero n-vector $\mathbf{z}$ is called an eigenvalue of $\mathbf{A}$ and $\mathbf{z}$ is an eigenvector associated with λ. We have seen earlier that a real matrix can have complex eigenvalues.

However, real symmetric matrices always have real eigenvalues. This fact is a consequence of a more general result, which is now proved.

Theorem 8.8 **Eigenvalues of Hermitian Matrices**

The eigenvalues of an Hermitian matrix $\mathbf{A}$ *are all real.*

Proof Assume that $\mathbf{A}$ is $n \times n$ and Hermitian. There are two steps.

(a) We first show that if $\mathbf{z}$ is any n-vector, then $\mathbf{z}^{\mathsf{H}}\mathbf{A}\mathbf{z}$ is real. The product $x = \mathbf{z}^{\mathsf{H}}\mathbf{A}\mathbf{z}$ is a 1×1 matrix (a scalar) and x is real if and only if $x = \overline{x}$. Using properties of the Hermitian operator (Exercises 8.5, 18), we have

$$x^{\mathsf{H}} = (\mathbf{z}^{\mathsf{H}}\mathbf{A}\mathbf{z})^{\mathsf{H}} = \mathbf{z}^{\mathsf{H}}\mathbf{A}^{\mathsf{H}}(\mathbf{z}^{\mathsf{H}})^{\mathsf{H}} = \mathbf{z}^{\mathsf{H}}\mathbf{A}\mathbf{z} = x.$$

and so $x = x^{\mathsf{H}} = \overline{x}^{\mathsf{T}} = \overline{x}$, showing that x is real.

(b) Let λ be an eigenvalue of $\mathbf{A}$ and let z be an eigenvector associated with λ. Then, multiplying by $\mathbf{z}^{\mathsf{H}}$, we have

$$\mathbf{A}\mathbf{z} = \lambda\mathbf{z} \quad \Rightarrow \mathbf{z}^{\mathsf{H}}(\mathbf{A}\mathbf{z}) = \mathbf{z}^{\mathsf{H}}(\lambda\mathbf{z}) = \lambda\mathbf{z}^{\mathsf{H}}\mathbf{z} = \lambda\|\mathbf{z}\|^2.$$

But $\|\mathbf{z}\|^2$ is real and positive because $\mathbf{z} \neq \mathbf{0}$ and so $\lambda = \mathbf{z}^{\mathsf{H}}\mathbf{A}\mathbf{z}/\|\mathbf{z}\|^2$ is real in light of (a).

Note that Theorem 8.8 shows that the eigenvalues of a real symmetric matrix are all real numbers.

The next goal is to see how the eigenvectors of a Hermitian matrix $\mathbf{A}$ are related.

Theorem 8.9 **Orthogonal Eigenvectors**

Let $\mathbf{A}$ *be an* $n \times n$ *Hermitian matrix and let* $\mathbf{z}_1$ *and* $\mathbf{z}_2$ *be complex eigenvectors associated, respectively, with* distinct *eigenvalues* λ_1 *and* λ_2. *Then* $\mathbf{z}_1$ *is orthogonal to* $\mathbf{z}_2$ *and we write* $\mathbf{z}_1 \perp \mathbf{z}_2$.

Proof Consider the equation $\mathbf{A}\mathbf{z}_1 = \lambda_1\mathbf{z}_1$. Applying the Hermitian operator and using its properties, we have

$$\mathbf{A}\mathbf{z}_1 = \lambda_1\mathbf{z}_1 \Rightarrow (\mathbf{A}\mathbf{z}_1)^{\mathsf{H}} = (\lambda_1\mathbf{z}_1)^{\mathsf{H}} \Rightarrow \mathbf{z}_1^{\mathsf{H}}\mathbf{A}^{\mathsf{H}} = \overline{\lambda_1}\mathbf{z}_1^{\mathsf{H}}$$

$$\Rightarrow \mathbf{z}_1^{\mathsf{H}}\mathbf{A} = \lambda_1\mathbf{z}_1^{\mathsf{H}} \tag{8.51}$$

because $\mathbf{A}$ is Hermitian and λ_1 is both a scalar and real. Multiplying (8.51) on right by $\mathbf{z}_2$ and $\mathbf{A}\mathbf{z}_2 = \lambda_2\mathbf{z}_2$ on the left by $\mathbf{z}_1^{\mathsf{H}}$, and subtracting we have

$$\left.\begin{array}{ll} \mathbf{z}_1^{\mathsf{H}}\mathbf{A} = \lambda_1\mathbf{z}_1^{\mathsf{H}} & \Rightarrow \mathbf{z}_1^{\mathsf{H}}\mathbf{A}\mathbf{z}_2 = \lambda_1\mathbf{z}_1^{\mathsf{H}}\mathbf{z}_2 \\ \mathbf{A}\mathbf{z}_2 = \lambda_2\mathbf{z}_2 & \Rightarrow \mathbf{z}_1^{\mathsf{H}}\mathbf{A}\mathbf{z}_2 = \lambda_2\mathbf{z}_1^{\mathsf{H}}\mathbf{z}_2 \end{array}\right. \Rightarrow (\lambda_1 - \lambda_2)\mathbf{z}_1^{\mathsf{H}}\mathbf{z}_2 = 0.$$

But λ_1, λ_2 are distinct and so $\mathbf{z}_1^{\mathsf{H}}\mathbf{z}_2 = \mathbf{z}_1 \cdot \mathbf{z}_2 = 0$. Hence $\mathbf{z}_1 \perp \mathbf{z}_2$.

■ ILLUSTRATION 8.2

Hermitian Matrices, Eigenvalues, Eigenvectors

Consider the Hermitian matrix shown in (8.52).

$$\mathbf{A} = \begin{bmatrix} 2 & 1+i \\ 1-i & 3 \end{bmatrix}. \tag{8.52}$$

The eigenvalues of **A** and convenient normalized eigenvectors associated with these values are:

$$\lambda_1 = 1, \quad \mathbf{z}_1 = \frac{1}{\sqrt{3}}\begin{bmatrix} 1+i \\ -1 \end{bmatrix}, \quad \lambda_2 = 4, \quad \mathbf{z}_2 = \frac{1}{\sqrt{6}}\begin{bmatrix} 1+i \\ 2 \end{bmatrix}.$$

The eigenvalues are distinct, and by Theorem 8.9 we know that $\mathbf{z}_1 \perp \mathbf{z}_2$ (verify). The set $\{\mathbf{z}_1, \mathbf{z}_2\}$ is orthonormal: The vectors are of unit length and each pair of vectors (there is only one pair in this case) are orthogonal. The next observation illustrates a general result.

Form a matrix $\mathbf{U} = [\,\mathbf{z}_1 \;\; \mathbf{z}_2\,]$ and compute $\mathbf{U}^\mathsf{H}\mathbf{U}$, as follows:

$$\mathbf{U}^\mathsf{H}\mathbf{U} = \begin{bmatrix} (1-i)/\sqrt{3} & -1/\sqrt{3} \\ (1-i)/\sqrt{6} & 2/\sqrt{6} \end{bmatrix}\begin{bmatrix} (1+i)/\sqrt{3} & (1+i)/\sqrt{6} \\ -1/\sqrt{3} & 2/\sqrt{6} \end{bmatrix} = \begin{bmatrix} 1 & 0 \\ 0 & 1 \end{bmatrix} = \mathbf{I}_2,$$

which shows that $\mathbf{U}^\mathsf{H} = \mathbf{U}^{-1}$. A matrix with this property is called *unitary*.

In parallel with the theory of diagonalization in the real case, we have

$$\mathbf{U}^{-1}\mathbf{A}\mathbf{U} = \mathbf{U}^\mathsf{H}\mathbf{A}\mathbf{U} = \begin{bmatrix} 1 & 0 \\ 0 & 4 \end{bmatrix} = \mathbf{D},$$

where **D** is diagonal that has the eigenvalues of **A** on its main diagonal in the same order that the corresponding eigenvectors appear in **U**. Thus

$$\mathbf{A}\mathbf{U} = \mathbf{U}\mathbf{D}, \quad \text{or} \quad \mathbf{A} = \mathbf{U}\mathbf{D}\mathbf{U}^\mathsf{H} = \lambda_1\mathbf{z}_1\mathbf{z}_1^\mathsf{H} + \lambda_2\mathbf{z}_2\mathbf{z}_2^\mathsf{H}. \tag{8.53}$$

Using (8.53), **A** can be *decomposed* as follows:

$$\mathbf{A} = \begin{bmatrix} 2 & 1+i \\ 1-i & 3 \end{bmatrix} = \frac{1}{3}\begin{bmatrix} 1+i \\ -1 \end{bmatrix}[\,1-i \;\; -1\,] + \frac{4}{6}\begin{bmatrix} 1+i \\ 2 \end{bmatrix}[\,1-i \;\; 2\,] \quad ■$$

Note that a decomposition $\mathbf{A} = \mathbf{U}\mathbf{D}\mathbf{U}^\mathsf{H}$ can be found for all Hermitian matrices **A**.

Cofactor, Adjugate, Inverse

In Chapter 5, following popular usage, we used the term *adjoint* for the transpose of the matrix of cofactors of a square matrix. In this section we will use the term *adjugate* for the latter matrix to avoid confusion with $\mathbf{A}^\mathsf{H}$ which is often referred to as the adjoint of **A**.

EXAMPLE 4 **Cofactors, Adjugate and Inverse**

We will find the cofactors, adjugate, and inverse of the matrix

$$\mathbf{A} = \begin{bmatrix} 0 & i & i \\ i & 0 & i \\ i & i & 0 \end{bmatrix}.$$

The cofactors of $\mathbf{A}$ are $c_{11} = 1$, $c_{12} = -1$, $c_{13} = -1$, $c_{21} = -1$, $c_{22} = 1$, $c_{23} = -1$, $c_{31} = -1$, $c_{32} = -1$, and $c_{33} = 1$.

The determinant $\det \mathbf{A} = 0 \cdot c_{11} + ic_{12} + ic_{13} = 0 \cdot 1 + i(-1) + i(-1) = -2i$.

The adjugate of $\mathbf{A}$ is $\operatorname{adj}(\mathbf{A}) = \begin{bmatrix} 1 & -1 & -1 \\ -1 & 1 & -1 \\ -1 & -1 & 1 \end{bmatrix}^{\mathsf{T}} = \begin{bmatrix} 1 & -1 & -1 \\ -1 & 1 & -1 \\ -1 & -1 & 1 \end{bmatrix}.$

The inverse of $\mathbf{A}$ is $\mathbf{A}^{-1} = \dfrac{1}{\det(\mathbf{A})} \operatorname{adj}(\mathbf{A})$ and so

$$\mathbf{A}^{-1} = \frac{1}{-2i} \begin{bmatrix} 1 & -1 & -1 \\ -1 & 1 & -1 \\ -1 & -1 & 1 \end{bmatrix} = \frac{i}{2} \begin{bmatrix} 1 & -1 & -1 \\ -1 & 1 & -1 \\ -1 & -1 & 1 \end{bmatrix}.$$ ■

Historical Notes

Charles Hermite (1822–1901) French mathematician. Hermite was perhaps the most influential mathematician in France in the latter half of the 19th century. He held positions at various Parisian universities during his working life and communicated with the well-known mathematicians of his day. He made important contributions to real and complex analysis, algebra, and number theory. In 1873, Hermite published the first proof that the exponential number e is *transcendental*, meaning that e is not a root of any polynomial with rational coefficients. Hermite's name is today connected with a number of mathematical concepts, including *Hermitian matrices*.

EXERCISES 8.5

1. Find the values of the variables that make the following equation valid.

$$\begin{bmatrix} a & b \\ 1+i & 2+2i \end{bmatrix} = i \begin{bmatrix} 1-i & 2i \\ x & 2y \end{bmatrix}$$

Exercises 2–5. Solve each linear system (S) by forward elimination followed by back-substitution when (S) is consistent.

2. $\begin{cases} iz_1 + 2iz_2 = 1 + 3i \\ (1+i)z_1 + iz_2 = 1 - i \end{cases}$

3. $\begin{cases} 2iz_1 + (1+i)z_2 = -4 + 6i \\ (1-i)z_1 + 2z_2 = 3 + 3i \end{cases}$

4. $\begin{cases} z_1 + z_2 + z_3 = 3 + 2i \\ iz_1 - iz_2 + (1+i)z_3 = -1 \end{cases}$

5. $\begin{cases} (2-i)z_1 + (1+i)z_2 = 2 + i \\ (1+2i)z_1 - (1-i)z_2 = i \end{cases}$

Exercises 6–7. Solve each linear system by Gauss–Jordan elimination.

6. $\begin{cases} (1+i)z_1 + (1-i)z_2 = 4 \\ (-1+i)z_1 + (1+i)z_2 = 4i \end{cases}$

7. $\begin{cases} iz_1 + iz_2 - z_3 = 2+2i \\ (1+i)z_1 + (1-i)z_2 + iz_3 = -3+2i \\ z_1 + 2z_2 + 3z_3 = 0 \end{cases}$

Exercises 8–9. Find the reduced row echelon form of each matrix **A** *and determine if* **A** *has linearly independent columns. Is* **A** *invertible?*

8. $\begin{bmatrix} 1+i & -2i & -1-3i \\ i & 2+i & 2 \\ 2+3i & 1 & -1-3i \end{bmatrix}$

9. $\begin{bmatrix} i & 1 & -i \\ 1 & -i & i \\ 0 & 1 & i \end{bmatrix}$

Exercises 10–11. Represent each linear system in the form $\mathbf{Ax} = \mathbf{b}$ *and solve by finding* $\mathbf{A}^{-1}$*. Cross-check the solution using Cramer's rule.*

10. $\begin{cases} (2+i)z_1 + (4+3i)z_2 = 2-2i \\ (4-3i)z_1 + (2-i)z_2 = 2+2i \end{cases}$

11. $\begin{cases} 2iz_1 + (2-i)z_2 = -1-i \\ (3+i)z_1 + (1+3i)z_2 = 6+6i \end{cases}$

Exercises 12–15. Compute the inverse of each matrix **A** *and verify that* $\det(\mathbf{A}^{-1}) = \dfrac{1}{\det(\mathbf{A})}$.

12. $\begin{bmatrix} i & 0 \\ 0 & i \end{bmatrix}$ **13.** $\begin{bmatrix} i & i \\ i & 0 \end{bmatrix}$

14. $\begin{bmatrix} 2 & 1+i \\ 1-i & i \end{bmatrix}$ **15.** $\begin{bmatrix} i & 1-i \\ 1+i & -i \end{bmatrix}$

Exercises 16–17. Solve the determinant equation.

16. $\begin{vmatrix} 1-z & 2 \\ -3 & -1-z \end{vmatrix} = 0$

17. $\begin{vmatrix} 0 & 2i & i \\ 4 & -2 & 1+i \\ i & 2i & 3i \end{vmatrix} + \begin{vmatrix} z & 3 \\ 4 & z \end{vmatrix} = 0$

18. *Hermitian Operation.* Let **A** and **B** be complex matrices and let z be a complex scalar. Prove each property.

$$(\mathbf{A}^{\mathsf{H}})^{\mathsf{H}} = \mathbf{A} \tag{8.54}$$

$$(\mathbf{A}+\mathbf{B})^{\mathsf{H}} = \mathbf{A}^{\mathsf{H}} + \mathbf{B}^{\mathsf{H}} \tag{8.55}$$

$$(z\mathbf{A})^{\mathsf{H}} = \bar{z}\mathbf{A}^{\mathsf{H}} \tag{8.56}$$

$$(\mathbf{AB})^{\mathsf{H}} = \mathbf{B}^{\mathsf{H}}\mathbf{A}^{\mathsf{H}} \tag{8.57}$$

19. Let **u** and **v** be complex vectors in $\mathbb{C}^n$. Use the properties of the Hermitian operator H to show that $\mathbf{u}\cdot\mathbf{v} = \overline{\mathbf{v}\cdot\mathbf{u}}$ and that $\|\mathbf{u}\|$ is real and nonnegative.

Exercises 20–21. Find the dot product $\mathbf{u}\cdot\mathbf{v}$ *and the norm of each vector. Confirm that* $\mathbf{u}\cdot\mathbf{v} = \overline{\mathbf{v}\cdot\mathbf{u}}$.

20. $\mathbf{u} = \begin{bmatrix} 1+i \\ i \end{bmatrix}$, $\mathbf{v} = \begin{bmatrix} 2+i \\ 1-i \end{bmatrix}$

21. $\mathbf{u} = \begin{bmatrix} 2 \\ -i \end{bmatrix}$, $\mathbf{v} = \begin{bmatrix} 1-i \\ 1+i \end{bmatrix}$

22. *The Pauli Spin Matrices.* The complex 2×2 matrices σ_x, σ_y, σ_z are used in the study of electron spin in quantum mechanics. A pair of $n \times n$ matrices **A** and **B** are called *anticommutative* when $\mathbf{AB} = -\mathbf{BA}$. Show that the Pauli spin matrices are anticommutative (there are three pairs to consider).

$$\sigma_x = \begin{bmatrix} 0 & 1 \\ 1 & 0 \end{bmatrix}, \quad \sigma_y = \begin{bmatrix} 0 & -i \\ i & 0 \end{bmatrix}, \quad \sigma_z = \begin{bmatrix} 1 & 0 \\ 0 & -1 \end{bmatrix}$$

23. Given an $m \times n$ complex matrix **A**, show that the matrices $\mathbf{F} = \mathbf{A}^{\mathsf{H}}\mathbf{A}$ and $\mathbf{G} = \mathbf{A}\mathbf{A}^{\mathsf{H}}$ are both Hermitian. Construct **F** and **G** when

$$\mathbf{A} = \begin{bmatrix} i & 0 & 1-i \\ 0 & 1+i & -i \end{bmatrix}.$$

24. Verify that the eigenvalues of the Hermitian matrix $\mathbf{A} = \begin{bmatrix} 4 & -i \\ i & 6 \end{bmatrix}$ are real.

25. Use eigenvalues to show that $\det \mathbf{A}$ is real when **A** is Hermitian. *Hint*: Consider the characteristic equation of **A**.

Exercises 26–28. Find the inverse of each matrix.

26. $\begin{bmatrix} 1+i & 1-i \\ 1-i & 1+i \end{bmatrix}$

27. $\begin{bmatrix} 0 & i & i \\ i & 0 & i \\ i & i & 0 \end{bmatrix}$

28. $\begin{bmatrix} 1 & i & 1-i \\ 1 & 1+i & 1 \\ 1 & i & 2-i \end{bmatrix}$

Exercises 29–34. Find the cofactors, the adjugate, the determinant, and the inverse (if it exists) of each matrix **A**.

29. $\begin{bmatrix} 2+3i & 1+i \\ 4-i & 1+i \end{bmatrix}$

30. $\begin{bmatrix} 1-i & 10+9i \\ 1+i & 10-9i \end{bmatrix}$

31. $\begin{bmatrix} 8+i & \sqrt{2}+i \\ \sqrt{2}-i & 8-i \end{bmatrix}$

32. $\begin{bmatrix} 3-2i & 1+2i \\ 2+i & 2+3i \end{bmatrix}$

33. $\begin{bmatrix} 2i & 0 & -1 \\ 0 & 2i & 1 \\ -i & 1 & -1 \end{bmatrix}$

34. $\begin{bmatrix} 0 & -2i & i \\ 4 & -2 & 1+i \\ i & 2i & i \end{bmatrix}$

Exercises 35–40. Find the eigenvalues, an eigenvector corresponding to each eigenvalue, and a diagonalizing matrix of each matrix.

35. $\begin{bmatrix} 0 & i \\ i & 0 \end{bmatrix}$

36. $\begin{bmatrix} 1 & 2 \\ -3 & 1 \end{bmatrix}$

37. $\begin{bmatrix} 2 & 1 \\ -1 & 2 \end{bmatrix}$

38. $\begin{bmatrix} 1-i & 1+i \\ 0 & 0 \end{bmatrix}$

39. $\begin{bmatrix} i & 0 \\ i & 0 \end{bmatrix}$

40. $\begin{bmatrix} 1 & 3 \\ 4 & 1 \end{bmatrix}$

USING MATLAB

The prime symbol $\mathbf{A}'$ is the transpose operator for real matrices. The same symbol acts as the Hermitian operator for complex matrices. If u is a *column* vector, then the command **norm(u)** returns the same value as **sqrt(u′*u)**.

Exercises 41–42. Find the norm of the given vector in two different ways.

41. $[\,1+i, 1-2i, 2-i, i\,]$

42. $\begin{bmatrix} i \\ 2+5i \\ 1+i \end{bmatrix}$

Exercises 43–44. Use hand calculation to find the determinant of each matrix by expanding on row 1. Cross-check using MATLAB.

43. $\begin{bmatrix} i & 0 & -2i \\ -1 & 2i & 1 \\ 2 & -i & 0 \end{bmatrix}$

44. $\begin{bmatrix} 1+i & 1-i & 1+i \\ -1+i & -1+i & -1+i \\ 1-i & -1-i & -1-i \end{bmatrix}$

45. Use MATLAB to check hand calculations in this exercise set.

CHAPTER 8 REVIEW

State whether each statement is true or false as stated. Provide a clear reason for your answer. If a false statement can be modified to make it true, then do so.

1. $\mathrm{Re}\,(8-i) = 8$ and $\mathrm{Im}\,(8-i) = -i$.
2. $(2-3i)(3+2i) = 6-6i$.
3. $\dfrac{(2+i)}{(1-2i)} = i$.
4. $(1+i)^{10} = 32i$
5. $1+i < 2+i$
6. $\sqrt{i} = \pm\,\mathrm{cis}(\pi/4)$
7. $1/i^{202} = 1$
8. The set of complex numbers $z = a+bi$ such that $2a - b \le 3$ is an unbounded region in the complex plane.
9. The complex number $1+i$ lies on the unit circle in the complex plane.

10. The domain of the complex function $f(z) = \dfrac{1}{|z| + 1}$ is the complex plane with the point $z = -1$ deleted.

11. $\text{Arg}\,(-1 - i) = 5\pi/4$

12. $\text{cis}^5\,\theta = \text{cis}(5\theta)$.

13. The roots of $z^2 + 2z + 2$ are $-1 + i,\ 1 - i$.

14. The polynomial $p(z) = z^4 - 1$ has four complex roots.

15. $z = e^{2\pi i/3}$ is a cube root of unity.

16. $\begin{bmatrix} i & 1 \\ 1 - i & -i \end{bmatrix}^{-1} = \begin{bmatrix} -1 & i \\ 1 + i & 1 \end{bmatrix}$.

17. If $\mathbf{u} = \begin{bmatrix} 1 + i \\ 1 - i \end{bmatrix}$ and $\mathbf{v} = \begin{bmatrix} 4i \\ 2 + i \end{bmatrix}$, then $\mathbf{u} \cdot \mathbf{v} = 5 + 7i$.

18. The set of all 2×2 matrices having purely imaginary diagonal entries is a subspace of $\mathbb{C}^{2\times 2}$.

19. The eigenvectors of the matrix $\begin{bmatrix} 1 & 1 + i \\ 1 - i & 1 \end{bmatrix}$ form an orthogonal set.

20. The matrix $\begin{bmatrix} 3 - i & -1 + i \\ -1 - i & 4 - i \end{bmatrix}$ is Hermitian.

George Bernard Dantzig (1914–)

During his work with the U.S. Air Force that began in 1941, George Dantzig and his research team invented a new mathematical theory called *linear programming* (LP for short), which was applied to streamline military planning operations. Other applications quickly followed. Computations for the early LP applications of the 1940s were carried out using mechanical calculating machines. By the 1950s, the development of computers made it possible to handle calculations for larger LP problems arising in both industry and government. The first industrial users of LP were the petroleum and food industries, whose goal was to minimize cost of production and transportation.

Dantzig graduated with a doctorate from the University of California at Berkeley in 1946 and was appointed Professor and Chair of the Operations Research Center there in 1960. He is currently Professor Emeritus of Operations Research and Computer Science at Stanford University. He is a member of the National Academy of Engineering, National Academy of Science, and the American Academy of Arts and Sciences. He is a recipient of the National Medal of Science and numerous honorary degrees. His work laid the foundation for the field of systems engineering and is widely used in network and component design in computer, mechanical, and electrical engineering.

Today, linear programming forms part of a vibrant field of mathematics and computer science called *mathematical programming*. The George B. Dantzig Prize, created in 1979 jointly by the Mathematical Programming Society (MPS) and the Society for Industrial and Applied Mathematics (SIAM), is awarded for original and significant research in the field of mathematical programming.

Chapter 9

LINEAR PROGRAMMING

Introduction

Some important modern advancements in engineering, science, medicine and other fields can be traced back to innovations made during World War II (1939–1945) that were further developed in peacetime. One such advancement in mathematics arose out of the military's need to plan and schedule operations and to organize processes, such as the movement and training of personnel and the acquisition and deployment of food and equipment. After the war it became standard practice for each U.S. military operation to have a clearly stated goal or *objective* and an associated logical plan or *program* that, when carried out, would meet the objective.

During their work for the U.S. Department of Air Force, George Dantzig, Marshall Wood, and others developed the idea that some military objectives could be expressed in terms of linear functions of many variables and that a program to meet a particular objective could be formulated after applying a mathematical process that we now call *linear programming*.

Two fundamental goals of commerce are the maximization of profits and the minimization of costs. Beginning in the 1950s, major petroleum companies and companies from other sectors of the economy started to apply LP methods to minimize the cost of blending new products (output) from a mix of ingredients (input). Such applications are called *blending models* and they arise when blending, for example, cattle feed, fertilizer, processed food, metal, grain, wine, balance sheets, and investment portfolios.

We will use the following simplified blending model for gasoline to illustrate a typical LP problem and to introduce some standard terminology. Focus on the global picture rather than details at this stage.

■ ILLUSTRATION 9.1

Blending Gasoline

An oil company wishes to minimize the cost of producing 14,000 gallons[1] of a new mid-grade gasoline with minimum octane rating[2] of 89 using three ingredients (petroleum products) labeled A, B, C. The octane rating, volume available, and cost of each ingredient are shown in the following table.

Ingredients	Octane Rating	Gallons	Cost per Gal
A	86	Unlimited	$0.40
B	92	3906	$0.50
C	100	4100	$0.65

Define *decision variables* x_1, x_2, x_3, which denote, respectively, the number of gallons used of each ingredient A, B, C in order to blend 14,000 gallons of the new gasoline. Define a linear *objective function* f given by

$$f = 0.4x_1 + 0.5x_2 + 0.65x_3 \quad \text{(dollars)}, \tag{9.1}$$

which represents the total cost of blending 14,000 gals. The objective is to minimize f subject to a set of *constraints* (conditions) that are expressed as linear inequalities or equations. Here are the constraints for this problem.

The quantities x_1, x_2, x_3 cannot be negative and so the inequalities

$$x_1 \geq 0, \quad x_2 \geq 0, \quad x_3 \geq 0$$

must all be satisfied. We call these *nonnegativity constraints*.

The *demand constraint* for this problem is expressed by the equation

$$x_1 + x_2 + x_3 = 14{,}000. \tag{9.2}$$

The *availability constraints* give upper bounds on the number of gallons available of each ingredient. These are the inequalities

$$x_2 \leq 3906, \quad \text{(for } B\text{)} \qquad x_3 \leq 4100 \quad \text{(for } C\text{)}. \tag{9.3}$$

The *blending constraint* is a mathematical statement that says that the new product must have an octane rating of at least 89. The condition is

$$\frac{86x_1 + 92x_2 + 100x_3}{x_1 + x_2 + x_3} \geq 89, \tag{9.4}$$

noting that $x_1 + x_2 + x_3 > 0$ by (9.2). The left side of (9.4) is called a *weighted average*. Multiplying the left and right sides of (9.4) by the positive quantity $x_1 + x_2 + x_3$ preserves the inequality and, after transferring variable terms to the left side, we obtain the blending constraint

$$-3x_1 + 3x_2 + 11x_3 \geq 0. \tag{9.5}$$

The objective is to find the minimum cost f that will satisfy all the constraints. ■

---

[1] *Gallon*, a unit of capacity. 1 U.S. gallon = 231 cubic inches or 3.79 liters. 1 imperial (British) gallon = 1.2 U.S. gallons.

[2] *Octane number or rating*, a measure of the antiknock quality of automobile fuel expressed as a percentage.

The *simplex algorithm*, discovered by George Dantzig in the 1940s, became the standard mathematical tool for solving LP problems. The algorithm has been refined over the years, and alternative algorithms[3] have been developed that give faster solutions in certain cases. The basic ideas of the simplex algorithm are explained in Section 9.2. The algorithm is applied at the end of the section to show that the minimum solution to the gasoline blending problem stated previously is

$$f = \$6322.10, \quad x_1 = 8768, \quad x_2 = 3906, \quad x_3 = 1326 \quad \text{gal.} \tag{9.6}$$

The oil company can now implement a program (or plan) to blend the new gasoline, which costs \$0.4516 per gallon and has an octane rating of 89 exactly. Real gasoline blending models are much more complicated and are constrained by other properties of petroleum products (volatility and vapor pressure, for example).

LP problems are commonly grouped into generic classes of problems according to their type. Blending models form one such class. Other classes include production scheduling, transportation and network analysis, decision making under uncertainty (using probabilities), and economic planning.

LP methods are today basic tools in planning and organization. Large corporations and governments devote considerable resources to solving LP problems by computer. The development of algorithms and commercial software in the field of *mathematical programming*[4] which began in the 1950s, has grown in step with advances in computer technology. Today, mathematical programming is an important field of pure research with widespread applications.

Historical Notes

The Nobel Prize in Economics for 1975 was awarded jointly to professors Leonid Kantorovich of the U.S.S.R. and Tjalling C. Koopmans of the United States for *contributions to the theory of optimum allocation of resources*. Kantorovich applied LP methods to study economic planning in Russia. Koopmans developed LP methods to optimize shipping transportation across the Atlantic during World War II and later applied these methods in peacetime to the input–output production models in economics.

9.1 Standard Forms, Geometrical Methods

The goal of an LP problem is to maximize or minimize (*optimize*) a linear objective function f in n decision variables $x_1, x_2, \ldots, x_n$ subject to a set of m constraints. The problem is called a *standard maximization problem* if it has the form

MAX

$$f = c_1x_1 + c_2x_2 + \cdots + c_nx_n, \tag{9.7}$$

[3] For example, *interior* and *barrier* methods are used in solving certain large-scale LP problems.

[4] *Mathematical programming*, a field of optimization that includes LP as well as applications where the objective function is nonlinear (*nonlinear programming*) and applications requiring integer solutions (*integer programming*).

ST

$$
\begin{aligned}
a_{11}x_1 + a_{12}x_2 + \cdots + a_{1n}x_n &\le b_1 \\
a_{21}x_1 + a_{22}x_2 + \cdots + a_{2n}x_n &\le b_2 \\
&\vdots \\
a_{m1}x_1 + a_{m2}x_2 + \cdots + a_{mn}x_n &\le b_m
\end{aligned}
\tag{9.8}
$$

where the decision variables $x_1, x_2, \ldots, x_n$ are nonnegative.

MAX means *maximize* the objective function f and **ST** means *subject to* the given constraints. The coefficients $c_1, c_2, \ldots, c_n$ in f are given real numbers; the m constraints appear as "less than or equal to" inequalities in which the coefficients a_{ij}, $1 \le i \le n$, $1 \le j \le m$; and constant terms b_j, $1 \le j \le m$, are given real numbers.

When an LP problem has only two or three decision variables, the method of solution can be visualized geometrically in $\mathbb{R}^2$, or $\mathbb{R}^3$, respectively. We begin by considering the case when $n = m = 2$ in (9.7) and (9.8).

■ ILLUSTRATION 9.2

A Simple Production Model

A company makes two products P_1 and P_2 using two machines M_1 and M_2. Producing one unit of P_1 requires one hour on each of M_1 and M_2. Producing one unit of P_2 requires 2 hours on M_1 and one hour on M_2. Due to the need for regular servicing, M_1 and M_2 can only run for 40 and 30 hours per week, respectively. We call these *time constraints*.

The preceding data are organized in the following table.

	P_1	P_2	Time Constraints (hr.)
M_1	1	2	≤ 40
M_2	1	1	≤ 30

The company expects net profits of \$20 and \$30, respectively, on each unit of P_1 and P_2 sold. How should the company schedule production in order to maximize net weekly profit?

The model translates directly into an LP problem in standard form with $n = m = 2$ in (9.7) and (9.8). Let x_1, x_2 represent, respectively, the number of units of products P_1 and P_2 produced per week. The company's goal is
MAX

$$f = 20x_1 + 30x_2 \quad \text{(net profit per week)} \tag{9.9}$$

ST

$$x_1 + 2x_2 \le 40 \quad \text{(hours, } M_1\text{)} \tag{9.10}$$

$$x_1 + x_2 \le 30 \quad \text{(hours, } M_2\text{)} \tag{9.11}$$

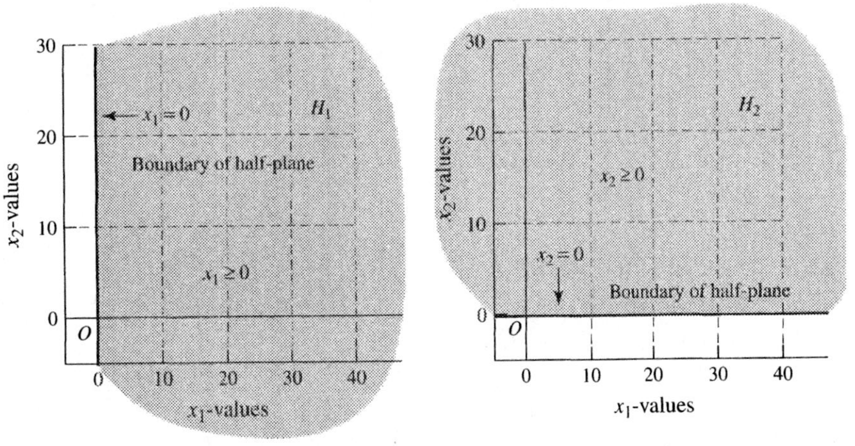

Figure 9.1 Half-planes H_1, H_2.

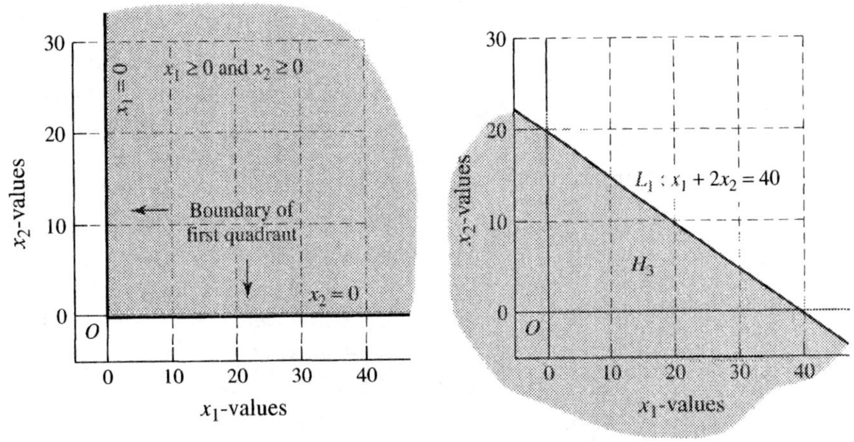

Figure 9.2 First quadrant and half-plane H_3.

with x_1, x_2 nonnegative. Moreover, the decision variables must be positive integers in this illustration, and so we call this model an *integer* LP problem.

As a first step toward visualization we identify the set of points (x_1, x_2) in $\mathbb{R}^2$ that satisfy *all* the constraints.

Refer to Figure 9.1. The points (x_1, x_2) in $\mathbb{R}^2$ that satisfy the nonnegativity constraint $x_1 \geq 0$ lie in the *half-plane* H_1, on and to the right of the line with equation $x_1 = 0$ that is the *boundary* of H_1. Similarly, the points (x_1, x_2) that satisfy the nonnegativity constraint $x_2 \geq 0$ lie in the half-plane H_2, on and above the line with equation $x_2 = 0$ that is the boundary of H_2.

Refer to Figure 9.2. Points (x_1, x_2) that satisfy *both* nonnegativity constraints lie in the *first quadrant* in $\mathbb{R}^2$ (positive axes included). Points (x_1, x_2) satisfying constraint (9.10) lie in the half-plane H_3, on and below the line $L_1 : x_1 + 2x_2 = 40$ that is the boundary of H_3. Similarly, the constraint (9.11) defines a half-plane H_4 with boundary line $L_2 : x_1 + x_2 = 30$.

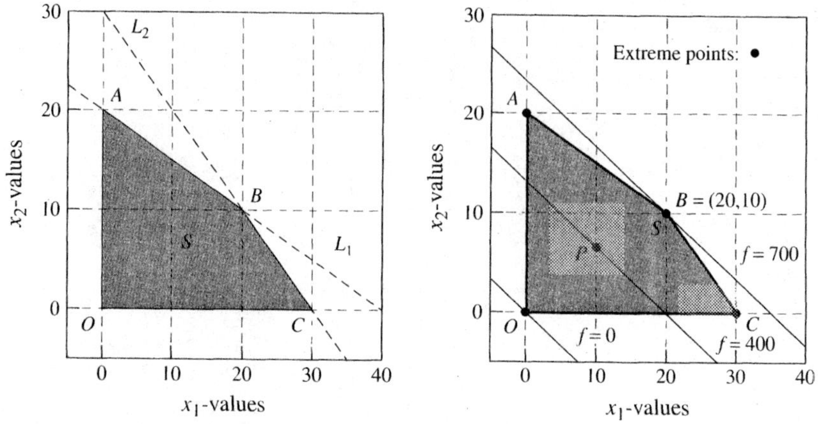

Figure 9.3 The feasible set S.

Refer to Figure 9.3. The set of all points in $\mathbb{R}^2$ common to all four half-planes defines a *polygon*[5] in $\mathbb{R}^2$, denoted by S, which is bounded by four line segments. We write $S = H_1 \cap H_2 \cap H_3 \cap H_4$ (the intersection of four half-planes).

Any point (x_1, x_2) lying in the interior or on the boundary of S is called a *feasible solution* for the problem because the values x_1, x_2 satisfy all the constraints. Hence, S is called the *feasible set* for the problem. More generally, the feasible set S for an LP problem is the *domain* of the objective function f.

If the values from any feasible solution (x_1, x_2) are substituted into (9.7), we obtain a *feasible value* for f. For example, the point $P = (10, \frac{20}{3})$ in Figure 9.3 is a feasible solution and $f = 20(10) + 30(\frac{20}{3}) = 400$ is the corresponding feasible value for f. A value for f is called *optimal* if f attains a minimum or maximum value at a feasible solution (x_1, x_2) and then (x_1, x_2) is called an *optimal feasible solution* for the problem. Keep in mind that optimal feasible solutions may not be unique in every case.

Extreme Points

We now identify some important feasible solutions. An *extreme point* or *vertex* in S is a point that does not lie on a line segment joining two other points in S. The points $O = (0, 0)$, $A = (0, 20)$, $B = (20, 10)$, $C = (30, 0)$ shown in Figure 9.3 are the only extreme points in S. These points occur at the intersection of some, but not necessary all, pairs of boundary lines. In this illustration, the extreme points are found by solving 2×2 linear systems that represent pairs of boundary lines.

Isovalues

The lines in $\mathbb{R}^2$ such that $f = k$, where k is a constant, are of special interest. We will call these lines *isovalues*.[6] The isovalues for the production model are parallel lines $(f =)\ 20x_1 + 30x_2 = k$, for $k \geq 0$. Three particular isovalues

---

[5] *Polygon*, a closed plane figure bounded by three or more straight lines.

[6] *Iso*, from the Greek meaning "equal." An isobar (isotherm) is a line on a map that connects points of equal pressure (temperature).

corresponding to $k = 0, 400, 700$ are shown in Figure 9.3. Looking at the set of isovalues, it is clear that the optimal (minimum or maximum) values of f occur at extreme points in the feasible set, namely O and B, respectively. Thus, our production problem could be solved geometrically by evaluating the objective function f at all extreme points in S, taking the maximum of these values. We find that $B = (20, 10)$ gives the maximum value $f = 700$ and $(x_1, x_2) = (20, 10)$ is the unique optimal feasible solution. ■

■ ILLUSTRATION 9.3

An LP Problem Visualized in $\mathbb{R}^3$

Suppose that a practical application translates into the following LP problem that is in standard form. Our goal is

MAX

$$f = -x_1 + 2x_2 + x_3 \tag{9.12}$$

ST

$$5x_1 + 4x_2 + 4x_3 \leq 24 \tag{9.13}$$

$$2x_1 + 4x_2 + x_3 \leq 12 \tag{9.14}$$

with x_1, x_2, x_3 nonnegative. This LP problem can be visualized in $\mathbb{R}^3$. The set of points (x_1, x_2, x_3) satisfying (9.13) is called a *half-space* H in $\mathbb{R}^3$. The boundary of H is a *plane* with equation $5x_1+4x_2+4x_3 = 24$. Similarly, the set of points satisfying (9.14) is a half-space with bounding plane $2x_1+4x_2+x_3 = 12$. The nonnegativity constraints also define half-spaces:

$$\begin{array}{llll} x_1 \geq 0 & \text{Bounding plane} & x_1 = 0 & (\text{the } x_2x_3\text{-plane}) \\ x_2 \geq 0 & \text{Bounding plane} & x_2 = 0 & (\text{the } x_1x_3\text{-plane}) \\ x_3 \geq 0 & \text{Bounding plane} & x_3 = 0 & (\text{the } x_1x_2\text{-plane}). \end{array}$$

The intersection of all five half-spaces defines the solid *polyhedron OABCDE* shown in Figure 9.4 that is the feasible set S for this problem. The extreme

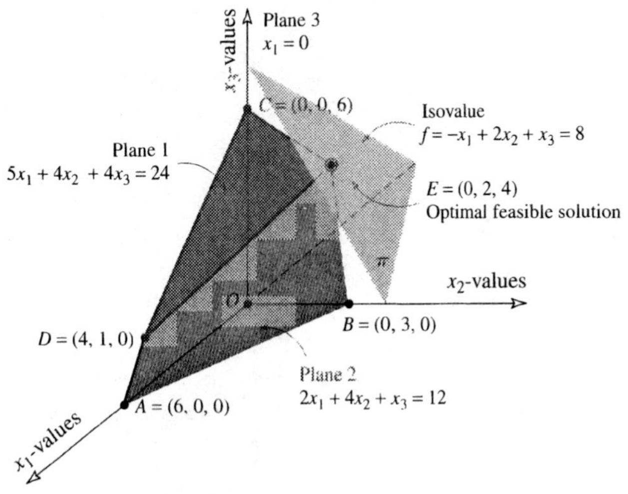

Figure 9.4 Solid polyhedron S in $\mathbb{R}^3$.

points in S are found by solving 3×3 linear systems. For example, $E = (0, 2, 4)$ is the unique solution to the system

$$\begin{cases} 5x_1 + 4x_2 + 4x_3 = 24 & \text{Plane 1} \\ 2x_1 + 4x_2 + x_3 = 12 & \text{Plane 2} \\ x_1 = 0 & \text{Plane 3.} \end{cases}$$

The point E lies at the intersection of Plane 1, Plane 2, and Plane 3. Isovalues $f = k$ in $\mathbb{R}^3$ are a family of parallel planes with equations $-x_1 + 2x_2 + x_3 = k$ for $k \geq 0$. Evaluating the objective function at each extreme point shows that the optimal feasible solution is $(0, 2, 4)$ and $f = 8$ is the optimal feasible value. The plane π with equation $-x_1 + 2x_2 + x_3 = 8$ passes through E. In Figure 9.4, the shaded triangle lies in π. ■

The preceding geometric illustrations are instructive but only of passing interest. The idea of finding all extreme points of the feasible set is computationally flawed because even a few constraints can lead to a large number of extreme points and the whole process becomes too labor intensive. More than this, any serious method for solving LP problems must accommodate any number of decision variables, not just two or three. The simplex method, explained in the next section, provides an algebraic method of solution that is programmable and extremely efficient.

Convex Sets

The feasible set S for any LP problem has a special property called *convexity*.

Definition 9.1

Convexity

A set S in $\mathbb{R}^n$ is called *convex* if for all pairs of vectors $\mathbf{u}$, $\mathbf{v}$ in S, the vector $\mathbf{x} = t\mathbf{u} + (1 - t)\mathbf{v}$ lies in S for all t with $0 \leq t \leq 1$.

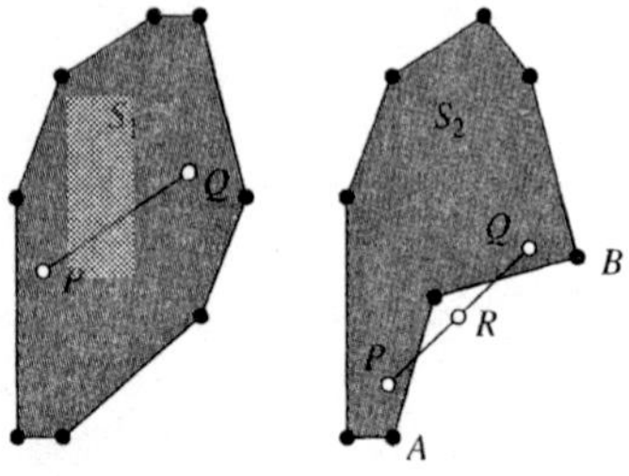

Figure 9.5 Convex and nonconvex sets.

Let us illustrate the definition in $\mathbb{R}^2$. Figure 9.5 shows a polygon S_1 in $\mathbb{R}^2$. Suppose the vectors $\mathbf{u}$ and $\mathbf{v}$ represent any pair of points P and Q, respectively, that lie inside S_1 or on its boundary. The set of vectors $\mathbf{x} = t\mathbf{u} + (1 - t)\mathbf{v}$ as t varies $0 \leq t \leq 1$ represent all points on the line segment PQ and consequently every such line segment PQ lies entirely in S_1. By definition, S_1 is convex. The feasible set S shown in Figure 9.3 is convex for the same reason. The polygon S_2 in Figure 9.5 is not convex due to the shape of the boundary between A and B. The point R on the line segment joining P and Q falls outside S_2 and the definition of convexity fails for the points P, Q.

The following important facts are stated without proof and hold for LP problems in general.

(a) The feasible set S is a convex set in $\mathbb{R}^n$.
(b) The objective function f attains its optimal (minimum or maximum) value at an extreme point in S.

LP problems can be classified according to certain geometric properties of the feasible set S and we will state these properties when S is in $\mathbb{R}^2$ and $\mathbb{R}^3$.

(a) S may be *bounded* (the points in S are all contained inside a sufficiently large circle [sphere], centered at the origin).
(b) S may be the *empty set* (there are no points in S).
(c) S may be *unbounded* (there are points in S outside any circle [sphere], centered at the origin).

Case (a) applies to Illustrations 9.2 and 9.3. Refer to Figure 9.6. Illustrating case (b), the constraints

$$x_1 \geq 0, \quad x_2 \geq 0, \quad L_1 : x_1 + x_2 \leq 2, \quad L_2 : 3x_1 + 4x_2 \geq 12$$

define two *disjoint* regions S_1 and S_2 in $\mathbb{R}^2$ with no point in common. The feasible set is empty in this case and no objective function with these constraints can be optimized. Illustrating case (c), the feasible set S_3 is *bounded* by the line $L_3 : -x_1 + x_2 = 1$ but is unbounded in $\mathbb{R}^3$ because there are points in S_3 outside any circle centered at the origin. The objective function $f = x_1 + x_2$, for example, has no maximum value on S_3.

LP Problems: Matrix Form

The standard maximization problem defined by (9.7) and (9.8) can be expressed compactly using matrix notation. Define matrices

$$\mathbf{c} = \begin{bmatrix} c_1 \\ c_2 \\ \vdots \\ c_n \end{bmatrix}, \quad \mathbf{x} = \begin{bmatrix} x_1 \\ x_2 \\ \vdots \\ x_n \end{bmatrix}, \quad \mathbf{A} = [a_{ij}]_{m \times n}, \quad \mathbf{b} = \begin{bmatrix} b_1 \\ b_2 \\ \vdots \\ b_m \end{bmatrix}.$$

Then a standard maximization problem in n decision variables and m constraints has the matrix form

$$\begin{array}{ll} \textbf{MAX} & f = \mathbf{c} \cdot \mathbf{x} = \mathbf{c}^{\mathsf{T}} \mathbf{x} \\ \textbf{ST} & \mathbf{A}\mathbf{x} \leq \mathbf{b}, \quad \mathbf{0} \leq \mathbf{x}. \end{array} \tag{9.15}$$

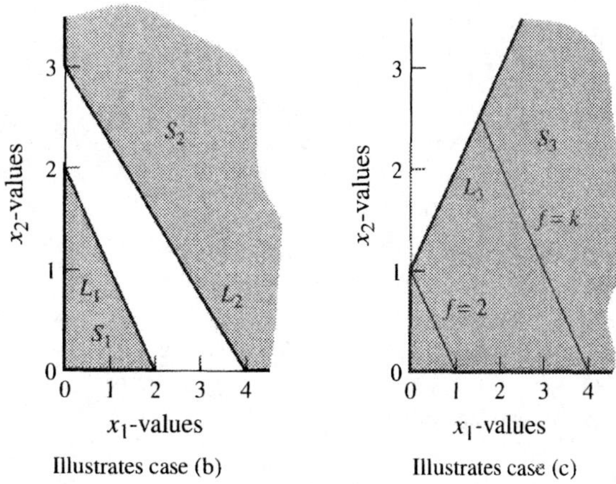

Figure 9.6 Empty and unbounded feasible sets.

Note that any constraint involving $\geq$ can be rewritten as a constraint involving $\leq$ by multiplying throughout by -1. For example,

$$x_1 + 4x_2 - x_3 \geq 2 \quad \Rightarrow \quad -x_1 - 4x_2 + x_3 \leq -2.$$

Any constraint equation can be replaced by two inequalities, both of which must be satisfied. For example, using (9.2), we have

$$x_1 + x_2 + x_3 = 1400 \quad \Rightarrow \quad \begin{cases} x_1 + x_2 + x_3 \leq 1400 \\ x_1 + x_2 + x_3 \geq 1400 \end{cases}. \tag{9.16}$$

The preceding two comments indicate that any LP problem can be written as a standard maximization problem.

Standard Minimization Problems

An LP problem is a *standard minimization problem* if it can be written in the form

$$\begin{array}{ll} \textbf{MIN} & f = \mathbf{c} \cdot \mathbf{x} = \mathbf{c}^{\mathsf{T}}\mathbf{x} \\ \textbf{ST} & \mathbf{Ax} \geq \mathbf{b}, \quad \mathbf{0} \leq \mathbf{x}. \end{array} \tag{9.17}$$

The constraints now appear as "greater than or equal to" inequalities and the objective function f is minimized subject to these constraints. Finding a minimum value for f is equivalent to finding a maximum value for the function $F = -f$, and vice versa. Hence, any standard minimization LP problem can be viewed as a standard maximization LP problem, and vice versa. However, for technical reasons, it is customary to focus on minimizing an objective function (this is true of many computer packages designed for solving LP problems).

EXERCISES 9.1

1. Draw the half-plane that corresponds to each given constraint. Shade the region in the half-plane for which $x_1 \geq 0$ and $x_2 \geq 0$.

   (a) $x_1 - x_2 \leq 1$ **(b)** $4x_1 + 6x_2 \geq 12$

   (c) $-x_1 + x_2 \leq 1$

   Suppose the intersection of the sets defined by (a), (b), (c) in the first quadrant is the feasible set S for an LP problem. Find the extreme points for S.

2. Refer to Exercise 1. Suppose we wish to maximize $f = 3x_1 + 5x_2$ on S. Draw the isovalue $f = 15$ and find the optimal value of f geometrically.

3. Consider the objective function $f = x_1 + 2x_2$ subject to constraints $-x_1 + x_2 \leq 0$, $x_1 - x_2 \leq -2$, $x_1, x_2 \geq 0$. Draw the feasible set S and draw the isovalue $f = 2$. Can f be maximized on S?

4. Describe the feasible set S for the objective function $f = 4x_1 + x_2$ subject to the constraints $x_1 + 2x_2 \geq 6$, $x_1, x_2 \geq 0$. Find the optimal (minimum, maximum) values for f when they exist.

5. Give the constraints for an LP problem with objective function $f = 5x_1 + 2x_2$ and convex feasible set S with extreme points $(0, 0)$, $(2, 2)$, $(4, 1)$. Find the optimal feasible solutions and the optimal (minimum, maximum) values for f.

6. Give the constraints for an LP problem that has objective function $f = x_1 + x_2$ and convex feasible set S with extreme points $(1, 2)$, $(3, 1)$, $(4, 4)$, $(6, 2)$. Find the optimal feasible solutions and the optimal (minimum, maximum) values for f.

7. Refer to Figure 9.4. Find the intercepts of the plane $\pi - x_1 + 2x_2 + x_3 = 8$ with the principal axes. Give a 3-vector that is normal to π. Check that the value of f at other extreme points is less than $f = 8$.

Exercises 8–11. Use geometrical methods to find the optimal (maximum) value of the objective function f subject to the given constraints. All decision variables are nonnegative. Draw the feasible set S and indicate the extreme points. Find the optimal feasible solution and show the isovalue when f is optimal.

8. $f = 2x_1 + 3x_2$, $\quad x_1 + x_2 \geq 2$, $\quad 4x_1 + 5x_2 \leq 20$

9. $f = 2x_1 + 3x_2$, $\quad x_1 + x_2 \leq 10$, $\quad 4x_1 - x_2 \geq 8$

10. $f = x_1 + x_2$, $\quad x_1 + 2x_2 \geq 12$, $\quad x_1 - x_2 \geq 5$, $\quad x_1 \leq 14$

11. $f = x_1 + 10x_2$, $\quad x_1 + x_2 \geq 1$, $\quad 3x_1 + 4x_2 \leq 12$, $\quad -x_1 + x_2 \leq 1$

Exercises 12–13. Find the minimum value of the objective function f subject to the given constraints.

12. $f = 6x_1 + 4x_2$, $\quad x_1 + x_2 \geq 4$, $\quad 5x_1 + 3x_2 \geq 15$

13. $f = -x_1 + 5x_2$, $\quad 1 \leq x_1 + x_2 \leq 2$, $\quad -4 \leq x_1 - x_2 \leq 0$

Exercises 14–17. Set up the following applications as LP problems. You will be asked for solutions in Section 9.2. Define the decision variables accurately. State the objective function and the system of constraints in matrix form.

14. *Production*. The Zen Bicycle Company in Shanghai, China makes two types of bicycles: 5-speed regular (R) and 10-speed deluxe (D).

The contribution margins (financial contribution that the product makes to fixed business expenses, factory overhead, wages, profit, etc.) are, respectively, \$10 and \$15 for each unit of R and D sold.

The factory constraint (available of floor space, machinery, etc.) dictates that not more than 50 units of both types can be made per week.

Stock constraints: The main components of each bicycle (labeled a, b, c, d) are in short supply. The number of these items received by the company per week and the number required for each bicycle type are shown in the following table.

Component	Stock	Regular	Deluxe
a	220	4	2
b	160	2	4
c	370	2	10
d	300	5	6

What program should the company follow in order to maximize total contribution margin?

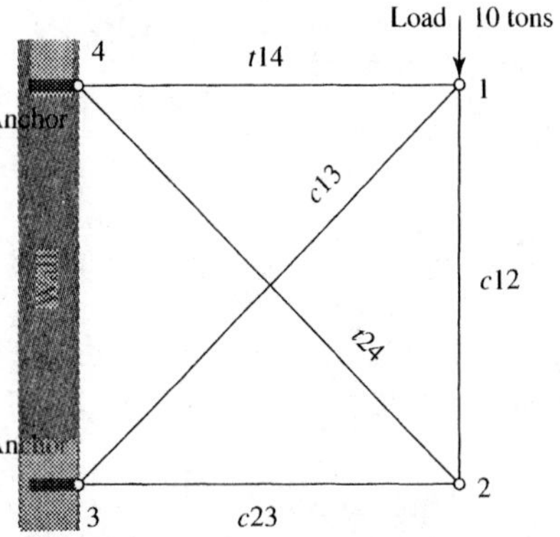

Figure 9.7 A cantilever structure.

15. *Structures*. A cantilever beam is shown in Figure 9.7. The square structure with four nodes, labeled 1, 2, 3, 4, and five structural *links* is attached to a wall at nodes 3, 4. A load of 10 tons is applied at node 1. The cost of each link depends on whether it is in tension or compression. Links that are only under tension generally cost \$100 per ton of load, but the specific link between nodes 2 and 4 that is under tension costs \$141 per ton load. Links that are under compression generally cost \$120 per ton of load, but the specific link between nodes 1 and 3 costs \$170 per ton of load. How much load should be carried by each link so that the cost of the structure is minimized and the structure will support the given load?

The notation $c13$, for example, denotes a compression load in the link between nodes 1 and 3 and t denotes tension. *Constraints*: Vertical and horizontal forces at nodes 1 and 2 must balance.

16. *Transportation*. Kraft Foods, Inc. was an early user of LP methods to make transportation decisions in order to minimize cost. Here is a simplified model. A cheese company has certain amounts of product in storage at supply points that can be transported to major cities at given costs (supply). Each week, customers (supermarkets, distributors) in the major cities request certain amounts of cheese (demand). The data for a simplified supply–demand model are given in the following table.

Supply (tons)		Demand (tons)	
Fort Atkinson	20	Chicago	13
Springfield	15	Indianapolis	9
		Kansas City	10

The dollar cost per ton of shipping from supply points is given next.

	Chicago	Indianapolis	Kansas City
Fort Atkinson	100	120	142
Springfield	111	113	127

How should the company program its trucks for delivery in order to minimize cost?

17. *Blending Metal.* Recycled iron and steel are available in the amounts specified and at the given costs.

	Carbon	Magnesium	Silicon	Amount	Cost/lb
Iron	3.0%	0.9%	3.0%	unlimited	\$0.2
Steel	0.4%	0.3%	0%	1500 lb	\$0.6

The goal is to produce 2000 lb (one short ton) of metal blended from recycled iron and steel at minimum cost and with the following properties:

Specifications		
	At Least	At Most
Carbon content	2.0%	2.6%
Magnesium content	0.8%	1.2%
Silicon content	2.5%	3.0%

18. Give a geometric argument that proves that the intersection of convex sets is convex. Explain the significance in terms of feasible sets in $\mathbb{R}^2$.

USING MATLAB

Consult online help for the commands **fill**, **patch**, *and related commands.*

19. Plot the polygon P defined by the intersection of the half-planes

$$x_1 - x_2 \leq 0, \quad -x_1 + 2x_2 \leq 4, \quad x_1 + x_2 \leq 2.5.$$

with x_1, x_2 nonnegative. Color P (a color of your choice) using the command **fill(x,y,[r,g,b])**

The vectors **x** and **y** carry the coordinates (x, y) of the vertices of the polygon in the plane and **[r,g,b]** is a probability vector that defines color having a mix of red (r), green (g) and blue (b) values, where **[0,0,0]** is black and **[1,1,1]** is white. Plot the optimal (maximum) isovalue for the objective function $f = 3x_1 + x_2$ defined on P.

20. Plot the feasible sets and optimal isovalues for Exercises 8–11.

9.2 The Simplex Algorithm

We now describe an algebraic, step-by-step procedure, called the *simplex algorithm*, that can be applied to solve LP problems with any number of decision variables and constraints. The first illustration uses a standard maximization LP problem in two decision variables x_1, x_2 so that the steps in the algorithm can be visualized geometrically in $\mathbb{R}^2$.

■ ILLUSTRATION 9.4

The simplex algorithm

Our goal is

MAX

$$f = 2x_1 + 3x_2 \tag{9.18}$$

ST

$$\begin{aligned} x_1 - x_2 &\leq 2 \\ 2x_1 - x_2 &\leq 5 \\ x_1 + 2x_2 &\leq 10 \end{aligned} \tag{9.19}$$

with x_1, x_2 nonnegative.

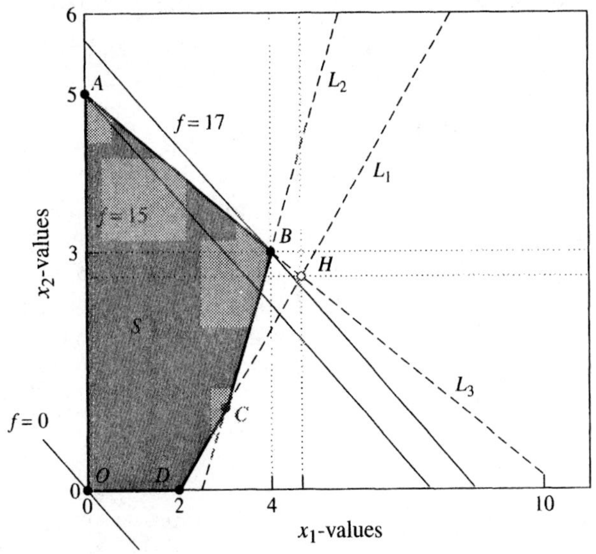

Figure 9.8 Feasible set S.

The feasible set S shown in Figure 9.8 is bounded and convex. The lines bounding S are the axes $x_1 = 0$ and $x_2 = 0$ and lines

$$L_1 \ : \ x_1 - x_2 = 2, \qquad L_2 \ : \ 2x_1 - x_2 = 5, \qquad L_3 \ : \ x_1 + 2x_2 = 10$$

defined by equality in (9.19).

Slack Variables

The first step is to transform the constraints in (9.19) into equations by introducing a new nonnegative variable for each constraint. Rearranging the first constraint for example, we have $0 \le 2-(x_1-x_2)$ and so defining $u_1 = 2-(x_1-x_2)$ results in the equation

$$x_1 - x_2 + u_1 = 2,$$

where u_1 is nonnegative. We call u_1 a *slack variable* because it represents the difference or "slack" between the values 2 and $x_1 - x_2$. Introducing nonnegative slack variables u_2 and u_3 in a similar way, the set of constraints (9.19) can be rewritten as a 3×5 linear system:

$$\left\{ \begin{array}{rrrrrrl} x_1 - & x_2 + u_1 & & & = & 2 & E_1 \\ 2x_1 - & x_2 & + u_2 & & = & 5 & E_2 \\ x_1 + & 2x_2 & & + u_3 & = & 10 & E_3 \end{array} \right. \tag{9.20}$$

where all variables are nonnegative. The effect of introducing slack variables is to extend the original problem from $\mathbb{R}^2$ to $\mathbb{R}^5$. A 5-vector $\mathbf{v} = (x_1, x_2, u_1, u_2, u_3)$ with nonnegative entries is a feasible solution to the extended problem if and only if $\mathbf{v}$ satisfies (9.20).

Boundary Points, Extreme Points

How do we identify boundary points and extreme points algebraically? Equation E_1 in (9.20) shows that a point (x_1, x_2) in $\mathbb{R}^2$ lies on the boundary line L_1 if and only if $u_1 = 0$ and lies in the half-plane above L_1 if and only if $u_1 > 0$. The conditions $u_2 \geq 0$ and $u_3 \geq 0$ are interpreted similarly using equations E_2 and E_3, respectively.

At each extreme point exactly two of the variables in the set x_1, x_2, u_1, u_2, u_3 are zero. The variables having value zero are called *nonbasic* variables and the three remaining nonzero variables are called *basic* variables. The set of basic variables at each extreme point is called the *basis*.[7] The following table describes the extreme points of S in Figure 9.8 in terms of basic and nonbasic variables.

Extreme point		Basic Variables		Nonbasic Variables
$O = (0, 0)$	$u_1 = 2$	$u_2 = 5$	$u_3 = 10$	$x_1 = x_2 = 0$
$A = (0, 5)$	$x_2 = 5$	$u_1 = 7$	$u_2 = 10$	$x_1 = u_3 = 0$
$B = (4, 3)$	$x_1 = 4$	$x_2 = 3$	$u_1 = 1$	$u_2 = u_3 = 0$
$C = (3, 1)$	$x_1 = 3$	$x_2 = 1$	$u_3 = 5$	$u_1 = u_2 = 0$
$D = (2, 0)$	$x_1 = 2$	$u_2 = 1$	$u_3 = 8$	$x_2 = u_1 = 0$

(9.21)

The values of the basic variables are found by setting the nonbasic variables in (9.20) to zero and solving the resulting 3×3 linear system for the basic variables. Each 3×3 linear system so formed will have a unique solution if and only if the equations in (9.20) are linearly independent. When this is the case we call the problem *nondegenerate* and otherwise *degenerate*.

Caution! Setting two variables from the set x_1, x_2, u_1, u_2, u_3 equal to zero does not necessarily signal an extreme point. For example, setting $u_1 = u_3 = 0$ in (9.20) gives $x_1 = 14/3$, $x_2 = 8/3$, $u_2 = -5/30$ and the negative value of u_2 indicates that $H = (14/3, 8/3)$ at the intersection of L_1 and L_3 is not an extreme point in S.

As mentioned Section 9.1, an optimal (minimum or maximum) value of the objective function f is attained at an extreme point in S. The simplex algorithm depends on locating an initial extreme point in S from which to start the solution process. This is called *initialization*. We then move step by step along the boundary of S from current extreme point to an adjacent extreme point, in such a way that the value of f is increased at each step, ending at a final extreme point when f cannot be increased further. The vector $\mathbf{v} = (x_1, x_2, u_1, u_2, u_3)$ associated with an extreme point is called a *basic feasible solution* to the problem. The vector $\mathbf{v}$ associated with the final extreme point is called the *optimal feasible solution*. At each extreme point two components in $\mathbf{v}$ are nonbasic and three are basic. In passing from current extreme point to adjacent extreme point, one basic variable moves out of the basis (becomes nonbasic) and one nonbasic variable enters the basis (becomes basic). This process is evident in table (9.21), moving clockwise around S in Figure 9.8.

[7] Not to be confused with a basis of vectors for a vector space.

Tableaux[8]

Each step of the simplex algorithm is recorded using an array (or table) that is traditionally called a *tableau*. A tableau, such as (9.23), displays numerical data, together with other labels: variables, current basis, indicator and pivot for the next step, and so on. The last row, called the *objective row*, is used to compute the current value of the objective function.

The first step in solving any LP problem is to identify an initial basic feasible solution to the extended problem. Let $x_1 = x_2 = 0$ and then $\mathbf{v} = (0, 0, 2, 5, 10)$ is such a solution, where x_1, x_2 are nonbasic variables. The solution corresponds to the origin $O = (0, 0)$ in $\mathbb{R}^2$, where $f = 0$. The basis at O is $(u_1, u_2, u_3) = (2, 5, 10)$ shown in (9.21) and in (9.23).

The *initial tableau* for our problem is shown in (9.23). It contains the augmented matrix of the system (9.20) and the objective row records the coefficients and constant term from equation (9.22), which is obtained by rearranging (9.18) and introducing slack variables.

$$f - 2x_1 - 3x_2 + 0u_1 + 0u_2 + 0u_3 = 0 \tag{9.22}$$

Initial Tableau (9.23)

f	x_1	x_2	u_1	u_2	u_3	b	Basis	Pivot Row (*)
	1	−1	1	0	0	2	u_1	
	2	−1	0	1	0	5	u_2	
	1	(2)	0	0	1	10	u_3	$10 \div 2 = 5^*$
1	−2	-3^*	0	0	0	0		

Notice that the basis variables correspond to the binary columns in (9.23).

We can now move to the next stage in the solution process. We will try to increase f by moving from O along the boundary of S to an adjacent extreme point. The possible directions from O are along the lines $x_1 = 0$ or $x_2 = 0$ (the axes). Equation (9.18) implies that increasing x_2 keeping $x_1 = 0$ is best because f will increase by $3x_2$ (3 units in f for every unit increase in x_2) as opposed to increasing x_1 keeping $x_2 = 0$, which will only increase f by $2x_1$. In practice, we locate the negative coefficient in the objective row, left of column b, which has the largest absolute value and call this number the *indicator* [starred in (9.23)]. The column in which the indicator appears is the *pivot column* (column 2).

If x_2 is increased while $x_1 = 0$, how much can x_2 be increased? The *positive* entries in the pivot column above the indicator will answer this question. In this case there is only one such positive entry (circled) and this is the *pivot* for this step. The third row of (9.23) corresponds to the equation $x_1 + 2x_2 + u_3 = 10$, and with $x_1 = 0$ we must have $x_2 \leq 10/2 = 5$ because u_3 is nonnegative. Hence x_2 cannot exceed 5, as seen in Figure 9.8.

To obtain a second tableau we apply elementary row operations (EROs) to the initial tableau, as in Gauss elimination: Scale the pivot to 1 and reduce

[8] In French, *tableau* means "list" or "table." *Tableaux* is plural of *tableau*.

all other entries in the pivot column to zero. The EROs are

$$\frac{1}{2}\mathbf{r}_3 \to \mathbf{r}_3, \quad \mathbf{r}_1 + \mathbf{r}_3 \to \mathbf{r}_1, \quad \mathbf{r}_2 + \mathbf{r}_3 \to \mathbf{r}_2, \quad \mathbf{r}_4 + 3\mathbf{r}_3 \to \mathbf{r}_4$$

and the next tableau is shown in (9.24).

Second Tableau (9.24)

f	x_1	x_2	u_1	u_2	u_3	b	Basis	Pivot Row (*)
	$\frac{3}{2}$	0	1	0	$\frac{1}{2}$	7	u_1	$7 \div (3/2) = 14/3$
	$\textcircled{\frac{5}{2}}$	0	0	1	$\frac{1}{2}$	10	u_2	$10 \div (5/2) = 4^*$
	$\frac{1}{2}$	1	0	0	$\frac{1}{2}$	5	x_2	$5 \div (1/2) = 10$
1	$-\frac{1}{2}^*$	0	0	0	$\frac{3}{2}$	15		

The second tableau corresponds to the point $A = (0, 5)$, where x_1, u_3 are nonbasic variables and the basic values are $x_2 = 5$, $u_1 = 7$, $u_2 = 10$, which can be read off using the binary columns in (9.24). Now u_3 has moved out of the basis and x_2 has moved into the basis. The objective row in (9.24) can be rearranged to indicate the current value of f that contains only nonbasic variables. We have

$$f - \frac{1}{2}x_1 + \frac{3}{2}u_3 = 15 \quad \Rightarrow \quad f = 15 + \frac{1}{2}x_1 - \frac{3}{2}u_3. \tag{9.25}$$

The procedure just described is repeated again and again, increasing f at each stage by moving to an adjacent extreme point until the optimal value is found.

We ask whether f can be increased from the current value $f = 15$ at A? From (9.25), the answer is to increase x_1 while keeping $u_3 = 0$ (increasing u_3 while keeping $x_1 = 0$ would only reduce f).

The indicator for the next step is $-\frac{1}{2}$ and column 1 now becomes the pivot column. How much can x_1 be increased from its current value $x_1 = 0$? The rows in (9.24) that contain a *positive* entry in the pivot column above the indicator hold the key. The equations corresponding to these rows are obtained from (9.24) and appear on the left in (9.26).

$$\begin{aligned}
\tfrac{3}{2}x_1 + u_1 + \tfrac{1}{2}u_3 &= 7 && \Rightarrow x_1 \le 7 \div (3/2) = 14/3 \\
\tfrac{5}{2}x_1 + u_2 + \tfrac{1}{2}u_3 &= 10 && \Rightarrow x_1 \le 10 \div (5/2) = 4 \\
\tfrac{1}{2}x_1 + x_2 + \tfrac{1}{2}u_3 &= 5 && \Rightarrow x_1 \le 5 \div (1/2) = 10
\end{aligned} \tag{9.26}$$

At present $u_3 = 0$ and all other variables must be nonnegative. Hence x_1 can only increase to the upper bounds shown on the right in (9.26). But all three equations must be satisfied and so $x_1 \le min\{\frac{14}{3}, 4, 10\} = 4$ and this identifies row 2 as the pivot row for the next stage with the pivot $p = \frac{5}{2}$ circled in (9.24). Geometrically, we are going to move along L_3 from A toward B. See Figure 9.8.

As before, scale the pivot to 1 and use EROs to reduce all other entries in the pivot column to zero. The required EROs are

$$\frac{2}{5}\mathbf{r}_2 \to \mathbf{r}_2, \quad \mathbf{r}_1 - \frac{3}{2}\mathbf{r}_2 \to \mathbf{r}_1, \quad \mathbf{r}_3 - \frac{1}{2}\mathbf{r}_2 \to \mathbf{r}_3, \quad \mathbf{r}_4 + \frac{1}{2}\mathbf{r}_2 \to \mathbf{r}_4$$

and the next tableau is shown in (9.27).

Third Tableau (9.27)

f	x_1	x_2	u_1	u_2	u_3	b	Basis	Pivot Row (*)
	0	0	1	$-\frac{3}{5}$	$\frac{1}{5}$	1	u_1	
	1	0	0	$\frac{2}{5}$	$\frac{1}{5}$	4	x_1	
	0	1	0	$-\frac{1}{5}$	$\frac{2}{5}$	3	x_2	
1	0	0	0	$\frac{1}{5}$	$\frac{8}{5}$	17		

The third tableau corresponds to vertex B with nonbasic variables $u_2 = u_3 = 0$ and the basis values $x_1 = 4$, $x_2 = 3$, $u_1 = 1$, which are evident from the binary columns in (9.27). Now u_2 has moved out of the basis, x_1 has entered the basis, and the objective row in (9.27) shows that

$$f + \frac{1}{5}u_2 + \frac{8}{5}u_3 = 17 \quad \Rightarrow \quad f = 17 - \frac{1}{5}u_2 - \frac{8}{5}u_3,$$

with f once again expressed only in terms of nonbasic variables. The process stops here because increasing the value of u_2 or u_3 will only decrease the value of f. Note that there are no negative entries in the objective row left of column b. Geometrically, $B = (4, 3)$ is the final extreme point in the sequence of steps and $f = 17$ is the optimal (maximum) value of f. The point $(4, 3, 1, 0, 0)$ in $\mathbb{R}^5$ is the optimal feasible solution that maximizes f. Also f has an optimal (minimum) value $f = 0$ attained at $O = (0, 0)$. ■

The General Problem

Consider the standard form of an LP problem given by (9.7) and (9.8). Introducing an m-vector (column) of slack variables $\mathbf{u}$ and letting $\mathbf{0}$ be the $m \times 1$ zero matrix, we obtain an $m \times (n+m)$ linear system of constraints (9.28) and a modified objective function (9.29) as follows:

$$\mathbf{A}\mathbf{x} + \mathbf{I}_m\mathbf{u} = \mathbf{b} \tag{9.28}$$

$$f - \mathbf{c}^\mathsf{T}\mathbf{x} + \mathbf{0}^\mathsf{T}\mathbf{u} = 0. \tag{9.29}$$

The following matrix provides the numerical data in the initial tableau:

$$\left[\begin{array}{cc|c} \mathbf{A} & \mathbf{I}_m & \mathbf{b} \\ -\mathbf{c}^\mathsf{T} & \mathbf{0}^\mathsf{T} & 0 \end{array}\right]. \tag{9.30}$$

At each extreme point of the feasible set S for the extended problem there are n nonbasic variables in the set $x_1, x_2, \ldots, x_n, u_1, \ldots, u_m$, and the remaining m basic values are found from (9.28) by solving the resulting $m \times m$ linear system. The problem is called *nondegenerate* when a solution to each such square system exists. This happens when the equations in (9.28) are linearly independent; otherwise the problem is called *degenerate*.

The Simplex Algorithm

Consider a standard maximization LP problem given in matrix form by

$$\begin{aligned} &\textbf{MAX} \quad f = \mathbf{c} \cdot \mathbf{x} = \mathbf{c}^{\mathsf{T}}\mathbf{x} \\ &\textbf{ST} \qquad \mathbf{Ax} \le \mathbf{b}, \quad \mathbf{0} \le \mathbf{x}. \end{aligned}$$

We shall impose the extra assumption that $\mathbf{0} \le \mathbf{b}$. The case when $\mathbf{b}$ has some negative entries is addressed briefly in Section 9.4.

Form the initial tableau as in (9.30). The top rows correspond to constraints expressed as equations using slack variables. The bottom row is the objective row.

Step 1 The tableau is optimal if all entries in the objective row left of column b are nonnegative. Otherwise identify a negative entry (the indicator) in column j that is largest in absolute value. If there are two or more equal entries with this property, choose one at random. Column j becomes the pivot column.

Step 2 For each positive entry p_{ij} in the pivot column above the indicator, form the nonnegative ratio b_i/p_{ij}, where b_i is the coefficient from column b in row i. The minimum positive quotient $q = b_i/p_{ij}$ identifies row i as the pivot row and the pivot is p_{ij}. If there are no positive entries above the indicator, the problem has no solution.

Step 3 Scale the pivot to 1 and use EROs to clear the pivot column, reducing all other entries to zero. The objective row shows the value of f in terms of the current nonbasic variables.

Step 4 Repeat from Step 1 until all entries in the objective row (excluding column b) are nonnegative.

EXAMPLE 1

Applying the Simplex Algorithm

Our goal is
MAX

$$f = x_1 + 3x_2 + 2x_3$$

ST

$$\begin{aligned} x_1 + 2x_2 - x_3 &\le 2 \\ 2x_1 + x_2 - x_3 &\le 4 \\ 2x_1 - 3x_2 + 2x_3 &\le 6 \end{aligned}$$

with x_1, x_2, x_3 nonnegative. Introducing slack variables u_1, u_2, u_3, the associated matrices in (9.28) and (9.29) are

$$\mathbf{c} = \begin{bmatrix} 1 \\ 3 \\ 2 \end{bmatrix}, \quad \mathbf{x} = \begin{bmatrix} x_1 \\ x_2 \\ x_3 \end{bmatrix}, \quad \mathbf{b} = \begin{bmatrix} 2 \\ 4 \\ 6 \end{bmatrix}, \quad \mathbf{u} = \begin{bmatrix} u_1 \\ u_2 \\ u_3 \end{bmatrix}, \quad \mathbf{A} = \begin{bmatrix} 1 & 2 & -1 \\ 2 & 1 & -1 \\ 2 & -3 & 2 \end{bmatrix}.$$

The initial tableau constructed using (9.30) is shown next.

Initial Tableau (9.31)

f	x_1	x_2	x_3	u_1	u_2	u_3	b	Basis	Pivot Row (*)
	1	(2)	−1	1	0	0	2	u_1	$2/2 = 2^*$
	2	1	−1	0	1	0	4	u_2	$4/1 = 4$
	2	−3	2	0	0	1	6	u_3	
1	−1	-3^*	−2	0	0	0	0		

An initial feasible value is $f = 0$ when $x_1 = x_2 = x_3 = 0$. The indicator -3 identifies the pivot column. We increase the value of f by increasing x_2. The positive entries above the indicator are used to form the positive quotients shown on the extreme right in (9.31). The minimum value 2 tells us that the pivot is $p_{12} = 2$ (circled). The EROs used to form the next tableau are

$$\frac{1}{2}\mathbf{r}_1 \to \mathbf{r}_1, \quad \mathbf{r}_2 - \mathbf{r}_1 \to \mathbf{r}_2, \quad \mathbf{r}_3 + 3\mathbf{r}_1 \to \mathbf{r}_3, \quad \mathbf{r}_4 + 3\mathbf{r}_1 \to \mathbf{r}_4,$$

Second Tableau

f	x_1	x_2	x_3	u_1	u_2	u_3	b	Basis	Pivot Row (*)
	$\frac{1}{2}$	1	$-\frac{1}{2}$	$\frac{1}{2}$	0	0	1	x_2	
	$\frac{3}{2}$	0	$-\frac{1}{2}$	$-\frac{1}{2}$	1	0	3	u_2	
	$\frac{7}{2}$	0	($\frac{1}{2}$)	$\frac{3}{2}$	0	1	9	u_3	$9/(1/2) = 18^*$
1	$\frac{1}{2}$	0	$-\frac{7}{2}^*$	$\frac{3}{2}$	0	0	3		

The objective row gives the new value for f, namely

$$f + \frac{1}{2}x_1 - \frac{7}{2}x_3 + \frac{3}{2}u_1 = 3 \quad \Rightarrow \quad f = 3 - \frac{1}{2}x_1 + \frac{7}{2}x_3 - \frac{3}{2}u_1$$

given as usual in terms of nonbasic variables x_1, x_3, u_1. Note that u_1 has moved out of the basis and x_2 has moved into the basis.

The new indicator is $-\frac{7}{2}$ and column 3 is the pivot column. The only positive entry above the indicator is $p = \frac{1}{2}$, which by default becomes the pivot. The EROs (leaving scaling to last for convenience) are

$$\mathbf{r}_1 + \mathbf{r}_3 \to \mathbf{r}_1, \quad \mathbf{r}_2 + \mathbf{r}_3 \to \mathbf{r}_2, \quad \mathbf{r}_4 + 7\mathbf{r}_3 \to \mathbf{r}_4, \quad 2\mathbf{r}_1 \to \mathbf{r}_1,$$

and the third tableau is

Third Tableau									
f	x_1	x_2	x_3	u_1	u_2	u_3	b	Basis	Pivot Row (*)
	4	1	0	2	0	1	10	x_2	
	5	0	0	1	1	1	12	u_2	
	7	0	1	3	0	2	18	x_3	
1	25	0	0	12	0	7	66		

The new value for f is now given by

$$f = 66 - 25x_1 - 12u_1 - 7u_3,$$

giving f in terms of nonbasic variables x_1, u_1, u_3. There is no new indicator, and $f = 66$ is optimal (maximum) for f attained when $x_1 = 0$, $x_2 = 10$, $x_3 = 18$. ■

Remarks

If it happens that there is no positive entry in the pivot column above the indicator, then the algorithm stops short of finding a solution. The feasible set is *unbounded* in this case.

EXAMPLE 2

An Unbounded Feasible Set

Our goal is
MAX

$$f = x_1 + 2x_2$$

ST

$$-x_1 + x_2 \leq 2, \quad x_1 \geq 0, \quad x_2 \geq 0.$$

The feasible set S is shown in Figure 9.9.

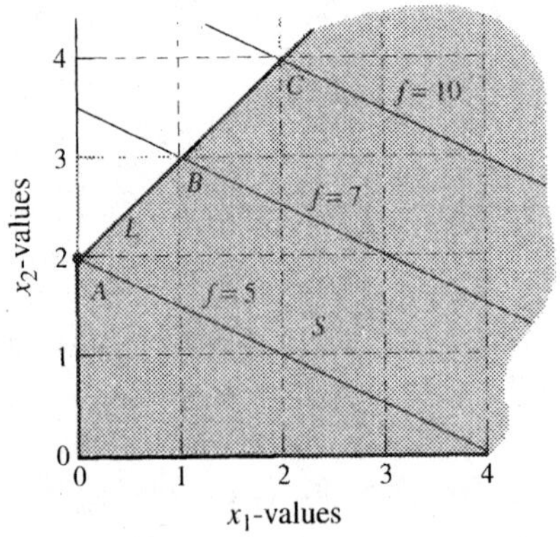

Figure 9.9 An unbounded feasible set.

The initial tableau is

Initial Tableau						
f	x_1	x_2	u_1	b	Basis	Pivot Row (*)
	-1	①	1	2	u_1	$2 \div 1 = 2^*$
1	-1	-2^*	0	0		

The initial feasible solution has nonbasic values $x_1 = x_2 = 0$, the indicator is -2 and the pivot is $p_{12} = 1$. Clearing column 2 using the ERO $\mathbf{r}_2 + 2\mathbf{r}_1 \to \mathbf{r}_2$, we have

Second Tableau						
f	x_1	x_2	u_1	b	Basis	Pivot Row (*)
	-1	1	1	2	x_2	
1	-3^*	0	2	4		

The indicator for the next step should be -3 in column 1, but there is no positive entry above the indicator and the algorithm terminates. The objective row gives $f = 4 + 3x_1 - 2u_1$ and the value $f = 4$ is attained at A when $x_1 = u_1 = 0$. Increasing x_1 will increase f. In fact, $B = (1, 3)$ gives $f = 7$, a feasible value for f, and each point along the boundary line $L: -x_1 + x_2 = 2$ (C, for example) gives larger values for f. Thus f increases *without bound* because the shaded region in Figure 9.9 is unbounded (above) and the solution is said to be *unbounded* in this case. ■

Minimization Models

The next illustration shows how the simplex algorithm can be applied (after some initial adjustments) to solve minimization problems. We will also see how constraint inequalities involving $\geq$ signs are handled.

■ ILLUSTRATION 9.5

Minimization

Our goal is
MIN

$$f = 2x_1 + 3x_2 \tag{9.32}$$

ST

$$x_1 + x_2 \geq 4 \tag{9.33}$$

$$x_1 + 3x_2 \geq 6 \tag{9.34}$$

with x_1, x_2 nonnegative.

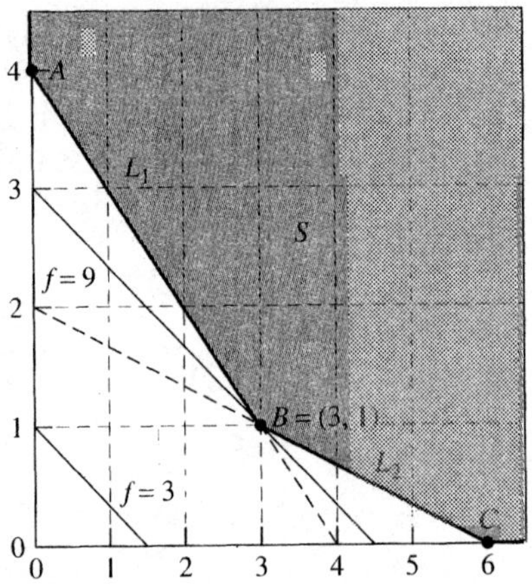

Figure 9.10 Minimization.

As mentioned at the beginning of Section 9.1, the minimization of the function f is equivalent to finding a maximum for the function $F = -f = -2x_1 - 3x_2$. The constraints define the feasible set S shown in Figure 9.10, which is *unbounded above* in the first quadrant.

Surplus Variables

Inequalities (9.33) and (9.34) are converted into equations by introducing terms $-u_1$ and $-u_2$ (note the minus signs), where u_1 and u_2 are nonnegative *surplus variables*. For example, the nonnegative value $u_1 = (x_1 + x_2) - 4$ is the difference or "surplus" between the values $x_1 + x_2$ and 4. The equations associated with (9.33) and (9.34) are

$$\begin{cases} x_1 + \ \ x_2 - u_1 \qquad\quad = 4 & E_1 \\ x_1 + 3x_2 \qquad\quad - u_2 = 6 & E_2 \end{cases}. \tag{9.35}$$

The boundary lines of S are given by

$$L_1 : x_1 + x_2 = 4, \qquad L_2 : x_1 + 3x_2 = 6, \qquad x_1 = 0, \qquad x_2 = 0 \quad \text{(the axes)}.$$

A point (x_1, x_2) lying on L_1 satisfies E_1 with $u_1 = 0$ and any value $u_1 > 0$ defines a set of points on a line $x_1 + x_2 = 4 + u_1$ lying above and parallel to L_1. A similar statement holds for u_2 and E_2. The points (x_1, x_2) lying inside (not on the boundary) of S are those points which satisfy (9.35) with u_1 and u_2 both positive, and conversely.

Artificial Variables

The simplex algorithm depends on finding an initial basic feasible solution. For standard maximization problems this is obtained by making all decision variables nonbasic. However, in this illustration $x_1 = x_2 = 0$ does not satisfy (9.33) and (9.34). Consequently, we introduce new nonnegative variables a_1 and a_2 (one for each constraint), which are called *artificial variables* because their only role is to force an initial basic feasible solution. Artificial variables have no physical meaning. We obtain the 2×6 linear system

$$\begin{cases} x_1 + \ \ x_2 - u_1 \qquad +a_1 \qquad = 4 \\ x_1 + 3x_2 \qquad -u_2 \qquad + a_2 = 6 \end{cases} . \tag{9.36}$$

The original LP problem is extended to an "artificial problem," which includes the artificial variables. The problem is nondegenerate because if we set four of the variables $x_1, x_2, u_1, u_2, a_1, a_2$ to zero we can solve (9.36) for the remaining two variables. For example, a basic feasible solution to the extended problem is obtained by setting $x_1 = x_2 = u_1 = u_2 = 0$ in (9.36), which gives $a_1 = 4$ and $a_2 = 6$. However, not all solutions $(x_1, x_2, u_1, u_2, a_1, a_2)$ to (9.36) provide a basic feasible solution for the original problem unless $a_1 = a_2 = 0$. For example, consider the solution (2, 1, 1, 1, 2, 2) to (9.36).

In order to force $a_1 = a_2 = 0$, we modify the objective function $F = -f = -2x_1 - 3x_2$ into the form

$$F = -2x_1 - 3x_3 + 0u_1 + 0u_2 - Ma_1 - Ma_2 \tag{9.37}$$

or

$$F + 2x_1 + 3x_3 + 0u_1 + 0u_2 + Ma_1 + Ma_2 = 0,$$

where M is a sufficiently large constant. Observe from (9.37) that F cannot attain its maximum value unless $a_1 = a_2 = 0$ because positive values for the artificial variables will only decrease F. Choosing $M = 20$, for example, we obtain the initial tableau as follows.

Initial Tableau

F	x_1	x_2	u_1	u_2	a_1	a_2	b	Basis	Pivot Row (*)
	1	1	−1	0	1	0	4	a_1	
	1	3	0	−1	0	1	6	a_2	
1	2	3	0	0	20	20	0		

As a preliminary step, the terms $M = 20$ in the objective row are reduced to zero using the EROs: $\mathbf{r}_3 - 20\mathbf{r}_1 \to \mathbf{r}_3$ and $\mathbf{r}_3 - 20\mathbf{r}_2 \to \mathbf{r}_3$. Note that M is chosen large enough for the coefficients of x_1 and x_2 in the objective row to become negative after applying the row replacements. We can now apply the simplex algorithm as before.

Second Tableau

F	x_1	x_2	u_1	u_2	a_1	a_2	b	Basis	Pivot Row (*)
	1	1	−1	0	1	0	4	a_1	$4 \div 1 = 4$
	1	(3)	0	−1	0	1	6	a_2	$6 \div 3 = 2^*$
1	−38	-77^*	20	20	0	0	−200		

The objective row in the second tableau gives

$$F = -200 + 38x_1 + 77x_2 - 20u_1 - 20u_2.$$

The indicator is -77 and the pivot is $p_{22} = 3$ (circled). Using EROs to clear column 2, we have

Third Tableau

F	x_1	x_2	u_1	u_2	a_1	a_2	b	Basis	Pivot Row (*)
	$\frac{2}{3}$ (circled)	0	-1	$\frac{1}{3}$	1	$-\frac{1}{3}$	2	a_1	$2 \div (2/3) = 3^*$
	$\frac{1}{3}$	1	0	$-\frac{1}{3}$	0	$\frac{1}{3}$	2	x_2	$2 \div (1/3) = 6$
1	$-\frac{37}{3}$*	0	20	$-\frac{17}{3}$	0	$\frac{77}{3}$	-46		

The objective row in the third tableau gives

$$F = -46 + \frac{37}{3}x_1 - 20u_1 + \frac{17}{3}u_2 - \frac{77}{3}a_2.$$

The indicator is $-\frac{37}{3}$ and the pivot is $p_{11} = \frac{2}{3}$ (circled). Using EROs to clear column 1, we have

Fourth Tableau

F	x_1	x_2	u_1	u_2	a_1	a_2	b	Basis	Pivot Row (*)
	1	0	$-\frac{3}{2}$	$\frac{1}{2}$	$\frac{3}{2}$	$-\frac{1}{2}$	3	x_1	
	0	1	$\frac{1}{2}$	$-\frac{1}{2}$	$-\frac{1}{2}$	$\frac{1}{2}$	1	x_2	
1	0	0	$\frac{3}{2}$	$\frac{1}{2}$	$\frac{37}{2}$	$\frac{39}{2}$	-9		

The objective row in the fourth tableau gives

$$F = -9 - \frac{3}{2}u_1 - \frac{1}{2}u_2 - \frac{37}{2}a_1 - \frac{39}{2}a_2.$$

There are no negative coefficients in the objective row (left of column b) and the simplex algorithm ends here. $F = -f = -9$ is the maximum value of F and $f = -F = 9$ is the minimum value of f. The optimal feasible solution to the original problem is (3, 1). ■

Blending Gasoline Revisited

Refer to the blending model in the introduction to this chapter. Minimizing f in (9.1) is equivalent to maximizing F, where

$$F = -0.4x_1 - 0.5x_2 - 0.65x_3 \quad (= -f) \tag{9.38}$$

subject to the same given constraints. Instead of replacing the demand constraint equation (9.2) by two inequalities, as in (9.16), we may introduce a nonnegative artificial variable a_1 into the equation, as follows:

$$x_1 + x_2 + x_3 + a_1 = 14{,}000$$

and so (9.2) is true if and only if $a_1 = 0$. Slack variables u_1, u_2 are introduced, one for each of the availability constraints (9.3). Multiplying the blending constraint (9.5) by -1, we obtain

$$3x_1 - 3x_2 - 11x_3 \leq 0$$

and a slack variable u_3 is introduced for this constraint. The objective function F in (9.38) is modified into the form required in the objective row of the initial tableau, namely

$$F + 0.4x_1 + 0.5x_2 + 0.65x_3 + 0u_1 + 0u_2 + 0u_3 + Ma_1 = 0,$$

where M is a sufficiently large constant (the choice $M = 10$ will suffice). We have

Initial Tableau

F	x_1	x_2	x_3	u_1	u_2	u_3	a_1	b	Basis	Pivot Row (*)
	1	1	1	0	0	0	1*	14000	a_1	
	0	1	0	1	0	0	0	3906	u_1	
	0	0	1	0	1	0	0	4100	u_2	
	3	−3	−11	0	0	1	0	0	u_3	
1	0.4	0.5	0.65	0	0	0	10	0	0	

The first step is to eliminate the constant M $(= 10)$ in the objective row using the replacement $\mathbf{r}_5 - 10\mathbf{r}_1 \to \mathbf{r}_5$.

Second Tableau

F	x_1	x_2	x_3	u_1	u_2	u_3	a_1	b	Basis	Pivot Row (*)
	1	1	1	0	0	0	1	14000	a_1	
	0	1	0	1	0	0	0	3906	u_1	
	0	0	1	0	1	0	0	4100	u_2	$4100 \div 1 = 4100$
	(3)	−3	−11	0	0	1	0	0	u_3	$0 \div 3 = 0^*$
1	$-\frac{48}{5}$*	$-\frac{19}{2}$	$-\frac{187}{20}$	0	0	0	0	−140000		

The indicator is $-\frac{48}{5}$ and the pivot is 3. Clear column 1 using EROs $\frac{1}{3}\mathbf{r}_4 \to \mathbf{r}_4$, $\mathbf{r}_1 - \mathbf{r}_4 \to \mathbf{r}_1$, $\mathbf{r}_5 + \frac{5}{48}\mathbf{r}_4 \to \mathbf{r}_5$.

Third Tableau

F	x_1	x_2	x_3	u_1	u_2	u_3	a_1	b	Basis	Pivot Row (*)
	0	2	$\textcircled{\frac{14}{3}}$	0	0	$-\frac{1}{3}$	1	14000	a_1	$14000 \div \frac{14}{3} = \frac{1000}{3}$*
	0	1	0	1	0	0	0	3906	u_1	
	0	0	1	0	1	0	0	4100	u_2	$14000 \div 1 = 14000$
	1	-1	$-\frac{11}{3}$	0	0	$\frac{1}{3}$	0	0	x_1	
1	0	$-\frac{191}{10}$	$-\frac{891}{20}$*	0	0	$\frac{16}{5}$	0	-140000		

The indicator is $-\frac{891}{20}$ and the pivot is $\frac{14}{3}$. Clear column 3 using $\frac{3}{14}\mathbf{r}_1 \to \mathbf{r}_1$, $\mathbf{r}_4 + \frac{11}{3}\mathbf{r}_1 \to \mathbf{r}_4$, $\mathbf{r}_5 + \frac{20}{891}\mathbf{r}_1 \to \mathbf{r}_5$.

Fourth Tableau

F	x_1	x_2	x_3	u_1	u_2	u_3	a_1	b	Basis	Pivot Row (*)
	0	$\textcircled{\frac{3}{7}}$	1	0	0	$-\frac{1}{14}$	$\frac{3}{14}$	3000	a_1	$3000 \div \frac{3}{7} = 142.85$*
	0	1	0	1	0	0	0	3906	u_1	$3906 \div 1 = 3906$
	0	$-\frac{3}{7}$	1	0	1	$\frac{1}{14}$	$-\frac{3}{14}$	1100	u_2	
	1	$\frac{4}{7}$	0	0	0	$\frac{1}{14}$	$\frac{11}{14}$	11000	x_1	$11000 \div \frac{4}{7} = 392.85$
1	0	$-\frac{1}{140}$*	0	0	0	$\frac{1}{56}$	$\frac{2673}{280}$	-6350		

Fifth Tableau

F	x_1	x_2	x_3	u_1	u_2	u_3	a_1	b	Basis	Pivot Row (*)
	0	0	1	$-\frac{3}{7}$	0	$-\frac{1}{14}$	$\frac{3}{14}$	1326	x_3	
	0	1	0	1	0	0	0	3906	x_2	
	0	0	0	$\frac{3}{7}$	1	$\frac{1}{14}$	$\frac{3}{14}$	2744	u_2	
	1	0	0	$-\frac{4}{7}$	0	$\frac{1}{14}$	$\frac{11}{14}$	8768	x_1	
1	0	0	0	$\frac{1}{140}$	0	$\frac{1}{56}$	$\frac{2673}{280}$	$-\frac{63221}{10}$		

The objective row in the last tableau gives

$$F = -6322.1 - \frac{1}{7}u_1 - \frac{1}{56}u_3 - \frac{2673}{280}a_1,$$

which is expressed in terms of nonbasic variables and the optimal (maximum) is $f = \$6322.10$, $(x_1, x_2, x_3) = (8768, 3906, 1326)$, as in (9.6).

EXERCISES 9.2

Exercises 1–8. Write each LP problem as a standard maximization problem using matrix notation. Use the simplex algorithm to maximize the objective function f subject to the given constraints. Draw an accurate diagram indicating the feasible set S, the optimal feasible solution, and the optimal isovalue. All decision variables are nonnegative.

1. $f = 4x_1 + 3x_2, \quad \begin{cases} -x_1 + x_2 \le 3 \\ 4x_1 - x_2 \le 1 \end{cases}$

2. $f = 4x_1 - 3x_2, \quad \begin{cases} -x_1 + x_2 \le 3 \\ 4x_1 - x_2 \le 1 \end{cases}$

3. $f = 2x_1 + x_2, \quad \begin{cases} x_1 + x_2 \le 4 \\ 5x_1 - 4x_2 \le 3 \end{cases}$

4. $f = x_1 - x_2, \quad \begin{cases} 2x_1 + 6x_2 \le 6 \\ 4x_1 + 3x_2 \le 24 \end{cases}$

5. $f = x_1 + x_2, \quad \begin{cases} x_1 - x_2 \le 0 \\ x_1 + x_2 \le 10 \\ -x_1 + x_2 \le 40 \end{cases}$

6. $f = 2x_1 + 3x_2, \quad \begin{cases} x_1 - x_2 \le 2 \\ 4x_1 - x_2 \le 11 \\ x_1 + x_2 \le 9 \end{cases}$

7. $f = 3x_1 + 2x_2, \quad \begin{cases} x_1 - x_2 \le 0 \\ 5x_1 - 3x_2 \le 10 \\ -5x_1 + 8x_2 \le 40 \end{cases}$

8. $f = 2x_1 + x_2, \quad \begin{cases} -x_1 + 6x_2 \le 6 \\ 4x_1 + 3x_2 \le 24 \\ 2x_1 - x_2 \le 2 \end{cases}$

9. Draw the feasible set for the objective function $f = x_1 + x_2$ subject to the constraints

$$\begin{cases} -x_1 + x_2 \le 1 \\ x_1 + x_2 \le 4 \end{cases}, \quad 0 \le x_1 \le 3, \quad 0 \le x_2 \le 2.$$

Maximize f using the simplex algorithm. Use the diagram to comment on the result.

Exercises 10–15. Use the simplex algorithm to find the maximum value of f subject to the given constraints. All decision variables are nonnegative.

10. $f = x_1 + x_2 + x_3, \quad \{ x_1 + 2x_2 + x_3 \le 4$

11. $f = 2x_1 + 2x_2 + x_3, \quad \begin{cases} 4x_1 - 2x_2 + x_3 \le 3 \\ x_1 + x_2 + x_3 \le 6 \end{cases}$

12. $f = x_1 + x_2 + x_3, \quad \begin{cases} x_1 + x_2 + x_3 \le 8 \\ x_1 + x_2 \le 4 \\ x_2 + x_3 \le 4 \end{cases}$

13. $f = 2x_1 + x_2 + 2x_3, \quad \begin{cases} 4x_1 + x_2 - x_3 \ge -3 \\ x_1 - x_2 + 2x_3 \le 5 \\ 3x_1 + 2x_2 + x_3 \le 10 \end{cases}$

14. $f = x_1 + x_2 + x_3, \quad \begin{cases} 2x_1 + 3x_2 + 4x_3 \le 9 \\ 3x_1 + 4x_2 + x_3 \le 8 \\ x_1 + 2x_2 + 3x_3 \le 6 \end{cases}$

15. $f = 2x_1 + x_2 - x_3, \quad \begin{cases} -2x_1 + 2x_2 + 2x_3 \le 2 \\ x_1 - x_2 + x_3 \le 1 \\ -x_1 - x_2 + x_3 \le 1 \\ x_1 + x_2 + x_3 \le 5 \end{cases}$

USING MATLAB

16. Use the M-file **gauss.m** to check the calculations in Example 1.

17. Refer to Exercises 8–11 in Section 9.1. Solve each LP problem using the simplex algorithm.

Exercises 18–21. Refer to Exercises 14–17 in Section 9.1. Solve each LP problem using the simplex algorithm.

18. Exercises 14.

19. Exercises 15.

20. Exercises 16.

21. Exercises 17.

22. Refer to Exercises 1–8 in this exercise set. For each LP problem, write an M-file to draw the feasible set $\mathcal{S}$. Use the **patch** command to color $\mathcal{S}$. Draw the isovalue f for the optimal (maximum) value of f.

9.3 Duality

Every LP problem has associated with it a second LP problem which is called its *dual*.[9] The coefficients and variables in the original or *primal problem* are

[9] *Dual*, meaning "two-fold" or "double."

used to formulate the *dual problem* and when the same dualization process is applied to the dual problem we arrive back at the primal problem once again. Schematically,

$$\textbf{PRIMAL} \quad \Rightarrow \quad \textbf{DUAL} \quad \Rightarrow \quad \textbf{PRIMAL}.$$

■ ILLUSTRATION 9.6

Duality

The interplay between primal and dual LP problems in real applications is illustrated by the following simplified gasoline blending model.

Primal Problem

The Penn Oil company blends two grades of gasoline, labeled A and B, from three ingredients, labeled C_1, C_2, C_3. The composition and net profit on each gallon of the new gasolines are shown in the following table, together with the available quantity of each ingredient.

Blend	C_1	C_2	C_3	Profit per Gallon
A	20%	50%	30%	\$0.86
B	30%	50%	20%	\$0.74
Quantity	6000	4500	2000	(gallons)

Introduce nonnegative decision variables x_1, x_2, which represent the quantity (gallons) produced of grade A and B, respectively. Then Penn Oil has the following standard maximization problem. We denote the objective function by p (for primal).

MAX

$$p = 0.86x_1 + 0.74x_2 \qquad \text{(total net profit)}$$

ST

$$\begin{cases} 0.2x_1 + 0.3x_2 \le 6000 \\ 0.5x_1 + 0.5x_2 \le 4500 \\ 0.3x_1 + 0.2x_2 \le 2000 \end{cases}$$

The Dual Problem

Suppose that a second oil company, Denver Oil, wishes to purchase ingredients C_1, C_2, C_3 from Penn Oil for their own use. We introduce nonnegative decision variables y_1, y_2, y_3, which represent the dollar cost per gallon that Denver Oil is willing to pay for C_1, C_2, C_3, respectively. Naturally Penn Oil would not sell to Denver Oil unless the total revenue received from Denver Oil for the ingredients is at least the total net profit that would be realized from sale of products A and B. Thus Denver Oil has the following minimization problem. We denote the objective function by d (for dual).

MIN

$$d = 6000y_1 + 4500y_2 + 2000y_3 \quad (= \text{total cost to Denver Oil})$$

ST

$$\begin{cases} 0.2y_1 + 0.5y_2 + 0.3y_3 \geq 0.86 \\ 0.3y_1 + 0.5y_2 + 0.2y_3 \geq 0.74 \end{cases}$$

Matrix Formulation of Primal and Dual Problems

The matrix form of the primal and dual problems shows clearly how they interact. We have

$$\textbf{PRIMAL}\quad \mathbf{c} = \begin{bmatrix} 0.86 \\ 0.74 \end{bmatrix}, \quad \mathbf{x} = \begin{bmatrix} x_1 \\ x_2 \end{bmatrix}, \quad \mathbf{A} = \begin{bmatrix} 0.2 & 0.3 \\ 0.5 & 0.5 \\ 0.3 & 0.2 \end{bmatrix}, \quad \mathbf{b} = \begin{bmatrix} 6000 \\ 4500 \\ 2000 \end{bmatrix}$$

$$\textbf{MAX}\quad p = \mathbf{c}^\mathsf{T}\mathbf{x} \qquad \textbf{ST}\quad \mathbf{Ax} \leq \mathbf{b}, \quad \mathbf{0} \leq \mathbf{x}.$$

Transposing **A** in the primal problem and using the costs to Denver Oil as components of a 3-vector **y**, we have

$$\textbf{DUAL}\quad \mathbf{b} = \begin{bmatrix} 6000 \\ 4500 \\ 2000 \end{bmatrix}, \quad \mathbf{y} = \begin{bmatrix} y_1 \\ y_2 \\ y_3 \end{bmatrix}, \quad \mathbf{A}^\mathsf{T} = \begin{bmatrix} 0.2 & 0.5 & 0.3 \\ 0.3 & 0.5 & 0.2 \end{bmatrix}, \quad \mathbf{c} = \begin{bmatrix} 0.86 \\ 0.74 \end{bmatrix}$$

$$\textbf{MIN}\quad d = \mathbf{b}^\mathsf{T}\mathbf{y} \qquad \textbf{ST}\quad \mathbf{A}^\mathsf{T}\mathbf{y} \geq \mathbf{c}, \quad \mathbf{y} \geq \mathbf{0}.$$

Dualizing the dual problem clearly results in the primal problem, and we say that the Penn Oil and Denver Oil problems are *dual* to each other. Using machine computation,[10] we find that the optimal values are

PRIMAL $\quad p = \$6900.00 \text{ (max)} \quad x_1 = 2000, \quad x_2 = 7000 \quad \text{(gal)}$

DUAL $\quad d = \$6900.00 \text{ (min)} \quad y_1 = \$0.00, \quad y_2 = \$1.00, \quad y_3 = \$1.20.$

The profit for Penn Oil is maximized in the primal problem by producing 2000 gallons of grade A and 7000 gallons of grade B, and the cost to Denver Oil is minimized in the dual problem by buying 4500 gallons of C_2 at \$1.00 per gallon and 2000 gallons of C_3 at \$1.20 per gallon. Note that the primal problem can be visualized in $\mathbb{R}^2$ and the dual problem in $\mathbb{R}^3$. Also note that $p = \mathbf{c}^\mathsf{T}\mathbf{x} = \mathbf{b}^\mathsf{T}\mathbf{y} = d$ (max = min) for the vectors $\mathbf{x} = (2000, 7000)$ and $\mathbf{y} = (0, 1, 1.2)$. ■

The following important facts on duality are stated without proof.

(a) If the primal and dual problems both have feasible solutions, then both problems have optimal solutions.

(b) If **x** is a feasible solution to a primal problem and **y** is a feasible solution to the dual problem, then $p = \mathbf{c}^\mathsf{T}\mathbf{x} \leq \mathbf{b}^\mathsf{T}\mathbf{y} = d$. If $p = \mathbf{c}^\mathsf{T}\mathbf{x} = \mathbf{b}^\mathsf{T}\mathbf{y} = d$, then **x** and **y** are optimal solutions to the primal and dual problems, respectively.

EXAMPLE 1

Duality, a Numerical Example

Refer to Figure 9.11. Consider the following primal standard maximization problem:

$$\textbf{MAX}\quad p = 4x_1 + 3x_2 \qquad \textbf{ST}\quad \begin{matrix} -x_1 + x_2 \leq 3 \\ 4x_1 - x_2 \leq 1 \end{matrix}$$

[10] Solved using LINDO® software.

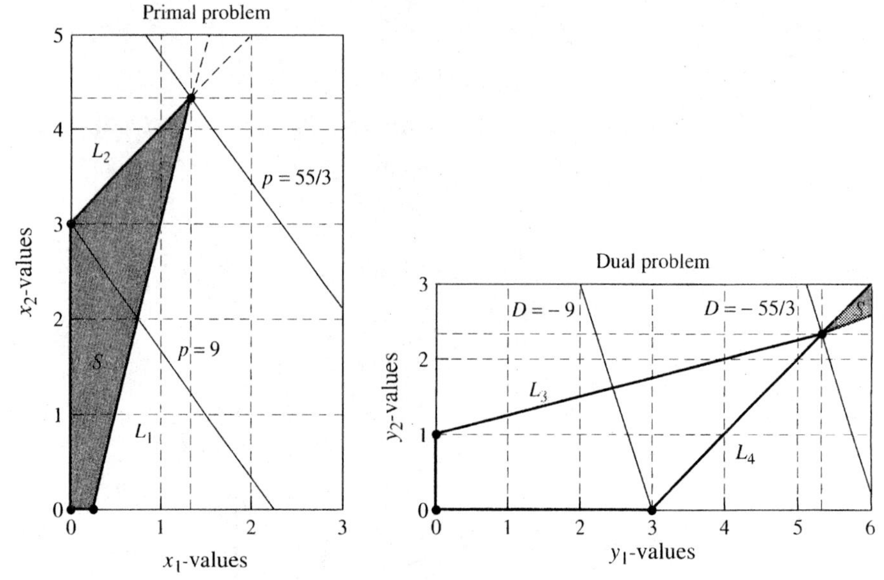

Figure 9.11 Feasible sets for primal and dual problems.

with x_1, x_2 nonnegative. The dual problem is

$$\textbf{MIN}\ d = 3y_1 + y_2 \qquad \textbf{ST} \qquad \begin{aligned} -y_1 + 4y_2 &\geq 4 \\ y_1 - \ y_2 &\geq 3 \end{aligned}$$

with y_1, y_2 nonnegative. To solve the primal problem, introduce nonnegative slack variables u_1, u_2 and rewrite the constraints and objective function as follows:

$$\begin{aligned} -x_1 + \ x_2 + \ u_1 \qquad\quad &= 3 \\ 4x_1 - \ x_2 \qquad + \ u_2 &= 1 \\ p - 4x_1 - 3x_3 + 0u_1 + 0u_2 &= 0. \end{aligned}$$

Applying the simplex algorithm, we have

Initial Tableau

p	x_1	x_2	u_1	u_2	b
	−1	1	1	0	3
	4	−1	0	1	1
1	−4	−3	0	0	0

⇒

Final Tableau

p	x_1	x_2	u_1	u_2	b
	0	1	$\frac{4}{3}$	$\frac{1}{3}$	$\frac{13}{3}$
	1	0	$\frac{1}{3}$	$\frac{1}{3}$	$\frac{4}{3}$
1	0	0	$\frac{16}{3}$	$\frac{7}{3}$	$\frac{55}{3}$

which gives the maximum $p = \frac{55}{3}$ when $\mathbf{x} = (x_1, x_2) = (\frac{4}{3}, \frac{13}{3})$.

The dual problem is first transformed into a standard maximization problem with the same solution, namely

$$\textbf{MAX}\ D = -d = -3y_1 - y_2 \qquad \textbf{ST} \qquad \begin{aligned} y_1 - 4y_2 &\leq -4 \\ -y_1 + \ y_2 &\leq -3 \end{aligned}$$

with y_1, y_2 nonnegative. Multiplying each constraint by -1, introducing two nonnegative slack variables u_1, u_2 and two nonnegative artificial variables a_1, a_2, the constraint equations and objective function become

$$\begin{aligned} -y_1 + 4y_2 - u_1 \qquad + a_1 \qquad &= 4 \\ y_1 - y_2 \qquad - u_2 \qquad + a_2 &= 3 \end{aligned}$$

$$D = -3y_1 - y_2 + 0u_1 + 0u_2 - Ma_1 - Ma_2,$$

where M is a sufficiently large positive constant. Choosing $M = 10$, for example, and applying the simplex algorithm, we have

Initial Tableau								
D	y_1	y_2	u_1	u_2	a_1	a_2	b
	-1	4	-1	0	1	0	4
	1	-1	0	-1	0	1	3
1	3	1	0	0	10	10	0

Final Tableau							
D	y_1	y_2	u_1	u_2	a_1	a_2	b
	0	1	$-\frac{1}{3}$	$-\frac{1}{3}$	$\frac{1}{3}$	$\frac{1}{3}$	$\frac{7}{3}$
	1	0	$-\frac{1}{3}$	$-\frac{4}{3}$	$\frac{1}{3}$	$\frac{4}{3}$	$\frac{16}{3}$
1	0	0	$\frac{4}{3}$	$\frac{13}{3}$	$\frac{26}{3}$	$\frac{17}{3}$	$-\frac{55}{3}$

The maximum value $D = -d = -\frac{55}{3}$ gives the minimum value $d = \frac{55}{3}$ when $\mathbf{y} = (y_1, y_2) = (\frac{16}{3}, \frac{7}{3})$. ■

Computational Note

Suppose the coefficient matrix $\mathbf{A}$ of the system of constraints in a primal LP problem is $m \times n$. Then the cost of computing the solution is $C_p = km^2n$, where k is a constant. Because the coefficient matrix $\mathbf{A}^\mathsf{T}$ for the dual problem is $n \times m$, the cost of computing the dual solution is $C_d = ln^2m$, where l is a constant. Choosing the smaller cost C_p or C_d (deciding to use the primal or dual formulation of the problem) may reduce computing time in large-scale problems.

Historical Notes

John von Neumann (1903–1957) One of the most influential mathematicians and scientists of the 20th century. Von Neumann introduced the concept of duality for LP models, and his discussions with Dantzig in the mid-1940s led to his conjecture that LP theory is equivalent to the theory of two-person, zero-sum games. This conjecture was eventually proved by Dantzig, Gale, Kuhn, and Tucker. The equivalence is used to give the most elementary proof of Von Neumann's famous *Minimax Theorem.*

EXERCISES 9.3

1. Draw the feasible set S corresponding to the system of constraints

$$x_1 \geq 0, \quad x_2 \geq 0, \quad 2x_1 - 2x_2 \leq 3, \quad -x_1 + x_2 \leq 1$$

(a) Is S empty or nonempty, bounded, or unbounded?

(b) Show that the objective function $f = 2x_1 + 3x_2$ has no feasible optimal solution with respect to the given constraints.

(c) Using f in part (b), formulate the dual problem and comment on whether an optimal feasible solution exists.

2. Solve the primal LP problem

MAX

$$f = 3x_1 + 4x_2$$

ST

$$\begin{cases} x_1 + x_2 \leq 5 \\ x_2 = 1 \end{cases}$$

with x_1, x_2 nonnegative. Draw a diagram in $\mathbb{R}^2$. Formulate the dual and solve it. Draw a diagram in $\mathbb{R}^2$. State the connection between the two solutions.

3. Solve the primal LP problem

MAX

$$f = x_1 + x_2 + x_3$$

ST

$$\begin{cases} x_1 + 2x_2 + x_3 \leq 4 \\ x_1 + x_2 = 1 \end{cases}$$

with x_1, x_2, x_3 nonnegative. Draw a diagram of the feasible set S. Formulate the dual problem and solve it. Draw a diagram of the feasible set S. State the connection between the two solutions.

Exercises 4–7. Refer to Exercises 9.2. Formulate the dual problem in each case and use the simplex algorithm to minimize the objective function.

4. Problem 1

5. Problem 3

6. Problem 5

7. Problem 7

8. *Production*. A major perfume company makes three brands of perfume P_1, P_2, P_3 using two processes Q_1, Q_2 (in any order). The time required for each product to pass through each process and the expected net profit on each ounce of product sold are given in the following table.

Product	Q_1	Q_2	Net profit
P_1	2	1	\$ 0.5
P_2	2	3	\$ 0.6
P_3	3	4	\$ 0.7

The company wishes to maximize total net profit. Formulate the LP problem and solve it. A rival company is contemplating renting the processes Q_1 and Q_2 to make its own products. Formulate the dual problem using decision variables y_1 and y_2. Solve the dual problem. Explain your solution in terms of a program for the rival company.

USING MATLAB

9. Use **gauss.m** to carry out all the computations in Illustration 9.6.

10. Use **gauss.m** to carry out all the computations in Example 1.

9.4 Mixed Constraints

The version of the simplex algorithm which appears in Section 9.2 applies to standard maximization LP problems in which the system of constraints is given by $\mathbf{Ax} \leq \mathbf{b}$, where $\mathbf{0} \leq \mathbf{b}$. We now briefly address the case when $\mathbf{b}$ has some negative entries. Such cases may arise from systems of *mixed constraints* that contain inequalities involving both $\leq$ and $\geq$ signs.

Consider, for example, the following system of mixed constraints

$$\begin{aligned} x_1 - 6x_2 &\geq 3 \\ 2x_1 + x_2 &\leq 4 \\ -4x_1 + 7x_2 &\geq 8 \end{aligned} \quad .$$

Rearranging the system into the standard maximization form $\mathbf{Ax} \le \mathbf{b}$ results in a vector $\mathbf{b}$ that has some negative entries. Multiplying the first and third equations by -1, we have

$$\begin{aligned} -x_1 + 6x_2 &\le -3 \\ 2x_1 + x_2 &\le 4 \\ 4x_1 - 7x_2 &\le -8 \end{aligned} \qquad \mathbf{A} = \begin{bmatrix} -1 & 6 \\ 2 & 1 \\ 4 & -7 \end{bmatrix}, \quad \mathbf{b} = \begin{bmatrix} -3 \\ 4 \\ -8 \end{bmatrix}.$$

Certain cases in which the vector $\mathbf{b}$ has at least one negative entry can be handled by modifying Step 2 in the simplex algorithm. The modification may not yield a solution in every case.

New Step 2

Replace Step 2 in the simplex algorithm (see page 422) with the following process.

Let k, $1 \le k \le m$, be the largest integer with $b_k < 0$ (b_k is the last negative entry in the column $\mathbf{b}$). Choose a negative entry $a_{kj} < 0$ in column j of row k (a_{kj} exists because the left side of constraint k is negative). Column j is the pivot column for this step. Compute the quotient b_k/a_{kj} and the quotient b_p/a_{pj} for each positive entry (if any) a_{pj}, $k < p \le m$. Let b_i/a_{ij} be the smallest among these quotients. Then a_{ij} is the pivot for this step which is used to clear column j in the usual way.

■ ILLUSTRATION 9.7

Mixed Constraints, Modified Simplex Algorithm

Our goal is

MAX

$$f = 2x_1 - 4x_2 + 3x_3 \tag{9.39}$$

ST

$$\begin{aligned} x_1 + 2x_2 - 3x_3 &\ge 3 \quad E_1 \\ 2x_1 - x_2 + x_3 &\ge -10 \quad E_2 \\ x_1 - x_2 + 2x_3 &\le 8 \quad E_3 \end{aligned} \tag{9.40}$$

with x_1, x_2, x_3 nonnegative. Multiply E_1 and E_2 throughout by -1 to form a uniform system of constraints $\mathbf{Ax} \le \mathbf{b}$, where

$$\mathbf{A} = \begin{bmatrix} -1 & -2 & 3 \\ -2 & 1 & -1 \\ 1 & -1 & 2 \end{bmatrix}, \quad \mathbf{x} = \begin{bmatrix} x_1 \\ x_2 \\ x_3 \end{bmatrix}, \quad \mathbf{b} = \begin{bmatrix} -3 \\ 10 \\ 8 \end{bmatrix}.$$

Note that not all entries in $\mathbf{b}$ are positive and so the modified version of the simplex algorithm is applicable. Introduce slack variables s_1, s_2, s_3 and arrive at the initial tableau.

Initial Tableau

f	x_1	x_2	x_3	u_1	u_2	u_3	b	Basis	Pivot Row (*)
	(−1)	−2	3	1	0	0	−3	u_1	$(-3)\div(-1)=3^*$
	−2	1	−1	0	1	0	10	u_2	
	1	−1	2	0	0	1	8	u_3	$8\div 1=8$
1	-2^*	4	−3	0	0	0	0		

Applying the new step 2 in the simplex algorithm, we have $k=1$ and $b_1=-3$. Choose $a_{11}=-1<0$ and so $j=1$. The required quotients are $b_1/a_{11}=3$ and $b_3/a_{31}=8$. The smallest quotient gives pivot $a_{11}=-1$. Clearing column 1, the EROs are

$$-\mathbf{r}_1\to\mathbf{r}_1,\quad \mathbf{r}_2+2\mathbf{r}_1\to\mathbf{r}_2,\quad \mathbf{r}_3-\mathbf{r}_1\to\mathbf{r}_3,\quad \mathbf{r}_4-2\mathbf{r}_1\to\mathbf{r}_4.$$

Second Tableau

f	x_1	x_2	x_3	u_1	u_2	u_3	b	Basis	Pivot Row (*)
	1	2	−3	−1	0	0	3	x_1	
	0	5	−7	−2	1	0	16	u_2	
	0	−3	(5)	1	0	1	5	u_3	$5\div 5=1^*$
1	0	8	-9^*	2	0	0	6		

There are no negative entries in column b of the second tableau, and so we use the usual Step 2. The largest negative entry in the objective row (the indicator) is -9, and so column 3 becomes the pivot column for this step. The only positive entry in column 3 above the objective row is 5, which becomes the pivot. Clearing column 3, the EROs are

$$\frac{1}{5}\mathbf{r}_3\to\mathbf{r}_3,\quad \mathbf{r}_1+3\mathbf{r}_3\to\mathbf{r}_1,\quad \mathbf{r}_2+7\mathbf{r}_3\to\mathbf{r}_2,\quad \mathbf{r}_4-9\mathbf{r}_3\to\mathbf{r}_4.$$

Third Tableau

f	x_1	x_2	x_3	u_1	u_2	u_3	b	Basis	Pivot Row (*)
	1	$\frac{1}{5}$	0	$-\frac{2}{5}$	0	$\frac{3}{5}$	6	x_1	
	0	$\frac{4}{5}$	0	$-\frac{3}{5}$	1	$\frac{7}{5}$	23	u_2	
	0	$-\frac{3}{5}$	1	($\frac{1}{5}$)	0	$\frac{1}{5}$	1	x_3	$1\div\frac{1}{5}=5^*$
1	0	$\frac{13}{5}$	0	$-\frac{1}{5}^*$	0	$\frac{9}{5}$	15		

The only negative entry in the objective row $-\frac{1}{5}$, which defines Column 4 as the pivot column and $\frac{1}{5}$ as the pivot, being the only positive entry in the pivot column above the indicator. Clearing column 4, the EROs are

$$5\mathbf{r}_3 \to \mathbf{r}_3, \quad \mathbf{r}_1 + \frac{2}{3}\mathbf{r}_3 \to \mathbf{r}_1, \quad \mathbf{r}_2 + \frac{3}{5}\mathbf{r}_3 \to \mathbf{r}_2, \quad \mathbf{r}_4 - \frac{1}{5}\mathbf{r}_3 \to \mathbf{r}_4.$$

Fourth Tableau

f	x_1	x_2	x_3	u_1	u_2	u_3	b	Basis	Pivot Row (*)
	1	−1	2	0	0	1	8	x_1	
	0	−1	3	0	1	2	26	u_2	
	0	−3	5	1	0	1	5	x_3	
1	0	2	1	0	0	2	16		

The objective row in the Fourth Tableau corresponds to

$$f + 2x_2 + x_3 + 2u_3 = 16 \quad \text{or} \quad f = 16 - 2x_2 - x_3 - 2u_3,$$

showing that $f = 16$ is the maximum for f, attained when $x_1 = 8$, $x_2 = 0$, $x_3 = 0$, $u_1 = 5$, $u_2 = 26$, $u_3 = 0$. ■

The modification to Step 2 breaks down in row k when $b_k < 0$ and there is no corresponding negative entry $a_{kj} < 0$, $1 \le j \le n$. No feasible solution exists because the inequality corresponding to row k has positive left side and negative right side. For example, the constraints $0 \le x_1, 0 \le x_2, 2x_1 + x_2 \le -5$ are inconsistent. Any problem containing such inconsistencies has no solution.

EXERCISES 9.4

Exercises 1–4. Use the (modified) simplex algorithm to determine the maximum (if it exists) of the objective function f subject to the given constraints.

1. $f = 3x_1 - 2x_2, \quad x_1 \ge 0, \quad x_2 \ge 0, \quad \begin{cases} x_1 + x_2 \le 9 \\ x_1 - x_2 \le 5 \\ x_1 + 2x_2 \ge 4 \end{cases}$

2. $f = x_1 + 2x_2 - x_3, \; x_1 \ge 0, \; x_2 \ge 0, \; x_3 \ge 0,$

$$\begin{cases} -x_1 + x_2 + x_3 \ge 1 \\ x_1 - x_2 + x_3 \le 1 \\ -x_1 - x_2 + x_3 \le 1 \\ x_1 + x_2 + x_3 \le 5 \end{cases}$$

3. $f = 3x_1 + x_2 + 2x_3, \quad x_1 \ge 0, \; x_2 \ge 0, \; x_3 \ge 0,$

$$\begin{cases} 2x_1 + 2x_2 + 3x_3 \le 14 \\ -2x_1 + x_2 - x_3 \ge 1 \\ 3x_1 + x_2 + 2x_3 \le 12 \\ 3x_1 + x_2 + x_3 \le 9 \end{cases}$$

4. $f = x_1 + x_2 + x_3, \; x_1 \ge 0, \; x_2 \ge 0, \; x_3 \ge 0,$

$$\begin{cases} x_1 + x_2 + x_3 \le 1 \\ x_1 + x_2 \ge 0.5 \\ x_2 + x_3 \ge 0.5 \end{cases}$$

USING MATLAB

5. Use the M-file **gauss.m** to carry out the row operations in Exercises 1–4.

6. Refer to Exercises 1–4. Write an M-file to draw each region of feasibility. Use the **patch** command to color the region red. Draw the objective function for its attained maximum value.

CHAPTER 9 REVIEW

State whether each statement is true or false as stated. Provide a clear reason for your answer. If a false statement can be modified to make it true, then do so.

1. Any LP problem can be written as a standard maximization problem.

2. Some LP problems cannot be written as standard minimization problems.
3. A feasible set for an LP problem in $\mathbb{R}^2$ is a polygon with at most four sides.
4. An isovalue is an equation formed by assigning a constant value to the objective function.
5. Every constraint equation in an LP program can be replaced by a pair of inequalities.
6. The feasible set in an LP program is the domain of the objective function.
7. The objective function has no minimum value when the feasible set is unbounded.
8. If a region in $\mathbb{R}^2$ is bounded, then it is convex.
9. If a region in $\mathbb{R}^2$ is convex, then it is bounded.
10. Assuming x_1, x_2 nonnegative, some objective function can have a maximum value when the feasible set in $\mathbb{R}^2$ is unbounded.
11. Every extreme point in $\mathbb{R}^2$ lies at the intersection of a pair of lines.
12. If a feasible set in $\mathbb{R}^2$ has extreme points $(0, 0)$, $(0, 1)$, $(2, 0)$, then the objective function $f = x_1 + (2 - \epsilon)x_2$, where $0 < \epsilon$, attains a maximum at the optimal feasible solution $(0, 1)$.
13. An optimal feasible solution is an optimal value for the objective function f.
14. Slack and artificial variables are not decision variables.
15. The LP problem **MAX** $x_1 + x_2$, **ST** $x_1 - x_2 \leq -1$, $-x_1 + x_2 \leq 0$, with x_1, x_2 nonnegative, has no feasible solution.
16. In any tableau, the indicator is the largest value in the objective row.
17. If an LP problem has n decision variables and m slack or artificial variables, then at each extreme point the values of the m basic variables can be found by solving a linear system.
18. If the primal problem has n decision variables, then the dual problem has n decision variables.
19. Any system of mixed constraints can be modified so that the problem becomes a standard minimization problem.
20. The simplex algorithm can be applied to solve any LP problem with mixed constraints.

Appendix A

MATLAB®

MATLAB[1] is a technical computing environment whose basic data structure is an array that does not require resizing—the array can grow or shrink to accommodate the data. MATLAB provides core mathematics and advanced graphical tools for data analysis, visualization, and other application development. Calculations and graphics are implemented through either *interactive computation* or *M-file programming*.

Demos

A tour of MATLAB is obtained by following the path Help → Demos in the MATLAB Command Window. This tour gives a brief overview of some of the software features, including how to enter matrices, and commands associated with basic matrix operations.

Interactive Computation

Commands known to MATLAB are entered at the prompt:

```
>>
```

Online help is obtained by typing **help** *topic* at the prompt, where *topic* is a command name known to MATLAB. For example, **help size** prints documentation on the M-file **size.m** that is used to determine the size of a matrix. The documentation also may indicate related commands, in this case **length**, and **ndims**.

MATLAB interprets and evaluates input at the prompt and will normally print a response unless the response is suppressed by using a semicolon after input.

[1] The name is an acronym for **MAT**trix **LAB**oratory.

The MATLAB Workspace

All keyboard input (variables and their current values, commands, symbolic expressions, and so on) is stored in the MATLAB *workspace*. For example, the 2×2 matrix **A** shown is entered into the workspace using either the command

```
A = [1 2; -3 4]   or   A = [1, 2; -3, 4]
```

where matrix entries are separated by at least one space or by commas, and a semicolon ends each row. The response is

```
A      =
            1     2
           -3     4
```

Individual matrix entries can be changed after the matrix has been created. For example, the command $\mathbf{A}(1,2) = 9.2$ changes the $(1,2)$-entry of **A** from 2 to 9.2. The insertion of a floating point number into the array causes the display mode to change from integers to **format short** mode (default), which displays four places of decimals. The response is

```
A =        1.0000     9.2000
          -3.0000     4.0000
```

There are various formats for numerical data—consult **help format**.

Submatrices

Consider a 3×3 matrix **A**, which is entered as follows:

```
A = [1, 2, 3; 4, 5, 6; 7, 8, 9]

  A =     1     2     3
          4     5     6
          7     8     9
```

The special character colon (:) denotes all rows (columns). Thus $\mathbf{B} = \mathbf{A}(:, 1:2)$ defines a submatrix **B** of **A** consisting of entries from all rows in **A** and only columns 1 and 2 (deleting the third column). The response is

```
B =     1     2
        4     5
        7     8
```

If the row vector [1 2 3] refers to the rows (columns) of **A** in natural order, then [3 1 2] refers to the rows (columns) of **A** in permuted order. The vector [2 1] refers to rows (columns) 1 and 2 in **A** reversed. Using vectors, a new array (submatrix) can be defined using the entries in **A** in any desired pattern. For example, the command **B** = **A**([3 1 2], [2 1]) gives the response

```
B =     8     7
        2     1     [Rows of A permuted, columns 1 and 2 reversed]
        5     4
```

Note that the identity matrix of order n is generated using the command **eye**(n) and a zero $m \times n$ matrix using the command **zeros**(m,n). The command **ones**(m,n) generates an $m \times n$ matrix with all entries 1.

The symbol [ ] denotes the empty matrix (no entries) and acts as a matrix concatenation operator. By this we mean that if **x** and **y** are matrices of the same size, then **z** = [**x**, **y**] (respectively, **z** = [**x**; **y**]) concatenates the rows (respectively, columns) of **x** to the rows (respectively, columns) of **y**. Note that the entries in a row or column vector are indexed by the positive integers $1, 2, 3, \ldots$ (not beginning with 0).

Saving Data

Commands and data in the workspace will normally be lost when the current session is ended. However, the command **diary** asks MATLAB to open a file called **diary** (a reserved name) and begin echoing all keyboard input and response to this file. The command **diary off** will close the file, and **diary on** will open the file again and append new data. As an alternative, use the command **diary** *filename* (your choice), with or without an extension. Diary files can be edited, printed, and incorporated into reports or other documents.

To check which variables are currently in the workspace and their properties, use the command **who**, or for more details, **whos**. Consult **help which** for related commands.

The command **save** *filename* (without extension) saves all variable names and their current values to the file *filename.mat*, adding the default extension *.mat*. The command **save** (without file name) saves the workspace data to a special file **matlab.mat**. Note that **save** *filename* **x p r** saves specific variables **x, p, r** and their current values. The command **load** *filename.mat* retrieves the saved data into the current workspace. The command **clear** deletes all variables and their values from the current workspace—**clear x y** deletes only the variables **x**, **y** and their values.

Smart Recall: Arrow Keys

All keyboard input from the current session is stored in a MATLAB buffer. The Up-Arrow ↑ and Down-Arrow ↓ keys are extremely useful for smart recall of previous commands. These keys scroll (at the MATLAB prompt) through the buffer and save much time by giving you the option of resubmitting or editing and submitting a previous command. To edit the command line, use the Right-Arrow → and the Left-Arrow ← keys, and the Home and End keys, which take you, respectively, to the beginning and end of the command line. Typing only the first few characters of a command followed by use of the up or down arrow keys displays only those commands starting with those characters.

M-File Programming

Interactive computing is the most direct use of the software. An alternative approach is to create a file containing MATLAB language code—this (program) is called an *M-file* because it always has extension **.m**. The name of an M-file (without extension) is a command that can be entered at the prompt in order to execute the contents of the file. Many (but not all) built-in commands of MATLAB are M-files. Creating M-files is a good way to save your work, automate tasks, share ideas with colleagues, and develop more complex algorithms.

■ ILLUSTRATION A.1

Solving Linear Systems, Reduced Form, Partial Pivoting

The small-scale linear system

$$(S)\quad \begin{cases} 2x_1 + x_2 - x_3 = 1 \\ 3x_1 - 2x_2 + 2x_3 = -3 \\ 5x_1 + x_2 - x_3 = 2 \end{cases}$$

will be used to illustrate the creation and use of an M-file. However, in practice, imagine that (S) is a large-scale system that needs to be updated and solved on a regular basis so that interactive computation is an inefficient option.

Open a new file using the path File → New → M-file and save the file under the name **myfirst** (the extension **.m** is added automatically). Enter the following program and documentation. Note that the symbol % is the MATLAB comment delimiter (all characters after it on the line will be ignored), which is used for adding comments and documentation.

```
% File name: myfirst.m
% Augmented matrix M of a 3 x 3 linear system
% Computes the reduced form of M and the step-by-step reduction of M
% using partial pivoting

M = [2, 1, -1, 1; . . .     % Where . . . means continue input on next line
3, -2, 2 -3; . . .          % Continue input on next line
5, 1, -1, 2];               % The semicolon suppresses printed response
M                           % Display the matrix
Mstar=rref(M)               % Compute the reduced echelon form Mstar of M
pause;                      % Pause before running the movie.
                            % Hit any key to continue
rrefmovie(M)                % Visualization: step-by-step reduction of M
```

Once the file is saved, you can run it in the Command Window using the command **myfirst**. Note that any information or comments written at the beginning of an M-file can be displayed in the Command Window using the command **help myfirst**.

The columns of **M** can be "pulled out" of M using the following command (many commands can be put on one line, separated by commas)

$$\mathbf{u1} = \mathbf{M}(:,1)\ ,\quad \mathbf{u2} = \mathbf{M}(:,2)\ ,\quad \mathbf{u3} = \mathbf{M}(:,3)\ ,\quad \mathbf{b} = \mathbf{M}(:,4)$$

The coefficient matrix of (S) is defined by the command $\mathbf{A} = [\mathbf{u1}, \mathbf{u2}, \mathbf{u3}]$. The commands **rank(A)** and **rank(M)** return values of 2 and 3, respectively, showing that (S) is inconsistent. The linear dependence of the columns of **A** is seen from the response to **rref (A)** or alternatively, the command **det(A)** returns a value 0. Thus **A** is singular—an attempt to take the inverse using the command **inv(A)** does not result in an error, but with a response (matrix) together with a warning that **A** is close to singular or badly scaled. ■

The M-file **myfirst** in Illustration A.1 is called a *script* file because it executes, line by line, a sequence of commands, requires no input arguments and returns no output arguments. Another type of M-file is described next.

Functions

An M-file that can accept input arguments and return output arguments is called a *function* file. MATLAB has its own set of built-in functions that take one or more parameters. The function **rand**(m,n), for example, generates a random $m \times n$ matrix with entries that are numbers (probabilities) r with $0 \leq r \leq 1$. For example, **r** = **rand**(1) returns a single random number r, while the command **P** = **rand**(2), or **rand**(2,2), generates a 2×2 probability matrix (random array).

A random matrix generated by the **rand** command is not column stochastic (Section 6.4) because each column sum is not generally 1. A stochastic matrix can be formed by scaling its columns. The M-file shown next, called **stochos.m**, defines a *function* that takes a parameter. Note that the coding conventions used for function M-files are as follows:

- The function line is the first line.
- The first comment line is a short summary of what the function does.
- The rest of the comment block follows the summary line.

```
function y = stochos(x)
% STOCHOS Computes a stochastic matrix
% Parameter x is an m x n matrix.
% y is x with normalized columns
[m , n] = size(x)                 % m = row dim, n = col dim
s = sum(x)                        % vector of column sums
for j = 1:n                       % for each column
y(:,j) = x(:,j)/s(j)              % scale jth column
end                               % end for
```

The functions **size** and **sum** are known to MATLAB. The first returns the size of the $m \times n$ matrix and **sum(x)** returns a row vector whose entries are the column sums for the matrix **x**. All other variables in the function declaration are local. For example, normalizing the columns of **P** shown, we have

P = [0.9347, 0.5194; 0.3835, 0.8310]
Q = **stochos(P)**

```
Q =
        0.7091      0.3846
        0.2909      0.6154
```

Comment lines are useful in finding files and printing M-file documentation. Consult **help lookfor**.

Characteristic Polynomial, Cayley–Hamilton

Refer to Section 6.1. Consider the matrix **A** = [2, 1; 3, −4]. The command [**E**, **D**] = **eig**(**A**) gives the response

```
E =
        0.9071      -0.1529
        0.4210       0.9882
D=
        2.4641            0
             0      -4.4641
```

The columns of **E** are eigenvectors of **A** and the corresponding eigenvalues of **A** lie on the main diagonal of the diagonal matrix **D**.

The command **p** = **poly**(**A**) returns the row vector **p** = [1.0000 2.0000 − 11.0000] corresponding to the characteristic polynomial $p(\lambda) = \lambda^2 + 2\lambda - 11$ of **A**. The command **nearzero** = **polyvalm**(**p**, **A**) evaluates $p(\lambda)$ with λ replaced by **A** and the response is

```
nearzero= 1.0e−014 *
            0.1776          0
                 0     0.1776
```

The Cayley–Hamilton theorem tells us that the 2 × 2 matrix **nearzero** should actually be the zero matrix **zeros**(2), the error occurring due to finite-precision machine calculations.

Arithmetic Operators

The arithmetic operators for matrix (array) algebra are

+	addition	−	subtraction	*	multiplication
\	left division	/	right division	^	power

The special symbol ′ (prime) is used to take the transpose of a matrix.

Caution! If **C** is a complex matrix, then **C**′ is the Hermitian transpose $\mathbf{C}^H$ (Section 8.5). For an unconjugated transpose, use **C**.′ (with a dot).

Matrix Products

The matrix product **A** ∗ **B** is defined whenever **A** and **B** are conformable for multiplication. A distinction between array operations and element-by-element operations must be made. The operations + and − are both array and element operations. MATLAB uses a period (.) in conjunction with other operators to perform element-by-element operations. We will give a number of examples using the row vectors

```
x = [2, 3]; y = [5, 6];
```

For multiplication, we have

```
z = x. * y        [* acts on corresponding entries]
  z =
          10       18
```

For powers, we have

```
w = x.^y        [entries in x to powers of entries in y]
  z =
          32       729
z = x.^3         [scalar power]
  z =
          8       27
z = 2.^[x, y]       [a vector of powers of 2, where [x, y] = [2 3 5 6]]
  z =
          4       8       32       64
```

If **u** and **v** are row vectors, each with n components, then **dot(u,v)** returns the dot product of **u** and **v**. The equivalent command is **w** = **u** * **v** ′.

An understanding of how left and right division works for elements will help when these operations are applied to arrays.

```
z = x.\y        [x divided into y, y divided by x]
  z =
          2.5000       2.0000
z = x./y        [x divided by y, y divided into x ]
  z =
          0.4000       0.5000
```

Note that MATLAB has both left division **A\B** and right division **B/A** for matrices. Think of **B** as being fixed and **A** as being divided into **B** from the left and from the right.

Suppose **A** is an invertible square matrix. Then left division formally corresponds to **inv**(**A**) ∗ **B**, whereas right division corresponds to **B** ∗ **inv**(**A**). Note that MATLAB performs left and right division without computing **inv**(**A**).

Matrix Equations

Suppose **A** is an $m \times n$ matrix.

The command **x** = **A\b** solves **A** ∗ **x** = **b**, where **x** is $n \times 1$ and **b** is $m \times 1$.

The command **x** = **A/b** solves **x** ∗ **A** = **b**, where **x** is $1 \times m$ and **b** is $1 \times n$.

The linear systems defined by $\mathbf{A} * \mathbf{x} = \mathbf{b}$ and $\mathbf{x} * \mathbf{A} = \mathbf{b}$ need not be consistent because MATLAB finds the closest solution in least squares (Section 4.4). Note that left division $\mathbf{A}\backslash\mathbf{B}$ is defined whenever **A** and **B** have the same number of rows, with **A** not necessarily square. Right division is then defined by $\mathbf{B}/\mathbf{A} = (\mathbf{A}'\backslash\mathbf{B}')'$.

Complex Numbers

Complex matrices (entries are complex numbers) can be entered using the command

```
C = [ 1 + 5i,  2 + 6i;  3 + 7i  4 + 8i]

C =       1.0000      +5.0000i      2.0000      +6.0000i
          3.0000      +7.0000i      4.0000      +8.0000i
```

Caution! Use the format $1 + 5i$ to enter complex numbers, without spaces between 5 and i (5*i is optional).

Alternatively, enter real and imaginary parts separately **Re** = [1 2; 3 4]; and **Im** = [5 6; 7 8]; and combine these using the command **C** = **Re** + **Im**$*i$.

Generating Arrays

Consult help for the commands **colon** and **fix**. The special character : (colon) can be used to generate vectors of any length, with varying equal increments between entries. The command **x** = a:b:c is the same as $[a,\ a+b,\ \ldots\ ,a+k*b]$, where $k =$ **fix**$((c-a)/b$. For example,

```
x = 2 : pi/3 : 7   [increment of π/3]

  x =
          2.0000      3.0472      4.0944      5.1416      6.1888
```

and the command **fix**((7 − 2)/(pi/3)) returns the value 4 (= k).

Alternatively, the command **linspace**(a, b, n) generates a vector with $n-1$ equally spaced points between a and b. For example,

```
linspace(1, 5, 4)      [response is ans = because no variable name is
                                 given]

  ans =
            1.0000      2.3333      3.6667      5.0000
```

Tables of independent and dependent values for a given function are easily produced. Choosing some values for x, we can compute corresponding values y given by the function $y = e^x \sin(x) + e^{-x}\cos(x)$ as follows.

```
x = [0 : 0.2 : 0.6]';     [define a vector of x-values, transposed]
y = exp(x). * sin(x) + exp(-x). * cos(x);
[x, y]  % Concatenation
  ans =
            0           1.0000
            0.2000      1.0451
            0.4000      1.1983
            0.6000      1.4818
```

Graphics

MATLAB provides excellent graphics capabilities for presenting and visualizing data and plotting graphs in two and three dimensions. The most basic commands for 2-D graphics are

plot	creates a plot of vectors or columns of matrices
title	adds a title to the graph
xlabel	adds label for x-axis
ylabel	adds label to y-axis
text	displays a text string at specified location

Consult **help subplot** and refer to Figure A.1.

The M-file shown next creates a figure with two subplots, labeled Plot 1 and Plot 2.

```
clf;                % Clear figure window
x=[1,3,8,5,1];      % Set of discrete points for x
subplot(1,2,1)      % First subplot
plot(x)             % Plot x against index
title('Plot 1')     % Title
hold on             % Hold axes for second plot
grid on             % Grid lines
subplot(1,2,2)      % Second subplot
y=[1,2,4,0,1];      % Set of discrete points for y
plot(x,y)           % Plot vector x against y
fill(x,y,'r')       % Fill closed polygon
grid on             % Grid lines
title('Plot 2')     % Title
```

For any vector **x**, the command **plot**(**x**) produces a linear graph of the entries in **x** plotted against the *index* (row position) of the entries in **x**. As mentioned previously, indexing begins with 1. The vector **x** defined draws a sequence of line segments between points in the plane, from (1, 1) to (2, 3), to (3, 8) to (4, 5),

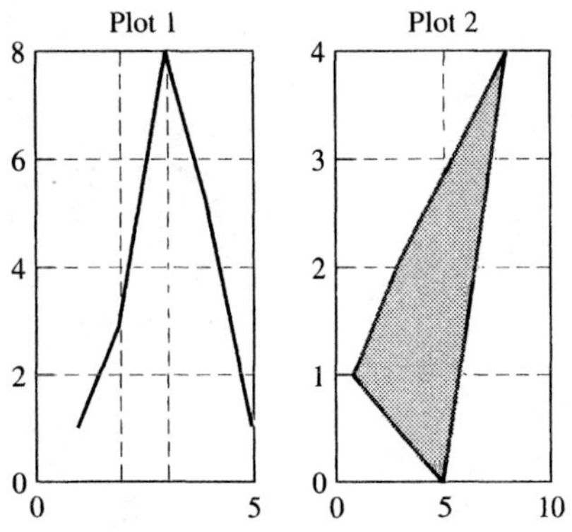

Figure A.1 Two subplots.

to (5, 1). The graph is called *piecewise linear*. The graph is **subplot**(1,2,1) in the figure.

If two vectors **x** and **y** of the same length n are specified, then **plot**(**x**, **y**) draws line segments between the successive points (x_i, y_i), $1 \leq i \leq n$. The command **fill(x,y,**'r') fills the closed polygon and colors it red.

Refer to Figure A.2. The next M-file plots three polynomials of best fit through a set of given data points. The polynomial of degree 5 is an exact fit. In order to draw a smooth curve, we must generate a vector of x-values with small increments, as shown on line 6 of the M-file. The line characteristics are specified by the third argument in the plot command: '-' is solid, '- -' is broken, and so on.

```
x=[-2 0 2 5 7 9];                          % Define x-data
y=[5 3 -3 1.5 4.5 4];                      % Define y-data
p3=polyfit(x,y,3);                         % Define degree 3 polynomial
p4=polyfit(x,y,4);                         % Define degree 4 polynomial
p5=polyfit(x,y,5);                         % Define degree 5 polynomial
xval=-2.5:0.1:9.5;                         % Range of x-values
p3plot=polyval(p3,xval);                   % Evaluate p3 at x-values
p4plot=polyval(p4,xval);                   % Evaluate p4 at x-values
p5plot=polyval(p5,xval);                   % Evaluate p5 at x-values
plot(xval,p3plot,'--', ...                 % Plot
     xval,p4plot,'--', ...                 % the three curves
     xval,p5plot,'-',x,y,'o');             % and six data points
grid on                                    % Grid lines
xlabel('x-values')                         % Label x-axis
ylabel('y-values')                         % Label y-axis
title('Approximation in least squares')    % Title
```

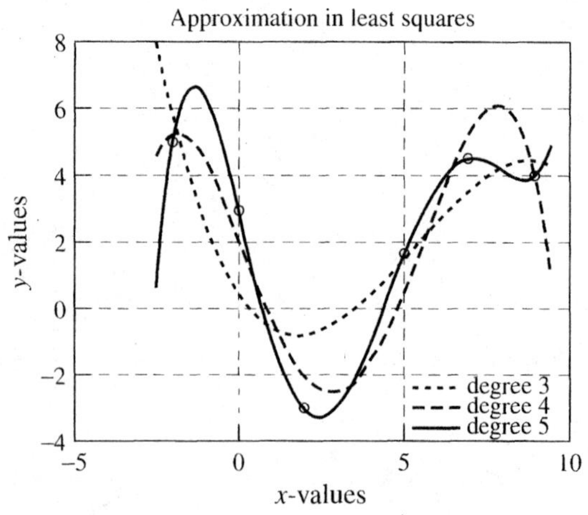

Figure A.2 Least squares.

Refer to Figure A.3. The code shown next is an M-file that is used to plot the complex sixth roots of unity (Section 8.4).

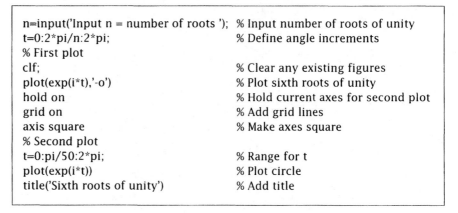

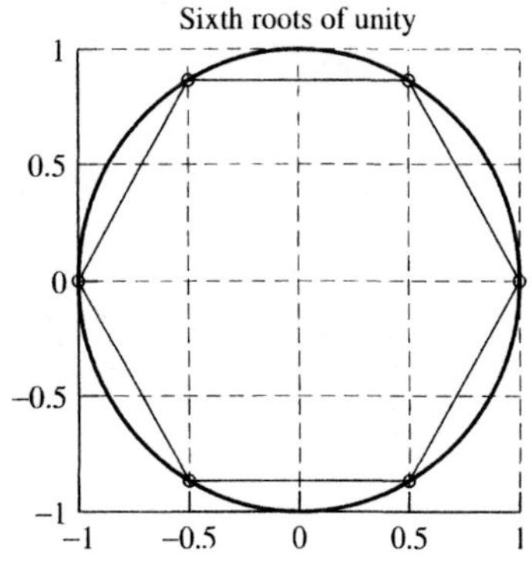

Figure A.3 Roots of unity.

```
n=input('Input n = number of roots ');  % Input number of roots of unity
t=0:2*pi/n:2*pi;                        % Define angle increments
% First plot
clf;                                    % Clear any existing figures
plot(exp(i*t),'-o')                     % Plot sixth roots of unity
hold on                                 % Hold current axes for second plot
grid on                                 % Add grid lines
axis square                             % Make axes square
% Second plot
t=0:pi/50:2*pi;                         % Range for t
plot(exp(i*t))                          % Plot circle
title('Sixth roots of unity')           % Add title
```

The Core Software

MATLAB has more than 600 mathematical, statistical, and engineering built-in functions that can be used to write customized programs to solve specific problems. The software is extended by a variety of other packages, such as Simulink (a simulation and prototyping environment for modeling, simulating, and analyzing dynamical systems), Stateflow (a graphical simulation environment), Blocksets, and Core Generation Tools. A *toolbox* is a collection of M-files that are written with a specific application in mind. A large set of application-specific toolboxes are available that extend the software environment into areas such as automotive data analysis and design, communications systems design, image processing, optimization, signal processing, control systems design, dynamic systems simulation, neural networks, finance, so on. In particular, the Symbolic/Extended Math Toolbox performs symbolic math computations.

REFERENCES

1. Pärt-Enander, et al., *The MATLAB Handbook*, Addison-Wesley, 1996.

Appendix B

TOOLBOX

Methods of Proof

If **(P)** and **(Q)** are statements, then

$$(\mathbf{P}) \Rightarrow (\mathbf{Q}) \tag{B.1}$$

is called a *logical implication*, which means that the truth of statement **(Q)** follows logically from the truth of statement **(P)**. In mathematics, we establish the truth of (B.1) in one of three ways.

I Direct Proof

We say that the implication (B.1) is proved *directly* if we assume the truth of **(P)** and apply correct mathematical reasoning to arrive at statement **(Q)**.

Theorem B.1 **Illustration of Direct Proof**

Let $\mathbf{A}$ *be an* $m \times m$ *matrix and* $\mathbf{B}$ *be an* $m \times n$ *matrix. If* $\mathbf{O}$ *is the* $m \times n$ *zero matrix, then*

$$(\mathbf{P}): \quad \mathbf{AB} = \mathbf{O} \textit{ and } \mathbf{A} \textit{ is invertible} \quad \Rightarrow \quad (\mathbf{Q}): \quad \mathbf{B} = \mathbf{O}. \tag{B.2}$$

Proof Assuming **(P)** is true we can premultiply $\mathbf{AB} = \mathbf{O}$ by $\mathbf{A}^{-1}$ and use associativity of matrix multiplication to obtain

$$\mathbf{A}^{-1}(\mathbf{AB}) = \mathbf{A}^{-1}\mathbf{O} \quad \Rightarrow \quad (\mathbf{AA}^{-1})\mathbf{B} = \mathbf{O} \quad \Rightarrow \quad \mathbf{I}_m\mathbf{B} = \mathbf{O},$$

where $\mathbf{I}_m$ is the identity matrix of order m. Then $\mathbf{I}_m\mathbf{B} = \mathbf{B}$ and so $\mathbf{B} = \mathbf{O}$.

II The Contrapositive

The *negation* of a statement (**P**) is its logical opposite, denoted by (not **P**). The implication (B.1) is *logically equivalent* to the implication

$$(\text{not } \mathbf{Q}) \quad \Rightarrow \quad (\text{not } \mathbf{P}), \tag{B.3}$$

which is called the *contrapositive* (reverse negated implication) of (B.1). For example, the contrapositive of the statement: *If this is Monday, it can't be Italy* is: *If this is Italy, it can't be Monday*. In general, the truth of (B.3) implies the truth of (B.1), and conversely. Proving the contrapositive is sometimes easier than finding a direct proof.

Some care should taken in negating a statement. For example, the statement: **A** *is a zero matrix* means that *all* the entries in **A** are equal to zero. The statement: **A** *is not a zero matrix* means that *at least one* entry in **A** (but not necessarily all) is nonzero.

Theorem B.2 **Illustration of Proof Using the Contrapositive**

Let **A** *be an* $n \times n$ *matrix. If* $\det(\mathbf{A})$ *is nonzero, then the rows of* **A** *are linearly independent.*

Proof Define statements:

(**P**) : det(**A**) is nonzero
(**Q**) : The rows of **A** are linearly independent.

Assume (not **Q**) is true. Then some row (say, row i) of **A** is a linear combination of other rows in **A**. By performing a sequence of row replacements (EROs) on **A**, we can obtain a matrix **B** in which row i is a zero row. Recall (Section 5.1) that row replacements do not change the value of the determinant. Thus, expanding the determinant on row i, we have $\det(\mathbf{B}) = 0 = \det(\mathbf{A})$, and so (not **P**) is true. Hence (**P**) implies (**Q**).

III Indirect Proof (Proof by Contradiction)

In order to prove (B.1), suppose we assume that (**P**) is true and make the temporary assumption that (**Q**) is false; that is, (not **Q**) is true. The method is then to apply correct deductive reasoning using the statements **P** and (not **Q**) in order to arrive at a statement that either contradicts (**P**), or some other true statement in mathematics. The contradiction can only mean that our assumption that (**Q**) is false must be false; that is, (**Q**) must be true. This method, called *indirect proof*, or *proof by contradiction*, depends on the fact that any mathematical statement is logically either true or false.

Theorem B.3 **Illustration of Indirect Proof**

Let $\mathbf{A} = [a_{ij}]$ *be an* $m \times n$ *matrix. If* $\mathbf{Ax} = \mathbf{0}$ *for every column vector* $\mathbf{x}$ *in* $\mathbb{R}^n$, *then* $\mathbf{A} = \mathbf{O}$, *where* $\mathbf{O}$ *is the* $m \times n$ *zero matrix.*

Proof We assume the following statements are true:

(**P**) : $\mathbf{Ax} = \mathbf{0}$, for every $\mathbf{x}$ in $\mathbb{R}^n$ and (not **Q**) : **A** is not the zero matrix.

Let $\mathbf{A} = [\, \mathbf{v}_1 \ \cdots \ \mathbf{v}_j \ \cdots \ \mathbf{v}_n \,]$. If (not **Q**) is true, then some entry a_{ij} in **A** is nonzero. Consider the vector $\mathbf{x} = [\, 0 \ 0 \ \cdots \ 1 \ \cdots \ 0]^{\mathsf{T}}$ having all entries zero except for an entry 1 in the jth position. Then $\mathbf{Ax} = \mathbf{0}$ for this particular **x** and so $\mathbf{v}_j = \mathbf{0}$. Using the same argument for any other nonzero entries, we have $\mathbf{A} = \mathbf{O}$, which contradicts the temporary assumption (not **Q**). Hence **Q** must be true, namely $\mathbf{A} = \mathbf{O}$.

The Converse

The *converse* of the implication (B.1) is the implication

$$(\mathbf{Q}) \;\Rightarrow\; (\mathbf{P}) \tag{B.4}$$

A common error is to assume that if (B.1) is true, then (B.4) is automatically true. For example, the implication *dog* $\Rightarrow$ *animal* is true, but the converse, *animal* $\Rightarrow$ *dog*, is certainly false! More mathematically, it is shown in Section 4.2 that if a set of nonzero vectors in $\mathbb{R}^n$ is orthogonal, then the set is linearly independent, but a linearly independent set in $\mathbb{R}^n$ is not necessarily orthogonal.

Regarding (B.1), we say that (**P**) is a *sufficient condition* for **Q** to be true, and that (**Q**) is a *necessary condition* for (**P**) to be true. The statements (**P**) and (**Q**) are *logically equivalent* if (B.1) and (B.4) are *both* true (or both false) together, and we express this fact by saying (**P**) *if and only if* (**Q**); that is, (**Q**) is true if (**P**) is true and (**P**) is true only if (**Q**) is true.

Mathematical Induction

Consider a mathematical statement, denoted by $S(n)$, that involves the nonnegative integer n. In some cases $S(n)$ is only true for certain values of n. For example, the statement

$$S(n): \quad p = n^2 + n + 17 \quad \text{is a prime number} \tag{B.5}$$

is true for the positive integers $n = 0, 1, 2, 3, \ldots, 15$, but is false when $n = 16$ because $p = 289 = 16^2 + 16 + 17 = 16(16+1) + 17 = 17^2$ is *composite* (a prime number has no factors other than 1 and itself).

In some cases a statement $S(n)$ is true for every positive integer $n \geq n_0$, where n_0 is some initial nonnegative integer, but how can we prove this? Testing the truth of $S(n)$ for specific values of n is computationally flawed because the testing process will never end. This problem is avoided by applying the *principle of mathematical induction*, stated as follows:

Step 1 Suppose that a statement $S(n)$ be true for some fixed positive integer $n = n_0$. This step is called *initialization* and the value n_0 is called the *initializing value* for n.

Step 2 Assume that $S(n)$ is true for some nonnegative integer n, with $n \geq n_0$. This assumption is called the *inductive hypothesis*.

Step 3 If the truth of $S(n+1)$ can be proved using the inductive hypothesis, then $S(n)$ is true for *all* positive integers $n \geq n_0$.

The principle works like this: $S(n_0)$ is true from Step 1, then setting $n = n_0$, $S(n_0 + 1)$ is true from Steps 2 and 3, and setting $n = n_0 + 1$, $S(n_0 + 2)$ is true from Steps 2 and 3, and so on.

Proofs by induction may look deceptively simple. Step 1 is usually routine and Step 2 requires no action. Step 3 is the crucial step—if $S(n + 1)$ cannot be proved, making use of the inductive hypothesis, then something is wrong and the proof by induction fails—perhaps the result is actually false for some value of n, as is the case with (B.5).

■ ILLUSTRATION B.1

Formulating a Conjecture, Proof by Induction

There are many situations in mathematics when we wish to find a *closed form expression* (formula) involving a positive integer n to describe a general pattern. In such cases no statement $S(n)$ is given to us and we need to create it! Consider finding, if possible, a formula for the powers of the matrix

$$\mathbf{A} = \begin{bmatrix} 1 & 2 \\ 0 & 3 \end{bmatrix}.$$

One approach is to find a few powers of $\mathbf{A}$ (perhaps by machine) to see if a general pattern will emerge. We find that

$$\mathbf{A}^2 = \begin{bmatrix} 1 & 8 \\ 0 & 9 \end{bmatrix}, \quad \mathbf{A}^3 = \begin{bmatrix} 1 & 26 \\ 0 & 27 \end{bmatrix}, \quad \mathbf{A}^4 = \begin{bmatrix} 1 & 80 \\ 0 & 81 \end{bmatrix}, \ldots.$$

The pattern suggests that a reasonable *conjecture* (guess) might be

$$S(n): \qquad \mathbf{A}^n = \begin{bmatrix} 1 & 3^n - 1 \\ 0 & 3^n \end{bmatrix}. \tag{B.6}$$

A conjecture requires proof and in this case we attempt a proof by mathematical induction.

Step 1 Let $n_0 = 1$. Then $S(n_0)$ is true.

Step 2 Assume $\quad S(n)$: $\quad \mathbf{A}^n = \begin{bmatrix} 1 & 3^n - 1 \\ 0 & 3^n \end{bmatrix}$ is true.

Step 3 Use $S(n)$ to show that $S(n + 1)$ is true as follows:

$$\mathbf{A}^{n+1} = \mathbf{A}^n\mathbf{A} = \begin{bmatrix} 1 & 2 \\ 0 & 3 \end{bmatrix}^n \begin{bmatrix} 1 & 2 \\ 0 & 3 \end{bmatrix}$$

$$= \begin{bmatrix} 1 & 3^n - 1 \\ 0 & 3^n \end{bmatrix} \begin{bmatrix} 1 & 2 \\ 0 & 3 \end{bmatrix} = \begin{bmatrix} 1 & 3^{n+1} - 1 \\ 0 & 3^{n+1} \end{bmatrix}.$$

Hence by the principle of mathematical induction, (B.6) is true for all integers $n \geq 1$. ■

A Geometric Sequence

For a fixed real number r with $0 < r < 1$, the sequence $\{r^n\}$ given by

$$r,\ r^2,\ r^3,\ \ldots\ , r^n,\ \ldots\ ,$$

tends to zero. We write $\{r^n\} \to 0$ as n becomes large. The meaning is this: If you are given *any* positive number ϵ no matter how small (say, $\epsilon = 0.0001$), then there is a value of n for which $0 < r^n < \epsilon$.

Proof Let $y = 1/r$. Then $y > 1$ so that $y = 1 + p$ with $p > 0$. Next, $y^2 = 1+2p+p^2 > 1+2p$ because $p^2 > 0$, and using mathematical induction, we have

$$y^n > 1 + np, \quad n = 1, 2, 3, \ldots .$$

Hence

$$\frac{1}{y^n} = r^n < \frac{1}{1+np} \to 0 \quad \text{as } n \text{ becomes large.}$$

Equivalence Relations

Let $\mathcal{S}$ be a set. A *relation*, denoted by $\sim$, defined on $\mathcal{S}$ is called an *equivalence relation* if the following three conditions are met for any members a, b, c of $\mathcal{S}$.

$E1$	$a \sim a$	[Reflexivity]
$E2$	$a \sim b \Rightarrow b \sim a$	[Symmetry]
$E3$	$a \sim b$ and $b \sim c \Rightarrow a \sim c$	[Transitivity]

EXAMPLE 1

Equivalent Linear Systems

Recall that two $m \times n$ linear systems (S_1) and (S_2) are equivalent if either system can be changed into the other system by applying a sequence of elementary equation operations (EROs) and we write $(S_1) \sim (S_2)$ when this is the case. The relation $\sim$ is an equivalence relation on the set $\mathcal{S}$ of all $m \times n$ linear systems. Checking the conditions, we have

Reflexivity $(S) \sim (S)$ for any (S) in $\mathcal{S}$. No EROs are applied.

Symmetry $(S_1) \sim (S_2)$ implies $(S_2) \sim (S_1)$. Each ERO used to change (S_1) into (S_2) has an inverse. The sequence of inverse EROs (applied in reverse order) changes (S_2) back into (S_1).

Transitivity Suppose $(S_1) \sim (S_2)$ and $(S_2) \sim (S_3)$. Start with (S_1) and apply the sequence of EROs that changes (S_1) into (S_2), then apply the sequence of EROs that changes (S_2) into (S_3). The combined sequence of EROs changes (S_1) into (S_3). ■

An equivalence relation on a set $\mathcal{S}$ *partitions* $\mathcal{S}$ into *equivalence classes* (disjoint subsets) in such a way that two members of the same class are equivalent and members from different classes are not equivalent.

Looking back at Section 1.1, any pair of $m \times n$ linear systems are either equivalent (in the same equivalence class) or not equivalent (in different equivalence classes). Moreover, pairs of linear systems in the same equivalence class have the same solution sets because the application of EROs do not change solution sets.

Another example comes from Section 7.3. Two $n \times n$ matrices $\mathbf{A}$ and $\mathbf{B}$ are similar if there is an $n \times n$ matrix $\mathbf{P}$ such that $\mathbf{B} = \mathbf{P}^{-1}\mathbf{AP}$. Similarity is an equivalence relation on the set of all $n \times n$ matrices and so a pair of $n \times n$ matrices are in the same equivalence class if and only if they represent the same linear operator on $\mathbb{R}^n$ relative to appropriately defined bases.

Recursive Functions

Some functions are defined by repeated application of the function itself. We call this a *recursive definition*. For example, $5! = 5 \cdot 4!$ defines $5!$ in terms of $4!$ and likewise $4! = 4 \cdot 3!$, and so on. Hence $n!$ is computed *recursively* by the formula

$$n! = n\ (n-1)! \quad \text{for} \quad n = 1, 2, 3 \ldots .$$

Recursive definitions are commonly required in computer programming. A function M-file written in MATLAB to calculate $n!$ might have the form

```
function   y = factorial(n)   % computes factorial n
factorial = 1;
    for k = 2:n
    factorial = factorial * k;
end
y = factorial
```

The determinant *function* is a rule that assigns a unique scalar $\det(\mathbf{A})$ to each square matrix $\mathbf{A}$. The definition of $\det(\mathbf{A})$, given in Chapter 5, is recursive—a determinant of order n is defined in terms of determinants of order $n - 1$.

Permutations

The next concept appears in many applied areas, including coding theory, probability, statistics, and the analysis of computer algorithms.

The number of ways of *permuting* (or *arranging*) r objects chosen from a set of n objects is

$$P(n, r) = n(n-1)(n-2)\cdots(n-r+1). \tag{B.7}$$

Refer to the right side of (B.7). Because the first term is $n - 0$ and the last $n - (r - 1)$, there are r terms corresponding to r "boxes"

box 1	**box 2**		**box r**
□	□	. . .	□
n ways	$(n-1)$ ways		$(n-r+1)$ ways

ordered from left to right. The number of ways to place an object in box 1 is n, leaving $n-1$ possibilities for box 2. The number of ways to place an object in box 1 and box 2 is $n(n-1)$. Continue this process for all r boxes.

There are $n!$ ways of arranging n objects in a row. When $n = 3$, for example, we obtain the $3! = 6$ terms *abc*, *acb*, *bac*, *bca*, *cab*, *cba*.

Selections

The number of ways of *selecting* (choosing) r objects from n objects is

$$C(n, r) = \frac{n(n-1)(n-2)\cdots(n-r+1)}{1 \cdot 2 \cdots r} \tag{B.8}$$

$$= \frac{n!}{r!(n-r)!} \tag{B.9}$$

Proof Each set of r objects chosen from n objects can be arranged in $r!$ ways. Hence $C(n, r)r! = P(n, r)$, which is equation (B.8). Multiply numerator and denominator of (B.8) by $(n-r)!$ to obtain (B.9).

We define $C(n, 0) = C(n, n) = 1$. Alternative notation for $C(n, r)$ is $\binom{n}{r}$.

EXAMPLE 2

The Number of Binary Matrices

The number of binary (entries 0 or 1) 2×2 matrices having exactly two entries equal to 1 is equal to the number of ways of selecting two entries from four entries, namely $C(4, 2) = 6$. The number of binary $m \times n$ matrices having exactly k entries equal to 1 is $C(mn, k)$.

Probability

The *probability* of obtaining a particular *outcome* of an event is measured by a real number p with $0 \le p \le 1$. The value $p = 0$ indicates *impossibility* (the outcome cannot occur) and $p = 1$ indicates absolute *certainty* (the outcome will occur). A value of p strictly between 0 and 1 is an indicator of how likely a particular outcome will be.

Sequential Events

Consider a sequence of two events E_1 and E_2 defined by tossing a coin twice in succession. These events are independent of each other. The probability of obtaining a head ($= H$) on each occasion is $p = \frac{1}{2}$. The probability of obtaining a particular sequence such as HH is obtained by writing out all possible outcomes of the events.

$$\begin{matrix} HH & HT \\ TH & TT \end{matrix} \quad \Rightarrow \quad p(HH) = \frac{1}{4} = \left(\frac{1}{2}\right)\left(\frac{1}{2}\right)$$

In general, if the events are independent we multiply probabilities.

A General Power Algorithm

Consider the problem of computing a power x^k, $k = 1, 2, \ldots$, where x denotes a number or an $n \times n$ matrix. Finding the power x^9, for example, requires eight separate multiplications by x. A more efficient squaring technique computes x^2, then $(x^2)^2 = x^4$, then $(x^4)^2 = x^8$ and finally multiplies by x, requiring only four multiplications to achieve the same result. A squaring algorithm, known for over 1200 years (see [1], pp. 441-443), that will compute x^k efficiently, is stated as follows.

S-X Binary Method for Powers

Step 1 Write k as a binary string.

Step 2 Reading the string from left to right, replace each 1 by the letters SX and each 0 by the letter S. The result is a string of X's and S's.

Step 3 Delete the pair SX at the extreme left of the string in Step 2. Reading from left to right, interpreting each S as squaring and each X as multiplication by x.

The S-X algorithm is extremely efficient and requires no temporary storage, except for x and the current partial product.

EXAMPLE 3

Applying the S-X Algorithm for Multiplication

The index $k = 19$ has binary representation 10011 and corresponds to the string SX S S SX SX. Delete SX at the extreme left to get S S S X S X. Reading the string left to right, compute

$$\begin{array}{ccccccc} x & x^2 & x^4 & x^8 & x^9 & x^{18} & x^{19} \\ & \text{S} & \text{S} & \text{S} & \text{X} & \text{S} & \text{X} \end{array}$$

which requires a mere six multiplications to find x^{19}. ▩

REFERENCES

1. Knuth, D. E., *The Art of Computer Programming* (Vol. 1–3, 2nd ed.) Reading, Mass., Addison-Wesley (1973–1981).

ANSWERS TO SELECTED ODD-NUMBERED EXERCISES

Exercises 1.1

1. $2x - y = -1$.

3. $y = 0$.

5. Pivot: x_1, x_2, x_3. Free: x_4. $(x_1, x_2, x_3, x_4) = (4 + 2t, 2 - 4t, 3 - 2t, t)$, where t is a real parameter. None.

7. Pivot: x_1, x_3. Free: x_2, x_4. $(x_1, x_2, x_3, x_4) = (4 - 0.5s, s, -1.5 + 0.5t, t)$, where s, t are real parameters. None.

9. $x_1 = 2$, $x_2 = -3.5$.

11. $(x_1, x_2, x_3) = (4, -2, 1)$. Three planes with a single point in common.

13. Inconsistent. Three planes with no point in common.

15. Inconsistent. Four planes with no point in common.

17. Zero or infinitely many. If the system is consistent, there will be free variables. $(x_1, x_2, x_3, x_4) = (1, 1 + s - t, s, t)$, where s and t are real parameters.

19. Zero or infinitely many. If the system is consistent, there will be free variables. $(x_1, x_2, x_3, x_4) = (5.5 + 0.5s + t, 2.5 + 1.5s, s, t)$, where s and t are real parameters.

21. $(x_1, x_2, x_3, x_4) = (2.8 + 0.6t, -0.6 + 0.2t, 2.4 + t, t)$, where t is a real parameter. With $t = -2.4$ we obtain $(1.36, -1.08, 0, -2.4)$.

23. $\mathbf{M} = \left[\begin{array}{ccc|c} 2 & 1 & 13 & 28 \\ -1 & 1 & -2 & -8 \\ 1 & 1 & 8 & 16 \end{array}\right]$, 3×3, 3×1, 3×4, $(x_1, x_2, x_3) = (12 - 5t, 4 - 3t, t)$. Particular solutions: $(12, 4, 0)$, $(7, 1, 1)$.

25. (a) $\alpha = -2$, (b) $\alpha \neq 1$ and $\alpha \neq -2$, (c) $\alpha = 1$

27. $(x_1, x_2, x_3, x_4) = (1 - t, t, 3, 1)$, where t is a real parameter. None.

29. $\mathbf{M}$ is 4×4, $\mathbf{A}$ is 4×3, $(x_1, x_2, x_3) = (6 - t, -1.5 - 1.5t, t)$, where t is a real parameter.

31. $\mathbf{M}$ is 4×5, $\mathbf{A}$ is 4×4, $(x_1, x_2, x_3, x_4) = (0, -3, 3, -2)$.

33. $(x, y) = (-8, 2)$.

35. $(x, y) = (1.5, 2)$.

37. $c = 4.5$, $(x, y) = (0.5, 0.5)$.

39. $y = 0.75x + 1.25$.

41. $(1.2, -1)$ is a particular solution.

43. $(a, b, c) = (7, -7, -8)$, $(p, q) = (-3.5, 3.5)$, $r = \sqrt{65/2} \simeq 5.7$.

45. $(x, y, z) = \pm(\frac{1}{\sqrt{2}}, \sqrt{2}, \frac{3}{\sqrt{2}})$.

47. $\frac{\pi}{3} \pm 2k\pi$, $\frac{5\pi}{3} \pm 2k\pi$, $k = 0, 1, \ldots$.

49. (a) $(0, \frac{1}{3}, \frac{5}{3})$, (b) $(-\frac{1}{2}, 0, \frac{5}{2})$.

51. $x_1 = -7 + 4t$, $x_2 = 5 - 3t$, $x_3 = t$, where t is a real parameter.

53. T.

55. F.

57. F. **59.** T. **61.** F.

63. T. **65.** T.

Exercises 1.2

1. Yes. $\mathbf{A}^* = \mathbf{A}$, pivot position $(1, 1)$, col 1, $r = 1$.

3. No, $\mathbf{A}^* = \begin{bmatrix} 1 & 0 \\ 0 & 1 \end{bmatrix}$, pivot positions $(1, 1)$, $(2, 2)$, cols 1, 2, $r = 2$.

5. Yes, $\mathbf{A}^* = \begin{bmatrix} 1 & 1 \\ 0 & 0 \end{bmatrix}$, pivot position $(1, 1)$, col 1, $r = 1$.

7. No. $\mathbf{A}^* = \begin{bmatrix} 1 & 1 & 1 & 0 \\ 0 & 0 & 0 & 1 \\ 0 & 0 & 0 & 0 \end{bmatrix}$, pivot positions $(1, 1)$, $(2, 4)$, cols 1, 4 $r = 2$

9. $\begin{bmatrix} 1 & 1 \\ 1 & 1 \end{bmatrix}$.

11. $\mathbf{A} \sim \mathbf{I}_2 \sim \mathbf{B}$, $r = 2$.

13. (a) Use sequence: $\mathbf{r}_2 \leftrightarrow \mathbf{r}_3$, $\frac{1}{2}\mathbf{r}_3 \rightarrow \mathbf{r}_3$, $\mathbf{r}_3 + \mathbf{r}_1 \rightarrow \mathbf{r}_3$. (b) Use sequence of inverses: $\mathbf{r}_3 - \mathbf{r}_1 \rightarrow \mathbf{r}_3$, $2\mathbf{r}_3 \rightarrow \mathbf{r}_3$, $\mathbf{r}_2 \leftrightarrow \mathbf{r}_3$.

15. $(x_1, x_2, x_3) = (5, 11, 7)$.

17. (a) $\alpha = 11$ and $\beta \neq 36$, (b) $\alpha \neq 11$, any β, (c) $\alpha = 11$ and $\beta = 36$.

19. (a) $p = 1$, (b) $p \neq 1$, (c) no value of p.

21. $x_1 = -5.4$, $x_2 = 3.2$, $x_3 = -0.6$, $x_4 = 0.6$.

23. The solutions are the columns of the matrix $\begin{bmatrix} -1 & 23 & -37 \\ 1 & -13 & 21 \end{bmatrix}$.

25. The solutions are the columns of the matrix $\begin{bmatrix} -7 & -15 & -23 \\ 5 & 11 & 17 \end{bmatrix}$.

27. $(x_1, x_2, x_3) = (t, -2t, t)$, where t is a real parameter, $r = 2$, $n - r = 3 - 2 = 1$ free variable.

29. $(x_1, x_2, x_3, x_4) = (0, 0, 0, 0)$, $r = 4$, $n - r = 4 - 4 = 0$ free variables.

31. $\begin{bmatrix} 1 & 0 \\ 0 & 1 \\ 0 & 0 \end{bmatrix}$. **33.** $\begin{bmatrix} 1 & 0 \\ 0 & 1 \end{bmatrix}$.

35. $r = 1$. **37.** $r = 1$. **39.** $r = 1$.

41. (a) $\alpha = -1$ or $\alpha = 3$, (b) $\alpha \neq -1$ and $\alpha \neq 3$.

43. (a) $\alpha = 0$, (b) $\alpha \neq 0$.

45. $r = \text{rank}\,\mathbf{A} = 3 = \text{rank}\,\mathbf{M}$. Case 2: $n = r = 3$, unique solution, $x_1 = -\frac{5}{6}$, $x_2 = \frac{7}{6}$, $x_3 = -\frac{5}{6}$.

47. $\text{rank}\,\mathbf{A} = 2 < 3 = \text{rank}\,\mathbf{M}$. Case 1: inconsistent.

49. $r = \text{rank}\,\mathbf{A} = 3 = \text{rank}\,\mathbf{M}$. Case 2: $n = r = 3$, unique solution, $(x_1, x_2, x_3) = (2, -1, 0)$.

51. $r = \text{rank}\,\mathbf{A} = \text{rank}\,\mathbf{M} = 2 = m$ and there is $n - r = 3 - 2 = 1$ free variable. Infinitely many solutions for every $\mathbf{b}$: If $x_2 = t$, a real parameter, then $x_1 = 7\alpha - 3\beta - t$, $x_3 = \beta - 2\alpha$.

53. Let $d = c - b - a$. We have $\text{rank}\,\mathbf{A} = 2$. The system is inconsistent if and only if $d \neq 0$, and $\text{rank}\,\mathbf{A} = 2 < 3 = \text{rank}\,\mathbf{M}$ in this case. We have $\text{rank}\,\mathbf{M} = 2$ if and only if $d = 0$ and there are infinitely many solutions in this case. No possibility of unique solution.

55. $\text{rank}\,\mathbf{A} = 2$. Infinitely many solutions: $(x_1, x_2, x_3) = (t, -7t, t)$, where t is a real parameter.

57. $(x_1, x_2) = (-t, t)$, for $\lambda = -2$, $(-5t, t)$ for $\lambda = 2$, where t is a nonzero real parameter, $(x_1, x_2) = (0, 0)$, for $\lambda = 0$.

59. $(x_1, x_2, x_3) = (s, t, 0)$, where s and t are nonzero real parameters, $(x_1, x_2, x_3) = (0, 0, 0)$, for $\lambda = 0$.

61. One row is a linear combination of other rows. $\mathbf{r} = \text{row } 4$, $\mathbf{r}_1 = \text{row } 1$, $\mathbf{r}_2 = \text{row } 2$, $\mathbf{r}_3 = \text{row } 3$, $s_1 = 0$, $s_2 = \frac{2}{7}$, $s_3 = \frac{1}{7}$.

63. Use the sequence $\mathbf{r}_1 + \mathbf{r}_2 \rightarrow \mathbf{r}_1$, $\mathbf{r}_2 - \mathbf{r}_1 \rightarrow \mathbf{r}_2$, $\mathbf{r}_1 + \mathbf{r}_2 \rightarrow \mathbf{r}_1$, $-\mathbf{r}_2 \rightarrow \mathbf{r}_2$.

65. F. **67.** T.

Exercises 1.3

1. P: \$ 0.6, D: \$ 0.5245 (million).

3. $(x_1, x_2) = (180, 360)$, where x_1, x_2 are, respectively, the number of units of P1, P2 produced. Yes.

5. Let x_i, $1 \leq i \leq 3$, be the respective number of servings of Skim Milk, Whole Wheat Bread, and Hamburger consumed per day. $(x_1, x_2, x_3) = (2, 4, 2)$.

7. B broke the law, D did not, A may or may not have done so because $21 \leq x_4 \leq 30$ and so $40 \leq x_1 \leq 58$, where x_1 and x_4 are the number of units of coal burned (daily) by A and D.

9. Let x_i, $1 \leq i \leq 3$ denote the number of bundles of type A, B, C, respectively. $(x_1, x_2, x_3) = (5, 9, 0)$, $(4, 8, 1)$, $(3, 7, 2)$, $(2, 6, 3)$, $(1, 5, 4)$, $(0, 4, 5)$. Two ways. The last solution is cheapest at \$169.00.

11. Maximum batches of rolls is 540.

13. Minimum $x_1 = 20$. $(x_1, x_2, x_3) = \frac{1}{k+1}(100k, 100, 20(4k - 1))$.

15. (a) At nodes A, B, C, D, E, F, we have $x_1 + x_4 = 700$, $x_2 + x_4 - x_6 = 200$, $x_3 - x_6 = -400$, $x_3 - x_7 = -200$, $x_2 + x_5 - x_7 = 250$, $x_1 + x_5 = 550$. (b) Min $x_7 = 200$. (c) $x_3 = 0$. (d) $x_4 = 150$.

17. $(I_1, I_2, I_3) = \frac{1}{15}(26, 22, 4)$.

19. $(I_1, I_2, I_3, I_4, I_5, I_6) = \frac{1}{13}(16, 10, 6, 4, 2, 6)$.

21. $(6, 6, 6, 1)$.

23. Suppose x_1 moles of cellulose and x_2 moles of oxygen produces x_3 moles of carbon dioxide, x_4 moles of water, plus x_5 moles of carbon. $(x_1, x_2, x_3, x_4, x_5) = (1, 6, 6, 5, 0)$, $(1, 5, 5, 5, 1)$, $(1, 4, 4, 5, 2)$.

25. $(m, c) = (0.7, 1.8)$

27. $y = 1 + x - x^2$.

29. $y = (-\frac{1}{4} + \frac{t}{8})x^2 + (\frac{3}{2} - \frac{3t}{4})x + t$, where t is a real parameter. The curve has equation $y = -4x^2 + 24x - 30$.

Exercises 2.1

1. $\begin{bmatrix} 2 & 3 \\ 3 & 4 \end{bmatrix}$

3. $\begin{bmatrix} 0 & -1 & -2 \\ 1 & 0 & -1 \\ 2 & 1 & 0 \end{bmatrix}$.

5. $\begin{bmatrix} 2 & 0 & 10 & 0 \\ 0 & 8 & 0 & 20 \end{bmatrix}$.

7. $x = y = 0, 2$

9. $x = 2$, $y = 16$, $z = -6$

11. $x = y = 1$, $z = 0$.

13. $\mathbf{X}$ is 2×2, $x_{12} = 1$.

15. $\mathbf{A} = \begin{bmatrix} 7 & -21 & 14 \\ 4 & -12 & 8 \end{bmatrix}$. A is 2×3.

17. $c_{12} = -2$ and $c_{33} = -8$.

19. $w_{13} = 10$, $w_{21} = -7$.

21. The $(2, 2)$-entry is 2.

23. $\mathbf{AB} = \begin{bmatrix} 7 & 2 & 1 \\ 3 & -7 & -1 \end{bmatrix}$. $\mathbf{BA}$ undefined because $\mathbf{B}$ has three columns while $\mathbf{A}$ has two rows.

25. $\mathbf{BC} = \begin{bmatrix} 5 \\ 15 \end{bmatrix}$. $\mathbf{CB}$ undefined because $\mathbf{C}$ has one column while $\mathbf{B}$ has two rows.

27. $\mathbf{BB}^{\mathsf{T}} = \begin{bmatrix} 5 & 2 \\ 2 & 26 \end{bmatrix}$, $\mathbf{B}^{\mathsf{T}}\mathbf{B} = \begin{bmatrix} 13 & 10 & 3 \\ 10 & 17 & 4 \\ 3 & 4 & 1 \end{bmatrix}$.

29. $(\mathbf{BC})\mathbf{A}$ is undefined because $\mathbf{BC}$ has one column while $\mathbf{A}$ has two rows. $(\mathbf{BC})^{\mathsf{T}}(\mathbf{BC}) = \begin{bmatrix} 250 \end{bmatrix}$.

31. Premultplication scales rows of $\mathbf{A}$ and postmultiplication scales columns of $\mathbf{A}$.

33. The $(1, 3)$-entry is 9 and the $(3, 2)$-entry is -6.

35. $\begin{cases} x_1 + 3x_2 - 2x_3 = 2 \\ -2x_1 \qquad\qquad x_3 = 3 \end{cases}$, $x_1 = -\frac{3}{2} + \frac{1}{2}t$, $x_2 = \frac{7}{6} + \frac{1}{2}t$, $x_3 = t$, where t is a real parameter.

37. When $\mathbf{A}$ and $\mathbf{B}$ commute, for example.

39. $2\mathbf{I}_2$.

41. $\mathbf{I}_3$.

43. $2\mathbf{A}^2 + 2\mathbf{B}^2$.

45. $22, \dfrac{m}{2}(2n - m + 1)$.

47. 1×12, 12×1, 2×6, 6×2, 3×4, 4×3.

49. 1×13, 13×1.

51. $x_1^2 + 5x_1x_2 + 4x_2^2$.

53. $k = 2$.

55. $k = 3$.

57. $\begin{bmatrix} 2 & 4 \\ -1 & -2 \end{bmatrix}$, $\mathbf{A} = \begin{bmatrix} 1 & 1 \\ -1 & -1 \end{bmatrix}$, $\mathbf{A}^{\mathsf{T}}$.

59. $\mathbf{A}^{\mathsf{T}} = (\mathbf{A}^2)^{\mathsf{T}} = (\mathbf{A}^{\mathsf{T}})^2$. $\mathbf{O}$, $\mathbf{I}$, $\mathbf{A} = \begin{bmatrix} -1 & 2 \\ -1 & 2 \end{bmatrix}$, $\mathbf{A}^{\mathsf{T}}$.

63. Equate corresponding entries in $\mathbf{A}^2$ and $\mathbf{J}$. The $(1, 2)$-entries imply that $a + d \neq 0$ and the $(2, 1)$-entries imply that $c = 0$. But then $a = d = 0$ from the $(1, 1)$- and $(2, 2)$-entries. Contradiction.

65. $\mathbf{AB} = \mathbf{CD}_1\mathbf{CCD}_2\mathbf{C} = \mathbf{CD}_1\mathbf{D}_2\mathbf{C}$, since $\mathbf{C}^2 = \mathbf{I}$. Similarly $\mathbf{BA} = \mathbf{CD}_2\mathbf{D}_1\mathbf{C}$. But the diagonal matrices $\mathbf{D}_1$ and $\mathbf{D}_2$ commute and hence $\mathbf{AB} = \mathbf{BA}$.

71. $\mathbf{S} = \begin{bmatrix} 1 & 2.5 \\ 2.5 & 4 \end{bmatrix}$, $\mathbf{K} = \begin{bmatrix} 0 & -0.5 \\ 0.5 & 0 \end{bmatrix}$.

73. $\mathbf{B}^{\mathsf{T}} = -\mathbf{B}$, $\mathbf{A} = \mathbf{I} + \mathbf{B}$, $\mathbf{AA}^{\mathsf{T}} = (\mathbf{I} + \mathbf{B})(\mathbf{I} + \mathbf{B})^{\mathsf{T}} = (\mathbf{I} + \mathbf{B})(\mathbf{I} + \mathbf{B}^{\mathsf{T}}) = (\mathbf{I} + \mathbf{B})(\mathbf{I} - \mathbf{B})$. Hence $\mathbf{AA}^{\mathsf{T}} = \mathbf{I}^2 - \mathbf{B}^2 = (\mathbf{I} - \mathbf{B})(\mathbf{I} + \mathbf{B}) = \mathbf{A}^{\mathsf{T}}\mathbf{A}$.

79. $\begin{bmatrix} p & 0 \\ 0 & q \end{bmatrix}$, where $p = x_0 + x_1 + \cdots + x_k$ and $q = x_0 - x_1 + \cdots + (-1)^k x_k$.

81. (a) $\mathbf{AB} = \begin{bmatrix} 2 & 2 & -1 \\ 3 & 9 & 5 \\ -1 & 8 & -4 \end{bmatrix}$.

(b) $\mathbf{AB} = \begin{bmatrix} 3 & -3 & 8 \\ 24 & 14 & 9 \end{bmatrix}$.

Exercises 2.2

1. $\begin{bmatrix} 3 & -1 \\ -5 & 2 \end{bmatrix}$.

3. $\frac{1}{a^2-b^2}\begin{bmatrix} a & -b \\ -b & a \end{bmatrix}$, where $a^2 - b^2 \neq 0$.

5. Singular.

7. $\begin{bmatrix} 0.5 & 0 & 0 \\ 0 & 0.25 & 0 \\ 0 & 0 & -0.5 \end{bmatrix}$.

9. $k = 0, \pm 4$.

11. $\mathbf{AXB} = \mathbf{C}$ implies $\mathbf{A}^{-1}(\mathbf{AXB}) = \mathbf{A}^{-1}\mathbf{C}$ implies $\mathbf{XB} = \mathbf{A}^{-1}\mathbf{C}$, using associativity and $\mathbf{A}^{-1}\mathbf{A} = \mathbf{I}$, and similarly postmultiply by $\mathbf{B}^{-1}$ to obtain the result. $\mathbf{B} = \mathbf{A}^{-1}$. Row 1: $[1\ 0\ 0]$, in fact, $\mathbf{X} = \mathbf{I}_3$.

13. (a) $\mathbf{A} = \begin{bmatrix} 2 & 1 \\ 5 & 3 \end{bmatrix}$, $\mathbf{B} = \mathbf{I}_2$, (b) $\mathbf{A} = \begin{bmatrix} 2 & 1 \\ 5 & 3 \end{bmatrix}$, $\mathbf{B} = -\mathbf{A}$.

15. $\mathbf{X} = (\mathbf{A}^{-1})^{\mathsf{T}}\mathbf{A}^{-1}$. $x_{11} = 49$, $x_{22} = 6$, $x_{33} = 11$.

19. All diagonal entries must be nonzero. The inverse is a diagonal matrix with entries $1/d_{ii}$, $1 \le i \le n$.

27. They are the identities $\mathbf{I}_n$.

29. Yes, because $\operatorname{rank}\mathbf{A} = \operatorname{rank}\mathbf{B} = n$.

31. $(x_1, x_2) = (1, -1)$.

33. $(x_1, x_2, x_3) = (66, -22, 4)$.

35. $(x_1, x_2, x_3, x_4) = (-2, 4, 3.5, 1.5)$.

39. $\mathbf{X} = \begin{bmatrix} 1 & 2 & 3 \\ -2 & -4 & -6 \\ 1 & 2 & 3 \end{bmatrix}$.

41. Form $(\mathbf{A} - \mathbf{B})\mathbf{x} = \mathbf{0}$ and use the previous exercise.

43. $\mathbf{A}^{-1} = [c_{ij}]$, where $c_{ii} = c_{i,i+1} = 1$, $1 \le i \le n$, and $c_{ij} = 0$ otherwise.

45. $\mathbf{A}$ is singular if and only if $\det(\mathbf{A}) = ad - bc = 0$ if and only if $\frac{a}{c} = \frac{b}{d}$.

47. Not elementary.

49. Elementary, $\mathbf{r}_1 \leftrightarrow \mathbf{r}_2$, $\mathbf{A}^{\mathsf{T}} = \mathbf{A}$.

51. Elementary, $\mathbf{r}_2 \leftrightarrow \mathbf{r}_3$, $\mathbf{A}^{\mathsf{T}} = \mathbf{A}$.

53. Not elementary.

55. $\mathbf{E}_1 = \mathbf{E}_1^{-1} = \begin{bmatrix} 0 & 1 \\ 1 & 0 \end{bmatrix}$, $\mathbf{E}_2 = \begin{bmatrix} 1 & -2 \\ 0 & 1 \end{bmatrix}$, $\mathbf{E}_2^{-1} = \begin{bmatrix} 1 & 2 \\ 0 & 1 \end{bmatrix}$, $\mathbf{E}_3 = \begin{bmatrix} 1 & 0 \\ 0 & 0.5 \end{bmatrix}$, $\mathbf{E}_3^{-1} = \begin{bmatrix} 1 & 0 \\ 0 & 2 \end{bmatrix}$,

$\mathbf{F} = \begin{bmatrix} -2 & 1 \\ 0.5 & 0 \end{bmatrix}$.

57. $\mathbf{F} = \begin{bmatrix} 0.5 & -1 \\ 0 & 0.5 \end{bmatrix}$.

59. $\mathbf{F} = \dfrac{1}{3}\begin{bmatrix} 2 & -1 & 0 \\ 1 & 1 & 0 \\ -11 & 4 & 3 \end{bmatrix}$.

61. $\mathbf{F} = \dfrac{1}{7}\begin{bmatrix} -2 & 6 & 1 \\ 1 & -3 & 3 \\ -2 & -1 & 1 \end{bmatrix}$.

63. $\mathbf{F} = \begin{bmatrix} 31 & -3 & -6 \\ -5 & 0 & 1 \\ -10 & 1 & 2 \end{bmatrix}$.

65. $\mathbf{F} = \begin{bmatrix} 1 & -2 \\ 0 & 1 \end{bmatrix}$. $\mathbf{F}$ corresponds the performing $\mathbf{r}_1 - 2\mathbf{r}_2 \to \mathbf{r}_1$ on $\mathbf{E}$.

67. $\mathbf{F} = \begin{bmatrix} 1 & 0 & 0 \\ 0 & 1 & 0 \\ 0 & 0 & \frac{1}{3} \end{bmatrix}$. $\mathbf{F}$ corresponds to performing $\frac{1}{3}\mathbf{r}_3 \to \mathbf{r}_3$ on $\mathbf{E}$.

77. $\mathbf{XA}$ is 2×2 and $\mathbf{XA}$ is 3×3. The $(1, 1)$ entry in $\mathbf{AX}$ is 8. The 3×3 linear system defined by $\mathbf{AX} = \mathbf{I}_3$ is inconsistent.

79. $\mathbf{X} = \begin{bmatrix} 1-s & -t \\ -2+s & 1+t \\ s & t \end{bmatrix}$, where s and t are real parameters. We have $\operatorname{rank}\mathbf{A} = 2 = m < n = 3$.

81. $\begin{bmatrix} 2 & -1 & 1 & 1 \\ -1 & 1 & 1 & 0 \\ 0 & 0 & 0 & -1 \\ 0 & 0 & -1 & 0 \end{bmatrix}$.

Exercises 2.3

1. $\begin{bmatrix} 1 & -1 \\ 0 & 1/2 \end{bmatrix}$.

3. $\begin{bmatrix} 1/2 & -1/6 & 1/30 \\ 0 & 1/3 & 2/15 \\ 0 & 0 & 1/5 \end{bmatrix}$.

5. $\begin{bmatrix} 1 & -2 & 3 \\ 0 & 1 & -1 \\ 0 & 0 & 1 \end{bmatrix}$.

9. $\mathbf{P}^{\mathsf{T}}$ is a right–inverse for $\mathbf{P}$ and so by Theorem 2.9 we have $\mathbf{P}^{-1} = \mathbf{P}^{\mathsf{T}}$. The matrix $\mathbf{P}^{\mathsf{T}}$ has one (and only on) entry equal to 1 in every row and every column. The columns of $\mathbf{P}$ are the rows of $\mathbf{I}_n$ in some order.

Hence, the rows of $\mathbf{P}^{\mathsf{T}}$ are the rows of $\mathbf{I}_n$ in some order.

11. $\begin{bmatrix} 1 & 0 \\ 3 & 1 \end{bmatrix}, \begin{bmatrix} 3 & -4 \\ 0 & 5 \end{bmatrix}.$

13. $\begin{bmatrix} 1 & 0 & 0 \\ 2 & 1 & 0 \\ 4 & 1 & 1 \end{bmatrix}, \begin{bmatrix} 1 & 5 & 1 \\ 0 & -6 & -8 \\ 0 & 0 & -9 \end{bmatrix}.$

15. $\mathbf{I}_3\mathbf{A}$.

17. Yes, take $\mathbf{L} = \mathbf{I}_2$ and $\mathbf{U} = \mathbf{A}$.

19. $\mathbf{L} = \begin{bmatrix} 1 & 0 & 0 \\ -3 & 1 & 0 \\ 0 & \frac{1}{6} & 1 \end{bmatrix}, \mathbf{U} = \begin{bmatrix} 1 & -2 & 2 \\ 0 & 6 & 0 \\ 0 & 0 & 1 \end{bmatrix},$

$\mathbf{x} = \begin{bmatrix} 1 \\ -2 \\ 1 \end{bmatrix}, \begin{bmatrix} -6 \\ \frac{1}{2} \\ \frac{7}{2} \end{bmatrix}, \begin{bmatrix} \frac{5}{3} \\ \frac{2}{3} \\ \frac{1}{3} \end{bmatrix}.$

21. $\mathbf{L} = \begin{bmatrix} 1 & 0 \\ \frac{3}{2} & 1 \end{bmatrix}, \mathbf{U} = \begin{bmatrix} 2 & 0 \\ 0 & 4 \end{bmatrix}.$

23. $\mathbf{L} = \begin{bmatrix} 1 & 0 & 0 \\ 4 & 1 & 0 \\ 7 & 2 & 1 \end{bmatrix}, \mathbf{U} = \begin{bmatrix} 1 & 2 & 3 \\ 0 & -3 & -6 \\ 0 & 0 & -8 \end{bmatrix}.$

25. $\mathbf{L} = \begin{bmatrix} 4 & 0 \\ 10 & 11 \end{bmatrix}, \mathbf{U} = \begin{bmatrix} 1 & -\frac{1}{2} \\ 0 & 1 \end{bmatrix}.$

27. $\mathbf{L} = \begin{bmatrix} 1 & 0 \\ 3 & 1 \end{bmatrix}, \mathbf{U} = \begin{bmatrix} 1 & 2 \\ 0 & -2 \end{bmatrix}, \mathbf{A}^{-1} =$

$\mathbf{U}^{-1}\mathbf{L}^{-1} = \begin{bmatrix} -2 & 1 \\ \frac{3}{2} & -\frac{1}{2} \end{bmatrix}.$

29. $\mathbf{L} = \begin{bmatrix} 1 & 0 & 0 \\ -1 & 1 & 0 \\ 1 & \frac{1}{3} & 1 \end{bmatrix}, \mathbf{U} = \begin{bmatrix} 1 & 2 \\ 0 & 9 \\ 0 & 0 \end{bmatrix}.$

Exercises 2.4

1. $\mathbf{A} = \mathbf{O}_{3\times 3}$.

3. $\mathbf{A} = \begin{bmatrix} \mathbf{O}_{3\times 3} & \mathbf{B} \\ \mathbf{B}^{\mathsf{T}} & \mathbf{O}_{3\times 3} \end{bmatrix}$, where $\mathbf{B} = \begin{bmatrix} 1 & 1 & 1 \\ 1 & 0 & 1 \\ 0 & 1 & 1 \end{bmatrix}.$

5. The graph on two vertices with no edges.

7. See Figure.

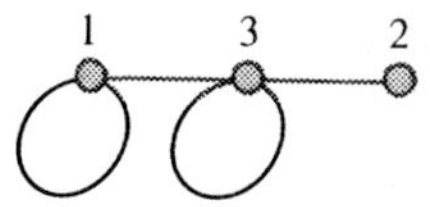

9. Walks: $2 \to 3 \to 2 \to 3$; $2 \to 3 \to 4 \to 3$; $2 \to 3 \to 5 \to 3$; $2 \to 1 \to 5 \to 3$ (path); $2 \to 1 \to 2 \to 3$. Walks: $4 \to 4 \to 3 \to 4$; $4 \to 3 \to 4 \to 4$; $4 \to 4 \to 4 \to 4$. No.

15. $\mathbf{B} = \begin{bmatrix} 1 & 2 & 2 \\ 1 & 1 & 1 \\ 1 & 2 & 2 \end{bmatrix}$. For example, there is only one directed walk (it is a cycle) $1 \to 3 \to 2 \to 1$ of length 3 from vertex 1 to itself.

19. See answer for Exercise 3.

21. (a) Leaving u and arriving at v accounts for one edge each. Visiting any vertex during the walk counts 2 edges for each visit. No edges are used twice. (b) The degrees of all vertices in G_8 are odd. (c) If an Eulerian walk begins and ends at the same vertex, all vertices have even degree.

23. (a) $\mathbf{A} = \begin{bmatrix} 0.9 & 0.2 \\ 0.1 & 0.8 \end{bmatrix}$. (b) $\mathbf{u}_1 = \begin{bmatrix} 1100 \\ 900 \end{bmatrix}$, $\mathbf{u}_2 = \begin{bmatrix} 1170 \\ 830 \end{bmatrix}$, $\mathbf{u}_3 = \begin{bmatrix} 1219 \\ 781 \end{bmatrix}$. (c) $\mathbf{u} = \begin{bmatrix} 1333.3 \\ 666.7 \end{bmatrix}$.

25. (a) $\mathbf{A} = \begin{bmatrix} 0.8 & 0.05 & 0.05 \\ 0.15 & 0.9 & 0.05 \\ 0.05 & 0.05 & 0.9 \end{bmatrix}$. (b) $\mathbf{u}_1 = \begin{bmatrix} 12.50 \\ 40.00 \\ 47.50 \end{bmatrix}$,

$\mathbf{u}_2 = \begin{bmatrix} 14.38 \\ 40.25 \\ 45.38 \end{bmatrix}, \mathbf{u}_3 = \begin{bmatrix} 15.78 \\ 40.65 \\ 43.57 \end{bmatrix}.$

(c) Steady state: $\mathbf{u} = \begin{bmatrix} 20.00 \\ 46.67 \\ 33.33 \end{bmatrix}$.

27. Steady state: $\mathbf{u} = \begin{bmatrix} 600 \\ 0 \\ 0 \end{bmatrix}$.

29. $\mathbf{u} = t\begin{bmatrix} 6 \\ 1 \end{bmatrix}$, where t is a parameter.

31. $\mathbf{u} = t\begin{bmatrix} 20 \\ 5 \\ 1 \end{bmatrix}$, where t is a parameter.

33. $\mathbf{u}_1 = \begin{bmatrix} 270 \\ 80 \\ 25 \end{bmatrix}$, $\mathbf{u}_2 = \begin{bmatrix} 195 \\ 108 \\ 20 \end{bmatrix}$, $\mathbf{u}_3 = \begin{bmatrix} 244 \\ 78 \\ 27 \end{bmatrix}$, ... No, the equation $\mathbf{Pu} = \mathbf{u}$ has no nonzero solution. The population is in slow decline. For example,
$$\mathbf{P}^{110}\mathbf{u}_0 = \begin{bmatrix} 9.1496 \\ 3.7667 \\ 0.9692 \end{bmatrix}.$$

37. $\mathbf{x} = \begin{bmatrix} 2 \\ 1 \end{bmatrix}$.

39. $\mathbf{x} = \begin{bmatrix} 168 \\ 85 \\ 70 \end{bmatrix}$.

41. Summing the rows of $\mathbf{A}$ produces a zero row. Yes.

43. $\mathbf{A} = \mathbf{I} - \mathbf{C} =$
$$\begin{bmatrix} -0.8 & 0 & 0.2 \\ 0.4 & -0.5 & 0.2 \\ 0.4 & 0.5 & -0.4 \end{bmatrix} \sim \begin{bmatrix} 1 & 0 & -0.25 \\ 0 & 1 & -0.60 \\ 0 & 0 & 0 \end{bmatrix}.$$
$\mathbf{x} = (x_1, x_2, x_3) = t(0.25, 0.60, 1)$, where $t > 0$ is a real parameter. A unique solution can be obtained if a value for the total production (sum of entries in $\mathbf{x}$) is known. If $x_1 + x_2 + x_3 = 37$, then $t = 20$ and $(x_1, x_2, x_3) = (5, 12, 20)$ (million \$).

45. $x_1 = \frac{8}{3}$, $x_2 = 3.5$.

47. $\begin{bmatrix} 4 \\ 5 \\ 10 \end{bmatrix}$.

49. $\mathbf{x} = \begin{bmatrix} 1.32 \\ 1.6 \\ 1 \end{bmatrix}$.

Exercises 3.1

1. $(r, s) = (3, 1)$

5. $[11\ \ -2]^{\mathsf{T}}$.

7. $[2\ \ -5]^{\mathsf{T}}$.

9. $[2\ \ 7\ \ 5]^{\mathsf{T}}$.

11. $\mathbf{x} = [1\ \ 1.5]^{\mathsf{T}}$.

13. Not a subspace, (**S1**) fails.

15. Not a subspace, (**S1**) fails.

17. Subspace.

19. Not a subspace, (**S1**), (**S2**) fail.

23. Let $x_1 = s$ and $x_3 = t$, where s and t are real parameters, then $\begin{bmatrix} x_1 \\ x_2 \\ x_3 \end{bmatrix} = s\begin{bmatrix} 1 \\ 2 \\ 0 \end{bmatrix} + t\begin{bmatrix} 0 \\ 4 \\ 1 \end{bmatrix} = s\mathbf{v}_1 + t\mathbf{v}_2$. We have $\mathcal{U} = \text{span}\{\mathbf{v}_1, \mathbf{v}_2\}$.

25. T.

27. T.

29. F. True if *finitely* is changed to *infinitely*. Also true if $\mathbf{u} = \mathbf{v} = \mathbf{0}$.

31. $(x_1, x_2, x_3) = (-\frac{5}{19}, -\frac{4}{19}, t)$, where t is a real parameter.

33. $(x_1, x_2, x_3) = (a, b, c)$.

35. $\mathbf{b} = 2\mathbf{v}_1 + 6\mathbf{v}_2$.

37. $\mathbf{b} = t\mathbf{v}_1 - 2t\mathbf{v}_2 + t\mathbf{v}_3$, for any scalar t.

39. $(x_1, x_2) = (0.4, 0.1)$.

41. $\mathcal{U}$ is a subspace spanned by $\begin{bmatrix} 1 \\ 0 \end{bmatrix}$.

43. $\mathcal{U}$ is a subspace spanned by $\begin{bmatrix} 1 \\ 0 \\ 0 \end{bmatrix}, \begin{bmatrix} -1 \\ 1 \\ 0 \end{bmatrix}, \begin{bmatrix} 0 \\ 2 \\ 1 \end{bmatrix}$.

45. $\mathcal{U}$ is a subspace spanned by $\begin{bmatrix} 1 \\ -1 \\ 0 \end{bmatrix}, \begin{bmatrix} 1 \\ 0 \\ -1 \end{bmatrix}$.

47. The points lie in the plane $y = 1$ which is not a subspace of $\mathbb{R}^3$.

49. $x_1\begin{bmatrix} 4 \\ -1 \end{bmatrix} + x_2\begin{bmatrix} 0 \\ 2 \end{bmatrix} + x_3\begin{bmatrix} 8 \\ 0 \end{bmatrix} = \begin{bmatrix} 0 \\ 0 \end{bmatrix}$, $(x_1, x_2, x_3) = t(2, 1, -1)$, for any scalar t.

51. The system $\begin{cases} 3x_1 + 2x_2 = -1 \\ x_1 - x_2 = 0 \\ 4x_1 = 2 \end{cases}$ is inconsistent. Hence $\mathbf{b}$ is not in $\text{span}\{\mathbf{v}_1, \mathbf{v}_2\}$.

53. (a) The unique solution is $(x_1, x_2, x_3) = (0, 0, 0)$. (b) The unique solution is $(x_1, x_2, x_3) = (-3, 5, 0)$. The reduced form of $\mathbf{A} = [\,\mathbf{v}_1\ \mathbf{v}_2\ \mathbf{v}_3\,]$ is $\mathbf{I}_3$ in both cases.

55. When rank $\mathbf{A} <$ rank $\mathbf{M}$, the linear system $\mathbf{Ax} = \mathbf{b}$ is inconsistent.

57. There will be $k - \operatorname{rank} \mathbf{A}$ parameters in the solution set to $\mathbf{Ax} = \mathbf{b}$.

59. When $\operatorname{rank} \mathbf{A} = n$ the linear system $\mathbf{Ax} = \mathbf{b}$ is consistent for every $\mathbf{b}$ (see Theorems 1.4 and 1.5 in Section 1.2).

Exercises 3.2

1. Linearly independent.

3. Linearly dependent. $t\mathbf{v}_1 - 2t\mathbf{v}_2 + t\mathbf{v}_3 = \mathbf{0}$, where t is a real parameter. Every vector.

5. Linearly independent.

7. Linearly dependent. $-2t\mathbf{v}_1 + t\mathbf{v}_2 + 0\mathbf{v}_3 + 0\mathbf{v}_4 = \mathbf{0}$, where t is a real parameter. $\mathbf{v}_1, \mathbf{v}_2$.

9. Linearly dependent. $2t\mathbf{v}_1 + t\mathbf{v}_2 - 8t\mathbf{v}_3 = \mathbf{0}$, where t is a real parameter. Every vector.

11. No value of k.

13. $k = -1.5$.

15. $k = -2 \pm \sqrt{3}$.

17. A nonzero vector is linearly independent. Pairs of vectors are linearly independent if neither vector is a scalar multiple of the other. $t\mathbf{v}_1 - t\mathbf{v}_2 + t\mathbf{v}_3 = \mathbf{0}$, where t is a real parameter. Every vector.

19. $\operatorname{rank} \mathbf{A} = 3$, linearly independent.

21. $\operatorname{rank} \mathbf{A} = 3$, linearly independent

23. If $\mathbf{A} = [\,\mathbf{v}_1\ \mathbf{v}_2\ \mathbf{v}_3\,]$, then $2t\mathbf{v}_1 - t\mathbf{v}_2 + 0\mathbf{v}_3 + t\mathbf{v}_4 = \mathbf{0}$, where t is a real parameter. $\mathcal{B} = \{\mathbf{v}_1, \mathbf{v}_2, \mathbf{v}_3\}$.

25. $\mathcal{S}: \begin{bmatrix} 1 \\ 1 \\ 0 \end{bmatrix}, \begin{bmatrix} -1 \\ 0 \\ 1 \end{bmatrix}$. $\mathcal{S}$ is linearly independent because neither vector is a scalar multiple of the other.

27. The set $\mathcal{S}$ contains 10 vectors in $\mathbb{R}^5$. Appeal to fact (3.20). The columns of the matrix $\begin{bmatrix} 1 & 0 & 0 & 1 & 0 \\ 1 & 1 & 0 & 0 & 1 \\ 1 & 1 & 1 & 0 & 0 \\ 0 & 1 & 1 & 1 & 1 \\ 0 & 0 & 1 & 1 & 1 \end{bmatrix}$ form a basis $\mathcal{B}$ for $\mathbb{R}^5$.

29. Rearranging $x\mathbf{u} + y(\mathbf{u} + \mathbf{v}) + z(\mathbf{u} + \mathbf{v} + \mathbf{w}) = \mathbf{0}$ gives $(x + y + z)\mathbf{u} + (y + z)\mathbf{v} + z\mathbf{w} = \mathbf{0}$, which has only the zero solution, because $\{\mathbf{u}, \mathbf{v}, \mathbf{w}\}$ is linearly independent. Hence, $x + y + z = 0$, $y + z = 0$, $z = 0$, which implies that $x = y = z = 0$.

31. Denoting the vectors by $\mathbf{v}_1, \mathbf{v}_2, \mathbf{v}_3$, we have $\mathbf{v}_3 = 8\mathbf{v}_1 - 2\mathbf{v}_2$. $\{\mathbf{v}_1, \mathbf{v}_2\}$, $\{\mathbf{v}_1, \mathbf{v}_3\}$, $\{\mathbf{v}_2, \mathbf{v}_3\}$.

33. Equation (3.17), with $k = 1$ and $\mathbf{v}_1 = \mathbf{0}$, is $x_1\mathbf{0} = \mathbf{0}$ which is true for all x_1, including nonzero values.

35. If $\mathcal{S}$ is linearly dependent, then $x_1\mathbf{v}_1 + x_2\mathbf{v}_2 = \mathbf{0}$, where at least one of x_1, x_2 are nonzero. If $x_1 \neq 0$, then $\mathbf{v}_1 = -(x_2/x_1)\mathbf{v}_2$. If $\mathbf{v}_1 = s\mathbf{v}_2$, for some scalar s, then $\mathbf{v}_1 - s\mathbf{v}_2 = \mathbf{0}$ has a nonzero solution and so $\mathcal{S}$ is linearly dependent.

37. Observe that $x_1\mathbf{v}_1 + \cdots + x_k\mathbf{v}_k + 0\mathbf{u}_1 + \cdots + 0\mathbf{u}_p = \mathbf{0}$, where some coefficient x_i is nonzero.

39. There are scalars, not all zero, such that $x_1\mathbf{v}_1 + x_2\mathbf{v}_2 + x_3\mathbf{v}_3 = \mathbf{0}$ and $x_3 \neq 0$, for otherwise $\{\mathbf{v}_1, \mathbf{v}_2\}$ would be linearly dependent. Hence $\mathbf{v}_3 = -(x_1/x_3)\mathbf{v}_1 - (x_2/x_3)\mathbf{v}_2$. If $\mathbf{v}_4 = y_1\mathbf{v}_1 + y_2\mathbf{v}_2 + y_3\mathbf{v}_3$, substitute for $\mathbf{v}_3$ to get $\mathbf{v}_4 = (y_1 - x_1/x_3)\mathbf{v}_1 + (y_2 - x_2/x_3)\mathbf{v}_2$.

41. $2t\mathbf{v}_1 - 3t\mathbf{v}_2 - t\mathbf{v}_3 + 0\mathbf{v}_4 = \mathbf{0}$, where t is a real parameter. $\mathbf{v}_1, \mathbf{v}_2, \mathbf{v}_3$.

43. If $\mathbf{A}$ is either upper or lower triangular, then the equation $\mathbf{Ax} = \mathbf{0}$ has only the zero solution because the reduced form of $\mathbf{A}$ is $\mathbf{I}_n$. Consider $\mathbf{A}^\mathsf{T}$ to show that the rows of $\mathbf{A}$ are linearly independent.

45. Yes, for example, $\begin{bmatrix} 1 \\ 0 \\ 0 \end{bmatrix}, \begin{bmatrix} 0 \\ 1 \\ 0 \end{bmatrix}, \begin{bmatrix} 1 \\ 1 \\ 0 \end{bmatrix}, \begin{bmatrix} 1 \\ -1 \\ 0 \end{bmatrix}$.

47. $d = r = 1$, basis $\mathcal{B}: \begin{bmatrix} 1 \\ 4 \end{bmatrix}$.

49. $d = r = 2$, basis $\mathcal{B}: \begin{bmatrix} 1 \\ 4 \\ 7 \end{bmatrix}, \begin{bmatrix} 2 \\ 5 \\ 8 \end{bmatrix}$.

51. $\mathbf{A} = \begin{bmatrix} 1 & 2 \\ 2 & 1 \end{bmatrix}$, $\operatorname{rank} \mathbf{A} = 2$ implies the columns of $\mathbf{A}$ are linearly independent and form a basis $\mathcal{B}$ for $\mathcal{U} = \mathbb{R}^2$ and $[\,\mathbf{v}\,]_\mathcal{B} = \begin{bmatrix} (2b - a)/3 \\ (2a - b)/3 \end{bmatrix}$.

53. $\mathbf{A} = \begin{bmatrix} 1 & 2 \\ 0 & 4 \end{bmatrix}$, $\operatorname{rank} \mathbf{A} = 2$ implies the columns of $\mathbf{A}$ are linearly independent and form a basis $\mathcal{B}$ for $\mathcal{U} = \mathbb{R}^2$ and $[\,\mathbf{v}\,]_\mathcal{B} = \begin{bmatrix} a - b/2 \\ b/4 \end{bmatrix}$.

55. If $\mathbf{A}$ is the given 3×3 matrix, then rank $\mathbf{A} = 3$ implies the columns of $\mathbf{A}$ are linearly independent and form a basis $\mathcal{B}$ for $\mathcal{U} = \mathbb{R}^3$ and $[\mathbf{v}]_{\mathcal{B}} = \begin{bmatrix} a+c \\ a+b+2c \\ a+b+c \end{bmatrix}$.

57. The rank of $[\mathbf{b}_1\ \mathbf{b}_2]$ is 2 and so $\mathcal{B} = \{\mathbf{b}_1, \mathbf{b}_2\}$ is a basis for $\mathbb{R}^2$ and $[\mathbf{v}]_{\mathcal{B}} = \begin{bmatrix} (c+d)/4 \\ (d-3c)/4 \end{bmatrix}$.

59. Let $\mathcal{U}$ be the set of all vectors in $\mathbb{R}^3$ with equal components and let $\mathbf{u}$ and $\mathbf{v}$ be in $\mathcal{U}$. Then $x_1\mathbf{u}+x_2\mathbf{v}$ is in $\mathcal{U}$ for any scalars x_1, x_2, showing that $\mathcal{U}$ is a subspace of $\mathbb{R}^3$. A basis for $\mathcal{U}$ is $[1\ 1\ 1]^{\mathsf{T}}$.

61. $\mathcal{B}$: $[1\ 1\ 0\ 0]^{\mathsf{T}}$, $[-1\ 0\ 1\ 0]^{\mathsf{T}}$, $[1\ 0\ 0\ 1]^{\mathsf{T}}$, $\dim \mathcal{U} = 3$.

63. Let $\mathcal{B}$ consist of the columns of $\mathbf{A}$. Then $[\mathbf{v}]_{\mathcal{B}} = \begin{bmatrix} a+b+0.75c \\ 0.5b \\ -0.25c \end{bmatrix}$.

65. $\mathbf{A} = \begin{bmatrix} 0 & 0 & 1 \\ 4 & 2 & 1 \\ -1 & -1 & 0 \end{bmatrix}$, $[\mathbf{v}]_{\mathcal{C}} = \begin{bmatrix} 2.5 \\ -0.5 \\ -5 \end{bmatrix}$,

$[\mathbf{v}]_{\mathcal{B}} = \begin{bmatrix} -5 \\ 4 \\ -2 \end{bmatrix}$.

67. T.

69. F. True if *independent* is replaced by *dependent*.

Exercises 3.3

1. The standard basis for $\mathbb{R}^3$ is a basis for null $\mathbf{A} = \mathbb{R}^3$. col $\mathbf{A} = \{\mathbf{0}\}$, row $\mathbf{A} = \{\mathbf{0}\}$. $n = 3 =$ nullity, rank $\mathbf{A} = 0$.

3. null $\mathbf{A} = \{\mathbf{0}\}$. The columns of $\mathbf{A}$ form a basis for col $\mathbf{A}$. The first two rows in $\mathbf{A}$ form a basis for row $\mathbf{A}$. $n = 2 =$ rank $\mathbf{A}$ and nullity $= 0$.

5. null $\mathbf{A}$ has a basis $\{\mathbf{v}\}$, where $\mathbf{v} = \begin{bmatrix}0 & 0 & 1\end{bmatrix}^{\mathsf{T}}$. The first two columns in $\mathbf{A}$ form a basis for col $\mathbf{A}$. Rows 1 and 3 form a basis for row $\mathbf{A}$. $n = 3$, rank $\mathbf{A} = 2$, nullity $= 1$.

7. null $\mathbf{A}$ has a basis $\{\mathbf{v}_1, \mathbf{v}_2\}$, where $\mathbf{v}_1 = \begin{bmatrix}-1 & 0 & 1 & 0\end{bmatrix}^{\mathsf{T}}$, $\mathbf{v}_2 = \begin{bmatrix}0 & -1 & 0 & 1\end{bmatrix}^{\mathsf{T}}$. The first two columns (rows) of $\mathbf{A}$ form a basis for col $\mathbf{A}$ (row $\mathbf{A}$). $n = 4$, rank $\mathbf{A} = 2$, nullity $= 2$.

9. null $\mathbf{A}$ has a basis $\{\mathbf{v}_1, \mathbf{v}_2\}$, where $\mathbf{v}_1 = \begin{bmatrix}-2 & 1 & 0 & 0 & 0\end{bmatrix}^{\mathsf{T}}$, $\mathbf{v}_2 = \begin{bmatrix}1 & 0 & 0 & 0 & 1\end{bmatrix}^{\mathsf{T}}$. Columns 1, 3, 4 in $\mathbf{A}$ form a basis for col $\mathbf{A}$ and the rows of $\mathbf{A}$ form a basis for row $\mathbf{A}$. $n = 5$, rank $\mathbf{A} = 3$, nullity $= 2$.

11. rank $\mathbf{A} = 3 = r$ giving col $\mathbf{A} =$ row $\mathbf{A} = \mathbb{R}^3$ and null $\mathbf{A} = \{\mathbf{0}\}$. We have $n = 3 = r + 0$ showing nullity $= 0$.

13. $\mathbf{A}$ is invertible if and only if rank $\mathbf{A} = n$ if and only if nullity $= 0$.

15. rank $\mathbf{A} = 3$ implies col $\mathbf{A} = \mathbb{R}^3$.

17. All entries on the main diagonal of $\mathbf{A}$ are nonzero.

19. Maximum dimension is 6.

21. Maximum rank $\mathbf{A} = 6$ and $8 =$ rank + nullity implies minimum nullity $= 2$.

23. $\mathcal{B} = \{\mathbf{v}_1, \mathbf{v}_2, \mathbf{v}_3\}$, where $\mathbf{v}_1, \mathbf{v}_2, \mathbf{v}_3$ denote, respectively, the first three rows of $\mathbf{A}^* = \begin{bmatrix} 1 & 0 & 0 & 0 \\ 0 & 1 & 0 & -1 \\ 0 & 0 & 1 & 1 \\ 0 & 0 & 0 & 0 \end{bmatrix}$.

The reduced form of $\mathbf{A}^{\mathsf{T}}$ is $\begin{bmatrix} 1 & 0 & 0 & 0 \\ 0 & 1 & 0 & 2 \\ 0 & 0 & 1 & -1 \\ 0 & 0 & 0 & 0 \end{bmatrix}$, which shows that $\mathcal{B}' = \{\mathbf{r}_1, \mathbf{r}_2, \mathbf{r}_3\}$, where $\mathbf{r}_1, \mathbf{r}_2, \mathbf{r}_3$ are the first three rows in $\mathbf{A}$. $\mathbf{r}_1 = 4\mathbf{v}_1 + \mathbf{v}_2 + 3\mathbf{v}_3$, $\mathbf{r}_2 = 2\mathbf{v}_1 + 4\mathbf{v}_2 + 5\mathbf{v}_3$, $\mathbf{r}_3 = 2\mathbf{v}_1 + 5\mathbf{v}_2 + 8\mathbf{v}_3$.

25. Let $\mathbf{A} = [\mathbf{v}_1\ \mathbf{v}_2\ \mathbf{v}_3\ \mathbf{v}_4\ \mathbf{v}_5]$, then $\mathcal{B} = \{\mathbf{v}_1, \mathbf{v}_3\}$ and $\mathbf{v}_2 = 2\mathbf{v}_1$, $\mathbf{v}_4 = \mathbf{v}_1 - 3\mathbf{v}_3$, $\mathbf{v}_5 = -\mathbf{v}_1 + 2\mathbf{v}_3$. A basis for null $\mathbf{A}$ is $\begin{bmatrix}-2 & 1 & 0 & 0 & 0\end{bmatrix}^{\mathsf{T}}$, $\begin{bmatrix}-1 & 0 & 3 & 1 & 0\end{bmatrix}^{\mathsf{T}}$, $\begin{bmatrix}1 & 0 & -2 & 0 & 1\end{bmatrix}^{\mathsf{T}}$.

27. $\begin{bmatrix} x_1 \\ x_2 \\ x_3 \end{bmatrix} = \begin{bmatrix} 2+3t \\ s \\ t \end{bmatrix} = \begin{bmatrix} 2 \\ 0 \\ 0 \end{bmatrix} + s\mathbf{v}_1 + t\mathbf{v}_2$, where

$\mathbf{v}_1 = \begin{bmatrix} 0 \\ 1 \\ 0 \end{bmatrix}$, $\mathbf{v}_2 = \begin{bmatrix} 3 \\ 0 \\ 1 \end{bmatrix}$ and s, t are real parameters.

$\mathcal{B} = \{\mathbf{v}_1, \mathbf{v}_2\}$ is a basis for null $\mathbf{A}$.

29. $\begin{bmatrix} x_1 \\ x_2 \end{bmatrix} = \begin{bmatrix} -0.5+t \\ t \end{bmatrix} = \begin{bmatrix} -0.5 \\ 0 \end{bmatrix} + t\mathbf{v}$, where $\mathbf{v} = \begin{bmatrix} 1 \\ 1 \end{bmatrix}$ and t is a real parameter. $\mathcal{B} = \{\mathbf{v}\}$ is a basis for null $\mathbf{A}$.

31. F. dim col $\mathbf{A}$ = rank $\mathbf{A} = r$ implies $3 = 2r$, which is impossible.

33. F. For any $m \times n$ matrix, null $\mathbf{A}$ is a subspace of $\mathbb{R}^n$ and col $\mathbf{A}$ is a subspace of $\mathbb{R}^m$.

35. F. If $\mathbf{A} = \begin{bmatrix} 1 & 1 \\ 1 & 1 \end{bmatrix}$, then $\mathbf{A}^* = \begin{bmatrix} 1 & 1 \\ 0 & 0 \end{bmatrix}$ and $\begin{bmatrix} 1 \\ 0 \end{bmatrix}$ is not in col $\mathbf{A}$.

37. T.

39. T.

41. T. $\mathbf{A}$ is $m \times 4$ and null $\mathbf{A}$ is one-dimensional implies the first three columns in $\mathbf{A}$ are pivot columns.

43. $\begin{bmatrix} 1 & 0 & -1 \\ 2 & 0 & -1 \\ 0 & -1 & 0.5 \end{bmatrix}$.

45. $\mathbf{A} = \begin{bmatrix} 1 & 0 & a & -1 \\ 0 & 1 & b & -2 \\ 0 & 1 & c & -2 \\ 2 & 0 & d & -2 \end{bmatrix}$.

47. (a) null $\mathbf{A}$: $\begin{bmatrix} t \\ 0 \end{bmatrix}$, col $\mathbf{A}$: $\begin{bmatrix} t \\ t \end{bmatrix}$, row $\mathbf{A}$: $[0\ t]^\mathsf{T}$, for all t. (b) null $\mathbf{A}$: $\begin{bmatrix} t \\ -t \end{bmatrix}$, col $\mathbf{A}$: $\begin{bmatrix} 0 \\ t \end{bmatrix}$, row $\mathbf{A}$: $[t\ t]^\mathsf{T}$, for all t.

Exercises 3.4

1. (a) Domain and codomain: $\mathbb{R}^2$, (b) ker $T = \{\mathbf{0}\}$, (c) T is one-to-one, (d) T is onto.

3. (a) Domain: $\mathbb{R}^3$, codomain : $\mathbb{R}^2$, (b) ker $T = \left\{ \begin{bmatrix} -5t \\ t \\ 3t \end{bmatrix} \right\}$, ran $T = \mathbb{R}^2$, (c) T is not one-to-one, (d) T is onto.

5. $\mathbf{A} = \begin{bmatrix} 1 & -1 \\ 1 & 1 \end{bmatrix}$. (a) Domain and codomain : $\mathbb{R}^2$, (b) ker $T = \{\mathbf{0}\}$, ran $T = \mathbb{R}^2$, (c) T is one-to-one, (d) T is onto.

7. $\mathbf{A} = \begin{bmatrix} 2 & -3 & 1 \\ 1 & 0 & 1 \end{bmatrix}$. (a) Domain : $\mathbb{R}^3$, codomain : $\mathbb{R}^2$, (b) ker $T = \left\{ \begin{bmatrix} 3t \\ t \\ -3t \end{bmatrix} \right\}$, ran $T = \mathbb{R}^2$, (c) T is not one-to-one, (d) T is onto.

9. rank $\mathbf{A} = 2$ implies dim null $\mathbf{A} = 0$. (a) T is one-to-one, (b) T is onto. null $\mathbf{A}$ is the zero subspace and the columns of $\mathbf{A}$ form a basis for col $\mathbf{A}$.

11. rank $\mathbf{A} = 2$ implies dim null $\mathbf{A} = 1$. (a) T is not one-to-one, (b) T is not onto. null $\mathbf{A}$ has a basis $[-2\ 4\ 3]^\mathsf{T}$ and the first two columns of $\mathbf{A}$ form a basis for col $\mathbf{A}$.

13. $\mathbf{x} = t\begin{bmatrix} 3 \\ 1 \end{bmatrix}$ for all scalars t.

15. $\mathbf{x} = t\begin{bmatrix} 1 \\ 0 \end{bmatrix}$ for all scalars t.

17. $\begin{bmatrix} -1 & 0 \\ 0 & 0 \end{bmatrix}$.

19. $\begin{bmatrix} 0 & 1 \\ 1 & 0 \end{bmatrix}$.

21. $\begin{bmatrix} 0 & 1 \\ 1 & 0 \end{bmatrix}$.

23. $\mathbf{A} = \frac{1}{6}\begin{bmatrix} -2 & -25 \\ 10 & 26 \end{bmatrix}$.

25. The equation of L' is $y' = x' - 6$.

27. rank $\mathbf{B} = 2$ implies col $\mathbf{B} = \mathbb{R}^2$. Hence we can solve $\mathbf{B}\mathbf{v}_1 = \mathbf{e}_1$ and $\mathbf{B}\mathbf{v}_2 = \mathbf{e}_2$ to find $\mathbf{A} = [\,\mathbf{v}_1\ \mathbf{v}_2\,]$ such that $\mathbf{BA} = \mathbf{I}_2$. We have $\mathbf{v}_1 = \begin{bmatrix} 1-t \\ t-2 \\ t \end{bmatrix}$ and $\mathbf{v}_2 = \begin{bmatrix} -s \\ 1+s \\ s \end{bmatrix}$, for any real scalars s, t.

Then $T_{\mathbf{B}} \circ T_{\mathbf{A}}$ is the identity mapping on $\mathbb{R}^2$ and so is one-to-one and onto. We have $\mathbf{C} = \mathbf{AB} = \begin{bmatrix} 1-t & -s & 1-t+s \\ t-2 & 1+s & -1-2s \\ t & s & t-s \end{bmatrix}$. The domain and codomain of $T_{\mathbf{C}}$ are both $\mathbb{R}^3$ and rank $\mathbf{C} = 2$ implies that $T_{\mathbf{C}}$ is neither one-to-one nor onto.

31. F, when n is odd, T by adding the phrase *when n is even*.

33. F.

35. F. T by adding the phrase *when $T_{\mathbf{A}}$ is onto*.

37. T. These are represented by idempotent matrices ($\mathbf{A}^2 = \mathbf{A}$).

Exercises 4.1

1. 2, (**DP1**).

3. -3, (**DP1**), (**DP2**).

5. 6, (**DP3**).

7. $\sqrt{38} \simeq 6.1644$, $\sqrt{9} + \sqrt{25} = 8$, (**N3**).

9. $2\sqrt{2} \simeq 2.8284$, 0.

11. 78°.

13. 57°.

15. (a) $(4t, 3t)$, for any scalar t, (b) $\pm(0.8, 0.6)$.

17. $k = -19$.

19. $\frac{1}{2}\|\mathbf{u}\|^2$ $(= \frac{1}{2}\|\mathbf{v}\|^2)$.

21. 17.

23. Smallest interior angle is at C and $\angle ACB \simeq 35.1°$.

25. 9 units^2.

27. $|BC| = |AC| \cos C + |AB| \cos B$.

31. $(\mathbf{u} + \mathbf{v}) \cdot (\mathbf{u} - \mathbf{v}) = \mathbf{u} \cdot \mathbf{u} + \mathbf{v} \cdot \mathbf{u} - \mathbf{u} \cdot \mathbf{v} - \mathbf{v} \cdot \mathbf{v} = \|\mathbf{u}\|^2 - \|\mathbf{v}\|^2$ using (**DP1**).

33. $4(\mathbf{u} \cdot \mathbf{v})$.

35. $d(\mathbf{u}, \mathbf{v}) = \sqrt{38} \simeq 6.1644$.

37. $3\sqrt{2} - 2 \simeq 2.2426$.

39. $\mathbf{u} \cdot \mathbf{v} = 0$ implies $\mathbf{u} \perp \mathbf{v}$. $\|\mathbf{u} + \mathbf{v}\|^2 = 36 = 30 + 6 = \|\mathbf{u}\|^2 + \|\mathbf{v}\|^2$.

41. $\text{proj}_{\mathbf{v}}(\mathbf{u}) = \frac{10}{17}\mathbf{v}$, $\text{proj}_{\mathbf{u}}(\mathbf{v}) = \frac{10}{13}\mathbf{u}$, $\text{comp}_{\mathbf{v}}(\mathbf{u}) = \frac{11}{13}\begin{bmatrix} -3 \\ 2 \end{bmatrix}$, $\text{comp}_{\mathbf{v}}(\mathbf{u}) = \frac{11}{17}\begin{bmatrix} 4 \\ -1 \end{bmatrix}$.

43. Note that $u_i v_i = v_i u_i$ for $1 \le i \le n$.

45. Note that $c(u_i v_i) = (cv_i)u_i = u_i(cv_i)$ for $1 \le i \le n$.

47. A sum of squares is nonnegative and so is the square root. If $\|\mathbf{v}\| = 0$, then $v_i^2 = 0$ for $1 \le i \le n$ and so $v_i = 0$ for each i. If $v_i = 0$ for $1 \le i \le n$, then $\|\mathbf{v}\| = 0$.

49. $\|(-2)\mathbf{u}\| = \sqrt{(-2)^2 + (-4)^2 + 2^2} = 2\sqrt{6} = |-2|\,\|\mathbf{u}\|$.

53. $6x_1 + 5x_2 - 12x_3 + 19 = 0$.

55. $\text{proj}_L(\mathbf{v}) = \frac{6}{17}\begin{bmatrix} 4 \\ 1 \end{bmatrix}$, $\text{ref}_L(\mathbf{v}) = \frac{1}{17}\begin{bmatrix} 31 \\ -22 \end{bmatrix}$, $\text{proj}_L(\mathbf{u}) = \frac{6}{5}\begin{bmatrix} 1 \\ 2 \end{bmatrix}$, $\text{ref}_L(\mathbf{v}) = \frac{1}{5}\begin{bmatrix} -8 \\ 19 \end{bmatrix}$

59. $\text{proj}_{\mathcal{U}}(\mathbf{v}) = \frac{2}{3}\mathbf{u}$, $\text{comp}_{\mathcal{U}}(\mathbf{v}) = \frac{1}{3}[1\ 6\ -2\ -1]^{\mathsf{T}}$, $\text{ref}_{\mathcal{U}}(\mathbf{v}) = \frac{1}{3}[1\ -6\ 4\ -1]^{\mathsf{T}}$.

61. $\mathbf{P} = \frac{1}{4}\begin{bmatrix} 1 & \sqrt{3} \\ \sqrt{3} & 3 \end{bmatrix}$, $\mathbf{R} = \frac{1}{2}\begin{bmatrix} -1 & \sqrt{3} \\ \sqrt{3} & 1 \end{bmatrix}$. (a) $\mathbf{Pv} = \frac{1}{2}\begin{bmatrix} 2+\sqrt{3} \\ 2\sqrt{3}+3 \end{bmatrix}$, $\mathbf{Rv} = \begin{bmatrix} -2+\sqrt{3} \\ 2\sqrt{3}+1 \end{bmatrix}$. (b) null $\mathbf{P}$ has a basis $\mathbf{b} = \begin{bmatrix} -\sqrt{3} & 1 \end{bmatrix}^{\mathsf{T}}$, null $\mathbf{R}$ is the zero subspace.

63. $\mathbf{P}^2 = \dfrac{\mathbf{u}\mathbf{u}^{\mathsf{T}}}{\mathbf{u}^{\mathsf{T}}\mathbf{u}}\dfrac{\mathbf{u}\mathbf{u}^{\mathsf{T}}}{\mathbf{u}^{\mathsf{T}}\mathbf{u}} = \dfrac{\mathbf{u}(\mathbf{u}^{\mathsf{T}}\mathbf{u})\mathbf{u}^{\mathsf{T}}}{(\mathbf{u}^{\mathsf{T}}\mathbf{u})^2} = \dfrac{\mathbf{u}\mathbf{u}^{\mathsf{T}}}{\mathbf{u}^{\mathsf{T}}\mathbf{u}} = \mathbf{P}$, cancelling the scalar $\mathbf{u}^{\mathsf{T}}\mathbf{u}$. For every $\mathbf{v}$ in $\mathbb{R}^n$, $\mathbf{Pv}$ is in the subspace $\mathcal{U}$ defined by $\mathbf{u}$ and so $\mathbf{P}(\mathbf{Pv}) = \mathbf{Pv}$ because vectors in $\mathcal{U}$ are unchanged by projection.

67. $\mathbf{A}$ defines an orthogonal projection, the other matrices are neither projections nor reflections.

69. 5, $\frac{5}{29}\vec{d}$.

71. $\mathbf{u} = \begin{bmatrix} -\cos\theta \\ -\sin\theta \end{bmatrix}$ $\mathbf{v} = \begin{bmatrix} \sin\theta \\ -\cos\theta \end{bmatrix}$ are unit vectors in the direction, respectively, of $\mathbf{b}$ and $\mathbf{a}$. Then $\mathbf{b} = \dfrac{\mathbf{w} \cdot \mathbf{u}}{\|\mathbf{u}\|^2} = 200 \sin\theta\, \mathbf{u}$ and $\mathbf{a} = \dfrac{\mathbf{w} \cdot \mathbf{v}}{\|\mathbf{v}\|^2} = 200 \cos\theta\, \mathbf{v}$. If $\theta = 0.45$ radians, then $\|\mathbf{b}\| \simeq 87$ lb.

Exercises 4.2

1. A set of two nonzero orthogonal vectors in $\mathbb{R}^3$ define a plane. Any third nonzero vector added to the set must be orthogonal to the plane.

3. The set $[0\ 0\ 0]^T$, $[1\ 0\ 0]^T$, $[0\ 1\ 0]^T$ is orthogonal.

5. $\left[-\frac{3}{5}\ 0\ \frac{4}{5}\right]^T, [0\ 1\ 0]^T$. **7.** $[-0.2\ 0.3]^T$.

9. $\left[\frac{2}{3}\ -2\ -1\right]^T$. **11.** Orthogonal

13. Orthogonal **15.** Not orthogonal

21. $x = \pm\frac{1}{\sqrt{2}}$, $y = \pm\frac{1}{\sqrt{6}}$, $z = \pm\frac{1}{\sqrt{3}}$.

23. If $\mathbf{B} = \mathbf{A}^{-1}$, then $\mathbf{B}^T = (\mathbf{A}^{-1})^T = (\mathbf{A}^T)^T = \mathbf{A} = \mathbf{B}^{-1}$.

25. Expand, noting that $\mathbf{A}\mathbf{A}^T = \mathbf{I}$ and $\mathbf{B}^T\mathbf{B} = \mathbf{I}$.

27. The rows form an orthonormal set and still do after rearranging.

29. Recall that $\cos 2\alpha = 2\cos^2\alpha - 1$ and $\sin 2\alpha = 2\sin\alpha\cos\alpha$ and let $\alpha = \theta/2$.

33. $\ker T = \{\mathbf{0}\}$ implies that T is one-to-one. $\operatorname{ran} T$ has a basis $\mathcal{B} = \{\mathbf{v}_1, \mathbf{v}_2\}$, where $\mathbf{v}_1 = [2\ 0\ 1]^T$, $\mathbf{v}_2 = [0\ 2\ 1]^T$. The matrix $\mathbf{B} = \begin{bmatrix} 1/3 & 1/3 & 1/3 \\ -1/2 & 1/2 & 0 \end{bmatrix}$ defines the inverse transformation on $\operatorname{ran} T$.

Exercises 4.3

1. Yes, because the basis vectors are orthogonal.

3. No, because $(2)(-1)+(-3)(1)+(3)(0) = -5 \neq 0$.

5. $\mathcal{B} = \{\mathbf{v}_1, \mathbf{v}_2\}$ is an orthogonal basis for the subspace $\mathcal{U}$ and $\operatorname{proj}_{\mathcal{U}}\mathbf{v} = \mathbf{v}_1 + 3\mathbf{v}_2$.

7. $\mathcal{B} = \{\mathbf{v}_1, \mathbf{v}_2\}$ is an orthogonal basis for the subspace $\mathcal{U}$ and $\operatorname{proj}_{\mathcal{U}}\mathbf{v} = \mathbf{v}_1 - \mathbf{v}_2$.

11. Let $\mathbf{A} = [\mathbf{v}_1\ \mathbf{v}_2\ \mathbf{v}_3]$. $\operatorname{null}\mathbf{A}^T$ has a basis $\mathbf{u} = [1\ -5\ 1]^T$ and $\mathbf{u}\cdot\mathbf{v}_i = 0$, for $1 \le i \le 3$, which shows that $\operatorname{null}\mathbf{A}^T$ is a subset of $(\operatorname{col}\mathbf{A})^\perp$. Conversely, if $\mathbf{w}$ belongs to $(\operatorname{col}\mathbf{A})^\perp$, then $\mathbf{w}\cdot\mathbf{v}_i = 0$, for $1 \le i \le 3$ and so $\mathbf{A}^T\mathbf{w} = \mathbf{0}$, which shows that $\mathbf{w}$ is in $\operatorname{null}\mathbf{A}^T$. Hence, the sets $(\operatorname{col}\mathbf{A})^\perp$ and $\operatorname{null}\mathbf{A}^T$ are identical.

13. For any $\mathbf{w}$ in $\mathbb{R}^n$, if $\mathbf{w}^T\mathbf{x} = 0$ for every $\mathbf{x}$ in $\mathbb{R}^n$, letting $\mathbf{x} = \mathbf{w}$ gives $0 = \mathbf{w}^T\mathbf{w} = ||\mathbf{w}||^2$, which implies $\mathbf{w} = \mathbf{0}$. Taking $\mathbf{w} = \mathbf{A}^T\mathbf{v}$ shows the implication. For the converse, if $\mathbf{A}^T\mathbf{v} = \mathbf{0}$, then $(\mathbf{A}^T\mathbf{v})^T\mathbf{x} = \mathbf{0}^T\mathbf{x} = 0$.

15. $\overline{\mathbf{v}} = \operatorname{proj}_{\mathcal{U}}(\mathbf{v}) = -\frac{5}{6}\mathbf{u}_1 + \frac{9}{210}\mathbf{u}_2$ and $\operatorname{comp}_{\mathcal{U}}\mathbf{v} = \mathbf{v} - \overline{\mathbf{v}}$.

17. Using the orthogonal basis $\mathbf{v}_1 = \mathbf{u}_1$, $\mathbf{v}_2 = 6\mathbf{u}_1 + 5\mathbf{u}_2 = [7\ 14\ 20]^T$, we have $\overline{\mathbf{v}} = -\frac{7}{5}\mathbf{v}_1 + \frac{16}{215}\mathbf{v}_2$ and $||\mathbf{v} - \overline{\mathbf{v}}|| = \frac{131}{395} \simeq 0.7924$.

19. In each part, the vectors $\{\mathbf{v}_1, \mathbf{v}_2 \ldots\}$ are an orthogonal basis for $\mathcal{U}$. (a) $\overline{\mathbf{v}} = \frac{17}{14}\mathbf{v}_1 - \frac{14}{13}\mathbf{v}_2 = \frac{1}{182}[271\ -221\ 1030]^T$ and $\mathbf{v} - \overline{\mathbf{v}} = \frac{1}{182}[93\ 403\ 62]^T$. (b) $\overline{\mathbf{v}} = \frac{9}{4}\mathbf{v}_1 - \frac{5}{4}\mathbf{v}_2 - \frac{3}{4}\mathbf{v}_3 = \frac{1}{4}[1\ 7\ 11\ 17]^T$ and $\mathbf{v} - \overline{\mathbf{v}} = \frac{7}{4}[1\ -1\ -1\ 1]$.

23. The orthonormal basis is $\frac{1}{5}[4\ 3]^T$, $\frac{1}{5}[-3\ 4]^T$.

25. The orthonormal basis is $[1\ 0\ 0\ 0]^T$, $\frac{1}{\sqrt{6}}[0\ 1\ 2\ 1]^T$, $\frac{1}{\sqrt{210}}[0\ -13\ 4\ 5]^T$.

27. $\operatorname{col}\mathbf{A}$ has a basis $\frac{1}{3}[2\ 1\ 2]^T$, $\frac{1}{\sqrt{18}}[-1\ 4\ -1]^T$. $\operatorname{col}\mathbf{A}^T$ has a basis $\frac{1}{5}[2\ 1]^T$, $\frac{1}{5}[-1\ 2]^T$. $\operatorname{null}\mathbf{A} = \{\mathbf{0}\}$.

Exercises 4.4

1. $\mathbf{A}$ is $m \times n$, $\mathbf{x}$ is $n \times 1$, $\mathbf{Ax}$ is $m \times 1$, as is $\mathbf{b}$. $\mathbf{A}^T$ is $n \times m$, $\mathbf{A}^T\mathbf{A}$ is $n \times n$ and $\mathbf{A}^T\mathbf{Ax}$ is $n \times 1$, as is $\mathbf{A}^T\mathbf{b}$.

3. If $\mathbf{A}$ is invertible, then $\mathbf{A}^T$ is also invertible and the steps in Exercise 2 are valid. The column space of $\mathbf{A}$ is $\mathbb{R}^n$ and so each vector $\mathbf{B}$ projects onto itself.

5. $\overline{\mathbf{x}} = \frac{2}{3}$.

7. $\mathbf{R}(\mathbf{x}_0) = [-2\ -5\ 1]^T$, $\overline{\mathbf{x}} = \frac{1}{16}[-2\ 3]^T$, $\mathbf{A}\overline{\mathbf{x}} = \left[-\frac{1}{2}\ 0\ \frac{1}{2}\right]^T$.

9. $\left[-\frac{1}{2}\ 0\ \frac{1}{2}\right]^T$.

11. $\overline{\mathbf{x}} = \begin{bmatrix} \frac{36}{86} + t \\ -\frac{3}{86} - 2t \\ t \end{bmatrix}$, where t is a real parameter.

15. Not necessarily. **19.** $y = 2x + 1$

29. $r = 0.86$, strong positive correlation.

31. $r = 0.67$, reasonable positive correlation.

Exercises 5.1

1. 4.08. **3.** $a^2 - b^2$.

5. $\mathbf{A}_{11} = \begin{bmatrix} 5 & 6 \\ 8 & 9 \end{bmatrix}$, $\mathbf{A}_{12} = \begin{bmatrix} 4 & 6 \\ 7 & 9 \end{bmatrix}$, $\mathbf{A}_{13} = \begin{bmatrix} 4 & 5 \\ 7 & 8 \end{bmatrix}$, $m_{11} = -3$, $m_{12} = -6$, $m_{13} = -3$, $\det(\mathbf{A}) = 0$.

7. -48. **9.** -15.

11. 44. **13.** 39.

15. 0. **17.** $\det(\mathbf{A}) = -80$.

21. $a = 0$ or $a = \pm\sqrt{2}b$.   **23.** 10^3.

25. -90.   **27.** 15.

29. 60.   **31.** 13.

35. $x = -6, -1$.   **37.** $x = 2, 3$.

39. $x = 0, 38$.

41. Perform $\mathbf{r}_3 - \mathbf{r}_2 \to \mathbf{r}_3$, pull out factors, perform $\mathbf{c}_2 + \mathbf{c}_3 \to \mathbf{c}_2$, and expand on row 3.

45. $|\mathbf{A}^T| = |-\mathbf{A}| \Rightarrow |\mathbf{A}| = (-1^n)|\mathbf{A}| \Rightarrow |\mathbf{A}|(1 - (-1)^n) = 0$ implies $2|\mathbf{A}| = 0$ when n is odd. None

of them. $\mathbf{A} = \begin{bmatrix} 0 & 0 & -1 & 1 \\ 0 & 0 & 0 & 0 \\ 1 & 0 & 0 & 1 \\ -1 & 0 & -1 & 0 \end{bmatrix}$,

$\mathbf{B} = \begin{bmatrix} 0 & 1 & -1 & 1 \\ -1 & 0 & 0 & 0 \\ 1 & 0 & 0 & 1 \\ -1 & 0 & -1 & 0 \end{bmatrix}$.

Exercises 5.2

1. 0, singular.   **3.** 58, invertible.

5. 81, $1/81$, 1125, $27/5^5$.

7. 2^7, $1/2^{10}$, $8/81$, 162, $16/81$.

9. $\begin{bmatrix} 4 & -2 \\ -3 & 1 \end{bmatrix}$.   **11.** $\begin{bmatrix} 1 & -1 & 0 \\ 0 & 1 & -1 \\ 0 & 0 & 1 \end{bmatrix}$

13. $\begin{bmatrix} -2 & 1 \\ 1.5 & -0.5 \end{bmatrix}$.

15. $\begin{bmatrix} 3/4 & -3/8 & -1/8 \\ -1/4 & 3/8 & 1/8 \\ -1 & 1/2 & 1/2 \end{bmatrix}$.

17. $|\mathbf{A}^T\mathbf{A}| = |\mathbf{A}^T||\mathbf{A}| = |\mathbf{A}|^2 \geq 0$.

25. $1/7 \begin{bmatrix} -4 & 5 \\ 3 & -2 \end{bmatrix}$.   **27.** $\begin{bmatrix} 15 & -14 \\ -16 & 15 \end{bmatrix}$.

29. $\begin{bmatrix} \sin\theta & -\cos\theta \\ \cos\theta & \sin\theta \end{bmatrix}$.

31. $\frac{1}{10} \begin{bmatrix} 16 & -26 & 11 \\ -6 & 6 & -1 \\ -2 & 12 & -7 \end{bmatrix}$.

41. $(x_1, x_2) = (-8, 2)$.   **43.** $x_2 = 2$, $x_3 = -2$.

45. $x_4 = 1/3$.

49. $\det(\mathbf{A}) = 2$, not unimodular.

51. $\det(\mathbf{A}) = -1$, unimodular.

$\mathbf{A}^{-1} = \begin{bmatrix} -1 & -6 & -3 \\ -2 & -13 & -6 \\ 3 & 16 & 8 \end{bmatrix}$.

53. $(-6, -11, 1)$.   **55.** $(22, 3, 15)$.

57. $(-12, -14, -2)$.

59. $2x - 4y - z + 3 = 0$. $a_T \simeq 2.2913$ units^2.

61. -2.   **63.** 2 units^3.

Exercises 6.1

1. No. $\mathbf{Ax} = \begin{bmatrix} -6 \\ 6 \end{bmatrix} \neq \lambda\mathbf{x}$ for any λ.

3. $\lambda = -2$, $\begin{bmatrix} 1 \\ 0 \end{bmatrix}$; $\lambda = 4$, $\begin{bmatrix} \frac{1}{3} \\ 1 \end{bmatrix}$.

5. $\lambda = 1$, $\begin{bmatrix} 1 \\ -1 \end{bmatrix}$; $\lambda = 9$, $\begin{bmatrix} 1 \\ 1 \end{bmatrix}$.

7. $\lambda = -1$, $\begin{bmatrix} 1 \\ 0 \\ -1 \end{bmatrix}$; $\lambda = 1$, $\begin{bmatrix} 0 \\ 1 \\ 0 \end{bmatrix}$; $\lambda = 3$, $\begin{bmatrix} 1 \\ 0 \\ 1 \end{bmatrix}$.

9. $\lambda = -1$, $\begin{bmatrix} 0 \\ -1 \\ 1 \end{bmatrix}$; $\lambda = 0$, $\begin{bmatrix} 1 \\ -2 \\ 1 \end{bmatrix}$; $\lambda = 1$, $\begin{bmatrix} 1 \\ 1 \\ -1 \end{bmatrix}$.

11. $\lambda = -1$, $\begin{bmatrix} 2 \\ 1 \\ 0 \end{bmatrix}$; $\lambda = 1$, $\begin{bmatrix} 7 \\ 5 \\ 2 \end{bmatrix}$.

13. $\mathbf{A} = \begin{bmatrix} -4 & -2 \\ 15 & 7 \end{bmatrix}$.

15. The eigenvalues are $-3, 4$.

17. $\mathbf{A} = \begin{bmatrix} 2 & 0.5 \\ -0.5 & 1 \end{bmatrix}$. **19.** $\mathbf{A} = \begin{bmatrix} 0 & 1 \\ 0 & 1 \end{bmatrix}$.

21. $\lambda = \pm\sqrt{a^2 + b^2}$.

23. $\lambda = 0, 0$. Eigenvectors $\mathbf{x} = \begin{bmatrix} t_1 \\ t_2 \end{bmatrix}$ in each case, where t_1 and t_2 are real parameters. $\lambda = 0$ and $a_0 = n$. Eigenvectors $\mathbf{x} = \begin{bmatrix} t_1 \\ \vdots \\ t_n \end{bmatrix}$, where $t_1, t_2, \ldots, t_n$ are real parameters. The standard basis for $\mathbb{R}^n$ is a basis of eigenvectors.

25. They have the same characteristic polynomial $\lambda^2 - 5\lambda - 2$.

27. $\begin{bmatrix} 1 & 2 \\ 2 & 1 \end{bmatrix}$.

29. $\mathbf{C} = \begin{bmatrix} 0.4 & 8.96 \\ 1 & 0 \end{bmatrix}$.

31. $\lambda = 2 : t\begin{bmatrix} 1 \\ 1 \end{bmatrix}$, $\lambda = 3 : t\begin{bmatrix} 1 \\ 2 \end{bmatrix}$, where t is a real parameter.

33. Show that the sum of the n row vectors in $\mathbf{A} - \alpha\mathbf{I}$ is zero. Hence $\text{rank}(\mathbf{A} - \alpha\mathbf{I}) < n$ implies $\det(\mathbf{A} - \alpha\mathbf{I}) = 0$, showing that α is an eigenvalue of $\mathbf{A}$.

35. $\lambda = 3, -0.4, -0.5$ and row sums are 3.

37. $|\mathbf{B} - \lambda\mathbf{I}| = |\mathbf{P}^{-1}\mathbf{A}\mathbf{P} - \lambda\mathbf{P}^{-1}\mathbf{P}| = |\mathbf{P}^{-1}||\mathbf{A} - \lambda\mathbf{I}||\mathbf{P}| = |\mathbf{A} - \lambda\mathbf{I}|$, because $|\mathbf{P}^{-1}| = |\mathbf{P}|^{-1}$. Hence $\mathbf{A}$ and $\mathbf{B}$ have the same characteristic polynomials. If $\mathbf{x}$ is an eigenvector of $\mathbf{B}$ associated with λ, then $\mathbf{P}^{-1}\mathbf{x}$ is an eigenvector of $\mathbf{A}$ associated with λ and if $\mathbf{x}$ is an eigenvector of $\mathbf{A}$ associated λ, then $\mathbf{P}^{-1}\mathbf{x}$ is an eigenvector of $\mathbf{B}$ associated with λ. The characteristic polynomials of $\mathbf{A}$ and $\mathbf{B}$ are $\lambda^2 - 4\lambda + 3$ and $\lambda^2 - 5\lambda + 3$, respectively.

39. We have $\mathbf{A}\mathbf{x} = \lambda\mathbf{x}$ if and only if $\mathbf{A}^{-1}\mathbf{x} = \frac{1}{\lambda}\mathbf{x}$, keeping in mind that the eigenvalues of $\mathbf{A}^{-1}$ are all nonzero.

41. $\mathbf{A}^k = \mathbf{O}$ for some positive integer k. If $\mathbf{A}\mathbf{v} = \lambda\mathbf{v}$, then by Exercise 38, $\mathbf{A}^k\mathbf{v} = \lambda^k\mathbf{v}$. Hence $\mathbf{O}\mathbf{v} = \mathbf{0} = \lambda^k\mathbf{v}$. But $\mathbf{v}$ is nonzero implies $\lambda^k = 0$ and so $\lambda = 0$.

43. Refer to equation (4.34) in Section 4.2. If $\mathbf{Q}$ is an orthogonal matrix, then $||\mathbf{x}|| = ||\mathbf{Q}\mathbf{x}|| = ||\lambda\mathbf{x}|| = |\lambda|||\mathbf{x}||$, which implies that $|\lambda| = 1$ because $||\mathbf{x}||$ is nonzero.

45. The entries on the main diagonal of $s\mathbf{A}$ are sa_{ii}, $1 \leq i \leq n$.

47. If $\mathbf{A} = \mathbf{C}\mathbf{B}\mathbf{C}^{-1}$, then (using Exercise 46) $\text{trace}(\mathbf{A}) = \text{trace}(\mathbf{C}\mathbf{B}\mathbf{C}^{-1}) = \text{trace}(\mathbf{C}\mathbf{C}^{-1}\mathbf{B}) = \text{trace}(\mathbf{B})$.

49. The characteristic polynomial is $\lambda^3 - 3\lambda^2 + 2\lambda = \lambda(\lambda - 1)(\lambda - 2)$. Show that $\mathbf{A}(\mathbf{A} - \mathbf{I})(\mathbf{A} - 2\mathbf{I}) = \mathbf{O}$.

Exercises 6.2

1. $\mathbf{A} = \begin{bmatrix} 2.5 & 0.5 \\ 0.5 & 2.5 \end{bmatrix}$, $\lambda = 2, 3$.

3. $\begin{bmatrix} -16 & -6 \\ 45 & 17 \end{bmatrix}$.

5. $\lambda = 0, 0$, yes, $a_0 = 2 = g_0$, any invertible $\mathbf{P}$.

7. $\lambda = 0, 4$, yes, distinct eigenvalues, $\mathbf{P} = \begin{bmatrix} 1 & -1 \\ 1 & 1 \end{bmatrix}$.

9. $\lambda = -1, 1, 1$, yes, $g_{-1} = a_{-1} = 1$, $g_1 = a_1 = 2$, $\mathbf{P} = \begin{bmatrix} 1 & 1 & 0 \\ -1 & 1 & 0 \\ 0 & 0 & 1 \end{bmatrix}$.

11. $(a - d)^2 + 4bc > 0$ is the discriminant of the quadratic characteristic polynomial $p(\lambda)$ of $\mathbf{A}$ and, being positive, shows there are distinct real roots for $p(\lambda)$. Hence $\mathbf{A}$ is diagonalizable by Theorem 6.4. The converse is false: Take $\mathbf{A} = \mathbf{I}_2$.

13. $\det(\mathbf{P}^{-1}\mathbf{A}\mathbf{P}) = \det(\mathbf{P}^{-1})\det(\mathbf{A})\det(\mathbf{P}) = \det(\mathbf{A}) = \det(\mathbf{D}) = \lambda_1 \cdots \lambda_n$. Yes (see p. 278).

15. $\mathbf{A}^{-1} = \mathbf{P}\mathbf{D}^{-1}\mathbf{P}^{-1}$, where the entries on the main diagonal of $\mathbf{D}^{-1}$ are the inverses of the nonzero eigenvalues of $\mathbf{A}$.

17. $\lambda\mathbf{I}_n$.

19. The eigenvalues $\lambda = 0, -1, 2$ are distinct and so $\mathbf{A}$ is diagonalizable. $\mathbf{P} = \begin{bmatrix} 1 & 1 & -1 \\ 1 & -1 & 1 \\ 0 & 1 & 2 \end{bmatrix}$.

21. $\mathbf{A}^k = \mathbf{A}$ when k is odd, $\mathbf{A}^k = \mathbf{I}_2$ when k is even, $\mathbf{I}_2$.

23. $\frac{1}{2}\begin{bmatrix} 3^k+(-1)^k & 0 & 3^k+(-1)^{k+1} \\ 0 & 2 & 0 \\ 3^k+(-1)^{k+1} & 0 & 3^k+(-1)^k \end{bmatrix}$, $\begin{bmatrix} 29525 & 0 & 29524 \\ 0 & 1 & 0 \\ 29524 & 0 & 29525 \end{bmatrix}$.

25. $\begin{bmatrix} 1 & 1-0.5^k & 1-0.5^{k-1} \\ 0 & 0.5^k & 0.5^{k-1} \\ 0 & 0 & 0 \end{bmatrix}$.

27. $\frac{1}{5}\begin{bmatrix} 2+(\frac{2}{5})^{k-1} & 2-(\frac{2}{5})^{k-1} & 2 \\ 2-(\frac{2}{5})^{k-1} & 2+(\frac{2}{5})^{k-1} & 2 \\ 1 & 1 & 1 \end{bmatrix}$

Exercises 6.3

1. $x_4 = y_4 = 810$.

3. Step matrix is $\mathbf{A} = \begin{bmatrix} 0.9 & 0.2 \\ 0.3 & 0.8 \end{bmatrix}$. $\mathbf{u}_2 = \begin{bmatrix} 798 \\ 618 \end{bmatrix}$, $\mathbf{u}_4 \simeq \begin{bmatrix} 904 \\ 839 \end{bmatrix}$. Possible error due to the size of the numbers.

7. $x_n = \frac{1}{3}(4-(-2)^n)$.

Exercises 6.4

1. $\mathbf{w} = \begin{bmatrix} \frac{9}{17} \\ \frac{8}{17} \end{bmatrix} \simeq \begin{bmatrix} 0.5294 \\ 0.4706 \end{bmatrix}$. $\mathbf{T}^k \simeq [\mathbf{w}\ \ \mathbf{w}]$ for $k \geq 27$.

3. If there are a_k cars at A and d_k cars at D on day k, then $\mathbf{u}_k = \begin{bmatrix} a_k \\ d_k \end{bmatrix}$ is the distribution vector for day k. A car rented from A is always returned to D. A car rented from D is returned to D 70% of the time and to A 30% of the time. $\mathbf{u}_3 = \begin{bmatrix} 132 \\ 168 \end{bmatrix}$. The steady state is $\mathbf{w} = \begin{bmatrix} 124 \\ 176 \end{bmatrix}$.

5. If there are, respectively, d_k, s_k people living downtown, in the suburbs, at the beginning of year k, then $\mathbf{u}_k = \begin{bmatrix} d_k \\ s_k \end{bmatrix}$ is the distribution vector for year k. $\mathbf{T} = \begin{bmatrix} 0.8 & 0.3 \\ 0.2 & 0.7 \end{bmatrix}$ $\mathbf{u}_1 = \begin{bmatrix} 0.8 \\ 0.2 \end{bmatrix}$, $\mathbf{u}_2 = \begin{bmatrix} 0.7 \\ 0.3 \end{bmatrix}$, $\mathbf{u}_3 = \begin{bmatrix} 0.65 \\ 0.35 \end{bmatrix}$. There is a 65% chance that the person will be in the city in year 3. $\mathbf{T}$ is regular. Steady state: $\mathbf{w} = \begin{bmatrix} 0.6 \\ 0.4 \end{bmatrix}$.

7. $\mathbf{u}_1 = \begin{bmatrix} 0 \\ 0.5 \\ 0 \\ 0.5 \end{bmatrix}$, $\mathbf{u}_2 = \begin{bmatrix} 0.25 \\ 0 \\ 0.25 \\ 0.5 \end{bmatrix}$, $\mathbf{u}_3 = \begin{bmatrix} 0.25 \\ 0.125 \\ 0 \\ 0.625 \end{bmatrix}$, $p = 0$.

9. (b) Regular for $0 < a, b < 1$. (c) If $\mathbf{u}_0 = \begin{bmatrix} x_0 \\ y_0 \end{bmatrix}$, then $c_1 = \dfrac{x_0+y_0}{a+b}$, $c_2 = \dfrac{ax_0-by_0}{a+b}$. $\mathbf{T}^k\mathbf{u}_0 = c_1\mathbf{v}_1 + c_2(1-a-b)^k\mathbf{v}_2$ and $0 < a+b < 2$ implies that $|1-a-b| < 1$ so that $(1-a-b)^k$ approaches zero as k becomes large.

13. $\mathbf{u}_2 = \begin{bmatrix} 0.381 \\ 0.183 \\ 0.436 \end{bmatrix}$, $\mathbf{w} = \begin{bmatrix} \frac{1}{3} \\ \frac{1}{6} \\ \frac{1}{2} \end{bmatrix}$.

15. $\mathbf{w} = \begin{bmatrix} \frac{11}{45} \\ \frac{13}{45} \\ \frac{21}{45} \end{bmatrix}$.

17. $\mathbf{P}$ is not regular, $\mathbf{w} = \begin{bmatrix} x_0 + \frac{1}{2}y_0 \\ 0 \\ \frac{1}{2}y_0 + z_0 \end{bmatrix}$.

19. First experiment: $\frac{1}{2}$ for drawing a red or a black. Second experiment: Suppose red (black) chosen on first experiment then $\frac{4}{9}$ for drawing a red (black) and $\frac{5}{9}$ for drawing a black (red). The probabilities are not independent of past experiments. A Markov chain results if the ball is replaced after each experiment.

Exercises 6.5

1. $y_1 = e^t + 1$, $y_2 = e^t - 1$.

3. $y_1 = e^t + e^{-t}$, $y_2 = -e^t + e^{-t}$.

5. $\mathbf{y} = \alpha\mathbf{v} + \beta(t\mathbf{v} + \mathbf{u})$, where $\mathbf{v} = (-3, 1)$, $\mathbf{u} = (-1, 0)$.

7. $y_1 = \alpha e^{2t} - (\beta + \gamma)e^{-t}$, $y_2 = \alpha e^{2t} + \gamma e^{-t}$ $y_3 = \alpha e^{2t} + \beta e^{-t}$. Particular solution: put $\alpha = 1$, $\beta = 0$, $\gamma = -2$.

9. $y_1 = \alpha e^{-t} + \beta e^{t}$, $y_2 = -\alpha e^{-t} + \beta e^{t}$, $y_3 = \gamma e^{t}$.

Exercises 7.1

1. In this case, the left side of (**A7**) is $(st)(x, y) = (2stx, 2sty)$, while the right side is $s(t(x, y)) = s(2tx, 2ty) = (4stx, 4sty)$ and taking $x = y = s = t = 1$ shows that (**A7**) fails.

3. Adding $-\mathbf{w}$ to the left side of the first equation in (**P1**), we have $(\mathbf{u}+\mathbf{w})+(-\mathbf{w}) = \mathbf{u}+(\mathbf{w}+(-\mathbf{w})) = \mathbf{u} + \mathbf{0} = \mathbf{u}$ and similarly for the right side, which gives $\mathbf{u} = \mathbf{v}$.

5. By (**A6**), $s(\mathbf{0} + \mathbf{0}) = s\mathbf{0} + s\mathbf{0}$. Also, $s(\mathbf{0} + \mathbf{0}) = s\mathbf{0}$ because $\mathbf{0} + \mathbf{0} = \mathbf{0}$ by (**A3**). But then $s\mathbf{0} + s\mathbf{0} = s\mathbf{0}$ implies (**P5**) using (**P2**).

7. If $s \neq 0$, let $t = 1/s$ so that $ts = 1$. Then (**A8**) gives $\mathbf{v} = 1\mathbf{v} = (ts)\mathbf{v} = t(s\mathbf{v})$ (using (**A7**)). But then $\mathbf{v} = t\mathbf{0} = \mathbf{0}$ by (**P5**). Thus either $s = 0$ or $\mathbf{v} = \mathbf{0}$.

9. Consider the equation $\mathbf{v} + \mathbf{w} = \mathbf{0}$. Using axioms (**A2**), (**A1**), (**A4**), (**A1**), (**A3**) in order, we have $-\mathbf{v} = (-\mathbf{v})+\mathbf{0} = (-\mathbf{v})+(\mathbf{v}+\mathbf{w}) = (-\mathbf{v}+\mathbf{v})+\mathbf{w} = (\mathbf{v} + (-\mathbf{v})) + \mathbf{w} = \mathbf{0} + \mathbf{w} = \mathbf{w} + \mathbf{0} = \mathbf{w}$, showing that $\mathbf{w} = -\mathbf{v}$.

11. We have $\alpha\begin{bmatrix} a_1 & 0 \\ 0 & b_1 \end{bmatrix} + \beta\begin{bmatrix} a_2 & 0 \\ 0 & b_2 \end{bmatrix} = \begin{bmatrix} \alpha a_1 + \beta a_2 & 0 \\ 0 & \alpha b_1 + \beta b_2 \end{bmatrix}$ is diagonal, for all scalars α, β. Use Exercise 2 to conclude that the set is a subspace. The result applies in $\mathbb{R}^{n\times n}$.

13. We have $\alpha\begin{bmatrix} a_1 & 0 \\ b_1 & c_1 \end{bmatrix} + \beta\begin{bmatrix} a_2 & 0 \\ b_2 & c_2 \end{bmatrix} = \begin{bmatrix} \alpha a_1 + \beta a_2 & 0 \\ \alpha b_1 + \beta b_2 & \alpha c_1 + \beta c_2 \end{bmatrix}$ is lower triangular, for all scalars α, β. Use Exercise 2 to conclude that the set is a subspace. The result applies in $\mathbb{R}^{n\times n}$.

15. The set is a subspace using Exercises 2 and the same result applies in $\mathbb{R}^{n\times n}$.

17. Axiom (**C1**) fails because, for example, $\begin{bmatrix} 1 & 0 \\ 0 & 0 \end{bmatrix} + \begin{bmatrix} 0 & 0 \\ 0 & 1 \end{bmatrix} = \mathbf{I}_2$, which is invertible. In general, the set of all singular matrices in $\mathbb{R}^{n\times n}$ is not a subspace.

19. If the $n \times n$ matrix $\mathbf{A} \neq \mathbf{O}$ and $\mathbf{A}^2 = \mathbf{A}$, then $(-\mathbf{A})^2 = \mathbf{A}^2 = \mathbf{A} \neq -\mathbf{A}$, showing that the object $(-1)\mathbf{A}$ is not idempotent and so (**S2**) fails.

21. Using Exercise 2, the set is a subspace.

23. Let $\mathbf{A}$ and $\mathbf{B}$ be $n \times n$ symmetric matrices. Then, using properties of the transpose, we have $(\alpha\mathbf{A} + \beta\mathbf{B})^{\mathsf{T}} = \alpha\mathbf{A}^{\mathsf{T}} + \beta\mathbf{B}^{\mathsf{T}} = \alpha\mathbf{A} + \beta\mathbf{B}$, showing that $\alpha\mathbf{A} + \beta\mathbf{B}$ is symmetric, for all scalars α, β. The set is a subspace.

25. The set is a subspace using Exercise 2. Note that the set of all $n \times n$ matrices of the form $a\mathbf{I}_n + \mathbf{S}$, where $\mathbf{S}^{\mathsf{T}} = -\mathbf{S}$, is a subspace of $\mathbb{R}^{n\times n}$.

27. Use Exercise 2. If $a_1+2b_1+3c_1 = 0$ and $a_2+2b_2+3c_2 = 0$, then $\alpha(a_1 + b_1t + c_1t^2) + \beta(a_2 + b_2t + c_2t^2) = (\alpha a_1+\beta a_2)+(\alpha b_1+\beta b_2)t+(\alpha c_1+\beta c_2)t^2$, and $\alpha a_1 + \beta a_2 + 2(\alpha b_1 + \beta b_2) + 3(\alpha c_1 + \beta c_2) = \alpha(a_1 + 2b_1 + 3c_1) + \beta(a_2 + 2b_2 + 3c_2) = 0$, for all scalars α, β. Hence, $\mathcal{U}$ is a subspace.

29. If $\mathbf{p}_1$, $\mathbf{p}_2$ are in $\mathcal{U}$, then $(\alpha\mathbf{p}_1+\beta\mathbf{p}_2)(0) = \alpha\mathbf{p}_1(0) + \beta\mathbf{p}_2(0) = 0 + 0 = 0$, for all scalars α, β. Hence, using Exercise 2, $\mathcal{U}$ is subspace.

31. Let $\mathbf{f}(t) = a$ and $\mathbf{g}(t) = b$ for all t, where a and b are constant, then $(\alpha\mathbf{f} + \beta\mathbf{g})(t) = \alpha\mathbf{f}(t) + \beta\mathbf{g}(t) = \alpha a + \beta b$, independent of t and therefore constant. Hence $\mathcal{U}$ is a subspace.

33. We have $(\alpha\mathbf{f} + \beta\mathbf{g})' = \alpha\mathbf{f}' + \beta\mathbf{g}'$, where $'$ denotes differentiation, and so $(\alpha\mathbf{f} + \beta\mathbf{g})$ is differentiable whenever $\mathbf{f}$ and $\mathbf{g}$ are. Hence $\mathcal{U}$ is a subspace.

35. If $\mathbf{v} = (x, y)$, where $y \neq 0$, then $1\mathbf{v} \neq \mathbf{v}$ and so (**A8**) fails (the other axioms are valid).

37. $(s + t)(x, y) = (x, y)$ and $s(x, y) + t(x, y) = (2x, 2y)$ showing that (**A5**) fails.

39. Consider objects $\mathbf{v}_1 = \mathbf{u}_1 + \mathbf{w}_1$ and $\mathbf{v}_2 = \mathbf{u}_2 + \mathbf{w}_2$, where $\mathbf{u}_1$, $\mathbf{u}_2$ are in $\mathcal{U}$ and $\mathbf{w}_1$, $\mathbf{w}_2$ are in $\mathcal{W}$. Then, for all scalars α, β, we have $\alpha\mathbf{v}_1 + \beta\mathbf{v}_2 = (\alpha\mathbf{u}_1 + \beta\mathbf{u}_2) + (\alpha\mathbf{w}_1 + \beta\mathbf{w}_2)$. But $\alpha\mathbf{u}_1 + \beta\mathbf{u}_2$ is in $\mathcal{U}$ and $\alpha\mathbf{w}_1 + \beta\mathbf{w}_2$ is in $\mathcal{W}$ because $\mathcal{U}$ and $\mathcal{W}$ are subspaces. Hence $\alpha\mathbf{v}_1 + \beta\mathbf{v}_2$ is the sum of an object from $\mathcal{U}$ and an object from $\mathcal{W}$. Thus $\mathcal{U} + \mathcal{W}$ is a subspace. If $\mathbf{v}_1$, $\mathbf{v}_2$ are in $\mathcal{U} \cap \mathcal{W}$, then for all scalars α, β, the object $\alpha\mathbf{v}_1 + \beta\mathbf{v}_2$ is in both $\mathcal{U}$ and $\mathcal{W}$ because these are subspaces. Hence $\mathcal{U} \cap \mathcal{W}$ is a subspace.

41. There are many possibilities. For example, let $\mathcal{U}$ be the set of matrices of the form $\begin{bmatrix} a & b \\ 0 & 0 \end{bmatrix}$ and $\mathcal{W}$ be the set of matrices of the form $\begin{bmatrix} 0 & 0 \\ c & d \end{bmatrix}$ for all a, b, c, d.

43. If $\mathbf{A}_1$ and $\mathbf{A}_2$ are matrices in $\mathcal{U}$, then $\text{trace}(\alpha\mathbf{A}_1 + \beta\mathbf{A}_2) = \alpha\,\text{trace}(\mathbf{A}_1) + \beta\,\text{trace}(\mathbf{A}_2) = \alpha 0 + \beta 0 = 0$, for all scalars α, β, showing that $\mathcal{U}$ is a subspace. The subspace is proper: For example, when $n = 2$, the matrix $\begin{bmatrix} 1 & 1 \\ 1 & -1 \end{bmatrix}$ is in $\mathcal{U}$ while $\begin{bmatrix} 1 & 1 \\ 1 & 1 \end{bmatrix}$ is not in $\mathcal{U}$.

47. T.

49. F. The objects in $\mathbf{F}[1, 2]$ are not objects in $\mathbf{F}$.

Exercises 7.2

1. No.

3. Yes, because $\sin^2 t + \cos^2 t = 1$, for all t.

5. Yes. $-2(-1+t+t^2)+3(1+2t+t^2) = 5+4t+t^2$.

7. The solution to $x_1\mathbf{M}_1 + x_2\mathbf{M}_2 + x_3\mathbf{M}_3 + x_4\mathbf{M}_4 = \begin{bmatrix} a & b \\ c & d \end{bmatrix}$ is $x_1 = \frac{1}{2}(b + c)$, $x_2 = -\frac{1}{2}(b + c) + d$, $x_3 = a - d$, $x_4 = a + \frac{1}{2}(b - c) - d$. Thus $\mathcal{S}$ spans $\mathbb{R}^{2\times 2}$.

9. No, because two objects in $\mathbf{P}_2$ cannot span $\mathbf{P}_2$.

11. The given set is linearly independent and contains 4 ($= \dim \mathbb{R}^{2\times 2}$) objects. Hence it is basis and spans $\mathbb{R}^{2\times 2}$.

13. No, because the equation $a(1+t)+b(1-t+t^2) = (3+t)$ is true for no a, b.

15. The equation $x_1\mathbf{p}_1+x_2\mathbf{p}_2+x_3\mathbf{p}_3 = a+bt+ct^2$ has solution $x_1 = \frac{1}{2}(a - b)$, $x_2 = b$, $x_3 = c$, showing that any object in $\mathbf{P}_2$ is a linear combination of the given objects.

17. The representation is $\mathbf{p} = (1 - \alpha)\mathbf{p}_1 + (3 - \alpha)\mathbf{p}_2 + \alpha\mathbf{p}_3$, for any scalar α. No. The object $\mathbf{p}$ can be written as a linear combination of $\mathbf{p}_1, \mathbf{p}_2, \mathbf{p}_3$ in infinitely many ways.

19. Linearly dependent. $\mathbf{M}_1 = -\mathbf{M}_2 - \mathbf{M}_3 + 2\mathbf{M}_4$.

21. Linearly dependent. $\mathbf{M}_4 = \mathbf{M}_1 + \mathbf{M}_2$.

23. Linearly dependent. $\mathbf{f}_3 = \cos 2t = \cos^2 t - \sin^2 t = \mathbf{f}_2 - \mathbf{f}_1$.

25. Linearly dependent. $\mathbf{f}_2 = \ln 3 + \ln t = \mathbf{f}_3 + \mathbf{f}_1$.

27. Linearly independent.

29. Linearly independent.

31. Adjoin $\mathbf{M}_4 = \begin{bmatrix} 1 & 0 \\ 0 & 0 \end{bmatrix}$.

33. Adjoin $1, t$.

35. $\mathcal{S}$ is linearly independent. Hence $\dim \mathcal{U} = 4$ and $\mathcal{U} = \mathbf{P}_3$.

37. $\mathcal{U}$ consists of all matrices of the form $\begin{bmatrix} a+b & c \\ b+c & a+c \end{bmatrix}$, for all scalars a, b, c. $\mathcal{S}$ is linearly independent and so it is a basis for $\mathcal{U}$ and $\dim \mathcal{U} = 3$. $\mathcal{U}$ is a proper subspace of $\mathbb{R}^{2\times 2}$.

39. A basis for $\mathcal{U}$ is $\mathbf{p}_1 = 1 - t$, $\mathbf{p}_2 = 1 - t^2$.

43. The set is linearly independent and contains five polynomials. Hence it is a basis because $\dim \mathbf{P}_4 = 5$.

Exercises 7.3

1. $\begin{bmatrix} \frac{9}{2} \\ -\frac{3}{2} \end{bmatrix}$.

3. $T(\alpha\mathbf{A}+\beta\mathbf{B}) = (\alpha\mathbf{A}+\beta\mathbf{B})\mathbf{P}+\mathbf{P}(\alpha\mathbf{A}+\beta\mathbf{B}) = \alpha\mathbf{AP}+\beta\mathbf{BP}+\alpha\mathbf{PA}+\beta\mathbf{PB} = \alpha(\mathbf{AP}+\mathbf{PA})+\beta(\mathbf{BP}+\mathbf{PB}) = \alpha T(\mathbf{A}) + \beta T(\mathbf{B})$, for all scalars α, β. Hence T is linear.

7. (a) The coordinate mapping defines an isomorphism G from $\mathcal{V}$ onto $\mathbb{R}^n$ and so the inverse transformation G^{-1} defines an isomorphism from $\mathbb{R}^n$ to $\mathcal{V}$. (b) If H denotes the isomorphism from $\mathcal{W}$ onto $\mathbb{R}^n$ defined by the coordinate mapping, then the composition mapping $H^{-1} \circ G$ defines an isomorphism from $\mathcal{V}$ onto $\mathcal{W}$.

11. If $\mathcal{S}$ is linearly dependent, then the equation $x_1\mathbf{v}_1 + x_2\mathbf{v}_2 + \cdots + x_k\mathbf{v}_k = \mathbf{0}$ has a nonzero solution. But using the linearity property (7.29), the equation $x_1T(\mathbf{v}_1) + x_2T(x_2) + \cdots + x_kT(\mathbf{v}_k) = \mathbf{0}$ also has a nonzero solution. If $\{T(\mathbf{v}_1), T(\mathbf{v}_2), \ldots, T(\mathbf{v}_k)\}$ is linearly independent, then $\mathcal{S}$ is linearly independent.

13. Suppose $\mathcal{S}$ is a basis for $\mathcal{V}$ and that $x_1T(\mathbf{v}_1) + x_2T(x_2) + \cdots + x_kT(\mathbf{v}_k) = \mathbf{0}$. By linearity of T, $T(x_1\mathbf{v}_1 + x_2\mathbf{v}_2 + \cdots + x_k\mathbf{v}_k) = \mathbf{0}$ and so $x_1\mathbf{v}_1 + x_2\mathbf{v}_2 + \cdots + x_k\mathbf{v}_k = \mathbf{0}$ because T is one-to-one. But then $x_1 = \cdots = x_k = 0$ because $\mathcal{S}$ is linearly independent, showing that $\{T(\mathbf{v}_1), T(\mathbf{v}_2), \ldots, T(\mathbf{v}_k)\}$ is linearly independent in $\mathcal{W}$. Hence $\dim \mathcal{W} \geq k = \dim \mathcal{V}$.

15. From the solution to Exercise 13, $\mathcal{S}' = \{T(\mathbf{v}_1), T(\mathbf{v}_2), \ldots, T(\mathbf{v}_k)\}$ is linearly independent. For any object $\mathbf{w}$ in $\mathcal{W}$ there is an object $\mathbf{v}$ in $\mathcal{V}$ such that $T(\mathbf{v}) = \mathbf{w}$ (T is onto) and $\mathcal{S}$ spans $\mathcal{V}$ implies $\mathbf{v} = x_1\mathbf{v}_1 + x_2\mathbf{v}_2 + \cdots + x_k\mathbf{v}_k$ for some scalars $x_1, x_2, \ldots, x_k$. But then $\mathbf{w} = T(\mathbf{v}) = x_1T(\mathbf{v}_1) + x_2T(x_2) + \cdots + x_kT(\mathbf{v}_k)$ and so $\mathcal{S}'$ spans $\mathcal{W}$.

17. The matrix equation $(\mathbf{P}[T]_{\mathcal{C}} - [T]_{\mathcal{B}}\mathbf{P})[\mathbf{v}]_{\mathcal{C}} = \mathbf{0}$ is true for all vectors $[\mathbf{v}]_{\mathcal{C}}$ in $\mathbb{R}^n$ so that $\mathbf{P}[T]_{\mathcal{C}} - [T]_{\mathcal{B}}\mathbf{P} = \mathbf{O}$, and the result follows.

19. $[T]_{\mathcal{B}} = \begin{bmatrix} 0 & 0 & 0 & 0 \\ 0 & 1 & 0 & 0 \\ 0 & 0 & 1 & 0 \\ 0 & 0 & 0 & 1 \end{bmatrix}$.

21. (a) $\mathbf{A} = \mathbf{PAP}^{-1}$, where $\mathbf{P} = \mathbf{I}_n$. (b) $\mathbf{A} = \mathbf{PBP}^{-1}$ implies $\mathbf{B} = \mathbf{QAQ}^{-1}$, where $\mathbf{Q} = \mathbf{P}^{-1}$. (c) If $\mathbf{A} = \mathbf{PBP}^{-1}$ and $\mathbf{B} = \mathbf{QCQ}^{-1}$, then $\mathbf{A} = (\mathbf{PQ})\mathbf{C}(\mathbf{PQ})^{-1}$.

23. Let λ be an eigenvalue of $\mathbf{B} = \mathbf{PCP}^{-1}$ with associated eigenvector $\mathbf{v}$. Then $\mathbf{Bv} = \lambda\mathbf{v}$ implies $\mathbf{PCP}^{-1} = \lambda\mathbf{v}$ implies $\mathbf{Cw} = \lambda\mathbf{w}$, where $\mathbf{w} = \mathbf{P}^{-1}\mathbf{v}$. Now $\mathbf{v}$ nonzero implies $\mathbf{w}$ nonzero and so λ is an eigenvalue of $\mathbf{C}$. By symmetry (see Exercise 21) every eigenvalue of $\mathbf{C}$ is an eigenvalue of $\mathbf{B}$.

25. Their traces are not equal.

27. Trace: 5, rank: 1, eigenvalues: 0,5. Take $\mathbf{P} = \begin{bmatrix} 1 & -1 \\ -2 & -1 \end{bmatrix}$.

29. Suppose $\mathbf{u}$, $\mathbf{v}$ are in $\ker T$. Then $T(\alpha\mathbf{u} + \beta\mathbf{v}) = \alpha T(\mathbf{u}) + \beta T(\mathbf{v}) = \alpha\mathbf{0} + \beta\mathbf{0} = \mathbf{0}$, showing that $\alpha\mathbf{u} + \beta\mathbf{v}$ is in $\ker T$ for all scalars α, β. Hence $\ker T$ is a subspace of $\mathcal{V}$ (see Exercises 7.1, 2). Suppose $\mathbf{w}_1$, $\mathbf{w}_2$ are in $\operatorname{ran} T$. Then $T(\mathbf{v}_1) = \mathbf{w}_1$ and $T(\mathbf{v}_2) = \mathbf{w}_2$ for some objects $\mathbf{v}_1$, $\mathbf{v}_2$ in $\mathcal{V}$. Then $T(\alpha\mathbf{v}_1 + \beta\mathbf{v}_2) = \alpha T(\mathbf{v}_1) + \beta T(\mathbf{v}_2) = \alpha\mathbf{w}_1 + \beta\mathbf{w}_2$, showing that $\alpha\mathbf{w}_1 + \beta\mathbf{w}_2$ is in $\operatorname{ran} T$ for all scalars α, β. Hence $\operatorname{ran} T$ is a subspace of $\mathcal{W}$.

31. (a) $\ker T = \{\mathbf{0}\}$, (b) $\operatorname{ran} T = \mathbf{P}_2$.

33. We have $T(\alpha\mathbf{p}_1 + \beta\mathbf{p}_2) = (\alpha\mathbf{p}_1 + \beta\mathbf{p}_2) + t(\alpha\mathbf{p}_1 + \beta\mathbf{p}_2)' = \alpha\mathbf{p}_1 + \beta\mathbf{p}_2 + \alpha\mathbf{p}_1' + \beta\mathbf{p}_2' = \alpha(\mathbf{p}_1 + \mathbf{p}_1') + \beta t(\mathbf{p}_2 + \mathbf{p}_2') = \alpha T(\mathbf{p}_1) + \beta T(\mathbf{p}_2)$. Hence T is linear. If $\mathbf{p} = a + bt + bt^2$, then $T(\mathbf{p}) = a + bt + ct^2 + t(b + 2ct) = a + 2bt + 3ct^2$ and so $T(\mathbf{p}_1) = T(\mathbf{p}_2)$ implies $\mathbf{p}_1 = \mathbf{p}_2$ (T is one-to-one) and $T(a + \frac{1}{2}t + \frac{1}{3}ct^2) = \mathbf{p}$ (T is onto).

35. F.

37. T.

Exercises 8.1

1. $13 - 3i$.

3. $41 - 13i$.

5. $14 - i\sqrt{2}$.

7. $-5 + 12i$.

9. $52 + 47i$.

11. $-95 + 63i$.

13. 0.

15. i.

17. $\frac{1}{2} + \frac{1}{2}i$.

19. $-1 + i$.

21. 0.

23. $\frac{37}{13} - \frac{36}{13}i$.

25. 0.

27. 2.

29. $-\frac{4}{25} + \frac{3}{25}i$.

31. $a = -\frac{6}{7}$, $b = -\frac{2}{7}$.

33. $a = 1$, $b = 1$.

35. $a = \frac{28}{19}$, $b = \frac{25}{19}$.

37. $x = -2$, $y = \frac{3}{2}$, $z = 2$, $w = -\frac{1}{2}$.

39. 5.

41. 1.

43. $\sqrt{5}$.

45. $z = \pm(c + i/c)$, where $c = \sqrt{1 + \sqrt{2}}$.

47. $z = \pm(2 + i)$.

49. $1 - \frac{3}{4}i$.

51. $\frac{3}{2}(1 + i)$, $-2 + 2i$.

53. $\sqrt{3} \pm i$.

55. Let $p = (1 - \sqrt{3})/2$, $q = (1 + \sqrt{3})/2$ then $z = p - iq, q - ip$.

59. $1 - 5x - 10x^2 + 10x^3 + 5x^4 - x^5$.

61. $5\sqrt{13}$.

63. 64.

65. 729/8.

67. $\sqrt{2}/4$.

69. $z = 1+3i, -1-3i$.

71. $8ixy(x^2-y^2)/(x^2+y^2)$.

Exercises 8.2

3. The values of $1/z$ are (i) $\frac{1}{2}$, (ii) $-i$, (iii) i, (iv) $\frac{1}{2}-\frac{1}{2}i$, (v) $\frac{3}{25}+\frac{4}{25}i$.

5. (i) The points on a circle of radius 3, center the origin. (ii) The points on a circle of radius 1, center the $z=2$. (iii) The points on a circle of radius 3, center $z=7$. (iv) The points on the unit circle.

9. The set of all points are $z = -\frac{1}{2}+yi$, $-\infty < y < \infty$.

Exercises 8.3

1. (i) $\operatorname{cis} 0$, (ii) $\operatorname{cis}\pi$, (iii) $3\operatorname{cis}\frac{\pi}{2}$, (iv) $\operatorname{cis}(-\frac{\pi}{2})$, (v) $\sqrt{2}\operatorname{cis}(-\frac{\pi}{4})$, (vi) $2\sqrt{2}\operatorname{cis}(\frac{\pi}{4})$, (vii) $5\operatorname{cis}(-\frac{\pi}{2})$, (viii) $\operatorname{cis}\frac{2\pi}{3}$, (ix) $\frac{1}{\sqrt{2}}\operatorname{cis}(-\frac{\pi}{4})$.

5. $\operatorname{cis}\pi$.

7. $2^8\operatorname{cis}\pi$.

9. $2^6\operatorname{cis}(-\frac{\pi}{3})$.

11. $\frac{2^9}{3^4}\cos\frac{4\pi}{3}$.

13. $2^4\operatorname{cis}\frac{2\pi}{3}$.

15. $2^3\operatorname{cis}(-\frac{\pi}{2})$.

17. $\frac{1}{16\sqrt{2}}\operatorname{cis}(-\frac{3\pi}{4})$.

19. $\operatorname{cis}\pi$.

21. $2^{-5}\operatorname{cis}\frac{\pi}{2}$.

23. $\operatorname{cis}(-3\theta)$.

25. $\operatorname{cis}(-2\beta)$.

27. $\operatorname{cis}(-3\theta)$.

29. $\operatorname{cis}(-4\theta)$.

31. $\operatorname{cis}(\frac{\pi}{2}+3\theta)$.

Exercises 8.4

1. $\pm\operatorname{cis}\theta$.

3. $\pm\operatorname{cis}(\pi/4) = \pm(1+i)/\sqrt{2}$.

5. $\pm\sqrt[4]{2}\operatorname{cis}-(\frac{\pi}{8})$.

7. $\operatorname{cis}\left(\theta+\frac{2k\pi}{3}\right)$ $k=0,1,2$.

9. $-i, \frac{1}{2}\left(\pm\sqrt{3}-i\right)$.

11. $\sqrt[6]{2}\operatorname{cis}\frac{3\pi}{4}$.

13. $\sqrt{2}\operatorname{cis}\left(\frac{8k+3}{12}\pi\right)$ $k=0,1,2$.

15. $\operatorname{cis}\left(\frac{4k-1}{12}\pi\right)$ $k=0,1,\ldots,5$.

17. $\sqrt[8]{2}\operatorname{cis}\left(\frac{8k+3}{16}\pi\right)$ $k=0,1,2,3$.

19. $\operatorname{cis}\left(\frac{8k+5}{12}\pi\right)$ $k=0,1,2$.

21. $2+i,\ 1-3i$.

23. $\sqrt[10]{2}\operatorname{cis}\frac{8k+1}{20}\pi$, $k=0,\ldots,4$.

25. $2\operatorname{cis}\left(\frac{3k-1}{6}\pi\right)$ $k=0,1,2,3$.

27. $0, \cos\left(\frac{4k-1}{12}\pi\right)$ $k=0,\ldots,5$.

29. $\operatorname{cis}\left(\frac{4k-1}{12}\pi\right)$ $k=0,\ldots,5$.

31. $(z+2)(z^2-2z+4)$.

33. $(z-2)(z+2)(z^2-2z+4)(z^2+2z+4)$.

35. $(z+i)(z-i)(z+2)(z-8)$.

37. $(z-\sqrt{2}-i\sqrt{3})(z-\sqrt{2}+i\sqrt{3})(z+i)(z-i)$.

Exercises 8.5

1. $a=1+i, b=-2, x=y=1-i$.

3. $(z_1,z_2)=(2+i,2i)$.

5. Inconsistent.

7. $(z_1,z_2,z_3)=(\frac{11}{3}+2i, -\frac{1}{3}-3i, -1+\frac{4}{3}i)$.

9. $\mathbf{I}_3$, linearly independent columns, yes.

11. $(z_1,z_2)=(1+i,1-i)$.

13. $\mathbf{A}^{-1}=\begin{bmatrix}0 & -i\\ -i & i\end{bmatrix}$, $\det(\mathbf{A}^{-1})=\det(\mathbf{A})=1$.

15. $\mathbf{A}^{-1}=\begin{bmatrix}i & -i\\ i & -i\end{bmatrix}$, $\det(\mathbf{A}^{-1})=\det(\mathbf{A})=-1$.

17. $z=1+i, -(1+i)$.

21. $\mathbf{u}\cdot\mathbf{v}=1-i$, $||\mathbf{u}||=\sqrt{5}\simeq 2.2361$, $||\mathbf{v}||=2$.

23. $\mathbf{F}=\begin{bmatrix}1 & 0 & -1-i\\ 0 & 2 & -1-i\\ -1+i & -1+i & 3\end{bmatrix}$,

$\mathbf{G}=\begin{bmatrix}3 & 1+i\\ 1-i & 3\end{bmatrix}$.

25. The constant term in the characteristic equation of **A** is det(**A**) and this is a product of the eigenvalues of **A**. Eigenvalues are real for Hermitian **A**.

27. $\frac{1}{2}\begin{bmatrix} i & -i & -i \\ -i & i & -i \\ -i & -i & i \end{bmatrix}$.

29. $\text{adj}\,(\mathbf{A}) = \begin{bmatrix} 1+i & -4+i \\ -1-i & 2+3i \end{bmatrix}$, $\det(\mathbf{A}) = -6+2i$,

$\mathbf{A}^{-1} = \frac{1}{20}\begin{bmatrix} -2-4i & 2+4i \\ 13+i & -3-11i \end{bmatrix}$.

31. $\text{adj}\,(\mathbf{A}) = \begin{bmatrix} 8-i & -\sqrt{2}+i \\ -\sqrt{2}-i & 8+i \end{bmatrix}$, $\det(\mathbf{A}) = 62$,

$\mathbf{A}^{-1} = \frac{1}{62}\,\text{adj}\,\mathbf{A}$.

33. $\text{adj}\,\mathbf{A} = \begin{bmatrix} -1-2i & -i & -2 \\ -1 & -3i & -2i \\ 2i & -2i & -4 \end{bmatrix}$, $\det(\mathbf{A}) = 6-2i$, $\mathbf{A}^{-1} = \begin{bmatrix} -1-7i & -3-i & -2+6i \\ 1-3i & 3-9i & 2-6i \\ -6-2i & 2-6i & -12-4i \end{bmatrix}$.

35. $\lambda_1 = i$, $\lambda_2 = -i$, $\mathbf{P} = \begin{bmatrix} 1 & -1 \\ 1 & 1 \end{bmatrix}$.

37. $\lambda_1 = 2+i$, $\lambda_2 = 2-i$, $\mathbf{P} = \begin{bmatrix} -i & i \\ 1 & 1 \end{bmatrix}$.

39. $\lambda_1 = 0$, $\lambda_2 = i$, $\mathbf{P} = \begin{bmatrix} 0 & 1 \\ 1 & 1 \end{bmatrix}$.

Exercises 9.1

1. Extreme points: (0, 0), (0, 1), (0.6, 1.6), (1.8, 0.8), (1, 0).

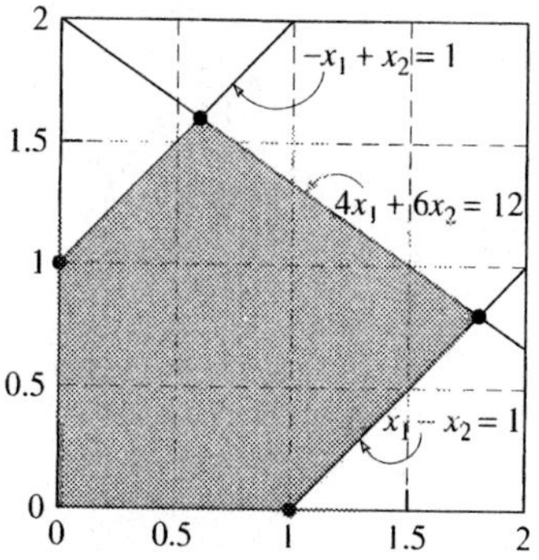

3. S is empty. No.

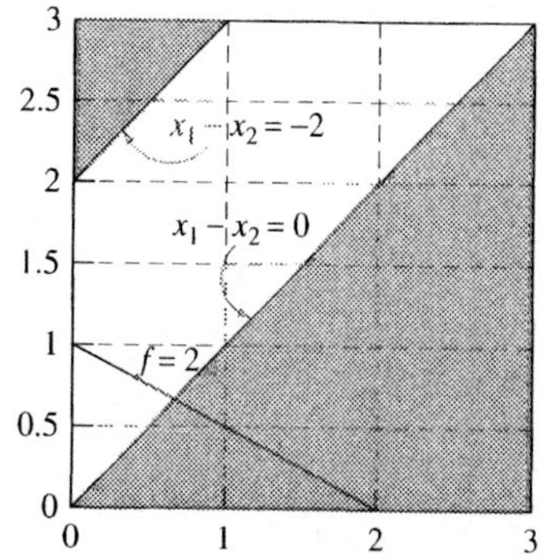

5. The constraints are: $x_1 - x_2 \geq 0$, $x_1 - 4x_2 \leq 0$, $x_1 + 2x_2 \leq 6$. The optimal feasible solutions and values are: (0, 0), $f = 0$ (minimum), (4, 1), $f = 22$ (maximum).

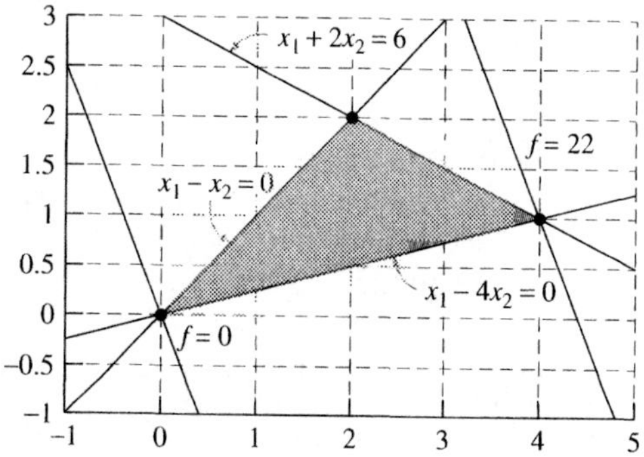

7. $(-8, 0, 0)$, $(0, 4, 0)$, $(0, 0, 8)$, $\mathbf{v} = (-1, 2, 1)$.

9. Extreme points: (28/9, 40/9), (3.6, 6.4), (8, 2). Optimal feasible solution: (3.6, 6.4). Maximum: $f = 26.4$.

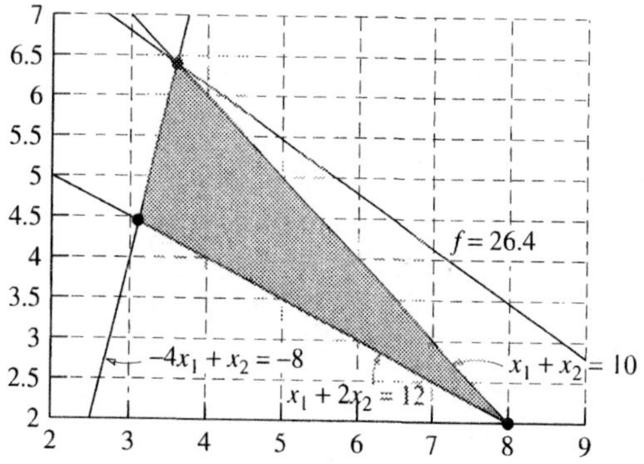

11. Extreme points: (0, 0), (4, 0), (8/7.15/7), (0, 1). Optimal feasible solution: (8/7, 15/7). Maximum: $f = \frac{158}{7} \simeq 22.57$.

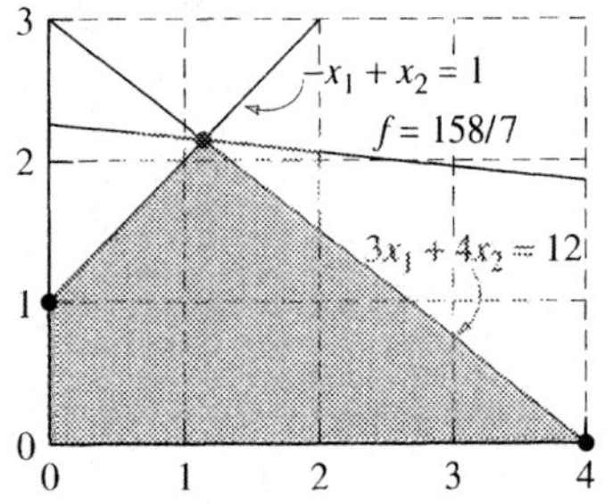

13. Optimal feasible solution: (0.5, 0.5). Minimum: $f = 2$.

15. Let $p = \sin(45°) = \cos(45°) \simeq 0.7071$. Minimize $170c13+120c12+120c23+100t14+141t24$ subject to the following constants: At node 1, $pc13+c12 = 10$ (vertically), $t14 - pc13 = 0$ (horizontally). At node 2, $pt24-c12 = 0$ (vertically), $pt24-c23 = 0$ (horizontally).

17. Let x_1, x_2 denote, respectively, the number of lb of recycled iron and steel to be used. Minimize $f = 0.2x_1 + 0.6x_2$ subject to $2000(0.02) \le 0.03x_1 + 0.004x_2 \le 2000(0.026)$, $2000(0.008) \le 0.009x_1 + 0.003x_2 \le 2000(0.012)$, $2000(0.025) \le 0.03x_1 \le 2000(0.03)$, $x_2 \le 1500$, $x_1 + x_2 = 2000$.

Exercises 9.2

1. $f = 55/3$, $x_1 = 4/3$, $x_2 = 13/3$.

3. $f = 55/9$.

5. $f = 10$.

7. $f = 44$.

9. $f = 5$. The maximum is attained for every pair (x_1, x_2) lying on $x_1 + x_2 = 4, 2 \le x_1 \le 3$.

11. $f = 12$.

13. $f = 11$.

15. $f = 8$.

19. 3404.186, $c13 = 14.14$, $t14 = 10$.

21. 523.08, $x_1 = 1692.30$, $x_2 = 307.70$.

Exercises 9.3

1. **(a)** S is nonempty and unbounded.

(b) Arguing geometrically, the values of the objective function f are unbounded above. Clearly, $f = 0$ is the (trivial) minimum solution.

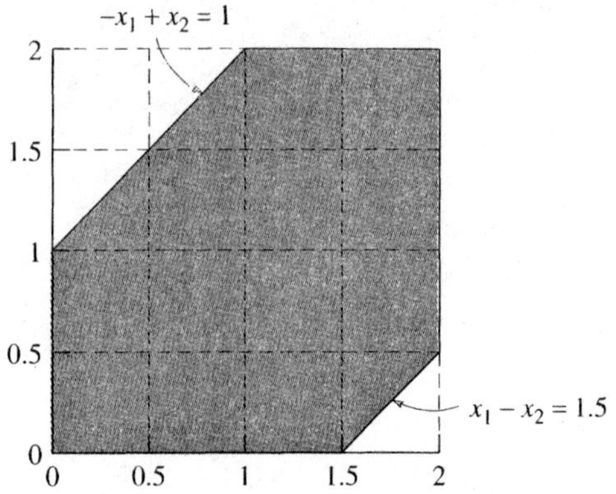

(c) The dual problem is: Minimize $d = 1.5y_1 + y_2$ subject to $y_1 - y_2 \ge 2$, $-y_1 + y_2 \ge 3$. The feasible set is empty. No optimal feasible solution.

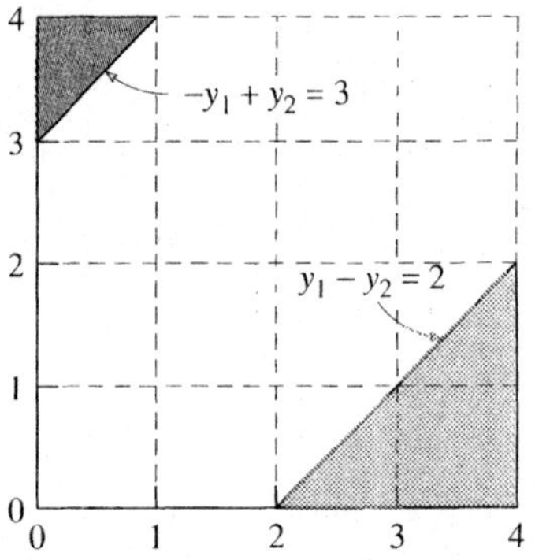

3. $f = 4$, $(x_1, x_2, x_3) = (1, 0, 3)$. Dual: $d = 2$, $(y_1, y_2) = (\frac{1}{3}, \frac{2}{3})$.

5. Minimize $d = 4y_1 + 3y_2$ subject to $y_1 + 5y_2 \ge 2$ and $y_1 - 4y_2 \ge 1$. $d \simeq 6.1$, $(y_1, y_2) \simeq (1.4, 0.1)$.

7. Minimize $d = 10y_2 + 40y_3$ subject to $y_1 + 5y_2 - 5y_3 \ge 3$ and $-y_1 - 3y_2 + 8y_3 \ge 2$. $d = 44$, $(y_1, y_2, y_3) = (0, 1.36, 0.76)$.

Exercises 9.4

1. $f = 17$.

3. $f = 8/5$.

INDEX

OTHER PH TEXTS IN INTRODUCTORY LINEAR ALGEBRA

GOODAIRE
Linear Algebra: A First Course Pure and Applied
0-13-047017-1

KOLMAN/HILL
Introductory Linear Algebra: An Applied First Course 8e
0-13-143740-2

SPENCE/INSEL/FRIEDBERG
Elementary Linear Algebra: A Matrix Approach
0-13-716722-9

BRETSCHER
Linear Algebra with Applications 3e
0-13-145334-3

KOLMAN/HILL
Elementary Linear Algebra 8e
0-13-045787-6

LEON
Linear Algebra with Applications 6e
0-13-033781-1

UHLIG
Transform Linear Algebra
0-13-041535-9

HILL/KOLMAN
Modern Matrix Algebra
0-13-948852-9

EDWARDS/PENNEY
Elementary Linear Algebra
0-13-258260-0

PH TEXTS IN LINEAR ALGEBRA/DIFFERENTIAL EQUATIONS

EDWARDS/PENNEY
Differential Equations & Linear Algebra 2e
0-13-148146-0

GOODE
Differential Equations and Linear Algebra 2e
0-13-263757-X

FARLOW/HALL/MCDILL/WEST
Differential Equations and Linear Algebra
0-13-086250-9

GREENBERG
Differential Equations and Linear Algebra
0-13-011118-X